Recent Optical and Photonic Technologies
Edited by Ki Young Kim

Republished by InTech
Janeza Trdine 9, 51000 Rijeka, Croatia

Additional hard copies can be obtained from orders@intechopen.com

Recent Optical and Photonic Technologies, Edited by Ki Young Kim
p. cm.
ISBN 978-953-7619-71-8

For EU product safety concerns:
IN TECH d.o.o., Prolaz Marije Krucifikse Kozulić 3, 51000 Rijeka, Croatia, info@intechopen.com or visit our website at intechopen.com.

Preface

Research and development in modern optical and photonic technologies have witnessed quite fast growing advancements in various fundamental and application areas due to availability of novel fabrication and measurement techniques, advanced numerical simulation tools and methods, as well as due to the increasing practical demands. The recent advancements have also been accompanied by the appearance of various interdisciplinary topics.

The book attempts to put together state-of-the-art research and development in optical and photonic technologies. It consists of 21 chapters that focus on interesting four topics of photonic crystals (first 5 chapters), THz techniques and applications (next 7 chapters), nanoscale optical techniques and applications (next 5 chapters), and optical trapping and manipulation (last 4 chapters), in which a fundamental theory, numerical simulation techniques, measurement techniques and methods, and various application examples are considered.

This book concerns itself with recent and advanced research results and comprehensive reviews on optical and photonic technologies covering the aforementioned topics. I believe that the advanced techniques and research described here may also be applicable to other contemporary research areas in optical and photonic technologies. Thus, I hope the readers will be inspired to start or to improve further their own research and technologies and to expand potential applications.

I would like to express my sincere gratitude to all the authors for their outstanding contributions to this book.

January 2010

Editor

Ki Young Kim
Department of Physics
National Cheng Kung University
Tainan, Taiwan
E-mail: kykim1994@gmail.com

Contents

PHOTONIC CRYSTALS

1

Dual-Periodic Photonic Crystal Structures

Alexey Yamilov and Mark Herrera[1]
Department of Physics, Missouri University of Science & Technology, Rolla, MO 65409, U.S.A.

1. Introduction

In this chapter we discuss optical properties of dual-periodic photonic (super-)structures. Conventional photonic crystal structures exhibit a periodic modulation of the dielectric constant in one, two or three spatial dimensions (Joannopoulos, 2008). In a dual-periodic structure, the dielectric constant is varied on two distinct scales $a_{1,2}$ along the same direction(s). An example of such a variation is given by the expression:

$$\varepsilon(x) = \varepsilon_0 + \Delta \times \left(1 + \gamma \cos\frac{2\pi}{a_1}x\right) \times \cos\frac{2\pi}{a_2}x. \tag{1}$$

In Sec. 2, after motivating our study, we describe one attractive possibility for a large-scale fabrication of the dual-periodic structures such as in Eq. (1) using the interference photo-lithorgraphy technique.
Sec. 3 presents the theory of slow-light effect in a dual-periodic photonic crystal. Here, four numerical and analytical techniques employed to study optical properties of the system. In the result, we obtain a physically transparent description based on the coupled-resonator optical waveguide (CROW) concept (Yariv et al., 1999).
Sec. 4 is devoted to discussion of a new type of optical waveguides - trench waveguide - in photonic crystal slabs. We demonstrate that this type of waveguide leads to an appearance of a second (super-) modulation in the slab, thus, slow-light devices / coupled-cavity micro-resonator arrays can be straightforwardly fabricated in the photonic crystal slab geometry. Importantly, the fabrication of such structures also does not require slow (serial) electron-beam lithography and can be accomplished with scalable (holographic) photolithography. The chapter concludes with a discussion and an outlook.

2. Dual-periodic structure as a photonic super-crystal

Optical pulse propagation in dielectrics is determined by the group velocity $v_g = d\omega(K)/dK$, where the dispersion $\omega(K)$ relates the frequency ω and the wave vector K inside the medium. One of the appealing features of photonic crystals has become a possibility to alter the dispersion of electromagnetic waves (Soukoulis, 1996) so that in a certain spectral region v_g becomes significantly smaller than the speed of light in vacuum. This "slowlight" effect (Milonni, 2005) attracted a great deal of practical interest because it can lead to low-threshold lasing (Nojima, 1998; Sakoda, 1999; Susa, 2001), pulse delay(Poon et al., 2004;

[1] Currently at department of Physics, University of Maryland

Vlasov et al., 2005), optical memories (Scheuer et al., 2005), and to enhanced nonlinear interactions (Soljacic et al, 2002; Xu et al., 2000; Jacobsen et al., 2006). Several approaches to obtaining low dispersion in photonic crystal structures have been exploited:

i. At frequencies close to the photonic band-edge, $\omega(K)$ becomes flat and group velocity approaches zero due to the Bragg effect at the Brillouin zone boundary. This property has been extensively studied and used in practice to control the spontaneous emission (Yablonovitch, 1987) and gain enhancement in lasers (Nojima, 1998; Sakoda, 1999; Susa, 2001). However, a large second order dispersion (i.e. dependence of v_g on frequency) in the vicinity of the bandedge leads to strong distortions in a pulsed signal that makes this approach unsuitable for, e.g., information processing applications.
ii. High order bands in two- and three-dimensional photonic crystals can have small dispersion not only at the Brillouin zone boundary but also throughout the band (Galisteo-López & López, 2004; Scharrer et al., 2006) where the second order dispersion can be significantly reduced. Nevertheless, these high-frequency photonic bands allow little control over v_g and are not spectrally isolated from other bands. These drawbacks and the increased sensitivity to fabrication errors (Dorado et al., 2007), limit the practical value of this approach.
iii. Based on the Coupled Resonator Optical Waveguide idea (CROW)(Stefanou & Modinos, 1998; Yariv et al., 1999; Poon et al., 2006; Scheuer et al., 2005), a low-dispersion photonic band can be purposefully created via hybridization of high-Q resonances arising from periodically positioned structural defects (Bayindir et al., 2001a;b; Altug & Vuckovic, 2005; Olivier et al., 2001; Karle et al., 2002; Happ et al., 2003; Yanik & Fan, 2004). This *spectrally isolated* defect-band is formed inside the photonic bandgap, with a dispersion relation given by

$$\omega(K) = \Omega[1 + \kappa \cos(KL)]. \qquad (2)$$

Here Ω is the resonance frequency for a single defect, κ is the coupling constant (assumed to be small) and L is the spacing between defects. These adjustable parameters allow one to control the dispersion in the band, and hence v_g, without significant detrimental effects associated with the second order dispersion.

A periodic arrangement of structural defects in the photonic crystal, described in (iii), creates a dual-periodic *photonic super-crystal* (PhSC) with short-range quasi-periodicity on the scale of the lattice constant and with long-range periodicity on the defect separation scale (Shimada et al., 2001; Kitahara et al., 2004; Shimada et al., 1998; Liu et al., 2002; Sipe et al., 1994; Benedickson et al., 1996; Bristow et al., 2003; Janner et al., 2005; Yagasaki et al., 2006). These structures usually need to be constructed with the layer-by-layer technique (or, more generally, serially) which is susceptible to the fabrication errors similarly to the other approaches (i,ii) above. We have recently proposed a PhSC with dual-*harmonic* modulation of the refractive index (Yamilov & Bertino, 2007), similar to Eq. (1), that can be fabricated by e.g. a single-step interference photolithography technique (Bertino et al., 2004; 2007). We considered four S-polarized laser beams defined by

$$\begin{bmatrix} \mathbf{q}_{L1}, E_{L1} \\ \mathbf{q}_{L2}, E_{L2} \\ \mathbf{q}_{R1}, E_{R1} \\ \mathbf{q}_{R2}, E_{R2} \end{bmatrix} = \begin{bmatrix} k_0\{-\sin(\theta_1), 0, \cos(\theta_1)\}, E_1 \\ k_0\{-\sin(\theta_2), 0, \cos(\theta_2)\}, E_2 \\ k_0\{\sin(\theta_1), 0, \cos(\theta_1)\}, E_1 \\ k_0\{\sin(\theta_2), 0, \cos(\theta_2)\}, E_2 \end{bmatrix}. \qquad (3)$$

Here **q** and E are the k-vector and amplitude of the beams respectively. Their interference pattern $E_{tot}(x) \propto \alpha\cos(k_1x) + \beta\cos(k_2x)$ leads to

$$n^2(x) = \varepsilon(x) = \varepsilon_0 + \Delta\varepsilon\left[\alpha\cos(k_1x) + \beta\cos(k_2x)\right]^2 \tag{4}$$

where $k_1 - k_2 \equiv \Delta k$, $(k_1 + k_2)/2 \equiv k$ and $\alpha + \beta = 1$. k and Δk are related to the short (a_S) and long range modulations of the refractive index: $a_S = 2\pi/\Delta k$, $a_L = \pi/k$. The parameters in Eqs. (3, 4) are related as $\alpha = E_1/(E_1 + E_2)$, $\beta = E_2/(E_1 + E_2)$ and $k_1 = k_0 \sin\theta_1$, $k_2 = k_0 \sin\theta_2$. Manipulation of the beams allows for an easy control over the structural properties of the resultant PhC: (i) fundamental periodicity a_S via k_0 and $\theta_{1,2}$; (ii) long-range modulation a_L via $\theta_1 - \theta_2$; and (iii) depth of the long-range modulation via relative intensity of the beams E_1/E_2. As we demonstrate in Sec. 3, the longer range modulation accomplishes the goal of creating the periodically positioned optical resonators. The condition of weak coupling $\kappa \ll 1$ between the states of the neighboring resonators requires sufficiently large barriers and therefore $a_S \ll a_L$, which we assume hereafter. Our approach to making dual-periodic structures has an advantage in that all resonators are produced at once and, therefore, it minimizes fabrication error margin and ensures the large-scale periodicity essential for hybridization of the resonances of individual cavities in an experiment.

Dual-periodic harmonic modulation of the refractive index can also be experimentally realized in optically-induced photorefractive crystals (Fleischer et al., 2003; Neshev et al., 2003; Efremidis et al., 2002). Although, the index contrast obtained is several orders of magnitude less than with QDPL (Bertino et al., 2004; 2007), the superlattices created in photorefractive materials offer a possibility of dynamical control - a feature lacking in the quantum dot system. While the study of dynamical and nonlinear phenomena in dual-periodic lattices is of significant interest, it goes beyond the scope of our study and will not be considered in this work.

3. Theory of slow-light effect in dual-periodic photonic lattices

In this section we theoretically investigate the optical properties of a one-dimensional PhSC using a combination of analytical and numerical techniques. We consider the dielectric function of the form given in Eq. (1) that can be produced with the interference photolithography method:

$$\varepsilon(x) = \varepsilon_0 + \frac{\Delta\varepsilon/2}{1+\gamma}\left[1 + \gamma\cos(2\pi x/L)\right]\left[1 + \cos(2\pi x/a)\right]. \tag{5}$$

Here ε_0 is the background dielectric constant. The amplitude of the short-range (on scale a) modulation gradually changes from $\Delta\varepsilon \times (1 - \gamma)/(1 + \gamma)$ to $\Delta\varepsilon$, c.f. Fig. 1a. $L = Na$ sets the scale of the long-range modulation, $N \gg 1$ is an integer.

The functional form in Eq. (5) was chosen to enable an analytic treatment and differs slightly from Eq. (4). Nonetheless, it shows the same spectral composition and modulation property. The discrepancy between the two forms is expected to cause only small deviations from the analytical results obtained in this section. Furthermore, the differences become insignificant in the limit $N \gg 1$.

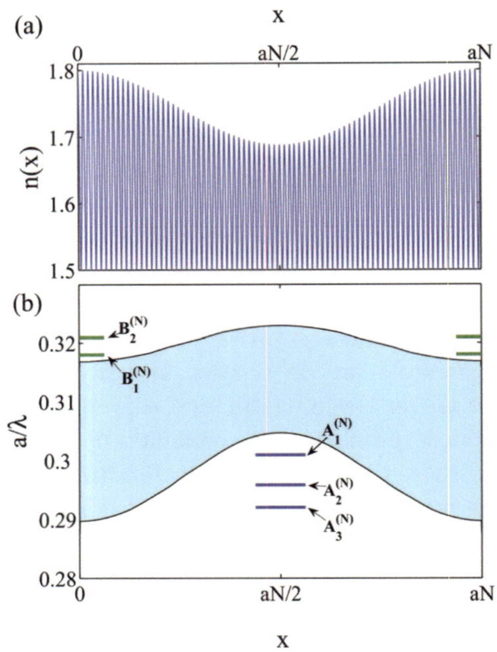

Fig. 1. (a) Dependence of the index of refraction in a dual-periodic photonic crystal as defined by Eq. (5). We used $\varepsilon_0 = 2.25$, $\Delta\varepsilon = 1$, $N = 80$ and the modulation parameter γ is equal to 0.25. (b) Local (position-dependent) photonic bandgap diagramfor $n(x)$ in (a). $A_i^{(N)}$ and $B_i^{(N)}$ mark the frequencies of the foremost photonic bands on the long- and short-wavelength sides of the photonic bandgap of the corresponding single-periodic crystal.

3.1 Transfer matrix analysis and coupled-resonators description of PhSC

The transmission/reflection spectrum of a one-dimensional PhSC of finite length, and the band structure of its infinite counterpart can be obtained numerically via the transfer matrix approach. Propagation of a field with wavevector $k = \omega/c$ through an infinitesimal segment of length dx is described by the transfer matrix (Yeh, 2005)

$$\widehat{M}(x,x+xd) = \begin{bmatrix} \cos(kn(x)dx) & n^{-1}(x)\sin(kn(x)dx) \\ -n(x)\sin(kn(x)dx) & \cos(kn(x)dx) \end{bmatrix} \tag{6}$$

where we have assumed that the refractive index $n(x)$ does not change appreciably over that distance. The matrix $\widehat{M}\,(x,\,x+dx)$ relates the electric field and its spatial derivative $\{E,1/k\ dE/dx\}$ at $x + dx$ and x. The total transfer matrix of a finite system is then given by the product of individual matrices

$$\widehat{M}_{tot} = \prod_{x=0}^{L} \widehat{M}(x,x+dx). \tag{7}$$

Since in our case the refractive index $n(x) = \varepsilon^{1/2}(x)$, Fig. 1(a), is not a piece-wise constant (in contrast to Refs. (Sipe et al., 1994; Benedickson et al., 1996)) but rather a continuous function of coordinate, one has to resort to numerical simulations. In what follows, we apply either scattering or periodic boundary conditions to obtain the transmission coefficient and Bloch number $K(\omega)$ respectively.

Figure 2(a) plots the transmission coefficient through one period of the dual-periodic system shown in Fig. 1. A series of progressively sharper resonances occur on the lower or upper edge of the spectral gap of the underlying single-periodic structure. Whether the peaks occur at the lower or upper band edge depends on the particular definition of the unit cell, as shown in the inset of Fig. 2(a). One can gain an insight into this effect by examining the modulation of the spectral position of the "local" photonic bandgap (PBG) with the position as shown in Fig. 1(b). This analysis is meaningful on the length scale of the order $a \ll \Delta x \ll L \equiv Na$. This condition can be satisfied in our case of slow modulation, with large N. At frequencies such as $A_i^{(N)}$ in Fig. 1(b), wave propagation is allowed in the vicinity of $x = aN \times (1/2 + m)$, whereas at the regions $x = aN \times m$, with m being an integer, it is locally forbidden. When considering a segment of the lattice with $0 < x < Na$, resonant tunneling via electromagnetic states $A_i^{(N)}$ of the cavity at the geometrical center leads to low-frequency peaks in the transmission coefficient, indicated by the solid line in Fig. 2(a). On the other hand, transmission through the segment $-Na/2 < x < Na/2$ exhibits a series of sharp resonances. These correspond to tunneling via $B_i^{(N)}$ cavity states in the high-frequency region.

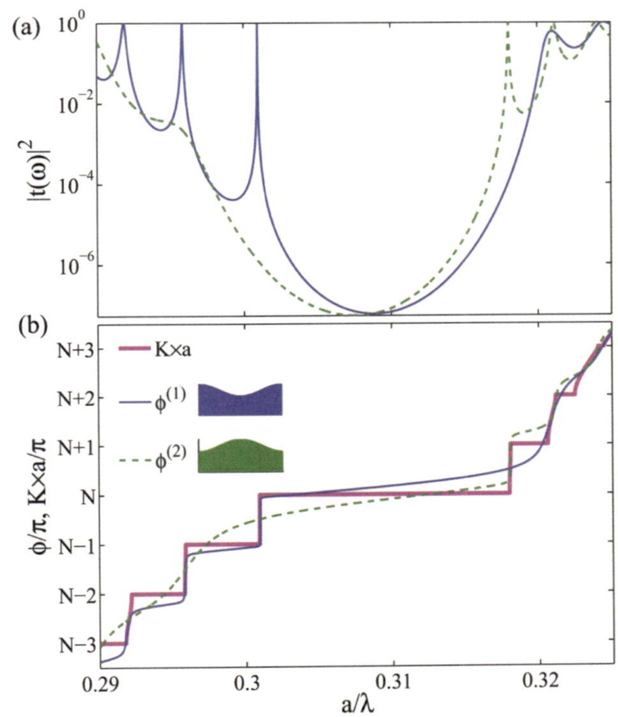

Fig. 2. (a) Transmission coefficient through a finite segment of length L (one period) of the periodic super-structure defined in Fig. 1. Solid and dashed lines correspond to $0 < x < Na$ and $-Na/2 < x < Na/2$ segments (shown in the inset of panel (b)) respectively. (b) Solid and dashed thin lines plot the corresponding phase of $t(\omega)$. Bold line depicts the Bloch number $K(\omega) \times a$ of the infinite crystal computed using Eq. 8.

The transmission coefficient through a finite segment of length L (equal to one period) can be related to the band structure of the corresponding periodic lattice (Benedickson et al., 1996) as

$$\cos(K(\omega)L) = Re[\frac{1}{t(\omega)}] \equiv \frac{1}{|t(\omega)|}\cos(\phi(\omega)), \tag{8}$$

where we have introduced the phase of the transmission coefficient $\phi(\omega)$ through $t = |t|\exp[i\phi]$. Fig. 3 shows that hybridization of the cavity resonances considered above leads to the formation of flat photonic bands. Their low dispersion and small group velocity may be exploited (Yamilov & Bertino, 2007) for practical applications.

In the vicinity of an isolated transmission resonance, $t(\omega)$ is given by the Lorentzian

$$t(\omega) = \frac{(-1)^N(\Gamma/2)}{i(\Gamma/2)-(\omega-\omega_0)} \tag{9}$$

where Γ is the full width at half maximum (FWHM) of the resonance and ω_0 is the resonant frequency. Substitution of Eq. 9 into Eq. 8 gives the flat band described by

$$\omega(K) = \omega_0[1 \pm \kappa\cos(KL)] \tag{10}$$

where $\kappa = \frac{\Gamma}{2\omega_0} = \frac{1}{Q} \ll 1$, and Q is the cavity Q-factor. Thus, the decrease of group velocity in the PhSC is directly related to the increase of confinement and the decrease of coupling between neighboring cavities. In our PhSC both these factors are described by the same parameter – the cavity Q-factor. In a single-periodic photonic crystal of finite length, the Q-factor of a band-edge mode depends on the system size. Comparing Fig. 1(b) and Fig. 3, one can see that $A_i^{(N)}$, $B_i^{(N)}$ modes are in fact band edge modes in their intervals of free propagation.

In our case L gives the characteristic length and as we demonstrate below, also determines the mode frequency. As N increases, the eigenfrequencies of the modes shift towards the bandgap. The associated decrease of the local group velocity contributes to an increase of the Q-factor of the resonators and to a further reduction of the group velocity in $A_i^{(N)}$, $B_i^{(N)}$ bands in the $N \to \infty$ limit.

Eq. (8) suggests that the dispersion relation $\omega(K)$ is independent of how the segment of length L (the period of our structure) is chosen. However, the transmission coefficient through the $0 < x < Na$ and $-Na/2 < x < Na/2$ segments of the crystal shows very different spectral composition, Fig. 2(a). In order to understand how these markedly different functions lead to the same $\omega(K)$, we analyze the phase of the transmission coefficient ϕ, shown in Fig. 2(b).

In a one-dimensional periodic system such as ours, the wave number $K(\omega)$ in Eq. 8 is equal to the integrated density of electromagnetic states. It is, by definition, a monotonically increasing function of frequency in the extended Brillouin zone scheme. In PhSC, $K \times L$ increases by π every time the frequency is increased through an allowed band, c.f. bold line in Fig. 2(b). At the frequency in the middle of the band, $\cos(KL) = 0$ because $K \times L = \pi \times (m + 1/2)$. From Eq. (8) one can see that ϕ should be equal to $\pi \times (m+1/2)$ at the same frequency. In the finite system, the mode counting phase $\tilde{\phi}$ defined (Lifshitz et al., 1998) as

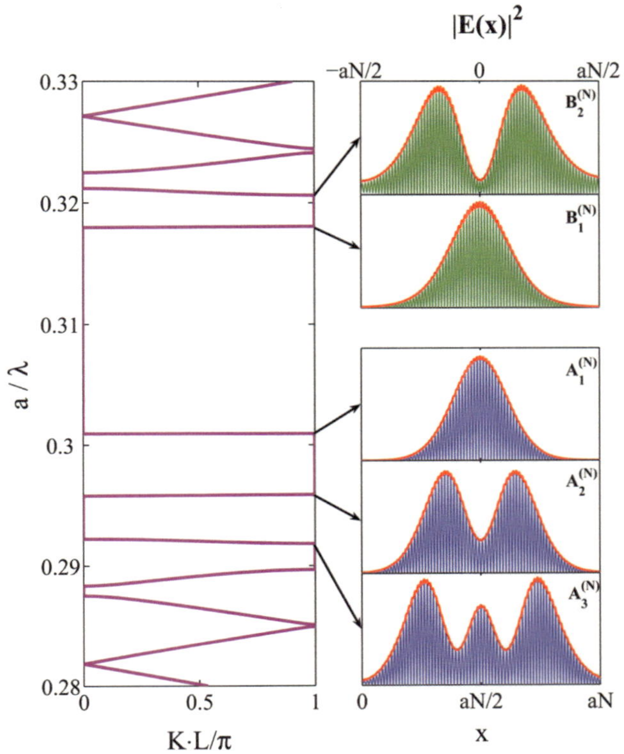

Fig. 3. The left panel shows dispersion of a PhSC $\omega(K)$ reduced to the first Brillouin zone. The eigenmodes which correspond to the series of flat bands in the vicinity of the parent-bandgap of the single periodic crystal are depicted on the right. Calculations were performed for the structure described in Fig. 1.

$\tan(\tilde{\phi}) = E'/E$ coincides with the phase of the transmission coefficient $\phi \equiv \tilde{\phi}$. This explains the monotonic behavior of $\phi(\omega)$. Eq. (8) leads to the fact that quasi-states of the finite system occur at the same place as the corresponding band center of the lattice, irrespective of the definition of the unit cell. Therefore, as can be also seen from Fig. 2(b), $\phi(\omega)$ and $K(\omega)L$ intersect at $\pi \times (m + 1/2)$.

Taylor expansion of the phase around the frequency ω_0 at the center of a pass band, where $K(\omega)L = \pi \times (m + 1/2)$ gives

$$\cos(K(\omega)L) = (\omega - \omega_0) \times \frac{(-1)^m}{|t(\omega_0)|} \frac{d\phi(\omega_0)}{d\omega}. \tag{11}$$

Here, the term that contained $d\ |t(\omega_0)|/d\omega$ dropped out because $\cos(K(\omega_0)L) = 0$. Comparing Eqs. (10) and (11) shows that it is $|t(\omega_0)|^{-1}\ d\phi(\omega_0)/d\omega$ that determines $Q = 1/\kappa$ and not just $|t(\omega_0)|$. Suppressed transmission compensates for a slow phase change (e.g. solid line in Fig. 2(b) in the high frequency spectral region) and leads to an identical $K(\omega)$ for two different definitions of the unit cell.

We also note that if the segment is chosen such that the corresponding "cavity" is located in the geometrical center ($|t(\omega_0)| = 1$), the FWHM of the resonance (Γ) in the transmission

coefficient is equal to the width of the pass band in the periodic lattice. This fact follows from Eqs. (8) and (9). It further emphasizes the analogy with CROW structures.
We conclude this section by reiterating that long-range refractive index modulation creates alternating spatial regions which serve as resonators separated by the tunneling barriers. Hybridization of the cavity resonances creates a series of photonic bands with low dispersion. The envelope of the eigenstates in these bands $A_i^{(N)}$, $B_i^{(N)}$ is a slowly varying function of the coordinate, c.f. Fig. 3. This effect stems from states proximity to the photonic band-edge of the underlying single-periodic lattice. The possibility of a separation into short (a of rapid field oscillations) and long (L of the slow amplitude variation) length scales will further inform analytical studies presented in the following sections.
In addition, the results of this section lead to somewhat counter-intuitive conclusion that the larger or even complete modulation in Eq. (5) would negatively affect (increase) the coupling between the resonators. This can also be seen from PBG diagram in Fig. 1b: in case of compete modulation of the refractive index ($\gamma = 1$), the local bandgap disappears at $x_m = aN \times (1/2 + m)$. Indeed, our photonic band structure calculations demonstrate that structures with 100% modulation are less advantageous and lead to significantly larger propagation speeds. The optimum value of γ depends on the experimental parameters (ε_0, $\Delta\varepsilon$ and N) and should be determined with the help of PBG diagram similar to Fig. 1b. The diagram also proves useful in explaining the advantage of A_N over B_N. In the latter case, the tunneling barriers are thinner and their localization length is longer (PBG is spectrally narrower at $x_m = aN \times (1/2 + m)$ than it is at $x_m = aN \times m$).

3.2 Resonant approximation

Forbidden gaps in the spectra of a periodic system arise due to a resonant interaction of the wave with its Bragg-scattered counterpart (Ashcroft & Mermin, 1976). The scattered wave appears due to the presence of Fourier harmonics in the spectrum of the periodic "potential", which in the case of the Helmholtz equation

$$E''(x) + \frac{\omega^2}{c^2}\delta\varepsilon(x)E(x) = \frac{\omega^2}{c^2}\bar{\varepsilon}E(x), \tag{12}$$

is represented by $(\omega^2/c^2)\delta\varepsilon(x) \equiv (\omega^2/c^2)\left[\varepsilon(x) - \bar{\varepsilon}\right]$. Here we have introduced the average value of the dielectric function $\bar{\varepsilon} = \overline{\varepsilon(x)} = \varepsilon_0 + \Delta\varepsilon/[2(1+\gamma)]$. When $\gamma = 0$, the condition $\Delta\varepsilon/\bar{\varepsilon} \ll 1$ is sufficient to obtain the position and width of spectral gaps. Otherwise, an additional condition $N \times \Delta\varepsilon/\bar{\varepsilon} \ll 1$ needs to be satisfied instead. We will discuss the physical meaning of this condition at the end of this section.
We begin by noticing that $\delta\varepsilon(x)$ of our choice (Eq. 5) contains only eight nonzero Fourier harmonics:

$$\varepsilon(x) = \sum_{m=-\infty}^{\infty} \varepsilon_m \exp\left[i\frac{2\pi}{L}mx\right], \tag{13}$$

where $m = \{\pm1, \pm(N-1), \pm N, \pm(N+1)\}$. This fact allows for an exhaustive study of all resonant interactions as follows. Expressing $E(x)$ in terms of its Fourier components

$$E(x) = \exp[iK(\omega)x] \sum_{m=-\infty}^{\infty} E_m \exp\left[i\frac{2\pi}{L}mx\right] \tag{14}$$

leads to an infinite system of linear coupled equations

$$\left[\frac{\omega^2}{c^2}\bar{\varepsilon} - \left(K(\omega) + \frac{2\pi}{L}m\right)^2\right]E_m + \frac{\omega^2}{c^2}\sum_{m'\neq 0}\varepsilon_{m'}E_{m-m'} = 0, \tag{15}$$

where K is the Bloch number that varies in the first Brillouin zone $[0, \pi/L]$. For the extreme values of K there exists a spectral range where the term in brackets in Eq. 15 can become simultaneously small for certain values of m and $-m$ at $K = 0$, and for m and $-m - 1$ at $K = \pi/L$. If $\varepsilon(x)$ contains a harmonic $\varepsilon_{m'}$ such that it couples these two Fourier components, the overall infinite system Eq. 15 can be reduced to two resonant equations.

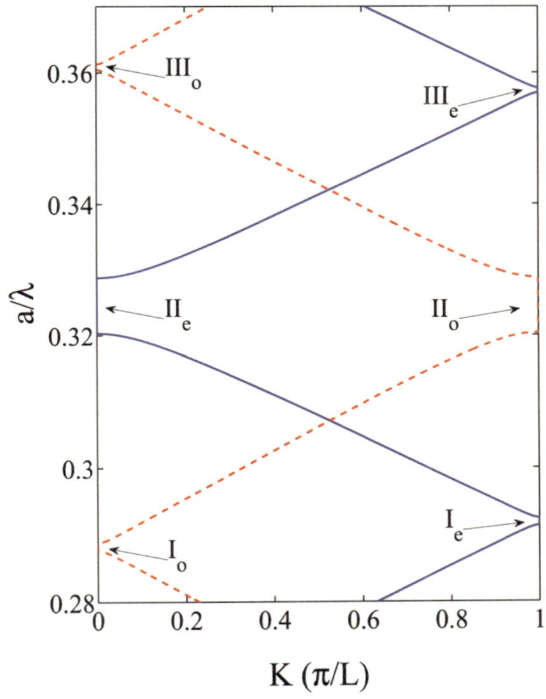

Fig. 4. Dispersion relation computed with the transfer matrix formalism for $\varepsilon_0 = 2.25$, $\Delta\varepsilon = 0.32$, $N = 9$ (dashed line) and $N = 10$ (solid line). The modulation parameter γ is equal to 0.25. For this set of parameters, the applicability condition Eq. 16 of the resonant approximation is satisfied.

The results of such an analysis are summarized in Table 1 and the corresponding band structure is shown in Fig. 4. Introduction of the long range modulation in the dielectric constant results in an expansion of the unit cell from a to $L = Na$ and, thus, to a reduction of the Brillouin zone, accompanied by the folding of photonic bands. The cases of even $N = 2s$ and odd $N = 2s + 1$ should be distinguished. In the former, the primary photonic bandgap (II_e) of the single-periodic lattice reappears at $K = 0$, whereas in the latter (II_o) it is located at

$K = \pi/L$. Our analysis shows that the nearest frequency gaps, $I_{o,e}$ and $III_{o,e}$, also become resonant, Fig. 4. For the refractive index modulation of Eq. (5), the normalized width $\varepsilon_{(N-1)} / \bar{\varepsilon} = \dfrac{\gamma\Delta\varepsilon / 4\bar{\varepsilon}}{1+\gamma}$ of the satellite gaps is smaller than that of the central gap by a factor of γ. By definition, this parameter is less than unity.

We can see that folding and the offset of the formation of flat $A_1^{(N)}$, $B_1^{(N)}$ bands is captured in this approximation. The criterion of its applicability can be found by considering the contributions of non resonant terms in Eq. (15). We find that for all three gaps the criteria are qualitatively the same. Therefore, we present the detailed analysis of only one particular resonance, III_e. The condition that the closest non-resonant Fourier components E_{-s-2}, E_{-s}, E_{s-1} and E_{s+1} be smaller than the resonant ones E_{-s-1} and E_s leads to the relation

$$\frac{(N+1)^2(\varepsilon_1+\varepsilon_N)}{4N\bar{\varepsilon}+2(N+1)^2\varepsilon_{N-1}} \ll 1. \tag{16}$$

In the limit of very large N the second term in the denominator becomes dominant and this condition cannot be satisfied for any value of $\Delta\varepsilon$. Thus, N should be finite. The condition that the first termin the denominator be dominant, is consistent with the entire inequality Eq. (16), and is equivalent to $N\varepsilon_i / \bar{\varepsilon} \ll 1$. Taking the most restrictive case for ε_i, we finally obtain

$$\frac{\Delta\varepsilon}{8\bar{\varepsilon}} \times N \ll 1, \tag{17}$$

where we have neglected γ for simplicity.

Equation (17) has a clear physical meaning. Indeed, from Table 1, it is clear that the frequency of bandgaps I and III approach the central gap inherited from the single periodic system as $1/N$. At some point, a bandgap of width $\Delta\omega/\omega_0 = \varepsilon_i/\bar{\varepsilon}$ begins to substantially perturb the pass band of width $K_{max} \times c \simeq \omega_0/N$ separating consecutive gaps. The resonant approximation breaks when these two scales become comparable. This condition results in Eq. (17). In other words, the approximation considered in this section can at most capture the onset of the flattening trend in the $A_1^{(N)}$, $B_1^{(N)}$ bands and fails when N is increased to the point where these states become abnormally flat, i.e., where $\Delta\omega/K_{max} \ll c/\bar{\varepsilon}$ throughout the band. More sophisticated approaches are considered below.

Resonant K, (even N)	π/L	0	π/L
(odd N)	0	π/L	0
Coupled components	$E_{-s}, E_{N-(s+1)}$	E_{s-N}, E_s	$E_{-(s+1)}, E_{N-s}$
Coupling harmonics	$\varepsilon_{-(N-1)}, \varepsilon_{N-1}$	$\varepsilon_{-N}, \varepsilon_N$	$\varepsilon_{-(N+1)}, \varepsilon_{N+1}$
Center frequency, ω_0	$\frac{c\pi}{\sqrt{\bar{\varepsilon}}L}(N-1)$	$\frac{c\pi}{\sqrt{\bar{\varepsilon}}L}N$	$\frac{c\pi}{\sqrt{\bar{\varepsilon}}L}(N+1)$
Normalized width, $\Delta\omega/\omega_0$	$\varepsilon_{(N-1)}/\bar{\varepsilon}$	$\varepsilon_N/\bar{\varepsilon}$	$\varepsilon_{(N+1)}/\bar{\varepsilon}$

Table 1. Results of resonant approximation analysis of Eq. (15) with dielectric function given by Eq. (5). Three columns correspond to the three resonant photonic band gaps that appear in the spectrum of the dual-periodic PhSC. The expressions hold for both even and odd N for the choice of parameter s: $N = 2s$ and $N = 2s + 1$ respectively.

3.3 Effective medium approximation

Gratings written in the core of photosensitive optical fibers are often analyzed with the help of coupled-mode theory (CMT) (Marcuse, 1991). In both shallow gratings with long-range modulation in fibers (Sipe et al., 1994; Janner et al., 2005) and in our PhSC, the forward and backward (locally) propagating waves continuously scatter into each other. The advantage of CMT is that it considers the amplitudes of the forward and backward waves directly. This tremendously simplifies Maxwell equations. Ref. (de Sterke, 1998) also considered fiber gratings with a deep piece-wise constant index modulation. In this section we employ the CMT-based method developed by Sipe, *et al.* (Sipe et al., 1994) to obtain the spectral positions of the flat photonic bands formed in a PhSC.

For *shallow modulation*, i.e., small $\Delta\varepsilon$, our Eq. (5) can be brought to resemble the model function considered in Ref. (Sipe et al., 1994)

$$n(x)/n_0 = 1+\sigma(x)+2\kappa(x)\cos\left[2k_0x+\varphi(x)\right] \tag{18}$$

with the following choice of parameters

$$\sigma(x) = \frac{\dfrac{\gamma\Delta\varepsilon/4}{1+\gamma}}{\varepsilon_0+\dfrac{\Delta\varepsilon/2}{1+\gamma}}\times\cos\frac{2\pi}{L}x;$$

$$\kappa(x) = \frac{\dfrac{\Delta\varepsilon/8}{1+\gamma}}{\varepsilon_0+\dfrac{\Delta\varepsilon/2}{1+\gamma}}\times\left(1+\gamma\cos\frac{2\pi}{L}x\right); \tag{19}$$

$$\varphi(x)\equiv 0; \quad n_0 = \left(\varepsilon_0+\frac{\Delta\varepsilon/2}{1+\gamma}\right)^{1/2}; \quad k_0 = \pi/a.$$

The CMT of Ref. (Sipe et al., 1994) is applicable as long as these functions have *a slow dependence on x*, on the scale much larger than k_0^{-1}. This condition is indeed satisfied in the PhSC with $N \gg 1$.

By introducing small detuning parameter

$$\delta = \frac{\omega-\omega_0}{\omega_0} \ll 1, \quad \omega_0 = \frac{k_0c}{n_0}$$

we can, following Ref. (Sipe et al., 1994), obtain the governing equation for the quantity E_{eff} related to the envelope of the electric field

$$\frac{d^2E_{eff}}{dx^2}+k_0^2n_{eff}^2(x,\omega)E_{eff}=0. \tag{20}$$

Frequency and position dependent effective refractive index

$$n_{eff} = \{(\sigma(x)+\Delta)^2-\kappa(x)^2\}^{1/2} \tag{21}$$

determines whether propagation is locally allowed (real n_{eff}) or forbidden (imaginary n_{eff}). This is similar to our definition of the local PBG diagram which we studied numerically in Section 3. Figure 5b compares CMT's region of evanescent propagation (solid lines) to the numerical calculation (dashed lines). We attribute the relatively small discrepancy observed there to the assumption of shallow modulation made in arriving at Eq. (21).

Eq. (20) is formally similar to the Schrödinger equation. Our previous analysis shows that the single-period states associated with photonic bands $A_i^{(N)}, B_i^{(N)}$ are confined to the region of classically allowed propagation, in the language of quantum mechanics. By analogy, the Wentzel-Kramers-Brillouin (WKB) approximation of quantum mechanics can be applied (Sipe et al., 1994) to determine the quantization of energies inside our optical equivalent of a quantum well

$$I(\omega) = k_0 \int_{x_L}^{x_R} n_{eff}(x,\omega)dx = (m+1/2)\pi \tag{22}$$

in which x_L and x_R are, respectively, the left and right turning points defined by the condition $n_{eff}(x_{L,R},\omega) = 0$, m is an integer. The solid line in Fig. 5a depicts the value of the integral in Eq. (22), as a function of ω, obtained numerically. The filled circles denote the frequencies at which quantization condition Eq. (22) is satisfied. In a system with the parameters which we used for illustration in previous sections, the obtained solutions are in fair agreement with numerical results obtained with the transfer matrix approach described in Section 3. This suggests that the index variation given by $\Delta\varepsilon = 1$, $\varepsilon_0 = 2.25$ was sufficiently small for this approach to still be qualitatively applicable.

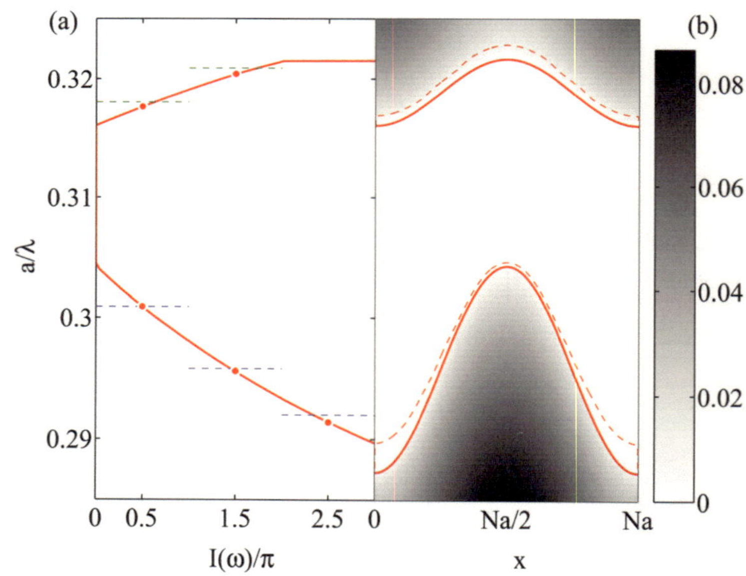

Fig. 5. (a) The value of the integral in Eq. (22), solid line, as a function of frequency is shown. For easy comparison with (b), the plot is transposed so that ω is plotted along the y-axis. The circles depict frequencies that satisfy the quantization condition of Eq. (22). The dashed lines denote the actual position of photonic states, as determined by direct numerical analysis of Section 3. (b) Gray-scale plot of $Re[n_{eff}(x,\omega)]$ given by Eq (21). The solid line shows the boundary of the regionwhere $Im[n_{eff}(x,\omega)] \neq 0$. For comparisonwe also show the local PBG of Fig. 1(b), dashed line. In both (a) and (b), the parameters of Fig. 1 are adopted.

We finish the current section by noting that it would be desirable to retain the attractive property of the CMT envelope approach without being constrained by the condition of small refractive index modulation. The latter may not always be justified in the experimental situation of interest (Bertino et al., 2004; 2007). In the following section we develop such an approach.

3.4 Bogolyubov-Mitropolsky approach

In this section we will consider the standing-wave solutions of Eq. (12). In this case, the corresponding $E(x)$ can be chosen to be a real function by an appropriate choice of normalization. Then, we make the Bogolyubov anzatz (Landa, 2001; Bogolyubov & Mitropolsky, 1974):

$$\begin{aligned} E(x) &= A(x)\cos\left(k_0 x + \phi(x)\right) \\ dE(x)/dx &= -k_0 A(x)\sin\left(k_0 x + \phi(x)\right), \end{aligned} \tag{23}$$

where, as in the preceding section, $k_0 = \pi/a$. The above equations define the amplitude and phase functions. Their substitution into Eq. (15) gives the so-called Bogolyubov equations in standard form (Landa, 2001; Bogolyubov & Mitropolsky, 1974)

$$\begin{aligned} \frac{d\phi(x)}{dx} &= \frac{1}{k_0}\left[\frac{\omega^2}{c^2}\varepsilon(x) - k_0^2\right]\cos^2\left(k_0 x + \phi(x)\right) \\ \frac{dA(x)}{dx} &= \frac{A(x)}{2k_0}\left[\frac{\omega^2}{c^2}\varepsilon(x) - k_0^2\right]\sin 2\left(k_0 x + \phi(x)\right), \end{aligned} \tag{24}$$

No approximations have been made so far. The structure of the above equation suggests that conditions $dA/dx \ll k_0 A$ and $d\phi/dx \ll k_0\phi$ can be satisfied in the vicinity of the spectral region where $(1/k_0)\left[\omega^2/c^2\overline{\varepsilon(x)} - k_0^2\right] \ll k_0$. Here, the overbar denotes an average over one period. Comparison with the analysis in the previous sections shows that this condition is satisfied in the vicinity of the primary photonic bandgap. In the system of interest, for which $N \gg 1$, this observation justifies the "averaging-out" of the fast spectral components, which is the Mitropolsky technique (Bogolyubov & Mitropolsky, 1974). This averaging procedure leads to the following system of nonlinear equations for the slow-varying amplitude and phase

$$\frac{d\phi(x)}{dx} = \frac{1}{2k_0}\left[\frac{\omega^2}{c^2}\varepsilon_0 - k_0^2 + \frac{\omega^2}{c^2}\frac{\Delta\varepsilon/2}{1+\gamma}\left(1+\gamma\cos\frac{2\pi}{L}x\right)\left(1+\frac{1}{2}\cos 2\phi(x)\right)\right] \tag{25}$$

$$\frac{d\log A(x)}{dx} = \frac{1}{2k_0}\frac{\omega^2}{c^2}\frac{\Delta\varepsilon/2}{1+\gamma}\left(1+\gamma\cos\frac{2\pi}{L}x\right)\sin 2\phi(x). \tag{26}$$

In deriving Eqs. (25) and (26) we have used the explicit form of $\varepsilon(x)$ given by Eq. (5).

We begin the analysis of Eqs. (25) and (26) with a discussion of the appropriate boundary conditions. In deriving these equations we have limited consideration to real-valued solutions of the original Eq. (15), which can be found only for a discrete set of frequencies. At these special frequencies, the corresponding amplitude function should reflect the periodicity of the dielectric function Eq. (5). This implies that

$$\phi(L) = \phi(0) + m\pi \tag{27}$$

$$\sin 2\phi(0, L/2, L) = 0. \tag{28}$$

The first condition is obtained by requiring $\sin 2\phi(x)$ in Eq. (26) to be periodic. Symmetry of the modulation profile $A(x)$, see Fig. 1(a), and continuity of its derivative lead to the condition $dA(x=0, L/2, L)/dx = 0$. This can only be satisfied by requiring Eq. (28), because other factors on the right hand side of Eq. (26) are positive functions.

Equation (25) which determines the evolution of the phase is self-contained. Hence, its solution together with the constraints given by Eqs. (27) and (28) is sufficient to obtain the spectrum of the system and $\phi(x)$. The amplitude is to be recovered in the second step by simple integration of Eq. (26) with the found phase $\phi(x)$.

Figure 6 shows the solutions of the Eqs. (25, 26, 27, 28) obtained by a fourth order Runge-Kutta numerical method. In accord with our expectation, for each band there exist two solutions $\phi(x)$, which correspond to standing-wave band-edge modes at $K = 0, \pi/L$, as seen in Fig. 6a,e. The corresponding solutions of the amplitude equation, Fig. 6b-d, f-g, agree with the envelopes extracted from direct solutions of the Helmholtz equation, Fig. 3. The eigenvalues of Eq. (25) also give the frequencies that correspond to band-edge states, and

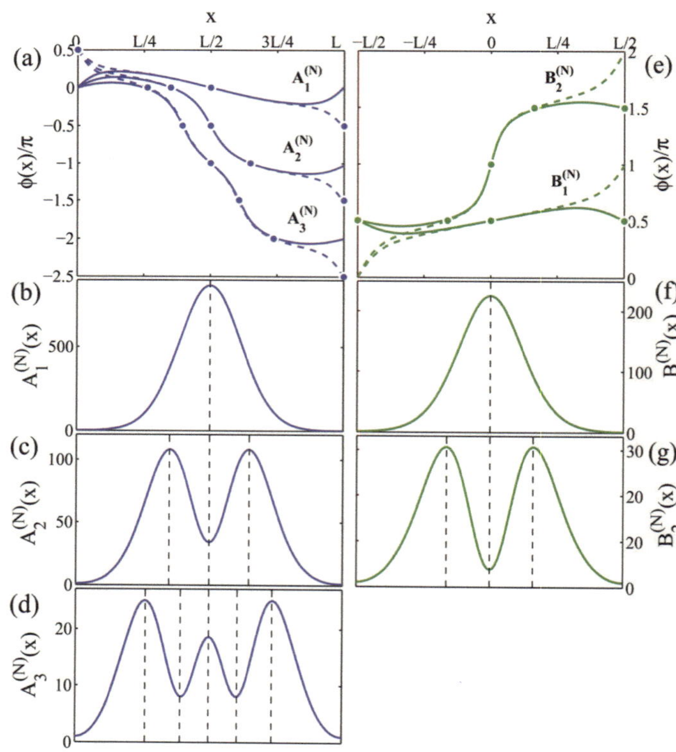

Fig. 6. Numerical solutions of Eqs. (25, 26, 27, 28) are shown. Filled circles in panels a,e denote the spatial position where the particular $\phi(x)$ is equal to $m\pi/2$. At these special points $dA(x)/dx = 0$ denoted by the vertical dashed lines in b-d and f-g panels. $A_i^{(N)}$ and $B_i^{(N)}$ denote the low-dispersion photonic bands as defined in Section 3.

are also in excellent agreement (the observed deviation is less than 0.1%), Fig. 3. Knowledge of the bandedge frequencies allows determination of all parameters of the tight-binding approximation for $\omega(K)$, Eq. (2). Therefore, the entire band structure in the spectral region of each of the flat bands can be obtained solely from solution of the amplitude-phase equation.

Filled circles in panels a,e of Fig. 6 denote the spatial position where the particular $\phi(x)$ is equal to $m\pi/2$. At these special points $dA(x)/dx = 0$, as denoted by the vertical dashed lines in b-d and f-g panels. Thus, the overall phase accumulated by $\phi(x)$ over one period is an important parameter indicative of the spatial structure of the amplitude. At the band-edge frequencies of the bands $A_1^{(N)}$ and $B_1^{(N)}$ (see Section 3 for notations), the phase is a bounded function $|\phi(x) - \phi(0)| \leq \pi/2$. Therefore, $x = 0, L/2, L$ are the only positions where the corresponding amplitude function takes minimum/maximum values. Thus, as seen in Fig. 6b, f $A(x)$ has only one "hump" for $A_1^{(N)}$ and $B_1^{(N)}$. A comparison of $\phi(x)$ computed at the $K = 0$ (solid lines) and $K = \pi/L$ (dashed lines) edges of each photonic band shows (Fig. 6) that the difference occurs in the spatial regions where electromagnetic waves propagate via the "tunneling mechanism" in the language of CROWs of Section 3. In these regions $A(x)$ is small, which explains the small spectral width of the corresponding photonic bands.

As the eigenfrequencies of the higher order states $A_2^{(N)}, B_2^{(N)}$... shift further away from the primary band-gap region, Fig. 1b, $\phi(x)$ becomes progressively steeper function, leading to a steady increase in the number of "humps" in $A(x)$, Fig. 6. This progression accelerates the spatial dependence of the amplitude and leads to an eventual breakdown of the scale separation approximation used in the derivation of Eqs. (25) and (26). Nevertheless, such a loss of applicability occurs well outside the spectral region of interest, demonstrating the robustness of the approach developed here.

3.5 Comparison of theoretical approaches

Although all of the methods considered above have their limitations, the results obtained with each technique complement each other:

While numerical simulations with transfer matrices in Sec. 3.1 allow one to compute the photonic band structure for an arbitrary refractive index modulations, this method, however, may not provide a complete physical insight into the nature of the photonic bands. The transfer matrix approach allowed us to compare the spectrum of the infinite (periodic) crystal with the transmission spectrum of a finite system with a length equal to one period of the superstructure. We also were able to identify the individual transmission resonances with the photonic bands and found a one-to-one correspondence. Furthermore, the spatial distribution of the fields at resonance demonstrated that in the $L \gg a$ limit the envelope (amplitude) of the state changes slowly - on the scale of L.

With a method commonly employed in condensed matter physics, in Sec. 3.2 we investigated the resonant interactions between Bloch waves when the second, longer-scale, modulation is introduced. It was shown that the flattening of photonic bands is related to, but goes beyond band folding. The reduction in the group speed (slow-light effect) arises due to increased coupling between Bloch waves with k-vectors at the boundaries of the Brillouin zone. The subsequent increase of the band-gap regions "squeezes" the bands making them progressively flatter as $N = L/a$ is increased. Although, this approach fails for very large N, it still provides an important insight into the origin of the anomalously small dispersion in the spectra of PhSCs.

Diffraction gratings introduced in optical fibers are often spatially modulated. Coupled-mode theory has been developed to reduce the problem to a study of the amplitudes of the forward and backward propagating waves and to avoid a direct solution of Maxwell's equations. Although the method had been initially developed for optical fibers where the induced refractive index contrast is small, the CMT-based approach of Sec. 3.3 provided a clear physical picture. It showed that the electromagnetic states of our optical resonators can be thought of as eigenstates of the photonic wells. This further reinforced the analogy with CROWs that we developed in Sec. 3.1.

Noting a formal similarity between the Helmholtz equation with the considered dual-periodic dielectric function Eq. (5) and the equation describing parametric resonance in oscillation theory, we adopted an amplitude-phase formalism, accompanied by a separation of scales (short a and long $L = N \times a$), Sec. 3.4. In the result we were able to derive a tractable set of equations for the envelope functions. This enabled us to study physically meaningful mode profiles directly, without assuming small modulations of the refractive index.

4. Dual periodicity in trench waveguide in photonic crystal slab

4.1 Slow-light effect in photonic crystal slab

Photonic crystals have provided a way to control light on a sub-wavelength scale to an unprecedented degree (Joannopoulos, 2008). Particularly, the ability of photonic crystal slab (PhCS) waveguides (Johnson et al., 1999; 2000; Lončar et al., 2000) to control the propagation of light through photonic confinement makes them a versatile tool for use in a wide variety of practical applications (Chutinan & Noda, 2000; Krauss, 2003). The planar geometry of the PhCS waveguides makes it easy to incorporate them into larger scale integrated optical devices.

By changing the size of structural units (usually a cylindrical holes in a dielectric slab) or omitting them altogether along a row in the PhCS lattice, a spatially confined region is created where the local effective dielectric constant differs from that in the surrounding regions. Thus, a band-edge mode splits off from either the upper or lower photonic band and moves into the spectral interval of the photonic bandgap. This allows for the propagation of light at such frequencies but only within the confines of the perturbed region (Johnson et al., 2000; Lončar et al., 2000). In order to make such a waveguide, it is required to use electron-beam lithography to fabricate the carefully designed waveguide region of the PhCS as well as the rest of the structure, including the photonic crystal slab itself (Lončar et al., 2000). This technique becomes time intensive because each individual structural element in the device must be created serially - one element at a time.

Here we consider a PhCS-based waveguide design which can be implemented with a combination of a scalable and cost-effective laser holography (Cho et al., 2005; 2007) and photolithography techniques. We show that efficient waveguiding can be achieved by creating a shallow trench in the pre-patterned (e.g. holographically) PhCS blank. Making the slab locally thinner accomplishes the same goal of perturbing the local effective dielectric constant as in more conventional designs. Thus, the waveguiding effect along the trench in the photonic slab is achieved *without* changing the radius of the holes, or making other small-feature adjustments (such as displacements) to the holes. This observation shows that with the proposed design, one is no longer restricted to the use of e-beam lithography. Our study shows that this design yields robust structures, which are expected to be almost insensitive to misalignment of the trench. At the same time, dispersion of the mode can be effectively tailored via adjustment of the trench width and its depth.

We note that the term "trench waveguide" was used by Yu. Vlasov and coworkers (Vlasov et al., 2004) in the context of a different structure - a uniform rectangular waveguide separated from PhCS's on both sides by small gaps (trenches).

We also make another observation, interesting from a conceptual point of view: unlike the guiding along a ridge which is also known in uniform (not PhCS) dielectrics, the guiding along a trench region is unique to the PhCS. In the latter system the vertical confinement does not originate from total internal reflection, but rather is related to the modifications of the optical dispersion due to the intrinsic periodicity in the PhCS.

Similar to other PhCS waveguides (when prepared on dielectric substrates), the trench waveguide suffers from propagation losses due to coupling between the guided mode and the bulk PhCS modes of the opposite parity (symmetry). In Sec. 4.2.3 we show that this effect is quite small in e.g. structures made of silicon. Thus, the trench waveguides could be used as an inexpensive alternative to carry optical signals over relatively short distances.

In Sec. 4.4 we demonstrate that when the trench waveguide is *rotated* with respect to the row of holes in the PhCS, the structure can be viewed as coupled-resonator optical waveguide (CROW) (Yariv et al., 1999) based on *dual-periodic photonic crystal* considered in Sec. 2,3. This makes our structures suitable for such applications as delay lines, optical storage, or coherently-coupled arrays of microlasers (Olivier et al., 2001; Karle et al., 2002; Yanik & Fan, 2004; Happ et al., 2003; Altug & Vučković, 2004).

Microlasers based on PhCS (Painter et al., 1999; Park et al., 2004) have attracted a great deal of attention due to low lasing thresholds and the possibility of on-chip integration. To increase the optical output and its efficiency, systems containing multiple coupled cavities were considered (Happ et al., 2003; Altug & Vučković, 2004). Usually, the efficiency of direct optical coupling between the microresonators is limited by the scalability of the fabrication process, as well as the ability to reliably reproduce cavities. In Sec. 4.4 we show that because different sections of the same (rotated) trench waveguide act as the optical cavities, the *resonator uniformity* is ensured. This feature of our design is expected to promote the optical coupling between individual resonators.

4.2 Formation of a guided mode in trench waveguide

4.2.1 Geometry

To investigate how the presence of a trench affects the optical properties of the PhCS, we computed the band structure $\omega_i(\vec{k})$ of several systems such as those shown in Fig. 7 with a plane-wave expansion method (Johnson & Joannopoulos, 2001). This method also provides us with the spatial distribution of the electric and magnetic fields at the eigen-frequency $\omega_i(\vec{k})$ found for the given wave vector $\vec{k}$. Because we are interested in the waveguiding properties of the structure, $\vec{k}$ will point in the direction of the trench.

In order to create a waveguiding channel, a line defect must be made in an existing photonic crystal slab. Rather than perturbing the shape/size of the holes, we alter the height of a linear region (stripe) of the material. Fig. 7 shows an example of one of the structures being considered. The system is a free-standing slab of silicon with a dielectric constant $\varepsilon_{slab} = 12.0$, and surrounding dielectric material of $\varepsilon_{air} = 1.0$. The dimensions of the structure are given in terms of the hexagonal lattice unit a, with the entire cell having dimensions of $2\sqrt{3}a \times 1a \times 4a$, which can be varied to achieve the desired level of accuracy. The radius of the holes in the PhCS is $r = 0.3a$. A parameter Δ controls the width of the linear defect, with $\Delta = 1.5 \times (\sqrt{3}a/2)$ corresponding to the distance between two rows of holes. The height of the slab h as well as its height in the waveguiding region, d, are the parameters the effects of which are to be investigated.

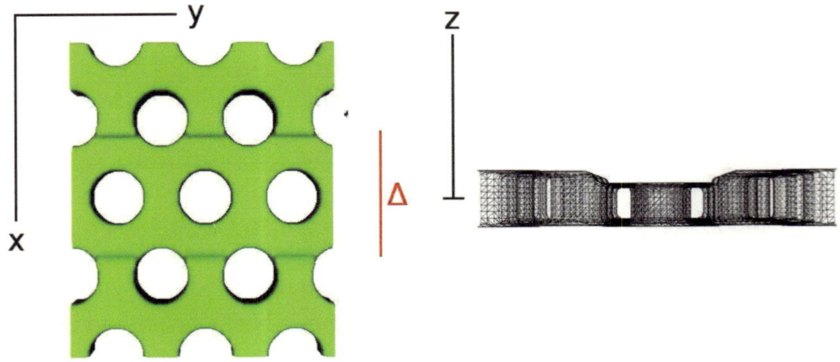

Fig. 7. An example of the computational super-cell being modeled (repeated by a factor of three in the y-direction for clarity). A linear defect, the trench, is created along y-axis in freestanding (membrane) PhCS structure with the dielectric constant of $\varepsilon_{slab} = 12.0$.

4.2.2 Symmetry considerations

In 2D or planar photonic structures such as a PhCS, the system parameters can be chosen such that a sizable photonic bangap can exist in the spectrum of either odd, TM-like, modes (e.g. high-index dielectric cylinders in air) or even, TE-like, modes (e.g. cylindrical air-holes in a high-index background) but not both simultaneously. The latter geometry, currently prevailing experimentally, is considered in this current work. Importantly, the very existence of the photonic bandgap relies on the possibility to separate TM- and TE-like modes into two *non-interacting* classes of modes. Our structure, c.f. Fig. 7, lacks the mirror-reflection symmetry with respect to the $z = 0$ plane dissecting the PhCS. The rest of this section will be devoted to the consequences of the interaction of the two classes of modes and the resulting detrimental effects of cross-talk between them.

While the systems we consider are not vertically symmetric, as we show below it is still possible to use an odd-like and even-like symmetry approach (with respect to the z-axis) to these systems. The solutions, while not having full odd or even symmetry, retain a large amount of their odd/even character, c.f. Fig. 8.

To illustrate the above point, we compare two systems schematically depicted on the inset of Fig. 9, both with the same parameters of $\Delta = 1.5$, $h = 0.5a$ and $d = 0.4a$ (thickness of PhCS in the trench region). The first system is vertically symmetric, with two trench (stripe) regions - one above and one below the PhCS - being removed. The second system is our original geometry, c.f. Fig. 7, which is not vertically symmetric. Fig. 9 plots the odd (red) and even (blue) modes of the symmetric system, as well as the full inseparable band structure for the non-symmetric system (black). The agreement between the non-symmetric and symmetric case is extremely high, except for the anti-crossing region highlighted with an arrow. This observation suggests that the non-symmetric modes still have a high degree of odd and even character. Fig. 8b displays both $\Re[H_z(x,y,\ z = 0)]$ of the guided mode at the Brillouin zone boundary $k = 0.5$, as well as its z-profile $|\int\int H_z(x,y,\ z)dxdy|$. The results demonstrate that the mode is indeed highly z-symmetric and confined to the trench.

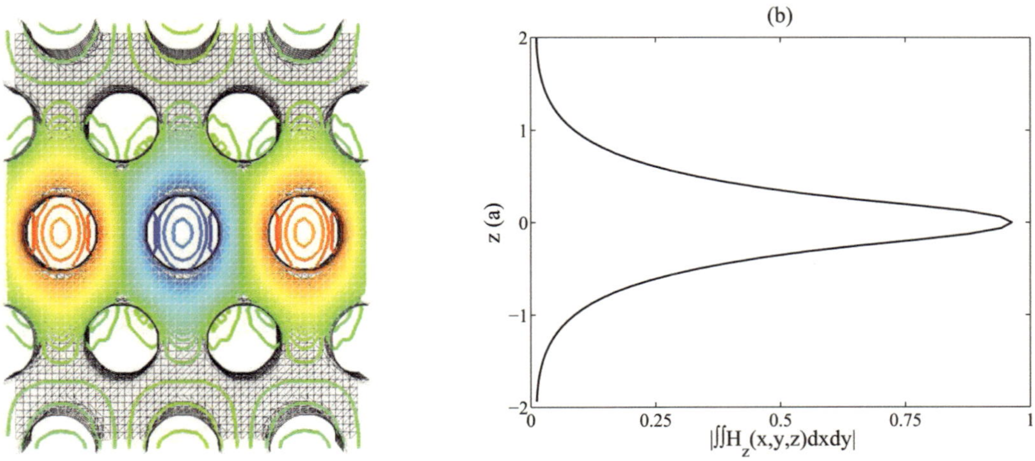

Fig. 8. (a)$\Re[H_z(x,y,\ z = 0)]$ for the waveguiding mode in the trench waveguide, see text. (b) $|\int\int H_z(x,y,\ z)dxdy|$ for the same mode, demonstrating its vertical confinement.

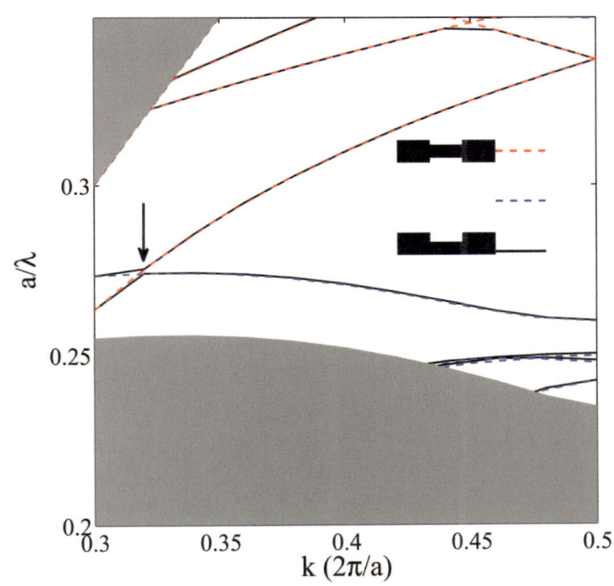

Fig. 9. Band structure diagram for even (blue) and odd (red) modes of the system symmetric about $z = 0$, and the inseparable band structure (black) of the system that is not symmetric about $z = 0$. The inset schematically shows xz cross-sections of both systems. The band structures of the symmetric and asymmetric waveguides are almost identical with exception of small deviations in the vicinity of the anti-crossing regions.

4.2.3 Cross talk between modes of different symmetries

The coupling between the waveguiding mode (which is, as seen in the above Sec. 4.2.2, predominantly even) and the odd modes leads to propagation loss. This is because the energy transfered to an odd mode is no longer spatially confined to the region of the waveguide and is irreversibly lost. To assess the efficacy of the waveguiding in PhCS with the trench, one needs to quantify the extent of the cross-talk.

In order to address this question, we compared magnetic field profiles of the waveguiding mode (even-like) with the odd bulk mode for the frequencies close to the anti-crossing, Fig. 9. We examined the overlap between two modes $\delta = \left| \int H_{z,1}(\mathbf{r}) H^{*}_{z,2}(\mathbf{r}) dV \right|^2$. Here, we assumed the H fields to be already normalized. Fig. 10(a,b) plots the band structure for $\Delta=1.5(\sqrt{3}a/2)$, $h = 0.5a$, $d = 0.4a$, and the values of δ for different branches of the dispersion curve. The frequency scales are aligned along the y-axis so the value of the overlap is plotted along the x-axis in Fig. 10b. The calculations indicate that the overlap between the bulk mode and the mode from a waveguiding branch is indeed small (no greater than ~ 2%). As expected, the degree of the overlap within the other branch gradually increases away from the anti-crossing. We argue that making the trench deeper (smaller d) and narrowing the width of the trench (smaller Δ) decreases the even- and odd- like character of the modes. The reasoning is the following: by decreasing the depth of the waveguiding region, one is introducing larger perturbations to the ideal, symmetric slab about z =0. Thus, the odd-like and even-like modes interact to a greater extent, and the odd-even symmetry is lost. Further, this should be seen in the overlap between the once even-like mode and the odd bulk mode. If odd-even symmetry has decreased, then one expects the overlap to be greater. Indeed, the calculations performed for a structure with $\Delta = 1.25(\sqrt{3}a/2)$, $h = 0.5a$, $d = 0.3a$ yield the results qualitatively similar to those in Fig. 10, but with greater degree of the overlap.

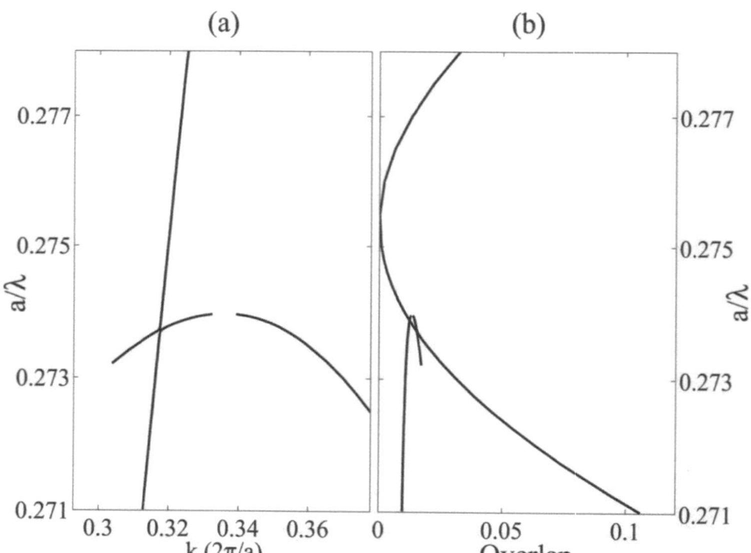

Fig. 10. (a) Band structure diagram for $h = 0.5a$, $d = 0.4a$ in the spectral vicinity of the region of the strongest leakage of the guided mode (low dispersion curve). (b) plots (on the x-axis) the overlap between the guided mode and the bulk mode of the opposite (odd) symmetry.

4.3 Control over the properties of the mode

4.3.1 The effect of trench depth

In Sec. 4.2.3 we have found that when the trench becomes too deep, the loss of the even symmetry of the guided mode may lead to increased propagation losses. Here, we investigate the possibility of tuning the optical properties of the trench waveguide while keeping it shallow $(h - d) \ll h$.

We varied the parameter d between $d = 0.36a$ and $d = 0.46a$ in steps of $\delta d = 0.02a$, while $\Delta = 1.5 \times (\sqrt{3}a/2)$ and $h = 0.5a$ were kept constant for all structures. The resulting dispersion relations are plotted in Fig. 11. One observes that for lower values of d, the frequency of the guided mode increases. This is to be expected, as the mode propagating in structures with a deeper trench (smaller d) should have more spatial extent in regions of air. The associated lowering of the effective index experienced by these modes leads to the increase of their frequency $\omega \propto n_{eff}^{-1}$.

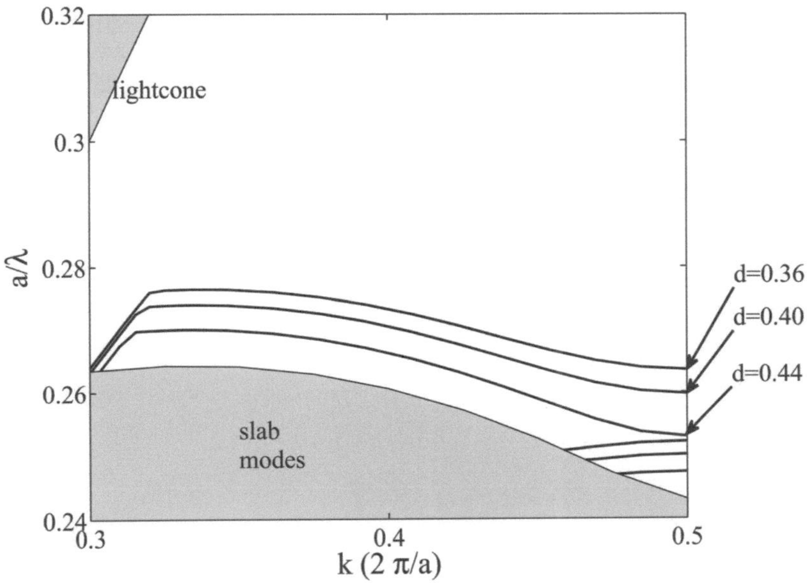

Fig. 11. Dispersion relations for the guided mode in the trench PhCS waveguide with parameters $h = 0.5a$, $\Delta = 1.5(\sqrt{3}a/2)$, and different values of d. The even bulk PhCS modes are superimposed as gray regions. A decrease in the depth of the trench $(h - d)$ leads to the decrease in the frequency of the guided mode in accordance with the effective index argument, see text.

4.3.2 Trench displacement

One of the structural parameters important from the experimental point of view is the alignment of the trench waveguide with the rows of cylindrical air-holes in the PhCS. To demonstrate the robustness of the waveguiding effect in our design, we studied the dependence of the band structure on the trench position. We introduce a displacement parameter t_d which measures the distance between the middle of the trench and the line containing the centers of the air-holes. By our definition, the maximum amount of trench displacement is $t_{d,max} = a \times (\sqrt{3}/4)$. By symmetry, any larger displacements are identical to one in $0 \leq t_d \leq t_{d,max}$ interval. The parameter t_d was varied in this range in steps of $\delta t_d = 0.2 \times t_{d,max}$, while $\Delta = 1.5\times(\sqrt{3}a/2)$ and $h = 0.5a$ were kept constant for all iterations. The dispersion relation plots are presented in Fig. 12.

As t_d approaches $t_{d,max}$, we note that the frequency of the waveguiding band shifts only slightly to lower frequencies. Thus, fabrication errors in the alignment of the trench with the background photonic crystal slab should have minimal effects on the frequency of the band. The most pronounced dependence on t_d appears at the edge of the Brillouin zone, $k = 1/2$. At $t_d = t_{d,max}$ a degeneracy created between the guided mode and the next highest-frequency even-like mode; the trench waveguide no longer operates in a single mode regime. This degeneracy can be explained by studying the z-component of the magnetic field, H_z. Fig. 12b plots $\Re[H_z(x_0, y, z)]$ for the guided mode with $t_d = 0$ (upper panel) and $t_d = t_{d,max}$ (lower panel). x_0 corresponds to the line containing the centers of the airholes. At $t_{d,max}$ displacement, an additional symmetry appears due to the fact that the trench is centered at the midpoint between two consecutive rows of air-holes. As highlighted by the structure of the mode in Fig. 12b, the combination of translation by $a/2$ along the direction of the trench (y-axis) and the $y-z$ mirror reflection leaves the structure invariant. Thus, the effective index sampled by two modes related by the above symmetry transformation, is identical. For the k-vectors other than $1/2$, the two modes remain spectrally separated for a large range of t_d, making the system robust against misalignment errors during fabrication.

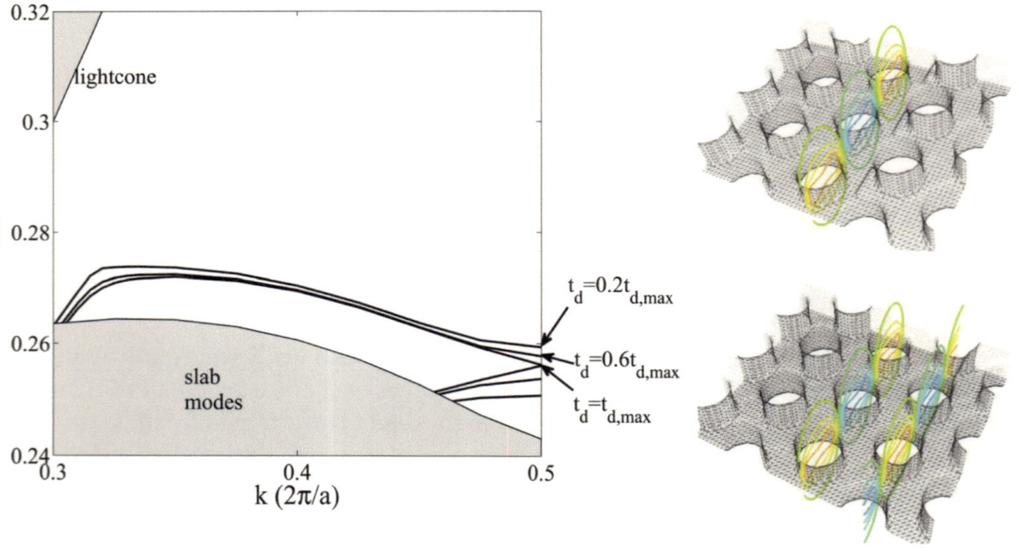

Fig. 12. Dispersion relations for $h = 0.5a$, $d = 0.4a$, and different values of the horizontal position t_d is shown in (a). The even bulk PhCS modes are superimposed as gray regions. One notes that as t_d approaches $t_{d,max}$ the bands become degenerate at the edge of the Brillouin zone. Panel (b) depicts the guided mode with $k = 1/2$ for the trench centered at (upper) or between (lower) rows of holes. Degeneracy of the lower mode for which $t_d = t_{d,max}$ is explained by the added symmetry of the trench for this particular value of t_d. This symmetry involves $a/2$ translation and mirror reflection, see text.

4.4 Rotated trench waveguide as an array of coupled micro-cavities

As previously discussed in Sec. 4.1, a wide range of new phenomena is expected when the direction of the trench waveguide is rotated with respect to the direction of the row of holes. Indeed, a rotation of the trench creates modulations along the waveguide - the trench alternates between the regions where it is centered on a hole and those between holes. We will see that these regions play the role of optical resonators which are optically coupled (by construction) to form a coupled resonator optical waveguide (CROW) (Yariv et al., 1999).

4.4.1 Effective index approximation analysis

In order to quantify the orientation of the trench, we use a parameter α, the angle between the trench and the row of holes in the nearest neighbor direction. The investigation of such structures can still be accomplished with the plane wave expansion method of Ref. (Johnson & Joannopoulos, 2001). The required super-cell, however, is greatly increased (c.f. Fig. 15 below). To allow the detailed qualitative study of the rotated trench structures, we first adopt an effective index approximation (Qiu, 2002), reducing the structures to two dimensions. The slab is now a 2D hexagonal lattice with the background dielectric constant $\varepsilon = 12.0$, with holes of radius r = 0.4a and $\varepsilon_{air} = 1.0$. The trench is represented by a stripe region with the reduced dielectric constant of $\varepsilon = 3.0$. A band gap is present in the spectrum of the TE-polarization modes propagating though this structure, with the guided mode of the same polarization. Similar to the original 3D system, the frequency of the mode is displaced up into the band gap due to the linear defect. An example of the super-cell of the 2D dielectric structure being modeled is depicted in the inset of Fig. 13a.

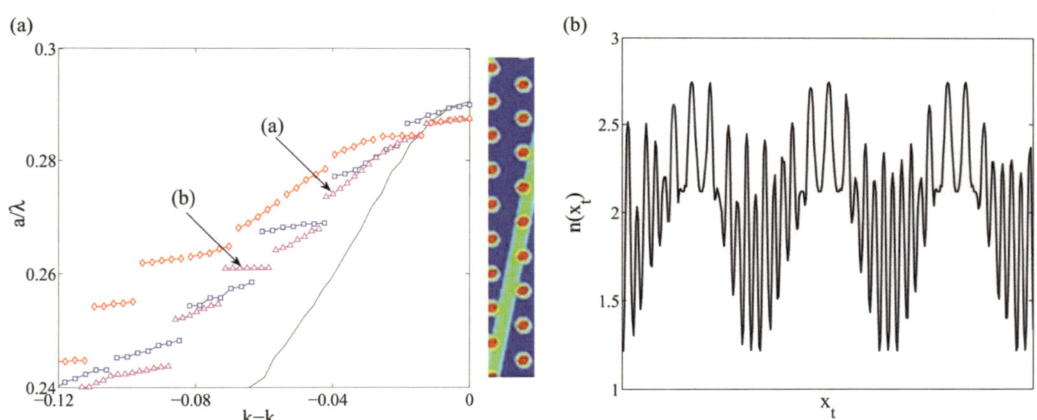

Fig. 13. (a) Band structure for the aligned trench waveguide $\alpha = 0°$ (solid line) is compared to the extended Brillouin zone band structures of the rotated trench waveguides with $\alpha = 9.8°$ (squares), $\alpha = 7.1°$ (triangles), and $\alpha = 4.9°$ (diamonds). The slowest group velocity (flattest band) occurs for the intermediate $\alpha = 7.1°$. The inset shows the 2D effective-medium approximation of the 3D trench. (b) $n(x_t)$ as a function of trench coordinate x_t. $n(x_t)$ is modulated in a periodic fashion, allowing for the 1D photonic crystal methods to be applied.

We consider trenches with a small rotation from the *M*-crystallographic direction of the hexagonal lattice. The smallness of the angle is determined in comparison to the other nonequivalent direction, *K*, which is separated by an angle of 30°. We studied the rotated trench waveguides with several values of α; here we report the results on $\alpha = 9.8°$, 7.1° and

4.9°. In order to model the structures with such small angles, a large (along the direction of the waveguide) computational super-cell is needed. As the result, the band structure of trench is folded due to reduction of the Brillouin zone (BZ)(Neff et al., 2007). Even when unfolded, the size of the BZ is reduced because a single period along the direction of the trench contains several lines of air-holes. Thus, to compare the dispersion of the rotated waveguide to that of the straight one, in Fig. 13a we show their band structures in the extended form. The obtained series of bands correspond to the different guided modes of the trench waveguide. Strikingly, we observe that the group velocity $v_g = d\omega(k)/dk$ associated with different bands varies markedly, c.f. bands (a,b) indicated by the arrows in Fig. 13a. The origin of such variations is discussed below.

4.4.2 Coupled resonator optical waveguide (CROW) description

As the trench defect crosses the lines of air-holes in the PhCS, the local effective index experienced by the propagating mode varies, c.f. inset in Fig. 13a. This creates a one-dimensional sequence of the periodically repeated segments with different modulations of the refractive index. Indeed, Fig. 13b shows the refractive index averaged over the cross-section of the trench and plotted along the waveguide direction. As it was shown Sec. 2, 3, this dual-periodic (1D) photonic super-crystal acts as a periodic sequence of coupled optical resonators. Furthermore, comparison of two modes in Fig. 14 demonstrates that at some frequency, a segment of the trench may play the role of the cavity, whereas at another, this particular section of the trench may serve as a tunneling barrier. This is similar to our results in Fig. 1b.

For applications such as optical storage or coupled laser resonators, small-dispersion modes (slow-light regime) are desired (Vlasov et al., 2005; Baba & Mori, 2007). Examining Fig. 13 we find that the band with the smallest group velocity (marked with (b) in the figure) occurs at $\alpha = 7.1°$. Comparison of the fast (a) and slow (b) modes, c.f. Figs. 13a, 14, provides a clue as to why there might exist such a dramatic variations in the dispersion. For mode (a) the resonator portion of the trench is long, whereas the barrier separating two subsequent resonator regions is short. The corresponding CROW mode is extended with weak confinement and high degree of coupling between the resonators. For mode (b) the resonator regions appear to be well separated, thus the cavities provide good confinement while the coupling is quite weak. This results in low dispersion of the CROW band (b).

(a) (b)

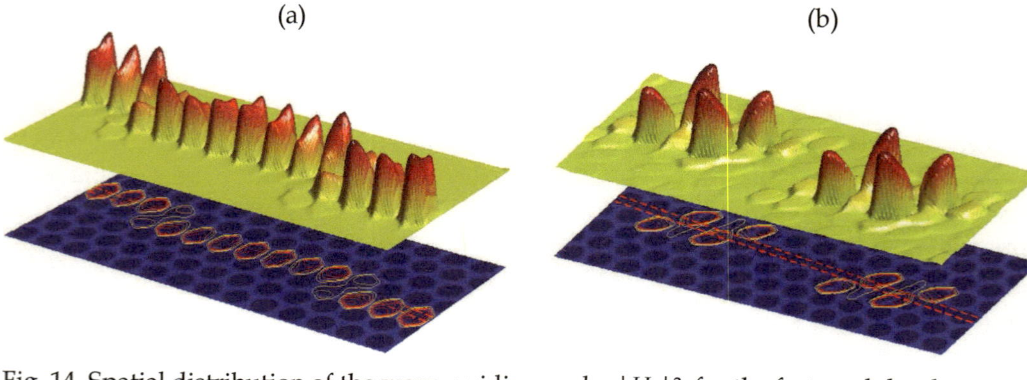

Fig. 14. Spatial distribution of the wave-guidingmode, $|H_z|^2$, for the fast- and the slow-bands denoted as (a) and (b) in Fig. 13a.

Our analysis of 1D structures in Ref. (Yamilov & Bertino, 2008) showed that increasing the period of the super-modulation monotonously leads to flatter bands - simultaneously enhancing confinement and weakening inter-cavity coupling. In the effective index approximation of our trench waveguide, an increase of the super-modulation corresponds to the decrease of the angle of rotation of the trench α. Lack of such a uniform reduction in the group velocity of the guided modes with the decrease of α (in the system considered, the minimum in v_g occurs for the intermediate value of $\alpha = 7.1^{\circ}$) shows that the reduction to 1D system (such as in Fig. 13b) may not be fully justified. In other words, the position of the trench with relation to the PhCS units is important in formation of the optical resonators, hence, simulation of a particular structure in hand is required.

4.5 Implementation of trench-waveguide

Although the band structure computations become significantly more challenging when one relaxes the effective index approximation employed in Sec. 4.4.1, 4.4.2, our CROW description of the guided modes in the rotated trench waveguide remains valid. Fig. 15 shows a representative mode found in the full 3D simulations. In the realistic 3D systems the CROW description is further complicated (Sanchis et al., 2005; Povinelli & Fan, 2006) due to the need to account not only the in-plane confinement $1/Q_{\parallel}$ but also the vertical confinement factor $1/Q_{\perp}$ even in a single cavity (a single-period section of the trench). Indeed, since the total cavity Q-factor contains both contributions $1/Q = 1/Q_{\parallel} + 1/Q_{\perp}$, the structures optimized in the 2D-approximation simulations which contain no $Q_{\perp}$, no longer appear optimized in 3D.

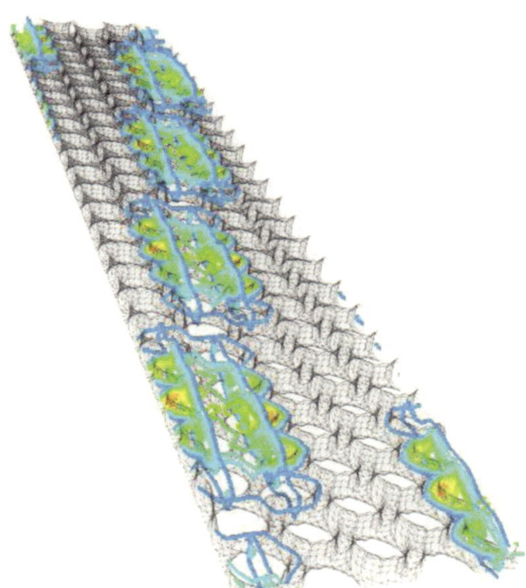

Fig. 15. A representative example of the guided mode obtained in full 3D simulation of the rotated trench waveguide. The system parameters are chosen as in Sec. 4.3, $\alpha = 9.8^{\circ}$.

Several designs aim at optimization of PhCS-based resonator cavities by balancing $Q_{\parallel}$ and $Q_{\perp}$ via "gentle localization" (Akahane et al., 2003), phase slip (Lončar et al., 2002; Apalkov &

Raikh, 2003) or double heterostructure (Song et al., 2005). In Ref. (Yamilov et al., 2006) we also demonstrated how random fluctuation of the thickness of PhCS give rise to *self-optimization* of the lasing modes. The results of Sec. 4.4.1, 4.4.2 suggest that by varying such structural parameters of the trench waveguide as its width, depth and the rotation angle, a variety of resonator cavities is created. Thus, we believe that, given a particular experimental realization, it would be possible to optimize the guided modes of the trench waveguide for the desired application. We stress that the adjustment of all three structural parameters of the considered design does not require the alteration of the structural unit of PhCS - the air-hole - and it should be possible to fabricate a trench waveguide in a PhCS "blank" prepared e.g. holographically. Therefore, the fabrication process of the finished device involving the trench waveguides may be accomplished without employing (serial) *e*-beam lithography opening up a possibility of parallel mass production of such devices.

5. Summary and outlook

In this contribution we presented the analytical and numerical studies of photonic super-crystals with short- and long-range harmonic modulations of the refractive index, c.f. Eq. (1). Such structures can be prepared experimentally with holographic photolithography, Sec. 2. We showed that a series of bands with anomalously small dispersion is formed in the spectral region of the photonic bandgap of the underlying single-periodic crystal. The related slow-light effect is attributed to the long-range modulations of the index, that leads to formation of an array of evanescently-coupled high-Q cavities, Sec. 3.1.

In Sec. 3, the band structure of the photonic super-crystal is studied with four techniques: (i) transfer matrix approach; (ii) an analysis of resonant coupling in the process of band folding; (iii) effective medium approach based on coupled-mode theory; and (iv) the Bogolyubov-Mitropolsky approach. The latter method, commonly used in the studies of nonlinear oscillators, was employed to investigate the behavior of eigenfunction envelopes and the band structure of the dual-periodic photonic lattice. We show that reliable results can be obtained even in the case of large refractive index modulation.

In Sec. 4 we discussed a practical implementation of a dual-periodic photonic super-crystal. We demonstrated that a linear trench defect in a photonic crystal slab creates a periodic array of coupled photonic crystal slab cavities.

The main message of our work is that *practical slow-light devices based on the coupled-cavity microresonator arrays can be fabricated with a combination of scalable holography and photo-lithography methods, avoiding laborious electron-beam lithography.* The intrinsic feature uniformity, crucial from the experimental point of view, should ensure that the resonances of the individual cavities efficiently couple to form flat photonic band and, thus, bring about the desired slow light effect. Furthermore, the reduction in fabrication costs associated with abandoning *e*-beam lithography in favor of the optical patterning, is expected to make them even more practical.

6. Acknowledgments

AY acknowledges support from Missouri University of Science & Technology. MH acknowledges the support of a Missouri University of Science & Technology Opportunities for Undergraduate Research Experiences (MST-OURE) scholarship and a Milton Chang Travel Award from the Optical Society of America.

7. References

Akahane, Y., Asano, T., Song, B.-S., and Noda, S. (2003). "High-Q photonic nanocavity in a two-dimensional photonic crystal," Nature (London) 425, 944-947.

Altug, H., and Vučković, J. (2004). "Two-dimensional coupled photonic crystal resonator arrays," Appl. Phys. Lett. 84, 161.

Altug, H. & Vukovic, J. (2005). "Experimental demonstration of the slow group velocity of light in two-dimensional coupled photonic crystal microcavity arrays," Appl. Phys. Lett. 86, 111102.

Apalkov, V. M., and Raikh, M. E. (2003). "Strongly localized mode at the intersection of the phase slips in a photonic crystal without band gap," Phys. Rev. Lett. 90, 253901.

Ashcroft, N. W., & Mermin, N. D. (1976). *Solid State Physics*, (Brooks Cole).

Baba, T., and Mori, D. (2007). "Slow light engineering in photonic crystals," J. Phys. D 40, 2659-2665.

Bayindir, M., Tanriseven, S., and Ozbay, E. (2001). "Propagation of light through localized coupled-cavity modes in one-dimensional photonic band-gap structures," Appl. Phys. A 72, 117-119.

Bayindir, M., Kural, C., and Ozbay, E. (2001). "Coupled optical microcavities in one-dimensional photonic bandgap structures," J. Opt. A 3, S184-S189.

Benedickson, J. M., Dowling, J. P., and Scalora, M. (1996). "Analytic expressions for the electromagnetic mode density in finite, one-dimensional, photonic band-gap structures," Phys. Rev. E 53, 4107-4121.

Bertino, M. F., Gadipalli, R. R., Story, J. G., Williams, C. G., Zhang, G., Sotiriou-Leventis, C., Tokuhiro, A. T., Guha, S., and Leventis, N. (2004). "Laser writing of semiconductor nanoparticles and quantum dots," Appl. Phys. Lett. 85, 6007-6009.

Bertino, M. F., Gadipalli, R. R.,Martin, L. A., Rich, L. E., Yamilov, A., Heckman, B. R., Leventis, N., Guha, S., Katsoudas, J., Divan, R., and Mancini, D. C. (2007). "Quantum dots by ultraviolet and X-ray lithography," Nanotechnology 18, 315603.

Bogolyubov, N. N. & Mitropolsky, Yu. A. (1974). *Asymptotic methods in theory of nonlinear oscillations*, (in Russian) (Moscow, Nauka).

Bristow, A. D., Whittaker, D. M., Astratov, V. N., Skolnick, M. S., Tahraoui, A., Krauss, T. F., Hopkinson, M., Croucher, M. P., and Gehring, G. A. (2003). "Defect states and commensurability in dual-period AlxGa1-xAs photonic crystal waveguides," Phys. Rev. B 68, 033303.

Chutinan, A., and Noda, S. (2000). "Waveguides and waveguide bends in two-dimensional photonic crystal slabs," Phys. Rev. B 62, 4488-4492.

Cho, C. O., Jeong, J., Lee, J., Kim, I., Jang, D. H., Park, Y. S., and Woo, J. C. (2005). "Photonic crystal band edge laser array with a holographically generated square-lattice pattern," Appl. Phys. Lett. 87, 161102.

Cho, C. O., Lee, J., Park, Y., Roh, Y.-G., Jeon, H., and Kim, I. (2007). "Photonic Crystal Cavity Lasers Patterned by a Combination of Holography and Photolithography," IEEE Photonics Tech. Lett. 19, 556-558.

de Sterke, C. M. (1998). "Superstructure gratings in the tight-binding approximation," Phys. Rev. E 57, 3502-3509.

Dorado, L. A., Depine, R. A., and Miguez, H. (2007). "Effect of extinction on the high-energy optical response of photonic crystals ," Phys. Rev. B. 75, 241101.

Fleischer, J. W., Carmon, T., Segev, M., Efremidis, N. K., and Christodoulides, D. N. (2003). "Observation of Discrete Solitons in Optically Induced Real Time Waveguide Arrays," Phys. Rev. Lett. 90, 023902.

Efremidis, N. K., Sears, S., Christodoulides, D. N., Fleischer, J. W., and Segev, M. (2002). "Discrete solitons in photorefractive optically induced photonic lattices," Phys. Rev. E 66, 046602.

Galisteo-López, J. F. & López, C. (2004). "High-energy optical response of artificial opals," Phys. Rev. B 70, 035108.

Happ, T. D., Kamp, M., Forchel, A. Gentner, J.-L., and Goldstein, L. (2003). "Two-dimensional photonic crystal coupled-defect laser diode," Appl. Phys. Lett. 82, 4.

Herrera, M., and Yamilov, A. (2009). "Trench waveguide in photonic crystal slab structures," J. Opt. Soc. Am. B (submitted).

Hofstadter, D. R. (1976). "Energy levels and wave functions of Bloch electrons in rational and irrational magnetic fields," Phys. Rev. B 14, 2239.

Jacobsen, R. S., Andersen, K. N., Borel, P. I., Fage-Pedersen, J., Frandsen, L. H., Hansen, O., Kristensen, M., Lavrinenko, A. V., Moulin, G., Ou, H., Peucheret, C., Zsigri B., and Bjarklev, A. (2006). "Strained silicon as a new electro-optic material," Nature 441, 199-202.

Janner, D., Galzerano, G., Della Valle, G., Laporta, P., Longhi, S., and Belmonte, M. (2005). "Slow light in periodic superstructure Bragg grating," Phys. Rev. E 72, 056605.

Joannopoulos, J.D., Johnson, S. G., Winn, J.N., and Meade, R. D., (2008). *Photonic Crystals: Molding the Flow of Light,* (2nd Ed., Princeton University Press, Princeton, NJ)

Johnson, S.G., Fan, S., Villeneuve, P.R., Joannopoulos, J.D., and Kolodziejski, L.A. (1999). "Guided modes in photonic crystal slabs," Phys. Rev. B 60, 5751-5758.

Johnson, S.G., Villeneuve, P.R., Fan, S., and Joannopoulos, J.D. (2000). "Linear waveguides in photonic-crystal slabs," Phys. Rev. B 62, 008212.

Johnson, S.G., and Joannopoulos, J.D. (2001). "Block-iterative frequency-domain methods for Maxwell's equations in a planewave basis," Optics Express 8, 173-190.

Karle, T.J., Brown, D.H., Wilson, R., Steer, M., and Krauss, T.E. (2002). "Planar photonic crystal coupled cavity waveguides," IEEE J. of Selected Topics in Quantum Electr., 8, 909 - 918.

Kitahara, H., Kawaguchi, T., Miyashita, J., Shimada, R., and Takeda, M. W. (2004). "Strongly Localized Singular Bloch Modes Created in Dual-Periodic Microstrip Lines," J. Phys. Soc. Jap. 73, 296-299.

Krauss, T. F. (2003). "Planar photonic crystal waveguide devices for integrated optics," Phys. Stat. Sol. (a) 197, 688702.

Landa, P. S. (2001). *Regular and Chaotic Oscillations,* (Springer, Berlin).

Lifshitz, I. M., Gredeskul, S. A., and Pastur, L. A. (1988). *Introduction to the Theory of Disordered Systems,* (Wiley, New York).

Liu, Z.-W., Du, Y., Liao, J., Zhu, S.-N., Zhu, Y.-Y., Qin, Y.-Q., Wang, H.-T., He, J.-L., Zhang, C., and Ming, N.-B., (2002). "Engineering of a dual-periodic optical superlattice used in a coupled optical parametric interaction," J. Opt. Soc. Am. B 19, 1676-1684.

Lončar, M., Doll, T., Vuckovic, J., and Scherer, A. (2000). "Design and fabrication of silicon photonic crystal optical waveguides," J. of Lightwave Technology 18, 1402-1411.

Lončar, M., Yoshie, T., Scherer, A., Gogna, P., and Qiu, Y. (2002). "Low-threshold photonic crystal laser," Appl. Phys. Lett. 81, 2680.

Marcuse, D. (1991). *Theory of Dielectric Optical Waveguides*, 2nd ed. (Academic, San Diego).

Milonni, P. W. (2005). *Fast light, slow light and left-handed light*, (Institute of Physics, Bristol).

Neff, C. W., Yamashita, T., and Summers, C. J. (2007). "Observation of Brillouin zone folding in photonic crystal slab waveguides possessing a superlattice pattern," Appl. Phys. Lett. 90, 021102.

Neshev, D., Ostrovskaya, E., Kivshar, Y., and Krolikovwski, W. (2003). "Spatial solitons in optically induced gratings," Opt. Lett. 28, 710-712.

Nojima, S. (1998). "Enhancement of optical gain in two-dimensional photonic crystals with active lattice points," Jpn. J. Appl. Phys. Part 2 37, L565-L567.

Olivier, S., Smith, C., Rattier, M., Benisty, H., Weisbuch, C., Krauss, T., Houdr´e, R., and Oesterl´e, U. (2001). "Miniband transmission in a photonic crystal coupled-resonator optical waveguide," Opt. Lett. 26, 1019-1021.

Painter, O., Lee, R. K., Scherer, A., Yariv, A., O'Brien, J. D., Dapkus, P. D., and Kim, I. (1999). "Two-Dimensional Photonic Band-Gap Defect Mode Laser," Science 284, 1819.

Park, H., Kim, S., Kwon, S., Ju, Y., Yang, J., Baek, J., Kim, S., and Lee, Y. (2004). "Electrically Driven Single-Cell Photonic Crystal Laser," Science 305, 1444.

Poon, J. K. S., Scheuer, J., Xu, Y., and Yariv, A. (2004). "Designing coupled-resonator optical waveguide delay lines," J. Opt. Soc. Am. B 21, 1665-1673.

Poon, J. K. S., Zhu, L., DeRose, G. A., and Yariv, A. (2006). "Polymer Microring Coupled-Resonator Optical Waveguides," J. Lightwave Technol. 24, 1843-1849.

Povinelli, M. L. and Fan, S. (2006). "Radiation loss of coupled-resonator waveguides in photonic-crystal slabs," Appl. Phys. Lett. 89, 191114.

Qiu, M. (2002). "Effective index method for heterostructure-slab-waveguide-based twodimensional photonic crystals," Appl. Phys. Lett. 81, 1163.

Sakoda, K. (1999). "Enhanced light amplification due to group-velocity anomaly peculiar to two- and three-dimensional photonic crystals," Opt. Express 4, 167-176.

Sanchis, P., Marti, J., Bogaerts, W., Dumon, P., Van Thourhout, D., and Baets, R. (2005). "Experimental results on adiabatic coupling into SOI photonic Crystal coupled-cavity waveguides," IEEE Photonics Tech. Lett. 17, 1199 - 1201.

Scharrer, M., Yamilov, A., Wu, X., Cao, H., and Chang, R. P. H. (2006). "Ultraviolet lasing in high-order bands of three-dimensional ZnO photonic crystals," Appl. Phys. Lett. 88, 201103.

Scheuer, J., Paloczi, G., Poon, J., and Yariv, A. (2005). "Toward the Slowing and Storage of Light," Opt. and Phot. News 16, 36.

Shimada, R., Koda, T., Ueta, T., and Ohtaka, K. (1998). "Energy Spectra in Dual-Periodic Multilayer Structures," J. Phys. Soc. Jap. 67, 3414-3419.

Shimada, R., Koda, T., Ueta, T., and Ohtaka, K. (2001). "Strong localization of Bloch photons in dual-periodic dielectric multilayer structures," J. Appl. Phys. 90, 3905-3909.

Sipe, J. E., Poladian, L., and de Sterke, C. M. (1994). "Propagation through noniniform grating structures," J. Opt. Soc. Am. A 11, 1307-1320.

Soljacic, M., Johnson, S. G., Fan, S., Ibanescu, M., Ippen, E., and Joannopoulos, J. D. (2002). "Photonic-crystal slow-light enhancement of nonlinear phase sensitivity," J. Opt. Soc. Am. B 19, 2052-2059.

Song, B.-S., Noda, S., Asano, T., and Akahane, Y. (2005). "Ultra-high-Q photonic double-heterostructure nanocavity," Nat. Mater. 4, 207-210.

Soukoulis, C. M., ed. (1996). *Photonic band gap materials*, (Kluwer, Dordrecht).

Stefanou, N. & Modinos, A. (1998). "Impurity bands in photonic insulators," Phys. Rev. B 57, 12127-12133.

Susa, N. (2001). "Threshold gain and gain-enhancement due to distributed-feedback in two-dimensional photonic-crystal lasers," J. Appl. Phys. 89, 815-823.

Vlasov, Y. A., Moll, N., and McNab, S. J. (2004). "Mode mixing in asymmetric double-trench photonic crystal waveguides," J. Appl. Phys. 95, 4538-4544.

Vlasov, Yu. A., O'Boyle, M., Hamann H. F., and McNab, S. J. (2005). "Active control of slow light on a chip with photonic crystal waveguides," Nature 438, 65-69.

Xu, Y., Lee, R. K., and Yariv, A., (2000). "Propagation and second-harmonic generation of electromagnetic waves in a coupled-resonator optical waveguide," J. Opt. Soc. Am. B 17, 387-400.

Yablonovitch, E. (1987). "Inhibited Spontaneous Emission in Solid-State Physics and Electronics," Phys. Rev. Lett. 58, 2059-2062.

Yagasaki, K., Merhasin, I.M., Malomed, B. A.,Wagenknecht, T., and Champneys, A. R. (2006). "Gap solitons in Bragg gratings with a harmonic superlattice," Europhys. Lett. 74, 1006-1012.

Yamilov, A., Wu, X., Liu, X., Chang, R. P. H., and Cao, H. (2006). "Self-optimization of optical confinement in ultra-violet photonic crystal slab laser," Phys. Rev. Lett. 96, 083905.

Yamilov A., & Bertino, M. (2007). "Disorder-immune coupled resonator optical waveguide," Opt. Lett. 32, 283-285.

Yamilov, A., Herrera, M. R., and Bertino, M. F. (2008). "Slow light effect in dual-periodic photonic lattice," J. Opt. Soc. Am. B 25, 599-608.

Yanik, M. F., and Fan, S. (2004). "Stopping Light All Optically," Phys. Rev. Lett. 92, 083901.

Yariv, A., Xu, Y., Lee, R. K., and Scherer, A. (1999). "Coupled-resonator optical waveguide: a proposal and analysis," Opt. Lett. 24, 711-713.

Yeh, P. (2005). "Optical waves in layered media," (Wiley, Hoboken).

2

Two-Dimensional Photonic Crystal Micro-cavities for Chip-scale Laser Applications

Adam Mock[1] and Ling Lu[2]
[1]*Central Michigan University*
[2]*University of Southern California*
USA

1. Introduction

1.1 Photonic crystals

In this chapter we will focus on two-dimensional photonic crystal devices and emphasize their use as building blocks in photonic integrated circuits with applications in high bandwidth optical communication systems. In particular we will discuss recent progress in designing high quality (Q) factor resonant cavities for building efficient micro- and nano-cavity lasers. The first section will provide a brief overview of two-dimensional photonic crystals and motivate their use in photonic integrated circuits. This will be followed by a first principles derivation of the role of the Q factor in estimating laser threshold. We will then focus on the photonic crystal heterostructure cavity due to its exceptionally large Q factor. Its spectral and modal properties will be discussed, and its use as a high output power edge-emitting laser will be presented. We conclude with remarks on continuous wave laser operation via heat sinking lower substrates and the issue of out-of-plane loss.

The term photonic crystal refers to any structure with a periodic variation in its refractive index (John, 1987; Yablonovitch et al., 1991; Joannopoulos et al., 1995). The periodicity can be in one, two or three spatial dimensions and can introduce a photonic bandgap (a range of frequencies for which electromagnetic radiation is non-propagating) with the same dimensionality. The bandgap arises due to Bragg reflection and occurs when the spatial periodicity has a length scale approximately one half that of the incident electromagnetic radiation. This same phenomenon gives rise to the electronic bandgap in semiconducting materials. Examples of photonic crystal structures with periodicity in varying spatial dimensions are shown in Figure 1. One dimensional photonic crystals have found many technology applications in the form of Bragg reflectors which are part of the optical feedback mechanism in distributed feedback lasers (Kogelnik & Shank, 1971; Nakamura et al., 1973) and vertical cavity surface emitting lasers (Soda et al., 1979). Two and three dimensional photonic crystals have been the subject of intense research recently in areas related to sensing (Lončar et al., 2003; Chow et al., 2004; Smith et al., 2007), telecommunications (Noda et al., 2000; McNab et al., 2003; Bogaerts et al., 2004; Notomi et al., 2004; Noda et al., 2000; Jiang et al., 2005; Aoki et al., 2008), slow light (Vlasov et al., 2005; Krauss, 2007; Baba & Mori, 2007; Baba, 2008) and quantum optics (Yoshie et al., 2004; Lodahl et al., 2004; Englund et al., 2005).

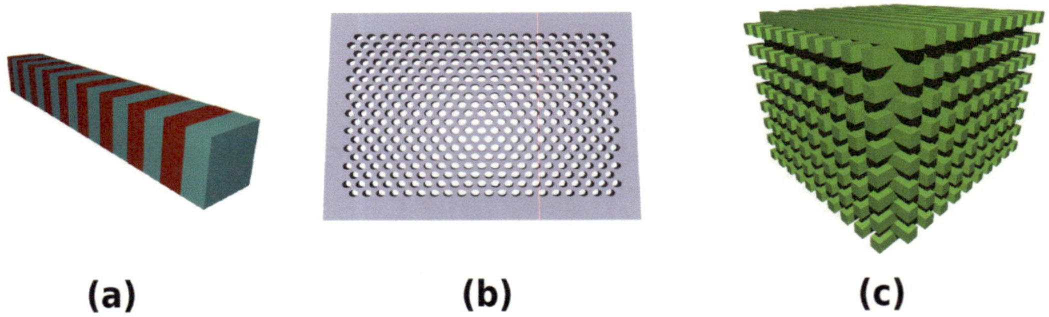

Fig. 1. Images depicting photonic crystals with periodicity in (a) one dimension, (b) two dimensions and (c) three dimensions.

Figure 1(b) displays a semiconductor slab perforated with a two-dimensional triangular array of air holes. Because of the periodic refractive index, the in-plane propagating modes of the slab can be characterized using Bloch's theorem. In the vertical, out-of-plane direction, the modes are confined via index guiding, and Figure 2(a) illustrates typical guided and radiation modes. These modes are peaked near the center of the slab and are either evanescent (guided) or propagating (radiation) out-of-plane. Figure 2(b) is a photonic band diagram corresponding to a photonic crystal structure similar to that shown in Figure 1(b) and Figure 2(a). The left vertical axis is written in terms of normalized frequency where a corresponds to the lattice constant of the photonic crystal, and c is the vacuum speed of light. The right vertical axis is denormalized and written in terms of free space wavelength using a lattice constant of a = 400nm. The photonic bandgap corresponds to the normalized frequency range 0.25-0.32 where there are no propagating modes in this structure. Using a lattice constant of a = 400nm places the near infrared fiber optic communication wavelengths of 1.3μm (low-dispersion) and 1.5μm (low-loss) within the bandgap making this geometry amenable to applications in fiber optic communication systems. The shaded regions on the left and right sides of Figure 2(b) represent the projection of the light cone onto the various propagation directions which is a result of the vertical confinement mechanism being due to index guiding. Photonic crystal modes that overlap the shaded regions in Figure 2(b) correspond to the radiation modes in Figure 2(a). Figure 2(b) shows the dispersion for the three lowest frequency bands with transverse electric polarization (out-of-plane magnetic field has even vertical symmetry). Figure 2(c) illustrates a unit cell corresponding to a triangular lattice photonic crystal. The band diagram in Figure 2(b) was calculated using the three-dimensional finite-difference time-domain method (Taflove & Hagness, 2000). The computational domain is similar to that shown in Figure 2(c) where the in-plane boundaries are terminated using Bloch boundary conditions. More details about this approach can be found in (Kuang et al., 2007). Photonic crystal geometries represent complicated electromagnetic problems and almost always demand a numerical approach for their analysis. Several numerical methods for solving Maxwell's equations exist. Some examples include the finite-element method (Kim 2004), transmission line method (Benson et al., 2005), scatterning based methods (Peterson et al., 1998; Sadiku 2000) and plane wave expansion methods for periodic structures (Joannopoulos et al., 1995; Sakoda 2001). In this work, we will be using the finite-difference time-domain method due to its generality, simplicity and linear scaling with problem size. For the band structure in Figure 2(b), the refractive index of the slab was set to n = 3.4, the hole radius to lattice constant ratio was set to r/a = 0.29 and the slab thickness to lattice constant ratio was set to d/a = 0.6. These photonic crystal geometry parameters will hold for the rest of the devices analyzed in this chapter.

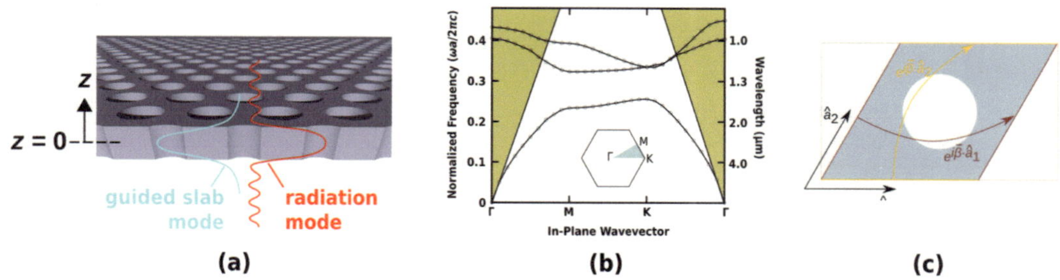

Fig. 2. (a) Cross section of a two-dimensional photonic crystal defined in a dielectric slab of finite thickness. The field distribution in the vertical direction for guided and radiation modes is shown. (b) Photonic band diagram for a two-dimensional photonic crystal defined in a single-mode slab. c denotes free space speed of light. a denotes the lattice cosntant. The diagram depicts the lowest three bands for the TE-like modes of the slab. The inset shows the region of the first Brillouin zone described by the dispersion diagram. (c) A unit cell of a triangular photonic crystal lattice and the phase relationships between the boundaries determined by Bloch's theorem.

1.2 Defects in two-dimensional photonic crystals

Much of the versatility and device applications of two-dimensional photonic crystal structures are associated with the introduction of defects into the periodic lattice. Figure 3(a) displays the out-of-plane component of the magnetic field of a typical mode associated with a photonic crystal waveguide formed by removing a single row of holes along the $\Gamma - K$ direction. For the TE-like modes of the slab, only the E_x, E_y and H_z fields are nonzero at the midplane, and H_z is displayed due to its scalar nature. It is clear that the mode is localized to the defect region along the y-direction due to the photonic crystal bandgap, and confinement along the z-direction is due to index guiding as discussed with regard to Figure 2(a). Figure 3(b) displays the unit cell used in the computation of the field shown in Figure 3(a). The finite-difference time-domain method was used with Bloch boundary conditions along the x-direction (Kuang et al., 2006). Figures 6(b) and 7 depict photonic crystal waveguide dispersion diagrams. The mode depicted in Figure 3(a) is associated with the lowest frequency band in the bandgap and a propagation constant of βa = 1.9. It has been shown that photonic crystal waveguides are capable of low loss optical guiding (McNab et al., 2003) and have the ability to redirect light along different directions in-plane with low loss waveguide bends (Shih et al., 2004).

Figure 3(c) displays the z-component of the magnetic field corresponding to a typical resonant mode of an L3 cavity (Akahane et al., 2003, 2005). The L3 cavity is formed by removing three adjacent holes along the $\Gamma - K$ direction in a triangular photonic crystal lattice. The two dimensional in-plane confinement due to the photonic crystal bandgap is apparent. In the case of photonic crystal defect cavities, a single unit cell with Bloch boundary conditions is no longer applicable, and large three dimensional computational domains must be analyzed. Such a cavity can be used as an optical filter, an optical buffer or a laser.

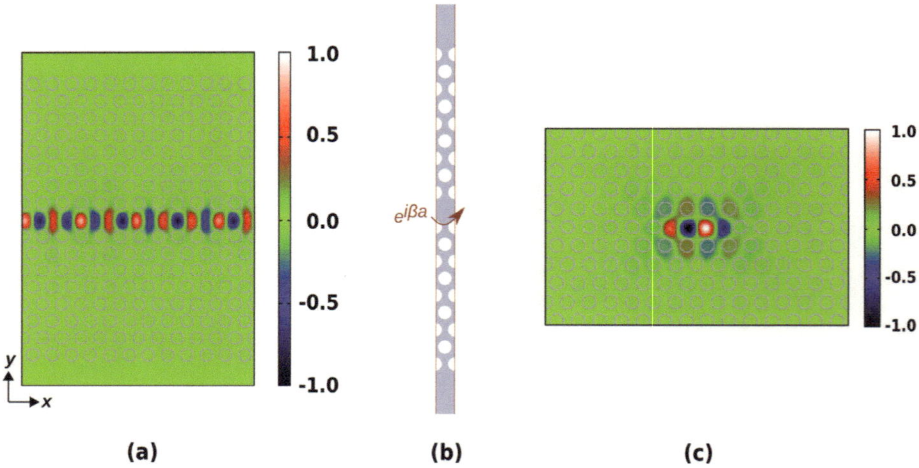

Fig. 3. (a) $H_z(x,y, z = 0)$ for a typical photonic crystal waveguide mode. $z = 0$ corresponds to the midplane of the slab as shown in Figure 2(a). (b) The unit cell of a photonic crystal waveguide and the phase relationship between the boundaries determined by Bloch's theorem. (c) $H_z(x,y, z = 0)$ for a typical L3 cavity mode.

1.3 Photonic integrated circuits

Photonic integratration is analogous to the integration of electronic devices on a silicon chip. Typical microprocessors contain on the order of 10^9 transistors in an area on the order of $1cm^2$. Such dense device integration has resulted in microprocessors with exceptional functionality. And because the devices share a common substrate and metal wiring network, they can be mass produced with limited overhead costs.

Similar to electronic integrated circuits, photonic integrated circuits are useful for any application in which a large number of devices need to be contained in a confined space. Photonic integrated circuits have a variety of applications including telecommunications, sensing and imaging. In telecommunication systems photonic integrated circuits have the potential for lower cost systems due to reduced packaging costs, improved reliability due to reduced alignment errors and improved bandwidth through all optical signal processing. Another application of photonic integrated circuits is in optical buses in multicore computer architectures. The inter-core communication and off-chip memory access can be a performance bottleneck for applications with heavy memory access. Optics has the potential to improve memory access bandwidth due in part to its ability to transmit signals at multiple wavelengths through a single waveguide. It also has the advantage of operating at a lower temperature due to the absence of resistive heating.

Figure 4 shows a schematic diagram of a photonic crystal based photonic integrated circuit that includes sources, modulators, filters and detectors integrated on a single chip. This particular structure consists of a bus waveguide passing from left to right carrying modulated optical signals at wavelengths λ_1 and λ_2. First, the signals encounter frequency selective filters which couple the filtered signal to an optical detector. On-chip lasers operating at λ_1 and λ_2 generate a new carrier beam which is modulated and rerouted to the bus waveguide. The input and output ports could lead to other on-chip processing or to coupled optical fibers. From this simple example, it is clear that photonic crystals offer a versatile platform for realizing a variety of different devices by local rearrangements of the

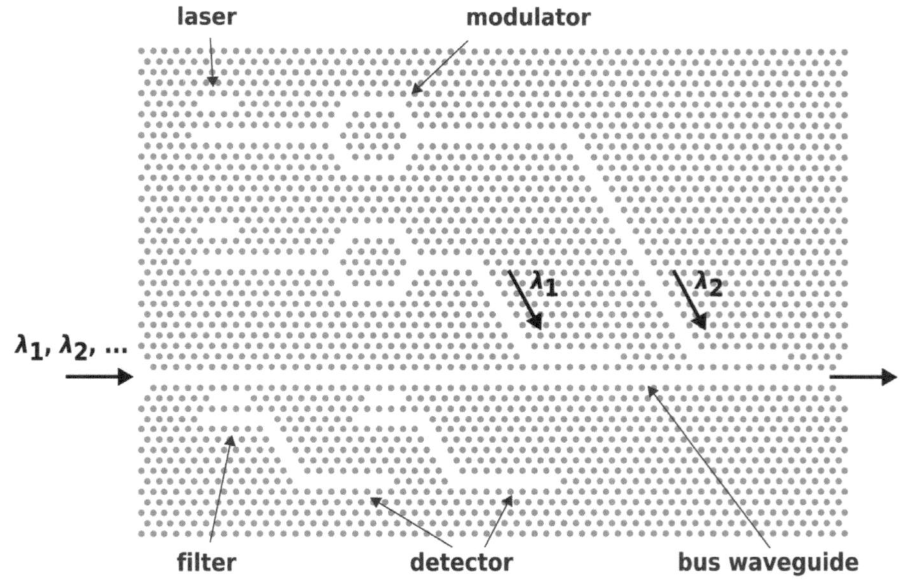

Fig. 4. Schematic diagram depicting a simple photonic crystal based photonic integrated circuit.

hole pattern. It is this flexibility along with their dispersive properties and high Q factor cavities that make photonic crystals an interesting candidate for photonic integration. Furthermore, devices based on photonic crystals have the potential to occupy smaller spaces than devices based on index guiding. In the remainder of this chapter we will focus our attention to aspects of designing photonic crystal cavities for on-chip laser sources.

2. High quality factor photonic crystal cavities

2.1 Role of the quality factor in determining laser threshold

The Q factor is a figure of merit used to quantify the radiation losses of an optical resonator. Formally it is defined by

$$Q = \omega_0 \frac{\langle U \rangle}{-\langle \frac{dU}{dt} \rangle}. \tag{1}$$

where ω_0 is the resonance frequency, and the angled brackets denote a time-average over an integer number of optical periods. U represents the electromagnetic energy and is given by

$$U = \int [\frac{1}{2} \epsilon \vec{E} \cdot \vec{E} + \frac{1}{2} \mu \vec{H} \cdot \vec{H}] dV. \tag{2}$$

where ϵ represents the electric permittivity, $\vec{E}$ represents the electric field, μ represents the magnetic permeability and $\vec{H}$ represents the magnetic field. Equation 1 can be considered a first order ordinary differential equation in $\langle U \rangle$. Its solution is

$$\langle U(t) \rangle = \langle U(0) \rangle \exp(-\omega_0 t / Q). \tag{3}$$

Equation 3 tells us that the temporal decay of the electromagnetic energy stored in a large Q factor cavity will be slower than that of a low Q factor cavity. From Eq. 3 a photon lifetime can be defined as $\tau_p = Q/\omega_0$. This quantity is related to the average length of time that a photon spends inside a cavity. It is clear that a large Q factor results in a long photon lifetime.

In order to illustrate the precise role that the Q factor plays in designing efficient, low-threshold lasers, we will derive the laser threshold condition applicable to photonic crystal resonant cavities made from semiconductor active material (Mock & O'Brien, 2009a). Equation 4 is the Poynting theorem in its time averaged form and is a statement of electromagnetic energy conservation. $\vec{S}$ is the Poynting vector and is given by $\vec{S} = \vec{E} \times \vec{H}$. Its closed surface integral represents the power radiated through the surface. P_a represents absorbed power which in the case of a semiconductor active material would occur in regions of the structure in which the carrier population is not inverted. P_s represents supplied power coming in the form of optical gain resulting from an external energy source.

$$\oint \langle \vec{S} \rangle \cdot d\vec{A} = -\langle \frac{\partial U}{\partial t} \rangle - \langle P_a \rangle + \langle P_s \rangle \tag{4}$$

If one considers a passive cavity in which $P_s = P_a = 0$ and substitutes dU/dt in Equation 1 into Equation 4 one gets

$$\oint \langle \vec{S}_m \rangle \cdot d\vec{A} = \omega_0 \frac{\langle U_m \rangle}{Q_p}. \tag{5}$$

where the subscript m is used to specify that the Poynting vector and electromagnetic energy correspond to a specific mode m whose passive Q factor is given by Q_p. The basic idea behind laser threshold is that the optical loss mechanisms should be exactly compensated by an optical power source which comes in the form of optical gain (Schawlow & Townes, 1958). At threshold when the loss just equals the gain, the temporal rate of change of the energy in the cavity is zero, and we can set $dU/dt = 0$ in Equation 4. If we then use Equation 5 in Equation 4, the result is

$$\omega_0 \frac{\langle U_m \rangle}{Q_p} + \langle P_a \rangle = \langle P_s \rangle. \tag{6}$$

Equation 6 is the laser threshold condition. The first term on the left side represents passive cavity (radiative) losses. The second term represents active cavity (absorptive) losses. The right side represents the supplied power required to offset the optical losses. From the first term in Equation 6, it is apparent that the passive Q factor should be as large as possible so as to reduce the radiative losses and thus the power required to reach threshold. It should be noted, however, that a high Q factor cavity often results in reduced output power, and tradeoffs between low threshold and sufficient output power should be considered when designing a prospective cavity for chip-scale laser applications.

2.2 Two-dimensional photonic crystal cavities

Figure 5 displays four cavity designs as well as the evolution of their Q factors over the passed decade. Early photonic crystal cavities were formed by removing a single hole from a uniform lattice (Painter et al., 1999; Ryu et al., 2002). More recently, linear defects have been shown to have higher Q factors than single missing hole cavities (Akahane et al., 2003, 2005), and the photonic crystal double heterostructure cavity has been shown to have the largest Q factor among two-dimensional photonic crystal cavities (Song et al., 2005; Tanaka et al., 2008). Because of its exceptionally high Q factor and small mode volume, the photonic crystal double heterostructure has been the subject of intense research for building efficient chip scale optical sources and will be highlighted in what follows.

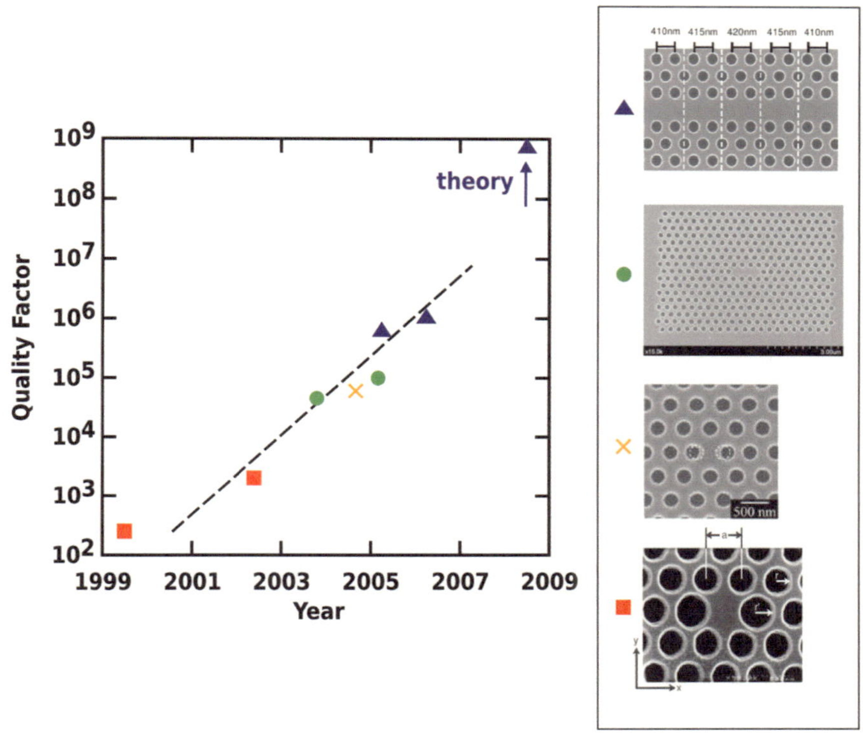

Fig. 5. Quality factor for different two-dimensional photonic crystal cavities as a function of time. Data points and figures come from (Painter et al., 1999; Ryu et al., 2002; Akahane et al., 2003; Zhang & Qiu, 2004; Nozaki & Baba, 2006; Akahane et al., 2005; Song et al., 2005; Asano et al., 2006; Tanaka et al., 2008)

3. Photonic crystal double heterostructure resonant cavities

3.1 Introduction

In 2005 Song et al. showed that a two-dimensional photonic crystal waveguide with a small, localized perturbation can form an ultra high Q factor cavity with a Q factor greater than 10^5 and a mode volume on the order of one cubic wavelength (Song et al., 2005, 2007). Such a cavity was termed a photonic crystal double heterostructure and is depicted in Figure 6. The cavity is formed from an otherwise uniform photonic crystal waveguide by enlarging the lattice constant along the x-direction of the light colored air holes. Below the schematic diagram, the resulting photonic band structure is shown illustrating the formation of a photonic well along the x-direction. Further experimental work on these cavities has demonstrated devices with passive Q factors as large as 10^6 (Asano et al., 2006). Since these initial reports, several groups have reported forming high Q factor photonic crystal double heterostructure cavities through a variety of methods including local modulation of a photonic crystal line defect width (Kuramochi et al., 2006), local air-hole infiltration (Smith et al., 2007), photosensitive materials (Tomljenovic-Hanic et al., 2007), effective index change through micro-fiber coupling (Kim et al., 2007) and local modulation of the hole radii (Kwon et al., 2008b). A numerical analysis showed that Q factors as high as 10^9 are possible with a tapered perturbation (Tanaka et al., 2008). The ultra high Q factors and cubic wavelength mode volumes along with the waveguide-like shape of the cavities have made them attractive for a variety of applications including chemical sensing (Kwon et al., 2008a), slow light (Tanabe et al., 2007; Takahashi et al., 2007), elements of coupled resonator optical waveguides (O'Brien et al., 2007) and edge-emitting lasers (Shih, Kuang, Mock, Bagheri, Hwang, O'Brien & Dapkus, 2006; Shih, Mock, Hwang, Kuang, O'Brien & Dapkus, 2006; Yang et al., 2007; Lu et al., 2007, 2008; Lu, Mock, Shih, Hwang, Bagheri, Stapleton, Farrell, O'Brien & Dapkus, 2009).

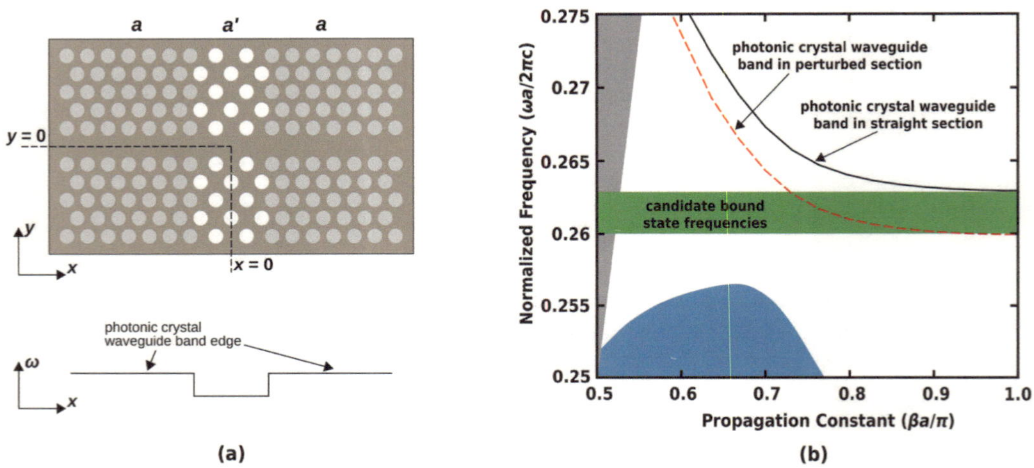

Fig. 6. (a) Schematic diagram of a photonic crystal double heterostructure resonant cavity formed in a uniform single line defect waveguide by increasing the lattice constant of the light colored holes along the x-direction. The resulting photonic well diagram is illustrated below. (b) Photonic crystal waveguide dispersion diagram depicting the photonic crystal waveguide bands associated with the straight (black, solid) and perturbed (red, dashed) portions of the waveguide. The waveguide frequencies of the perturbed section that fall into the mode gap of the straight waveguide are labeled "candidate bound state frequencies." The blue region denotes the photonic crystal cladding modes, and the gray region denotes the light cone.

3.2 Spectral and modal properties

When the lattice constant is locally increased, it shifts the frequencies of the waveguide band associated with the perturbed region to lower frequencies as shown in Figure 6(b). The bound state will oscillate near frequencies of the perturbed waveguide section that fall into the mode gap of the uniform waveguide sections. Candidate frequencies for bound state resonances are labeled in Figure 6(b). Only below the minima of the dispersion relation in the uniform photonic crystal waveguide regions is there a possibility for a mode to exist in the central region without the possibility of there simultaneously being a mode in the cladding at the same frequency a small distance in wavevector away. In other words, only in these cases is there no mode that is nearby (in the wavevector sense) in the cladding at the same frequency (Mock et al., 2006, 2008). This mode formation is analogous to the formation of bound states in electronic heterostructures at the extrema of the electronic dispersion relations.

Figure 7 is a comparison between the spectral features of a photonic crystal double heterostructure resonance spectrum and the frequency axis of the photonic crystal waveguide dispersion diagram corresponding to the underlying straight waveguide. The red dotted lines illustrate that the bound state resonance frequencies occur just below the waveguide dispersion minima. The resonance spectrum was obtained by taking a discrete Frourier transformation of a 2×10^5 element time sequence. The time sequence was calculated via the three dimensional finite-difference time-domain method. The computational domain included 20 uniform photonic crystal cladding periods on either side of the central defect region along the x-direction and 8 photonic crystal layers above and below the central waveguide core along the y-direction. This geometry was discretized with 950 × 340 × 200 discretization points along the $x \times y \times z$ directions and parallelized using 11 × 4 × 3 processors (132 total processors) along the $x \times y \times z$ directions. The geometry was discretized using 20 points per lattice constant (a). The lattice constant of the perturbed region was increased by 5% along the x-direction.

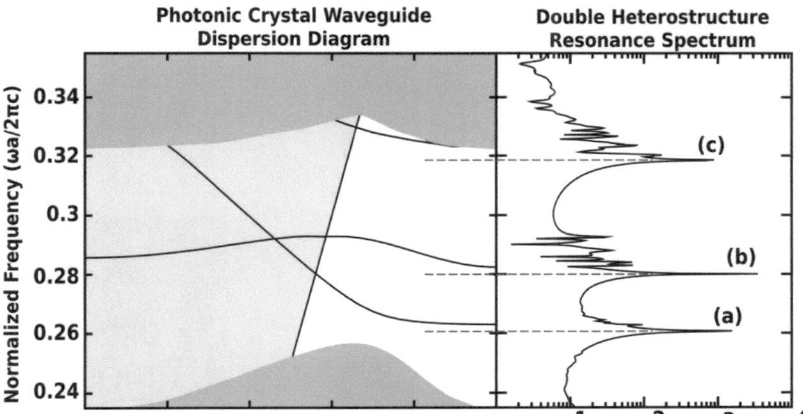

Fig. 7. Left: photonic crystal waveguide dispersion diagram. Black lines correspond to the photonic crystal waveguide dispersion bands. Blue regions denote photonic crystal cladding modes. The gray region denotes the light cone projection. Right: photonic crystal double heterostructure resonant spectrum. Dashed lines illustrate correspondance between heterostructure bound state frequencies and waveguide dispersion extrema.

Figure 8 depicts the z-component of the magnetic field for the bound state resonances labeled in Figure 7. These mode profiles may be interpreted as consisting of the waveguide mode of the underlying straight waveguide multiplied by a confining envelope function centered at the perturbation. It is interesting to point out that mode (c) exhibits significant extension into the photonic crystal cladding. This can be attributed to the close proximity of the corresponding photonic crystal waveguide band to the photonic crystal cladding modes in Figure 7. To the right of each $H_z(x,y,\, z = 0)$ mode profile is the corresponding spatial Fourier transform. Specifically, $\log(|FT(E_x)|^2 + |FT(E_y)|^2)$ is plotted where FT stands for Fourier transform. The two-dimensional spatial Fourier transform yields the spatial wavevector components that make up the bound state resonance. The spatial wavevector distributions are centered at $\beta_x = \pm\pi/a$. This is consistent with the observation that bound state resonance frequencies occur near the waveguide dispersion minima which coincide with the Brillouin zone boundary at $\beta_x = \pm\pi/a$ for this particular waveguide.

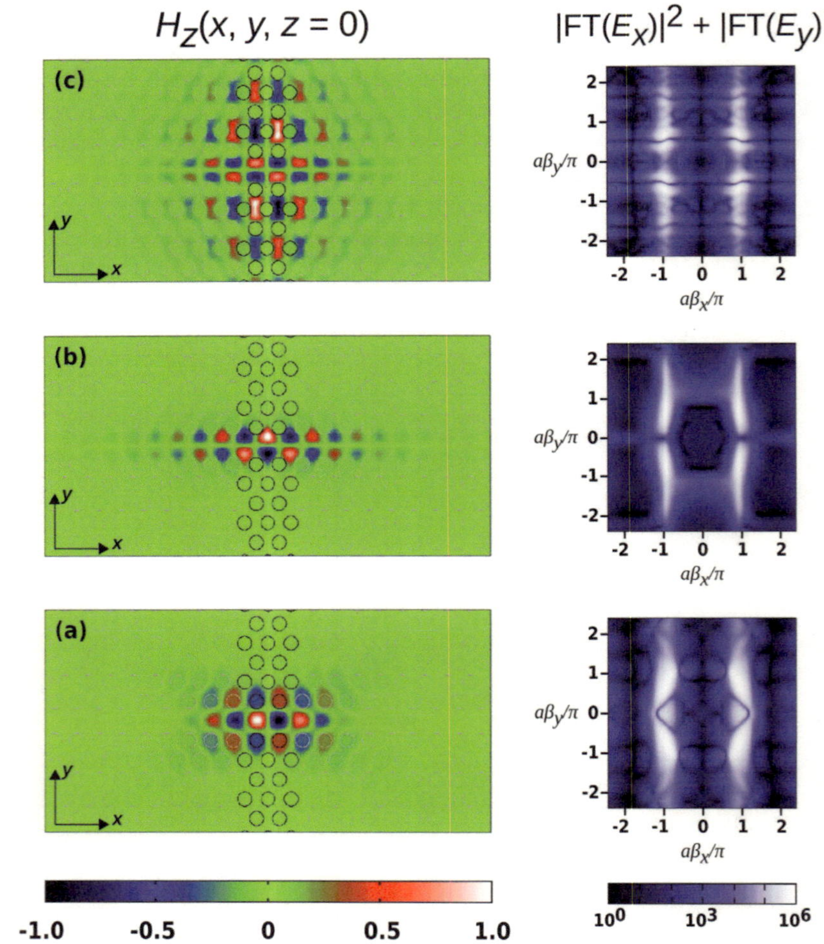

Fig. 8. Left: $H_z(x, y, z = 0)$ for the three modes summarized in Table 1. Right: spatial Fourier transform $\log(|FT(E_x)|^2 + |FT(E_y)|^2)$ of each mode illustrating the distribution of spatial wavevectors making up the different resonant modes.

Table 1 summarizes the normalized resonance frequencies of the three modes shown in Figure 8. The resonance frequencies and Q factors were obtained using the Padé interpolation method (Mock & O'Brien, 2008). It is clear that the bound state resonance frequencies fall just below the photonic crystal waveguide band edge. Mode (a) has the largest Q factor and will be featured in the remainder of this chapter.

Mode	Resonance Frequency	Bandedge Frequency	Q factor
(a)	0.2606	0.2629	336,700
(b)	0.2800	0.2824	10,800
(c)	0.3184	0.3227	8,250

Table 1. Summary of Q factors and resonant frequencies for the resonant modes associated with a photonic crystal heterostructure cavity.

3.3 Higher-order bound states

The previous section discussed a photonic crystal double heterostructure cavity resulting from a 5% lattice constant stretching along the x-direction. Figure 9 shows several interesting features of the high Q factor mode in Table 1 as the degree of perturbation is varied. First, the Q factor exhibits a strong dependence on the percent lattice constant increase. For very shallow perturbations (<3%), Q factors in excess of one million are predicted. Whereas for perturbations exceeding 20%, the Q factor dips below one thousand. Intuitively, one would expect that by increasing the lattice constant perturbation, the photonic well is deepened which would lead to improved confinement. It turns out that deepening the well makes the transition between the straight waveguide and the perturbation region more abrupt and introduces high spatial frequencies into the envelope function of the mode along the x-direction (Akahane et al., 2003). Because these modes have Fourier space distributions centered near $\beta_x = \pi/a$, large spatial frequencies in the envelope function get shifted to regions in Fourier space near $\beta_x = 0$. Fourier components inside the light cone centered at the origin in Fourier space radiate out-of-plane, and this loss mechanism dominates the overall loss properties of the mode. Researchers have investigated designs that smoothen the transition between the straight waveguide regions and the perturbation region and have obtained improved Q factors as a result (Akahane et al., 2005; Song et al., 2005).

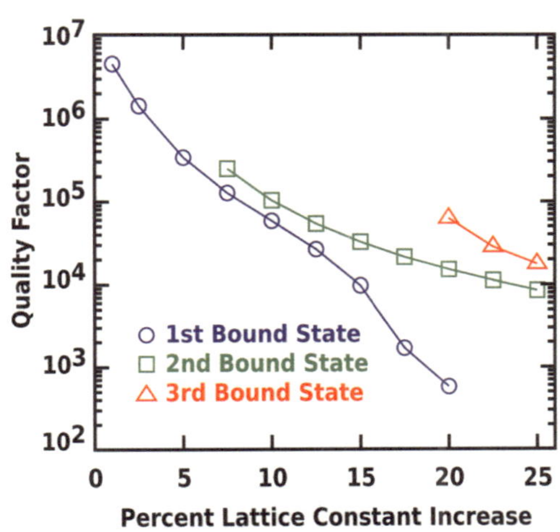

Fig. 9. Q factor versus perturbation depth for the first, second and third order bound states.

The second interesting feature of Figure 9 is the presence of higher order bound states. For perturbations greater than 7.5%, the cavity supports both a first order and a second order bound state. For perturbations greater than 20%, the cavity supports three bound states. The Q factors of the higher order bound state resonances exhibit a similar dependence on percent lattice constant increase as the first order bound state. It should also be pointed out that for a given perturbation, the highest order bound state has the largest Q factor. Figure 10 displays the z-component of the magnetic field of the first three bound states associated with a heterostructure cavity. Also shown are the envelope functions obtained by extracting

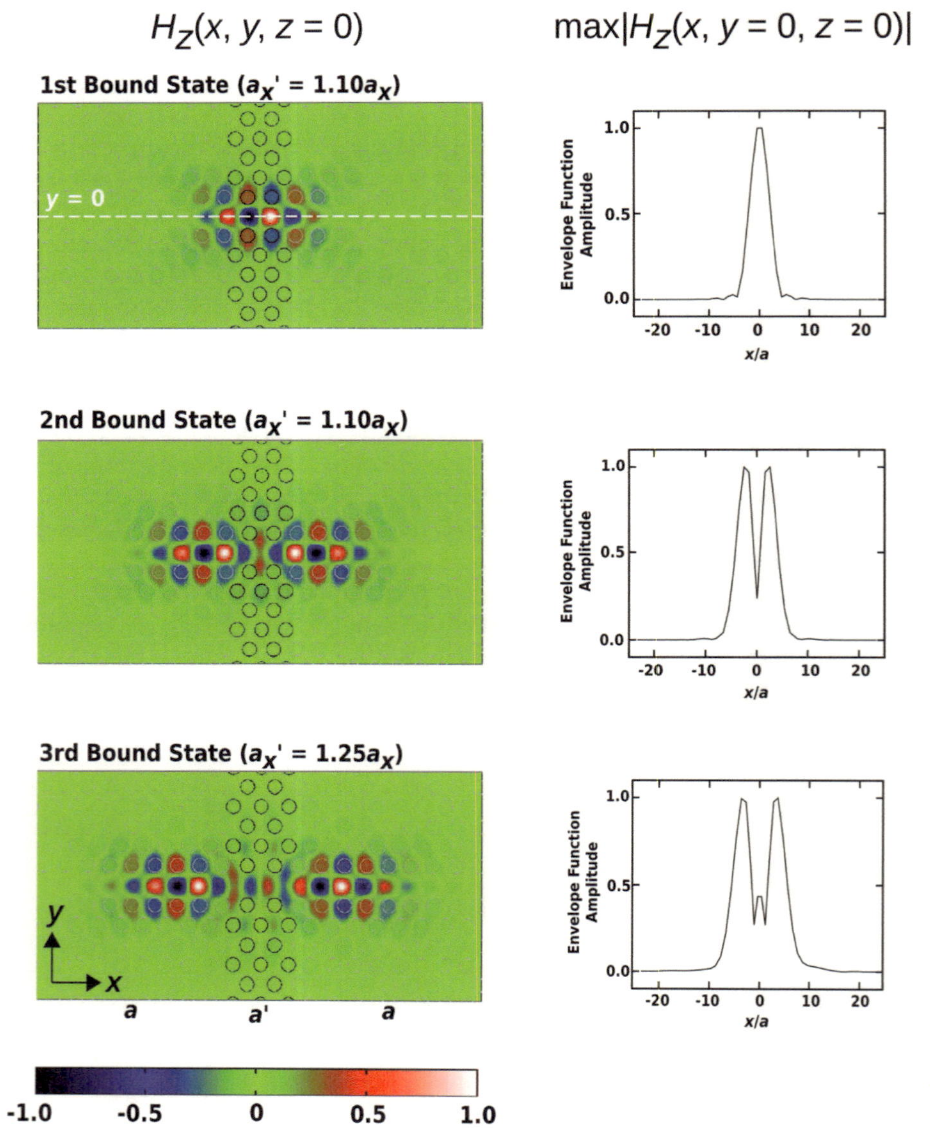

Fig. 10. Left: $H_z(x, y, z = 0)$ field distributions for the first, second and third order bound states. Right: field envelopes extracted from $|H_z(x, y = 0, z = 0)|$.

the maxima of $|H_z(x,y = 0, z = 0)|$ (the dotted white line in the top image of Figure 10 depicts the contour along which the maxima were measured). The envelope functions show a similar behavior to the wavefunction solutions corresponding to a quantum well problem in quantum mechanics.

From a technology perspective, if one chooses to work with a cavity that supports multiple bound states, then it is useful to be able to enhance one mode relative to the others. From Equation 6, one sees that the mode with the largest Q factor will be the first to reach threshold if the cavity is multimoded. However, from Figure 10, the various bound states have different spatial mode distributions and thus different overlap integrals with the spatial gain distribution. For instance, an optical pump beam directly centered on the heterostructure cavity will preferentially pump the first order bound state, and this mode could reach threshold first even though it has a smaller Q factor than the second order bound state. In order to get around this issue one can introduce cavity modifications that significantly reduce the Q factors of the unwanted modes while leaving the Q factor of the featured mode intact. Such a mode discrimination scheme improves side mode suppression as well.

One strategy to perform mode discrimination is to place extra holes in the cavity near the maxima of the electric field corresponding to the mode we wish to suppress. This enhances out-of-plane radiation and lowers the Q factor (Kuang et al., 2005). Figure 11 displays modified cavities that were fabricated in a 240-nm-thick suspended InGaAsP membrane containing four compressively strained quantum wells. The semiconductor dry-etch was done in an inductively coupled plasma etcher using BCl_3 chemistry at 165°C. The rest of the fabrication processes are the same as those in (Shih, Kuang, Mock, Bagheri, Hwang, O'Brien & Dapkus, 2006). The inset of Figure 11(a) displays a scanning electron micrograph of a cavity with a 10% perturbation which supports both the first order and the second order bound states. The inset of Figure 11(b) illustrates a cavity with holes placed at $x = \pm 2.4a$ to suppress the second order bound state, and Figure 11(c) illustrates a cavity with a hole placed at $x = 0$ to suppress the first order bound state (Mock et al., 2009).

The devices were optically pumped at room temperature by an 850 nm diode laser at normal incidence with an 8 ns pulse width and 1% duty cycle. The size of the pump spot was about 2 μm in diameter. The lower spectrum in Figure 11(a) is the single-mode lasing spectrum operating in the first bound state, while the upper multimode lasing spectrum shows the existence of the second bound state approximately 20 nm away from the first one when the pump spot is slightly moved off the device center along the waveguide core. The two modified structures in Figure 11(b) and (c) both operate in stable single-mode operation with respect to pump beam location. Their lasing wavelengths line up with the first and second bound states lasing in the unmodied structure. All four lasing spectra were taken at the peak incident power of 1.7mW. The broad resonance peak between 1.40 and 1.45 μm corresponds to a higher order waveguide dispersion band. Figure 11(d) depicts the light-in-light-out (L-L) curves of the three lasers shown in (a)-(c). They have almost identical thresholds but different slopes, indicating the same amount of total optical loss but different portions of collected laser power. It is possible that the extra holes are causing some excess vertical scattering that is collected by the collection setup.

3.4 Edge-emitting lasers

As mentioned earlier in this chapter, large Q factor cavities lead to reduced laser thresholds resulting from reduced radiative losses. However, reduced radiative losses also reduce the

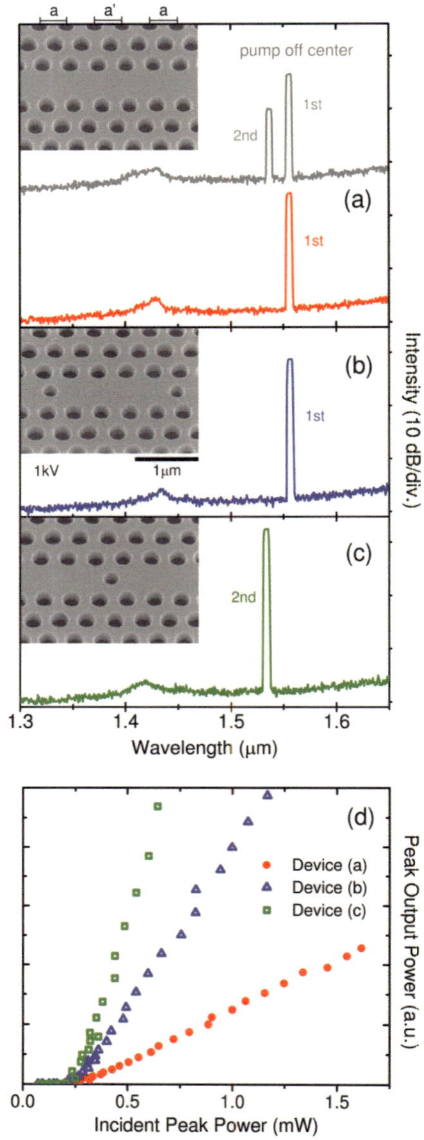

Fig. 11. (a)-(c) Lasing spectra of three double-heterostructure lasers with 10% perturbation ($a_{x'} = 1.10a_x$). Their SEM images are shown as insets. (d) Light-in-light-out curves of the lasers in (a)-(c).

output power of a laser for a given pump level. For applications related to integrated photonics, it is important to have sufficient output power for efficient on-chip detection with a large signal-to-noise ratio and low bit error rate. Fortunately, high Q factor cavities give the designer freedom to lower the Q factor intentionally by introducing losses in a preferential radiation direction. For photonic integrated circuits, the preferential output direction is in-plane. Because the photonic crystal double heterostructure cavity is formed from a perturbation of a straight waveguide, the in-plane losses can be enhanced along the waveguide direction by reducing the number of uniform waveguide periods cladding the perturbation region. Figure 12 displays a finite-difference time-domain simulation of a fabricated heterostructure cavity in which several uniform photonic crystal waveguide cladding periods on one side of the structure have been removed. The geometry was defined in the numerical simulation from a scanning electron micrograph of a fabricated device. For this structure, only five uniform waveguide periods remain between the etched facet and the perturbation region. In excess of 100 microWatts of peak power was measured from this cavity (Lu, Mock, Shih, Hwang, Bagheri, Stapleton, Farrell, O'Brien & Dapkus, 2009). One issue with this structure is that the output power is diffracting at relatively large angles, so that our measurement setup did not collect all of the edge-emitted output power. More recent cavity designs have improved the directionality and have demonstrated output powers in excess of 500 microWatts (Lu, Mock, Hwang, O'Brien & Dapkus, 2009). It should be noted that in a real integrated photonics application, these cavities would be coupled to in-plane waveguides. Coupling efficiencies between defect cavities and photonic crystal waveguides as high as 90% have been reported (Nozaki & Watanabe, 2008) which suggests that over a milliWatt of power could be coupled to an in-plane waveguide. These results strengthen the technological viability of photonic crystal lasers as commercial on-chip sources.

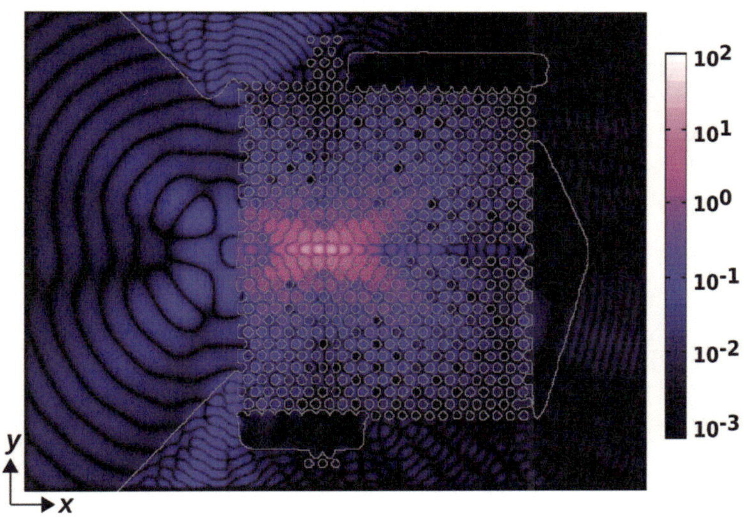

Fig. 12. Finite-difference time-domain calculation of the lasing mode in a fabricated edge-emitting heterostructure laser. $|H_z(x,y, z=0)|$ is plotted.

4. Heat sinking dielectric substrates

In addition to demonstrating that photonic crystal lasers are capable of outputting sufficient output power, showing that these lasers are capable of continuous wave (CW) operation at room temperature and above is another important step in advancing the commercial viability of these sources. The experimental lasing results described thus far correspond to a pulsed optical pump. As mentioned previously, these cavities are formed in thin, suspended membranes which makes it difficult for the material to dissipate heat buildup in the vertical direction. Furthermore, the lattice of airholes reduces the effective thermal conductivity in-plane further reducing the ability of these lasers to dissipate heat. Pulsed pumping allows sufficient carrier excitation for the laser to reach threshold, but the pump beam is on for short enough time periods that the gain properties of the active material are not degraded due to heating.

One approach to achieving CW operation is to grow or bond the semiconductor active material to a thermally conductive lower substrate to aid in heat dissipation (Cao et al., 2005). Two candidate materials include silicon dioxide whose thermal conductivity is 0.014W/cm-K and sapphire whose thermal conductivity is 0.34W/cm-K. These thermal conductivities are significantly larger than that of air alone (0.00024W/cm-K). The trade-off associated with introducing a thermally conductive lower substrate is that the index contrast between the semiconductor slab and the vertical cladding structure is reduced which increases the optical leakage into the substrate. Figure 13 depicts the total Q factor of the heterostructure cavity as a function of the index of the lower substrate. Also depicted are the directional Q factors characterizing radiative losses into the waveguide direction (WG), photonic crystal cladding (PC), air and substrate. Because the directional Q factors add as inverses to express the total Q factor, the lowest directional Q factor will represent the dominant loss mechanism. In this case, it is confirmed that radiation into the substrate is the dominant loss mechanism as the substrate index is increased. The refractive indices of silicon dioxide and sapphire are 1.45 and 1.74, respectively. From Figure 13, the total Q factor has dropped to 4600 for n=1.45 and 610 for n=1.74. Previous studies have found that the minimum Q factor required for CW lasing in sapphire bonded cavities is 1000 (Shih et al., 2006). One sees that the photonic crystal heterostructure cavity is predicted to have a Q factor below 1000 when its substrate refractive index is consistent with that of sapphire suggesting that this configuration is not capable of achieving CW threshold when bonded to sapphire.

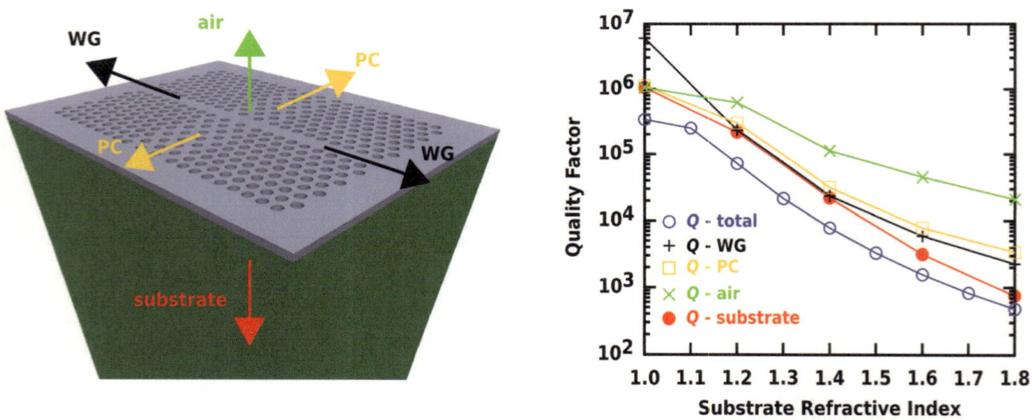

Fig. 13. (a) Depiction of a photonic crystal heterostructure cavity bonded or grown on a lower substrate. Radiation directions are indicated. (b) Q factor versus substrate refractive index for the total Q factor as well as the directional Q factors.

In order to reduce optical radiation into the substrate, an alternative cavity design based on introducing a glide-plane along the waveguide direction has been proposed (Kuang & O'Brien, 2004; Mock & O'Brien, 2009b). Intuitively, the glide-plane introduces a phase shift between the fields on either side of the waveguide core, so that when the two fields combine on-axis, they interfere destructively, and the out-of-plane radiation is reduced. This configuration is known as a type B photonic crystal heterostructure. A schematic of the cavity is shown in Figure 14(a) along with a directional loss anlaysis in Figure 14(b). It should be noted that the cavity analysed in Figure 14(b) included a tapered perturbation in which intermediate stretchings of 2.5% were included on either side of the 5% perturbation. The motivation for this was to reduce the glide-plane symmetry breaking which reduced in-plane losses. It is apparent from the directional loss analysis that for substrate refractive indices between 1.0 and 1.5, the dominant loss mechanism is in-plane along the waveguide direction. The total Q factor remains flat over this range of substrate refractive indicies due to the improved susceptibility to out-of-plane radiation. For substrate refractive indicies greater than 1.5, the substrate losses dominate. It should be pointed out that the type B heterostructure has a total Q factor in excess of 1000 when its substrate refractive index is consistent with that of sapphire making this geometry a promising candidate for edge-emitting photonic crystal lasers operated under CW conditions.

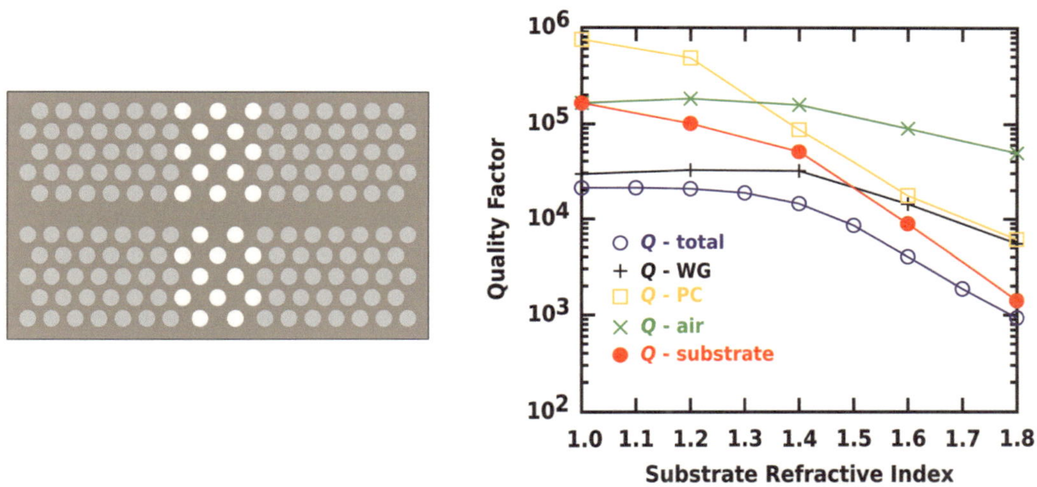

(a) (b)

Fig. 14. (a) Schematic diagram of a type B heterostructure cavity. (b) Q factor versus substrate refractive index for the total Q factor as well as the directional Q factors. Radiation directions are the same as those in Figure 13(a).

5. Conclusion

In this chapter we have highlighted recent progress in the design of high Q factor two dimensional photonic crystal cavities for on-chip lasing. The photonic crystal double heterostructure was featured due to its ultra high Q factor and small mode volume. A first principles derivation of the laser threshold condition was presented which motivated using a large Q factor cavity to achieve low threshold. This was contrasted with the issue of output power in which the cavity Q factor was intensionally lowered in order to extract higher output power. The ability to discriminate between modes when these cavities supported multiple bound states was demonstrated. Finally issues of heat sinking were discussed, and the out- of-plane loss properties associated with dielectric lower substrates were quantified.

A novel cavity based on introducing a glide-plane symmetry was shown to have reduced out-of-plane losses.

6. References

Akahane, Y., Asano, T., Song, B.-S. & Noda, S. (2003). High-*Q* photonic nanocavity in a two-dimensional photonic crystal, *Nature* 425: 944–947.

Akahane, Y., Asano, T., Song, B.-S. & Noda, S. (2005). Fine-tuned high-*Q* photonic-crystal nanocavity, *Optics Express* 13(4): 1202–1214.

Aoki, K., Guimard, D., Nishioka, M., Nomura, M., Iwamoto, S. & Arakawa, Y. (2008). Coupling of quantum dot light emission with three-dimensional photonic crystal nanocavity, *Nature Photonics* 2: 688–692.

Asano, T., Song, B.-S. & Noda, S. (2006). Analysis of the experimental *Q* factors (~1 million) of photonic crystal nanocavities, *Optics Express* 14(5): 1996–2006.

Baba, T. (2008). Slow light in photonic crystals, *Nature Photonics* 2: 465–473.

Baba, T. & Mori, D. (2007). Slow light engineering in photonic crystals, *Journal of Physics D* 40: 2659–2665.

Benson, T.M., Sewell, P., Boriskina, S. V., Janyani, V., Al-Jarro, A., Vukovic, A., Sakhnenko, N., Smotrova, E. I., Nosich, A. I. & Nerukh, A. G. (2005). Microcavities: An inspiration for advanced modelling techniques, *7th International Conference on Transparent Optical Networks*, Barcelona, Spain, pp. 272–275.

Bogaerts, W., Taillaert, D., Luyssaert, B., Dumon, P., Van Campenhout, J., Bienstman, P., Van Thourhout, D. & Baets, R. (2004). Basic structures for photonic integrated circuits in silicon-on-insulator, *Optics Express* 12(8): 1583–1591.

Cao, J. R., Kuang, W., Wei, Z.-J., Choi, S.-J., Yu, H., Bagheri, M., O'Brien, J. D. & Dapkus, P. D. (2005). Sapphire-bonded photonic crystal microcavity lasers and their far-field radiation patterns, *Photonics Technology Letters* 17(1): 4–6.

Chow, E., Grot, A., Mirkarimi, L. W., Sigalas, M. & Girolami, G. (2004). Ultracompact biochemical sensor built with two-dimensional photonic crystal microcavity, *Optics Letters* 29(10): 1093–1095.

Englund, D., Fattal, D., Waks, E., Solomon, G., Zhang, B., Nakaoka, T., Arakawa, Y., Yamamoto, Y. & Vučković, J. (2005). Controlling the spontaneous emission rate of single quantum dots in a two-dimensional photonic crystal, *Physical Review Letters* 95: 013904 1–4.

Jiang, Y., Jiang, W., Gu, L., Chen, X. & Chen, R. T. (2005). 80-micron interaction length silicon photonic crystal waveguide modulator, *Applied Physics Letters* 87: 221105.

Joannopoulos, J. D., Meade, R. D. & Winn, J. N. (1995). *Photonic Crystals*, Princeton University Press, Princeton, NJ.

John, S. (1987). Strong localization of photons in certain disordered dielectric superlattices, *Physical Review Letters* 58(23): 2486–2489.

Kim, M.-K., Hwang, I.-K., Seo, M.-K. & Lee, Y.-H. (2007). Reconfigurable microfiber-coupled photonic crystal resonator, *Optics Express* 15(25): 17241–17247.

Kim, W. J. (2004). *Ph. D Thesis*, University of Southern California, Los Angeles, CA.

Kogelnik, H. & Shank, C. V. (1971). Stimulated emission in a periodic structure, *Applied Physics Letters* 18(4): 152–154.

Krauss, T. F. (2007). Slow light in photonic crystal waveguides, *Journal of Physics D* 40: 2666–2670.

Kuang, W., Cao, J. R., Choi, S.-J., O'Brien, J. D. & Dapkus, P. D. (2005). Modied suspended membrane photonic crystal d3 laser cavity with improved sidemode suppression ratio, *Optics Express* 17(5): 941–943.

Kuang, W., Kim, W. J., Mock, A. & O'Brien, J. D. (2006). Propagation loss of line-defect photonic crystal slab waveguides, *IEEE Journal of Selected Topics in Quantum Electronics* 12(6): 1183–1195.

Kuang, W., Kim, W. J. & O'Brien, J. D. (2007). Finite-difference time domain method for nonorthogonal unit-cell two-dimensional photonic crystals, *Journal of Lightwave Technology* 25(9): 2612–2617.

Kuang, W. & O'Brien, J. (2004). Reducing the out-of-plane radiation loss of photonic crystal waveguides on high-index substrates, *Optics Letters* 29(8): 860–862.

Kuramochi, E., Notomi, M., Mitsugi, S., Shinya, A. & Tanabe, T. (2006). Ultrahigh-*Q* photonic crystal nanocavities realized by the local width modulation of a line defect, *Applied Physics Letters* 88: 041112-1–041112-3.

Kwon, S.-H., Sünner, T., Kamp, M. & Forchel, A. (2008a). Optimization of photonic crystal cavity for chemical sensing, *Optics Express* 16(16): 11709–11717.

Kwon, S.-H., Sünner, T., Kamp, M. & Forchel, A. (2008b). Ultrahigh-*Q* photonic crystal cavity created by modulating air hole radius of a waveguide, *Optics Express* 16(7): 4605– 4614.

Lodahl, P., van Driel, A. F., Nikolaev, I. S., Irman, A., Overgaag, K., Vanmaekelbergh, D. & Vos, W. L. (2004). Controlling the dynamics of spontaneous emission from quantum dots by photonic crystals, *Nature* 430: 654–657.

Lončar, M., Scherer, A. & Qiu, Y. (2003). Photonic crystal laser sources for chemical detection, *Applied Physics Letters* 82(26): 4648–4650.

Lu, L., Mock, A., Bagheri, M., Hwang, E. H., O'Brien, J. D. & Dapkus, P. D. (2008). Double heterostructure photonic crystal lasers with reduced threshold pump power and increased slope efficiency obtained by quantum well intermixing, *Optics Express* 16(22): 17342–17347.

Lu, L., Mock, A., Hwang, E. H., O'Brien, J. D. & Dapkus, P. D. (2009). High peak power efficient edge-emitting photonic crystal nanocavity lasers, *Integrated Photonics and Nanophotonics Research and Applications Topical Meeting*, Optical Society of America, Honolulu, Hawaii, USA, p. ITuB3.

Lu, L., Mock, A. Yang, T., Shih, M. H., Hwang, E. H., Bagheri, M., Stapleton, A., Farrell, S., O'Brien, J. D. & Dapkus, P. D. (2009). 120 μw peak output power from edge-emitting photonic crystal double heterostructure nanocavity lasers, *Applied Physics Letters* 94(13): 111101.

Lu, L., Yang, T., Mock, A., Shih, M. H., Hwang, E. H., Bagheri, M., Stapleton, A., Farrell, S., O'Brien, J. D. & Dapkus, P. D. (2007). 100 μw edge-emitting peak power from a photonic crystal double-heterostructure laser, *Conference on Lasers and Electro-Optics Technical Digest*, Baltimore,MD, p. CMV3.

McNab, S. J., Moll, N. & Vlasov, Y. A. (2003). Ultra-low loss photonic integrated circuit with membrane-type photonic crystal waveguides, *Optics Express* 11(22): 2927–2939.

Mock, A., Kuang, W., Shih, M. H., Hwang, E. H., O'Brien, J. D. & Dapkus, P. D. (2006). Spectral properties of photonic crystal double-heterostructure resonant cavities, *Laser and Electro-Optics Society Annual Meeting Technical Digest*, Montreal, Canada, p. ML4.

Mock, A., Lu, L., Hwang, E. H., O'Brien, J. & Dapkus, P. D. (2009). Modal analysis of photonic crystal double-heterostructure laser cavities, *Journal of Selected Topics in Quantum Electronics* 15(3): 892–900.

Mock, A., Lu, L. & O'Brien, J. D. (2008). Spectral properties of photonic crystal double heterostructure resonant cavities, *Optics Express* 16(13): 9391–9397.

Mock, A. & O'Brien, J. D. (2008). Convergence analysis of padé interpolation for extracting large quality factors in photonic crystal double heterostructure resonant cavities, *Conference on Numerical Simulation of Optoelectronic Devices Technical Digest*, Nottingham, England, p. TuB3.

Mock, A. & O'Brien, J. D. (2009a). Photonic crystal laser threshold analysis using 3-d fdtd with a material gainmodel, *Integrated Photonics and Nanophotonics Research and Applications Topical Meeting*, Optical Society of America, Honolulu, Hawaii, USA, p. ITuD6.

Mock, A. & O'Brien, J. D. (2009b). Quality factor dependence on vertical slab structure in photonic crystal double heterostructure resonant cavities, *Integrated Photonics and Nanophotonics Research and Applications Technical Digest*, Honolulu, HI, p. IMF2.

Nakamura, M., Yariv, A., Yen, H. W., Somekh, S. & Garvin, H. L. (1973). Optically pumped gaas surface laser with corregation feedback, *Applied Physics Letters* 22(10): 515–516.

Noda, S., Tomoda, K., Yamamoto, N. & Chutinan, A. (2000). Full three-dimensional photonic bandgap crystals at near-infrared wavelengths, *Science* 289(5479): 604–606.

Notomi, M., Shinya, A., Mitsugi, S., Kuramochi, E. & Ryu, H.-Y. (2004). Waveguides, resonators and their coupled elements in photonic crystal slabs, *Optics Express* 12(8): 1551–1561.

Nozaki, K. & Baba, T. (2006). Laser characteristics with ultimate-small modal volume in photonic crystal slab point-shift nanolasers, *Applied Physics Letters* 88: 211101-1–211101- 3.

Nozaki, K. & Watanabe, HidekiBaba, T. (2008). Hundred micro-watts peak output power from an edge-emitting photonic crystal double-heterostructure laser, *Applied Physics Letters* 92(2): 021108.

O'Brien, D., Settle, M. D., Karle, T., Michaeli, A., Salib, M. & Krauss, T. F. (2007). Coupled photonic crystal heterostructure nanocavities, *Optics Express* 15(3): 1228–1233.

Painter, O., Lee, R. K., Scherer, A., Yariv, A., O'Brien, J. D., Dapkus, P. D. & Kim, I. (1999). Two-dimensional photonic band-gap defect mode laser, *Science* 284(11): 1819–1821.

Peterson, A. F., Ray, S. L. & Mittra, R. (1998). *Computational Methods for Electromagnetics*, IEEE Press, New York.

Ryu, H.-Y., Kim, S.-H., Park, H.-G., Hwang, J.-K. & Lee, Y.-H. (2002). Square-lattice photonic band-gap single-cell laser operating in the lowest-order whispering gallery mode, *Applied Physics Letters* 80(21): 3883–3885.

Sadiku, M. N. O. (2000). *Numerical Techniques in Electromagnetics*, CRC Press, Boca Raton.

Sakoda, K. (2001). *Optical Properties of Photonic Crystals*, Springer, Germany.

Schawlow, A. L. & Townes, C. H. (1958). Infrared and optical masers, *Physical Review* 112(6): 1940–1949.

Shih, M. H., Kim, W. J., Wuang, W., Cao, J. R., Yukawa, H., Choi, S. J., O'Brien, J. D. & Marshall, W. K. (2004). Two-dimensional photonic crystal machzehnder interferometers, *Applied Physics Letters* 84(4): 460–462.

Shih, M. H., Kuang, W., Mock, A., Bagheri, M., Hwang, E. H., O'Brien, J. D. & Dapkus, P. D. (2006). High-quality-factor photonic crystal heterostructure laser, *Applied Physics Letters* 89: 101104-1–101104-3.

Shih, M. H., Kuang, W., Yang, T., Bagheri, M., Wei, Z.-J., Choi, S.-J., Lu, L., O'Brien, J. D. & Dapkus, P. D. (2006). Experimental characterization of the optical loss of sapphire-bonded photonic crystal laser cavities, *Photonics Technology Letters* 18(3): 535–537.

Shih, M. H., Mock, A., Hwang, E. H., Kuang, W., O'Brien, J. D. & Dapkus, P. D. (2006). Photonic crystal heterostructure laser with lattice-shifted cavity, *Conference on Lasers and Electro-Optics Technical Digest*, Baltimore, MD, p. paper CMKK3.

Smith, C. L. C., Wu, D. K. C., Lee, M. W., Monat, C., Tomljenovic-Hanic, S., Grillet, C., Eggleton, B. J., Freeman, D., Ruan, Y., Madden, S., Luther-Davies, B., Giessen, H. & Lee, Y.-H. (2007). Microfluidic photonic crystal double heterostructures, *Applied Physics Letters* 91: 121103-1–121103-3.

Soda, H., Iga, K., Kitahara, C. & Suematsu, Y. (1979). Gainasp/inp surface emitting injection lasers, *Japan Journal of Applied Physics* 18(12): 2329–2330.

Song, B.-S., Asano, T. &Noda, S. (2007). Heterostructures in two-dimensional photonic-crystal slabs and their application to nanocavities, *Journal of Physics D* 40: 2629–2634.

Song, B.-S., Noda, S., Asano, T. & Akahane, Y. (2005). Ultra-high-*Q* photonic double heterostructure nanocavity, *Nature Materials* 4: 207–210.

Taflove, A. & Hagness, S. C. (2000). *Computational electrodynamics*, Artech House, Massachusetts.

Takahashi, Y., Hagino, H., Yoshinori, T., Song, B.-S., Asano, T. & Noda, S. (2007). High-*Q* nanocavity with a 2-ns photon lifetime, *Optics Express* 15(25): 17206 17213.

Tanabe, T., Notomi,M., Kuramochi, E., Shinya, A. & Taniyama, H. (2007). Trapping and delaying photons for one nanosecond in an ultrasmall high-*Q* photonic-crystal nanocavity, *Nature Photonics* 1: 49–52.

Tanaka, Y., Asano, T. & Noda, S. (2008). Design of photonic crystal nanocavity with *Q*-factor of ~ 10^9, *Journal of Lightwave Technology* 26(11): 1532–1539.

Tomljenovic-Hanic, S., Steel, M. J., de Sterke, C. M. & Moss, D. J. (2007). High-*Q* cavities in photosensitive photonic crystals, *Optics Letters* 32(5): 542–544.

Vlasov, Y. A., O'Boyle, M., Hamann, H. F. & McNab, S. J. (2005). Active control of slow light on a chip with photonic crystal waveguides, *Nature* 438: 65–69.

Yablonovitch, E., Gmitter, T. J. & Leung, K. M. (1991). Photonic band structure: The face-centered-cubic case employing nonspherical atoms, *Physical Review Letters* 67(17): 2295–2298.

Yang, T., Lipson, S., Mock, A., O'Brien, J. D. & Deppe, D. G. (2007). Edge-emitting photonic crystal double-heterostructure nanocavity lasers with inas quantum dot material, *Optics Letters* 32(9): 1153–1155.

Yoshie, T., Scherer, A., Hendrickson, J., Khitrova, G., Gibbs, H. M., Rupper, G., Ell, C., Shchekin, O. B. & Deppe, D. G. (2004). Vacuum rabi splitting with a single quantum dot in a photonic crystal nanocavity, *Nature* 432: 200–203.

Zhang, Z. & Qiu, M. (2004). Small-volume waveguide-section high *Q* microcavities in 2d photonic crystal slabs, *Optics Express* 12(17): 3988–3995.

3

Anisotropy of Light Extraction Emission with High Polarization Ratio from GaN-based Photonic Crystal Light-emitting Diodes

Chun-Feng Lai[1], Chia-Hsin Chao[2], and Hao-Chung Kuo[1]

[1]*Department of Photonics and Institute of Electro-Optical Engineering, Nation Chiao-Tung University*

[2]*Electronics and Opto-Electronics Research Laboratories, Industrial Technology Research Institute*

Hsinchu, Taiwan, Republic of China

1. Introduction

1.1 General background

GaN-based materials have been attracted a great deal of attention due to the large direct band gap and the promising potential for the optoelectronic devices, such as light emitting diodes (LEDs) and laser diodes (LDs). LEDs have the advantages of small size, conserve energy, and have a long lifespan. LEDs of solid-state lighting will be in a position to replace conventional lighting sources within years. At present, the efficiency of LEDs is still lower than that of fluorescence lamps in general lighting applications. Therefore, the ultimate optimization of all aspects of LED efficiency is necessary in solid-state lighting development. Several factors are likely to limit the light extraction efficiency of LEDs. One may think that the main limiting factor is internal light generation as internal quantum efficiency (IQE). Nevertheless, this is not the case in a variety of material where the conversion from carriers to photons reaches 50% to 90% if the material's quality is high enough. In this case, the strongest limiting factor is that of external extraction efficiency, i.e. the ability for photons generated inside the semiconductor material to escape into air. Unfortunately, most of the light emitted inside the LED is trapped by total internal reflection (TIR) at the material's interface with air. Although many efficient light extraction strategies have already been applied, they are mostly based on the principle of randomizing the paths followed by the light, such as surface roughening [1-2], flip-chip [3-4], and photonic crystals (PhCs) [5-6].

1.2 Research niche

Light-emitting diodes (LEDs) have become ubiquitous in illumination and signal applications as their efficiency and power level improve. While the improvement of the basic characteristics will benefit the replacement of the conventional light sources, further improvement in other characteristics can bring about unique applications. One notable example is the polarized light emission which is highly desirable for many applications [7], e.g. back-lighting for liquid crystal displays and projectors. For the application of next-generation LEDs, such in projector displays, backlight displays, and automobile headlights, further improvements the light extraction efficiency, the polarized emission, and the directional far-field patterns of light sources are required. Recently, PhC has attracted much attention for the possibility to improve the extraction efficiency [8-9], polarization [10], and directional far-field patterns [11-12] from GaN-based LEDs and GaN-based film-transferred LEDs, respectively. In order to optimize the PhC LED performance for a specific system, detailed knowledge of the light extraction and polarization, especially the angular distribution, is required. The light wave propagating in the PhC LED waveguide, with its propagation partially confined by the TIR, can interact with the reciprocal lattice vectors of the two-dimensional (2D) PhC lattice to exhibit a variety of novel behaviors from the light localization. On the other hand, through the Bragg diffraction with the PhC which fabricated on LEDs can scatter the guided light into the escaping cone to circumvent the deleterious effects due to TIR, which traps the majority of the emitted light in LED chips. In this study, the GaN-based LEDs with PhCs were demonstrated and investigated in the light extraction, and polarization.

In this chapter, we first introduce the theory analysis and design method of GaN-based PhC LED structures in section 2. Then, in section 3, we exhibit the direct imaging of the azimuthal angular distribution of the 2D PhC light extraction using a specially designed waveguide structure. The optical images of the light extraction patterns from the guided electroluminescence (EL) light are obtained with a current injected into the center of the annular structure made on the GaN multilayer. With increasing lattice constant, symmetric patterns with varying number of petals according to the symmetry of the PhC are observed. The observed anisotropy is charted using the Ewald construction according to the lattice constant and the numerical aperture of the observation system. The appearance and disappearance of the petals can be explained using the Ewald construction in the reciprocal space. In addition, several novel features of light propagations associated with the PhC can also be directly observed including the focusing and collimating behavior. These results can be used for the optimization of LED devices with PhC extraction. Next, in section 4, polarization characteristics of the GaN-based PhC LEDs using an annular structure with square PhC lattice have been studied experimentally and theoretically. The observed a strong polarization dependence of the lattice constant and orientation of the PhC. It is found that the PhC can be as a polarizer to improve the *P/S* ratio of the extracted EL emission. The results of the *P/S* ratio for light propagating in different lattice orientation were found to be consistent with the results obtained using the PhC Bloch mode coupling theory. This polarization behavior suggests an efficient means to design and control the GaN blue PhC LEDs for polarized light emission. Finally, conclusions are provided in section 5.

2. Fundamental and modelling of photonic crystal LEDs

2.1 Waveguide properties of LED structures

Although the IQE of GaN-based LEDs have reached up to 90%, the light emission from a multi-quantum well (MQW) into the air is fundamentally limited by TIR. LEDs have such low external extraction efficiency that most of the light generated in a high-index material is trapped by TIR. Due to the GaN-based LED layer behaving as a waveguide, trapped light is distributed in a series of so-called guided modes. The propagation properties, including electromagnetic field distributions and wave vectors of guided modes, affect PhC light extraction behavior. In general, the high order guided modes interact strongly with PhC to have high extraction efficiency. By contrast, the low order guided modes have weak light extraction efficiency due to the poor overlap with the PhC regions. But the light of energy distribution coupling to the low order guided modes is larger. Therefore, our discussion begins with the guided mode properties in a waveguide structure of LED semiconductor layers, which is helpful to optimize the design of PhC structure on LEDs with high light extraction efficiency.

A large number of waveguide modes exist in a typical GaN-based LED structure as asymmetric slab waveguide in geometry. For example, GaN-based blue LED structure is grown by metal-organic chemical vapor deposition (MOCVD) on *c*-sapphire substrate. The GaN blue LED structure consists of a 2 μm-thick un-GaN buffer layer, a 2-μm-thick n-GaN layer, a 100 nm InGaN/GaN MQW region, and a 200 nm-thick p-GaN layer, as shown in Fig. 1(a). In order to study the guided modes in the LED structures, the guided mode distributions were calculated in the asymmetric slab waveguide with the vertical effective refractive index profile, as shown in Fig. 1(b). Since the emitted light from the MQW is predominantly TE polarized in the waveguide plane [13], only TE modes are analyzed. In this case, thirty-two TE guided modes with effective refractive index distribution are obtained by using waveguide theory [14]. The first three and the last of the thirty-two guided modes of electric field distributions are plotted in Fig. 2, respectively. Each guided mode has different electromagnetic field distribution and wave vector. In a planar GaN-based LED on a sapphire substrate, 66% of the total emitted light is wave guided within the GaN layer, while the remainder is found in the delocalized modes in the sapphire, as shown in Fig. 3(a). Only 8.7% of the light generated can directly escape from both top and bottom surfaces of the GaN medium into the air. Further, when the MQW emitter position was be considered in the LED structure, that the guided modes excited a percentage of relative intensity as shown in Fig. 3(b). In the fundamental mode (TE_{00}), the excited percentage is 19.5%; in the other guided modes, the excited percentages are 14.1%, 9.6%, 6.6%, 5.1%, and 3.5%, respectively. The relative intensity ratio of the higher-order modes becomes weak due to the poor field overlap with the MQW emission regions. Therefore, the guided mode energy distribution is mainly in the lower-order modes.

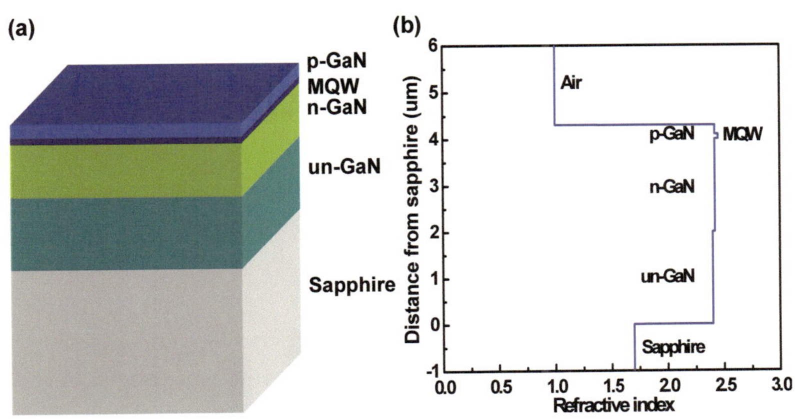

Fig. 1. (a) Schematic diagram of the MOCVD-grown GaN-based blue LED structure (dominant λ = 470 nm). (b) Vertical effective refractive index profile of the characterized GaN-based LED.

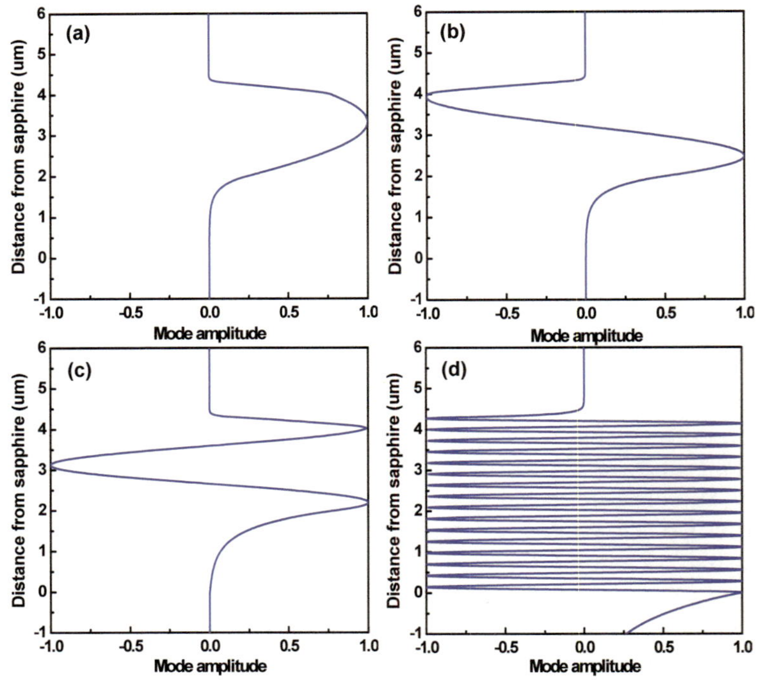

Fig. 2. Electric field distributions of the asymmetric slab waveguide for TE mode are (a) TE_{00} (fundamental mode), (b) TE_{01}, (c) TE_{02}, and (d) TE_{31}.

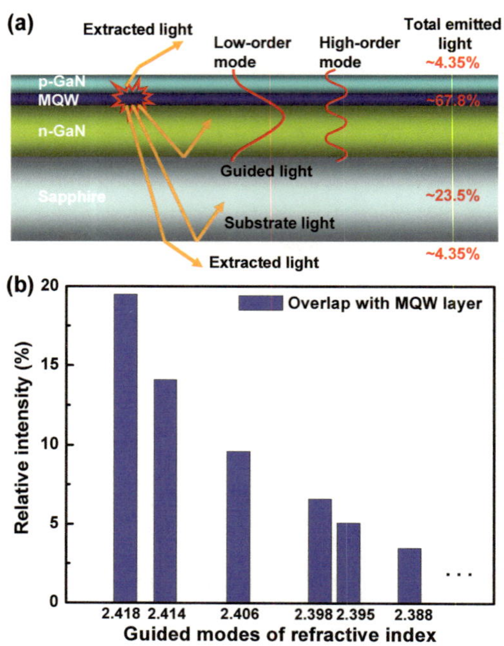

Fig. 3. (a) Possible paths for emitted light in a GaN-based blue LED structure. (b) The guided modes excited percentage of relative intensity indicates overlap with MQW.

2.2 Ewald construction of Bragg's diffraction theoretical analysis methods for photonic crystals

Photonic crystals (PhCs) are artificial structures containing periodic arrangements of dielectric materials which exhibit unique dispersion properties (e.g. such as photonic bandgap (PBG) [15]) and that manipulate light emission behaviors. In this chapter, we will concentrate on the extraction of waveguide light from GaN-based LED structures. There are several schemes to obtain light extraction through PhC nanostructures [16], as shown in Fig. 4, such as (a) inhibition of guided modes emission by PBG, (b) spontaneous emission enhanced in a small cavity by Purcell effect, and (c) emission extraction on the whole surface by leaky mode coupling. Accordingly, the emission region can be deeply etched with a pattern to forbid propagation of guided modes, as shown in Fig. 4(a), and thus force the emitted light to be redirected towards the outside. Defects in PhCs behave as microcavities, as shown in Fig. 4(b), such that the Purcell effect can be excited for spontaneous emission enhancement. Then, light can only escape through leaky modes coupling, as shown in Fig. 4(c). In addition, PhCs can also act as 2D diffraction gratings in slabs or waveguides to extract guided modes to the air and to redirect the emission directions.

The optimal design of PhC structures for high extraction efficiency is promising, which is strongly dependent on various parameters such as lattice constant (*a*), the type of lattice (square, triangular…), filling factor (*f*), and etch depth (*t*). Among parameters described here, we paid special attention to the effect of the lattice constant *a*. In order to discuss the effect of the lattice constant, we use the Ewald construction of Bragg's diffraction theory. In addition, the plane-wave expansion method (PWE) and the finite-difference time-domain method (FDTD) are implemented to investigate the optical properties of PhC numerically.

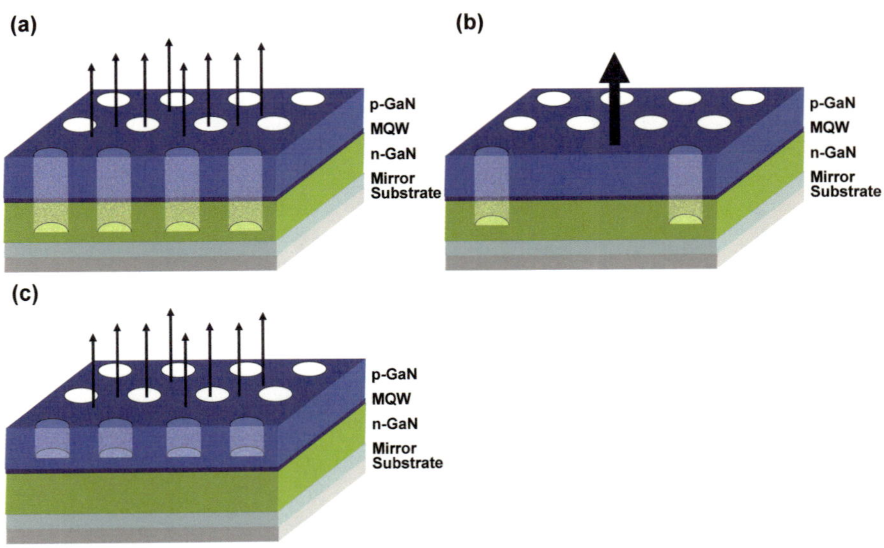

Fig. 4. Schematic the various extraction methods relying on PhCs are (a) PBG, (b) Purcell effect, and (c) leaky mode coupling.

Figure 4(c) is a schematic of the surface grating devices that can be discussed in relation to the light extraction of the lattice constant of PhCs by using the Ewald construction of Bragg's diffraction theorem. The light extraction of guided waves through diffraction by PhC is discussed. According to Bragg's diffraction law, $k_g\sin\theta_1+mG= k_0\sin\theta_2$, the phase-matching diagrams in the wave number space are shown in the Fig. 5(a). The two circles in the Fig. 5 correspond to 1.) the waveguide mode circle with radius $k_g = 2n\pi/\lambda$ at the outside, where n is the effective refractive index of the guided mode; 2.) the air cone with radius $k_o = 2\pi/\lambda$ at the inner circle. The light extraction from PhC also can be quantitatively analyzed using the Ewald construction in the reciprocal space. The extraction of waveguide light into air can be described by the relation $|k_g + G| < k_0$, where G is the diffraction vectors. Such a relation can be represented graphically with the Ewald construction commonly used in the X-ray crystallography. In the present case, for reasons of simplicity, PhC is treated as a 2D in an overall 3D structure as is commonly done. In such case, the reciprocal lattice of the 2D PhC will be represented as the rods protruding perpendicular to the waveguide plane. Figure 5(b) depicts the Ewald spheres for a square lattice with the k vector of the incident light pointing directly at a reciprocal lattice point. The center of the sphere is at the end of the vector and the radius is the magnitude of k_g. The intersection points of the sphere with the protruding rods define the extraction direction of the diffracted light. For simplicity, only the in-plane propagation needs to be treated and a consideration of the projection on the waveguide plane is sufficient. When the in-plane component of the resultant wavevector after the coupling to a reciprocal lattice vector falls inside the air circle, the diffracted light can escape into air, as shown in Fig. 5(c).

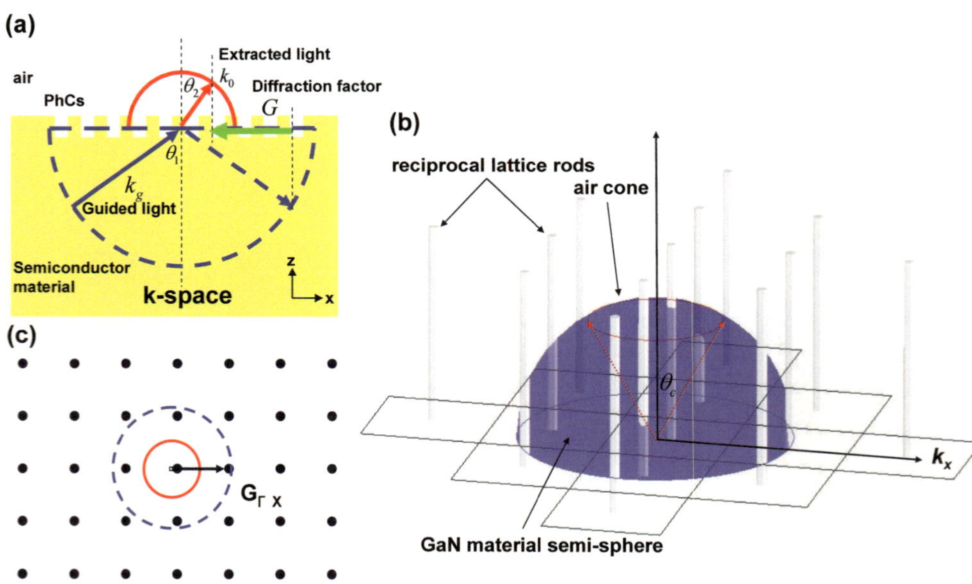

Fig. 5. (a) A schematic of the 2D PhC structure of the Bragg diffraction phase matching diagrams. (b) The Ewald construction for square lattice PhC. (c) The projection of the Ewald sphere construction on the waveguide plane. Thick red circle is air cone and dashed blue circle is waveguide mode cone.

Further, an actual 2D square lattice of PhC as grating has the anisotropy of the diffraction vector [23]. Figure 6 shows the diffraction vector for various lattices constant a, dispersion circles for the in-plane wavevector in air, k_0, and in the semiconductor material, k_g. For example, in the square lattice of PhC, $G_{\Gamma X}$ and $G_{\Gamma M}$ are $2\pi/a$ and $2\sqrt{2}\pi/a$, respectively. When $G_{\Gamma X} > k_0 + k_g$ $[a/\lambda < 1/(n+1)]$, the zone-folded curve does not enter the air curve, so the diffraction does not occur, as shown in Fig. 6(a). When *a* is larger than this value, some amount of diffraction occurs, as shown in Fig. 6(b). When *a* is large enough to satisfy $G_{\Gamma M} < k_0$ ($a/\lambda > \sqrt{2}$), the diffraction vector is wholly included in the air curve, and this gives the maximum light diffraction efficiency. However, the diffraction efficiency cannot be unity for such larger *a*, since light can find not only the extracted light cone but also another solid angle not extracted by the diffraction. Even in light diffracted into the extracted light cone, half goes downward.

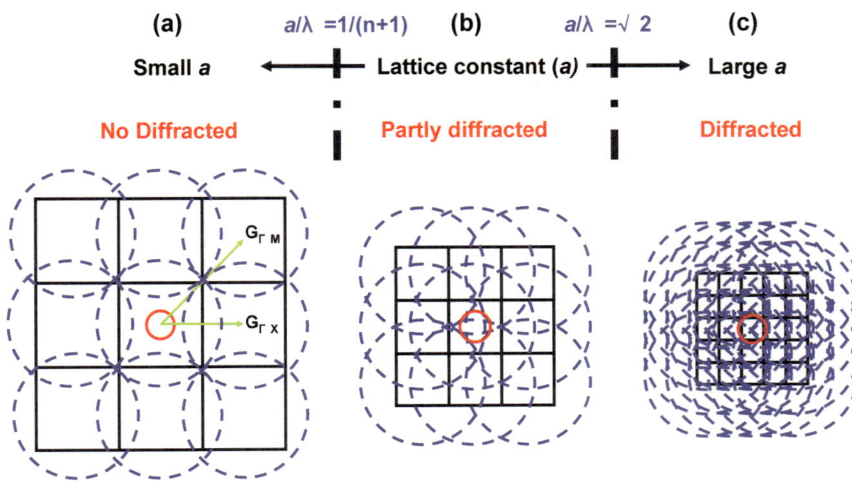

Fig. 6. Brillouin zones for 2D square lattice, dispersion curves of k_0 (center thick red circle) and k_g (dashed blue circle).

3. Anisotropy light extraction properties of GaN-based photonic crystal LEDs

3.1 Sample prepared and measurement results

In order to optimize the PhC LED performance for high light extraction efficiency, detailed knowledge of light extraction is required especially the angular distribution [9, 26]. Therefore, we present the direct imaging of the azimuthal angular distribution of the extracted light using a specially designed annual PhC structure, as shown in Fig. 7(a). The GaN-based LED samples used in this study were grown by metal-organic chemical vapor deposition (MOCVD) on a *c*-axis sapphire (0001) substrate. The LED structure (dominant wavelength λ at 470 nm) was composed of a 1-µm-thick GaN bulk buffer layer, a 2-µm-thick n-GaN layer, a 100-nm-thick InGaN/GaN MQW, and a 130-nm-thick top p-GaN layer. An annular region of square PhC lattice with an inner/outer diameter of 100/200 µm was patterned by holographic lithography. Two different periods of the lattice constant are used by 260 and 410 nm. A scanning electron microscopy (SEM) image of the square-lattice PhC structure is shown inset in Fig. 7(b). The holes were then etched into the top p-GaN layer using inductively coupled plasmon (ICP) dry etching to a depth of t =120 mm. The electron-beam-evaporated Ni/Au film was used as the transparent ohmic contact layer (TCL) to p-GaN, and a 200-nm-thick SiO_2 layer was used for passivation. Finally, Ti/Al/Ti/Au layer was deposited on the n-GaN as an n-type electrode and onto TCL as a p-type electrode on LEDs, respectively. In addition, the schematics for the experimental setup are shown in Fig. 7(b). An electroluminescence (EL) probe station system was utilized for the experiment after fabrication, which included a continuous wave (CW) current source and a 15x microscope objective with numerical aperture (NA)=0.32. A 15x UV objective with NA of 0.32 was used to collect the on-axis emission signal from the sample, which formed a high-resolution image on a charge-coupled device (CCD); this was recorded with a digital camera. The experiment of the observed image is shown inset in Fig. 7(b).

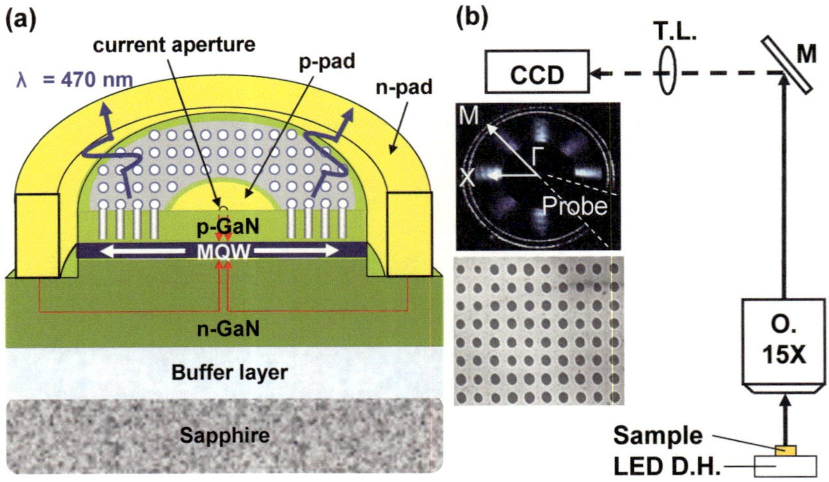

Fig. 7. (a) Schematic diagram of the GaN-based blue LED structure with annular PhC region. (b) EL probe station and CCD imaging system setup, where D.H.:driver holder; M.:mirror; T.L.: tube lens; O.: objective.

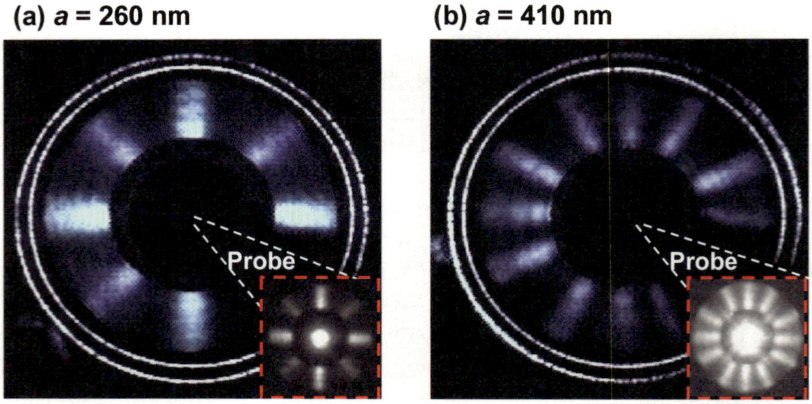

Fig. 8. CCD images taken with square lattices with a = (a) 260 nm and (b) 410 nm. Inset of the photoluminescence (PL) CCD images.

Figure 8 depicts the CCD images for the square PhC structures with lattice constant a of 260 and 410 nm corresponding to a/λ of 0.553 and 0.872, respectively. The EL light was partially guided toward the surrounding PhC region by the waveguide formed by GaN epitaxial layers. This guided light was then coupled into the PhC region and diffracted by the PhC lattice while propagating inside the PhC region. Depending on the lattice constant of the PhC, some of the diffracted light left the wafer and formed the images shown in Fig. 11. It can be seen that a varying number of petals appears as the lattice constant increases. Under certain conditions, some of the petals may become weaker or disappeared altogether. The observed anisotropy, therefore, primarily arises from the diffraction of guided EL light into the air, which is picked up by the microscope objective.

3.2 Bragg diffraction theoretical discussion

The appearance and disappearance of the petals observed in Fig. 8 can be qualitatively analyzed using the Ewald construction in the reciprocal space. The above observation established that the use of 2D Ewald construction explains the observed images. It can be invoked to determine the boundaries between regions with varying numbers of petals. As shown in Fig. 9, as a/λ increases above the cutoff, the resultant wave vector will start to couple to the shortest lattice vector $G_{\Gamma X}$. The resultant wave vector falls inside the NA circle as shown in Fig. 9(a), where the NA circle with radius NA=0.32k_0 at the inside corresponds to the acceptance angle of the objective lens with NA numerical aperture. For the ΓM direction, the resultant wave vector falls outside the NA circle and will not be seen by the NA=0.32 objective lens as shown in Fig. 9(b). Therefore, a pattern with four petals pointing in the ΓX direction is observed. As a/λ increases further, the resultant wave vector after coupling to $G_{\Gamma X}$ may fall short of the NA circle and therefore it will not be observed, as shown in Fig. 9(c). Thus, there is a range of a/λ within which the resultant wave vector can fall into the NA circle for a particular propagation direction. The boundary for when this range with four petals pointing in the ΓX direction starts to appear can be determined by the relation $k = |G_{\Gamma X} - NA|$ to be $a/\lambda = 1/(n+NA)$. For further increase of a/λ, the resultant wavevector will leave the NA circle as shown Fig. 9(c).

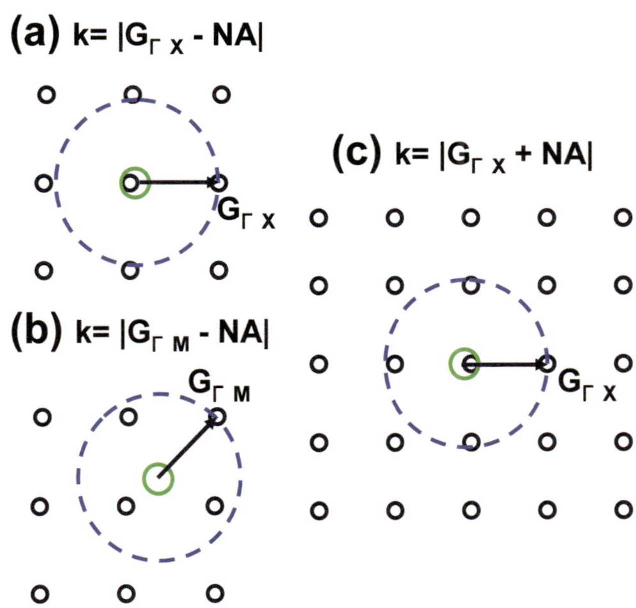

Fig. 9. Ewald constructions for a/λ increases above the cutoff and just start to couple with the shortest lattice vector $G_{\Gamma X}$ (a) in the ΓX directions. (b) ΓM direction with the resultant wave vector falling outside the NA circle and will not be seen by the NA=0.32 objective. (c) a/λ increases further as nk_0 just starts to leave the NA circle to disappear from the CCD image.

For larger lattice constants, the escape cone and the guided mode circle become larger relative to the reciprocal lattice. For $a/\lambda > \sqrt{2}/n$, the coupling to $G_{\Gamma M}$ becomes possible and four more petals appears representing four equivalent ΓM directions. For even larger lattice constants, coupling to the third nearest wave vectors is possible and the number of petals increases to 16. These increased coupling possibilities are observed as the increased number of petals in the images. The boundaries separating these regions can be readily derived using the Ewald construction as shown in Fig. 10 along with our observations.

The above discussion considers the simple case of single mode propagation in the waveguide plane. Since the thickness of the epitaxial layer used for the present study is 3 um, the waveguide is multimode. Every mode can couple with different reciprocal vectors to form their own boundaries for a given number of pedals. When plotted on the map, these boundaries will appear as a band of lines. To present these multimode extractions clearly, only the first and the last mode with modes number '*m*' are shown on Fig. 10. The two outermost lines, $G^{+}_{\Gamma X}$ and $G^{m-}_{\Gamma X}$, define the boundary of the possible a/λ's for all the modes that can fall into NA circle after coupling to $G_{\Gamma M}$. The a/λ values shown on the right side of Fig. 10 correspond to the boundaries for NA=1.

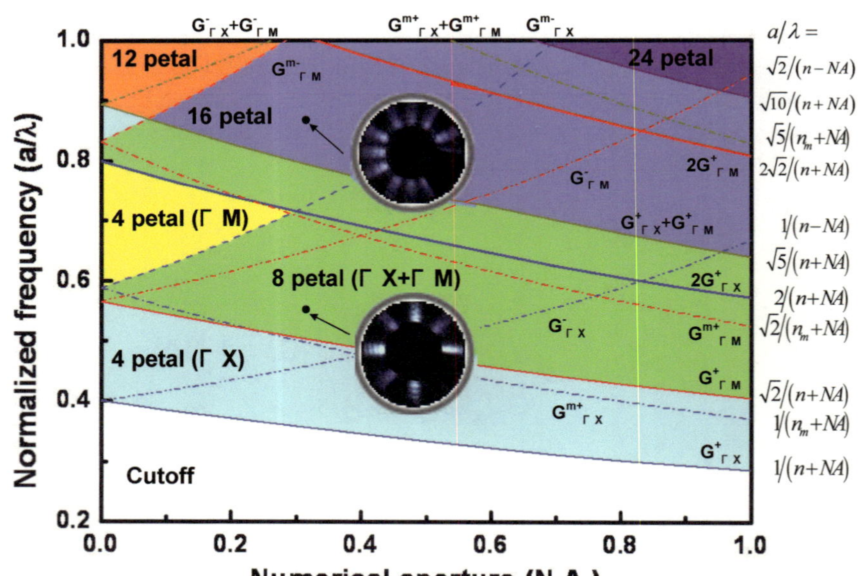

Fig. 10. Map showing regions with different number of petals. The formulas on the right of the figure are the boundary for regions for NA=1. The insets showed the observed 8-fold (*a* = 260 nm) and 16-fold (*a* = 410nm) symmetry patterns. The regions of various petals are shown with different colors. The directions of the petals are shown in the parenthesis. The "+" and "-" signs indicate the lower and upper boundary for the regions. The highest mode order number is designated as '*m*' with n_m=1.7 (Sapphire) and the maximum index is n=2.5 (GaN).

In addition, we also observed that the intensity of the light propagating inside the PhC is found to decrease with a decay length of 70-90 μm, depending on the orientation and the size of the holes. The decay length is determined using the data in the middle dynamic range of the CCD camera where the intensity decay appears as a linear line on the log linear plot. This value is in the same range of that reported in David et al.[17]. Such a parameter is needed for the design of the PhC light extractors.

4. Polarized light emission properties of GaN-based photonic crystal LEDs

Due to valence band intermixing, the side emission of light from quantum well structure is predominantly polarized in the TE direction (along the wafer plane). The observed polarization ratio has been reported to be as high as 7:1 for GaN/InGaN QWs [18]. For common GaN LED structures grown along the *c* axis, access to this polarized light can only be gained by measurements taken from the edge of the sample [19-20]. Several authors have reported polarized light emission for LED structures grown on nonpolar or semipolar GaN substrates [21-22]. In the present study, we investigate the approach employing photonic crystals (PhCs) which do not require the growth on different orientation of sapphire or GaN substrates nor using specific wafer orientations. PhC has been widely studied in recent years [9, 23-26] for the enhancement of LED efficiency, but polarized light emission using PhC has not been investigated. In this section, we use the PhC structure to access the polarized emission and measured their orientation dependence using a specially designed PhC structure to extract the waveguided light. It is found that the PhC can behave as a polarizer to improve the P/S ratio of the extracted EL emission. The results of the P/S ratio for light propagating in different lattice orientation was found to be consistent with the results obtained using the PhC Bloch mode coupling theory [10, 27-28].

4.1 Measurement results

The GaN-based PhC LED samples used in the present work are the same as described before section 3. The polarization properties of the GaN blue PhC LEDs were measured at room temperature using a scanning optical microscopic system, which included continuous wave (CW) current source (Keithley 238), a 20× microscope objective with numerical aperture (NA) = 0.45, a 40× microscope objective with NA = 0.6, and charge-coupled device (CCD) spectrometer with spectral resolution of 0.1 nm. A 20× objective is used to collect the on-axis emission signal from the sample and formed a high-resolution image with a digital camera CCD. Figure 11(a) shows EL CCD image for the sample with square lattice constant a = 260 nm corresponding to a/λ = 0.553. Inset in Fig. 11(a) are the PL CCD image and the reduced Brillioun zone. The observed light emission is from the light propagation along the ΓM and ΓX directions as reported before section 3. Further, the extraction enhancement of the PhC LED chips was determined to be above 100% by mounting the dies on TO packages and using an integration sphere with Si photodiode, when compared to the GaN-based LED chips without PhC. A polarizer (Newport, 10LP-VIS-B) was placed on the GaN blue PhC LEDs for the EL measurements. Figure 11(b) presents CCD image of room temperature EL for samples biased at a drive current of 20 mA. The red dash line indicates the polarization axis for the polarizer. Since the polarization direction of the light is perpendicular to its propagation directions, the light propagated in the direction align with the axis of the polarizer will be blocked. The luminescent signal emitted by the sample was collected by the

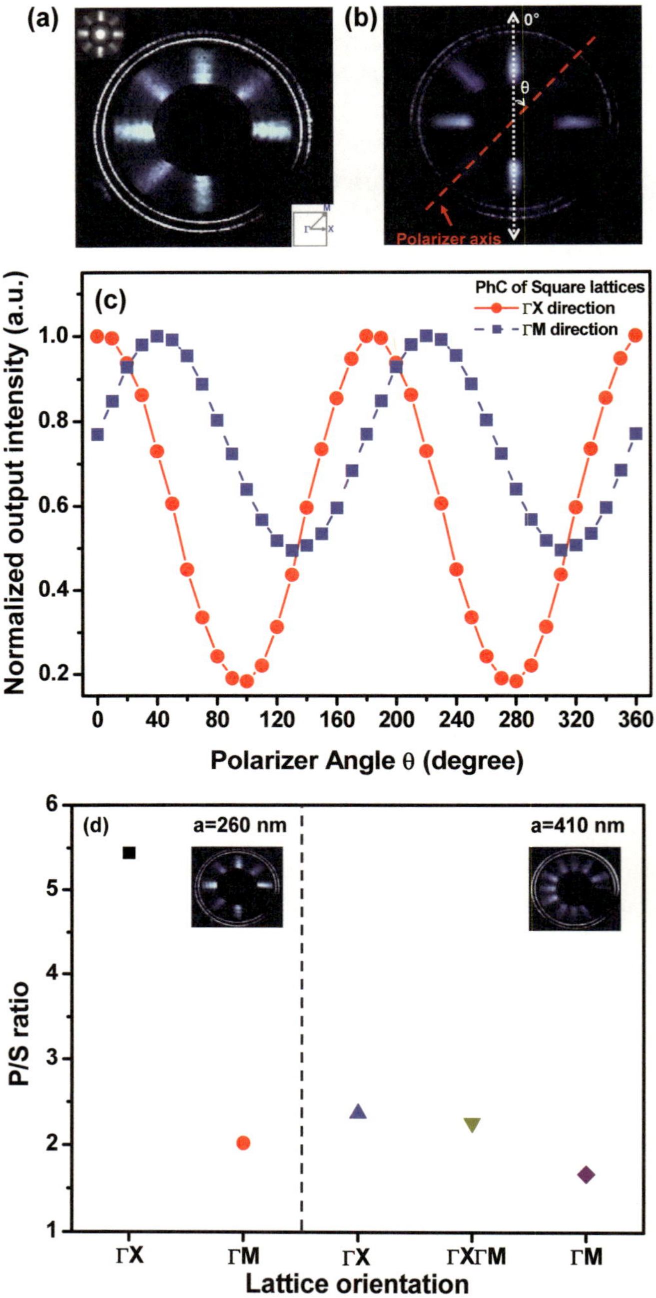

Fig. 11. (a) CCD EL images for lattice constants a = 260 nm, inset of the PL CCD image, and the reduced Brillouin zone. (b) CCD EL images show polarization properties; the red line indicates the polarization axis of the polarizer. (c) Spectrally integrated EL intensity of the GaN PhC LED as a function of polarizer angle θ. (d) P/S ratio of different lattice constant as a function of orientation direction.

40× objective lens of the confocal microscope and was transferred to a monochromator for μ-PL measurement through an optical fiber with core diameter of 600 μm. Figure 11(c) shows the EL intensity as a function of the orientation of the polarizing filter placed between the GaN blue PhC LED and the spectrometer, at a drive current of 20 mA. The intensity at various angles was determined from image taken under the same bias condition. Thus the polarization for different propagation direction can be determined as shown in Fig. 11(c). It can be seen that there is a periodic variation of the EL intensity with angular orientation of the polarizer. This indicates that the light collected from the PhC LED is partially polarized, and the P/S ratio [defined as $P/S=I_{max}/I_{min}$] were 5.5 and 2.1 for square lattice (a = 260 nm) in ΓX and ΓM direction, respectively, as shown in Fig. 11(d). Fig. 11(d) also shows the P/S ratio measured in other samples with different period. For square-lattice PhC LEDs, P/S ratio in ΓX orientation is larger than those in other orientations despite the lattice constants. In addition, for the same orientations, PhC LEDs with shorter lattice constant have higher P/S ratio.

4.2 Coupled mode theoretical discussion

The experimental results described above can be explained by examining the electromagnetic field distributions of PhC Bloch modes. Field distributions of Bloch modes were calculated by plane wave expansion (PWE) method using the structure with PhC sandwiched in between air and GaN materials. Figure 12(a) schematically shows the device structure where light is generated and extracted through PhCs. Due to the valence band mixing effects in MQW, guided light propagating in the GaN slab is nearly linear polarized in transverse direction as shown in Fig. 12(b). For PhC a/λ = 0.553, the field distribution for propagation in ΓX and ΓM directions are shown schematically in Fig. 12(c) and Fig. 12(d), respectively, where the arrows indicate the electric field vectors in the plane, and black circles indicate the locations of holes. The individual electric field distributions are complicated, resulting in complicated polarization pattern. It can be seen that the field distribution in ΓX orientation has linear-like polarization behavior, and those in ΓM orientation has circular-like polarization [29]. This behavior can be inferred from the arrangement of the atoms relative to the propagation direction. For ΓX direction, the propagating wave sees the same atom arrangement in the planes perpendicular to the propagating direction from one lattice plane to plane, while in the ΓM direction, the field distribution sees an alternately displaced atom arrangements from plane to plane. Such a staggered atom arrangement will tend to generate the field components normal to the polarization plane. Based on the couple mode theory, the polarization behavior of extracted light can follow the Bloch modes in PhCs and reveal the similar polarization characteristics. Therefore P/S ratio of light extracted through ΓX orientation would be higher than through ΓM orientation. From the Bloch mode patterns in Fig. 12, the experimental polarization results can be realized and consistent with the above discussion.

At a/λ = 0.872, the field distribution in ΓX orientation also has more linear-like than circular-like behavior, and those in ΓM orientation have stronger circular-like polarization as shown in Fig. 12(e) and 12(f). The degree of the polarization appears to be much weaker than that for a/λ = 0.553. In order to discuss this observation, P/S ratio as a function of normalized frequency was calculated. We employ the plane-wave expansion method to calculate the

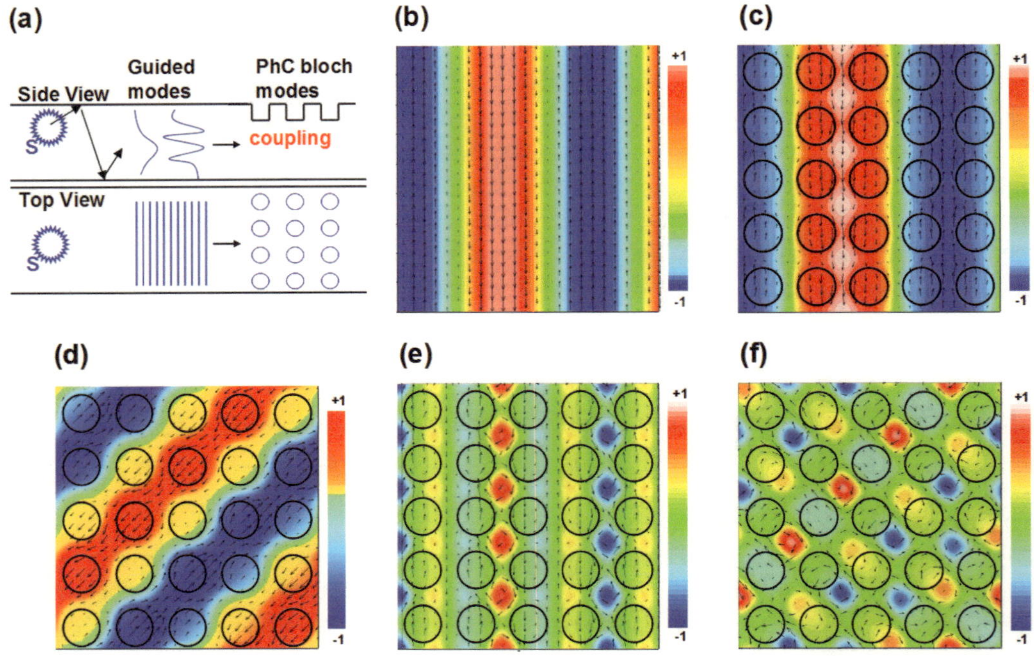

Fig. 12. (a) Schematic of the light generating, propagating, and coupling to PhC Bloch modes. Electromagnetic field distributions for a waveguiding mode in the (b) plane slab guide mode and PhC Bloch modes in the (c) ΓX and (d) ΓM directions of the frequency a/λ = 0.553 and in the (e) ΓX and (f) ΓM directions of the frequency a/λ = 0.872, respectively. Arrows indicate the electric field vectors in the plane, and black circles indicate the locations of lattice points.

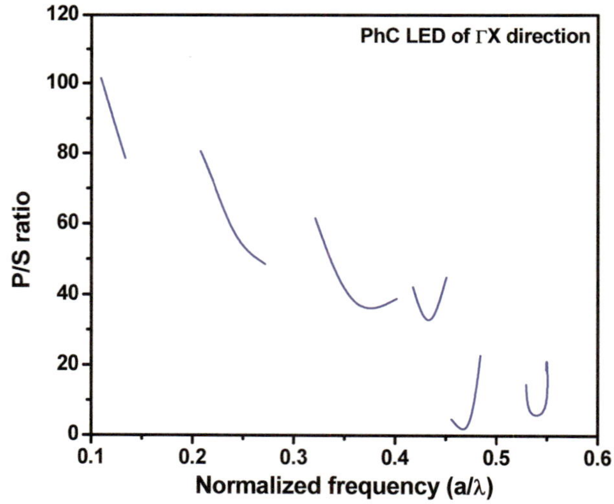

Fig. 13. P/S ratio of PhC Bloch leaky modes in ΓX direction as a function of normalized frequency.

polarization properties (P/S ratio) of the leaky modes in the ΓX directions as a function of normalized frequency. In the calculation, the polarization of the generated light is assumed to be TE polarized. The calculation was carried for each band alone the ΓX direction up to the light line where the light becomes guided and its polarization is then the same as they were generated. As can be seen in Fig. 13, the trend of P/S ratio is decreasing with normalized frequency although the trend within each band is not uniform depending on the filed distribution. Details of this discussion will appear in later publication. It can also be seen from Fig. 13 that by varying the fill factor the lattice constant, the PhC can be designed to have higher extraction efficiency for TE polarization while discriminating the TM polarization. In such case, very high P/S ratio (>10^2) can be achieved. The maximum efficiency for the polarized emission that can be obtained in such case is equal to the TE portion of the total emission which be as high as 88% for a 7:1 P/S ratio.

5. Conclusion

In conclusion, we have experimentally and theoretically demonstrated that surface emitted anisotropic light extraction and polarized light from GaN-based LEDs. The EL images of the anisotropy light extraction distribution in the azimuthal direction were obtained with specially designed annual GaN PhC LED structures, which is dependent on the orientations of the PhC lattice and lattice constants and shows a four-fold symmetric light extraction patterns with varying numbers of petals in the plane of the waveguide. The regions corresponding to the various numbers of petals are determined for increasing lattice constant. More petals appear in the observed image with increasing lattice constant, and some of the petals may disappear. The regions for the appearance and disappearance of the petals are determined by the Bragg diffraction analysis using Ewald construction. In addition the angular dependence of the light extraction for waveguided light incidents to plane with various lattice orientations is also determined. The results show that the light extraction for the square lattices can only occur for certain crystal directions according to the lattice symmetry. Further, a P/S ratio of 5.5 (~85% polarization light) has been observed. The polarization characteristics are theoretically discussed by couple mode theory. At lower normalized frequency, PhC LED has better polarization property, and lattice orientation not only affects the extraction efficiency but also P/S ratio of radiative light. This polarization behavior suggests an efficient means to design and control the GaN blue PhC LEDs for polarized light emission.

6. References

[1] Fujii, T., Gao, Y., Sharma, R., Hu, E. L., DenBaars, S. P., and Nakamura, S. (2004). Increase in the extraction efficiency of GaN-based light-emitting diodes via surface roughening. *Applied Physics Letters*, Vol. 84, pp. 855–857.

[2] Stocker, D. A., Schubert, E. F., and Redwing, J. M. (1998). Crystallorgaphic wet chemical etching of GaN. *Applied Physics Lett.*, Vol. 73, pp. 2654–2656.

[3] Chang, S. J., Chang, C. S., Su, Y. K., Lee, C. T., Chen, W. S., Shen, C. F., Hsu, Y. P., Shei, S. C., and Lo, H. M. (2005). Nitride-based flip-chip ITO LEDs. *IEEE Transactions on Advanced Packaging*, Vol. 28, pp. 273–277.

[4] Wierer, J. J., Steigerwald, D. A., Krames, M. R., O'Shea, J. J., Ludowise, M. J., Christenson, G., Shen, Y. C., Lowery, C., Martin, P. S., Subramanya, S., Gotz, W., Gardner, N. F., Kern, R. S., and Stockman, S. A. (2001). High-power AlGaInN flip-chip light-emitting diodes. *Applied Physics Letters*, Vol. 78, pp. 3379-3381.

[5] Oder, T. N., Kim, K. H., Lin, J. Y., and Jiang, H. X. (2004). III-nitride blue and ultraviolet photonic crystal light emitting diodes. *Applied Physics Letters*, Vol. 84, pp. 466-468.

[6] Wierer, J. J., Krames, M. R., Epoer, J. E., Gardner, N. F., Craford, M. G., Wendt, J. R., Simmons, J. A., and Sigalas, M. M. (2004). InGaN/GaN quantum-well heterostructure light-emitting diodes employing photonic crystal structure. *Applied Physics Letters*, Vol. 84, pp. 3885-3887.

[7] Kim, K., Schubert, E. F., and Cho, J., (2007). Linearly polarized emission from GaInN lightemitting diodes with polarization-enhancing reflector. *Opt. Express*, Vol. 15, pp. 11213-11218.

[8] David, A., Meier, C., Sharma, R., Diana, F. S., DenBaars, S. P., Hu, E., Nakamura, S., and Weisbuch, C. (2005). Photonic bands in two-dimensionally patterned multimode GaN waveguides for light etraction. *Appl. Phys. Lett.*, Vol. 87, pp. 101107-1-101107-3.

[9] Lai, C. F., Kuo, H. C., Chao, C. H., Hsueh, H. T., Wang, J.F. T., Yeh, W. Y., and Chi, J. Y. (2007). Anisotropy of light extraction from two-dimensional photonic crystal light-emitting diodes. *Appl. Phys. Lett.*, Vol. 91, pp. 123117-1-123117-3.

[10] Lai, C. F., Chi, J. Y., Yen, H. H., Kuo, H. C., Chao, C. H., Hsueh, H. T., Wang, J. F. T., Huang, C. Y., and Yeh, W. Y. (2008). Polarized light emission from photonic crystal light-emitting diodes. *Applied Physics Letters*, Vol. 92 No. 24, pp. 243118-1-243118-3.

[11] McGroddy, K., David, A., Matioli, E., Iza, M., Nakamura, S., DenBaars, S., Speck, J. S., Weisbuch, C., and Hu, E. L. (2008). Directional emission control and increased light extraction in GaN photonic crystal light emitting diodes. *Applied Physics Letters*, Vol. 93, pp. 103502-1-103502-3.

[12] Lai, C. F., Chao, C. H., Kuo, H. C., Yen, H. H., Lee, C. E., and Yen, W. Y. (2009). Directional light extraction enhancement from GaN-based film-transferred photonic crystal light-emitting diodes. *Appl. Phys. Lett.*, Vol. 94, pp. 123106-1-123106-3.

[13] Schubert, M.F., Chhajed, S., Kim, J.K., Schubert, E.F. and Cho, J. (2007) Polarization of light emission by 460 nm GaInN/GaN light-emitting diodes grown on (0001) oriented sapphire substrates. *Applied Physics Letters*, Vol. 91, pp. 051117-1-051117-3.

[14] Yeh, P. (1991), *Optical Waves in Layered Media*, Wiley.

[15] Yablonovitch, E. (1987). Inhibited Spontaneous Emission in Solid-State Physics and Electronics. *Physical Review Letters*, Vol. 58, pp. 2059-2062.

[16] Weisbuch, C., David, A., Fujii, T., Schwach, C., Denbaars, S., Nakamura, S., Rattier, M., Bensity, H., Houdre, R., Stanley, R., Carlin, J.F., Krauss, T.F. and Smith, C.J.M. (2004). Recent results and latest views on microcavity LEDs. Proceedings of SPIE, Vol. 5366, pp. 1-19.

[17] David, A., Meier, C., Sharma, R., Diana, F.S., DenBaars, S.P., Hu, E., Nakamura, S. and Weisbuch, C. (2005). Photonic bands in two-dimensionally patterned multimode GaN waveguides for light extraction. *Applied Physics Letters,* Vol. 87, pp. 101107-1-101107-3.

[18] Schubert, M. F., Chhajed, S., Kim, J. K., Schubert, E. F., and Cho, J. (2007). Polarization of light emission by 460 nm GaInN/GaN light-emitting diodes grown on (0001) oriented sapphire substrates. *Appl. Phys. Lett.,* Vol. 91, pp. 051117-1-051117-3.

[19] Shakya, J., Knabe, K., Kim, K. H., Li, J., Lin, J. Y., and Jiang, H. X. (2005). Polarization of III-nitride blue and ultraviolet light-emitting diodes. *Appl. Phys. Lett.*, Vol. 86, pp. 091107-1-091107-3.

[20] Jia, C., Yu, T., Mu, S., Pan, Y., Yang, Z., Chen, Z., Qin, Z., and Zhang, G. (2007). Polarization of edge emission from III-nitride light emitting diodes of emission wavelength from 395 to 455 nm. *Appl. Phys. Lett.,* Vol. 90, pp. 211112-1-211112-3.

[21] Sharma, R., Pattison, M., Masui, H., Farrell, R. M., Baker, T. J., Haskell, B. A., Wu, F., DenBaars, S. P., Speck, J. S., and Nakamura, S. (2005). Demonstration of a semipolar $(10\overline{13})$ InGaN/GaN green light emitting diode. *Appl. Phys. Lett.,* Vol. 87, pp. 231110-1-231110-3.

[22] T. Koyama, T. Onuma, H. Masui, A. Chakraborty, B. A. Haskell, S. Keller, U. K. Mishra, J. S. Speck, S. Nakamura, and S. P. DenBaars, *Appl. Phys. Lett.,* Vol. 89, pp. 091906, 2006.

[23] Ichikawa, H., and Baba, T. (2004). Efficiency enhancement in a light-emitting diode with a two-dimensional surface grating photonic crystal. *Appl. Phys. Lett.,* Vol. 84, pp. 457-459.

[24] Oder, T. N. , Kim, K. H., Lin, J. Y., and Jiang, H. X., (2004). III-nitride blue and ultraviolet photonic crystal light emitting diodes. *Appl. Phys. Lett.,* Vol. 84, pp. 466-468.

[25] Chen, L., and Nurmikko, A. V. (2004). Fabrication and performance of efficient blue light emitting III-nitride photonic crystals. *Appl. Phys. Lett.,* Vol. 85, pp. 3663-3665.

[26] Lai, C. F., Chi, J. Y., Kuo, H. C., Chao, C. H., Hsueh, H. T., Wang, J.F. T., and Yeh, W. Y. (2008). Anisotropy of light extraction from GaN two-dimensional photonic crystals. *Optics Express,* Vol. 16, pp. 7285-7294.

[27] Lai, C. F., Chi, J. Y., Kuo, H. C., Yen, H. H., Lee, C. E., Chao, C. H., Hsueh, H. T., and Yeh, W. Y. (2009). Far-field of GaN film-transferred green light-emitting diodes with two-dimensional photonic crystals. *Optics Express,* Vol. 17, pp. 8795-8804.

[28] Lai, C. F., Chi, J. Y., Kuo, H. C., Yen, H. H., Lee, C. E., Chao, C. H., Yeh, W. Y., and Lu, T. C. (2009). Far-Field and Near-Field Distribution of GaN-Based Photonic Crystal

LEDs With Guided Mode Extraction. *IEEE. J. Sel. Top. Quant. Electron.*, Vol. 15, pp. 1234-1241.

[29] Imada, M., Chutinan, A., Noda, S., and Mochizuki, M. (2002). Multidirectionally distributed feedback photonic crystal lasers. *Phys. Rev. B*, Vol. 65, pp. 195306-1-195306-8.

4

Holographic Fabrication of Three-Dimensional Woodpile-type Photonic Crystal Templates Using Phase Mask Technique

Di Xu[1], Kevin P. Chen[1], Kris Ohlinger[2] and Yuankun Lin[2]
[1]*University of Pittsburgh,*
[2]*University of Texas-Pan American*
U.S.A.

1. Introduction

The telecommunication and computing industries are currently facing increasing challenges to transfer data at a faster rate. Researchers believe that it might be possible to engineer a device operate at optical frequencies. Photonic technology using photon instead of electron as a vehicle for information transfer paves the way for a new technological revolution in this field. Photons used for communication has several advantages over electrons which are currently being used in electronic circuits. For example, photonic devices made of a specific material can provide a greater bandwidth than the conventional electronic devices and can also carry large amount of information per second without interference.

Photonic crystals are such kind of material. They are periodic structures that allow us to control the flow of photons. (John, 1987; Yablonovitch, 1987) To some extent it is analogous to the way in which semiconductors control the flow of electrons: Electrons transport in a piece of silicon (periodic arrangement of Si atoms in diamond-lattice), and interact with the nuclei through the Coulomb force. Consequently they see a periodic potential which brings forth allowed and forbidden electronic energy bands. The careful control of this electronic band allowed the realization of the first transistor. Now, we change our perspective from atom scale to wavelength scale and imagine a slab of dielectric material in which periodic arrays of air cylinders are placed. Photons propagating in this material will see a periodic change in the index of refraction. To a photon this looks like a periodic potential analogous to the way it did to an electron. The difference of the refractive index between the cylinders and the background material can be adjusted such that it confines light and therefore, allowed and forbidden regions for photon energies are formed. (Joannopoulos et al., 1995)

Nowadays, extensive theoretical and experimental studies have revealed many unique properties of photonic crystals useful in optical communication. Intrigued by their vast potential in photonics engineering, tremendous efforts have been invested into the fabrication of three dimensional (3D) photonic crystal structures. However, the fabrication of those photonic crystals with a complete photonic bandgap, i.e. can exhibit bandgaps for the incident lights from all directions, still proves to be a challenge. Considerable efforts have been dedicated to develop fabrication techniques to produce large area defect-free 3D photonic structures toward device applications. This part of research needs to develop a CMOS-compatible, fast and repeatable technique to produce 3D photonic crystal structures with complete bandgaps around the visible and near infrared telecommunication windows. (Ho et al., 1994; Blanco et al., 2000 ; Campbell et al., 2000 ; Deubel et al., 2004)

The Chapter is organized as follows: Section 2 recalls the definition of photonic crystals, its optical properties and the laser holographic lithography fabrication technique for 3D photonic crystal templates. After that, based on the related fundamentals of optics and the interference principle of light beams, Section 3 introduces the novel phase mask techniques for our laser holographic fabrication. The utilization of the phase masks simplifies the fabrication configuration of photonic crystals and is amendable for massive production and chip-scale integration of 3D photonic structures. In Section 4, we discuss specific cases for 3D photonic crystal template fabrication with phase masks techniques. The templates have woodpile symmetries constructed and synthesized at sub-micron scale by pattern rotation and superposition. Section 5 concludes the chapter.

2. Photonic crystal holographic lithography fabrication

2.1 3D photonic crystals

Photonic crystals are typically classified into three categories: 1D, 2D and 3D crystals according to the dimensionality of the stack. Depending on the refractive index contrast, structure geometry and the periodicity, photonic bandgaps are determined for specific frequency ranges in the electromagnetic (EM) spectra. (Joannopoulos et al., 1995) The band structure of a photonic crystal indicates the response of the crystal to certain wavelengths of the EM spectra for a certain propagation direction. It defines optical properties of the crystal such as transmission, reflection and their dependence on the direction of propagation of light. No EM waves can propagate inside the corresponding bandgap ranges. Using this property allows one to manipulate, guide and confine photons, which in turn makes it possible to produce an all optical integrated circuit.

Currently, the fabrication of photonic crystals is quite a hot topic; many groups with many different techniques have shown the formation of photonic crystals with different dimensionalities. Among them, 3D photonic crystals have attracted enormous interest in the last decade in both science and technology communities. Its unique capability to trap photons offers an interesting scientific perspective and can be useful for optical communication and sensing. It is now possible to produce 1D or 2D photonic crystal, at high volume and low cost, through use of deep ultraviolet photolithography, which is the standard tool of the electronics industry. But efficient micro-fabrication of 3D photonic bandgap microstructures, especially at a large-scale has been a scientific challenge over the past decade. So far, a number of fabrication techniques have been employed to produce sub-micron 3D photonic crystals or templates. They include: conventional multilayer stacking of woodpile structures by using semiconductor fabrication processes, (Ho et al., 1994) colloidal self-assembly, (Hynninen et al., 2007) and multi-photon direct laser writing, (Deubel et al., 2004). Each method posses some extent of success. However, we still need to find an economic and rapid way to produce defect-free nano/mrico-scale structure over uniform and large area. This mission has been accomplished by the application of the holographic lithography method. (Berger et al. 1997)

2.2 Holographic lithography method

Holographic lithography has recently been employed to fabricate 3D photonic crystals by exposing a photoresist or polymerizable resin to interference patterns of laser beams. (Campbell et al., 2000) This interference technique requires that multiple coherent beams converge on the same spatial region, which is also called multi-beam interference lithography. It is promising because it creates periodic microstructures in polymers without extensive lithography and etching steps. The monochromaticity and spatial volume of laser light has produced nearly defect-free structures, at submicron scale and over large substrate areas. Photonic structures are defined in photoresist by iso-intensity surfaces of interference patterns. In the case of negative photoresist, the underexposed material is then dissolved away in the post-exposure processing. The overexposed region forms a periodic network motif and acts as a 3D photonic crystal template. In the post processing step, the template can be infiltrated at room temperature with SiO_2 and burned away, leaving behind a daughter inverse template. Then, the daughter SiO_2 template is inverted by infiltration with silicon and selective etching of SiO_2. (Tétreault et al., 2006) The final structure has relative higher index contrast ratio (Si/Air holes) in 3D form, corresponding to relative larger photonic bandgap.

Holographic lithography allows complete control of the translational symmetry of the photonic crystal and provides considerable freedom for design of the unit cell. The electrical field of a laser beam can be described by

$$\vec{E}_i(\vec{r},t) = E_i \cos(\vec{k}_i \cdot \vec{r} - \omega t + \delta_i) \tag{1}$$

where k and ω are the wave vector and angular frequency, respectively, E is the electric field strength, and δ is the phase. When two or more coherent laser beams are presented simultaneously in the same region, the waves interfere with each other and generate a periodic spatial modulation of light. The intensity distribution of the interference field I for N laser beams can be described by a Fourier superposition,

$$I = <\sum_{i=1}^{N} E_i^2(\vec{r},t)> + \sum_{i<j}^{N} \vec{E}_i \bullet \vec{E}_j \cos[(\vec{k}_i - \vec{k}_j) \bullet \vec{r} + (\delta_i - \delta_j)] \tag{2}$$

The structure of the interference pattern can be designed by controlling beam properties such as electric field strength, polarization, wave vector, and phase. The photonic structure formed through holographic lithography has the translational periodicity determined by the difference between the wave vectors k_i-k_j of the interfering beams. Therefore, lattice constants of the photonic structure are proportional to the wavelength of the interfering laser beam. The polarization, represented by the electric field vector, determines the motif placed within the unit cell of the photonic lattice. The initial phase difference shifts the interference pattern and changes the motif within the unit cell. The laser intensity, exposure time, photoresist preparation, and post-exposure development condition will also contribute to the motif of the interference pattern. The photonic structure formed through holographic lithography should have good connectivity in both the dielectric and the air component so that the structure is self-supporting and the unwanted photoresist can be dissolved away.

The N coherent laser beams produce an intensity pattern with maximal (N-1) dimensional periodicity if the difference between the wave vectors is non-coplanar. For example, two interfering beams form a 1D fringe pattern and three crossed beams form a 2D hexagonal log-pile pattern. By using a 4-beam interference setup, a pattern with 3D symmetry can be designed. (Shoji et al., 2003; Lai et al., 2005) Hence, by selecting different beam combinations, and even performing some pattern translations, patterns with different lattice symmetries are possible to make.

To have successful interference lithography, coherence requirements must be met. It is preferred to use a monochromatic and coherent light source. This is readily achieved with a laser or filtered broadband sources. The monochromatic requirement can be reached if a diffraction element is used as a beam splitter, since different wavelengths would diffract into different angles but eventually recombine anyway. In this case, spatial coherence and normal incidence would still be necessary. The coherent length for our laser system requires that the path difference not exceed 10 cm.

3. Phase mask techniques

Fabrication strategies that rely on interference of multiple independent beams can introduce alignment complexity and inaccuracies due to differences in the optical path length and angles among the interfering beams as well as vibration instabilities in the optical setup. In order to improve the optical setup, diffractive optical elements or phase masks have been introduced to create the interference pattern for the holographic fabrication of photonic crystals. (Diviliansky et al., 2003; Lu et al., 2005; Lin et al., 2005) Other than the traditional bulk optical reflective/refractive elements such as mirrors, beam splitters and top-cut prisms (Campbell et al., 2000; Yang et al., 2002), a diffractive optical element is a promising alternative CMOS-compatible choice for 3D holographic lithography. It can be incorporated into phase/amplitude masks used in optoelectronic circuit fabrications to enable a full integration of 3D photonic structures on-chip. A phase mask, typical a phase grating with periodically index variation in height direction, can create multiple laser beams in various diffraction orders that are inherently phase-locked and stable for reproducible creation of 3D interference patterns from a single laser beam. Fig. 1 shows a schematic of the propagation of a laser beam through a 1D phase mask as a diffractive optical element. The phase mask will create three major stable co-plane output beams. These coherent beams then generate a pattern inside the overlap region below the phase mask, in the shape of 2D log-pile. The pattern is recorded in a photoresist to form a periodic template.

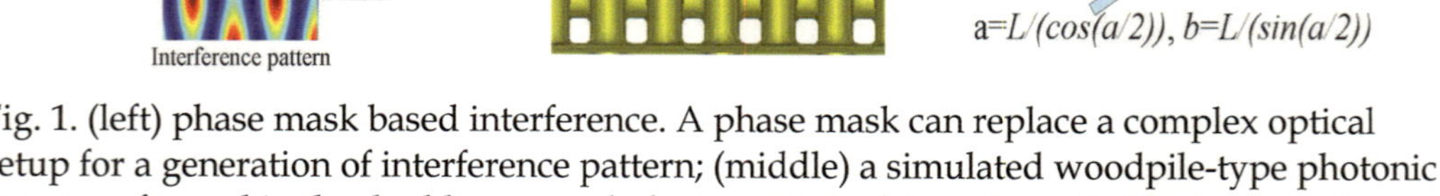

Fig. 1. (left) phase mask based interference. A phase mask can replace a complex optical setup for a generation of interference pattern; (middle) a simulated woodpile-type photonic structure formed in the doubly-exposed photoresist; (right) a schematic illustration of woodpile-type photonic structure with orthorhombic or tetragonal symmetry and its lattice constants.

Theoretically, when a single beam goes through a one-dimensional phase grating, the beam will be diffracted into three as shown in Fig. 1 (left). Beams 1 and 2 are from first order diffraction and beam 3 is from zero order diffraction. Beam 1 and 2 has a diffraction angle θ relative to beam 3. Mathematically these three beams are described by:

$$\vec{E}_1(\vec{r},t) = E_1 \cos[(k\cos\theta)z - (k\sin\theta)x - \omega t + \delta_1] \tag{3}$$

$$\vec{E}_2(\vec{r},t) = E_2 \cos[(k\cos\theta)z + (k\sin\theta)x - \omega t + \delta_2] \tag{4}$$

$$\vec{E}_3(\vec{r},t) = E_3 \cos(kz - \omega t + \delta_3) \tag{5}$$

These three beams will generate a two-dimensional interference pattern. The interference pattern is determined by the laser intensity distribution I in 3D space:

$$\begin{aligned} I = {} & \frac{1}{2}E_1^2 + \frac{1}{2}E_2^2 + \frac{1}{2}E_3^2 + (E_1E_2)\cos[(2k\sin\theta)x + (\delta_2 - \delta_1)] \\ & + (E_1E_3)\cos[(2k\sin^2(\frac{\theta}{2}))z + (k\sin\theta)x + (\delta_3 - \delta_1)] \\ & + (E_2E_3)\cos[(2k\sin^2(\frac{\theta}{2}))z - (k\sin\theta)x + (\delta_3 - \delta_2)] \end{aligned} \tag{6}$$

Fig. 1 (left) shows the interference pattern generated behind the phase mask, assuming the incident laser beam has polarization in y direction. The interference pattern is periodic in the z-direction as well as in x-direction. The periodicity of the interference pattern along x direction is λ/(sinθ) (where λ is the wavelength of laser beam generating the interference pattern). The periodicity c of the interference pattern along z direction is $\lambda/(2\sin^2(\theta/2))$. (Lin et al., 2006a) After the photoresist is exposed to such interference pattern, the sample is rotated by an angle of α along the propagation axis of the incident beam and its position is displaced along the laser propagation direction by 1/4 times $\lambda/(2\sin^2(\theta/2))$. Then the photoresist receives second exposure. The doubly-exposed photoresist is then developed to form a 3D woodpile-type photonic crystal template. Fig. 1 (middle) shows a simulated photonic structure formed with the rotation angle α=90° if a negative photoresist is used. After the photoresist development, the under-exposed area is dissolved away while the area exposed with above-threshold laser dosage is networked to form periodic structures. We illustrate in Fig. 1 (right) how we construct crystal lattices. We set a fundamental length scale L=λ/(sinθ) for such structure because the three beam interference pattern is determined by the laser wavelength and the interference angle θ. L is actually equal to the grating period of the phase mask. The lattice constants in xy plane depend on the angle. They are related by a=L/(cos(α/2)) and b=L/(sin(α/2)), respectively. The photonic crystal template has a lattice constant $c=\lambda/(2\sin^2(\theta/2))=L(\cot(\theta/2))$ along the z direction. If the sample rotation angle is 90°, we have a=b. Thus the 3D structure has a face-centered-tetragonal or face-centered-cubic symmetry. (Lin, 2006a) If the angle α is less than 90°, a face-centered-orthorhombic (a≠b≠c) or face-centered-tetragonal (a≠b=c) structure is formed in the photoresist.

Contrary to an intensity (amplitude) mask, the laser beam travels through a phase mask and accumulates an additional phase relative to light that travels through the air gap. However, a phase mask has much larger diffractive efficiency than an amplitude mask. This property enables those periodic structures have enough contrast ratio to the background in the photoresist polymerization.

The fabrication technique of 1D phase masks has been demonstrated by H. Jiang etc in 1999, (Jiang etc., 1999). E-beam lithography has been used to pattern fine gratings in Si substrate. Then an inversed elastomer mask is obtained by casting a layer of silicon-based organic Polydimethylsiloxane on the substrate. After curing and removal of the master mold, the elastomer mold is used as a mask again in the photolithography process to reproduce a photoresist phase grating on glass substrate. The phase grating fabricated has high quality surface and profile but the required processes are costly, laborious and time-consuming.

Here we demonstrate our holographic approach for phase mask fabrication. (Xu et al., 2008) The experimental setup is based on the principle of a Mach-Zehnder interferometer. As shown in Fig. 2, two coherent beams were cleaned, collimated, separated and focused back into the same photoresist region to produce interference patterns directly. The pattern recorded is a series of parallel fringes with sinusoidal profile. Thus we obtain the phase grating made of photoresist. This phase grating has refractive index difference between photoresist (n=1.67) and air gap, which can produce three coherent beams when it is used as a mask. This approach simplified the previous process and can be applied to more complicated fabrication. If we do multiple exposures and rotate the receiving photoresist between each exposure, we can get a phase mask with higher dimension. In the next section we will describe some concrete experimental steps to fabricate holographic phase masks and use them for the interference lithography fabrication of complex photonic crystal templates.

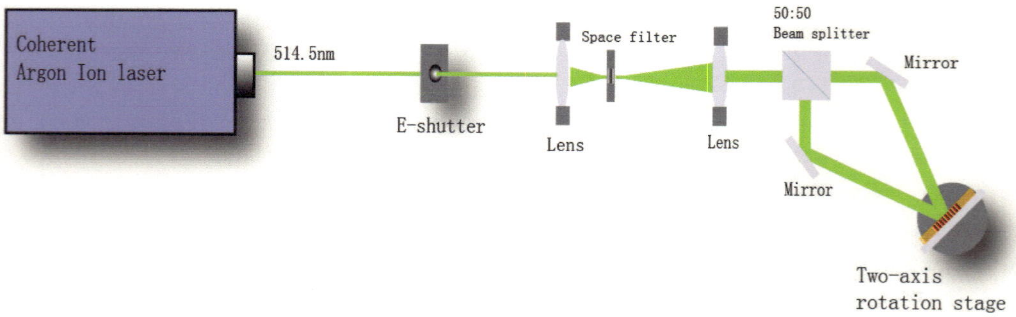

Fig. 2. Experimental setup for holographic fabrication of phase masks.

Overall, the optical diffractive elements are designed to avoid the alignment complexity and inaccuracies due to differences in the optical path length and angles among the interfering beams as well as vibration instabilities in the optical setup. They provide improved convenience in holographic lithography. It is useful to note that the current optical elements, such as beam splitter, mirrors, prisms and phase masks, can be used to generate pattern with all fourteen Bravais lattices in the space group. (Berger et al., 1997; Sharp et al., 2003) All that remains is a need to establish a better understanding of the relationship between the resulting symmetries and the beam parameters.

4. 3D woodpile photonic crystal fabrication by using 1D phase mask

It is well known that a 1D phase mask can generate three beam interference patterns, which has a 2D log-pile structure. The structure is polarization dependent of the incident EM wave thus lacks completeness of bandgap required for photonic communication. Previous researchers have proposed a method of building a 3D structure using two orthogonal 1D phase masks. (Chan, et al., 2006; Chanda et al., 2006) The beams that propagate through two phase masks will have two log-pile patterns recorded inside the photoresist. If well controlled, a 3D woodpile structure may be piled up by the log-pile structure. However, additional diffractions occur. The distance between two phase masks can also bring unwanted phase delay, which is difficult to adjust in practice. Our solution consists of multiple exposures through one 1D phase mask, which is spatially shifted between exposures, demonstrating an new approach for controllable 3D woodpile structure fabrication.

4.1 Pattern transformation

Here we demonstrate the fabrication process of 3D woodpile photonic crystals template, which can have orthorhombic or tetragonal structure depending on the rotational angle. (Lin et al., 2006a; Poole et al., 2007) Furthermore, the elongation in the z-direction can be compensated by rotating phase mask by an appropriate angle, which increases the lattice constant in the other direction. Theory predicts that the optimized rotation angle of a phase mask can achieve up to a 50% increase in photonic bandgap compared with those formed by two orthogonally oriented phase masks.

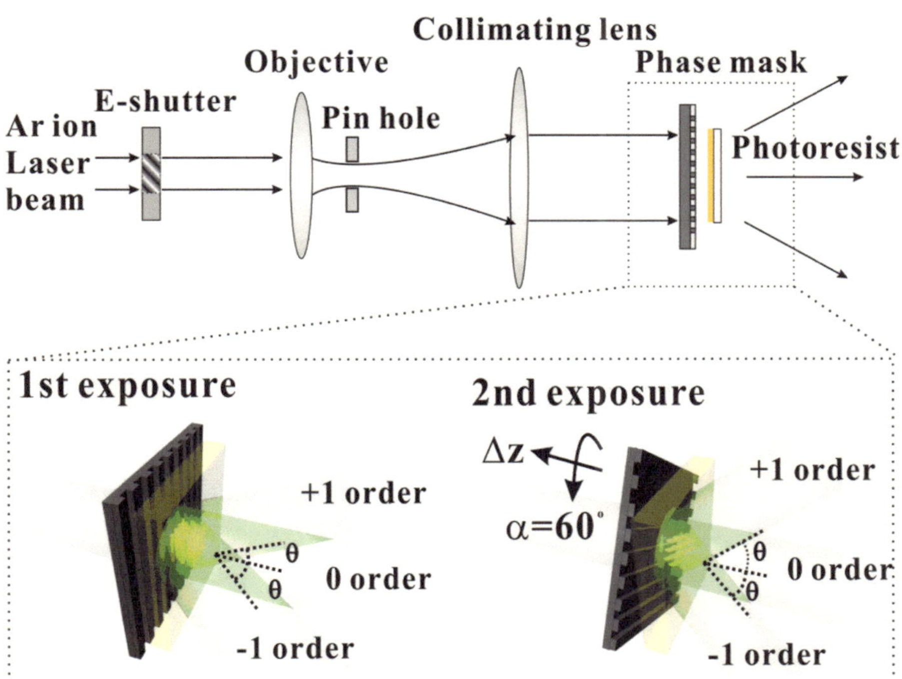

Fig. 3. Experimental setup for 3D photonic crystal template fabrication. Zoom in view is the schematic sketch of the double exposures procedure.

The interference pattern for a single exposure through a phase mask is in 2D log-pile structure, which is periodic in the z direction as well as in the x (or y) direction, as shown in Fig. 3. If we do a second exposure to record another log-pile structure on the same region, with appropriate relative rotation and shifting, we can have a 3D woodpile structure, which has periodic structure in all x, y and z directions, as shown in Fig. 4. It demonstrates a simulated structure from the dual-exposure procedure. Similar to how the photoresist reacts to illumination, the structure represents the receiving laser intensity distribution, i.e. the interference region, in the negative photoresist. The boundary of the 3D pattern is defined by setting a threshold value. The regions with intensity lower than the threshold value are removed and the regions with intensity equal and greater than the threshold value are sustained. Thus the photoresist records the interference pattern and can be visualized after development.

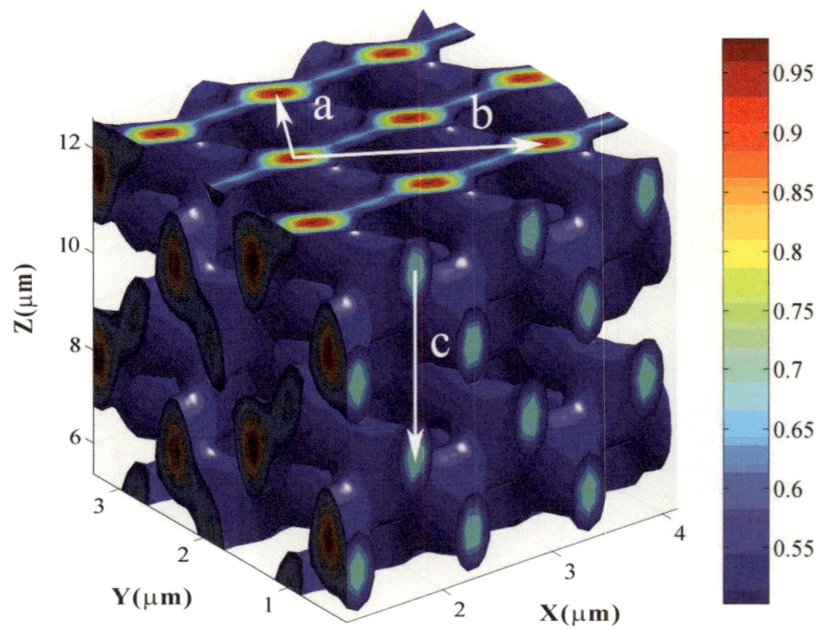

Fig. 4. Simulated 3D woodpile structure generated by double exposures. The rotational angle of phase mask is 60º. The scale bar shows the accumulated laser energy density upon two exposures.

Theoretically, the rotation of the interference pattern can be regarded as replacing the wave vector k of the diffractive beams, by a coordinate transform with rotation angle α;

$$\vec{k}_1 = (k\cos\theta)z - [(k\sin\theta)\cos\alpha]x - [(k\sin\theta)\sin\alpha]y \quad (7)$$

$$\vec{k}_2 = (k\cos\theta)z - [(k\sin\theta)\cos\alpha]x + [(k\sin\theta)\sin\alpha]y \quad (8)$$

$$\vec{k}_3 = (k\cos\theta)z \quad (9)$$

while the spatial movement of the pattern can be induced through the phase shift of interfering beams. When a phase difference (ρ_i-ρ_j) is introduced between interference beams, the interfering term I_{int} in Eq.(2) becomes

$$I_{\text{int}} = \sum_{i<j}^{3} \vec{E}_i \bullet \vec{E}_j \cos[(\vec{k}_i - \vec{k}_j)\bullet \vec{r} + (\rho_i - \rho_j) + (\delta_i - \delta_j)] \tag{10}$$

Such a phase difference between laser beams will translate the interference pattern by r_s as described by (Lin et al., 2006b)

$$I_{\text{int}} = \sum_{i<j}^{3} \vec{E}_i \bullet \vec{E}_j \cos[(\vec{k}_i - \vec{k}_j)\bullet (\vec{r} + \vec{r}_s) + (\delta_i - \delta_j)] \tag{11}$$

where the translation r_s is determined by (k_i-k_j) · r_s=(ρ_i-ρ_j). In general, the initial phase difference δ_i -δ_j is a constant if the laser beams are mutually coherent. It will shift the interference pattern relative to the one generated with (δ_i-δ_j)=0. But the initial phase difference will be the same for two exposures. The interference pattern generated by the second exposure needs to be shifted relative to the first one to fabricate the woodpile photonic crystal. The shifting is produced through the extra phase shift of (ρ_i-ρ_j). Specifically, the initial phase difference is zero if all diffracted beams are generated through a single diffractive optical element. Then the final 3D structure can be expressed by adding up the interfering terms I_{int} for two exposures, normalizing and setting proper threshold isosurface values.

Experimentally, the basic approach utilized to fabricate an interconnected periodic polymeric structure is the double-exposure of photosensitive material to three interfering laser beams generated by a 1D phase mask as shown in Fig. 3. A linearly polarized beam from an argon ion laser at 514.5nm is expanded, collimated, and passed through a phase mask to produce two 1st order and one 0th order diffracted beams (intensity ratio 1:5). A layer of photoresist on a silicon wafer is first exposed to the interference of the three laser beams. Thus, a spatially modulated chemical change in the photoresist is produced. A second rotated and translated phase mask is then used to induce a second set of spatially modulated chemical changes in the photoresist. The orientation of the second interference pattern is controlled by the orientation angle α of the second phase mask with respect to the first one. To form an interconnected 3D woodpile structure, the phase mask was shifted along the z direction (c-axis) by a distance r_s=(0, 0, Δz) for the second exposure. This shift has a significant impact on the size of overlap between the two interference patterns and consequently on the size of the bandgap formed in the final structure. A translation of Δz= 0.25c of the second interference pattern along the c-axis yields an optimized fully-interconnected woodpile structure as shown in Fig. 4. High-precision motion stages were used to control the movements of the phase masks with ±100nm accuracy. By controlling the rotational angle and the relative shift of the phase mask along the optic axis, both orthorhombic and tetragonal photonic crystal structures were formed. Fig. 4 shows a simulated face-centered orthorhombic photonic crystal structure formed by rotating the phase mask by α=60° between two exposures. The lattice constants (a, b, c) labelled in Fig. 4 are determined by the angle of diffraction θ of the 1st order beams in the photoresist and by the angular rotation of the phase mask α as $(L/(\cos(\alpha/2))$, $L/(\sin(\alpha/2))$, and $L(\cot(\theta/2))$, (Lin et al., 2006a) respectively. Where L is the grating period given by $L= \lambda/\sin\theta$, and λ is the laser wavelength in the photoresist material.

4.2 Band diagram of woodpile photonic crystal

The woodpile-type photonic crystal template will be converted into high refractive index materials using the approach of CVD infiltration (Miguez et al., 2002; Tétreault et al., 2005) in order to achieve a full bandgap photonic crystal. (Maldovan& Thomas, 2004) We calculated the photonic bandgap for converted silicon structures where 'logs' are in air while the background is in silicon. The calculation has been performed for photonic structures formed with various interference angles θ and rotation angles α. Fig. 5 (left) shows the first Brillouin surface of the face-centered-orthorhombic lattice. Coordinates of high symmetric points on the Brillouin surface varies with different structures. MIT Photonic-Bands Package (Johnson & Joannopoulos, 2005) was used to calculate the photonic bandgap of the converted silicon structure. Fig. 5 (right) shows the photonic band structure for the converted silicon woodpile-type structure with $c/L=2.4$ and $\alpha=51°$ (the dielectric constant of 11.9 was used for silicon in the calculation). (Toader et al., 2004) We would like to clarify that the λ_{photon} in the y-axis label of the Fig. 5 (right) is the wavelength of photons in the photonic band, not the wavelength of the exposure laser. The band structure shows that a photonic full bandgap exists between the 2nd and 3rd bands with a bandgap size of 8.7 % of the gap central frequency.

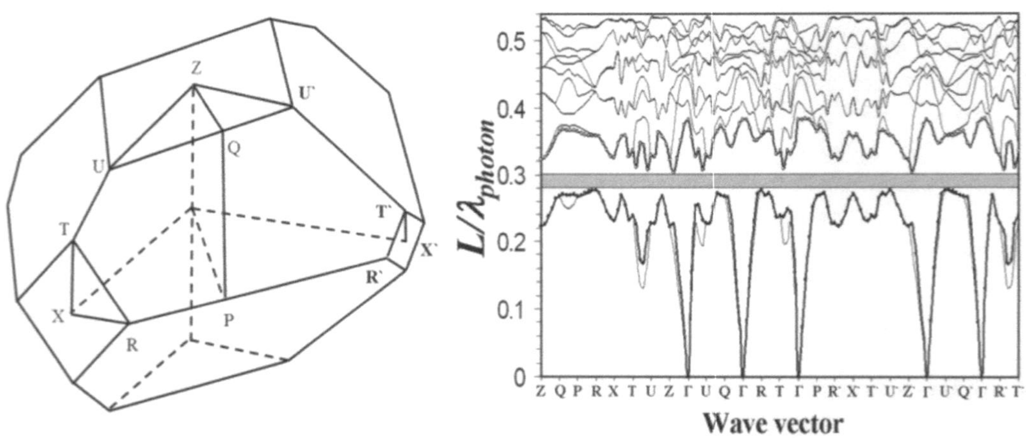

Fig. 5. (left) First Brillouin surface of face-centered-orthorhombic lattice; (right) photonic band structure for an orthorhombic photonic crystal. λ_{photon} is the wavelength of photons in the photonic band.

4.3 Bandgap size *vs* shifting Δz and rotation α

The significance of the overlap between the two alternating high-intensity stacks controlled by the translation Δz of the second phase mask along the optical axis is depicted in Fig. 6. The relative bandgap size is measured from the bandgap diagram as shown in Fig. 5 (right) and defined by the ratio of central frequency and the frequency range of the bandgap. From

Fig. 6 we can see that a global bandgap of 4% exists in structures with α=60° and Δz=0.03c. The maximum photonic bandgap appears at Δz=0.25c, where the 2nd log-pile pattern moves to a location closest to the 1st log-pile pattern, symmetrising the whole 3D woodpile structure. In structures where Δz≤0.03c, the width of the bandgap reduces rapidly and eventually vanishes. A maximum bandgap of 17% was achieved at a shift Δz=0.25c.

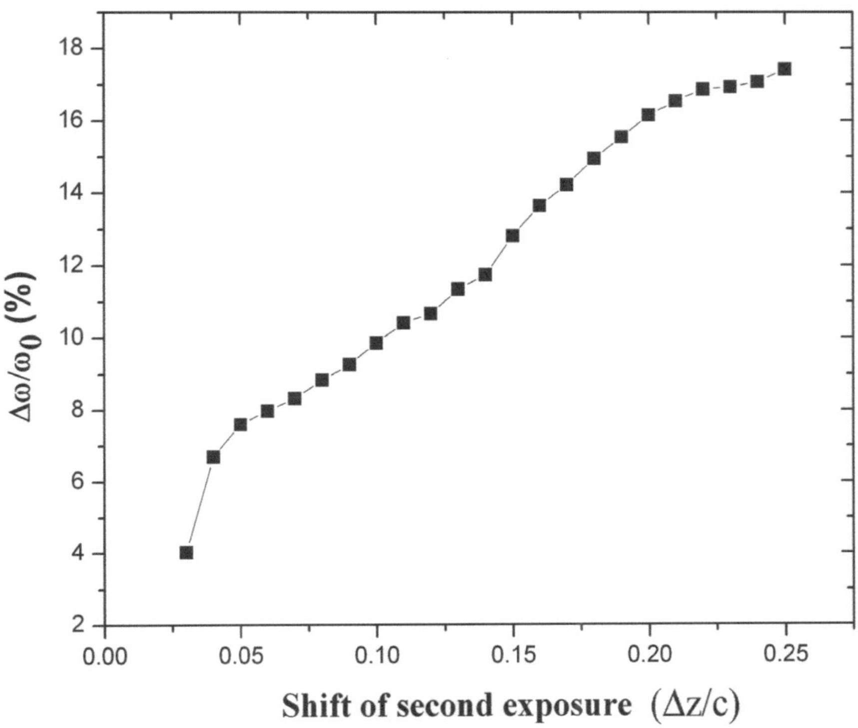

Fig. 6. Photonic bandgap as function of the phase mask displacement Δz between two exposures. The phase mask rotational angle α is 60°. Insets are the first Brillouin surface and photonic band diagram for the face-centered-orthorhombic structure.

To study the dependence of the size of the bandgap on α, photonic bandgap calculations were performed with various c/L ratios as shown in Fig. 7. Since all the laser beams come from the same half-space, the interference pattern generated will be elongated along the c-axis due to relatively small interference angles. This elongation, along with a rotational angle of 90°, causes the lattice constant c to be larger than a and b, yielding a face-centered tetragonal structure. When the rotation angle of phase mask decreases from 90°, the lattice constant b increases, while a decreases; in effect reducing the photonic crystal structure to a lattice with orthorhombic symmetry. A small phase mask rotational angle α can transfer the lattice back into tetragonal again when the lattice constant b is equal to c. When the value of b approaches that of c, the structure becomes more symmetric and the bandgap increases. From simulation, we found that the maximum bandgap occurs when the structure has the highest possible symmetry. For relatively small c/L ratios, where c approaches a and b, and

α=90°, the widest bandgap is produced. For larger c/L ratios, the maximum bandgap occurs at a rotational angle $\alpha \neq 90°$. Fig. 7 also illustrates the rotation angles α that maximize the bandgap for structures with a large c/L values. When c is larger than 1.9L, a small rotational angle of the phase mask is required to maximize the bandgap. For c/L=2.0, a 60° rotational angle maximizes the photonic bandgap. Maximizing the bandgap for structures with c/L ratios larger than 2 requires less than 60° angular displacements. For this c/L ratio, varying the rotation angle from 90° initially results in a drop in the width of the gap followed by an increase. This is consistent with the symmetry transformation of the photonic structure, changing from tetragonal symmetry to orthorhombic symmetry then back to tetragonal symmetry.

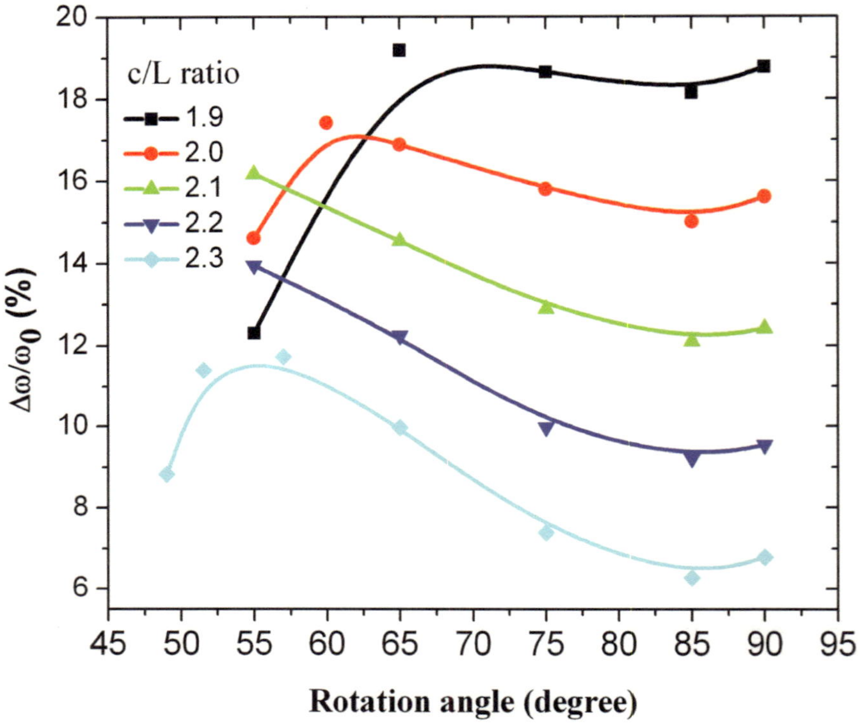

Fig. 7. Photonic bandgap as a function of the phase mask rotational angle α.

4.4 Bandgap size *vs* c/L ratio

Fig. 8 shows the optimum bandgap size in face-centered-tetragonal photonic structures which is formed with the rotation angle α=90° and in face-centered-orthorhombic structure where $\alpha \neq$ 90°, under different beam interference geometries. When c/L is small (beams have a larger interference angle), a rotation angle of 90° is preferred in order to have a larger bandgap. However if c/L is larger than 2.0, then the face-centered-orthorhombic structure is preferred for a larger bandgap. At c/L=2.3, the optimum bandgap size is 11.7% of the gap central frequency for a face-centered-orthorhombic structure formed with a rotation angle near 55°. While the face-centered-tetragonal structure formed with α=90° has a gap size of 6.7%.

To demonstrate the feasibility of the proposed fabrication technique, both orthorhombic and tetragonal structures were recorded into a modified SU-8 photoresist. Utilizing the phase mask method a number of photonic structures can be generated; however there are some practical issues in realizing a photonic structure with a full photonic bandgap. Fig. 8 shows that a photonic bandgap exists in structures with smaller c/L values. Because c/L=cot(θ/2), a bigger interference angle is required in order to generate an interference pattern for a structure with a full bandgap. When the photoresist is exposed into an interference pattern, the interference pattern recorded inside the photoresist will be different from that in air. In the case of c/L=2.5, an interference angle θ=43.6° is required, which is greater than the critical angle of most of photoresist.

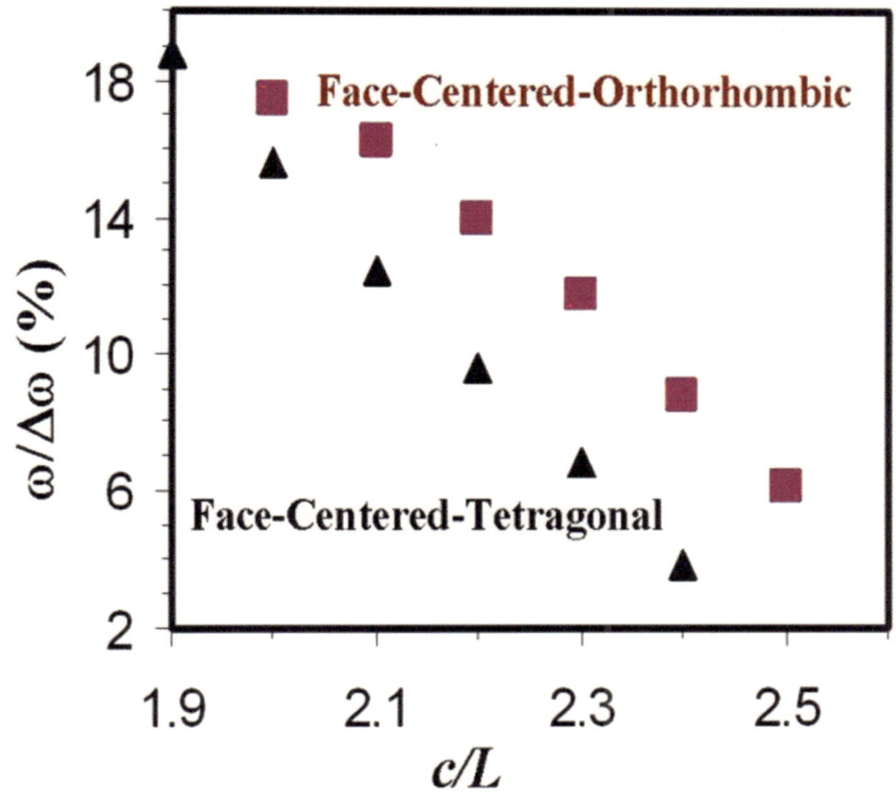

Fig. 8. Photonic bandgap size in face-centered-tetragonal structures (= 90°) and in face-centered-orthorhombic structures (< 90°) for various structures with a different c/L value.

4.5 Experimental results

In order to expose the photoresist to an interference pattern formed under a bigger interference angle, a special setup is arranged for the phase mask and the photoresist as shown in Fig. 9 (left). The photoresist is placed on the backside of the phase mask with the contact surface wetted with an index-match mineral oil. The design of the phase mask is modified correspondently. As a proof-of-principle, we show in Fig. 9 (right) scanning electron microscopy (SEM) of woodpile-type structures in SU-8 photoresist formed through the phase mask based holographic lithography. An Ar ion laser was used for the exposure of 10 μm thick SU-8 photoresist spin-coated on the glass slide substrate. The photoresist and phase mask were both mounted on high-precision Newport stages. Both the phase mask and photoresist were kept perpendicular to the propagation axis of the incident Ar laser beam.

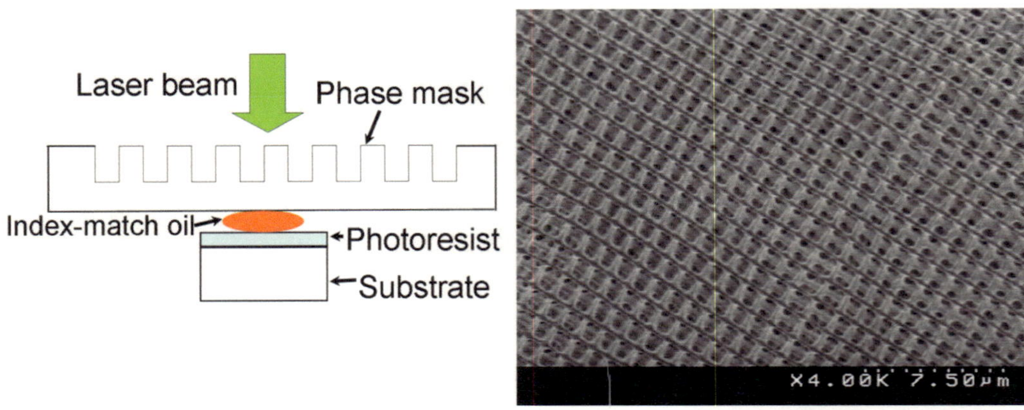

Fig. 9. (left) an arrangement of the phase mask and the photoresist. The interface between the backside of the phase mask and the photoresist is wetted with an index-match fluid; (right) SEM top-view of an orthogonal woodpile-type structure in SU-8 photoresist formed through the phase mask based holographic lithography.

The photoresist solution was prepared by mixing 40 gram SU-8 with 0.5 wt % (relative to SU-8) of 5,7-diiodo-3-butoxy-6-fluorone (H-Nu470), 2.5 wt% of iodonium salt co-initiator (OPPI) and 10 ml Propylene Carbonate to assist the dissolution. Due to the large background energy presented in the generated interference pattern (53% of 0th order), the photoresist solution was further modified by the addition of 20 mol percent Triethylamine. Subsequent exposure to light generates Lewis acids that are vital in the crosslinking process during post exposure bake. The addition of Triethylamine, acting as an acid scavenger, allowed the formation of an energy gap which prevented the polymerization process in locations exposed below the energy threshold. The substrates utilized for crystal fabrication were polished glass slides cleaned with Piranha solution and dehumidified by baking on a hot plate at 200 °C for 20 min. Each substrate was pre-coated with 1μm layer of Omnicoat to enhance adhesion. The SU-8 mixture was spin-coated onto the pre-treated substrate at speeds between 700 and 1500 rpm; resulting in a range of thicknesses from 25 to 5 μm. Pre-bake of SU-8 films was preformed at a temperature of 65 °C for about 30 min. The prepared samples were first exposed under 500mw illumination for 0.9 s using the first phase mask. A second phase mask, which was rotated by α about the optic axis and translated by Δz with respect to the first, was then used for an additional 0.9 s exposure. The samples were post-baked at 65 °C for 10 min and 95 °C for 5 min and immersed in SU-8-developer for 5 min.

Fig. 10(a) shows an SEM top view picture of a woodpile orthorhombic structure recorded in SU-8 with an α of 60°. The inset of the same figure details the predicted structure from simulation. The 3D span of the structure visible in Fig. 10(b) was also imaged by SEM. The layer-by-layer, woodpile nature of the structure is clearly demonstrated. The overlapping and cross-connection of neighbouring layers ensures a stable formation of 3D structures for further processing. From figure 10 (a) and (b), we measured in the SEM the lattice constants to be b=1.3 μm and c=3.4 μm. The elongation in the z-direction was thus compensated by the 60° rotation, compared with b=1.06 μm and c=6.13 μm in the structure generated by two orthogonally-oriented phase masks with similar period used in this work.

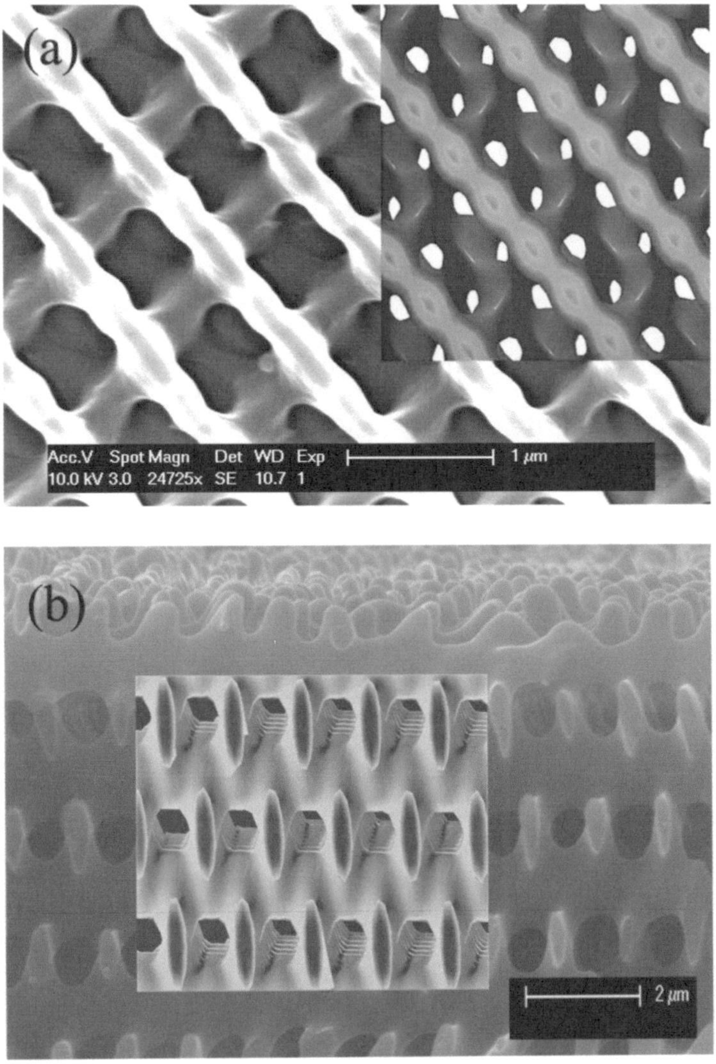

Fig. 10. (a) A SEM top view picture; and (b) a SEM side view picture of a woodpile orthorhombic structure recorded in SU-8 with α=60°. Simulated structures are inserted in Fig.s.

5. Conclusion

In summary, we demonstrate the fabrication of 3D photonic crystal templates in SU-8 using phase mask based holographic lithography technique. Both face-centered-orthorhombic and face-centered-tetragonal woodpile-type photonic crystals have been fabricated. The usage of phase mask dramatically simplified the optical setup and improved the sample quality. The structure and symmetry of the photonic crystals have been demonstrated by controlling the rotational angle of a phase mask to compensate the structural elongation in z-direction in order to enlarge the photonic bandgap. Photonic bandgap computations have been preformed optimally on those woodpile structures with α between 50° to 70° as well as traditional 90° rotation. Our simulation predicts that a full bandgap exists in both orthorhombic and tetragonal structures. The study not only leads to a possible fabrication of photonic crystals through holographic lithography for structures beyond intensively-studied cubic symmetry but also provides a blueprint defining the lattice parameter for an optimum bandgap in these orthorhombic or tetragonal structures.

6. References

V. Berger, O. Gauthier-Lafaye and E. Costard (1997). Photonic band gaps and holography, *J. Appl. Phys.* 82, 62-64.

A. Blanco, E. Chomski, S. Grabtchak, M. Ibisate, S. John, S. W. Leonard, C. Lopez, F. Meseguer, H.Miguez, J. P. Mondia, P. Jessica, G. A. Ozin, A. Geoffrey, O. Toader, and H. M. van Driel (2000). Large-scale synthesis of a silicon photonic crystal with a complete three-dimensional bandgap near 1.5 micrometers, *Nature,* 405, 437-440, London.

M. Campbell, D. N. Sharp, M. T. Harrison, R. G. Denning and A. J. Turberfield (2000). Fabrication of photonic crystals for the visible spectrum by holographic lithography, *Nature,* 404, 53, London.

T. Y. M. Chan, O. Toader and S. John (2006). Photonic band-gap formation by optical-phase-mask lithography, *Phys. Rev. E* 73, 046610.

D. Chanda, L. Abolghasemi and P. R. Herman (2006). One-dimensional diffractive optical element based fabrication and spectral characterization of three-dimensional photonic crystal templates, *Opt. Express,* 14, 8568-8577.

M. Deubel, G. V. Freymann, M. Wegener, S. Pereire, K. Busch, C. M. Soukoulis (2004). Direct laser writing of three-dimensional photonic crystal templates for telecommunications, *Nat. Mater.* 3, 444-447.

I. Diviliansky, T. S. Mayer, K. S. Holliday and V. H. Crespi (2003). Fabrication of three-dimensional polymer photonic crystal structures using single diffraction element interference lithography, *Appl. Phys. Lett.,* 82, 1667.

K. M. Ho, C. T. Chan, C. M. Soukoulis, R. Biswas and M. Sigalas (1994). Photonic band gaps in three dimensions: new layer-by-layer periodic structures, *Solid State Commun.,* 89, 413-416.

A. Hynninen, J. H. J. Thijssen, E. C. M. Vermolen, M. Dijkstra and A. Van Blaaderen (2007). Self-assembly route for photonic crystals with a bandgap in the visible region, *Nat. Mater.,* 6, 202.

H. Jiang, S. Maat, L. Shen and G. J. Mankey (1999). Magnetization reversal in submicron FeNi wires fabricated by phase mask lithography, *Spring MRS Meeting,* San Francisco, CA.

J. Joannopoulos, R. Meade & J. Winn (1995). Photonic Crystals, *Princeton Press,* Princeton, NJ.

S. John (1987). Strong localization of photons in certain disordered dielectric superlattices, *Phys. Rev. Lett.,* 58, 2486.

S. G. Johnson and J. D. Joannopoulos (2001). Block-iterative frequency-domain methods for Maxwell's equations in a planewave basis, *Opt. Express,* 8, 173.

N. D. Lai, W. P. Liang, J. H. Lin, C. C. Hsu , and C. H. Lin (2005). Fabrication of two- and three-dimensional periodic structures by multi-exposure of two-beam interference technique, *Opt. Express,* 13, 9605-9611.

Y. Lin, P. R. Herman and K. Darmawikarta (2005). Design and holographic fabrication of tetragonal and cubic photonic crystals with phase mask: toward the mass-production of three-dimensional photonic crystals, *Appl. Phys. Lett.,* 86, 071117.

Y. Lin, D. Rivera and K. P. Chen (2006a). Woodpile-type photonic crystals with orthorhombic or tetragonal symmetry formed through phase mask techniques, *Opt. Exp.* 14, 887.

Y. Lin, D. Rivera, Z. Poole and K. P. Chen (2006b). Five-beam interference pattern controlled through phases and wave vectors for diamondlike photonic crystals, Appl. Optics 45, 7971.

C. Lu, X. K. Hu, I. V. Mitchell, and R. H. Lipson (2005). Diffraction element assisted lithography: Pattern control for photonic crystal fabrication, *Appl. Phys. Lett.,* 86, 193110/1-3.

M. Maldovan and E. L. Thomas (2004). Diamond-structured photonic crystals, *Nat. Mater.,* 3, 593.

H. Miguez, N. Tétreault, B. Hatton, S. M.Yang, D. Perovic, G. A. Ozin (2002). Mechanical stability enhancement by pore size and connectivity control in colloidal crystals by layer-by-layer growth of oxide, *Chem. Commun.,* 22, 2736-2737, Cambridge, UK.

Y. V. Miklyaev, D. C. Meisel, A. Blanco, G. Von Freymann, K. Busch, W. Koch, C. Enkrich, M. Deubel, M. Wegener (2003). Three-dimensional face-centered-cubic photonic crystal templates by laser holography: fabrication, optical characterization, and band-structure calculations, *Appl. Phys. Lett.,* 82, 1284.

J. H. Moon, S. Yang, D. J. Pine and S. M. Yang (2005). Translation of interference pattern by phase shift for diamond photonic crystals, *Opt. Express,* 13, 9841.

Z. Poole, D. Xu, K. P. Chen, I. Olvera, K. Ohlinger and Y. Lin (2007). Holographic fabrication of three-dimensional orthorhombic and tetragonal photonic crustal templates using a diffractive optical element, *Appl. Phys. Lett.,* 91, 251101.

D. N. Sharp, A. J. Turberfield, and R. G. Denning (2003). Holographic photonic crystals with diamond symmetry, *Phys. Rev. B,* 68, 205102.

S. Shoji, H. Sun, and S. Kawata (2003). Photofabrication of wood-pile three-dimensional photonic crystals using four-beam laser interference, *Appl. Phys. Lett.,* 83, 608-710.

N. Tétreault, G. V. Freymann, M. Deubel, M. Hermatschweiler, F. P. Willard, S. John, M. Wegener, and G. A. Ozin (2006). New route to three-dimensional photonic bandgap materials: silicon double inversion of polymer templates, *Adv. Mater.,* 18, 457, Weinheim, Germany.

O. Toader, T. Y. M. Chan and S. John (2004). Photonic band gap architectures for holographic lithography, *Phys. Rev. Lett.,* 92, 043905.

V. P. Tondiglia, L. V. Natarajan, R. L. Sutherland, D. Tomlin, and T. J. Bunning (2002). Holographic formation of Electro-Optical Polymer-Liquid Crystal Photonic Crystals, *Adv. Mater.,* 14, 187-191.

D. Xu, K. P. Chen, K. Ohlinger, Y. Lin (2008). Holographic fabrication of diamondlike photonic crystal template using two-dimensional diffractive optical elements, *Appl. Phys. Lett.*, 93, 3, 031101.

E. Yablonovitch (1987). Inhibited Spontaneous Emission in Solid-State Physics and Electronics, *Phys. Rev. Lett.*, 58, 2059.

S. Yang, M. Megens, J. Aizenberg, J. Aizenberg, P. Wiltzius, P. M. Chaikin and W. B. Russel (2002). Creating Periodic Three-Dimensional Structures by Multibeam Interference of Visible Laser, *Chem. Mater.*, 14, 2831.

5

Quantum Electrodynamics in Photonic Crystal Nanocavities towards Quantum Information Processing

Yun-Feng Xiao[1,2], Xu-Bo Zou[3], Qihuang Gong[1], Guang-Can Guo[3], and Chee Wei Wong[2]

[1]*State Key Lab for Artificial Microstructure and Mesoscopic Physics, School of Physics, Peking University, Beijing 100871,*

[2]*Optical Nanostructures Laboratory, Center for Integrated Science and Engineering, Solid-State Science and Engineering, and Department of Mechanical Engineering, Columbia University, New York, NY 10027,*

[3]*Key Laboratory of Quantum Information, University of Science and Technology of China, Hefei 230026,*

[1,3]*People's Republic of China*

[2]*USA*

1. Introduction

Cavity quantum electrodynamics (cavity QED) describes few atoms coupling to quantized electromagnetic fields inside an optical cavity (Mabuchi & Doherty, 2002). The core of cavity QED is the strong coherent interaction between the single-mode electromagnetic field and the internal states of the atom. It is one of few experimentally realizable systems in which the intrinsic quantum mechanical coupling dominates losses that due to dissipation (Cirac et al., 1997). Furthermore, it represents an almost ideal and the simplest quantum system which allows quantitative studying of a dynamical open quantum system under continuous observation. Up to the present, three representative optical microcavities have been proposed for studying quantum optics and implementing quantum information (Vahala, 2004). The first one is the conventional Fabre-Perot (FP) type cavities consisting of two concave dielectric mirrors facing each other at a distance of the order of a few 100 μm, where single neutral atoms can be trapped through magneto-optical trap (MOT), optical dipole trap or magnetic trap for a long time (up to several seconds). The second is the microcavities supporting whispering gallery modes, including microspheres, microdisks, and microtoroids. The third type is the nanoscale cavities in photonic crystal (Foresi et al., 1997).

With FP-type microcavities, numerous theoretical schemes have been suggested for generating nonclassical states of cavity fields (Vogel et al., 1993; Parkins et al., 1993; Law et al., 1996) and entangled states of many atoms (Cabrillo et al., 1999), and realizing two-qubit logic gates (Pellizzari et al., 1995; Pachos and Walther, 2002) and universal gates for Fock-state qubits (Santos, 2005), which lead to experimental realization of the Einstein-Podolsky-Rosen (EPR) state of two atoms, Greenberger-Horne-Zeilinger (GHZ) states of three parties (two atoms plus one cavity mode), Schrödinger cat state (Brune et al., 1996), and single-photon state (Brattke et al., 2001) of a cavity field. However, FP-type microcavities have their inherent problems. For example, it is extremely difficult to realize a scalable quantum computation in experiment by integrating many microcavities, though theoretical protocols may be simple and elegant. Recently, whispering gallery microcavities have been studied for cavity QED toward quantum information processing (Xiao et al., 2006) due to their ultrahigh quality factors (Q, which is proportional to the confinement time in units of the optical period) and high physical scalability. Strong-coupling regime has been demonstrated when cold caesium atoms fall through the external evanescent field of a whispering gallery mode. Nevertheless, the cold atoms are ideal stationery qubits (quantum bits), but not suited for good flying qubits. Thus, a solid-state cavity QED (involved single quantum dot (QD), for example) system with whispering gallery microcavities seeks further advancements.

As a new resonant configuration, nanocavities in photonic crystal with high quality factors (Q) and ultrasmall mode volumes (V) are attracting increasing attention in the context of optical cavity QED (Faraon et al., 2008; Fushman et al., 2008; Hennessy et al., 2007; Badolato et al., 2006; Reithmaier et al., 2004; Yoshie et al., 2004). Combined with low loss and strong localization, they present a unique platform for highly integrated nanophotonic circuits on a silicon chip, which can also be regarded as quantum hardware for nanocavity-QED-based quantum computing. Toward this goal, strong interactions between a QD and a single photonic crystal cavity have been observed experimentally (Hennessy et al., 2007; Badolato et al., 2006; Reithmaier et al., 2004; Yoshie et al., 2004). Moreover, single photons from a QD coupled to a source cavity can be remarkably transferred to a target cavity via an integrated waveguide in an InAs/GaAs solid-state system (Englund et al., 2007a), which opens the door to construct the basic building blocks for future chip-based quantum information processing systems. Weak coupling nanocrystal ensemble measurements are reported in TiO2-SiO2 and AlGaAs cavity systems (below 1 μm wavelengths) recently (Guo et al., 2006; Fushman et al., 2005) and also independently in silicon nanocavities with lead chalcogenide nanocrystals (a special kind of QDs) at near 1.55 μm fibre communication wavelengths recently (Bose et al., 2007).

In this Chapter, we theoretically study the coherent interaction between single nanocrystals and nanocavities in photonic crystal. This Chapter is organized as follows. In section 2, our attention is focused on a single QD embedded in a single nanocavity. First, we introduce, derive, and demonstrate the explicit conditions toward realization of a spin-photon phase gate, and propose these interactions as a generalized quantum interface for quantum information processing. Second, we examine single-spin-induced reflections as direct evidence of intrinsic bare and dressed modes in our coupled nanocrystal-cavity system. In section 3, however, our attention is switched on the N coupled cavity-QD subsystems. We examine the spectral character and optical delay brought about by the coupled cavities interacting with single QDs, in an optical analogue to electromagnetically induced transparency (EIT) (Fleischhauer et al., 2005). Furthermore, we then examine the usability of this coupled cavity-QD system for QD-QD quantum phase gate operation and our numerical examples suggest that a two-qubit system with high fidelity and low photon loss.

2. Nanocrystals in silicon photonic crystal standing-wave cavities

In this section, we examine the single-photon pulse (or weak coherent light pulse) interactions of a single semiconductor nanocrystal in a system comprised of standing-wave high-Q/V silicon photonic crystal nanocavities (Xiao et al., 2007a). In contrast to earlier travelling-wave whispering gallery cavity studies (Xiao et al., 2006), we show here that a QED system based on coupled standing-wave nanocavities can realize a spin-photon phase gate even under the bad-cavity limit and provide a generalized quantum interface for quantum information processing. In addition, we demonstrate numerically a solid-state universal two-qubit phase gate operation with a single qubit rotation. This theoretical study is focused within the parameters of near 1.55 μm wavelength operation for direct integration with the fiber network, and in the silicon materials platform to work with the vast and powerful silicon processing infrastructure for large-array chip-based scalability.

2.1 Theoretical model

We begin by considering a combined system consisting of coupled point-defect high-*Q/V* photonic crystal cavities, a line-defect photonic crystal waveguide, and an isolated single semiconductor nanocrystal. We offer some brief remarks on this system before building our theoretical model. When a photon pulse is coupled into the cavity mode via a waveguide (Fig. 1(a)), photons can couple out of the cavity along both forward and backward propagating directions of the waveguide because the cavity supports standing-wave modes. While each cavity can each have a Faraday isolator to block the backward propagating photon, such implementation may not be easily scalable to a large-array of cavities. To obtain only forward transmission, here we examine theoretically a defect cavity system with accidental degeneracy (Fan et al., 1998; Xu et al., 2000; Min et al., 2004) as a generalized study of *cavity-dipole-cavity* systems, and which also provides close to 100% forward-only drop efficiency. This framework is also immediately applicable to non-reciprocal magneto-optic cavities which have larger fabrication tolerances. Both systems support two degenerate even $|e\rangle$ and odd $|o\rangle$ cavity modes (*h*-polarized, dominant in-plane *E*-field) that have opposite parity due to the mirror symmetry, as shown in Fig. 1(a). The waveguides can support both *v*-polarizations (dominant in-plane *H*-field) and *h*-polarizations for polarization diversity (Barwicz et al., 2007).

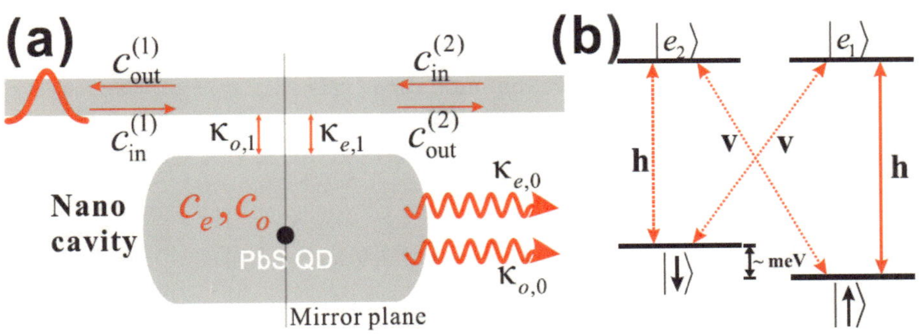

Fig. 1. (a) Sketch of a waveguide side coupled to a cavity which supports two degenerate modes c_e and c_o with opposite parity. (b) lead chalcogenide (e.g. lead sulphide) nanocrystal energy levels and the electron-exciton transitions in the presence of a strong magnetic field along the waveguide direction, which produces nondegenerate transitions from the electron spin states $|\uparrow\rangle$ and $|\downarrow\rangle$ to the charged exciton states $|e_1\rangle$ and $|e_2\rangle$ under the transition selection rules.

Fig. 1(b) shows the energy levels and electron-exciton transitions of our cavity-dipole-cavity system. In order to produce nondegenerate transitions from the electron spin states, a magnetic field is applied along the waveguide direction (Atatüre et al., 2006). $|\uparrow\rangle$ and $|\downarrow\rangle$ play the rule of a stationary qubit, which have shown much longer coherence time than an exciton (dipole or charge). The transition $|\uparrow\rangle \leftrightarrow |e_1\rangle$, with the descending operator $\sigma_- = |\uparrow\rangle\langle e_1|$, is especially chosen and coupled with the cavity modes with single-photon coupling strengths $g_e(\vec{r})$ and $g_o(\vec{r})$, while other transitions are decoupled with the cavity modes.

Now we construct our model by studying the interaction between the nanocrystal and the cavity modes. The Heisenberg equations of motion for the internal cavity fields and the nanocrystal are (Duan et al., 2003; Duan & Kimble et al., 2004; Sørensen & Mølmer, 2003)

$$\frac{dc_e}{dt} = -\mathrm{i}[c_e, H] - \kappa_e c_e + \mathrm{i}\sum_{j=1,2} \sqrt{\kappa_{e1}}\, c_{\mathrm{in}}^{(j)}, \tag{1}$$

$$\frac{dc_o}{dt} = -\mathrm{i}[c_o, H] - \kappa_o c_o + (-1)^{j+1}\sum_{j=1,2} \sqrt{\kappa_{o1}}\, c_{\mathrm{in}}^{(j)}, \tag{2}$$

$$\frac{d\sigma_-}{dt} = -\mathrm{i}[\sigma_-, H] - \gamma\sigma_-, \tag{3}$$

where the interaction Hamiltonian

$$H = \delta_{al}\sigma_+\sigma_- + \sum_{p=e,o} [\delta_{cl} c_p^{\dagger} c_p + g_p(\vec{r}) c_p \sigma_+ + h.c.] \tag{4}$$

is in a rotating frame at the input field frequency ω_l. In contrast to earlier work [Xiao et al., 2006; Waks & Vuckovic, 2004; Srinivasan & Painter, 2007], here we examine the case with the two $|e\rangle$ and $|o\rangle$ modes in the standing-wave cavities in order for forward-only propagation of the qubit. The cavity dissipation mechanism is accounted for by $\kappa_{e(o)} = \kappa_{e(o)0} + \kappa_{e(o)1}$, where $\kappa_{e(o)0}$ is intrinsic loss and $\kappa_{e(o)1}$ the external loss for the even (odd) mode. The nanocrystal dissipation is represented by $\gamma \equiv \gamma_s/2 + \gamma_p$ where γ_s is the spontaneous emission rate and γ_p the dephasing rate of the nanocrystal.

When the two degenerate modes have the same decay rate, i.e., $\kappa_{e0} = \kappa_{o0} = \kappa_0$, $\kappa_{e1} = \kappa_{o1} = \kappa_1$, and $\kappa \equiv \kappa_0 + \kappa_1$, two new states $|\pm\rangle = (|e\rangle \pm \mathrm{i}|o\rangle)/\sqrt{2}$ are suitable to describe this system, which can be thought as two traveling (or rotating) modes. In this regard, the interaction Hamiltonian is expressed as

$$H = \delta_{al}\sigma_+\sigma_- + \sum_{s=+,-} [\delta_{cl} c_s^{\dagger} c_s + g_s(\vec{r}) c_s \sigma_+ + h.c.] \tag{5}$$

where the effective single-photon coupling rates are $g_{\pm}(\vec{r}) = (g_e(\vec{r}) \mp \mathrm{i} g_o(\vec{r}))/\sqrt{2}$. In this case, Eqs. (1), (2), and (3) are rewritten into the corresponding forms with $c_{\pm}$.

The nanocrystal-cavity system is excited by a weak monochromatic field (e.g., single-photon pulse), so that we solve the above motion equations for the below explicit analytical expressions

$$\sigma_-(\omega) = -\mathrm{i}\sum_{s=+,-} g_s(\vec{r}) c_s(\omega)/(\mathrm{i}\delta_{al} + \gamma) \tag{6}$$

and $c_{\pm}(\omega)$ are given as

$$\mathrm{i}\sqrt{2\kappa_1}c_{\mathrm{in}}^{(1)}(\omega)-(\mathrm{i}\delta_{cl}+\kappa)c_{+}(\omega)-\mathrm{i}g_{+}^{*}(\vec{r})\sigma_{-}(\omega)=0\,, \tag{7}$$

$$\mathrm{i}\sqrt{2\kappa_1}c_{\mathrm{in}}^{(2)}(\omega)-(\mathrm{i}\delta_{cl}+\kappa)c_{-}(\omega)-\mathrm{i}g_{-}^{*}(\vec{r})\sigma_{-}(\omega)=0\,. \tag{8}$$

Note that orthogonality of the $|e\rangle$ and $|o\rangle$ basis modes (as shown in Fig. 1a) forces the nanocrystal to choose only either $g_e(\vec{r})=ig_o(\vec{r})$, or $g_e(\vec{r})=-ig_o(\vec{r})$, or both (in which case $|e\rangle$ and $|o\rangle$ are uniquely zero), but no other possibilities. Photon qubit input from only the left waveguide forces only one of the cavity states ($|e\rangle$ + i$|o\rangle$) to exist (Fan et al., 1998), and we assume this cavity environment from the existing photon qubit enhances the $g_e(\vec{r})=-ig_o(\vec{r})$ probability. Of course, with only the left waveguide qubit input in a non-reciprocal magneto-optic cavity, this condition is strictly enforced. Hence we can take $g_e(\vec{r})=-\mathrm{i}g_o(\vec{r})$, which implies $g_{-}(\vec{r})=0$, $g_{+}(\vec{r})=\sqrt{2}g_e(\vec{r})$, to further simplify Eqs. (7)-(8). Now note that the left output $c_{\mathrm{out}}^{(1)}$ remarkably vanishes, while the right output is given by $c_{\mathrm{out}}^{(2)}=c_{\mathrm{in}}^{(1)}(\kappa-2\kappa_1-\mathrm{i}\delta+\lambda)/(\kappa-\mathrm{i}\delta+\lambda)$, where $\lambda=2|g_e(\vec{r})|^2/[\mathrm{i}(\Delta-\delta)+\gamma]$, and $\Delta\equiv\delta_{al}-\delta_{cl}$ and $\delta\equiv-\delta_{cl}$ denote the nanocrystal-cavity and input-cavity detunings, respectively. Importantly, this implies that our quantum phase gate provides a *true* one-way transmission through the cavity-dipole-cavity system.

2.2 Spin-photon phase gate

To examine more of the underlying physics, we consider first the case of exact resonance ($\Delta=0,\delta=0$). When $g_e(\vec{r})|^2/\kappa\gamma>>1$ (the nanocrystal occupies the spin state $|\uparrow\rangle$), we obtain $c_{\mathrm{out}}^{(2)}\approx c_{\mathrm{in}}^{(1)}$. When $g_e(\vec{r})=0$ (the nanocrystal occupies the spin state $|\downarrow\rangle$), we obtain $c_{\mathrm{out}}^{(2)}\approx -c_{\mathrm{in}}^{(1)}$ for $\kappa_1>>\kappa_0$, which indicates that the system achieves a *global* phase change $e^{\mathrm{i}\pi}$. This distinct characteristic allows the implementation of a spin-photon phase gate. After the photon pulse passes though the cavity system, we easily obtain a gate operation

$$\begin{aligned}&|h\rangle|\uparrow\rangle\to|h\rangle|\uparrow\rangle,\ |h\rangle|\downarrow\rangle\to-|h\rangle|\downarrow\rangle,\\&|v\rangle|\uparrow\rangle\to|v\rangle|\uparrow\rangle,\ |v\rangle|\downarrow\rangle\to|v\rangle|\downarrow\rangle.\end{aligned} \tag{9}$$

This two-qubit phase gate combined with simple single-bit rotation is, in fact, universal for quantum computing. More importantly, this interacting system can be regarded as a quantum interface for quantum state sending, transferring, receiving, swapping, and processing.

To efficiently evaluate the quality of the gate operation, the gate fidelity is numerically calculated, as shown in Fig. 2. Considering specifically a lead chalcogenide (e.g. lead sulphide) nanocrystal and silicon photonic nanocavity system for experimental realization, we choose the spontaneous decay as γ_s ~ 2 MHz and all non-radiative dephasing γ_p ~ 1 GHz at cooled temperatures. Photonic crystal cavities have an ultrasmall mode volume V (~ 0.1 μm^3 at 1550 nm), with a resulting calculated single-photon coherent coupling rate $|g_e|$ of ~ 30 GHz. High Q of up to even ~10^6 experimentally and ~10^7 theoretically (Asano et al., 2006; Kuramochi et al., 2006) has been achieved in photonic crystal cavities.

With these parameters, as shown in Fig. 2a, the gate fidelity of the cavity-dipole-cavity system can reach 0.98 or more, even when photon loss is taken into account, and even when the vacuum Rabi frequency g_e is lower than the cavity decay rate κ (bad-cavity limit). The gate fidelity increases initially as the cavity approaches more into the over-coupling regime due to less photon loss and eventually decreases as the nanocrystal-cavity system moves away from the strong coupling regime. Secondly, we note that with non-zero detuning (Δ/κ_o=2; Case III and VI), the gate fidelity slightly decreases but is still adequate. With increasing nanocrystal dissipation rate (Fig. 2b), the fidelity decreases as expected and the system moves away from strong coupling (less nanocrystal interactions with the cavity). The physical essence behind such high fidelities is the true one-way transmission where the nanocrystal couples to $|+\rangle$ mode, with only forward propagation with no backward scattering of the qubit. In addition, accidental degeneracy mismatch may degrade the gate performance. To validate the feasibility of the present scheme, we perform a direct calculation of gate fidelity for different frequency and lifetime of the opposite-parity cavity modes. Even with degeneracy mismatch (ω_e - ω_l = $\delta_{el} \neq \delta_{ol}$ = ω_o - ω_l; in Case IV and VII) with some backward scattering of the qubit, the gate fidelity is shown to remain high. Moreover, with different lifetimes of the cavity modes (Case VII), the fidelity remains high as long as the $|g(\vec{r})|^2/\kappa\gamma >> 1$ condition is satisfied.

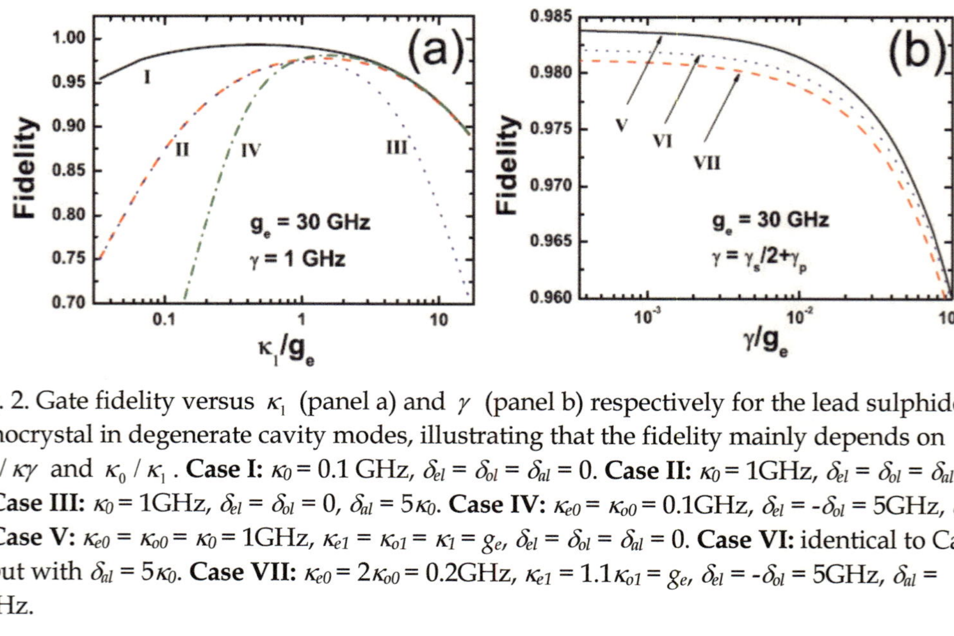

Fig. 2. Gate fidelity versus κ_1 (panel a) and γ (panel b) respectively for the lead sulphide nanocrystal in degenerate cavity modes, illustrating that the fidelity mainly depends on $g_e^2/\kappa\gamma$ and κ_0/κ_1. **Case I:** κ_0 = 0.1 GHz, δ_{el} = δ_{ol} = δ_{al} = 0. **Case II:** κ_0 = 1GHz, δ_{el} = δ_{ol} = δ_{al} = 0. **Case III:** κ_0 = 1GHz, δ_{el} = δ_{ol} = 0, δ_{al} = 5κ_0. **Case IV:** κ_{e0} = κ_{o0} = 0.1GHz, δ_{el} = -δ_{ol} = 5GHz, δ_{al} = 0. **Case V:** κ_{e0} = κ_{o0} = κ_0 = 1GHz, κ_{e1} = κ_{o1} = κ_1 = g_e, δ_{el} = δ_{ol} = δ_{al} = 0. **Case VI:** identical to Case V but with δ_{al} = 5κ_0. **Case VII:** κ_{e0} = 2κ_{o0} = 0.2GHz, κ_{e1} = 1.1κ_{o1} = g_e, δ_{el} = -δ_{ol} = 5GHz, δ_{al} = 1GHz.

2.3 Single-spin-induced reflections

Furthermore, we show that the above cavity-dipole-cavity interaction mechanism can result in interesting transmissions and reflections based on the presence or absence of dipole interaction, and with different detunings. We examine the case of $g_e(\vec{r}) = g$ and $g_o(\vec{r}) = 0$, such as when the nanocrystal is positioned at the cavity mirror plane. Some typical transmission and reflection spectra are shown in Fig. 3. In the absence of a dipole (i.e., the nanocrystal occupies the spin state $|\downarrow\rangle$), the cavity system has near-unity transmission, except when on-resonance. However, when the located nanocrystal is in state $|\uparrow\rangle$, the interacting system is transmission-free and remarkably reflects the cavity field strongly. We emphasize that this reflection is induced by a *single* spin state, and hence can be termed single-spin-induced reflection. The constructive interference of the cavity field can be considered as an optical-analog to electromagnetically induced absorption in the excited state of a 3-level atomic system. The three reflection peaks in Fig. 3(a) can be understood by considering the strong cavity-dipole-cavity interaction, where the input photon pulse experiences three modes: bare odd mode (central peak) and two dressed even modes (side peaks). When the total cavity decay increases, the three peaks overlap increasingly and form a new peak (Figs. 3b-3d) at zero detuning input. We note the high reflectivity for the cavity-dipole-cavity system at zero detuning, even under the bad-cavity limit. This high reflectivity is helpful to permit arrayed controlled phase flip operation with a single circulator at the input.

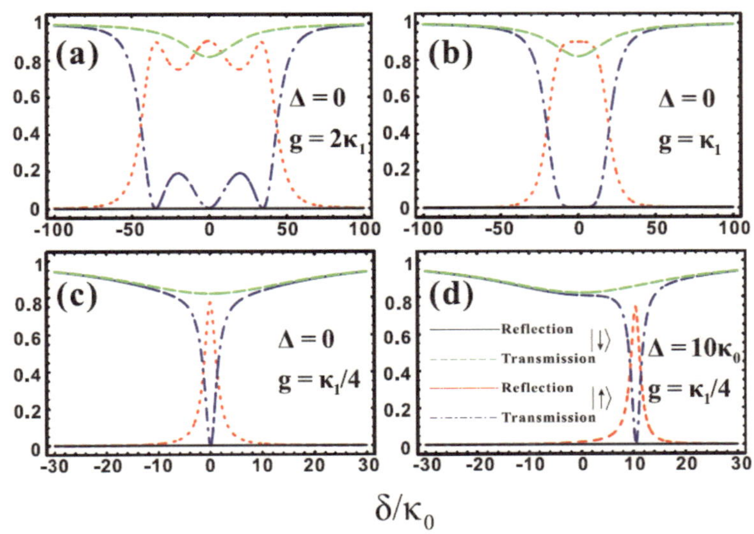

Fig. 3. Reflection and transmission of the spin-photon phase gate with an isolated semiconductor nanocrystal in the degenerate point-defect standing-wave cavity modes. Other conditions for this parameter set include: $\gamma = \kappa_0 / 10$ and $\kappa_1 = 20\kappa_0$, with the nanocrystal located at the cavity mirror plane ($g_e(\vec{r}) = g$ and $g_o(\vec{r}) = 0$). The black solid (green dashed) line is the reflection (transmission) in the absence of a dipole in the cavity. The red dotted (blue dashed-dot) line is the reflection (transmission) in the presence of a dipole in the cavity.

3. Coupled electrodynamics in photonic crystal cavities

Over the past few years, theoretical and experimental interests are mainly focused on a single cavity interacting with atoms, and tremendous successes have been made ranging from strongly trapping single atoms and deterministic generation of single-photon states, to observation of atom-photon quantum entanglement and implementation of quantum communication protocols. For more applications, current interest also lies in the coherent interaction among distant cavities. The coherent interaction of cavity arrays has been studied as an optical analogue to EIT in both theory (Smith et al., 2004; Xiao et al., 2007b) and experiment (Xu et al., 2006; Totsuka et al., 2007). Coupled cavities can be utilized for coherent optical information storage because they provide almost lossless guiding and coupling of light pulses at slow group velocities. When dopants such as atoms or QDs interact with these cavities, the spatially separated cavities have been proposed for implementing quantum logic and constructing quantum networks. Recent studies also show a strong photon-blockade regime and photonic Mott insulator state (Hartmann et al., 2006; Hartmann & Plenio, 2007), where the two-dimensional hybrid system undergoes a characteristic Mott insulator to superfluid quantum phase transition at zero temperature (Greentree et al., 2006; Angelakis et al., 2007). Recently, it has shown that coupled cavities can also model an anisotropic Heisenberg spin-1/2 lattice in an external magnetic field (Hartmann et al., 2007). The character of a coupled cavity configuration has also been studied using the photon Green function (Hughes, 2007).

3.1 Model of coupled N cavity–QD subsystems

Using transmission theory, we study coherent interactions in a cavity array that includes N cavity–QD subsystems (Xiao et al., 2008), with indirect coupling between adjacent cavities through a waveguide (Fig. 4). First, we investigate a subsystem in which a single cavity interacts with an isolated QD. Here for simplicity we suppose that only a single resonance mode (h-polarized) is present in the cavity, although two-mode cavity–QD interactions have been considered in the previous section. The cavity–QD–waveguide subsystem has mirror-plane symmetry, so that the mode is even with respect to the mirror plane. We can easily obtain the Heisenberg equations of motion

$$\frac{d\hat{c}_j}{dt} = -i\left[\hat{c}_j, H_j\right] - \Gamma_j \hat{c}_j + i\sqrt{\kappa_{1,j}}\left(\hat{a}_{\text{in}}^{(j)} + \hat{b}_{\text{in}}^{(j)}\right) \tag{10}$$

$$\frac{d\hat{\sigma}_{-,j}}{dt} = -i\left[\hat{\sigma}_{-,j}, H_j\right] - \gamma_j \hat{\sigma}_{-,j} + \sqrt{\gamma_j'}\hat{\sigma}_j' \tag{11}$$

where c_j is bosonic annihilation operator of the j-th cavity mode with resonant frequency $\omega_{c,j}$. $\hat{a}_{\text{in}}^{(j)}(\hat{b}_{\text{in}}^{(j)})$ and $\hat{a}_{\text{out}}^{(j)}(\hat{b}_{\text{out}}^{(j)})$ describe the input and output fields in the left (right) port respectively, with standard input-output relations $\hat{a}_{\text{out}}^{(j)} = \hat{b}_{\text{in}}^{(j)} + \sqrt{\kappa_{1,j}}\hat{c}_j$ and $\hat{b}_{\text{out}}^{(j)} = \hat{a}_{\text{in}}^{(j)} + \sqrt{\kappa_{1,j}}\hat{c}_j$. $2\Gamma_j$ represents total cavity decay with $\Gamma_j = \left(\kappa_{0,j} + 2\kappa_{1,j}\right)/2$, where $\kappa_{0,j}$ is the intrinsic cavity decay rate and $\kappa_{1,j}$ the external cavity decay rate. $\hat{\sigma}_{-(+),j}$ is the descending (ascending) operator of the interacting two-level QD with transition frequency $\omega_{r,j}$. γ_j is the total decay rate of the QD, including the spontaneous decay (at rate γ_s) and dephasing (at rate γ_p) in the excited state $|e\rangle$; H_j is the subsystem Hamiltonian $H_j = \omega_{c,j}\hat{c}_j^{\dagger}\hat{c}_j + \omega_{r,j}\hat{\sigma}_{+,j}\hat{\sigma}_{-,j} + \left[g_j(\vec{r})\hat{\sigma}_{+,j}\hat{c}_j + h.c.\right]$, where $g_j(\vec{r})$ is the coupling strength between the cavity mode and the dipolar transition $|g\rangle \leftrightarrow |e\rangle$. $\hat{\sigma}_j'$ is the vacuum noise operator associated with the decay rate γ_j'.

In the weak excitation limit (excited by a weak monochromatic field or a single photon pulse with frequency ω), and by omitting the term which concerns the Langevin noises, the motion equations can be solved, with the transport relation in the frequency domain

$$\begin{pmatrix} \hat{b}_{\text{in}}^{(j)}(\omega) \\ \hat{b}_{\text{out}}^{(j)}(\omega) \end{pmatrix} = T_j \begin{pmatrix} \hat{a}_{\text{in}}^{(j)}(\omega) \\ \hat{a}_{\text{out}}^{(j)}(\omega) \end{pmatrix} \tag{12}$$

Here the transmission matrix is

$$T_j = \frac{1}{\alpha_j + \kappa_{1,j} - \Gamma_j} \begin{pmatrix} -\kappa_{1,j} & \alpha_j - \Gamma_j \\ \alpha_j - \Gamma_j + 2\kappa_{1,j} & \kappa_{1,j} \end{pmatrix} \tag{13}$$

where $\alpha_j = i\Delta_{c,j} + \left|g_j(\vec{r})\right|^2 / \left(i\Delta_{r,j} - \gamma_j\right)$, $\Delta_{c,j} = \omega - \omega_{c,j}$ $(\Delta_{r,j} = \omega - \omega_{r,j})$ represents the detuning between the input field and the cavity mode (QD transition). For convenience, we also define the cavity-QD detuning $\delta_j \equiv \omega_{c,j} - \omega_{r,j}$. The transport matrix can be regarded as a basic cell in cascading subsystems and obtaining the whole transportation for the N-coupled cavity-QD system. The transport properties can thus be expressed as

$$\begin{pmatrix} \hat{b}_{\text{in}}^{(N)}(\omega) \\ \hat{b}_{\text{out}}^{(N)}(\omega) \end{pmatrix} = T_N T_0 \cdots T_0 T_2 T_0 T_1 \begin{pmatrix} \hat{a}_{\text{in}}^{(1)}(\omega) \\ \hat{a}_{\text{out}}^{(1)}(\omega) \end{pmatrix}. \tag{14}$$

Here T_0, the transmission matrix via the waveguide between the two adjacent nanocavities, can be expressed as

$$T_0 = \begin{pmatrix} 0 & e^{-i\theta} \\ e^{i\theta} & 0 \end{pmatrix} \tag{15}$$

where $\theta = \vec{k} \cdot \vec{L}$, and L is the distance between the adjacent nanocavities.

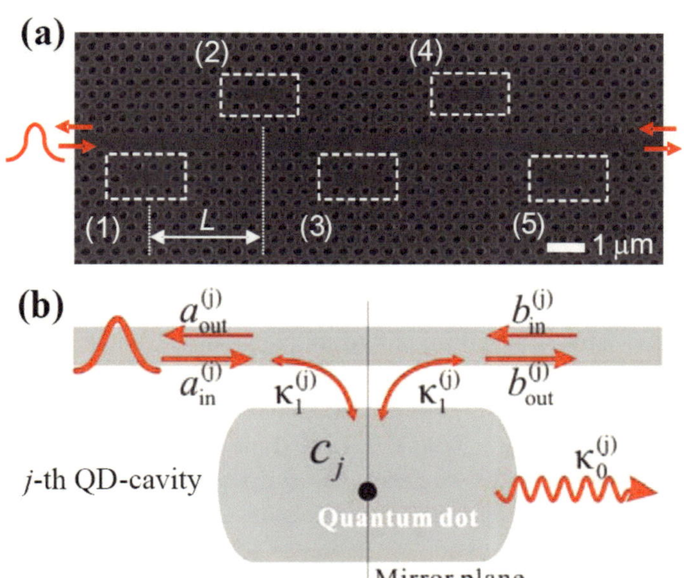

Fig. 4. (a) Example scanning electronic micrograph of periodic waveguide-resonator structure containing N side-coupled nanocavities (h-polarized) at a distance L. The nanocavities are side coupled through the integrated waveguide, with no direct coupling between any two nanocavities. (b) The j-th quantum dot - cavity subsystem. $\hat{a}_{\text{in}}^{(j)}(\hat{b}_{\text{in}}^{(j)})$ and $\hat{a}_{\text{out}}^{(j)}(\hat{b}_{\text{out}}^{(j)})$ describe the input and output fields in the left (right) port, respectively.

When studying only the spectral character of the coupled cavity-QD interaction (Section 3.2), we note that this is analogous to classical microwave circuit design, where the transmission and reflection characteristics from Eq. (14) can also be examined with coupled-mode theory with dipole terms inserted. Examining the spectral character first (Section 3.2) helps to understand the coupled cavity-QD controlled quantum phase gate operation and performance (Sections 3.3 and 3.4).

3.2 Spectral character of coupled cavity-QD arrays

To examine the physical essence, we need to first examine the spectral character of the coupled cavity-QD system. The reflection and transmission coefficients are defined as $r_{N1}(\omega) \equiv \hat{a}^{(1)}_{\mathtt{out}}(\omega)/\hat{a}^{(1)}_{\mathtt{in}}(\omega)$ and $t_{N1}(\omega) \equiv \hat{b}^{(N)}_{\mathtt{out}}(\omega)/\hat{a}^{(1)}_{\mathtt{in}}(\omega)$. In the following, we also assume that these cavities possess the same dissipation characteristic without loss of generality, *i.e.*, $\kappa_{0,j} = \kappa_0$, $\kappa_{1,j} = \kappa_1$, $\kappa_1 = 50\kappa_0$, and $\Gamma_j = \Gamma$.

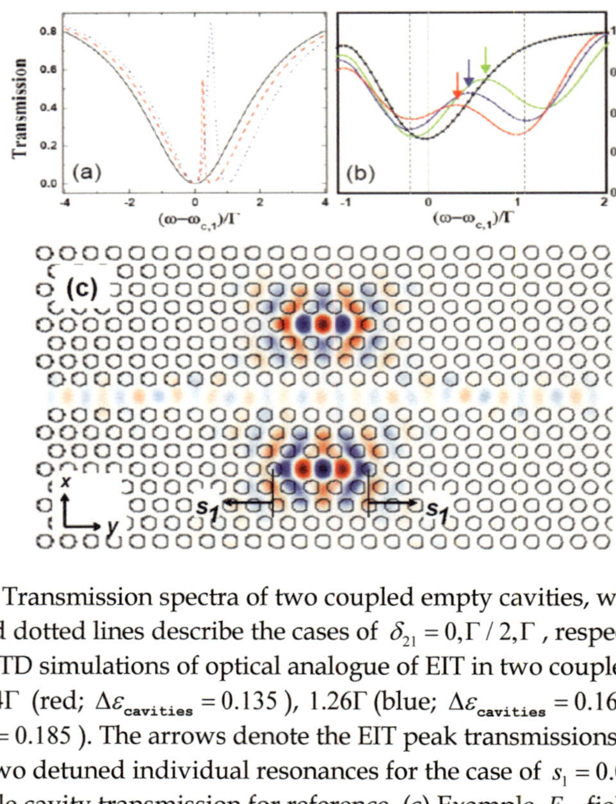

Fig. 5. (a) and (b): Transmission spectra of two coupled empty cavities, where $\theta = 20\pi$. Solid, dashed, and dotted lines describe the cases of $\delta_{21} = 0, \Gamma/2, \Gamma$, respectively. (b) Numerical 3D FDTD simulations of optical analogue of EIT in two coupled cavity ($\theta = 0$) for detunings 1.14Γ (red; $\Delta\varepsilon_{\mathtt{cavities}} = 0.135$), 1.26Γ (blue; $\Delta\varepsilon_{\mathtt{cavities}} = 0.160$), and 1.49Γ (green; $\Delta\varepsilon_{\mathtt{cavities}} = 0.185$). The arrows denote the EIT peak transmissions. The dashed grey lines denote the two detuned individual resonances for the case of $s_1 = 0.05a$. The black curve is for a single cavity transmission for reference. (c) Example E_x -field distribution of coupled empty photonic crystal cavities.

Fig. 5a describes the transmission spectra of two coupled *empty* cavities (without QD) with different detuning ($\delta_{21} \equiv \omega_{c,2} - \omega_{c,1}$). When the two cavities are exactly resonant, a transmission dip is observed; with increasing δ_{21}, a sharp peak exists at the center position between the two cavity modes. This is an optical analogue to the phenomenon of EIT in atomic vapors. We examine the classical optical analogue exactly through 3D finite-difference time-domain (FDTD) numerical simulations. Specifically, Fig. 5b and 5c show an example of the transmission and field distributions through the coherent interaction with two coupled *empty* cavities, where the resonance of one cavity is detuned by three cases: $\delta_{21} = 1.14\Gamma$, 1.26Γ, and 1.49Γ. The optical transparency peak from the FDTD is broader than in Fig. 5a due to the finite grid-size resolution, and is observed on top of a background Fabry-Perot oscillation (due to finite reflections at the waveguide facets). The analogy and difference between an all-optical analogue to EIT and atomic EIT are recently discussed in Ref. (Xiao et al., 2007).

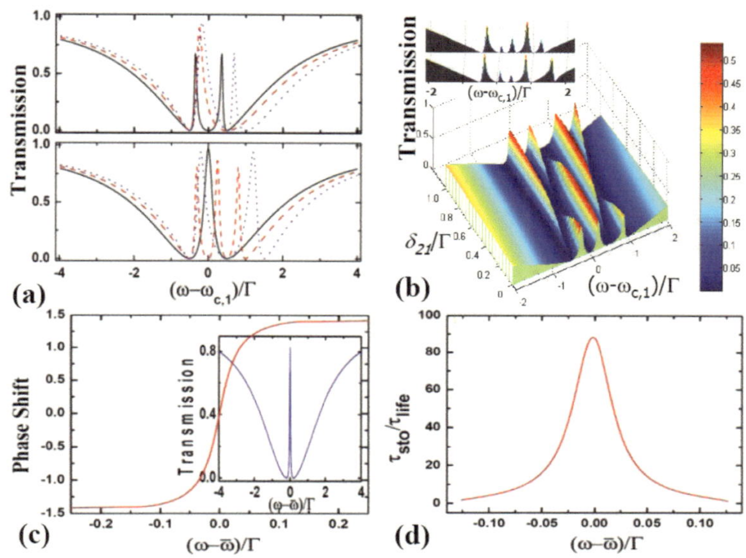

Fig. 6. (a) Transmission spectra of two coupled subsystems with one QD (top) and two QDs (bottom) where $g = \Gamma/2$, $\gamma = \gamma_1 = \gamma_2 = \kappa_0$. Other conditions are same as Fig. 5a. (b) Spectral character of three coupled subsystems with $\delta_{31} = \Gamma/2$ and $\delta_{1,2,3} = 0$. Inset: $\delta_{31} = \Gamma/2$, $\delta_{21} = 0$, $\delta_1 = \delta_3 = 0$, with $\delta_2 = \Gamma/2$ (top) and Γ (bottom). (c) and (d): Photon phase shift and delay (τ_{sto}) through two cavity-QD subsystems, where $\omega_{c(r),j} = \varpi$, $g = 0.2\Gamma$. Inset: transmission spectrum.

In the presence of QDs, Fig. 6a (top) shows the spectral characteristics in which a single QD resonantly interacts with the first cavity. When both cavities are resonant, there exist two obvious sharp peaks located symmetrically around $\omega = 0$ (for convenience, we define $\omega_{c,1} = 0$). This fact can be explained by dressed-mode theory. Resonant cavity-QD interaction results in two dressed modes, which are significantly detuned from the second *empty* cavity with the detuning $\pm|g_1(\vec{r})| = \pm\Gamma/2$. Both dressed modes non-resonantly couple with the empty mode, resulting in two transparency peaks located at frequencies $\omega \approx \pm\Gamma/4$. When $\delta_{21} = \Gamma/2$, one dressed cavity mode non-resonantly couples with the empty mode with a detuning Γ, which leads to a transparency peak located near $\omega \simeq 0$; while the other dressed mode resonantly couples with the empty mode, which does not result in a transparency peak. When δ_{21} continually increases, e.g., $\delta_{21} = \Gamma$, the vanished peak reappears since the two dressed modes are always non-resonant with the empty mode.

Fig. 6a (bottom) illustrates the case where both cavities resonantly interact with a single QD each. Similar to the above analysis, we can explain the number and locations of sharp peaks with respect to different δ_{21} by comparing the two pairs of dressed modes. For example, when $\delta_{21} = \Gamma$, the dressed modes in the first cavity is located at $\pm\Gamma/2$ while the second pair is at $\Gamma/2$ and $3\Gamma/2$, so the transparency peaks are located at $[-\Gamma/2,\Gamma/2]$ and $[\Gamma/2,3\Gamma/2]$, i.e., two peaks are near 0 and Γ. Fig. 6b shows the spectral character of three coupled cavity-QD subsystems, under various cavity-cavity and cavity-QD detunings. These transmission characteristics are helpful during experimental realization efforts to identify the required tuning and detunings when multiple QD transitions and cavity resonances are involved.

Phase shift and photon storage.- To further examine this coupled cavity-QD system, Fig. 6c shows the transmission phase shift for various detunings of the input photon central frequency, where the cavity and QD transition are resonant for both subsystems. The phase shift has a steep change as we expected intuitively, which corresponds to a strong reduction of the group velocity of the photon. As shown in Fig. 6d, the delay time (τ_{sto}) in this coupled system is almost hundreds of the cavity lifetime ($\tau_{life} = 1/2\Gamma$). This coupled cavity-QD system can essentially be applied to the storage of the photon. Moreover, our solid-state implementation has an achievable bandwidth of ~ 50 MHz in contrast to less than 100 kHz in atomic systems, although the delay-bandwidth product is comparable. To obtain longer photon storage, one can consider dynamical tuning (Yanik et al., 2004; Xu et al., 2007; Yanik & Fan, 2007) to tune the cavity resonances with respect to the QD dipolar transitions to break the delay-bandwidth product in a solid-state cavity-QD array system.

3.3 Quantum phase gate operation

In the section above, we have shown the novel transport character of the coupled cavity-QD system. Now we study the possibility of quantum phase gate operation of the QDs based on this transport character. The schematic to realize this multi-QD coupled cavity-cavity system is illustrated in Fig. 7. The QDs are represented by two ground states $|g\rangle$ and $|r\rangle$, where the state $|r\rangle$ is largely detuned with the respective cavity mode. The two ground states can be prepared via QD spin-states such as demonstrated remarkably in experiment in Ref. (Atatuer et al., 2006) with near-unity fidelity. The input weak photon pulse is assumed *h*-polarized, with an input pulse duration *D* (e.g. 1 ns) larger than the loaded cavity lifetime for the steady-state approximation. To remove the distinguishability of the two output photon spatial modes in the waveguide (transmitted and reflected), a reflecting element is inserted in the end of waveguide (such as a heterostructure interface), as shown in Fig. 7. This ensures that the photon always exits in the left-propagating mode $|L\rangle$ (from a right-propagating input mode $|R\rangle$) without any entanglement with the QD states. Alternatively, a Sagnac interferometer scheme such as introduced in Reference (Gao et al., 2008) can also be implemented to remove the spatial mode distinguishability and QD-photon entanglement. In this single input single output mode scheme, $|h\rangle$ and $|v\rangle$ represent the two polarization states of the input photon. We emphasize that in the below calculations we have considered the complete characteristics of the full system (including the end reflecting element and the resulting "standing wave" due to the long photon pulse width) where we examined the final left-propagating output mode $|L\rangle$ from a right-propagating input mode $|R\rangle$ (Fig. 7). The reflection interference is included where we force $\hat{b}_{\text{in}}^{(2)} = \hat{b}_{\text{out}}^{(2)}$ (Fig. 4b) from the reflection element, when calculating the temporal pulse delays for the different QD states.

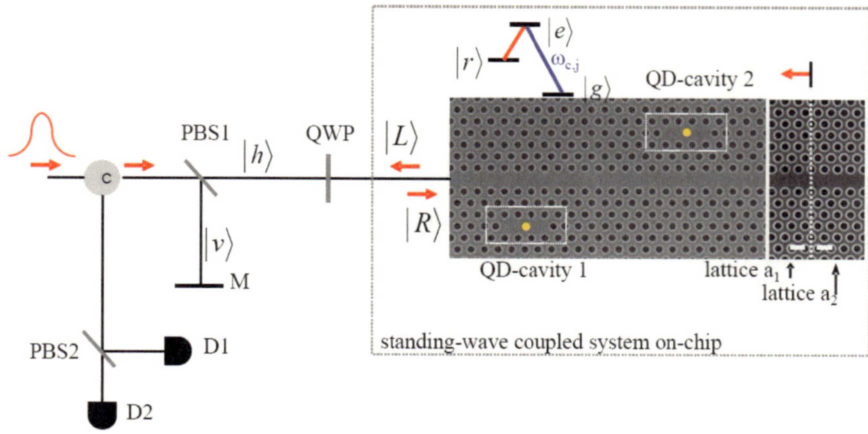

Fig. 7. Schematic to illustrate two-qubit quantum phase gate based on the coupled cavity-cavity multi-QD scheme. A heterostructure reflection element is introduced in the end of waveguide to remove spatial mode distinguishability, with only a single output mode $|L\rangle$ for an input photon mode $|R\rangle$. The QDs have a superposition of two ground states, $|g\rangle$ and $|r\rangle$. PBS1 and PBS2 represent the polarization beam splitters, D1 and D2 the single photon detectors, C the circulator, M the reflecting mirror. Here PBS1 and PBS2 are actually regarded as filters since only the h-polarized photon is required in our scheme. The response of detectors D1 and D2 provide an indication to show the gate operation success, and can also be used for measurement-induced entanglement in future.

To facilitate the discussion but without loss of generality, we describe the all resonance case (i.e., $\omega = \omega_{c(r),j}$) to describe the idea of the phase gate operation; in the subsequent numerical calculations, we will demonstrate the gate feasibility under non-ideal detunings. As an example, we focus on the realization of a two-qubit (two QDs) phase gate. Fig. 8 now shows the calculated reflection field (real and imaginary components) of the complete coupled cavity-cavity and two QD systems for the four superpositioned states $|\Psi\rangle_{\text{in}} = \alpha_1|r\rangle_1|r\rangle_2 + \alpha_2|g\rangle_1|r\rangle_2 + \alpha_3|r\rangle_1|g\rangle_2 + \alpha_4|g\rangle_1|g\rangle_2$. We address the following cases for the different QD states.

Case I: The two QDs are initially prepared in $|u\rangle_1|v\rangle_2$ ($u,v = g,r$) and at least one QD occupies the ground state $|r\rangle$. From Figs. 8a and 8b and for the all resonance case, we see that $\text{Re}[r_{21}] \simeq -1$ and $\text{Im}[r_{21}] \simeq 0$ under the over-coupling regime ($\kappa_0 \ll \kappa_1$) and with the large Purcell factor ($g^2/\Gamma\gamma \gg 1$). This fact can be understood by regarding the resonant condition ($\omega_{c,j} = \omega$). The input photon will be almost reflected by one *empty* cavity, in which the QD is in $|r\rangle$, resulting in a final state $|u\rangle_1|v\rangle_2|L\rangle$.

Case II: The QDs initially occupy in $|g\rangle_1|g\rangle_2$. In this case, note that the photon pulse interacts coherently with both cavities and two QDs, including the reflection element which is placed specifically to achieve $\hat{b}_{\text{in}}^{(2)} = \hat{b}_{\text{out}}^{(2)}$, before completely exiting the system. As demonstrated in Figs. 8a and 8b and for the all resonance case, the final output state is described by $\text{Re}[r_{21}] \simeq 1$ and $\text{Im}[r_{21}] \simeq 0$. The resulting state is $|g\rangle_1|g\rangle_2|L\rangle$. We note that the photon loss is small for all four cases when the input photon pulse is on resonance with the cavity resonances, as can be done experimentally by tuning the input photon.

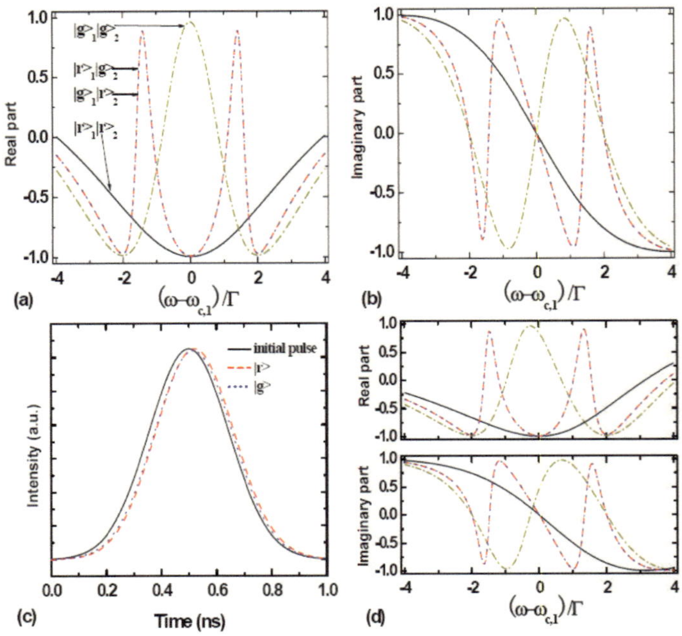

Fig. 8. Real (a) and imaginary (b) parts of the reflection coefficients for initial QD states, $|r\rangle_1|r\rangle_2$, $|r\rangle_1|g\rangle_2$, $|g\rangle_1|r\rangle_2$, and $|g\rangle_1|g\rangle_2$. Here we assume $|g_1(\vec{r})| = |g_2(\vec{r})| = 2\Gamma$, $\delta_{21} = \delta_1 = \delta_2 = 0$, and the propagation phase between the second cavity and the reflection element is adjusted as $\theta' = n\pi + \pi/2$ to compensate the phase shift induced by the mirror reflection (ideally, π). Other parameters are the same as Fig. 5a. (c) Shape function of the photon pulse for cases when the single QD is coupled ($|g\rangle$) or decoupled ($|r\rangle$) to the single cavity, and without the cavity. (d) Real and imaginary parts of the reflection coefficient when the reflection phase of the mirror deviates from ideal π with a deviation of 0.5.

Therefore, with the exit of the photon of the single input single output system, the state of the two QDs after the interaction is now described by $|\Psi\rangle_{out} = -\alpha_1|r\rangle_1|r\rangle_2 - \alpha_2|g\rangle_1|r\rangle_2 - \alpha_3|r\rangle_1|g\rangle_2 + \alpha_4|g\rangle_1|g\rangle_2$. Hence, after the above process and recombining at PBS1, the state of the QD-QD gate described by $U = e^{i\pi|g\rangle_{12}\langle g|}$ can be manipulated. Moreover, if $\alpha_i = 1/2 (i = 1,2,3,4)$, we have $|\Psi\rangle_{out} = (1/\sqrt{2})(|r\rangle_1|-\rangle_2 + |g\rangle_1|+\rangle_2)$ [where $|-\rangle_2 = 1/\sqrt{2}(|g\rangle_2 - |r\rangle_2)$ and $|+\rangle_2 = 1/\sqrt{2}(|g\rangle_2 + |r\rangle_2)$], which is the generation of the maximally entangled state in the coupled QDs. Most importantly, this idea can also be extended to realize an N-qubit gate with only one step, which is of importance for reducing the complexity of practical quantum computation and quantum algorithms for physical realization. In addition, using this configuration, some special entangled QD states (for e.g., the cluster state) can be generated (Cho and Lee, 2005).

We provide a few more notes on this designed coupled cavity-cavity multi-QD system. First, the temporal distinguishability is small for the single cavity-QD system, where in Fig. 8c we plot the shape function of the output photon pulse for cases when the QD is coupled ($|g\rangle$), decoupled ($|r\rangle$), or without the cavity, through numerical simulation of the dynamical evolution of the system. The pulse shape function overlaps very well. Secondly, the calculated temporal distinguishability in the coherently coupled cavity-cavity multi-QD system is also small compared to the pulse duration D. Specifically, with the parameters in Fig. 8a, the photon delays due to the coupling to the cavities are calculated as approximately τ_{life}, $2\tau_{\text{life}}$, $2\tau_{\text{life}}$, and $4\tau_{\text{life}}$ in case of the states $|r\rangle_1|r\rangle_2$, $|r\rangle_1|g\rangle_2$, $|g\rangle_1|r\rangle_2$, and $|g\rangle_1|g\rangle_2$, respectively, where the loaded cavity τ_{life} is about 0.02 ns. The photon delay of the complete system is therefore sizably smaller than the pulse duration (of 1 ns, for example). Furthermore, this cavity-induced delay will furthermore decrease with increasing the coupling rate g, furthering reducing temporal distinguishability. Of course, the size of the chip is also small (tens of microns) so that the propagation time ($2S/v$, where S denotes the distance between the first cavity and the reflector, and v the group velocity) in the waveguide is much smaller than the pulse duration D. Thirdly, we examine the dependence of the overall system reflection coefficient on the phase variation from the reflection element, when deviating from the ideal π phase shift. Fig. 8d shows the numerical results, where a slight dependence is observed when there is a phase deviation of 0.5 from π. Moreover, the phase shift from the reflection element can be externally controlled stably, such as with an external and focused pump beam to thermally tune the reflection region.

3.4 Gate fidelity and photon loss

To exemplify the coupled cavity system, isolated single semiconductor QDs in high-Q small modal volume (V) photonic crystal cavities are potential candidates, such as self-assembled InAs QDs in GaAs cavities, or PbS nanocrystals in silicon cavities at near 1550 nm wavelengths. For PbS nanocrystal and silicon cavity material system, we use the following parameters in our calculations: $\gamma_s \sim 2$ MHz, $\gamma_p \sim 1$ GHz at cooled temperatures, $V \sim 4$ μm^3 at 1550 nm, with resulting single-photon coherent coupling rate $g \sim 12.4$ GHz. Loaded cavity Q in the range of 10^4 and 10^5 are achievable experimentally, with intrinsic Q up to 10^6 reported recently (Noda et al., 2007; Tanabe et al., 2007).

To characterize the present gate operation, Figs. 9a and 9b present the two-qubit phase gate fidelity F and photon losses P for various g and the parameters described above, even under non-ideal detuning conditions and the bad cavity limit. It should be noted here that, with δ_{21} and $\delta_{1,2}$, we can know the detuning between two QDs. For example, in case of $\delta_{21} = 5\kappa_0$ and $\delta_{1,2} = 0$, we deduce the detuning between the two QDs is $5\kappa_0$. Based on the above parameters, F can reach to 0.99 or more, and P can be below 0.04. As shown in Fig. 9, for cavity-cavity detunings in the range of the intrinsic cavity decay rate, both F and does not degrade significantly but is strongly dependent on the cavity decay rate. Likewise, with cavity-QD detuning that is comparable with the intrinsic cavity decay rate, both F and P does not change significantly but is dependent on the cavity decay rate. We note that the on-resonant photon loss P can be larger than the non-resonant case when g is small. This can be explained by considering the decay of QDs. When the QDs resonantly interact with cavity modes, the decay of QDs becomes distinct, resulting in an increasing of photon loss. Moreover, we note the QD-QD detuning plays an important role in the quantum gate operations. Given the current large inhomogeneous distribution of QD transitions, however, active tuning methods such as Stark shifts would probably be needed to control the detuning within acceptable bounds to obtain strong quantum gate fidelity and low photon loss. Furthermore, we note that, with increasing g, the photon loss P exhibits an increase before a decrease, which can be understood by studying the photon loss when the QDs are in the state of $|g\rangle_1|g\rangle_2$. When $g \simeq \Gamma/2$, the absorption strength (resulting from κ_0 and γ) of the input photon by the coupled cavities reaches the maximum.

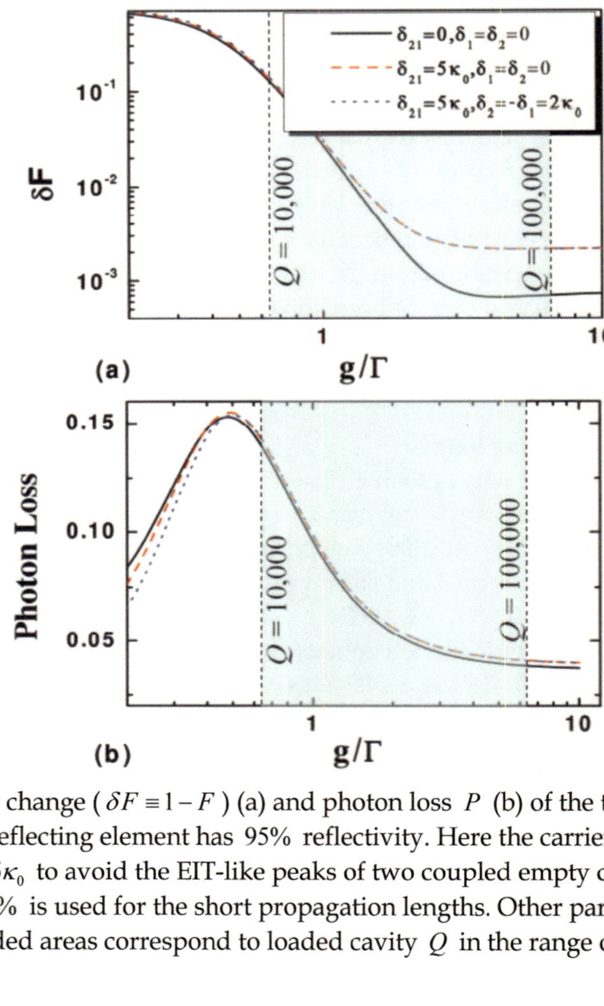

Fig. 9. Gate fidelity change ($\delta F \equiv 1 - F$) (a) and photon loss P (b) of the two-qubit gate versus g/Γ. The reflecting element has 95% reflectivity. Here the carrier frequency is assumed as $\varpi - 2.5\kappa_0$ to avoid the EIT-like peaks of two coupled empty cavities, and a scattering loss of 1% is used for the short propagation lengths. Other parameters are same as Fig. 6a. The shaded areas correspond to loaded cavity Q in the range of 10^4 to 10^5.

4. Summary

In this Chapter, with the nanocavities in photonic crystal, we theoretically introduce, derive, and demonstrate the robust implementation of a single spin-photon phase gate in a cavity-dipole-cavity system. The conditions of accidental degeneracy are examined to enforce complete transfer, either in the forward transmission or in reflection, of the qubit. In addition, we observe that a photon pulse is strikingly reflected by a cavity interacting with a single spin, even under the bad-cavity limit. This combined nanocrystal-cavity system, examined in a silicon materials platform with lead chalcogenide nanocrystals in the near infrared, can serve as a QD spin-photon two-qubit quantum phase gate and, indeed, as a general quantum interface for large-array chip-based quantum information processing. To further utilize the high-Q and small-V nanocavities of photonic crystal, we also investigate the operation and performance of a scalable cavity–QD array on a photonic crystal chip towards controlled QD-QD quantum gates. The coupling among single-QD emitters and quantized cavity modes in a coherent array results in unique transmission spectra, with an optical analogue of EIT-like resonances providing potential photon manipulation. In the quantum phase gate operation, we note that the gate fidelity can reach 0.99 or more and the photon loss can be below 0.04 in a realistic semiconductor system, provided the non-ideal detunings are kept within the cavity decay rates. Our study provides a potential for a chip-scale quantum gate towards a potential quantum computing network with the platform of silicon photonic crystal.

For future experimental quantum information processing in photonic crystal, we note that it is possible to realize the initial idea in a single nanocavity-QD coupled system, as experimentally demonstrated in the context of strong cavity-QD coupling (Hennessy et al., 2007; Badolato et al., 2006; Reithmaier et al., 2004; Yoshie et al., 2004; Faraon et al., 2008; Fushman et al., 2008) and the quantum state transfer between a single QD and a target cavity (Englund et al., 2007a). This opens the door to construct the basic building blocks for future chip-based quantum information processing systems. However, it is still a challenge to implement quantum information with a nanocavity array in photonic crystal. The challenge includes several main technique difficulties. First, it is necessary to precisely place a single two-level nanocrystal (or other QDs) with respect to the corresponding nanocavity mode in photonic crystal, for the largest Rabi frequency, and also to position across an array. Second, both the cavity resonances and QD transitions should spectrally overlap within approximately the cavity or exciton linewidths, although our theoretical model is still robust with small QD-QD, cavity-QD, and cavity-cavity detunings. The former challenge depends on careful nanofabrication techniques, while the latter condition can be relaxed through high-precision tunability of the cavity resonances or QD transitions. Moreover, we note that the ultrahigh-Q and low-V regime is desired to operate well into the strong coupling regime, suppressed chip-scale photon losses or improved collection to improve quantum state transfer, as well as control of dephasing especially at high temperatures in order for chip-level scalability in solid-state quantum information nanosciences.

5. References

Angelakis, D. G.; Santos, M. F. & Bose, S. (2007). Photon-blockade-induced Mott transitions and XY spin models in coupled cavity arrays. *Physical Review A*, Vol. 76, No. 3, 031805.

Asano, T.; Song, B.-S. & Noda, S. (2006). Analysis of the experimental Q factors (~1 million) of photonic crystal nanocavities. *Optics Express*, Vol. 14, No. 5, 1996-2002.

Atatüre, M.; Dreiser, J.; Badolato, A.; Högele, A.; Karrai, K. & Imamoglu, A. (2006). Quantum-dot spin-state preparation with near-unity fidelity. *Science*, Vol. 312, No. 5773, 551-553.

Badolato, A.; Hennessy, K.; Atatüre, M.; Dreiser, J. Hu, E.; Petroff, P. M. & Imamoglu, A. (2005). Deterministic coupling of single quantum dots to single nanocavity modes. *Science*, Vol. 308, No. 5725, 1158-1161.

BarWicz, T.; Watts, M. R.; Popović, M. A.; Rakich, P. T.; Socci, L.; Kärtner, F. X.; Ippen, E. P. & Smith, H. I. (2007). Polarization-transparent microphotonic devices in the strong confinement limit. *Nature Photonics*, Vol. 1, No. 1, 57-60.

Bose, R.; Yang, X.-D.; Chatterjee, R.; Gao, J. & Wong, C.-W. (2007). Weak coupling interactions of colloidal lead sulphide nanocrystals with silicon photonic crystal nanocavities near 1.55 mu m at room temperature. *Applied Physics Letters*, Vol. 90, No. 11, 111117.

Brattke, S.; Varcoe, B. T. H. & Walther, H. (2001). Generation of Photon Number States on Demand via Cavity Quantum Electrodynamics. *Physical Review Letters*, Vol 86, No. 16, 3534-4537.

Brune, M.; Hagley, E.; Dreyer, J.; Maitre, X.; Maali, A.; Wunderlich, C.; Raimond, J. M. & Haroche, S. (1996). Observing the Progressive Decoherence of the "Meter" in a Quantum Measurement. *Physical Review Letters*, Vol 77, No. 24, 4887-4890.

Cabrillo, C.; Cirac, J. I.; Garcia-Fernandez, P. & Zoller, P. (1999). Creation of entangled states of distant atoms by interference. *Physical Review A*, Vol 59, No.2, 1025-1033.

Cirac, J. I.; Zoller, P.; Kimble, H. J. & Mabuchi, H. (1997). Quantum state transfer and entanglement distribution among distant nodes in a quantum network. *Physical Review A*, Vol. 78, No. 16, 3221-3224.

Cho, J. & Lee, H.-W. (2005). Generation of atomic cluster states through the cavity input-output process. *Physical Review Letters*, Vol. 95, No. 16, 160501.

Duan, L.-M.; Kuzmich, A. & Kimble, H. J. (2003). Cavity QED and quantum-information processing with "hot" trapped atoms. *Physical Review A* Vol. 67, No. 3, 032305.

Duan, L.-M. & Kimble, H. J. (2004). Scalable photonic quantum computation through cavity-assisted interactions. *Physical Review Letters*, Vol. 92, No. 12, 127902.

Englund, D.; Faraon, A.; Fushman, I.; Stoltz, N.; Petroff, P. & Vuckovic, J. (2007a). Controlling cavity reflectivity with a single quantum dot. *Nature*, Vol. 450, 857-861.

Gao, J.; Sun, F.-W. & Wong, C.-W. (2008). Implementation scheme for quantum controlled phase-flip gate through quantum dot in slow-light photonic crystal waveguide. *Applied Physics Letters*, Vol. 93, (2008) 151108.

Fan, S.; Villeneuve, P. R.; Joannopoulos, J. D. & Haus, H. A. (1998). Channel drop tunneling through localized states. *Physical Review Letters*, Vol. 80, No. 5, 960-963.

Faraon, A.; Fushman, I.; Englund, D.; Stoltz, N.; Petroff, P. & Vuckovic, J. (2008). Coherent generation of nonclassical light on a chip via photon-induced tunneling and blockade. *Nature Physics*, Vol 4, 859-863.

Fleischhauer, M.; Imamoglu, A. & Marangos, J. P. (2005). Electromagnetically induced transparency: Optics in coherent media. *Reviews of Modern Physics*, Vol. 77, No. 2, 633-673.

Foresi, J. S.; Villeneuve, P. R.; Ferrera, J.; Thoen, E. R.; Steinmeyer, G.; Fan, S.; Joannopoulos, J. D.; Kimerling, L. C.; Smith, H. I. & Ippen, E. P. (1997). Photonic-bandgap microcavities in optical waveguides. *Nature*, Vol. 390, 143-145.

Fushman, I.; Englund, D. & Vučković, J. (2005). Coupling of PbS quantum dots to photonic crystal cavities at room temperature. *Applied Physics Letters*, Vol. 87, No. 24, 241102.

Fushman, I.; Englund, D.; Faraon, A.; Stoltz, N.; Petroff, P. & Vuckovic, J. (2008). Controlled Phase Shifts with a Single Quantum Dot, Science, Vol 320, No. 5877, 769-772.

Guo, L.; Krauss, T. D.; Poitras, C. B.; Lipson, M.; Teng, X. & Yang, H. (2006). Energy transfer between colloidal semiconductor nanocrystals in an optical microcavity. *Applied Physics Letters*, Vol. 89, No. 6, 061104.

Greentree, A. D.; Tahan, C.; Cole, J. H. & Hollenberg, L. C. L. (2006). Quantum phase transitions of light. *Nature Physics*, Vol. 2, 856-861.

Hartmann, M. J.; Brandão, F. G. S. L. & Plenio M. B. (2006). Strong interacting polaritons in coupled arrays of cavities. *Nature Physics*, Vol. 2, 849-855.

Hartmann, M. J. & Plenio M. B. (2007). Strong photon nonlinearities and photonic Mott insulators. *Phys. Rev. Lett.*, Vol. 99, No. 10, 103601.

Hartmann, M. J.; Brandão, F. G. S. L. & Plenio, M. B. (2007). Effective spin systems in coupled microcavities. *Physical Review Letters*, Vol. 99, No. 16, 160501.

Hennessy, K.; Badolato, A.; Winger, M.; Gerace, D.; Atatüre, M.; Gulde, S.; Fält, S.; Hu, E. L. & Imamoglu, A. (2007). Quantum nature of a strongly coupled single quantum dot-cavity System. *Nature*, Vol. 445, 896-899.

Hughes, S. (2007). Coupled-cavity QED using planar photonic crystals. *Physical Review Letters*, Vol. 98, No. 8, 083603.

Khitrova, G.; Gibbs, H. M.; Kira, M.; Koch, S. W. & Scherer, A. (2006). Vacuum Rabi splitting in semiconductors. *Nature Physics*, Vol. 2, No. 81, 81-90.

Law, C. K. & Eberly, J. H. (1996). Arbitrary Control of a Quantum Electromagnetic Field. *Physical Review Letters*, Vol. 76, No. 7, 1055-1058.

Mabuchi, H. & Doherty, A. C. (2002). Cavity quantum electrodynamics: coherence in context. *Science* Vol. 298, No. 5597, 1372-1377.

Min, B.-K.; Kim, J.-E. & Park, H. Y. (2004). Channel drop filters using resonant tunneling processes in two-dimensional triangular lattice photonic crystal slabs. *Optics Communications*, Vol. 273, No. 1-3, 59-63.

Noda, S.; Fujita, M. & Asano, T. (2007). Spontaneous-emission control by photonic crystals and nanocavities. *Nature Photonics*, Vol. 1, 449-458.

Pachos, J. & Walther, H. (2002). Quantum Computation with Trapped Ions in an Optical Cavity. *Physical Review Letters*, Vol 89, No. 18, 187903.

Parkins, A. S.; Marte, P.; Zoller, P. & Kimble, H. J. (1993). Synthesis of arbitrary quantum states via adiabatic transfer of Zeeman coherence. *Physical Review Letters*, Vol. 71, No. 19, 3095-3098.

Pellizzari, T.; Gardiner, S. A.; Cirac, J. I. & P. Zoller (1995). Decoherence, Continuous Observation, and Quantum Computing: A Cavity QED Model. *Physical Review Letters*, Vol 75, No. 21, 3788-3791.

Reithmaier, J. P.; SJk, G.; Löffler, A.; Hofmann, C.; Kuhn, S.; Reitzenstein, S.; Keldysh, L. V.; Kulakovskii, V. D.; Reinecke, T. L. & Forchel, A. (2004). Strong coupling in a single quantum dot-semiconductor microcavity system. *Nature*, Vol. 432, No. 7014, 197-200.

Santos, M. F. (2005). Universal and Deterministic Manipulation of the Quantum State of Harmonic Oscillators: A Route to Unitary Gates for Fock State Qubits. *Physical Review Letters*, Vol. 95, No. 1, 010504.

Smith, D.; Chang, H.; Fuller, K. A.; Rosenberger, A. T. & Boyd, R. W. (2004). Coupled-resonator-induced transparency. *Physical Review A*, Vol. 69, No. 6, 063804.

Srinivasan, K. & Painter, O. (2007). Mode coupling and cavity-quantum-dot interactions in a fiber-coupled microdisk cavity. *Physical Review A*, Vol. 75, No. 2, 023814.

Sørensen, A. S. & Mølmer, K. (2003). Measurement induced entanglement and quantum computation with atoms in optical cavities. *Physical Review Letters*, Vol. 91, No. 9, 097905.

Tanabe, T.; Notomi, M.; Kuramochi, E.; Shinya, A. & Taniyama, H. (2006). Trapping and delaying photons for one nanosecond in an ultrasmall high-Q photonic-crystal nanocavity. *Nature Photonics*, Vol. 1, 49-52.

Tosuka, K.; Kobayashi, N.; Tomita, M. (2007). Slow light in coupled-resonator-induced transparency. *Physical Review Letters*, Vol. 98, No. 21, 213904.

Vahala, H. J. (2004). Optical cavities. *Nature*, Vol 424, No. 7253, 839-846.

Vogel, K.; Akulin, V. M. & Schleich, W. P. (1993). Quantum state engineering of the radiation field. *Physical Review Letters*, Vol. 71, No. 12, 1816-1819.

Waks, E. & Vuckvic, J. (2006). Dispersive properties and large Kerr nonlinearities using dipole-induced transparency in a single-sided cavity. *Physical Review A.*, Vol. 73, No. 4, 041803.

Xiao, X.-F.; Han, Z.-F. & Guo G.-C. (2006). Quantum computation without strict strong coupling on a silicon chip. *Physical Review A*, Vol. 73, No. 5, 052324.

Xiao, Y.-F.; Gao, J.; Yang, X.; Bose, R. & Guo, G.-C. (2007a). Nanocrystals in silicon photonic crystal standing-wave cavities as spin-photon phase gates for quantum information processing. *Applied Physics Letters*, Vol. 91, 151105.

Xiao, Y.-F.; Zou, X.-B.; Jiang, W.; Chen, Y.-L. & Guo, G.-C. (2007b). Analog to multiple electromagnetically induced transparency in all-optical drop-filter systems. *Physical Review A*, Vol. 75, No. 6, 063833.

Xiao, Y.-F.; Gao, J.; Zou, X.-B.; McMillan, J.-F.; Yang, X.; Chen, Y.-L.; Han, Z.-F.; Guo, G.-C. & Wong, C. W. (2008). Coupled quantum electrodynamics in photonic crystal cavities towards controlled phase gate operations. *New Journal of Physics*, Vol. 10, 123013.

Xu, Q.; Sandhu, S.; Povinelli, M. L.; Shakya, J.; Fan, S. & Lipson, M. (2006). Experimental Realization of an on-chip all-optical analogue to electromagnetically induced transparency. *Physical Review Letters*, Vol. 96, No. 12, 123901.

Xu, Q.; Dong, P. & Lipson, M. (2007). Breaking the delay-bandwidth limit in a photonic structure. *Nature Physics*, Vol. 3, 406-410.

Xu, Y.; Li, Y.; Lee, R. K. & Yariv, A. (2000). Scattering-theory analysis of waveguide-resonator coupling. *Physical Review E*, Vol. 62, No. 5, 7389-7404

Yanik, M. F.; Suh, W.; Wang, Z. & Fan, S. (2004). Stopping light in a waveguide with an all-optical analog of electromagnetically induced transparency. *Physical Review Letters*, Vol. 93, No. 23, 233903.

Yoshie, T.; Scherer, A.; Hendrickson, J.; Khitrova, G.; Gibbs, H. M.; Rupper, G.; Ell, C.; Shchekin, O. B. & Deppe, D. G. (2004). Vacuum Rabi splitting with a single quantum dot in a photonic crystal nanocavity. *Nature*, Vol. 432, 200-203.

THz Techniques and Applications

6

Terahertz-wave Parametric Sources

Shin'ichiro Hayashi[1] and Kodo Kawase[1,2]
[1]*RIKEN,*
[2]*Nagoya University*
Japan

1. Introduction

Terahertz waves are the electromagnetic radiation whose frequency ranges from millimeter waves to the far infrared, shown in Figure 1. While both sides of this range have had a long history of research and development, leading to already commercially available sources, detectors, meters, and many additional devices, the terahertz wave range is still in its infancy, representing the last unexplored part of the electromagnetic spectrum between radio waves and visible light. This delayed development was mainly caused by the difficulty of producing reliable and strong terahertz wave generators, as well as the unavailability of sensors that can detect this unusual radiation. Technology extrapolation from both neighboring sides has been facing difficult problems: Up-conversion from the microwaves is inefficient due to the frequency being too high; down-conversion from the infrared is limited by the energy gaps.

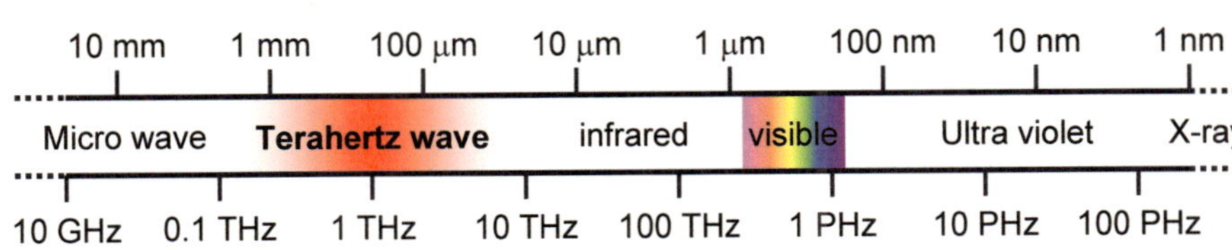

Fig. 1. A schematic showing the terahertz wave within the electromagnetic spectrum.

In recent years, terahertz wave sources have received considerable attention for use in many applications. Especially, recent research using terahertz waves, transmission imaging and fingerprint spectra have had an important contribution in the bioengineering and security fields, such as in material science, solid state physics, molecular analysis, atmospheric research, biology, chemistry, drug and food inspection, and gas tracing (Tonouchi, 2007).

There are several ways to generate terahertz waves. In the laboratory, one of the most widespread processes is the optical rectification or photoconductive switching produced using femtosecond laser pulses (Smith et al., 1988; Zhang et al., 1990). Applied research, such as time domain spectroscopy (THz-TDS), makes use of the good time resolution and the ultra broad bandwidth, up to the terahertz region. Novel tunable sources already exist in the sub-THz (several hundred GHz) frequency region, such as the backward-wave oscillator (BWO). However, the output power of a BWO rapidly decreases in the frequency region above 1 THz, and its tuning capability is relatively limited.

Only few sources bring together qualities such as room temperature operation, compactness, and ease of use. The terahertz wave parametric generation is based on an optical parametric process in a nonlinear crystal (Sussman, 1970; Pietrup et al., 1975). The principles of the terahertz wave parametric generator (TPG) (Shikata et al., 2000; Sato et al., 2001; Shikata et al., 2002) and the terahertz wave parametric oscillator (TPO) (Kawase et al., 1996; 1997; 2001) allow building systems that are not only compact but also operate at room temperature, making them suitable as practical sources. In principle, both a narrow linewidth and a wide tunability are possible in injection-seeded TPG/TPO (is-TPG/TPO) systems with single-longitudinal-mode near-infrared lasers as seeders (Kawase et al., 2001; 2002; Imai et al., 2001).

In basic research, these sources were pumped using flash lamp- or laser diode- pumped Q-switched Nd:YAG lasers which have Gaussian beam profile and long pulse widths (15 ~ 25 ns). The output energy of terahertz wave increases with the pump energy, but eventually the damage threshold of the crystals is reached. Recently, we demonstrated how the output energy/power was further enhanced and how the TPG was reduced to palmtop size by using a small pump source having a short pulse width and top-hat beam profile (Hayashi et al., 2007). These characteristics of the pump beam enable high intensity pumping especially close to the output surface of the terahertz wave without thermal damage of the crystal surface. The higher intensity pumping and smaller absorption of the terahertz wave inside the crystal enable higher output energy than previously reported. Further, we also demonstrated a compact and tunable terahertz wave parametric source pumped by a microchip Nd:YAG laser, seeded with the idler wave provided by an external cavity diode laser (ECDL) (Hayashi et al., 2009). We show how the output peak power and tunability were further enhanced and how the is-TPG was reduced to palmtop size by using a passively Q-switched small pump source having a short pulse width. These characteristics of the pump beam permit high intensity pumping close to the output surface of the terahertz wave without thermal damage to the crystal surface. The higher intensity pumping and smaller absorption of the terahertz wave inside the crystal allow a broader tuning range than that previously reported.

2. Principles of a Terahertz-wave parametric generation

When a strong laser beam propagates through a nonlinear crystal, photon and phonon transverse wave fields are coupled, behave as new mixed photon-phonon states, called polaritons. The generation of the terahertz wave results from the efficient parametric scattering of laser light via a polariton, that is, stimulated polariton scattering. The scattering process involves both second- and third-order nonlinear processes. Thus, strong interaction occurs among the pump beam, the idler wave, and the polariton (terahertz) wave.

One of the most suitable nonlinear crystal to generate terahertz wave is the lithium niobate ($LiNbO_3$) thanks to its large nonlinear coefficient (d_{33} = 25.2 pmV^{-1} at λ = 1064 nm) (Shoji et al., 1997) and its transparency over a wide wavelength range (0.4 – 5.5 μm). $LiNbO_3$ has four infrared- and Raman-active transverse optical (TO) phonon modes, called A_1-symmetry modes, and the lowest mode (ω_0 ~ 250 cm^{-1}) is useful for efficient terahertz wave generation because it has the largest parametric gain as well as the smallest absorption coefficient.

The principle of tunable terahertz wave generation is as follows. The polaritons exhibit phonon-like behavior in the resonant frequency region (near the TO-phonon frequency ω_{TO}). However, they behave like photons in the non resonant low-frequency region as shown in

Figure 2, where a signal photon at terahertz frequency (ω_T) and a near-infrared idler photon (ω_i) are created parametrically from a near-infrared pump photon (ω_p), according to the energy conservation law $\omega_p = \omega_T + \omega_i$ (p: pump beam, T: terahertz wave, i: idler wave). In the stimulated scattering process, the momentum conservation law $\boldsymbol{k_p} = \boldsymbol{k_i} + \boldsymbol{k_T}$ (noncollinear phase-matching condition, Figure 2) also holds. This leads to the angle-dispersive characteristics of the idler and terahertz waves. Thus, broadband terahertz waves are generated depending on the phase-matching angle. Generation of a coherent terahertz wave can be achieved by applying an optical resonator (in the case of the TPO) or injecting a "seed" for the idler wave (in the case of the is-TPG/TPO). Continuous and wide tunability is accomplished simply by changing the angle between the incident pump beam and the resonator axis or the seed beam.

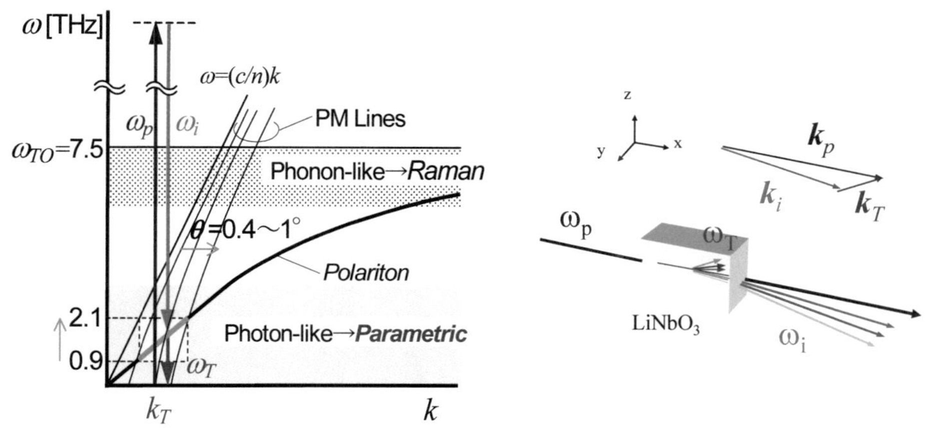

Fig. 2. Dispersion relation of the polariton. An elementary excitation is generated by the combination of a photon and a transverse optical phonon (ω_{TO}). The polariton in the low energy region behaves like a photon at terahertz frequency. Due to the phase-matching condition as well as the energy conservation law which hold in the stimulated parametric process, tunable terahertz wave is obtained by the control of the wavevector $\boldsymbol{k_i}$. The right figure shows the noncolinear phase-matching condition.

The bandwidth of the TPG is decided by the parametric gain and absorption coefficients in the terahertz region. According to a plane-wave approach, analytical expressions of the exponential gain for the terahertz and idler wave are given by (Shikata et al, 2000; 2002)

$$g_T = \frac{\alpha_T}{2}\left\{\sqrt{1+16\cos\phi\left(\frac{g_0}{\alpha_T}\right)^2}-1\right\}, \tag{1}$$

where a_T is the absorption coefficient of the terahertz wave in the nonlinear crystal. Parameter φ is the phase-matching angle between the pump beam and the terahertz wave; g_0 is the parametric gain in the low-loss limit, and takes the form

$$g_0 = \sqrt{\frac{\pi\omega_p\omega_i I_p}{2c^3 n_T n_i n_p}}\chi_p \propto \sqrt{\omega_i\omega_T I_p}, \tag{2}$$

$$\chi_p = d_E + \frac{S_0\omega_0^2}{\omega_0^2 - \omega^2} d_Q \,, \tag{3}$$

where I_p is the pump intensity, n_T, n_i, n_p are the crystal refractive indices at the wavelengths of the terahertz wave, idler wave, and pump beam, respectively, ω_0 is the resonance frequency of the lowest A_1-mode, and S_0 is the oscillator strength. The nonlinear coefficients d_E and d_Q represent second- and third-order nonlinear processes, respectively. The absorption coefficient α_T in the terahertz region is given by,

$$\alpha_T = \frac{2\omega}{c}\left|\mathrm{Im}\left(\sqrt{\varepsilon_T}\right)\right| \,, \tag{4}$$

where ε_T is the dielectric constant of the nonlinear crystal.

Figure 3 shows the calculated gain and the absorption coefficient at several pump intensities. The gain curve has a broad bandwidth of around 3 THz, with a dip appearing at around 2.6 THz. This is because the low frequency modes of doped MgO in the $MgO:LiNbO_3$ work as a crystal lattice defects for $LiNbO_3$.

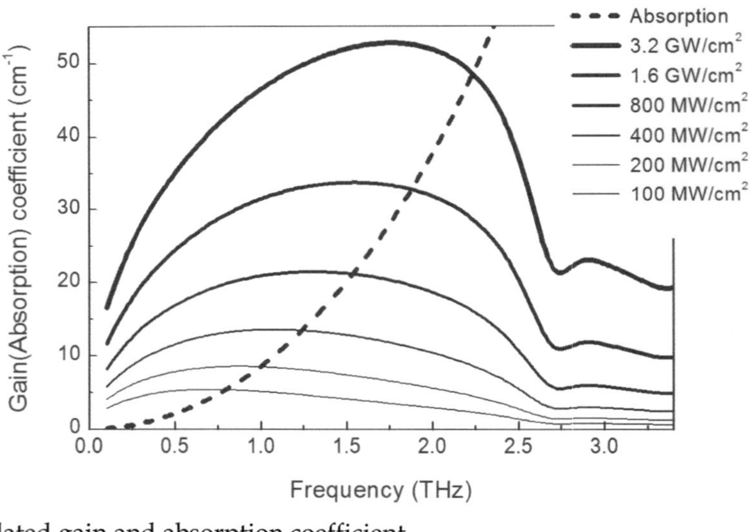

Fig. 3. Calculated gain and absorption coefficient.

3. Terahertz-wave parametric generator (TPG)

Broadband terahertz waves are generated by single-pass pumping, in a TPG. The linewidth of the terahertz wave emitted from the TPG is typically broad, about 1 THz. In addition, several applications are better suited to a broadband source (TPG) than to a nawwor linewidth source (TPO or is-TPG), such as tomographic imaging, interferometric spectroscopy, and diffuse reflection spectroscopy. Tomographic imaging and interferometric spectroscopy have to use a broadband source. The detection of scattered terahertz radiation strongly depends on the grain size of samples made of particles; using a broadband source reduces this effect. Also, the TPG is useful for many industrial applications such as transmission imaging for materials or food inspection.

A TPG uses a quite simple configuration since it needs no resonator or seeder, as shown in Figure 4. The MgO:LiNbO$_3$ crystal used in the experiment was cut to the size 65 (x) × 5 (y) × 4 (z) mm^3. The *x*-surfaces at both ends were mirror polished and antireflection coated. The *y*-surface was also mirror polished, in order to minimize the coupling gap between the prism base and the crystal surface, and to prevent scattering of the pump beam. The pump beam passed through the crystal close to the *y*-surface, to minimize the travel distance of the terahertz wave inside the crystal.

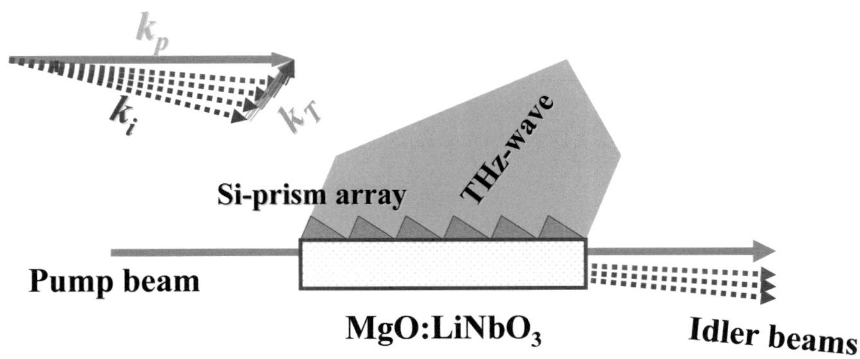

Fig. 4. A terahertz wave parametric generator with a Si-prism array. The Si-prism array was placed on the *y*-surface of the MgO:LiNbO$_3$ to increase the output and to reduce the diffraction angle of the terahertz wave by increasing the coupling area.

Most of the generated terahertz wave was absorbed or totally reflected inside the crystal due to the material's large absorption coefficient and large refractive index. Therefore, it was rather difficult to couple out the terahertz wave efficiently to the free space. We introduced a Si-prism coupler (n ≈ 3.4) to extract the terahertz wave generated inside a nonlinear crystal, thereby substantially improving the exit characteristics. The terahertz wave output energy, peak power and linewidth emitted from the TPG is typically 1 pJ/pulse, 300 μW, and 1 THz respectively.

4. Terahertz-wave parametric oscillator (TPO)

Coherent tunable terahertz waves can be generated by realizing a resonant cavity for the idler wave. This is the basic configuration of a TPO, and it consists of a Q-switched Nd:YAG laser, the nonlinear crystal placed inside the 15 cm long cavity, as shown in Figure 5. Both mirrors were half-area-coated, so that only the idler wave could resonate and the pump beam propagate through the uncoated area without scattering. The mirrors and a nonlinear crystal were mounted on a rotating stage, and tunability was obtained by rotating the stage slightly to vary the angle of the resonator with respect to the pump beam. The pump power and pulsewidth were 30 mJ/pulse and 25 ns, respectively. The pump beam entered the *x*-surface of the crystal and passed through the MgO:LiNbO$_3$ crystal close to the surface of the Si-prism coupler to minimize the absorption loss of the terahertz wave. A near-infrared idler oscillation around 1.07 μm was clearly recognized by its oscillating spot above a threshold pump power density of about 130 MW/cm^2. The idler wave is amplified in the resonator consisting of flat mirrors with a half-area HR coating. The mirrors and crystal are installed on a precise, computer-controlled rotating stage for precise tuning. When the incident angle of the pump beam into the MgO:LiNbO$_3$ is varied between 3.13 and 0.84 deg, the angle between the pump wave and the idler wave in the crystal changes from 1.45 down to 0.39 deg, whereas the angle between the terahertz wave and the idler wave changes from 67.3 down to 64.4 deg. With this slight variation in the phase-matching condition, the wavelength (frequency) of the terahertz wave could be tuned between 100 and 330 mm (3 – 0.9 THz); the corresponding idler wavelength changed from 1.075 down to 1.067 mm. The terahertz wave radiation was monitored with a 4K Si bolometer.

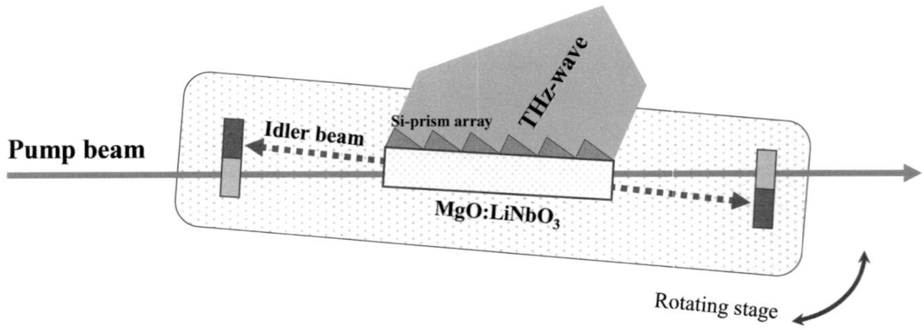

Fig. 5. TPO configuration. The TPO consists of a Q-switched Nd:YAG laser, a nonlinear crystal, and a parametric oscillator. The idler wave is amplified in the resonator consisting of flat mirrors with a half-area HR coating. The mirrors and crystal are installed on a precise, computer-controlled, rotating stage for fine tuning.

Typical input-output characteristics of a TPO are shown in Figure 6, in which the oscillation threshold was 18 mJ/pulse. With a pump power of 34 mJ/pulse, the output energy of terahertz wave from TPO was 192 pJ/pulse (≅ 19 mW at the peak), calibrated using the sensitivity of the bolometer. Since the Si-bolometer output becomes saturated at approximately 5 pJ/pulse, we used several sheets of thick paper as an attenuator after they were properly calibrated. The minimum sensitivity of the Si-bolometer is approximately 1 fJ/pulse, therefore, the dynamic range of measurement using the TPO as a source is 192 pJ / 1 fJ, which exceeds 50 dB.

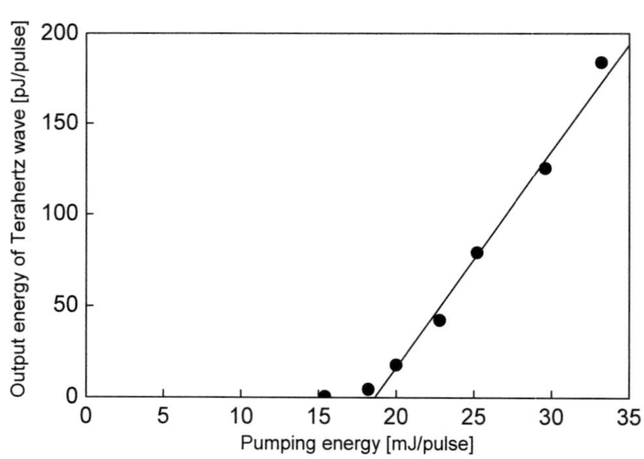

Fig. 6. Input-output characteristics of a terahertz wave parametric oscillator.

5. Injection-seeded Terahertz-wave parametric generator (is-TPG)

The TPG spectrum was narrowed to the Fourier Transform limit of the pulse width by introducing an injection seeding for the idler wave. Figure 7 shows our experimental setup of the is-TPG. An array of seven Si-prism couplers was placed on the *y*-surface of the secondary MgO:$LiNbO_3$ crystal for efficient coupling of the terahertz wave. The pumping laser was a single longitudinal mode Q-switched Nd:YAG laser (wavelength: 1.064 μm; energy: < 50 mJ/pulse; pulsewidth: 15 ns; beam profile: TEM_{00}). The pump beam diameter was 0.8mm. The pump beam was almost normal to the crystal surfaces as it entered the crystals and passed through the crystal close to the *y*-surface. A continuous wave tunable diode laser (wavelength: 1.066–1.074 μm; power: 50 mW) was used as an injection seeder for the idler. Observation of the intense idler beam easily confirmed the injection-seeded terahertz wave generation. The polarizations of the pump, seed, idler, and terahertz waves were all parallel to the *z*-axis of the crystals. The terahertz wave output was measured with a 4K Si bolometer.

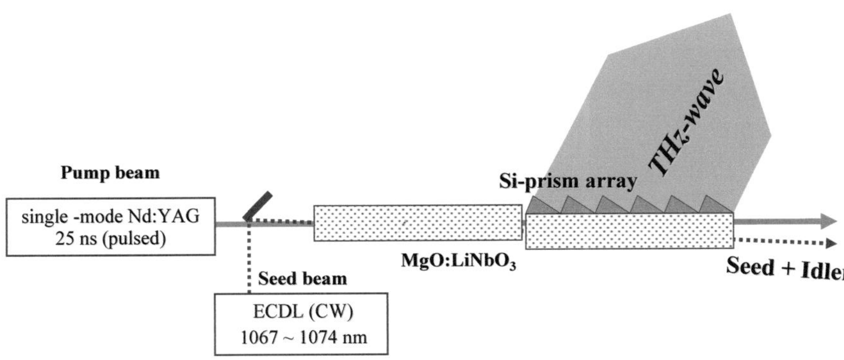

Fig. 7. Experimental setup of the is-TPG.

It was possible to tune the terahertz wavelength using an external cavity laser diode as a tunable seeder. A wide tunability, from 125 to 430 μm (frequency: 0.7 to 2.4 THz), was achieved as shown in Figure 8 by changing simultaneously the seed wavelength and the seed incident angle. Open squares and closed circles indicate the tunability of the terahertz and idler waves, respectively. Both crystals were MgO:$LiNbO_3$ in this experiment. The wavelength of 430 μm (0.7 THz) was the longest ever observed during our study of TPGs and TPOs. In the longer-wavelength region, the angle between the pump and idler becomes less than 1°; thus it is difficult for the TPO to oscillate only the idler inside the cavity without scattering the pump. In the shorter-wavelength region, the terahertz wave output is comparatively smaller than the idler wave output, due to the larger absorption loss inside the crystal.

The absorption spectrum of low-pressure (< 1 torr) water vapor was measured to demonstrate the continuous tunability and the high resolution of the is-TPG. The absorption gas cell used was an 87-cm-long stainless steel pipe with TPX windows at both ends. Figure 9 shows an example of measurements at around 1.92 THz, where two neighboring lines exist. Resolution of less than 100 MHz (0.003 cm^{-1}) was clearly shown. In fact, it is not easy for FTIR spectrometers in the terahertz wave region to demonstrate a resolution better than 0.003 cm^{-1} because of the instability of the scanning mirror for more than a meter. The system is capable of continuous tuning at high spectral resolution in 4 GHz segments

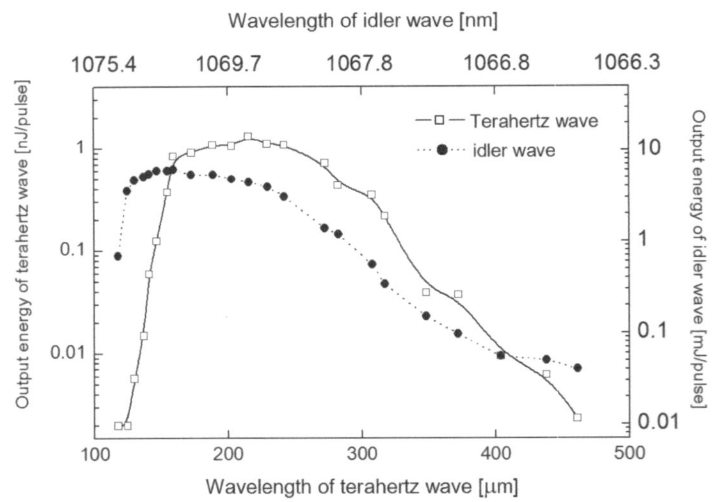

Fig. 8. Wide tunability of an is-TPG. Open squares and closed circles indicate the tunability of the THz and idler waves, respectively.

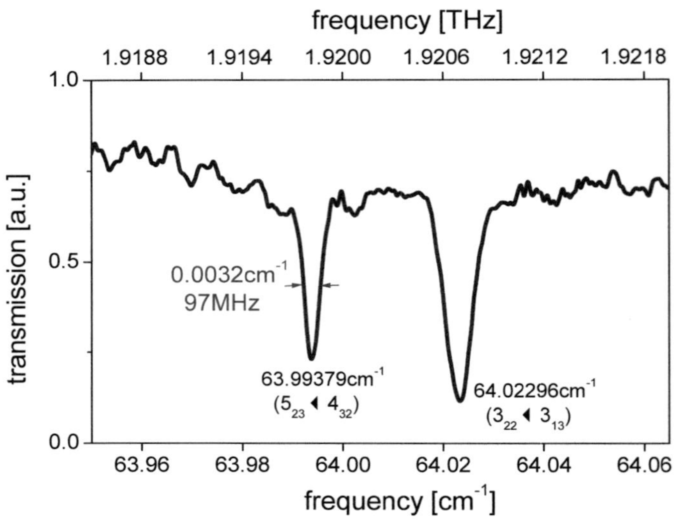

Fig. 9. An example of the absorption spectrum measurement of low-pressure (<1 torr) water vapor at around 1.919 THz. Resolution of less than 100 MHz (0.003 cm-1) was clearly demonstrated.

anywhere in the region from 0.7 to 2.4 THz. Since there is no cavity to be slaved, the continuous tuning is extendible, in principle, to the full tunability of the is-TPG by using a mode-hop-free seeder, such as a Littman-type external cavity diode laser.

The input-output characteristic of the terahertz wave from an is-TPG is shown in Figure10. The maximum conversion efficiency was achieved when the pump and seed beams almost fully overlapped at the incident surface of the first MgO:$LiNbO_3$ crystal. The maximum terahertz wave output of 1.3 nJ/pulse (peak power over 300 mW) was obtained with a single-mode pump beam of 34 mJ/pulse and a seed beam of 50 mW. To prevent saturating the Si bolometer, again we used several sheets of thick paper as an attenuator after calibrating them. In our previous studies, the maximum terahertz wave output from a conventional TPG and a TPO was 1 and 190 pJ/pulse, respectively. The Si-bolometer became saturated at about 5 pJ/pulse, so we used several thick calibrated sheets of paper as an attenuator. As the minimum sensitivity of the Si-bolometer is about 1 fJ/pulse, the dynamic range of the is-TPG system was from 1.2 nJ down to 1 fJ, that is, ~ 60 dB, which is sufficient for most applications. The dynamic range can be significantly increased using a lock-in amplifier.

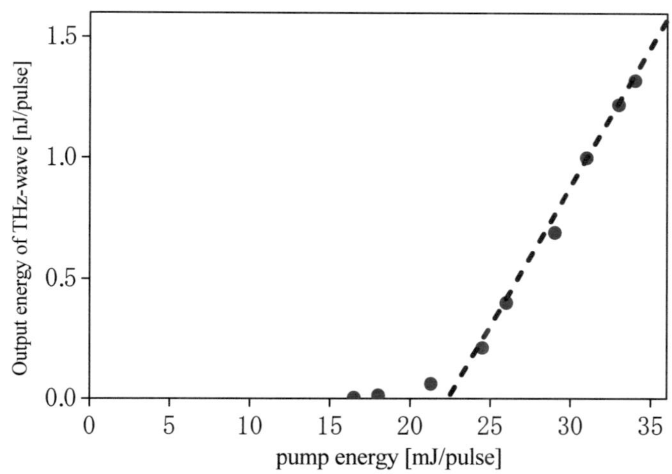

Fig. 10. Input-output characteristics of the is-TPG.

6. Recent progress

6.1 Energy-enhanced TPG

In this section, we report some recent developments of a TPG using a small pump source with a short pulse width and top-hat beam profile. In our experimental configuration, the output energy of the TPG is mainly limited by the damage threshold of the nonlinear crystal. We can generate high energy, broadband terahertz waves by using a short-pulsed pump beam with a top-hat beam profile which can provide high intensity pumping near the crystal surface without damaging the crystal.

The experimental apparatus, shown in figure 11, consists of a flash-lamp pumped Q-switched Nd:YAG laser, a lens, mirrors, and two nonlinear crystals. All components, except for the detector in figure 11, can be mounted on a 12 × 22 cm breadboard. The small pump source has a short pulse width, of around 5 ns. Its slight divergence is corrected by a lens placed at the output of the source. It has a top-hat profile with a beam diameter of 1.3 mm (full width at half maximum) on the first crystal. We used two 65-mm-long nonlinear MgO:$LiNbO_3$ crystals. Both crystal ends are antireflection coated for a wavelength of 1064 nm. The gap between the two crystals is about 100 μm in our experiment, which is short enough for the phase matching condition. A Si-prism array placed on the *y*-surface of the MgO:$LiNbO_3$ crystal acts as an efficient output coupler for the terahertz waves to avoid the total internal reflection of the terahertz waves on the output side crystal surface.

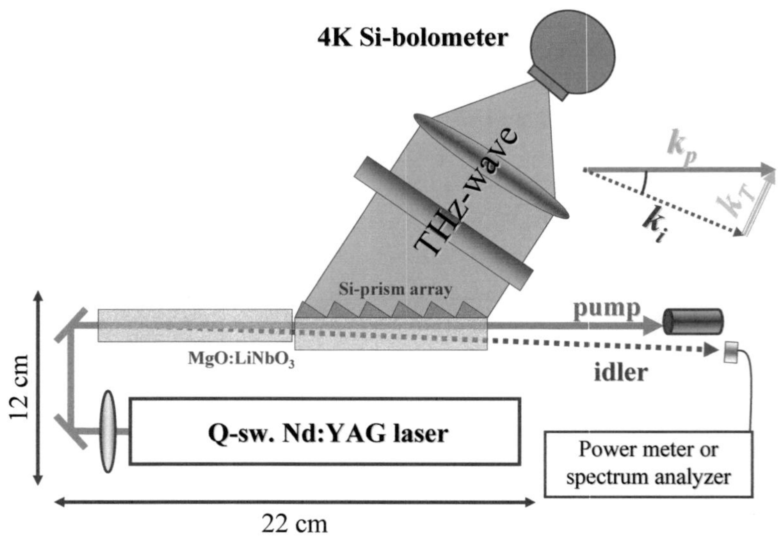

Fig. 11. Experimental setup of the energy-enhanced TPG.

For an efficient extraction of the terahertz waves, the pumped region inside the second crystal must be as close as possible to the Si-prism array, because of the large absorption coefficient of the $MgO:LiNbO_3$ crystal in the terahertz range. A top-hat beam profile is suitable for this purpose, since the high intensity region of the pump beam can be brought closer to the *y*-surface than in the case of a Gaussian beam. The distance between the *y*-surface and the beam center was precisely adjusted to obtain a maximum terahertz wave output, and it was approximately equal to the pump beam radius.

The terahertz wave output extracted through the Si-prism array was measured using a 4.2 K silicon bolometer, while the idler wave energy was measured using a pyroelectric detector. The minimum and maximum sensitivity levels of the bolometer are about 0.01 pJ and 10 pJ without any amplifier or attenuator. Attenuators were used when the detector was saturated; to cut diffused light from the pump and idler, a thick black polyethylene sheet was used.

Figure 12 shows the output energy/power (peak) of the terahertz wave as a function of the pump intensity. As the pump intensity is higher, the terahertz wave starts to be detected at a pump intensity of around 300 MW/cm^2 (25 mJ/pulse) then increase monotonically. The highest values obtained are 105 pJ/pulse (62 mW peak power) for the terahertz wave when the pump intensity is 830 MW/cm^2 (66 mJ/pulse), which corresponds to a pump energy of 66 mJ/pulse. The output of terahertz wave appears to saturate when the pump intensity exceeds 750 MW/cm^2 (60 mJ/pulse). Because higher intensity pumping leads to broader bandwidth as indicated by Eq. (1), however, the absorption coefficient for the terahertz wave rapidly increases in the high frequency range.

In previous TPG/TPO research, the crystal damage threshold was below the value of 200 MW/cm^2 for the pump beam intensity. With this report, by using a short-pulsewidth pump beam, the damage threshold is increased about 4 times. Moreover, the top-hat beam profile enables high intensity pumping especially close to the terahertz wave output surface, without any thermal damage to the crystal surface. These combined characteristics of pump beam yield 100 times more output energy of the terahertz wave.

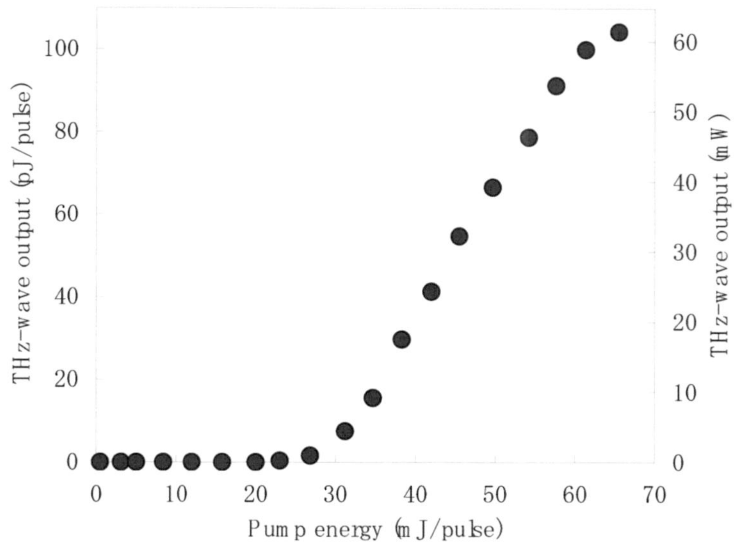

Fig. 12. The input-output characteristics of the energy-enhanced TPG.

Figure 13 shows the idler wave spectrum observed for varying pump energies. According to the noncollinear phase-matching condition, the propagating direction of the generated idler waves is slightly different from that of the pump beam, with an angle between them of around 1.5° outside the crystal. As the pump energy increases, the idler wave spectrum covers a broader spectral region, especially towards longer wavelengths. At the maximum pump energy, the idler wave spectrum was found to cover the range 1067 - 1079 nm. This spectrum corresponds to the terahertz wave frequency range 0.898 - 3.87 THz (77.6 - 334 μm). The measured spectrum is much broader than that observed in a previous report. The main reason for this broader spectrum might be the fact that the parametric gain could have broader bandwidth by higher pump energy as shown in Figure 3. The dip in the

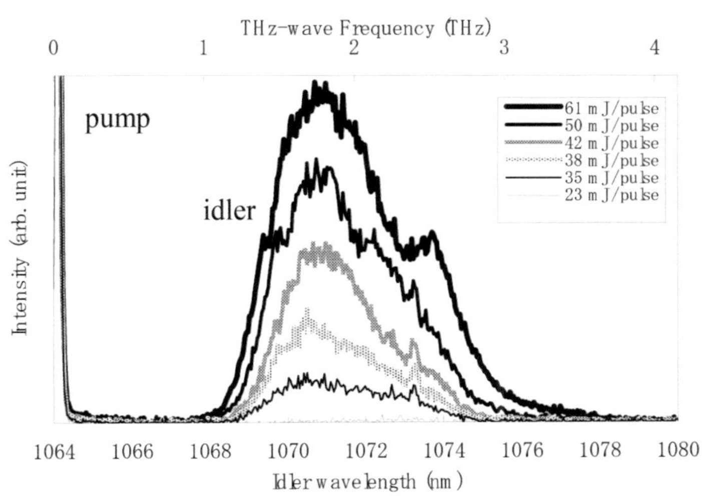

Fig. 13. Idler spectra at several pump energies.

spectra around 1073 nm appears due to the MgO doping of the $LiNbO_3$; the lattice defects produced by the MgO leads to additional peaks of the absorption coefficient α_T.

6.2 Tunability-enhanced is-TPG

We have enhanced the tunability of terahertz wave parametric generator using MgO:$LiNbO_3$ pumped by a sub-nanosecond, passively Q-switched, microchip Nd:YAG laser. This pump source allows high intensity pumping without damaging of the nonlinear crystal and generates a narrow linewidth and tunable terahertz wave with injection seeding by an external cavity diode laser for the idler wave. The high intensity pumping causes a gain curve broadening of the terahertz wave parametric generation, especially in the high frequency region.

The experimental setup, shown in Figure 14, consists of a pumping source (Microchip Nd:YAG laser), a seeding source (External Cavity Diode Laser) and the nonlinear crystal. The pump source is a diode end-pumped single-mode microchip Nd^{3+}:YAG laser, passively Q-switched by a Cr^{4+}:YAG saturable absorber. This microchip configuration enables the low order axial and transverse mode laser oscillation, whose linewidth is below 0.009 nm. The laser delivers 1.1 MW peak power pulses (530 μJ/pulse) with 430 ps pulse width at 100 Hz repetition rate with a M^2 factor of 1.09. This laser is free from the electric noise, unlike the active Q-switched lasers we used before. Additionally, this kind of fixed passively Q-switching allows us to obtain a stabilized peak power, with less than ±2 % power jitter (Pavel et al., 2001; Sakai et al., 2008). The pump beam diameter on the first crystal is 0.3 mm (full width at half maximum). We used two 65-mm-long nonlinear MgO:$LiNbO_3$ crystals. A silicon-prism array placed on the y surface of the second crystal acts as an efficient output coupler for the terahertz waves to avoid the total internal reflection of the terahertz waves on the crystal output side. For an efficient terahertz wave emission, the pumped region within the second crystal must be as close as possible to the silicon-prism array, because of the large absorption coefficient of the MgO:$LiNbO_3$ crystal in the terahertz range. The distance between the y-surface and the beam center was precisely adjusted to obtain a maximum terahertz wave output, and it was approximately equal to the pump beam radius. The terahertz wave output extracted through the silicon-prism array was measured using a 4.2 K silicon bolometer, while the idler-wave energy was measured using a pyroelectric detector.

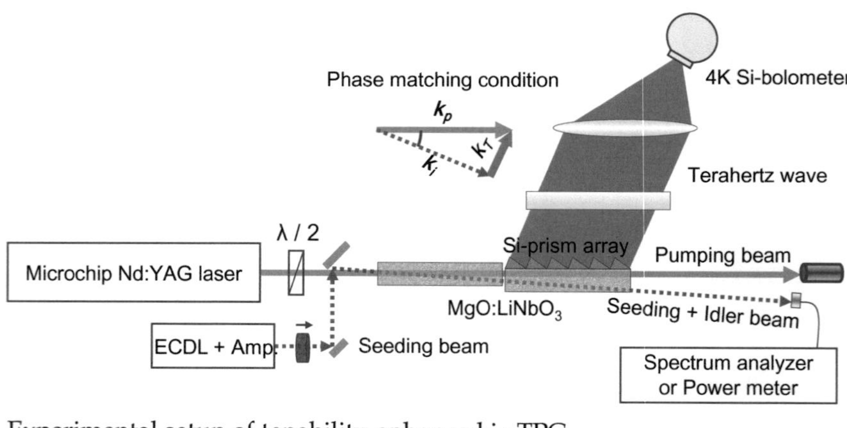

Fig. 14. Experimental setup of tenability-enhanced is-TPG.

Figure 15 shows the idler wave spectrum when the pump intensity is 2.9 GW/cm^2. According to the noncollinear phase matching condition, the propagating direction of the generated idler waves is slightly different from that of the pump beam, with an angle between them of around 1.5° outside the crystal. The idler wave spectrum was found to cover the range 1069 - 1075 nm, corresponding to the terahertz wave frequency range 1.4 - 2.9 THz without seeding beam. Compared with a previous report, this is a shift to higher frequency. The dip in the spectra around 1073 nm appears due to the MgO doping of the $LiNbO_3$; the lattice defects produced by the MgO leads to additional peaks of the absorption coefficient.

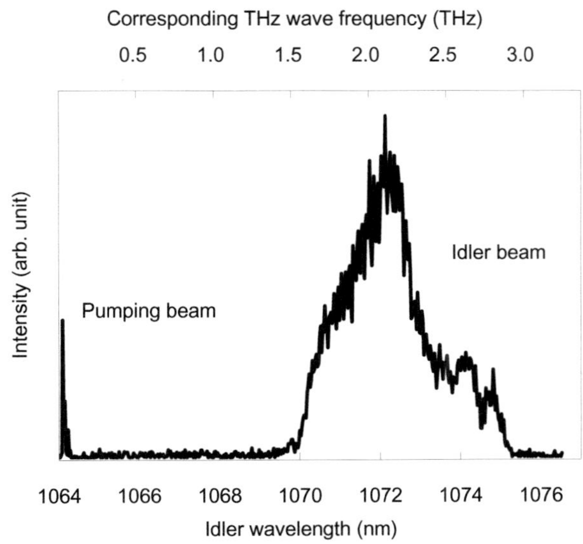

Fig. 15. Idler spectrum.

Figure 16 shows the time waveform of the terahertz wave output signals measured by the 4 K silicon bolometer. When we generate the terahertz wave without injection seeding to the idler wave, we observe a broadband terahertz wave with the peak power of about 1 mW (lower curve), however, after injection seeding, we observed a narrow linewidth terahertz wave with a peak power of about 20 mW (upper curve). This is about more than 100 times narrower and 20 times higher than when the seeding laser is cut off. In addition, the pulse width of this microchip laser is the shortest among our parametric sources.

It is possible to tune the terahertz frequency using an ECDL as a tunable seeder. When the pump intensity is 1.8 GW/cm^2 (peak, energy of 650 μJ/pulse) and the seeding power is 80 mW (CW), a wide tunability from 0.9 - 3 THz is observed, as shown in figure 6 by changing both the seed wavelength and the seed incident angle. The maximum output peak power of terahertz wave was about 100 mW at around 1.8 THz. The tuning curve has a broad bandwidth, with a dip appearing at around 2.7 THz. This is because the low frequency modes of doped MgO in the MgO:$LiNbO_3$ work as crystal lattice defects for $LiNbO_3$.[5]

Figure 18 shows an example of wavelength and linewidth measurement by a scanning Fabry–Perot etalon consisting of two Ni metal-mesh plates with a 65 μm grid. The displacement of one of the metal mesh plates corresponds directly to half of the wavelength. We observed a narrow linewidth terahertz wave with a wavelength of 140 μm and peak power of about 60 mW by the 4 K silicon bolometer. The free spectral range (FSR) of the etalon was about 100 GHz, and the linewidth was measured to be less than 10 GHz.

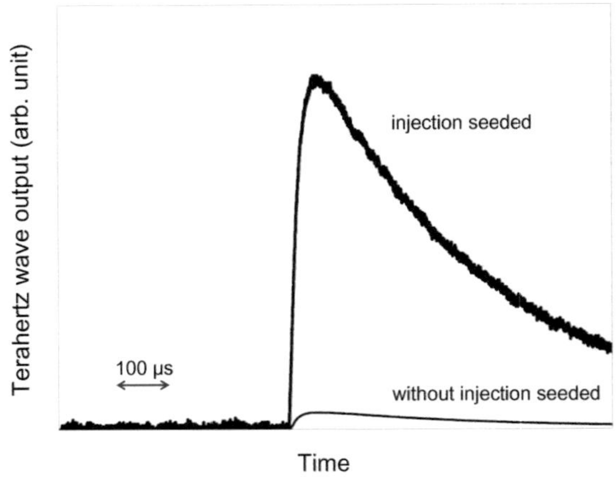

Fig. 16. Time dependent terahertz wave output signals.

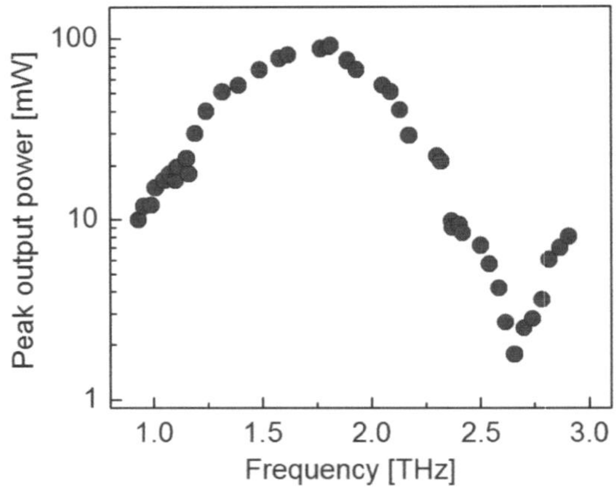

Fig. 17. Wide tunability of is-TPG with a dip at aroud 2.7 THz.

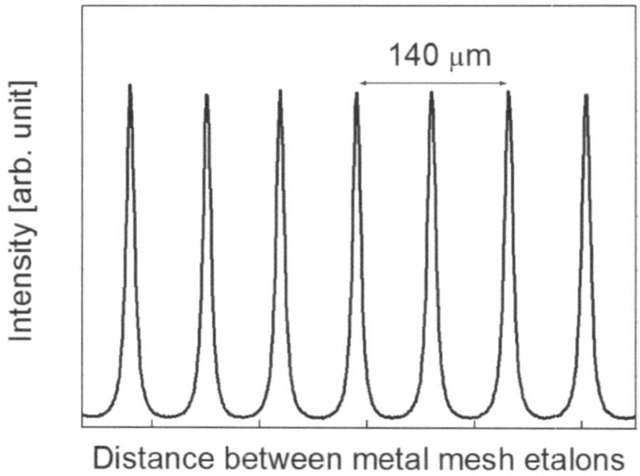

Fig. 18. Example of the wavelength and linewidth measurement using the scanning Fabry-Perot etalon consisting of metal mesh plates.

7. Conclusion

We reviewed terahertz wave parametric sources based on the optical parametric process. We have introduced several types of TPG, TPO, and is-TPG. Measurements on tunability, coherency, and power have been accomplished, proving this method to be suitable for many application fields. These include spectroscopy, communication, medical and biological applications, THz imaging, and so forth.

We also demonstrated output power enhancement of the TPG, while at the same time achieving a considerable downsizing of this terahertz source, all of which were realized by using a small pump source with a short pulse width and a top-hat beam profile. We measured a terahertz wave output energy of 105 pJ/pulse, with a power peak at 62 mW, and a broadband spectrum, extending from 0.9 to 3.8 THz. The new source is more than 100 times brighter and has a spectrum more than twice broader than previously reported.

In the next section, we demonstrated a compact and tunable terahertz wave source pumped by a microchip Nd:YAG laser. This source generates a narrow linewidth and high peak power terahertz wave by injection seeding for the idler wave. Using a microchip laser as the pumping source allowed high intensity pumping and the broadening of the tuning range towards the high frequency region. We could also observe a dip around 2.7 THz in the tuning curve, as expected from the calculation.

Further improvement of our system is possible. As OPGs and OPOs have improved tremendously in the last decade, the use of TPGs and TPOs shows great potential to move towards a lower threshold, higher efficiency, and wider tunability. A lower threshold and a narrower linewidth can be expected using a nonlinear optical waveguide and a longer pump pulsewidth, respectively. Operation in other wavelength regions, through proper crystal selection, should also be possible. Success in this will prove the practicality of a new widely tunable THz-wave source, the IS-TPG, that will compete with free-electron lasers and p-Ge lasers. For tunable THz-wave applications, the simplicity of the wave source is an essential requirement since cumbersome systems do not encourage new experimental thoughts and ideas. Compared with the available sources, the present parametric method has significant advantages in compactness, tunability, and ease of handling.

8. Acknowledgements

The authors to thank Dr. Takayuki Shibuya, Dr. Hiroaki Minamide, Dr. Tomofumi Ikari, Prof. Takanari Yasui, Prof. Yuichi Ogawa, Prof. Jun-ichi Shikata, and Prof. Hiromasa Ito for useful discussions, and Prof. Takunori Taira, and Dr. Hiroshi Sakai for providing the microchip laser, Mr. Choichi Takyu for his excellent work coating the crystal surface, and Mr. Tetsuo Shoji for superbly polishing the crystals.

9. References

Hayashi, S.; Minamide, H.; Ikari, T.; Ogawa, Y.; Shikata, J.; Ito, H; Otani, C. & Kawase, K. (2007). Output power enhancement of a palmtop terahertz-wave parametric generator. Appl. Opt., Vol. 46, 117 - 123, ISSN: 00036935.

Hayashi, S.; Shibuya, T.; Sakai, H.; Taira, T.; Otani, C.; Ogawa, Y.; & Kawase, K. (2009). Tunability enhancement of a terahertz-wave parametric generator pumped by a microchip Nd:YAG laser. Appl. Opt., Vol. 48, No. 15, 2899-2902, ISSN: 00036935.

Imai, K.; Kawase, K.; Shikata, J.; Minamide, H. & Ito, H. (2001). Injection-seeded terahertz-wave parametric oscillator. *Appl. Phys. Lett.*, Vol. 78, 1026–1028, ISSN: 00036951.

Kawase, K.; Sato, M.; Taniuchi, T. & Ito, H. (1996). Coherent tunable THz-wave generation from $LiNbO_3$ with monolithic grating coupler. *Appl. Phys. Lett.*, Vol. 68, 2483–2485, ISSN: 00036951.

Kawase, K.; Sato, M.; Nakamura, K.; Taniuchi, T. & Ito, H. (1997). Unidirectional radiation of widely tunable THz wave using a prism coupler under noncollinear phase matching condition, *Appl. Phys. Lett.*, Vol. 71, 753–755, ISSN: 00036951.

Kawase, K.; Shikata, J; Minamide, H.; Imai, K. & Ito, H. (2001). Arrayed silicon prism coupler for a terahertz-wave parametric oscillator. *Appl. Opt.*, Vol. 40, 1423–1426, ISSN: 00036935.

Kawase, K.; Shikata, J.; Imai, K. & Ito, H. (2001). Transformlimited, narrow-linewidth, terahertz-wave parametric generator. *Appl. Phys. Lett.*, Vol. 78, 2819–2821, ISSN: 00036951.

Kawase, K.; Minamide, H.; Imai, K.; Shikata, J. & Ito, H. (2002). Injection-seeded terahertz-wave parametric generator with wide tenability. *Appl. Phys. Lett.*, Vol. 80, 195–197, ISSN: 00036951.

Kawase, K.; Shikata, J. & Ito, H. (2002). Terahertz-wave parametric source. J. Phys. D, Vol. 35, R1–R14, ISSN 0022-3727.

Pavel, N.; Saikawa, J.; Kurimura, S. & Taira, T. (2001). High Average Power Diode End-Pumped Composite Nd:YAG Laser Passively Q-switched by Cr^{4+}:YAG Saturable Absorber. Jpn. J. Appl. Phys., Vol. 40, pt. 1, no. 3A, 1253-1259, ISSN:0021-4922.

Piestrup, M. A.; Fleming, R. N. & Pantell, R. H. (1975). Continuously tunable submillimeter wave source. Appl. Phys. Lett., Vol. 26, 418–421, ISSN: 00036951.

Sakai, H.; Kan, H. & Taira, T. (2008). > 1 MW peak power single-mode high-brightness passively Q-switched Nd^{3+}:YAG microchip laser. Opt. Exp., Vol. 16, 19891-19899, ISSN: 1094-4087.

Sato, A.; Kawase, K.; Minamide, H.; Wada, S. & Ito H. (2001). Tabletop terahertz-wave parametric generator using a compact, diode-pumped Nd:YAG laser. Rev. Sci. Instrum., Vol. 72, 3501–3504, ISSN: 0034-6748.

Shikata, J.; Kawase, K.; Karino, K.; Taniuchi, T. & Ito H. (2000). Tunable terahertz-wave parametric oscillators using $LiNbO_3$ and MgO:$LiNbO_3$ crystals. IEEE Trans. Microwave Theory Tech., Vol. 48, 653–661, ISSN: 0018-9480.

Shikata, J.; Kawase, K.; Taniuchi, T. & Ito, H. (2002). Fouriertransform spectrometer with a terahertz-wave parametric generator. Jpn. J. Appl. Phys., Vol. 41, 134–138, ISSN:0021-4922.

Shoji, I.; Kondo, T.; Kitamoto, A.; Shirane, M.& Ito,R. (1997). Absolute scale of second-order nonlinear-optical coefficients. *J. Opt. Soc. Am.* B, Vol. 14, 2268-2294, ISSN: 0740-3224.

Smith, P. R.; Auston, D. H. & Nuss, M. C. (1988). Subpicosecond photoconducting dipole antennas. *IEEE J. Quantum Electron.*, Vol. 24, 255–260, ISSN: 0018-9197.

Sussman, S. S. (1970). Tunable Scattering from Transverse Optical Modes in Lithium Niobate. *Microwave Laboratory Report*, No. 1851 (Stanford University).

Tonouchi, M. (2007), Cutting-edge terahertz technology, *Nature Photonics*, Vol. 1, 97-105, ISSN: 1749-4885.

Zhang, X. C.; Hu, B. B.; Darrow J. T. & Auston D. H. (1990). Generation of femtosecond electromagnetic pulses from semiconductor surfaces. Appl. Phys. Lett., Vol 56, 1011–1013, ISSN: 00036951.

7

Cherenkov Phase Matched Monochromatic Tunable Terahertz Wave Generation

Koji Suizu[1], Takayuki Shibuya[1,2] and Kodo Kawase[1,2]
[1]*Nagoya University,*
[2]*RIKEN*
Japan

1. Introduction

Terahertz (THz) waves present attractive possibilities in advanced applications including biomedical analysis and stand-off detection for hazardous materials. The development of monochromatic and tunable coherent THz-wave sources is of great interest for use in these applications. Recently, a parametric process based on second-order nonlinearities was used to generate tunable monochromatic coherent THz waves using nonlinear optical crystals (Boyd et al., 1972; Rice et al., 1994; Shi et al., 2002; Tanabe et al. 2003). In general, however, nonlinear optical materials have high absorption coefficients in the THz-wave region, which inhibits efficient THz-wave generation.

Avetisyan et al. proposed surface-emitting THz-wave generation using the difference frequency generation (DFG) technique in a periodically poled lithium niobate (PPLN) waveguide to overcome these problems (Avetisyan et al., 2002). A surface-emitted THz wave radiates from the surface of the PPLN and propagates perpendicular to the direction of the pump beam. The absorption loss is minimized because the THz wave is generated from the PPLN surface. Moreover, the phase-matching condition can be designed using PPLN with an appropriate grating period (Sasaki et al., 2002). Surface-emitted THz-wave devices have the potential for high conversion efficiency, and continuous wave THz-wave generation has been successfully demonstrated (Sasaki et al., 2005). Unfortunately, the tuning range of the THz waves is limited to about 100 GHz by the nature of PPLN, and a wide tuning range cannot be realized using the quasi-phase–matching method.

We developed a Cherenkov phase-matching method for monochromatic THz-wave generation using the DFG process with a lithium niobate crystal, which resulted in both high conversion efficiency and wide tunability. Although THz-wave generation by Cherenkov phase matching has been demonstrated using femtosecond pumping pulses (Auston et al., 1984; Kleinman et al., 1984; Hebling et al., 2002; Wahlstrand, 2003; Badrov et al., 2009), producing very high peak power (Yeh et al., 2007), these THz-wave sources are not monochromatic. Our method generates monochromatic and tunable THz waves using a nanosecond pulsed laser source.

2. Cherenkov phase matching

The Cherenkov phase-matching condition is satisfied when the velocity of the polarization wave inside the nonlinear crystal is greater than the velocity of the radiated wave outside.

The radiation angle θ is determined by the refractive index of the pumping wave in the crystal, n_{opt}, and that of THz-wave in the crystal, n_{THz} (Sutherland, 2003),

$$\cos\theta_{crystal} = \frac{\lambda_{THz}/n_{THz}}{2L_c} = \frac{\lambda_{THz}/n_{THz}}{\lambda_1\lambda_2/(n_1\lambda_2 - n_2\lambda_1)} \approx \frac{n_{opt}}{n_{THz}} \tag{1}$$

where λ is a wavelength of the contributing waves in the DFG process ($\omega_1 - \omega_2 = \omega_{THz}$), n_1, n_2 ($n_1=n_2\cong n_{opt}$) and n_{THz} are refractive index of the crystal at pump waves and THz-wave frequencies, respectively, and L_c is the coherence length of the surface-emitted process ($L_c = \pi/\Delta k$, where $\Delta k=k_1-k_2$ and k is the wave number). The Cherenkov angle, $\theta_{crystal}$, is determined by the refractive indices of the pumping wave and the THz-wave in the crystal, so the angle is strongly dependent on the choice of material. THz-frequency waves radiated at Cherenkov angles propagate to the crystal-air interface, and if the angle is greater than a critical angle (determined by the difference in refractive indices at the interface), the THz-frequency wave is totally reflected at the interface. To prevent total internal reflection, a clad material with a lower refractive index than that of the crystal in the THz range and a proper prism shape, is coupled in at the output. Figure 1 shows a schematic of Cherenkov radiation and output coupling of a THz-frequency wave.

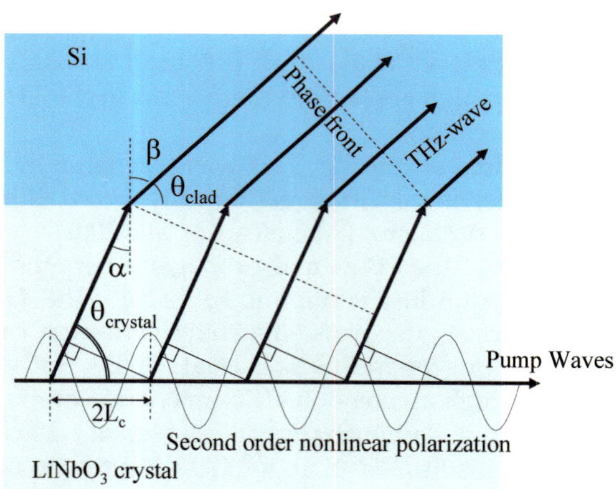

Fig. 1. Schematic of Cherenkov phase-matched monochromatic THz-wave generation.

Figure 2 shows relation of Cherenkov angle and critical angle of several clad materials. We choose polyethylene, diamond, Si and Ge as clad materials, because these materials have low absorbance and low dispersion character at THz frequency region. A total internal reflection occurs below the curve. For example, lithium niobate ($LiNbO_3$) has 2.2 and 5.2 of refractive index at near infrared and THz-wave region, results in 65 degree of Cherenkov angle in the crystal. On the other hands, critical angle of total internal reflection from the crystal to air, polyethylene, diamond, Si and Ge in a θ manner are 79, 76, 63, 49 and 40 degrees, respectively. The figure tells that diamond, Si and Ge prevent total internal reflection of Cherenkov radiation for lithium niobate crystal.

The angle in the clad material, θ_{clad}, is determined by Snell's law as shown in Fig. 1, using the refractive index of the clad material n_{clad}.

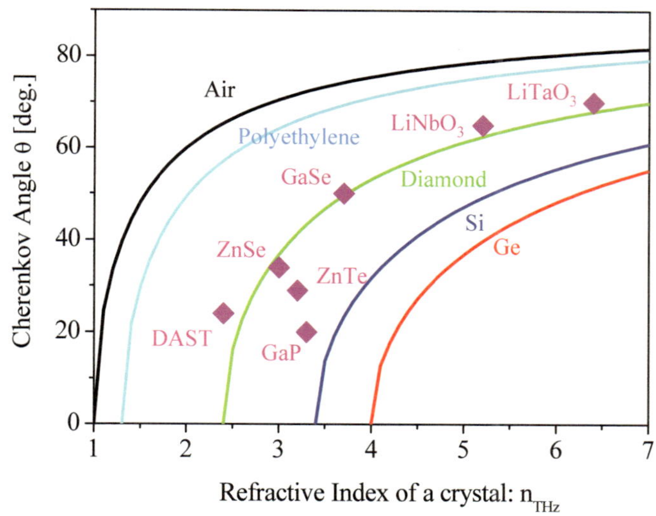

Fig. 2. Cherenkov angle for various nonlinear crystals (pink collared diamonds) and calculated critical angle between a crystal and a clad. Black, aqua, green, blue and red curve represent Air, polyethylene, diamond, Si and Ge as a clad material, respectively. A total internal reflection occurs below the curve.

$$\begin{aligned}\theta_{clad} &= \frac{\pi}{2} - \beta = \frac{\pi}{2} - \arcsin\left(\frac{n_{THz}}{n_{clad}}\sin(\alpha)\right) = \frac{\pi}{2} - \arcsin\left(\frac{n_{THz}}{n_{clad}}\sin\left(\frac{\pi}{2} - \theta_{crystal}\right)\right) \\ &= \frac{\pi}{2} - \arcsin\left(\frac{n_{THz}}{n_{clad}}\sin\left(\frac{\pi}{2} - \arccos\left(\frac{n_1\lambda_2 - n_2\lambda_1}{n_{THz}(\lambda_2 - \lambda_1)}\right)\right)\right) \\ &= \arccos\left(\frac{n_1\lambda_2 - n_2\lambda_1}{n_{clad}(\lambda_2 - \lambda_1)}\right)\end{aligned} \quad (2)$$

The radiation angle θ_{clad}, which is important for practical applications, is determined by the refractive indices of the pumping waves in the crystal and the THz-wave in the clad layer. Equation (2) is mathematically equivalent to a model in which the THz-wave is directly radiated to a clad layer. The equation tells us that n_{clad} should be larger than that of the nonlinear crystal in the pumping wave region. A comparison of the refractive indices of various nonlinear crystals with that of Si (about 3.4 in the THz-region) indicates that Si is an appropriate Cherenkov radiation output coupler for many crystals.

The radiation angle hardly changes during THz-frequency tuning because the silicon has low refractive index dispersion in the THz-wave region and the optical wavelength requires only slight tuning. The change in radiation angle is less than 0.01° for a fixed pumping wavelength. The actual angle change of the THz wave is significantly better than for the THz parametric oscillator (TPO) with a Si prism coupler (Kawase et al., 2001), which has an angle change of about 1.5° in the 0.7–3 THz tuning range.

3. Cherenkov phase-matched monochromatic THz-wave generation using difference frequency generation with a bulk lithium niobate crystal

3.1 Experimental setup

We demonstrated the method described above using the experimental setup shown in Fig. 3 (Suizu et al. 2008). The frequency-doubled Nd:YAG laser, which has pulse duration of 15 ns, a pulse energy of 12 mJ when operating at 532 nm, and a repetition rate of 50 Hz, was used as the pump source for a dual-wavelength potassium titanium oxide phosphate (KTP) optical parametric oscillator (OPO). The KTP-OPO, which consists of two KTP crystals with independently controlled angles, is capable of dual-wavelength operation with independent tuning of each wavelength (Ito et al., 2007). The OPO has a tunable range of 1300 to 1600 nm. The maximum output energy of 2 mJ was obtained for a pumping energy of less than 12 mJ. The 5 mol% MgO-doped lithium niobate crystal (MgO:$LiNbO_3$) used in the experiment was cut from a 5 × 65 × 6 mm wafer, and the x-surfaces at both ends were mirror-polished. An array of seven Si prism couplers was placed on the y-surface of the MgO:$LiNbO_3$ crystal. The y-surface was also mirror-polished to minimize the coupling gap between the prism base and the crystal surface, and to prevent scattering of the pump beam, which excites a free carrier at the Si prism base. To increase the power density, the pump beam diameter was reduced to 0.3 mm. The polarizations of the pump and THz waves were both parallel to the Z-axis of the crystals. The THz-wave output was measured with a fixed 4 K Si bolometer.

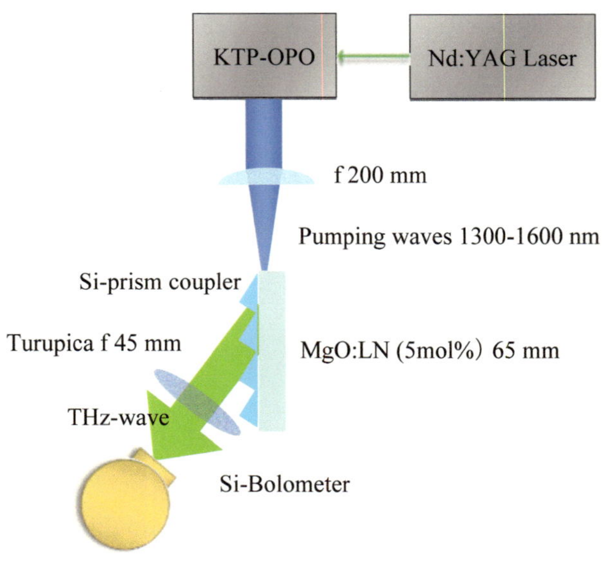

Fig. 3. Experimental setup for Cherenkov phase-matching monochromatic THz-wave generation with a bulk lithium niobate crystal.

3.2 Results and discussions

The THz-wave output map for various pumping wavelengths and corresponding THz-wave frequencies is shown in Fig. 4. The magnitude of the map denotes the output voltage of a Si bolometer with a gain of 200. The noise level of the bolometer was about 10 mV and is shown as the blue region in the figure. The regions where over 2 V of output voltage were obtained is red. As seen in the figure, wide tunability in the range 0.2–3.0 THz was obtained by choosing the proper pumping wavelength. Especially for lower frequency below 1.0 THz, this was very efficient compared to our previous TPO systems that used 1470 nm pumping.

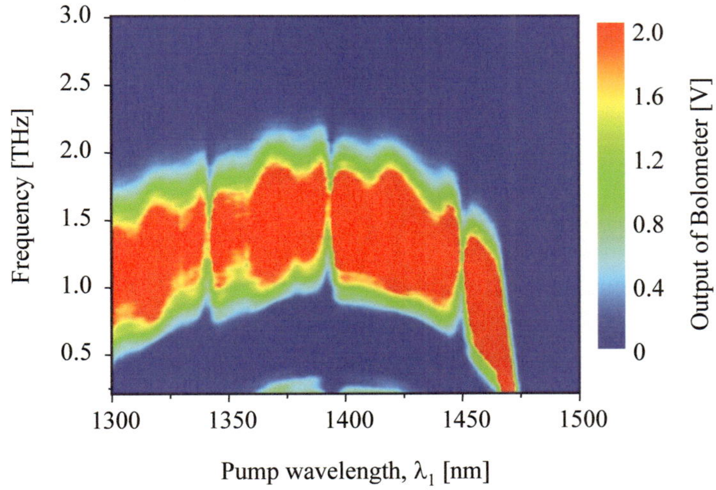

Fig. 4. THz-wave output mapping for various pumping wavelengths and corresponding THz-wave frequencies. The X-axis and Y-axis denote pumping wavelength λ_1 and THz-wave frequency, respectively. The magnitude of the map values indicates the output voltage of the detector.

Figure 5 (a) shows cross sections of the THz-wave output map of Fig. 4. The highest THz-wave energy obtained was about 800 pJ, using the fact that 1 V ≈ 101 pJ/pulse for low repetition rate detection, pulsed heating of the Si device, and an amplifier gain of 200 at the bolometer, and the energy conversion efficiency from the λ_1 wave (1 mJ/pulse) was about 10^{-4}%. This value is comparable to that obtained with our previous TPO systems, despite the low excitation energy of only 1 mJ. The figures clearly show the strong dependence of THz-wave output energy on the pumping wavelength. In the case of 0.8 THz generation, the output energy had a dip at a pumping wavelength of approximately 1400 nm as shown in Fig. 5(a). We obtained extremely high energy in the low-frequency region below 0.3 THz (millimeter wave region) using 1470 nm pumping. The reason for this is not clear, and the dispersion of pumping waves cannot explain the results; thus, an explanation is left for future research. The important result is that we could obtain a flat output spectrum in the range 0.2–2 THz by choosing proper pumping wavelength, as shown in Fig. 5(b).

Cherenkov phase matching inherently requires a waveguide structure for nonlinear polarization waves in the crystal to suppress phase mismatching in the direction perpendicular to the guiding mode (i.e., normal to the crystal surface). If we reduce the width of the pumping beams in the direction of THz-wave propagation to about one-half of the THz wavelength, (i.e., about 10 μm for 3 THz) by taking into account the refractive index of MgO:$LiNbO_3$ in the THz-wave region, no need exists to consider phase matching in that direction (Suizu et al., 2006). In our case, the waist of the pump beams in the MgO:$LiNbO_3$ was about 300 μm, which corresponds to about five cycles of THz waves at 1.0 THz, and one cycle of THz waves at 0.2 THz. Although the experimental conditions did not satisfy the requirement for Cherenkov phase matching, we did successfully detect Cherenkov-radiated

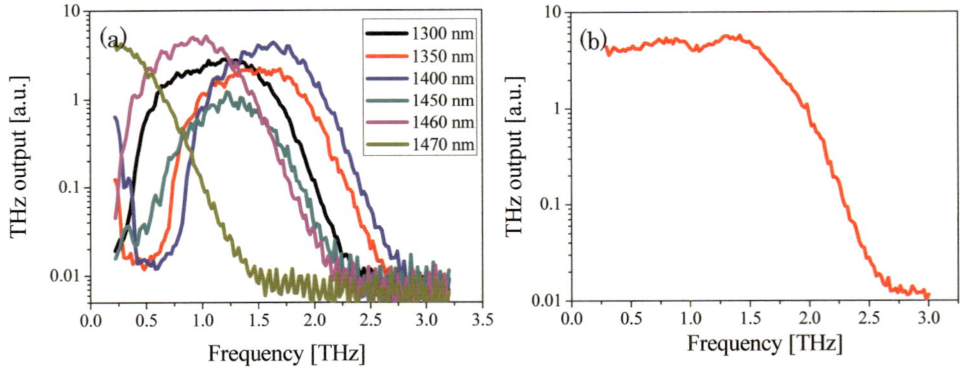

Fig. 5. THz-wave spectra (a) at various pumping wavelength and (b) under choosing proper pumping wavelengths.

THz waves, which originated in the higher absorbance area of the crystal at the THz-wave region. The THz waves generated far from the crystal surface would be attenuated and no significant phase mismatch would occur. This also remains an area for future study.

By shaping the pumping beams with a focused cylindrical lens or by adopting the waveguide structure of the crystal, we could neglect phase mismatches and obtain a higher power density of the pumping beams, resulting in higher conversion efficiency.

4. Efficient Cherenkov-type phase-matched widely tunable THz-wave generation via an optimized pump beam shape

We demonstrated the Cherenkov-type phase-matching method for monochromatic THz-wave generation via the DFG process using bulk lithium-niobate crystal. We successfully generated monochromatic, widely tunable THz waves in the 0.2- to 3.0-THz range. We obtained efficient energy conversion in the low-frequency region below 0.5 THz and achieved a flat tuning spectrum by varying the pumping wavelength during THz-wave tuning. The highest THz-wave energy was about 800 pJ pulse^{-1}, which was obtained for a broad spectral region in the range of 0.2 to 2.0 THz. However, obtaining high conversion efficiency in the frequency domain above 2 THz was difficult, and the output was almost zero at 3 THz. The output of the THz wave decreased in the high-frequency region due to a phase mismatch incurred by the finite size of the pumping beam diameter. As shown in Fig. 6(b), Cherenkov-type phase matching arises due to a superposition of spherical THz waves from the nonlinear polarization maxima created by pumping lights of two different frequencies in the NLO crystal, and thus, when the finite beam size is taken into account, the phase shift of the wave depends on the distance from the y-surface of the crystal. THz waves generated far from the crystal surface destructively interfere with those generated in the neighbourhood of a crystal surface. The beam diameter of the pumping wave in a lithium-niobate crystal in our previous work was about 300 μm, corresponding to about the wavelength of the THz wave at 0.2 THz, and ten cycles of THz waves at 2.0 THz, as the refractive index of lithium niobate is about 5.2. Since the 300-μm beam diameter is over 15 times the wavelength of a THz wave above the 3-THz region, a phase mismatch occurred and the THz-wave output decreased. In this experiment, we attempted to improve the THz-wave generation efficiency above 3 THz by optimizing the beam shape of the pumping wave to decrease the beam-diameter dependence effect (Shibuya et al., 2009).

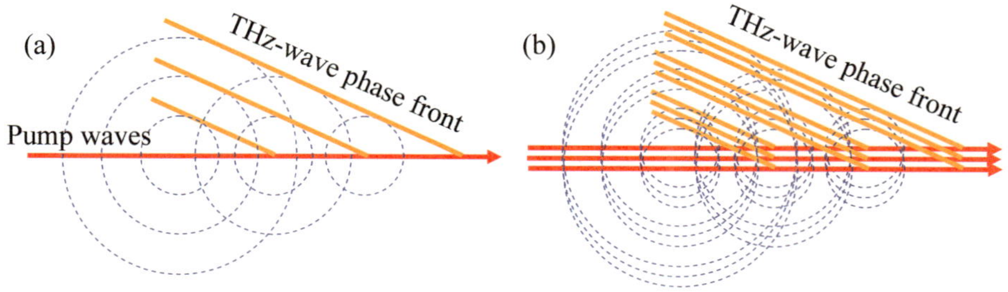

Fig. 6. (a) Ideal Cherenkov-type phase-matching condition; (b) Cherenkov-type phase-matching condition when the beam diameter of the exciting light is considered. In (b), the phase mismatch is caused by the finite size of the beam diameter.

4.1 Experimental setup

A dual-wavelength potassium titanium oxide phosphate (KTP) optical parametric oscillator (OPO) with a pulse duration of 15 ns, a pulse energy of 1.6 mJ, a 50-Hz repetition rate, and a tunable range of 1300 to 1600 nm was used for a DFG pumping source. The size of the MgO-doped lithium-niobate crystal was 5×65×6 mm^3. We used cylindrical lenses to reduce the pump beam diameter. The focal lengths of the cylindrical lenses were 20, 50, 100, and 150 mm, and the beam widths parallel to the crystal's y-axis were 35, 46, 83, and 127 μm (FWHM), respectively. The pump power was adjusted, and the power density on the focus position was made constant at 200 MW cm^{-2} for all lenses.

The obtained THz-wave output spectrum is shown in Fig. 7. The vertical axis is the THz-wave pulse energy calculated from the output voltage of a Si-bolometer detector. The horizontal axis is the THz-wave frequency. THz-wave output spectra were measured by selecting the excitation wavelength in which the maximum output was obtained for each THz-wave frequency. The output in the high-frequency region increased as the focal length of the cylindrical lens decreased. THz-wave generation was confirmed over the 3-THz region with the 20-mm and 50-mm cylindrical lenses. The tunable range for the 20-mm cylindrical lens was about 0.2 to 4 THz. This is the widest tuning range for the previous lithium-niobate crystal-generated THz-wave source. The pumping-wave beam diameter in the lithium-niobate crystal using the 20-mm cylindrical lens was about 35 μm, which corresponded to about 1.8-THz wave cycles at 3 THz. The phase mismatch is thought to have decreased as the beam diameter decreased, leading to an output improvement in the high-frequency region. Meanwhile, the conversion efficiency decreased because the pumping-wave beam diameter corresponded to over 2.3-THz wave cycles and the absorption coefficient increased rapidly above 4 THz. The absorption coefficient of the crystal at 4 THz was 425 cm^{-1}. When the pump beam moved 100 μm away from the y-surface of the crystal, 98.6% of the output was lost. Additionally, narrowing the beam diameter further was difficult due to diffraction. As the beam diameter narrowed, the confocal length shortened and the conversion efficiency decreased. The low-frequency region generation efficiency was expected to decrease for the 20-mm cylindrical lens case because the confocal length shortened. This problem can be prevented by using a waveguiding structure. By limiting the beam diameter of the pump wave to half of the wavelength using only the waveguide mode for THz-wave generation, the phase mismatch can be neglected and absorption loss reduced. This is because the distance from the y-surface to the pump beam drops to almost zero, causing a higher conversion efficiency and a wider spectrum.

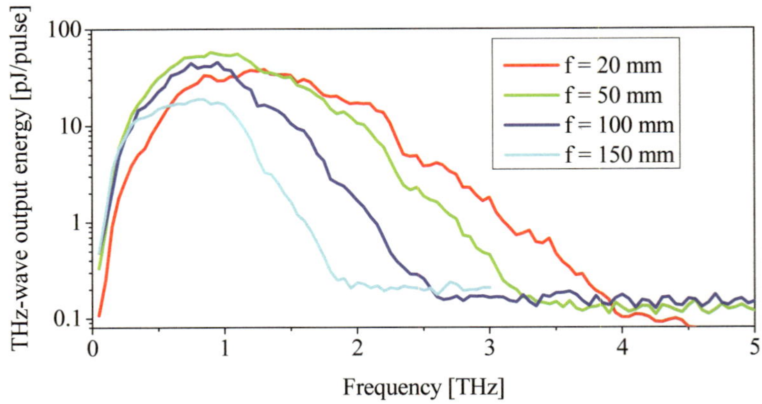

Fig. 7. THz-wave output spectra obtained using various cylindrical lenses, as measured by selecting the excitation wavelength in which the maximum output was obtained for each THz-wave frequency.

5. Extremely frequency-widened Cherenkov-Type Phase-Matched terahertz wave generation with a lithium niobate waveguide

Here, we show that Cherenkov radiation with waveguide structure is an effective strategy for achieving efficient and extremely wide tunable THz-wave source (Suizu et al., 2009). We fabricated MgO-doped lithium niobate slab waveguide with 3.8 μm of thickness and demonstrated difference frequency generation of THz-wave generation with Cherenkov phase matching. Extremely frequency-widened THz-wave generation, from 0.1 to 7.2 THz, without no structural dips successfully obtained. The tuning frequency range of waveguided Cherenkov radiation source was extremely widened compare to that of injection seeded-Terahertz Parametric Generator. The tuning range obtained in this work for THz-wave generation using lithium niobate crystal was the widest value in our knowledge. The highest THz-wave energy obtained was about 3.2 pJ, and the energy conversion efficiency was about 10^{-5} %. The method can be easily applied for many conventional nonlinear crystals, results in realizing simple, reasonable, compact, high efficient and ultra broad band THz-wave sources.

5.1 Experimental setup

Here, we prepared a slab waveguide of a lithium niobate crystal. A Y-cut 5 mol % MgO-doped lithium niobate crystal on a thick congruent lithium niobate substrate was polished down to 3.8 μm. A thin MgO-doped lithium niobate layer worked as an optical slab waveguide, because the refractive indexes of 5 mol % MgO-doped lithium niobate and congruent lithium niobate at 1300 nm are 2.22 and 2.15, respectively. The waveguide device was 5-mm wide and 70-mm long (X-axis direction). Each X-surface facet was mechanically polished to obtain an optical surface. We demonstrated difference-frequency generation using the experimental setup shown in Fig. 8(b). A dual-wavelength potassium titanium oxide phosphate (KTP) optical parametric oscillator (OPO) with a pulse duration of 15 ns, a pulse energy of 1 mJ and a 1300- to 1600-nm tunable range was used as a pumping source. A thin (3.4-μm thick) polyethylene terephthalate (PET) film was slipped between the array of Si prism couplers and the Y-surface of the MgO-doped lithium niobate crystal. Directly placing an array of Si prism couplers on the Y-surface of the MgO-doped lithium niobate will inhibit the function of the MgO-doped lithium niobate layer as a waveguide for pumping waves, because the refractive index of Si in the near-infrared region is higher (about 3.5) than that of lithium niobate (about 2.2). A PET, in contrast, has a lower refractive index in that region (about 1.3), so adding a thin PET film does not inhibit the function of the crystal as a waveguide. An array of Si prism couplers on a PET film can work as a coupler for THz-frequency waves, because the PET film is thin compared to the wavelength of a THz-frequency wave. A schematic of the coupling system of the pumping wave and THz-wave emitting system is shown in Fig. 8(a). To couple pumping waves, the pump beam was reduced to few micrometers in the X-axis direction by a 3-mm diameter glass rod lens. The width of the pumping beams in the Z-direction was about 1.9 mm. The waveguide power density was about 53 MW cm^{-2}, estimated from the pump wave pulse energy after waveguide propagation (about 60 μJ). We did not observe or calculate the waveguide mode of the structure in which a thin MgO-doped lithium niobate layer was sandwiched by a thick congruent lithium niobate layer and a thin PET film. It remains an area of future work to optimize the waveguide structure. The pump wave and THz-frequency wave polarizations were parallel to the crystal's Z-axis. The THz-wave output was measured with a fixed 4-K Si bolometer.

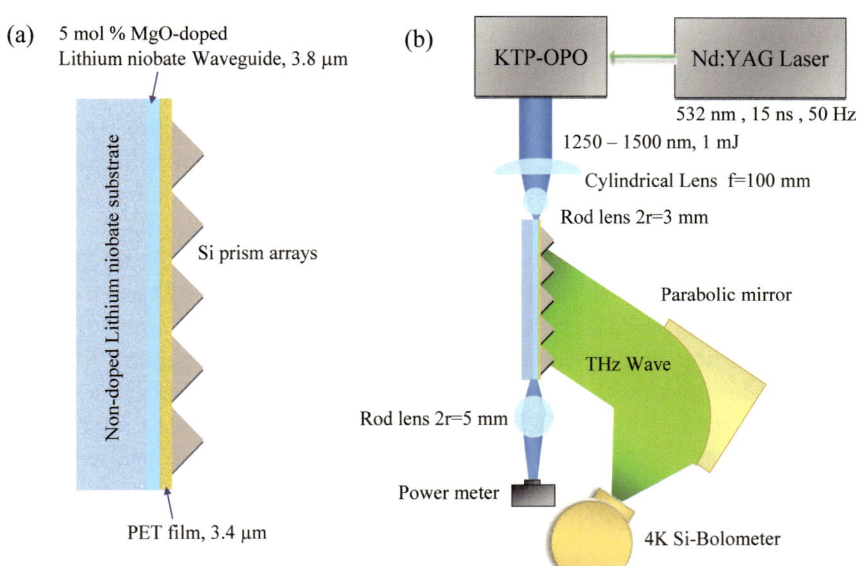

Fig. 8. (a) Schematic of the lithium niobate waveguide device with Si prism array coupler. (b) THz-wave detection experimental setup.

Figure 9 shows a THz-wave spectrum at various wavelength of λ_1 from 1250 to 1350 nm. The spectrum was obtained by varying λ_2 at fixed λ_1. As shown in Fig. 9, high-frequency THz-wave output ranging to about 7.2 THz was confirmed. We were unable to observe THz-wave generation around 7.2 THz due to very strong THz-wave absorption at 7.5 THz by the LO-phonon mode. The THz-wave spectrum does not depend on pumping wavelength because the near-infrared refractive index is almost constant in the 1250- to 1350-nm range.

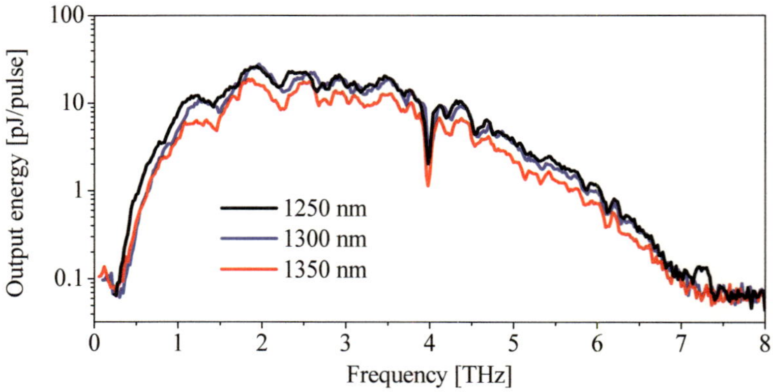

Fig. 9. THz-frequency spectrum of waveguided Cherenkov radiation. Black, red, blue and green curves represent pumping wavelengths of 1250, 1300, 1350 nm, respectively.

Figure 10 shows a comparison of normalized tuning spectrum of the waveguided Cherenkov radiation source and injection seeded terahertz parametric generator (is-TPG) (Kawase et al., 2002). Nevertheless each THz source were based on a same nonlinear crystal, MgO-doped lihitum niobate, a tuning frequency range of waveguided Cherenkov radiation source was extremely widened compare to that of is-TPG. We converted the output voltage of the Si bolometer to the actual THz-wave energy, using the fact that 1 V ≈ 20 pJ pulse^{-1} for low repetition rate detection, pulsed heating of the Si device, and an amplifier gain of 1000 at the bolometer. The highest THz-wave energy obtained was about 28 pJ, and the energy conversion efficiency from the λ_1 wave (30 μJ pulse^{-1}) was about 10^{-4}%. This value is comparable to our previous work on Cherenkov radiation using bulk crystal, despite the low excitation energy of only 30 μJ. The tuning range obtained in this work for THz-wave generation using lithium niobate crystal was the widest value in our knowledge.

The THz-wave emitting angle was absolutely constant, as Si dispersion in this range is almost flat. The device would be work well in an optical rectification process using a femtosecond laser. Such a range, free from structural dips between 0.1 and 7.2 THz, is suitable for ultra-short pulse generation. Also, the surface emission process used here is loss-less, permitting the generation of a continuous, widely-tunable THz-frequency range, and requiring only two easily commercially available diode lasers. Compact, robust and reasonable THz-wave sources can be realized by this method. Although we demonstrated this method using only a lithium niobate crystal, it can be adopted for other nonlinear crystals, such as $LiTaO_3$, GaSe, GaP, ZnSe, ZnTe, ZGP, DAST and so on. By choosing the best clad materials for the nonlinear crystals (in many case Si or Ge), the Cherenkov condition is easily satisfied, and control of crystal angles to satisfy phase-matching conditions, such as birefringence phase-matching, is not required. This method opens the door to simple, reasonable, compact, highly efficient and ultra-broadband THz-wave sources.

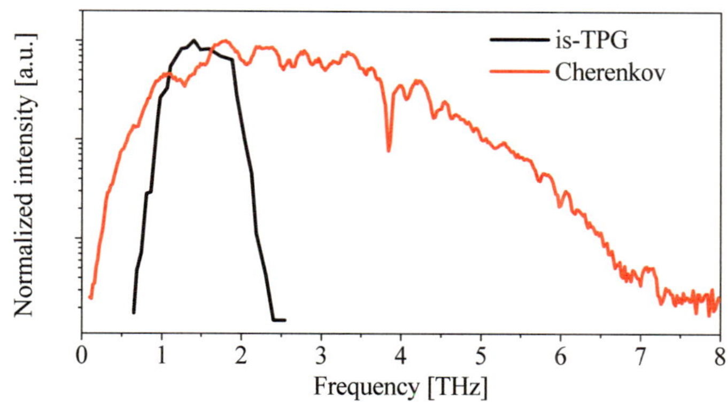

Fig. 10. A comparison of normalized tuning spectrum of the waveguided Cherenkov radiation source under 1250 nm pumping (red curve) and is-TPG (black curve).

6. Cherenkov phase matched THz-wave generation with surfing configuration for bulk Lithium Niobate crystal

We demonstrated a Cherenkov phase matched THz-wave generation with surfing configuration for bulk lithium niobate crystal (Suizu et al., 2009). THz-wave output was enhanced about 50 times by suppressing phase mismatching for THz-wave propagation direction. The suppression was achieved by combining two pumping waves with dual wavelength with finite angle, and THz-frequency was controllable by changing the angle within 2.5 degrees range. Higher frequency THz-wave generation at around 4.0 THz was successfully obtained by the method.

6.1 Cherenkov phase mating with surfing configuration

We demonstrated Cherenkov phase matching method for monochromatic THz-wave generation via DFG process using bulk lithium niobate crystal. We successfully generated monochromatic THz-waves with wide tunability in the range 0.2–2.5 THz. The highest THz-wave energy was about 800 pJ/pulse, and this energy could be obtained for the broad spectral region in the range around 0.2–2.0 THz. Although we successfully got wide tunable characteristics of THz-wave generation, conversion efficiency of a THz-wave generation at higher frequency region above 2.0 THz was slightly low. It would be caused by phase mismatch of generated THz-wave in a propagating direction of THz-wave. Beam diameter of pumping waves in a lithium niobate crystal in our previous work was about 300 μm, which corresponded to about ten cycles of THz waves at 2.0 THz because the refractive index of lithium niobate is about 5.2. THz-wave generated at far from a crystal surface interfered with that generated at neighborhood of a crystal surface, resulted in denying each other. By reducing the width of beam diameter in the crystal in the direction of THz-wave propagation to about one-half of the THz wavelength, there was no need to consider phase matching in that direction. We observed the effects by condensing a pump beam diameter to a THz-wave propagation direction by cylindrical lenses. Although higher THz-wave around

4.0 THz was successfully generated under tight focusing by the cylindrical lens with 20 mm of focus length, output of THz-wave at lower frequency region was reduced, because tight focusing resulted in reducing interaction length for pumping wave propagating direction.

In this study, we propose surfing configuration of Cherenkov type phase matching for THz-wave generation for bulk crystal to suppress a phase mismatching. Interference pattern of pumping waves in the crystal is induced by combining the dual wavelengths beams with finite angle. It provides a same spatial pattern of second order nonlinear polarization in THz-frequency. The interference pattern has not checkerboard one, which is a results of interference of tilted beams with same frequency, because dual wavelength beam courses other spatial interference pattern, corresponding to difference frequency, and the interference pattern is superimposed in checkerboard one.

Figure 11 shows electric field distribution of (a) pumping waves and (b) excited nonlinear polarization, with λ_1=1300 nm, λ_2=1317 nm (here, three waves in DFG interaction has a relation of $\omega_1=\omega_2-\omega_{THz}$, and corresponding THz frequency is 3 THz) and 3.7 degrees of angle between divided pumping beams, α. The periods of nonlinear polarization pattern of dual wavelengths beams, A for x-axis and B for y-axis are represented by following equations,

$$A = \frac{2\pi}{(k_1 - k_2)\cos\frac{\alpha}{2}},\ B = \frac{4\pi}{(k_1 + k_2)\sin\frac{\alpha}{2}} \tag{3}$$

where $k_1=2\pi n_1/\lambda_1$ and $k_2=2\pi n_2/\lambda_2$, here n_1 and n_2 are refractive index of λ_1 and λ_2, respectively. We used Sellmeier equation at near-infrared region for a lithium niobate crystal (Jundt, 1997). On the other hands, Cherenkov angle of the crystal, θ_c, is decided by relation of length A and THz-wavelength in the crystal, $C=\lambda_{THz}/n_{THz}$, here λ_{THz} and n_{THz} are THz-wavelength in vacuum and refractive index of the crystal at THz frequency. A phase matching condition for THz-wave propagation direction is satisfied by choosing an appropriate angle α of the pump beams for required THz-frequency. The angle α is formulated from geometric relation of A, B and C, $A^2C^2=B^2C^2=A^2B^2$, as shown in Fig.11(c).

$$\alpha = 2\arccos\left(\sqrt{\frac{-(k_1+k_2)^2 + \frac{16\pi^2}{(\lambda_{THz}/n_{THz})^2}}{4(k_1-k_2)^2 - (k_1+k_2)^2}}\right) \tag{4}$$

Generated THz-wave can propagate without influence of phase mismatching in the direction of propagating direction, just like as surf rider on nonlinear polarization waves, as shown in Fig.11 (b). The required angle for frequency tuning was shown in Fig.12 (a) internal and (b) external crystal. Phase matching condition is satisfied by changing the angle α for required THz-wave and pumping wave wavelength. And slightly narrow tunability (about 300 GHz at around 3 THz generation) is obtained at a fixed angle, α=4.0 degrees.

6.2 Experimental setup

Figure 13 shows the schematic of experimental setup. A pump source for DFG process was same as our previous works, and which has a tunable range of 1250 to 1500 nm, 15 ns of pulse duration and 0.88 mJ of pulse energy. An output of the source with dual wavelength

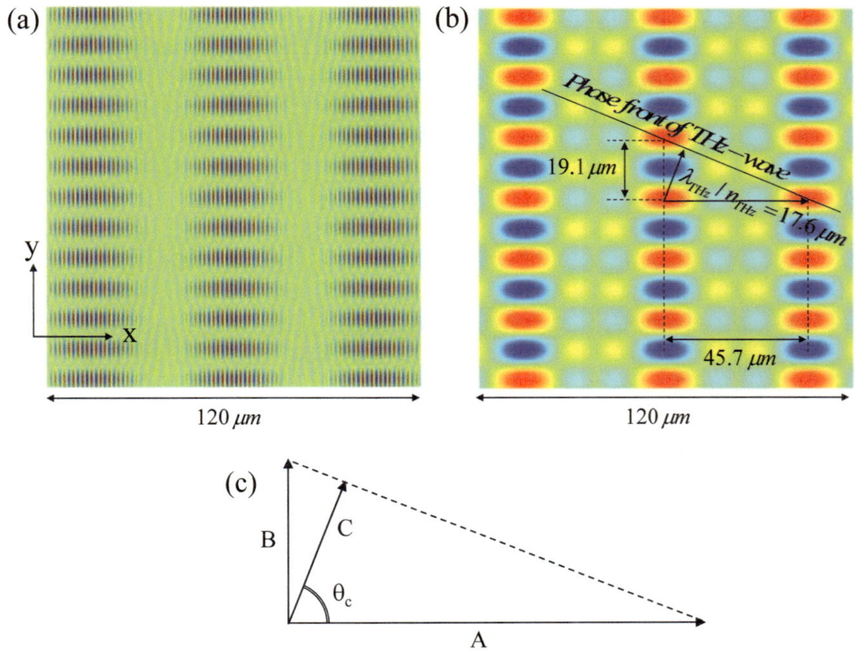

Fig. 11. Normalized electric field distribution of (a) combined dual wavelength pump beams with finite angle, and (b) exited second order nonlinear polarization of difference frequency. Here, λ_1=1300 nm, λ_2=1317 nm and 3 THz of difference frequency with 3.7 degrees of beam angle. (c) Geometric relation of A: excited nonlinear polarization for x-direction, B: interference period of pump beams for y-direction and C: THz-wavelength in the crystal.

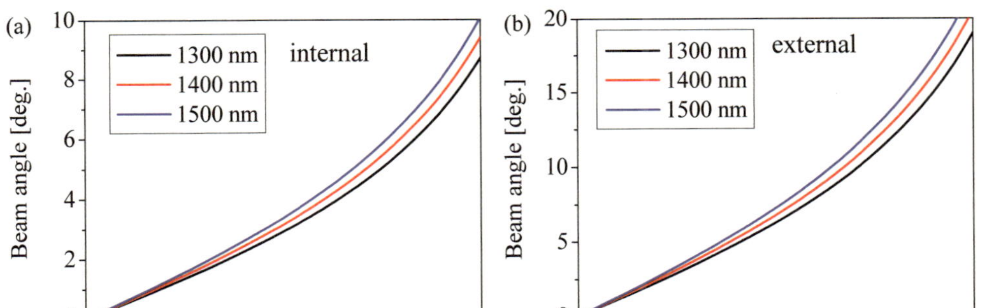

Fig. 12. Tuning angle (a) internal and (b) external of crystal under 1300, 1400 and 1500 nm of pumping wavelength of λ_1.

was focused by circular lens (f=500 mm) before divided by half beam splitter, and combined again with finite angle. The spot diameter of the combined beam was 0.45 mm. The 5 mol % MgO-doped lithium niobate crystal (MgO:$LiNbO_3$) used in the experiment was cut from a 5

× 65 × 6 mm wafer. The polarizations of the pump and THz waves were both parallel to the Z-axis of the crystals. The THz-wave output was measured with a fixed 4 K Si bolometer.

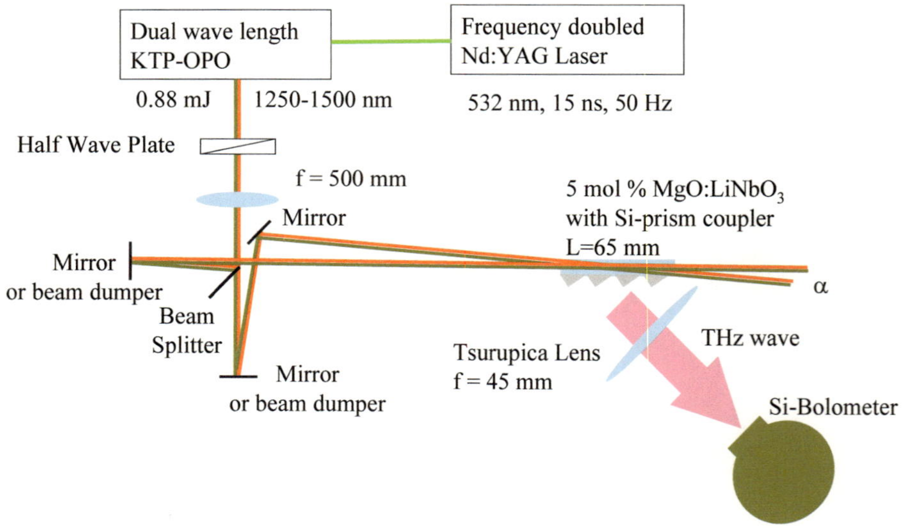

Fig. 13. Schematic of experimental setup for Cherenkov phase matching THz-wave generation with surfing configuration.

6.3 Results and discussions

Input-output properties of THz-wave for pumping energy are shown in Fig.14 at 1.0 THz generation with α=2.49 degrees. Circles and triangles denotes THz-wave output signal with combined beams and with single beam by dumping the other beam before entrance to the crystal, respectively. Maximum pumping energy of only 0.44 mJ was achieved at single beam pumping, because a half of whole pumping energy was dumped as shown in Fig.13. The vertical axis is the THz-wave pulse energy calculated from the output voltage of a Si-bolometer detector, a pulse energy of about 101 pJ/pulse corresponded to a Si-bolometer voltage output of 1 V when the repetition rate was less than 200 Hz. As shown in the figure, remarkable enhancement of THz-wave generation with surfing configuration, whose magnetic was about 50 times, was successfully observed. Inset of Fig.14 shows double logarithmic plot of input-output properties. Slope efficiency under combined beams and single beam pumping were almost same values. It means that enhancement factor of about 50 was a result of a suppression of phase miss-matching.

The generated THz-waves at different position in the crystal were in-phase each other, and outputted THz-wave was enhanced. Intensity of overlapping in-phase THz-waves in an absorptive media was calculated as shown in Fig.15. A 5 mol % MgO-doped Lithium Niobate crystal at THz-wave frequency region would has about 30 cm^{-1} of absorption coefficient (Palfalvi et al., 2005). The enhancement effect of in-phase interference would be effective for about 2 mm of traveling distance of THz-wave, this fact leads optimum pumping beam width in y-axis direction is about 1.8 mm. In this study, pumping beam width in y-axis was about 0.45 mm, results in a propagating length of a THz-wave was about 1.2 mm. Higher enhancement above 50 would be obtained with tight focused beam only for z-axis by cylindrical lens.

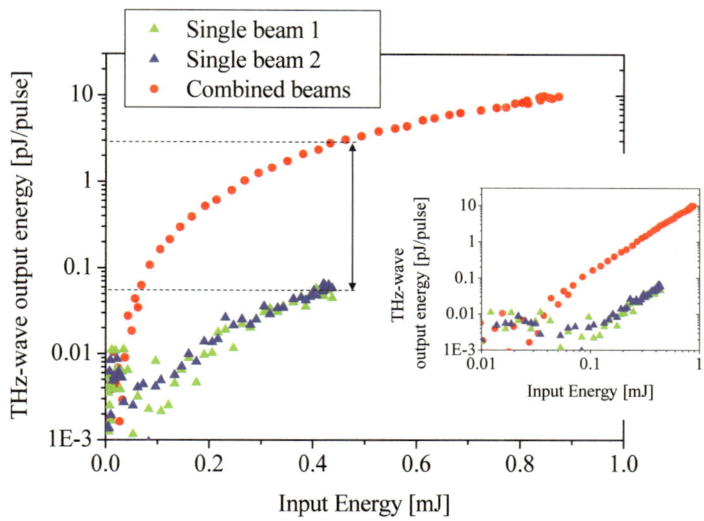

Fig. 14. Input-output property of THz-wave for pumping energy at 1.0 THz generation with α=2.49 degrees. Circles and triangles denotes THz-wave output signal with combined beams and with single beam. Inset shows double logarithmic plot of input-output properties.

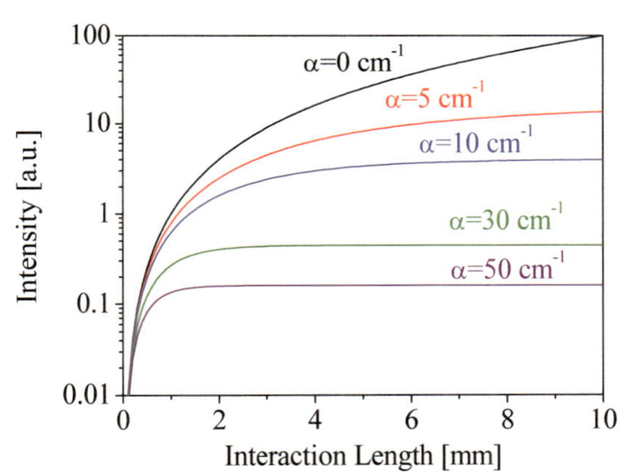

Fig. 15. Calculated intensity of overlapping in-phase THz-waves in an absorptive media.

Figure 16 shows THz-wave output characteristics under fixed pumping wavelength of 1300 nm and several fixed angle, 2.49, 3.80 and 5.03 degrees. Maximum THz-wave output at each angle was obtained at higher frequency in the bigger angle, α. Obtained peaks of THz-wave output were about 1.1, 1.6 and 1.9 THz, respectively. The relation between the angle and the frequency where maximum output was obtained agree well with Equation 4, 1.08, 1.61 and 2.07 THz under 1300 nm pumping respectively. Tuning range for higher frequency region was remarkably improved compare with our previous collinear and not tight focused configuration. THz-wave output at around 4 THz was successfully obtained.

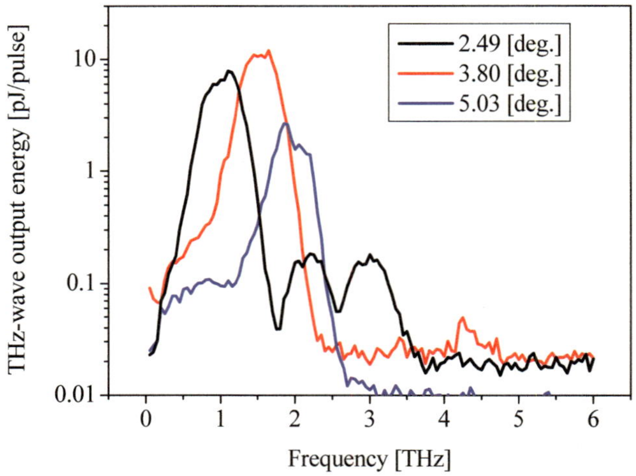

Fig. 16. THz-wave output spectra under fixed pumping wavelength of 1300 nm and several fixed angle, 2.49, 3.80 and 5.03 degrees.

As described in our previous work, because the linewidth of each pumping wave is about 60 GHz, the source linewidth is about 100 GHz, which is slightly broader than that obtained from sources such as injection-seeded terahertz parametric generator (Kawase et al., 2002) or DAST crystal-based difference-frequency generators (Powers et al., 2005). This occurs because the linewidth of the THz-wave depends on that of the pumping source.

The spectrum with α=2.49 degrees pumping had two dips at 1.8 and 2.6 THz. It coursed by perfect phase miss-matching of THz-wave propagation. Figure 17 shows calculated nonlinear polarization distributions at (a) 1.8 and (b) 2.6 THz generation with α=2.49. THz-wavelength in the crystal at 1.8 THz generation is 32.2 μm. Generated THz-wave at point "a" in Fig.17 interferes with that at point "b", which has a phase difference by π compare to that of point "a", results in destructive interference. Similarly, and adding higher order interference, generated THz-wave at point "c" has destructive interference with that at point "d". THz-wave generation was observed at around the dips, because perfect phase miss-matching was relaxed at these frequencies. We have not yet completed the analytical solution predicting the frequency due to destructive interference, and it remains an area of future work.

Broader tuning range would be obtained by controlling the angle α within about only 2.5 degrees range. Because lithium niobate is strongly absorbing at THz-frequencies, the beam-crossing position was set near the crystal surface to generate the THz-wave. In this configuration, the pumping beam passing through a Si prism yields an optical carrier excitation in Si that prevents THz-wave transmission, while the interaction length decreases at larger pumping angles, α. The interaction lengths,

$$l = 2D / \tan\alpha \tag{5}$$

where D is the beam diameter, are 21.4 and 10.7 mm for αs of 2.49° and 5.03°, respectively. If we use a shorter lithium niobate crystal, the optical carrier excitation can be avoided, and larger pumping angles can be employed to obtain higher-frequency generation. The method is very simple way to obtain higher frequency and efficient generation of THz-wave, because the method does not require a special device such as slab waveguide structure.

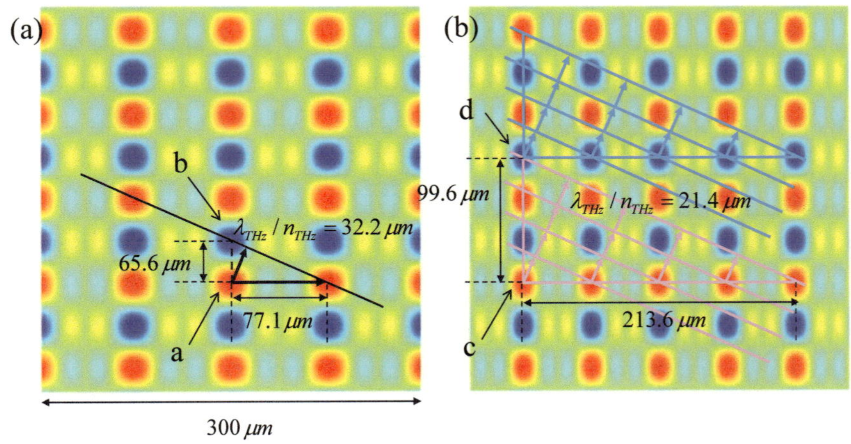

Fig. 17. Calculated nonlinear polarization distributions at (a) 1.8 and (b) 2.6 THz generation with α=2.49.

7. References

Auston, D. H.; Cheung, K. P.; Valdmanis, J. A. & Kleinman, D. A. (1984). Cherenkov radiation from femtosecond optical pulses in electro-optic media. *Phys. Rev. Lett.* 53, 1555–1558.

Avetisyan, Y.; Sasaki, Y. & Ito, H. (2001). Analysis of THz-wave surface-emitted difference-frequency generation in periodically poled lithium niobate waveguide. *Appl. Phys. B* 73, 511–514.

Bodrov, S. B.; Stepanov, A. N.; Bakunov, M. I.; Shishkin1, B. V.; Ilyakov, I. E. & Akhmedzhanov, R. A. (2009). Highly efficient optical-to-terahertz conversion in a sandwich structure with $LiNbO_3$ core. *Opt. Express* 17, 1871-1879.

Boyd, G. D.; Bridges, T. J.; Patel, C. K. N. & Buehler, E. (1972). Phase-matched submillimeter wave generation by difference-frequency mixing in $ZnGeP_2$. *Appl. Phys. Lett.* 21, 553–555.

Hebling, J.; Almasi, G.; Kozma, I. & Kuhl, J. (2002). Velocity matching by pulse front tilting for large area THz-pulse generation. *Opt. Express* 10, 1161–1166.

Ito, H.; Suizu, K.; Yamashita, T. & Sato, T. (2007). Random frequency accessible broad tunable terahertz-wave source using phase-matched 4-dimethylamino-N-methyl-4-stilbazolium tosylate (DAST) crystal. *Jpn. J. Appl. Phys.* 46, 7321–7324.

Jundt, D. H. (1997). Temperature-dependent sellmeier equation for the index of refraction, ne, in congruent lithium niobate. *Opt. Lett.* 22, 1553-1555.

Kawase, K.; Shikata, J.; Minamide, H.; Imai, K. & Ito, H. (2001). Arrayed silicon prism coupler for a terahertz-wave parametric oscillator. *Appl. Opt.* 40, 1423-1426.

Kawase, K.; Minamide, H.; Imai, K.; Shikata, J. & Ito, H. (2002). Injection-seeded terahertz-wave parametric generator with wide tenability. *Appl. Phys. Lett.* 80, 195-197.

Kleinman, D. A. & Auston, D. H. (1984). Theory of electro-optic shock radiation in nonlinear optical media. *IEEE J. Quantum Electron.* 20, 964–970.

Palfalvi, L.; Hebling, J.; Kuhl, J.; Peter, A. & Polgar, K. (2005). Temperature dependence of the absorption and refraction of Mg-doped congruent and stichiometric $LiNbO_3$ in the THz range. *J. Appl. Phys.* 97, 123505.

Powers, P. E.; Alkuwari, R. A.; Haus, J. W.; Suizu, K. & Ito, H. (2005). Terahertz generation with tandem seeded optical parametric generators. *Opt. Lett.* 30, pp. 640-642.

Rice, A.; Jin, Y.; Ma, X. F.; Zhang, X. C.; Bliss, D.; Larkin, J. & Alexander, M. (1994). Terahertz optical rectification from <110> zinc-blende crystals. *Appl. Phys. Lett.* 64, 1324–1326.

Sasaki, Y.; Avetisyan, Y.; Kawase, K. & Ito, H. (2002). Terahertz-wave surface-emitted difference frequency generation in slant-stripe-type periodically poled $LiNbO_3$ crystal. *Appl. Phys. Lett.* 81, 3323–3325.

Sasaki, Y.; Avetisyan, Y.; Yokoyama, H. & Ito, H. (2005). Surface-emitted terahertz-wave difference frequency generation in two-dimensional periodically poled lithium niobate. *Opt. Lett.* 30, 2927–2929.

Sasaki, Y.; Yokoyama, H. & Ito, H. (2005). Surface-emitted continuous-wave terahertz radiation using periodically poled lithium niobate. *Electron. Lett.* 41, 712–713.

Shi, W.; Ding, Y. J.; Fernelius, N. & Vodopyanov, K. (2002). Efficient, tunable, and coherent 0.18–5.27-THz source based on GaSe crystal. *Opt. Lett.* 27, 1454–1456.

Shibuya, T.; Tsutsui, T.; Suizu, K.; Akiba, T. & Kawase, K. (2009). Efficient Cherenkov-Type Phase-Matched Widely Tunable THz-Wave Generation via an Optimized Pump Beam Shape. *Appl. Phys. Express* 2, 032302.

Suizu, K.; Suzuki, Y.; Sasaki, Y.; Ito, H. & Avetisyan, Y. (2006). Surface-emitted terahertz-wave generation by ridged periodically poled lithium niobate and enhancement by mixing of two terahertz waves. *Opt. Lett.* 31, 957-959.

Suizu, K.; Tutui, T.; Shibuya, T.; Akiba, T. & Kawase, K. (2008). Cherenkov phase-matched monochromatic THz-wave generation using difference frequency generation with lithium niobate crystal. *Opt. Express* 16, 7493-7498.

Suizu, K.; Koketsu, K.; Shibuya, T.; Tsutsui, T.; Akiba, T. & Kawase, K. (2009). Extremely frequency-widened terahertz wave generation using Cherenkov-type radiation. *Optics Express*, 17, 6676-6681.

Suizu, K.; Tsutsui, T.; Shibuya, T.; Akiba, T. & Kawase, K. (2009). Cherenkov phase-matched THz-wave generation with surfing configuration for bulk lithium niobate crystal. *Optics Express*, 17, 7102-7109.

Sutherland, R. L. (2003). *Handbook of Nonlinear Optics*, Chap. 2. Marcel Dekker, New York.

Tanabe, T.; Suto, K.; Nishizawa, J.; Saito, K. & Kimura, T. (2003). Tunable terahertz wave generation in the 3- to 7-THz region from GaP. *Appl. Phys.* Lett. 83, 237–239.

Wahlstrand J. K. & Merlin, R. (2003). Cherenkov radiation emitted by ultrafast laser pulses and the generation of coherent polaritons. *Phys. Rev.* B 68, 054301.

Yeh, K.-L.; Hoffmann, M. C.; Hebling, J. & Nelson, K. A. (2007). Generation of 10 μJ ultrashort THz pulses by optical rectification. *Appl. Phys. Lett.* 90, 171121.

8

Nonreciprocal Phenomena on Reflection of Terahertz Radiation off Antiferromagnets

T. Dumelow[1], J. A. P. da Costa[1], F. Lima[2] and E. L. Albuquerque[2]
[1]*Universidade do Estado do Rio Grande do Norte*
[2]*Universidade Federal do Rio Grande do Norte*
Brazil

1. Introduction

There are a number of ways that reciprocity principles in optics may be affected by the presence of a static magnetic field (Potton, 2004). A classic example is Faraday rotation in which a plane polarised electromagnetic beam propagating through a suitable medium is rotated in the presence of a static magnetic field along the direction of propagation. The handedness of this rotation depends on the propagation direction, a nonreciprocal effect usefully applied to the construction of optical isolators (Dötsch et al., 2005). Nonreciprocal effects of this type are closely related to the idea that magnetic fields break time reversal symmetry. Similar nonreciprocal phenomena can occur, in various guises, on reflection off a semi-infinite sample. We discuss such behaviour in the present chapter, in the context of reflection off antiferromagnetic materials. In contrast to nonreciprocal phenomena based on the Faraday effect, our interest is in the Voigt geometry, in which the static magnetic field is perpendicular to the direction of propagation. We consider the well established phenomena of nonreciprocity in the intensity and phase of oblique incidence radiation, but concentrate mainly on recent developments on nonreciprocal power flow and finite beam effects.

We restrict discussion to the simple two dimensional geometry shown in Figure 1. Radiation is reflected, in the xy plane, off a semi-infinite sample, isotropic in this plane, in the presence of a static magnetic field $\mathbf{B}_0$ along z (into the page). Note that, in this configuration, we do not have to worry about polarisation effects, since there is no mixing between s-polarised (electromagnetic $\mathbf{E}$ field component along z) and p-polarised (electromagnetic $\mathbf{H}$ field component along z) radiation.

Now compare Figure 1(a) to Figure 1(b), in which the sign of the incident angle has been reversed. In the absence of the magnetic field ($\mathbf{B}_0$ = 0), we can consider Figure 1(b) as the mirror reflection of Figure 1(a) through the yz plane, so we expect no change in the reflection behaviour. In terms of the incident and reflected beam signals, this is a trivial example of the Helmholtz reciprocity principle, which, in the present context, can be interpreted as saying that, in the absence of magnetic fields, an interchange of source and detector should not affect the signal received by the detector (Born & Wolf, 1980). When $\mathbf{B}_0$ is nonzero, however, the mirror reflection of Figure 1(a) through the yz plane no longer leads to Figure 1(b), as one might expect. The essential point here is that the static magnetic field $\mathbf{B}_0$ is an axial vector, and a mirror symmetry operation through the yz plane would therefore involve reversing the direction of this field, so that it would come out of the page (Scott & Mills, 1977). In fact

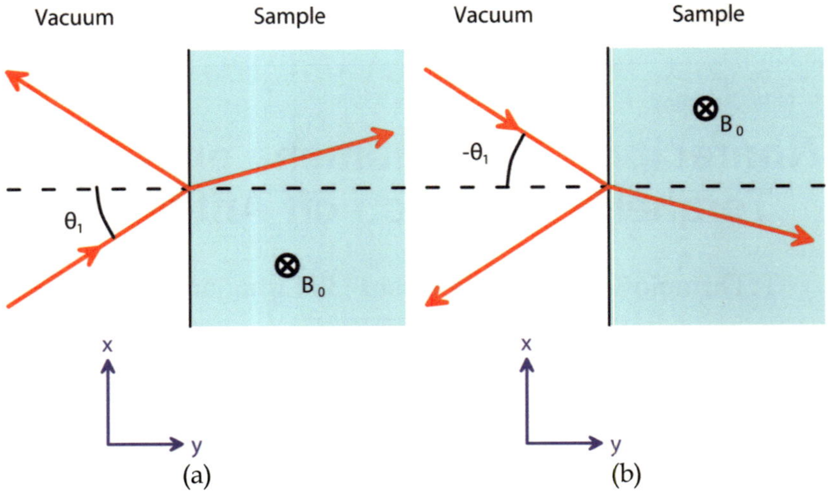

Fig. 1. Reflection geometry, showing interchange of incident and reflected beams.

there is no symmetry operation that leads us from Figure 1(a) to Figure 1(b), and the two figures are not equivalent. Nonreciprocal behaviour is thus, in principle, possible. Whether or not it occurs in practice, however, depends on the material properties of the sample.

In the present chapter we consider nonreciprocity associated with reflection off a simple uniaxial antiferromagnet. In this case the static field represented by $\mathbf{B}_0$ in Figure 1 is an external field, since an antiferromagnet has no intrinsic macroscopic magnetic field. We consider a geometry in which the anisotropy associated with the spin directions, along with the external field $\mathbf{B}_0$, is perpendicular to the plane of incidence. This is equivalent to putting the anisotropy along z in Figure 1, thus leaving the antiferromagnet isotropic in the xy plane. The electric component of the electromagnetic field is along z and the magnetic component is in the xy plane (s-polarisation).

In considering nonreciprocity in the intensity and phase of the reflected beam, it is sufficient to simply consider the effect of interchanging the incident and reflected beams (i.e. reversing the sign of θ_1). However, we note that a rotation of Figure 1(b) around the y axis brings us back to Figure 1(a), but with the field direction reversed. It is therefore possible to consider nonreciprocity in terms of a change in optical behaviour when the external field direction is reversed. This turns out to be more convenient when considering nonreciprocal effects inside the antiferromagnet and finite beam effects. It is notable that some of the new phenomena under investigation in this chapter occur at normal incidence, so such a test is simpler to visualise in such cases than a test based on the configurations of Figure 1. Thus our general test for nonreciprocity will be to see what happens when we reverse the sign of B_0.

2. Antiferromagnet permeability

The crucial parameter that determines the nonreciprocal optical properties of antiferromagnets is the magnetic permeability in region of the magnon (or spin wave) frequencies (Mills & Burstein, 1974), which typically lie in the terahertz range. We think of an antiferromagnet as two interpenetrating sublattices having opposite spin directions. Waves consisting of spins precessing in opposite directions in the two sublattices are then possible, and magnons of this type can interact with electromagnetic radiation. Their resonant frequencies are linked not only to the anisotropy field B_A that tends to align the spins along a preferred axis (the z axis in our coordinate system), but also to the interaction between the spins in the two sublattices. In the long wavelength limit (applicable to terahertz frequencies), and in the absence of any external field, the resonance frequency is given by

$$\omega_r = \gamma\left(2B_A B_E + B_A^2\right)^{1/2} \tag{1}$$

Here B_E is the exchange field representing the interaction between the spins of the opposing sublattices and γ is the gyromagnetic ratio. In the presence of an electromagnetic field whose **H** component lies in the xy plane, the induced magnetisation follows the direction of this field component, since the spins in the two sublattices precess in opposite directions with equal amplitudes. The permeability tensor $\boldsymbol{\mu}$ is thus diagonal and of the form

$$\boldsymbol{\mu} = \begin{pmatrix} \mu & 0 & 0 \\ 0 & \mu & 0 \\ 0 & 0 & 1 \end{pmatrix}. \tag{2}$$

The scalar quantity μ is given, at frequency ω, by

$$\mu = 1 + \frac{2\mu_0\gamma^2 B_A M_S}{\omega_r^2 - \omega^2 + i\omega\Gamma}. \tag{3}$$

where M_S is the sublattice magnetisation and Γ is a damping parameter.
In this study, we are interested in propagation of electromagnetic waves (strictly speaking polariton waves, since the waves include a contribution from the precessing spins in addition to that of the electromagnetic radiation) within the xy plane. We consider the electromagnetic **E** field component to be directed along z with the corresponding **H** field component in the xy plane. In this case, for plane waves of the form

$$\mathbf{E}(x,y,t) = \mathbf{E}_0 \exp\left[i\left(k_x x + k_{1y} y - \omega t\right)\right], \tag{4}$$

$$\mathbf{H}(x,y,t) = \mathbf{H}_0 \exp\left[i\left(k_x x + k_{1y} y - \omega t\right)\right], \tag{5}$$

the polaritons follow the familiar dispersion relation

$$k_x^2 + k_y^2 = \epsilon\,\mu k_0^2 \tag{6}$$

where k_x and k_y are wavevector components and ϵ is the dielectric constant of the medium. k_0 is the modulus of the free space wavevector, given by

$$k_0 = \frac{\omega}{c}. \tag{7}$$

In the presence of an external field $\mathbf{B}_0$ along the anisotropy axis, the two sublattices are no longer equivalent. This leads to two effects. Firstly, there are now two resonances instead of one and, secondly, the permeability tensor is no longer diagonal, but gyromagnetic. It thus takes the form (Mills & Burstein, 1974):

$$\boldsymbol{\mu} = \begin{pmatrix} \mu_1 & i\mu_2 & 0 \\ -i\mu_2 & \mu_1 & 0 \\ 0 & 0 & 1 \end{pmatrix}, \tag{8}$$

where

$$\mu_1 = 1 + \mu_0 \gamma^2 B_A M_S (Y^+ + Y^-), \tag{9}$$

$$\mu_2 = \mu_0 \gamma^2 B_A M_S (Y^+ - Y^-), \tag{10}$$

with

$$Y^{\pm} = \frac{1}{\omega_r^2 - (\omega \pm \gamma B_0 + i\Gamma)^2}. \tag{11}$$

The diagonal elements μ_1 do not depend on the sign of the applied field $\mathbf{B}_0$, but μ_2 changes sign when $\mathbf{B}_0$ is reversed. This is the basis of the nonreciprocal effects discussed in this chapter. The polariton dispersion relation (Equation 6) is now replaced by

$$k_x^2 + k_y^2 = \epsilon\, \mu_v k_0^2 \tag{12}$$

where μ_v is known as the Voigt permeability, and is given by

$$\mu_v = \frac{\mu_1^2 - \mu_2^2}{\mu_1}. \tag{13}$$

It is straightforward to see that μ_v does not depend on the sign of the external field B_0, so the polariton dispersion relation (Equation 12) is similarly unaffected. Thus polariton dispersion corresponding to propagation through an antiferromagnet (as bulk polaritons) is, in the present geometry, totally reciprocal. Nonreciprocal effects only occur in the presence of a surface, as in the case of reflection off an antiferromagnet (Camley, 1987).

3. Nonreciprocity in reflection of plane waves

3.1 Reflected intensity

As discussed in the introduction, we can regard reflectivity R as nonreciprocal if there is a change in reflected intensity when the incident and reflected beams are interchanged, i.e. $R(\theta_1) \neq R(-\theta_1)$ where θ_1 is the angle of incidence (see Figure 1), or, equivalently, when the applied field $\mathbf{B}_0$ is reversed, i.e. $R(\mathbf{B}_0) \neq R(-\mathbf{B}_0)$. The possibility of nonreciprocal reflectivity in the present geometry was first analysed using thermodynamic arguments (Remer et al., 1984; Camley, 1987; Stamps et al., 1991). This analysis shows that reflectivity should be reciprocal in the absence of absorption, but that it need not be in the presence of absorption. Here we demonstrate the same result explicitly in the case of reflection off a uniaxial antiferromagnet, using the arguments outlined by Abraha & Tilley (1996) and Dumelow et al. (1998).

We are interested in reflection from vacuum in s polarisation. The complex reflection coefficient r in this case can be easily worked out from the field continuity conditions at the vacuum/antiferromagnet interface. Written in terms of the $\mathbf{E}$ field component of the electromagnetic radiation, the complex reflection coefficient is given by

$$r = \frac{k_{1y}\mu_v - k_{2y} - ik_x(\mu_2 / \mu_1)}{k_{1y}\mu_v + k_{2y} + ik_x(\mu_2 / \mu_1)} \tag{14}$$

with a corresponding transmission coefficient

$$t = 1 + r = \frac{2k_{1y}\mu_v(\mu_2 / \mu_1)}{k_{1y}\mu_v + k_{2y} + ik_x(\mu_2 / \mu_1)}. \tag{15}$$

k_x is the in-plane component of the wavevector, which is continuous in both media and determined by the angle of incidence θ_1:

$$k_x = k_0 \sin\theta_1. \tag{16}$$

k_{1y} and k_{2y} are the normal components of the wavevector in vacuum and the antiferromagnet respectively, and are given by

$$k_{1y} = \sqrt{k_0^2 - k_x^2} \tag{17}$$

and

$$k_{2y} = \sqrt{\epsilon\mu_v k_0^2 - k_x^2}. \tag{18}$$

Since $k_x(\theta_1) = -k_x(-\theta_1)$ and $\mu_2(\mathbf{B}_0) = -\mu_2(-\mathbf{B}_0)$, the effect of either changing the sign of θ_1 or changing the sign of B_0 is to change the sign of the term $ik_x(\mu_2/\mu_1)$ in both the numerator and the denominator of Equation 14, all other terms in this equation being unaffected.

Let us first consider how this sign change affects the complex reflection coefficient r in the case of zero absorption ($\Gamma = 0$). In this case, all the parameters in Equation 14 are real, except for k_{2y}, which may be either real or imaginary depending on the frequency. When k_{2y} is real, there is propagation of radiation, as bulk polaritons, into the interior of the sample. When it is imaginary the field within the antiferromagnet is evanescent, decaying away from the interface. We refer to frequencies corresponding to k_{2y} real as bulk region frequencies and those corresponding to k_{2y} imaginary as reststrahl region frequencies.

At bulk region frequencies (k_{2y} real), separation of Equation 14 into real and imaginary parts leads to

$$r^*(\mathbf{B}_0) = r(-\mathbf{B}_0). \tag{19}$$

The overall reflectivity is given by

$$R = rr^*, \tag{20}$$

and is therefore reciprocal.

At reststrahl region frequencies (k_{2y} imaginary), the numerator of Equation 14 is the complex conjugate of the denominator. Thus one can see from Equation 20 that R must be equal to 1, regardless of the sign of B_0, so once again the reflectivity is reciprocal. The result $R = 1$ is of course what one should expect, since when k_{2y} is imaginary there is no propagation into the sample, leading to total reflection.

Simulated oblique incidence ($\theta_1 = 45°$) reflectivity spectra off MnF_2 at 4.2 K, in an external magnetic field $\mathbf{B}_0$ of magnitude 0.1 T, are shown in Figure 2(a), in which zero damping is assumed. The frequency scale is expressed in terms of wavenumbers $\omega/2\pi c$, and the MnF_2 parameters used in the calculation are (Dumelow & Oliveros, 1997) $\epsilon = 5.5$, $M_S = 6.0\times10^5$ A/m,

$B_A = 0.787$ T, $B_E = 53.0$ T and $\gamma = 0.975$ cm^{-1}/T, corresponding to $\omega_r = 8.94$ cm^{-1}. The curves for $B_0 = +0.1$ T and $B_0 = -0.1$ T are coincident at all frequencies, confirming that the reflectivity is reciprocal.

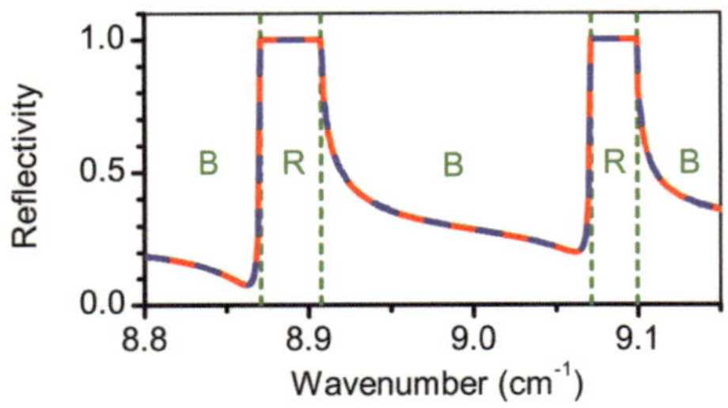

(a) Ignoring damping

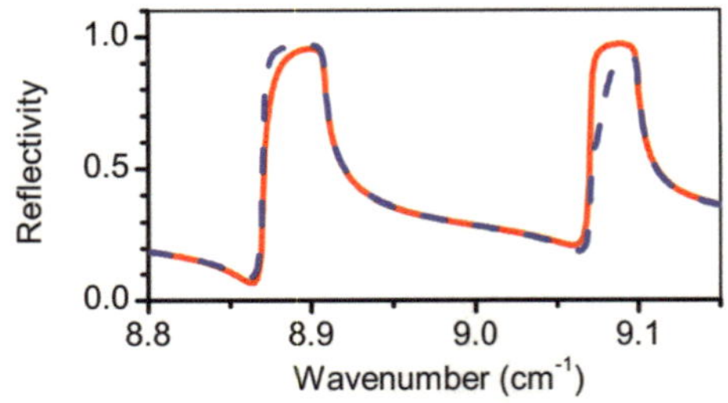

(b) Damping included

Fig. 2. Calculated oblique incidence ($\theta_1 = 45°$) reflectivity spectrum off MnF_2 in an external field of $B_0 = +0.1$ T (solid curves) and $B_0 = -0.1$ T (dashed curves). Note that in (a) the two curves are coincident. The symbols B and R in (a) represent the bulk and reststrahl frequency regions respectively.

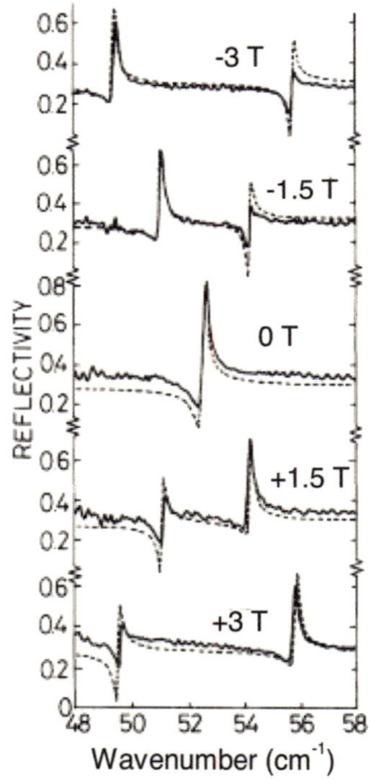

Fig. 3. Experimental (solid curves) and theoretical (dashed curves) oblique incidence reflectivity ($\theta_1 = 45°$) reflectivity spectra off FeF_2 at 4.2 K, in the presence of positive and negative external magnetic fields. After Brown et al. (1994).

When the damping is nonzero, the above symmetry arguments do not apply, and the reflectivity R is, in general, nonreciprocal. This can be seen from Figure 2(b), in which damping has been included in the calculation, using the experimental value of $\Gamma = 0.0007$

cm^{-1}. In this figure, the curves for B_0 = +0.1 T and B_0 = −0.1 T are not coincident. The difference in this case is quite small, since the damping in MnF_2 is small. Remer et al. (1986) was able to observe this small difference experimentally using a field scan at fixed frequency, but found a larger nonreciprocity at higher temperature, corresponding to larger damping. 4.2 K frequency scans of the type shown in Figure 2 have been performed on FeF_2 (Brown et al., 1994), which has a considerably higher damping parameter than MnF_2, using far infrared Fourier transform spectroscopy (Brown et al., 1998). In this case the nonreciprocity in the reflectivity is quite clear, as seen in Figure 3.

3.2 Reflected phase

The complex reflection coefficient r given by Equation 14 is commonly expressed in terms of a reflection amplitude ρ_r and phase ϕ_r:

$$r = \rho_r \exp(i\phi_r), \tag{21}$$

where

$$\rho_r = R^{1/2}, \tag{22}$$

and

$$\phi_r = \tan^{-1}\left[\frac{\mathrm{Im}(r)}{\mathrm{Re}(r)}\right]. \tag{23}$$

Although thermodynamic arguments show that, in the absence of damping, the reflected intensity R, and hence the amplitude ρ_r, should be reciprocal (Remer et al., 1984), such arguments cannot be applied to the reflected phase ϕ_r. A detailed discussion of nonreciprocity in the reflected phase on reflection off antiferromagnets is given in Dumelow et al. (1998). Here we summarise the main results.

We consider first the case of zero damping (Γ = 0). In the bulk regions, Equation 19 should apply. Thus, since the phase is given by Equation 23, we can see quite straightforwardly that

$$\phi_r(\mathbf{B}_0) = -\phi_r(-\mathbf{B}_0) + 2\pi m, \tag{24}$$

where m is an arbitrary integer. We include the term $2\pi m$ since it is convenient to plot the phase outside the range $-\pi < \phi_r < \pi$.

Equation 24 shows that, in the bulk regions, the reflected phase is nonreciprocal even in the absence of damping. This is also the case in the reststrahl regions, but the phase does not follow a simple symmetry relation of the type given by this equation.

The amplitude and phase for reflection off MnF_2 in the absence of damping are shown in Figures 4(a) and 4(c) respectively. The conditions are the same as those used in Figure 2. Note that we have shown the phase as varying within the range π to 3π in order to show that it changes continuously with frequency. The amplitude is reciprocal, in agreement with Figure 2, but nonreciprocity in the reflected phase is quite marked in both the bulk and reststrahl regions, obeying Equation 24 in the bulk regions.

Figures 4(b) and 4(d) show reflection amplitude and phase respectively in the presence of damping. In line with the reflectivity results in the previous subsection, the reflection amplitude now shows slight nonreciprocity. The phase shows the same type of nonreciprocity as seen without damping, although the bulk region symmetry arguments of Equation 24 no longer apply.

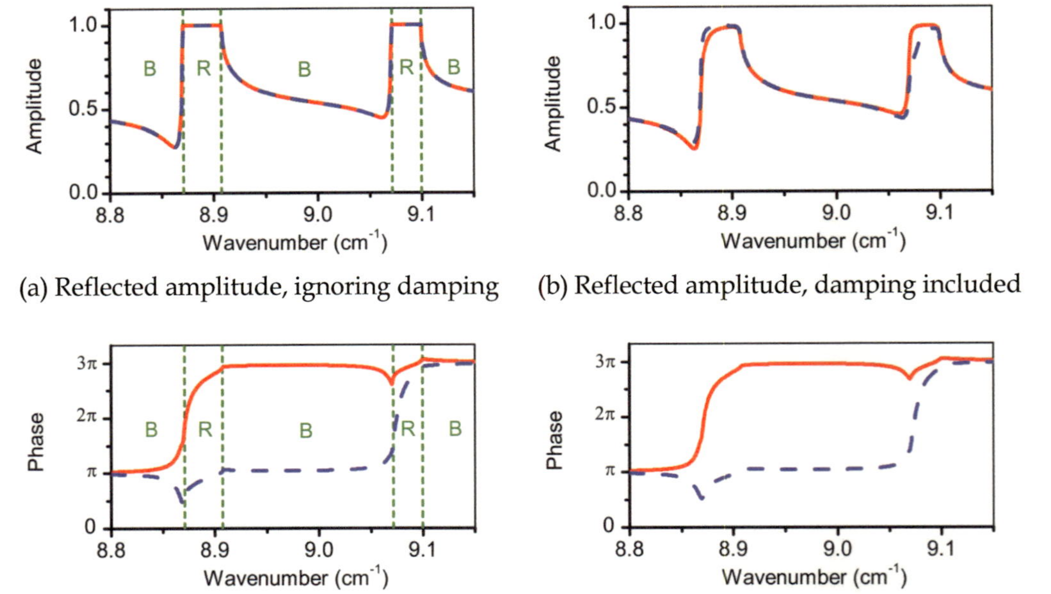

(a) Reflected amplitude, ignoring damping (b) Reflected amplitude, damping included

(c) Reflected phase, ignoring damping (d) Reflected phase, damping included

Fig. 4. Calculated reflected amplitude and phase spectra for oblique incidence (θ=45°) reflection off MnF_2 in an external field of $B_0 = +0.1$ T (solid curves) and $B_0 = -0.1$ T (dashed curves). Note that in (a) the two curves are coincident. The symbols B and R represent the bulk and reststrahl frequency regions respectively.

3.3 Power flow

The nonreciprocal phenomena described in the Subsections 3.1 and 3.2 were analysed ten or more years ago, and concern the behaviour of a reflected plane wave. Recently we have started studying nonreciprocal behaviour within the antiferromagnet itself, in particular with respect to the direction of the internal power flow (Lima et al., 2009), represented by the time-averaged Poynting vector (Landau & Lifshitz, 1984),

$$\langle \mathbf{S} \rangle = 1/2 \ \mathrm{Re}\left(\mathbf{E} \times \mathbf{H}^*\right). \tag{25}$$

We consider an angle of refraction in terms of the direction of the time-averaged Poynting vector $\langle \mathbf{S}_2 \rangle$ (which is not necessarily the same as the wavevector direction) in the antiferromagnet, as shown in Figure 5. The angle of refraction θ_2, defined in this way, is given by

$$\tan\theta_2 = \frac{\langle S_{2x} \rangle}{\langle S_{2z} \rangle}. \tag{26}$$

In s polarisation the **E** field is confined along z, so the Poynting vector is most easily represented in terms of the E_z field, making use of the conversion $\mathbf{k} \times \mathbf{E} = \omega\mu_0\mu\mathbf{H}$. The resulting time averaged Poynting vector has components

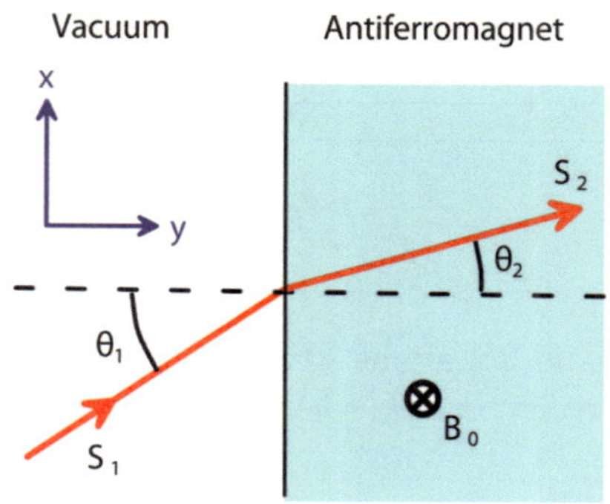

Fig. 5. Angle of refraction θ_2 defined by power flow direction.

$$\langle S_{2x}\rangle = \frac{|E_z|^2}{2\omega\mu_0}\,\mathrm{Re}\left[\frac{k_x - ik_{2y}(\mu_2/\mu_1)}{\mu_v}\right], \tag{27}$$

$$\langle S_{2y}\rangle = \frac{|E_z|^2}{2\omega\mu_0}\,\mathrm{Re}\left[\frac{k_{2y} + ik_x(\mu_2/\mu_1)}{\mu_v}\right]. \tag{28}$$

The direction of power flow can thus be obtained by substitution into Equation 26.
We now investigate the above expressions in order to search for possible nonreciprocity in the power flow direction in the antiferromagnet, taking power flow to be nonreciprocal if

$$\theta_2(-\mathbf{B}_0) \neq \theta_2(\mathbf{B}_0). \tag{29}$$

In order to consider power flow, we restrict ourselves initially to the case where there is no damping in the system ($\Gamma = 0$). The calculated values of θ_2 for oblique incident reflection off MnF_2 in this case are shown in Figure 6.

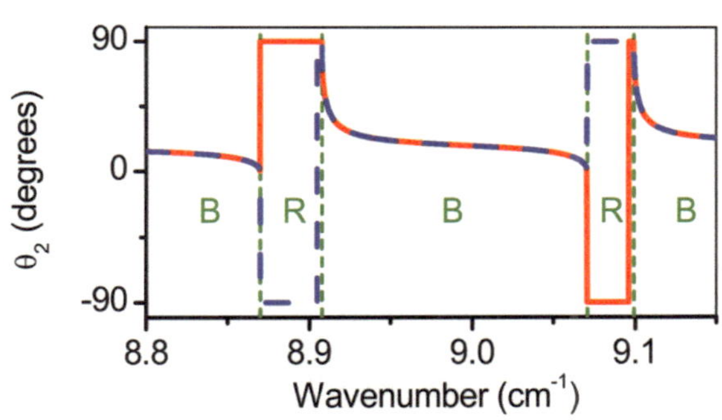

Fig. 6. Calculated θ_2 values on oblique incidence reflection ($\theta_1 = 45°$) off MnF_2 in a field of B_0 = +0.1 T (solid curve) and $B_0 = -0.1$ T (dashed curve), ignoring damping. Both curves are coincident in the bulk regions. The symbols B and R represent the bulk and reststrahl frequency regions (separated by dashed vertical lines) respectively.

In the case of zero damping, as discussed previously, μ_1, μ_2, and μ_v are all wholly real. k_{2y} is real in the bulk regions and imaginary in the reststrahl regions.
First we consider power flow for k_{2y} real (i.e. in the bulk regions). Equations 27 and 28 then give

$$\langle S_{2x}\rangle = \frac{|E_z|^2 k_x}{2\omega\mu_0\mu_v}, \tag{30}$$

$$\langle S_{2y}\rangle = \frac{|E_z|^2 k_{2y}}{2\omega\mu_0\mu_v}, \tag{31}$$

Thus, since none of the terms in Equations 30 and 31 depend on the sign of B_0, we see that the power flow direction is reciprocal ($\theta_2(-B_0) = \theta_2(B_0)$). We also see from Equations 26, 30 and 31 that

$$\tan\theta_2 = \frac{k_x}{k_{2y}}, \tag{32}$$

so that $\mathbf{S}_2$ is parallel to $\mathbf{k}_2$. Since k_x is continuous and k_{2y} positive it also follows that refraction must be positive, i.e. θ_2 always has the same sign as θ_1. Overall, therefore, power flow follows the type of behavior shown in Fig. 7(a). This is confirmed by the calculations of the θ_2 shown in Figure 6 for the bulk regions. Note that in these regions radiation may, in principle, flow an infinite distance into the antiferromagnet, and is thus unaffected by the sample surface. Our result that power flow is reciprocal in this case is thus consistent with the idea that radiation should display reciprocal behavior in the interior of a sample.

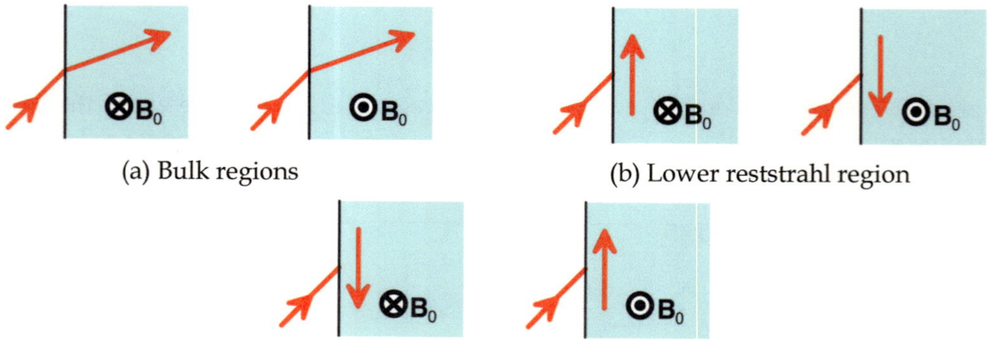

(a) Bulk regions (b) Lower reststrahl region

(c) Upper reststrahl region

Fig. 7. Power flow in the various spectral regions for oblique incidence radiation.

We now consider the case when k_{2y} is imaginary. In this case we get

$$\langle S_{2x}\rangle = \frac{|E_z|^2\left[k_x - ik_{2y}(\mu_2/\mu_1)\right]}{2\omega\mu_0\mu_v}, \tag{33}$$

$$\langle S_{2y}\rangle = 0. \tag{34}$$

There is thus no propagation into the antiferromagnet, but energy can travel along the surface in the form of an evanescent wave within the antiferromagnet. In such cases, which are characterized by $\theta_2 = \pm 90°$, the angle θ_2 does not describe refraction in the normal sense of the word, since the associated radiation has an evanescent component. Indeed there is no transfer of radiation across the interface, since power is wholly reflected. The direction of power flow may or may not be reciprocal, depending on the relative magnitudes of the two terms in the numerator of Equation 33. However, if the external field is sufficiently large, the second term dominates within both reststrahl regions. In the present example, for $B_0 = \pm 0.1$ T, this results in $\theta_2 = \pm 90°$ in most of the lower reststrahl region (Figure 7(b)) and $\theta_2 = \mp 90°$ in most of the upper reststrahl region (Figure 7(c)), as shown in the calculation in Figure 6, and nonreciprocity, as defined by Equation 29, occurs. There is, however, a narrow region of reciprocal power flow (in terms of direction if not of magnitude) near the upper frequency limit of each reststrahl region.

Possibly one of the most interesting aspects of the power flow is its behaviour in the case of normally incident radiation (Lima et al., 2009). The calculated θ_2 values are shown in Figure 8. In the bulk regions, as expected, power flow is normal to the surface, as shown in Figure 9(a). In the reststrahl regions, however, power flow parallel to the sample surface is induced in the same way as occurs at oblique incidence. The direction of this power flow depends on the sign of the external field, as shown in Figures 9(b) and 9(c). This is distinctly counter-intuitive considering that we are modelling plane waves normally incident on an infinite surface, so that there should be no spatial variation of the field along x. Nevertheless, the

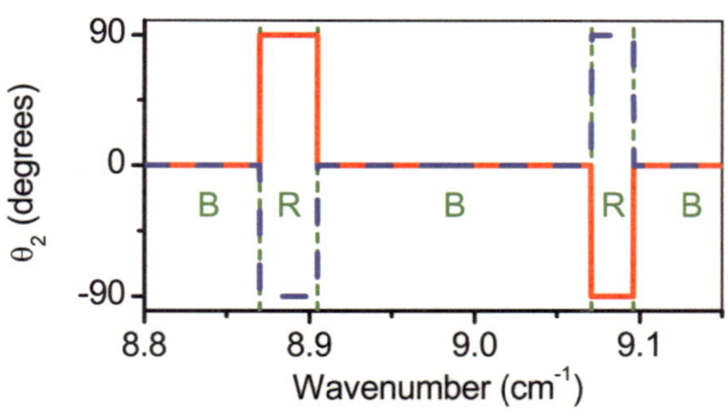

Fig. 8. Calculated θ_2 values on normal incidence reflection off MnF_2 in a field of $B_0 = +0.1$ T (solid curve) and $B_0 = -0.1$ T (dashed curve), ignoring damping. The symbols B and R represent the bulk and reststrahl frequency regions (separated by dashed vertical lines) respectively. After Lima et al. (2009).

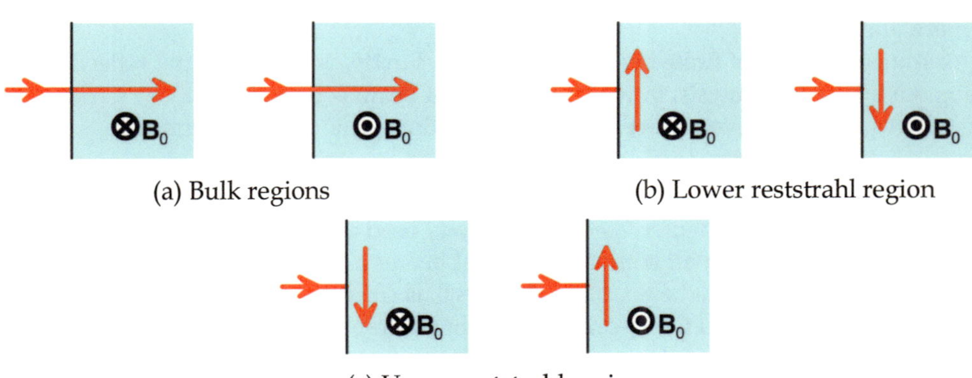

Fig. 9. Power flow in the various spectral regions for normal incidence radiation.

very fact that the field is assumed to extend over an infinite plane means that energy flow along the surface does not violate causality.

We now turn to the case where damping is present. Now k_{2y} should, in general, be complex, with the its imaginary part greater near the reststrahl regions. Since imaginary k_{2y} results in nonreciprocal power flow, we should also expect such nonreciprocity in the case of complex k_{2y}. In Fig. 10 we show the power flow directions both for normal and for oblique incidence, assuming $\Gamma = 0.0007\text{cm}^{-1}$. There is now no distinct division between reststrahl and bulk regions, and nonreciprocal power flow occurs both inside and outside the nominal reststrahl regions. At normal incidence the power flow directions are now no longer restricted to $\theta_2 = 0°$ and $\theta_2 = \pm 90°$. Since nonzero θ_2 implies nonreciprocity, the associated fields must be some extent be bound to the sample surface in all regions for which $\theta_2 \neq 0°$.

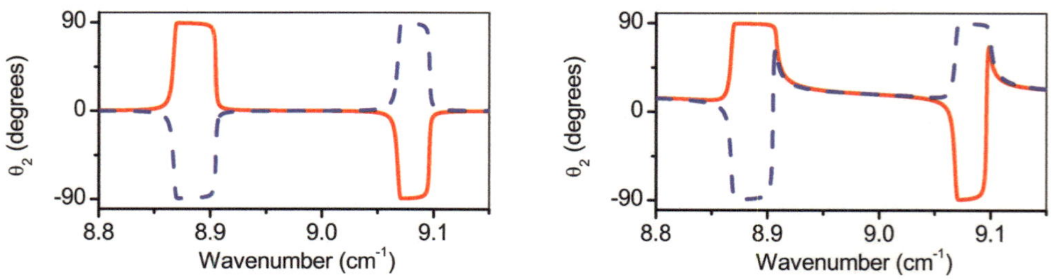

(a) $\theta_1 = 0$ (b) $\theta_1 = 45°$

Fig. 10. Calculated θ_2 values on reflection off MnF_2 in a field of $B_0 = +0.1$ T (solid curve) and $B_0 = -0.1$ T (dashed curve), assuming a damping parameter of $\Gamma = 0.0007\ \text{cm}^{-1}$.

4. Reflection of a finite beam

4.1 Introduction

The previous section discusses various phenomena associated with plane wave reflection off an antiferromagnet. However, when the plane wave is replaced by a finite beam, we predict additional effects concerning the profile and position of the reflected beam (Lima et al., 2008; 2009). Such effects are expected with either normal or oblique incidence radiation. However, we concentrate on normal incidence effects firstly for simplicity and secondly because they are more unexpected.

We examine reflection of finite beams in two ways. Firstly, we interpret the reflection using an angular spectrum analysis, in which the incident beam is considered as a Fourier sum of plane waves. We then give a power flow interpretation of the predicted effects.

4.2 Angular spectrum analysis

Here we summarise the angular spectrum analysis used in describing reflection of a finite beam normally incident on an antiferromagnet (Lima et al., 2008), using a two dimensional model in which the incident beam, centred at $x = 0$, is considered as an angular spectrum of plane waves propagating in the xy plane. It can thus be represented in the form

$$E_i(x,y) = \int_{-k_0}^{k_0} \psi(k_x)\exp\left[i\left(k_x x + k_{1y} y\right)\right] dk_x, \tag{35}$$

where $\psi(k_x)$ is a distribution function representing the shape of the beam. At the sample surface, which we place at $y = 0$, the electric field of the incident beam is

$$E_i(x,0) = \int_{-k_0}^{k_0} \psi(k_x)\exp(ik_x x)\,dk_x. \tag{36}$$

The electric field of the corresponding reflected beam is given, at the surface, by

$$E_r(x,0) = \int_{-k_0}^{k_0} r(k_x)\psi(k_x)\exp(ik_x x)\,dk_x. \tag{37}$$

Here $r(k_x)$ represents the complex reflection coefficient for the relevant plane wave component, and is given by Equation 14. It is convenient to consider this complex coefficient in terms of amplitude $\rho_r(k_x)$ and phase $\phi_r(k_x)$, in the form of Equation 21, i.e.,

$$r(k_x) = \rho_r(k_x)\exp[i\phi_r(k_x)]. \tag{38}$$

If the beam is sufficiently wide, there will be a narrow distribution of k_x values centred, at normal incidence, around $k_x = 0$. We can thus substitute Equation 38 into Equation 37 and expand $\rho_r(k_x)$ and $\phi_r(k_x)$ as a Taylor series around $k_x = 0$. If we ignore terms in k_x^2 and above, this leads to

$$E_r(x,0) = r(0)\int_{-k_0}^{k_0} \psi(k_x)\exp(ik_x X)\,dk_x + r'(0)\int_{-k_0}^{k_0} k_x\psi(k_x)\exp(ik_x X)\,dk_x, \tag{39}$$

where

$$r(0) = \rho_r(0)\exp[i\phi_r(0)], \tag{40}$$

$$r'(0) = \frac{d\rho_r}{dk_x}\Big|_{k_x=0} \exp[i\phi_r(0)], \tag{41}$$

$$X = x + \frac{d\phi_r}{dk_x}\Big|_{k_x=0}. \tag{42}$$

The second term on the right hand side of Equation 39 can, in practice, normally be ignored (Lima et al., 2008). In fact, in the absence of damping, it is identically zero since reciprocity in ρ_r implies $d\rho_r / dk_x|_{k_x=0} = 0$, and hence $r'(0) = 0$. Even in the presence of damping, however, the contribution of this term is negligible for a sufficiently narrow k_x distribution.

The reflected field is thus given, to a good approximation, by the first term on the right hand side of Equation 39. This term is simply the reflection coefficient $r(0)$ for a normally incident plane wave multiplied by an integral which gives the profile of the field along x. This integral is identical to that of the incident beam (Equation 36) except that x has been replaced by X, given by Equation 42. Thus the shape of the reflected beam is the same as that of the incident beam, but it has been shifted along the surface of the sample by a distance D_r equal to

$$D_r = -\frac{d\phi_r}{dk_x}\Big|_{k_x=0} \tag{43}$$

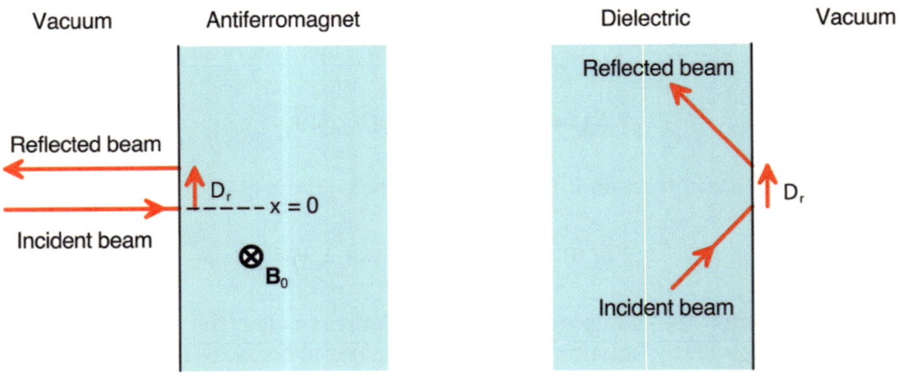

(a) Normal incidence reflection off AF (b) Conventional Goos-Hänchen shift

Fig. 11. Comparison of the normal incidence Goos-Hänchen shift predicted for external reflection off an antiferromagnet with the conventional Goos Hänchen shift for total internal reflection.

as shown in Figure 11(a). In the case of reciprocal reflected phase (i.e. $\phi(-k_x) = \phi(k_x)$), $d\phi_r / dk_x |_{k_x=0}$ must be zero, so there will be no lateral shift. In the case of reflection off antiferromagnets, however, the reflected phase is, in general, nonreciprocal, as discussed in Section 3.2, and a nonzero displacement of the reflected beam is expected.

Before discussing the lateral shift described by Equation 43 in any detail, we note that the behaviour of the reflected beam is in some ways similar to that of an oblique incidence finite beam which undergoes total internal reflection when passing from an optically denser to a less dense medium. Such a beam also suffers a lateral displacement D_r upon reflection, as shown in Figure 11(b). This displacement was observed experimentally by Goos & Hänchen (1947) and is normally referred to as a Goos-Hänchen shift (Lotsch, 1970). Goos-Hänchen shifts have also been reported in the case of external reflection in specific instances such as reflection off metals (Wild & Giles, 1982; Leung et al., 2007), and the lateral displacement discussed here can be considered as a type of a normal incidence Goos-Hänchen shift. Equation 43 is, indeed, a normal incidence version of the classical expression commonly used to describe Goos-Hänchen shifts (Artmann, 1948), and the angular spectrum analysis used above in deriving the equation is basically the same as that previously used to describe Goos-Hänchen shifts in the case of total internal reflection (Horowitz & Tamir, 1971; McGuirk & Carniglia, 1977).

We now apply Equation 43 to the specific case of reflection off an antiferromagnet. In the absence of damping, the reflection coefficient r can easily be resolved into its real and imaginary parts, allowing explicit evaluation of this equation. This gives

$$D_r = \frac{2(\mu_2/\mu_1)}{k_0(\mu_v - \epsilon)} \tag{44}$$

in both the bulk and the reststrahl regions. Since the sign of μ_2 depends on the sign of B_0, it is immediately obvious that the direction of the lateral displacement will be reversed if the external field direction is reversed. In the presence of damping, Equation 43 can be evaluated numerically, but the results are found to be almost identical to those with $\Gamma = 0$ (Lima et al., 2008).

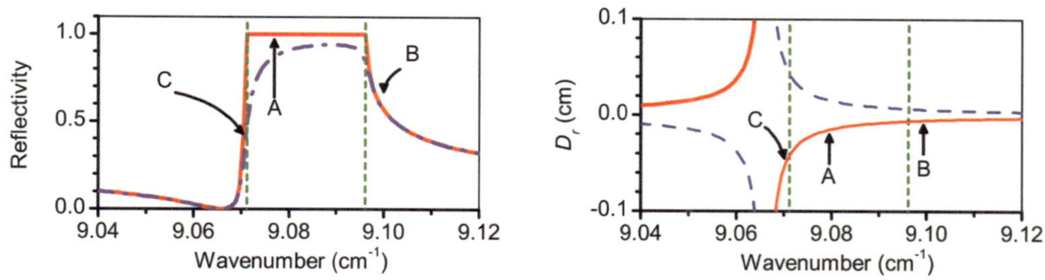

(a) Reflectivity (b) Lateral shift of reflected beam

Fig. 12. Calculated normal incidence reflectivity spectrum and 1ateral shift for reflection off MnF_2 in an external field of $B_0 = +0.1$ T. In (a) the solid curve is calculated without damping and the dot-dash curve with $\Gamma = 0.0007$ cm^{-1}. In (b) the solid curves are for an external field of $B_0 = \pm 0.1$ T and the dashed curves for an external field of $B_0 = -0.1$ T. The vertical lines separate the bulk and reststrahl regions. The marked frequencies are A: 9.0769 cm^{-1}; B: 9.0994 cm^{-1}; C: 9.0705 cm^{-1}. After Lima et al. (2009).

The calculated values of D_r, ignoring damping, are compared with the reflectivity spectrum at upper reststrahl region frequencies in Figure 12. A lateral shift is predicted in both the bulk and the reststrahl regions. Similar shifts are predicted around the lower reststrahl region, but of opposite sign (Lima et al., 2008). There is a divergence in D_r at the reflectivity minimum just below the reststrahl region. At this frequency, the assumptions made in deriving Equation D_r clearly do not apply.

In order ot verify the shifts shown in Figure 12, reflection of a particular incident beam profile may be modelled. For simplicity, we have considered a gaussian beam whose focal plane is the surface of the sample (Lima et al., 2008; 2009). The incident beam can thus be represented by Equation (35) with (Horowitz & Tamir, 1971)

$$\psi(k_x) = \frac{g}{2\sqrt{\pi}} \exp\left(-\frac{g^2 k_x^2}{4}\right), \tag{45}$$

where $2g$ represents the beam width at the focal plane. At the sample surface the incident beam profile is thus represented by Equation 36 and the reflected beam profile by Equation 37. The integrals in these two equations can be evaluated numerically, and the profiles of the corresponding **E** fields thus obtained.

The incident and reflected beam intensities can, in general, be well represented by $|E|^2$. In Figure 13 we show the resulting intensity profiles along x for the three frequencies marked in Figure 12. It is seen that, although the modelled incident beam is very narrow ($g = \lambda$, the free space wavelength), all the reflected beams are displaced along the surface in excellent agreement with Equation 44. It is also observed that damping does not noticeably affect this displacement. Explicit simulations have also confirmed the prediction of Equation 44 that beam displacement is, to a very good approximation, independent of beam width (Lima et al., 2008).

It is useful not only to consider the incident and reflected fields at the surface, but also the overall **E** field distribution in the xy plane. We conveniently consider the electric fields in terms of an incident field $E_i(x,y)$ and a reflected field $E_r(x,y)$ in the region $x < 0$ (vacuum) and a transmitted field $E_t(x,y)$ in the region $x > 0$ (antiferromagnet). $E_i(x,y)$ is given by Equation 35, with $E_r(x,y)$ and $E_t(x,y)$ given by

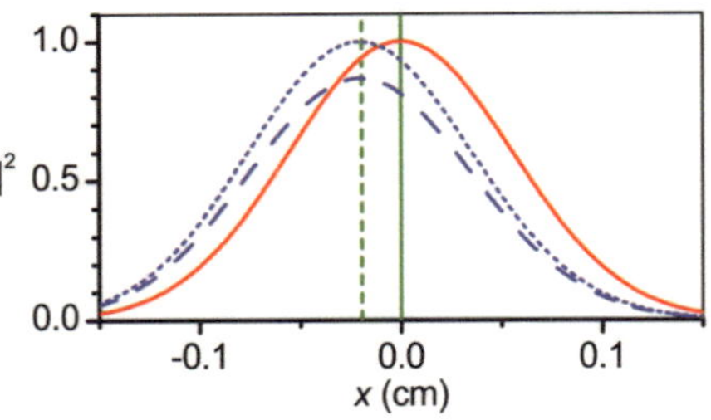

(a) Frequency A (reststrahl)

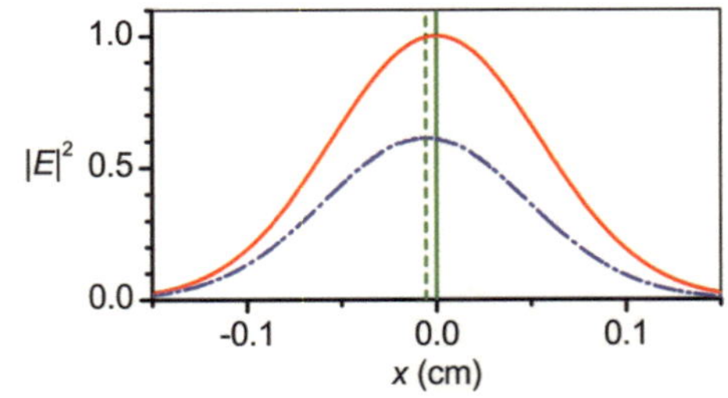

(b) Frequency B (bulk)

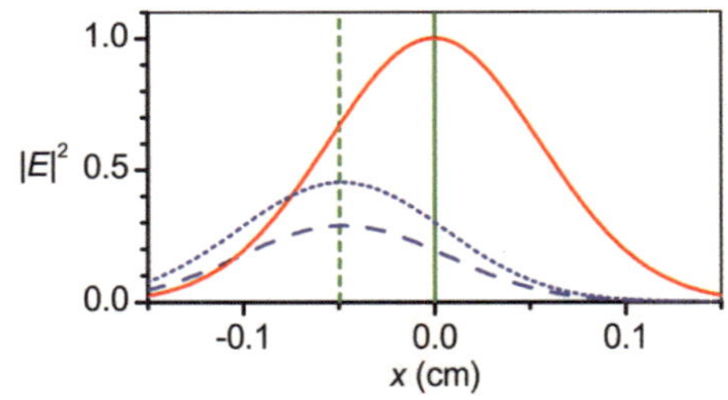

(c) Frequency C (bulk)

Fig. 13. Intensity profiles, along the sample surface, in the case of reflection of a normally incident gaussian beam off MnF_2 in an external magnetic field of 0.1 T. Solid curve: incident beam; dotted curve: reflected beam ignoring damping; dashed curve: reflected beam with $\Gamma = 0.0007\ cm^{-1}$; vertical solid line: centre of incident beam; vertical dashed line: centre of reflected beam predicted by Equation 44. Note that, in (b), the reflected intensities with and without damping are almost identical, and the curves cannot be separated. After Lima et al. (2009).

$$E_r(x,y) = \int_{-k_0}^{k_0} r(k_x)\psi(k_x)\exp\left[i\left(k_x x - k_{1y}y\right)\right]dk_x \tag{46}$$

and

$$E_t(x,y) = \int_{-k_0}^{k_0} t(k_x)\psi(k_x)\exp\left[i\left(k_x x + k_{2y}y\right)\right]dk_x \tag{47}$$

respectively. The resulting profiles at the three frequencies indicated in Figure 12 are shown in Figures 14 and 15. Figure 14 shows the profile in the absence of damping and Figure 15 shows the profile when damping is included. The left hand panels show the incident and transmitted fields, whilst the right hand panels show the reflected field.

In all cases the reflected fields are displaced along x in accordance with Figures 12 and 13. In addition, we see that the transmitted field is also displaced. This result can be anticipated from the fact that the phase ϕ_t of the complex transmission coefficient t is nonreciprocal. An analysis equivalent to that used in determining D_r then gives a lateral displacement D_t of the transmitted field profile given by

$$D_t = -\frac{d\phi_t}{dk_x}\Big|_{k_x=0}. \tag{48}$$

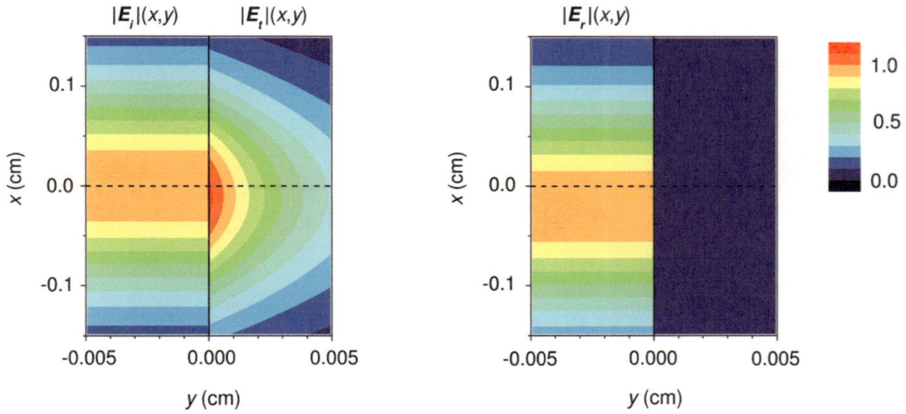

(a) Frequency A (reststrahl)

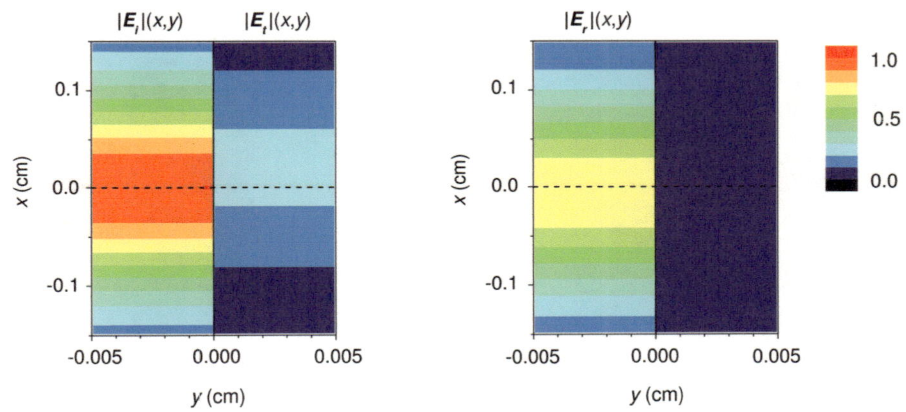

b) Frequency B (bulk)

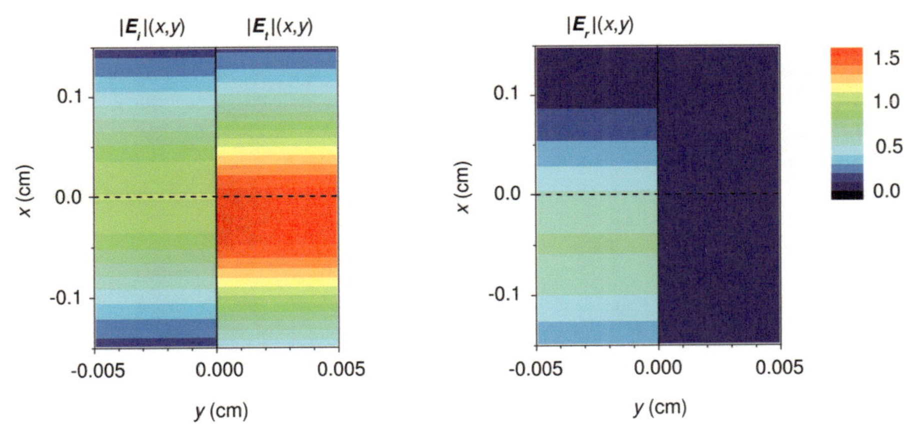

(c) Frequency C (bulk)

Fig. 14. Profiles of the amplitudes of the incident ($E_i(x,y)$), transmitted ($E_t(x,y)$), and reflected ($E_r(x,y)$) fields in the case of reflection of a normally incident gaussian beam off MnF_2 in an external magnetic field of 0.1 T, ignoring damping. After Lima et al. (2009).

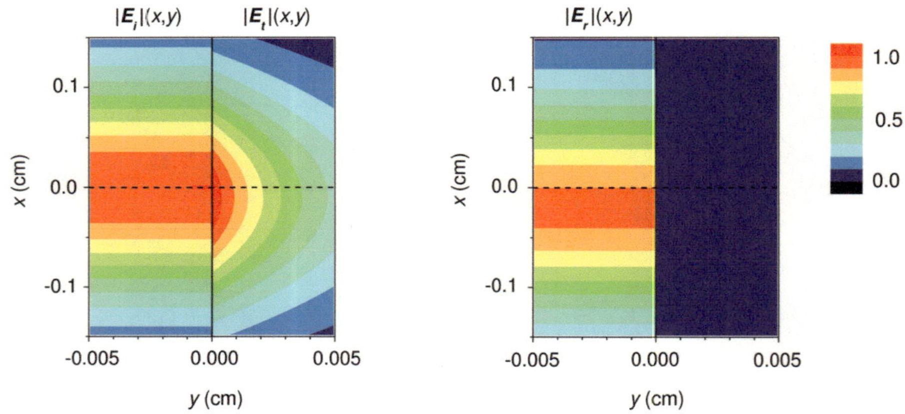

(a) Frequency A (reststrahl)

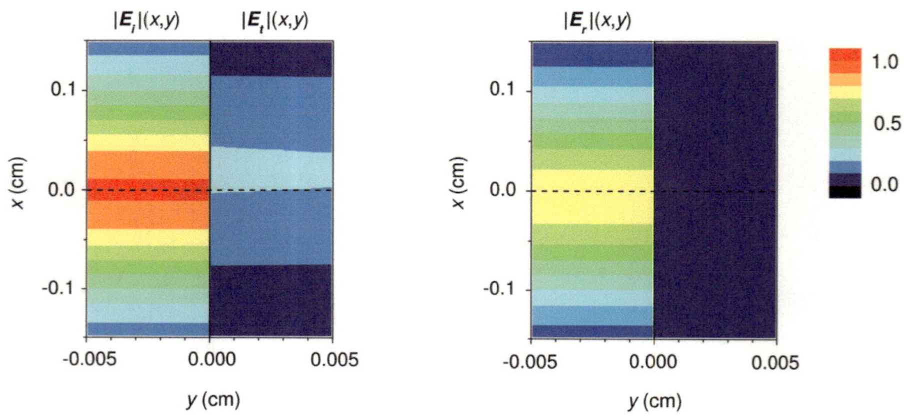

(b) Frequency B (bulk)

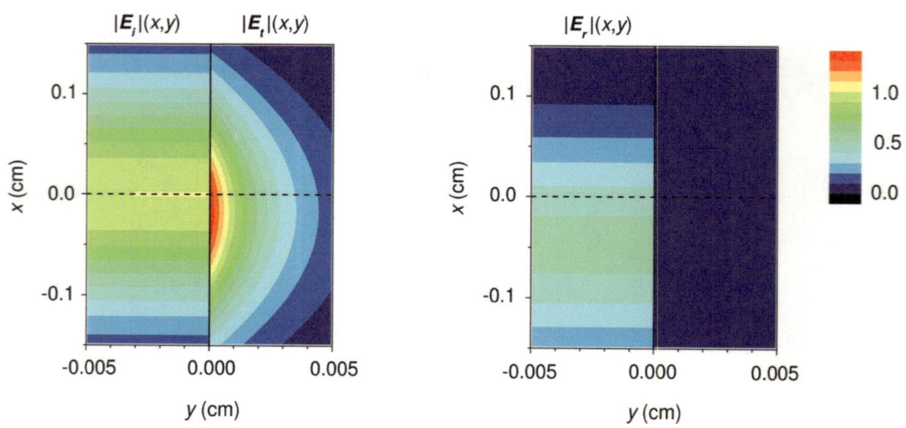

(c) Frequency C (bulk)

Fig. 15. Profiles of the amplitudes of the incident ($E_i(x,y)$), transmitted ($E_t(x,y)$), and reflected ($E_r(x,y)$) fields in the case of reflection of a normally incident gaussian beam off MnF_2 in an external magnetic field of 0.1 T, for $\Gamma = 0.0007$ cm^{-1}.

It is noticeable that, in the case of the bulk frequency B, near the top of the reststrahl band, there is a much larger displacement in the transmitted field than in that of the reflected beam. In the absence of damping, the field decays away from the interface in the reststrahl region (Figure 14(a)), but propagates into the sample, perpendicular to the surface, in the bulk regions (Figures 14(b) and 14(c)). When damping is included, the reststrahl region behaviour is largely unchanged, but there may now be significant decay in the bulk regions. At frequency B, just above the top of the reststrahl region, this decay is fairly small, but at frequency C, near the bottom of the reststrahl region, it is similar to that in the reststrahl region itself.

4.3 Power flow analysis

4.3.1 Reststrahl region

In the above analysis we have considered the displacement of the reflected beam as an interference effect. It is also useful, however, to consider this effect in terms of power flow. In the reststrahl regions, in the absence of absorption, energy conservation principles can be used in a straightforward way to analyse the lateral shift (Lima et al., 2009). The analysis is similar to that used by Renard (1964) in the case of the conventional Goos-Hänchen shift for total internal reflection.

We make use of the result of subsection 3.3 that, in the reststrahl regions and in the absence of damping, a normally incident plane wave reflected off an antiferromagnet induces power flow parallel to the surface within the antiferromagnet, as shown in Figures 9(b) and 9(c). We consider this to be the behaviour in the centre portion of a wide finite incident beam, such as that represented in Figure 16. The central portion of this incident beam lies between x_2 and x_3, and it gradually decays away to zero between x_2 and x_1 and between x_3 and x_4. The internal energy flux associated with plane wave reflection in the central portion is represented by P_2. Energy conservation therefore requires that there is a net flux P_1 entering the antiferromagnet near one edge of the beam and a net flux leaving near the other edge. This is equivalent to a lateral shift D_r of the reflected beam with respect to the incident beam. Thus P_1 enters in the region between x_1 and $x_2 + D_r$ and P_3 leaves in the region between x_3 and $x_4 + D_r$. We consider energy flow within a slice, of thickness Δz, in the xy plane. Within this slice we have $P_1 = P_2 = P_3$.

P_1 is the difference between the incident and reflected flux between X_2 and $x_2 + D_r$, and can be written as

$$P_1 = \int_{x_1}^{x_2} \langle S_i(x)\rangle \Delta z\, dx + \int_{x_2}^{x_2+D_r} \langle S_i(x)\rangle \Delta z\, dx + \int_{x_1+D_r}^{x_2+D_r} \langle S_r(x)\rangle \Delta z\, dx, \tag{49}$$

where $\langle S_i(x)\rangle$ and $\langle S_r(x)\rangle$ are the intensities, represented in terms of time averaged Poynting vectors along y, of the incident and reflected waves respectively. Since we expect the reflected beam to have the same shape as the incident beam, we can write

$$\langle S_i(x)\rangle = -\langle S_r(x + D_r)\rangle, \tag{50}$$

so that the first and last terms on the right hand side of Equation 49 cancel. Thus only the second term contributes to P_1, and the integral is only performed between x_2 and $x_2 + D_r$, in the central portion of the beam. Within this interval $\langle S_i(x)\rangle$ has a constant value, which we denote as $\langle S_{max}\rangle$. Equation 49 thus becomes

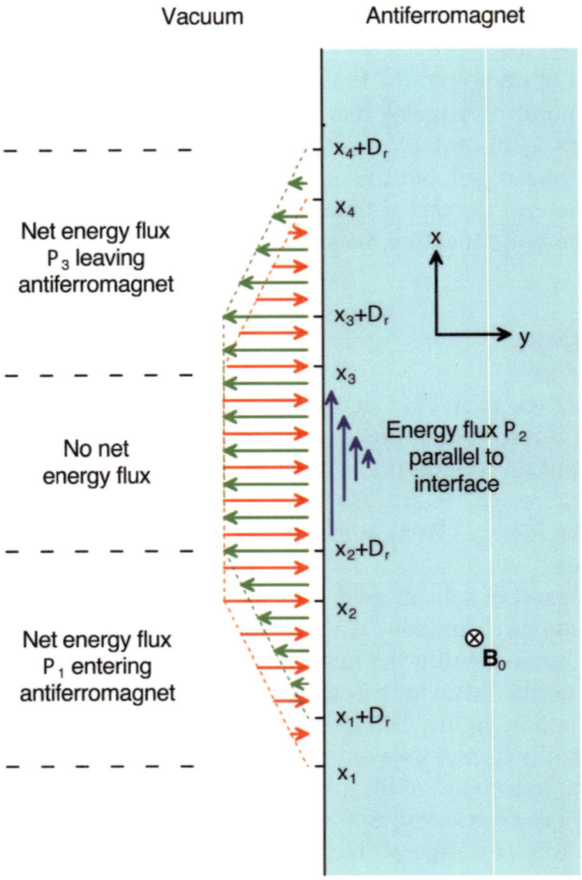

Fig. 16. Model of lateral displacement of reflected beam in the reststrahl regions in terms of power flow along the antiferromagnet surface.

$$P_1 = \langle S_{max} \rangle \Delta z D_r. \tag{51}$$

Using Equation 25 and a standard application of Maxwell's equations, $\langle S_{\max} \rangle$ can be expressed in terms of the incident field $E_{\max}$ within the central region of the beam as

$$\langle S_{max} \rangle = \frac{k_0 |E_{max}|^2}{2\omega\mu_0}. \tag{52}$$

The flux P_2 within the antiferromagnet is given by

$$P_2 = \int_0^\infty \langle S_{2x}(y) \rangle \Delta z \, dy. \tag{53}$$

$\langle S_{2x}(y) \rangle$ can be expressed in terms of the electric field $E_z(y)$ within the antiferromagnet using Equation 27. $E_z(y)$ itself can be related to the field $E_{\max}$ of the incident beam by the equation

$$E_z(y) = tE_{max} \exp\left(ik_{2y} y\right), \tag{54}$$

so P_2 may also be obtained as a function of E_{max}. On putting $P_1 = P_2$ in the above equations, and solving for D_r we get exactly the same result as obtained using the angular spectrum analysis (Equation 44). Thus simple energy conservation principles may be used to predict the Goos-Hänchen shift in this case.

We may examine explicitly the power flow behaviour in the xy plane using the type of calculation used in obtaining Figures 14 and 15. Figure 17(a) shows the overall power intensity and flux lines (Lai et al., 2000) for reflection off MnF_2 at frequency A, corresponding to the upper reststrahl region. The behaviour is similar to that predicted from Figure 16 (although in the present case the shift is negative). Thus, for $x > -0.01$ cm, the incident intensity is greater than the reflected intensity, so that the overall power flow is to the right. At $x \simeq -0.01$ cm, the incident and reflected beams cancel, whereas for $x < -0.01$ cm, the reflected beam dominates and the overall power flow is to the left. Flux continuity is thus preserved in the manner illustrated in Figure 16.

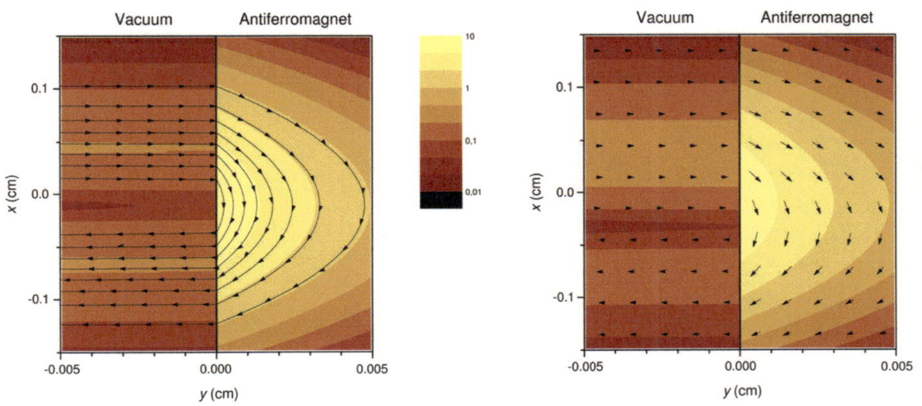

(a) Ignoring damping (b) Damping included

Fig. 17. Overall power intensity and flow for reflection off MnF_2 in an external magnetic field of 0.1 T in the upper reststrahl region (frequency A). After Lima et al. (2009).

Figure 17(b) shows power flow when damping is included. In this case the flux continuity argument no longer applies (hence we have not attempted to show continuous flux lines), but the overall behaviour is not significantly changed.

It is interesting to note that the power flow explanation of the normal incidence Goos Hänchen shift gives a very clear example of the breaking of time reversal symmetry by a static magnetic field. In the example shown in Figure 16, there is power flow along the positive x direction within the antiferromagnet, and the reflected beam is thus displaced in this direction. If, however, time reversal were applied to the reflected beam, so that it became a new incident beam, without reversing the direction of $\mathbf{B}_0$, the flow of energy within the antiferromagnet would not retrace its path along the negative x direction, but would once again flow along positive x, and there would be a further displacement of the reflected beam in this direction.

4.3.2 Bulk regions

The above power conservation principles used in analysing the normal incidence Goos-Hänchen shift in the reststrahl regions are based on the principle that all the incident energy will be reflected back from the antiferromagnet surface. This is clearly not the case in the bulk regions. Furthermore, in the absence of damping, the power flow resulting from normally incident plane waves is transmitted normal to the surface, as seen in Figure 9(a). The reststrahl region analysis requires power flow parallel to the interface, so one might expect that there would be no lateral shift of the reflected beam in the bulk regions. The angular spectrum analysis, however, predicts that such a shift should occur in both the reststrahl and the bulk regions, so it is important to understand, in terms of energy flow, how such a shift is possible in the bulk regions.

In order to analyse the power flow, we once again take the example of reflection of a Gaussian beam and examine the power flow profile in the xy plane. The resulting overall power intensity and flux at the two bulk frequencies B and C are shown in Figure 18. Before analysing the lateral shift in any detail, it is worth noting that the profile of the power intensity within the antiferromagnet is different from that of the electric field amplitudes

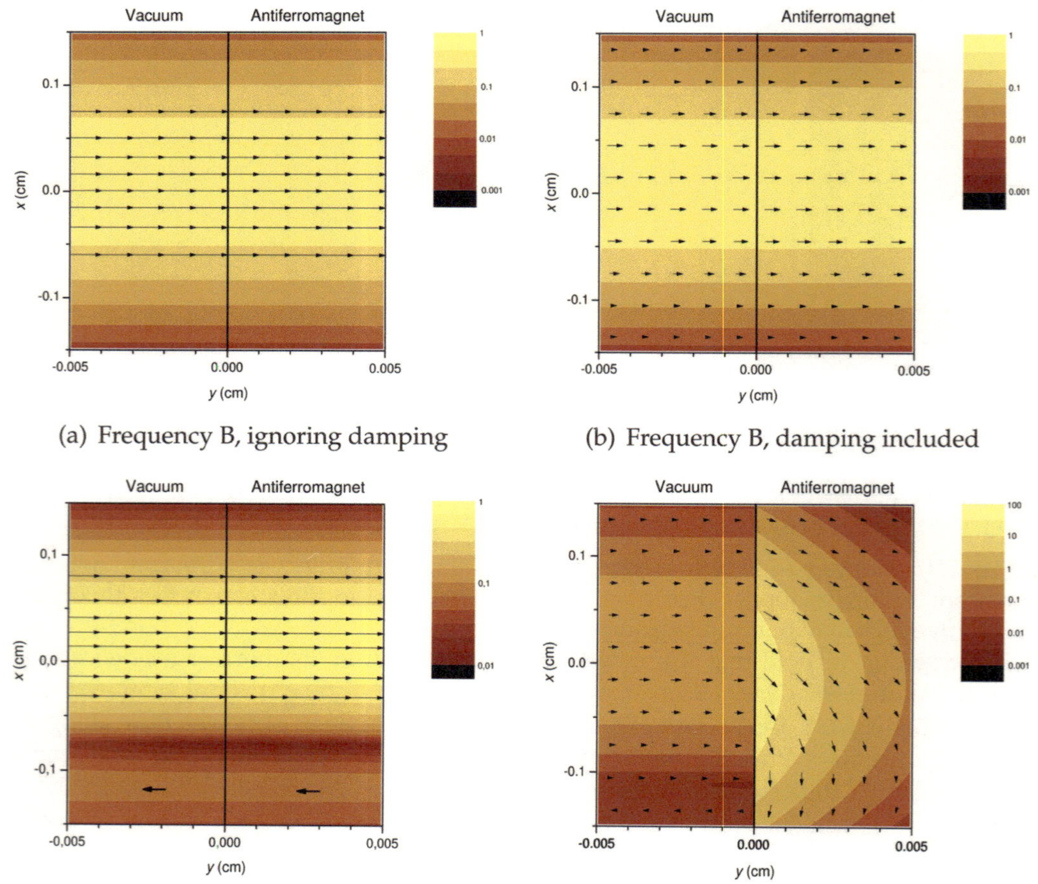

(a) Frequency B, ignoring damping (b) Frequency B, damping included

(c) Frequency C, ignoring damping (d) Frequency C, damping included

Fig. 18. Overall power intensity and flow for reflection off MnF_2 in an external magnetic field of 0.1 T in the bulk regions. Note that the arrows in the bottom half of (c) are simply intended to show the direction of power flow and are not a continuation of the flux lines in the upper part of the figure. After Lima et al. (2009).

shown in Figures 14 and 15. This is because the power intensity depends on the magnetic field component of the electromagnetic field as well as its electric field component. The magnetic field profile is in fact very different from that of the electric field (Lima et al., 2009), leading to a different power intensity profile.

At frequency B, ignoring damping (Figure 18(a)), overall power flow is to the right both in the vacuum region $y < 0$ and within the antiferromagnet $y > 0$, and remains perpendicular to the surface. There is flux continuity across the interface, as expected, and the overall profile is slightly displaced in the positive x direction with respect to the incident beam, which is centred at $x = 0$. We can understand this if we recall that the reflected beam is slightly displaced in the negative x direction and is less intense than the incident beam (see Figure 13(b)). The overall flux for $y < 0$, within the vacuum region, is the incident beam minus the reflected beam, so it is shifted in the positive x direction with respect to the pure incident beam. Flux continuity requires that this displacement is transferred to the transmitted beam, i.e. in the bulk regions the transmitted beam is displaced along x in the direction opposite to the displacement of the reflected beam. The behaviour in the presence of damping (Figure 18(b)) is not appreciably different.

At frequency C in the absence of damping (Figure 18(c)) the situation is similar to that at frequency B except that there is now a resultant power to the left for $x < -0.08$ cm^{-1}. This is related to the fact that the reflected beam is more intense than the incident beam at these values of x (see Figure 13(c)). This also means that there must be power flow to the left within the antiferromagnet. This presents an apparent problem since there is no obvious source for the associated energy. Nevertheless, detailed calculations of power flow several centimeters into the antiferromagnet show that a portion of the incident energy does indeed return to the left in this region (Lima et al., 2009). This is easier to see when damping is present (Figure 18(d)), in which case some of the incident energy returns to the left without penetrating a long distance into the antiferromagnet. In fact, in this case, the overall power flow is somewhat similar to that observed in the reststrahl regions. This is another illustration of the concept that, in the presence of damping, the bulk and reststrahl regions can be thought of as merging into one another.

In the above examples, frequency B (Figures 18(a) and 18(b)) can be regarded as representative of most of the bulk frequencies. Frequency C (Figures 18(c) and 18(d)), in contrast, displays rather peculiar behaviour particular to frequencies between the reststrahl region and the reflection minimum (see Figure 12).

5. Conclusions and future prospects

In this chapter we have examined various nonreciprocal effects associated with reflection of terahertz radiation off antiferromagnets. Of these effects, only nonreciprocity in the reflectivity has, to our knowledge, been investigated experimentally at the time of writing (Remer et al., 1986; Brown et al., 1994).

A simple, if slightly indirect, way of observing the nonreciprocal reflected phase has been suggested (Dumelow & Camley, 1996; Dumelow et al., 1998), and uses the configuration shown in Figure 19. Here, a dielectric layer is deposited onto the surface of an antiferromagnet and the overall reflectivity off the overall structure measured. In this setup, there is reciprocal reflection from the vacuum/dielectric interface, but the phase of the radiation reflected from the dielectric/antiferromagnet interface is nonreciprocal. Interference between these partial waves is thus nonreciprocal, leading to a nonreciprocal

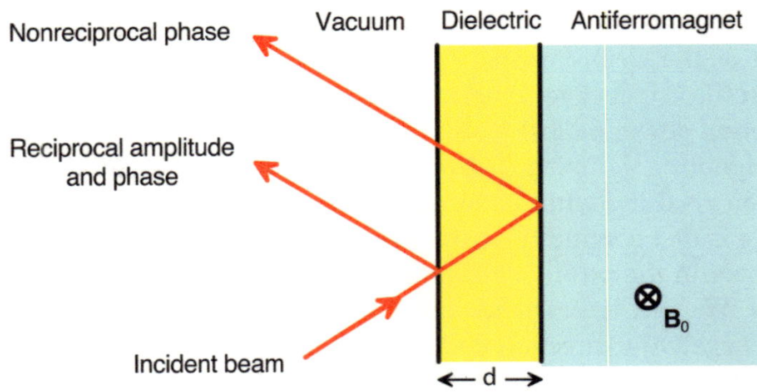

Fig. 19. Use of a dielectric layer for investigating nonreciprocal phase on reflection off an antiferromagnet.

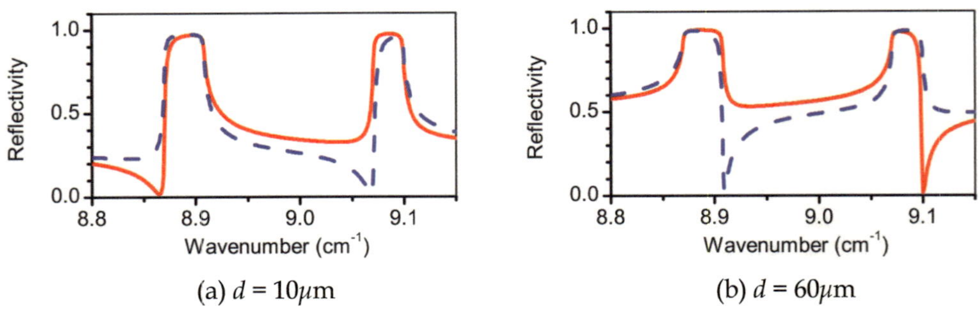

Fig. 20. Calculated oblique incidence reflectivity spectrum off a Si/MnF_2 structure of the type shown in Figure 19 in an external field of $B_0 = +0.1$ T (solid curves) and $B_0 = -0.1$ T (dashed curves).

overall reflectivity which depends on the dielectric layer thickness, as shown in Figure 20. This is true even when the reflectivity off the pure antiferromagnet is close to reciprocal, as is the case for MnF_2.

The discussion of nonreciprocity in the power flow is concerned with power flow behaviour within the interior of an antiferromagnet. Obviously it is not straightforward to measure this experimentally. It appears more reasonable to investigate the effect of this nonreciprocal power flow on the radiation interacting with a finite sized sample of a given shape. The analysis presented in this chapter does not extend to this type of system, since the antiferromagnet is considered to be infinite along x. However, other techniques such as the finite difference time domain (FDTD) method should help clarify the expected behaviour.

The lateral shift predicted in the case of reflection of a finite beam off an antiferromagnet should in principle be measurable given a suitable coherent source such as a far infrared laser (Rosenbluh et al., 1976), backward wave oscillator (Dobroiu et al., 2004), or YIG oscillator with frequency multiplied output (Kurtz et al., 2005). In order to observe the normal incidence shift, a beam splitting arrangement appears necessary. It is also, however, important to consider the effect at oblique incidence, both theoretically and experimentally. In this case the effect should be observable directly without the use of a beamsplitter.

In this chapter we have only discussed phenomena in the Voigt configuration with the external field aligned along the anisotropy axis, deliberately avoiding the more complicated configurations in which the external field makes an angle with the anisotropy field (Almeida

& Mills, 1988), or in which these axes are not perpendicular to the plane of incidence. However, we should point out that theoretical works on the reflected amplitude and phase do exist for more complex geometries (Stamps et al., 1991; Dumelow et al., 1998), and, in the case of reflectivity, there is some experimental work (Abraha et al., 1994; Brown et al., 1995). Finally, we stress that, although we have concentrated on reflection off antiferromagnets in this chapter, the basic priciples involved stem from the form of the permeability tensor given in Equation 8. However, there are other types of material, such as ferromagnets or ferrimagnets, that also have a gyromagnetic permeability of this form. We therefore expect similar phenomena for these materials, although some of the symmetry arguments have to be looked at in a slightly different way since, in general, such materials have their own internal macroscopic magnetic field. One can also have a dielectric tensor of this form, such as that associated with magnetoplasma excitations. In this case, p-polarisation radiation should give results similar to those presented here for s-polarisation reflection off antiferromagnets (Remer et al., 1984).

6. Acknowledgments

This work was partially financed by the Brazilian Research Agency CNPq (projects Universal 482238/2007-0, CT-ENERG 554889/2006-4, and CNPq-Rede NanoBioestruturas 555183/2005-0).

7. References

Abraha, K., Brown, D. E., Dumelow, T., Parker, T. J. & Tilley, D. R. (1994). Oblique incidence far-infrared reflectivity study of the uniaxial antiferromagnet fef$_2$, *Phys. Rev. B* 50(10): 6808–6816.

Abraha, K. & Tilley, D. R. (1996). Theory of far infrared properties of magnetic surfaces, films and superlattices, *Surf. Sci. Rep.* 24(5-6): 129-222.

Almeida, N. S. & Mills, D. L. (1988). Dynamical response of antiferromagnets in an oblique magnetic field: Application to surface magnons, *Phys. Rev. B* 37(7): 3400–3408.

Artmann, K. (1948). Berechnung der seitenversetzung des totalrelektierten strahles, *Ann. Physik* 437: 87–102.

Born, M. &Wolf, E. (1980). *Principles of Optics*, sixth edn, Pergamon Press, Oxford.

Brown, D. E., Dumelow, T., Parker, T. J., Abraha, K. & Tilley, D. R. (1994). Nonreciprocal reflection by magnons in FeF_2: a high resolution study, *Phys. Rev. B* 49(17): 12266–12269.

Brown, D. E., Dumelow, T., Parker, T. J., Abraha, K. & Tilley, D. R. (1995). Far infrared reflectivity off FeF_2, *J. Magn. Magn. Mater.* 140-144: 181–182.

Brown, D. E., Dumelow, T., Parker, T. J., Jensen, M. R. F. & Tilley, D. R. (1998). A high resolution fourier transform spectrometer for far infrared magneto-optic spectroscopy of magnetic materials, *Infrared Physics and Technology* 40: 219–230.

Camley, R. E. (1987). Nonreciprocal surface modes, *Surf. Sci. Rep.* 7(3-4): 103–188.

Dobroiu, A., Yamashita, M., Ohshima, Y. N., Morita, Y., Otani, C. & Kawase, K. (2004). Terahertz imaging system based on a backward-wave oscillator, *Appl. Opt.* 43(30): 5637- 5646.

Dötsch, H., Bahlmann, N., Zhuromskyy, O., Hammer, M., Wilkens, L., Gerhardt, R., Hertel, P. & Popkov, A. F. (2005). Applications of magneto-optical waveguides in integrated optics: review, *J. Opt. Soc. Am. B* 22(1): 240–253.

Dumelow, T. & Camley, R. E. (1996). Nonreciprocal reflection of infrared radiation from structures with antiferromagnets and dielectrics, *Phys. Rev. B* 54(17): 12232–12237.

Dumelow, T., Camley, R. E., Abraha, K. & Tilley, D. R. (1998). Nonreciprocal phase behavior in reflection of electromagnetic waves from magnetic materials, *Phys. Rev. B* 58(2): 897- 908.

Dumelow, T. & Oliveros, M. C. (1997). Continuum model of confined magnon polaritons in superlattices of antiferromagnets, *Phys. Rev.* B 55(2): 994-1005.

Goos, F. & Hänchen, H. (1947). Ein neuer und fundamentaler versuch zur totalreflexion, *Ann. Physik* 436(6): 333–346.

Horowitz, B. R. & Tamir, T. (1971). Lateral displacement of a light beamat a dielectric interface, *J. Opt. Soc. Am.* 61(5): 586–594.

Kurtz, D., Crowe, T., Hesier, J., Porterfield, D., Inc, V. & Charlottesville, V. (2005). Frequency domain terahertz spectroscopy, *Infrared and Millimeter Waves and 13th International Conference on Terahertz Electronics, 2005.*, Vol. 1, pp. 76–77.

Lai, H. M., Kwok, C. W., Loo, Y. W. & Xu, B. Y. (2000). Energy-flux pattern an the Goos-Hänchen effect, *Phys. Rev. E* 62(5): 7330–7339.

Landau, L. D. & Lifshitz, E. M. (1984). *Electrodynamics of Continuous Media*, second edn, Pergamon Press, Oxford.

Leung, P. T., Chen, Z. W. & Chiang, H.-P. (2007). Large negative Goos-Hänchen shift at metal surfaces, *Opt. Commun.* 276(2): 206–208.

Lima, F., Dumelow, T., Albuquerque, E. L. & da Costa, J. A. P. (2009). Power flow associated with the Goos-Hänchen shift of a normally incident electromagnetic beam reflected off an antiferromagnet, *Phys. Rev. B* 79(15): 155124.

Lima, F., Dumelow, T., da Costa, J. A. P. & Albuquerque, E. L. (2008). Lateral shift on normal incidence reflection off an antiferromagnet, *Europhys. Lett.* 83(1): 17003.

Lotsch, H. K. V. (1970). Beam displacement at total reflection: the Goos-Hänchen effect, *Optik* 32: 116,189,299,553.

McGuirk, M. & Carniglia, C. K. (1977). An angular spectrum approach to the Goos-Hänchen shift, *J. Opt. Soc. Am.* 67(1): 103–107.

Mills, D. L. & Burstein, E. (1974). Polaritons: the electromagnetic modes of media, *Rep. Prog. Phys.* 37(7): 817-926.

Potton, R. (2004). Reciprocity in optics, *Rep. Progr. Phys.* 67(5): 717–753.

Remer, L., Lüthi, B., Sauer, H., Geick, R. & Camley, R. E. (1986). Nonreciprocal optical reflection of the uniaxial antiferromagnet MnF_2, *Phys. Rev. Lett.* 56(25): 2752–2754.

Remer, L., Mohler, E., Grill, W. & Lüthi, B. (1984). Nonreciprocity in the optical reflection of magnetoplasmas, *Phys. Rev. B* 30(6): 3277-3282.

Renard, R. H. (1964). Total reflection: a new evaluation of the Goos-Hänchen shift, *J. Opt. Soc. Am.* 54(10): 1190–1197.

Rosenbluh, M., Temkin, R. J. & Button, K. J. (1976). Submillimeter laser wavelength tables, *Appl. Opt.* 15(11): 2635–2644.

Scott, R. Q. &Mills, D. L. (1977). Propagation of surface magnetostatic waves on ferromagnetic crystal structures, *Phys. Rev. B* 15(7): 3545–3557.

Stamps, R. L., Johnson, B. L. & Camley, R. E. (1991). Nonreciprocal reflection from semi-infinite antiferromagnets, *Phys. Rev. B* 43(4): 3626–3636.

Wild, W. J. & Giles, C. L. (1982). Goos-Hänchen shifts from absorbing media, *Phys. Rev. A* 25(4): 2099–2101.

9

Room Temperature Integrated Terahertz Emitters based on Three-Wave Mixing in Semiconductor Microcylinders

A. Taormina[1], A. Andronico[2], F. Ghiglieno[1], S. Ducci[1], I. Favero[1] and G. Leo[1]

[1]*Université Paris Diderot, Laboratoire MPQ – CNRS-UMR 7162,*
[2]*Institute for Genomics and Bioinformatics, University of California Irvine, CA 92697*
[1]*France*
[2]*USA*

1. Introduction

In the last years, the Terahertz (THz) domain has attracted an increasing interest in the scientific community due to the large number of applications that have been identified (Tanouchi, 2007).

Even if many different Terahertz sources - like photomixers, quantum cascade lasers, and photoconductive antennas (Mittleman, 2003) - have been investigated in the past, the fabrication of a compact device operating at room temperature and with an output power at least in the µW range still constitutes a challenge.

A very promising approach to this problem relies on the nonlinear optical process called Difference Frequency Generation (DFG) in materials like III-V semiconductors (Boyd, 2003).

In this chapter, we will propose an efficient, compact, and room-temperature THz emitter based on DFG in semiconductor microcylinders. These are whispering gallery mode (WGM) resonators capable to provide both strong spatial confinement and ultra-high quality factors. Nonlinear optics applications benefit from an ultra-high-Q cavity, since the fields involved in the nonlinear mixing interact for a long time, giving rise to an efficient conversion.

The structure we investigate is based on the technology of GaAs, owing to its wide transparency range (between about 0.9 and 17 µm), large refractive index for strong field confinement, and a huge nonlinear coefficient. Moreover, it offers attracting possibilities in terms of optoelectronic integration and electrical pumping.

After an introductory part about whispering gallery modes, we will present the study of the DFG inside GaAs microcylinders. The evanescent coupling with an external waveguide allows a selective excitation of the pump cavity modes.

At the end, on the theoretical premises of the first part, we will show that an appealingly simple structure can be used to confine both infrared and THz modes. Moreover, embedding self-assembled quantum dots in the cavity allows the integration of the pump sources into the device. With an appropriate choice of the cylinder radius, it is possible to phase match two WGMs with a THz mode, and have a compact, room-temperature THz emitter suitable for electrical pumping.

2. Whispering gallery modes

2.1 Microcylinder cavity modes

Whispering gallery modes are the optical modes of microcylinders, and, being the eigenmodes of a 3D structure, they cannot in general be derived analytically. However, the simple approximation we describe in the following (Heebner et al., 2007) can be used to reduce the 3D problem to a more manageable (2+1)D problem[1].

From Maxwell's equations in Fourier space and without source terms, we can easily obtain the familiar wave equation:

$$\nabla^2\vec{F} + \frac{n^2\omega^2}{c^2}\vec{F} = 0 \tag{1}$$

where $\vec{F}$ is either $\vec{E}$ or $\vec{H}$, and n is the refractive index (in general frequency dependent) of the medium.

Using the cylindrical coordinates (ρ, θ, x) shown in Fig. 1, equation (1) can be rewritten as:

$$\left(\frac{\partial^2}{\partial\rho^2} + \frac{1}{\rho}\frac{\partial}{\partial\rho} + \frac{1}{\rho^2}\frac{\partial^2}{\partial\rho^2} + \frac{\partial^2}{\partial x^2} + \frac{n^2\omega^2}{c^2}\right)\vec{F} = 0 \tag{2}$$

Let us assume that it is possible to classify the modes as purely TE or TM: in the first case, the non-vanishing components are H_x, E_ρ and E_θ, whereas, in the second, they are E_x, H_ρ and H_θ. This assumption, which echoes the optical slab waveguide case, is only approximate but greatly reduces the complexity of the problem: as we will see in the following, it is equivalent to decoupling the in-plane problem from the vertical problem, using the effective index method to take the latter into account (Tamir, 1990).

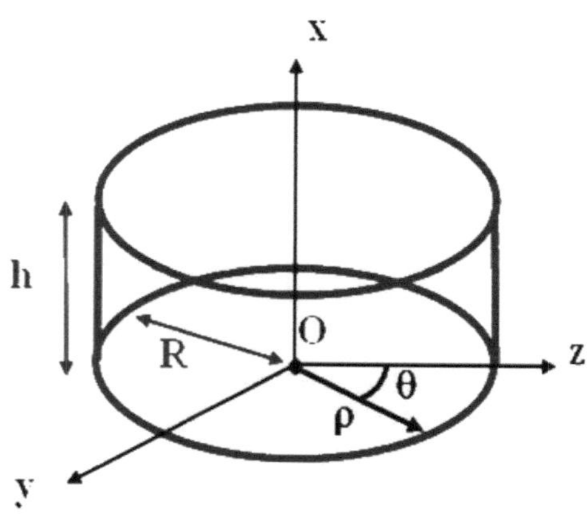

Fig. 1. General scheme of a microcylinder with radius R and thickness h. The cylindrical reference system used in the chapter is also shown.

Returning to Eq. (2) and writing the only independent field component F_x in the factorized form $F_x = \psi(\rho)\,\Theta(\theta)\,G(x)$, we find the following three equations:

---

[1] Recently, fully vectorial 3D approaches have also been proposed (Armaroli et al., 2008).

$$\begin{cases} \dfrac{d^2G}{dx^2} + \dfrac{\omega^2}{c^2}\left(n^2 - n_\xi^2\right)G = 0 \\ \dfrac{d^2\Theta}{d\theta^2} + m^2\Theta = 0 \\ \dfrac{d^2\Psi}{d\rho^2} + \dfrac{1}{\rho}\dfrac{d\Psi}{d\rho} + \left(\dfrac{n_\xi^2\omega^2}{c^2} - \dfrac{m^2}{\rho^2}\right)\Psi = 0 \end{cases} \tag{3}$$

The first equation tells us that G(x) is the eigenfunction of a slab waveguide with effective index n_ξ (ξ = TE/TM), whereas the second can be integrated to obtain $\Theta(\theta) = e^{-im\theta}$, m being the (integer) azimuthal number.

The radial mode dependence is obtained using the last equation in (3): if the microdisk radius is R, then $\psi(\rho)$ can be written in terms of first-kind Bessel functions (for $\rho \le R$) and second-kind Hankel functions (for $\rho > R$):

$$\Psi(\rho) = \begin{cases} NJ_m(kn_\xi\rho) & \rho \le R \\ NBH_m^{(2)}(k\rho) & \rho > R \end{cases} \tag{4}$$

where N is a normalization constant, $k = \omega/c$, $B = J_m(kn_\xi R)/H_m^{(2)}(kR)$, and we assume that the microcylinder is surrounded by air.

If we impose the continuity of tangential components $\vec{E}$ and $\vec{H}$, we find the following dispersion relations:

TM modes
$$n_\xi \frac{\dot{J}_m(kn_\xi R)}{J_m(kn_\xi R)} - \frac{\dot{H}_m^{(2)}(kR)}{H_m^{(2)}(kR)} = 0$$

TE modes
$$\frac{\dot{J}_m(kn_\xi R)}{J_m(kn_\xi R)} - n_\xi \frac{\dot{H}_m^{(2)}(kR)}{H_m^{(2)}(kR)} = 0 \tag{5}$$

Once these equations are numerically solved with respect to the variable k, we obtain the resonance eigenfrequencies $\tilde{\omega}$ of the cavity.

At this point it is worth stressing that, despite the formal analogy with the optical slab waveguide, the frequency of a WGM is a complex number, even if the effective index n_ξ that appears in equation (3) is real. This is due to the fact that the microcylinder walls are curved and then all its resonances are affected by radiation losses, which can be quantified by defining the WGM quality factor of a resonator mode:

$$Q^{WGM} = \frac{\mathrm{Re}(\tilde{\omega})}{2\mathrm{Im}(\tilde{\omega})} \tag{6}$$

Simply stated, the bent geometry of the microdisk gives rise to a continuous decay rate of the energy confined within the cavity, broadening the resonances linewidth.

Fig. 2 shows the square modulus of equations (5) versus the angular frequency for a microcylinder of radius R = 1 μm and effective index n = 2.2: it is evident that once we establish the structure under investigation (i.e. the disk radius and thickness, and subsequently the effective index n_ξ) and the azimuthal number m, multiple radial solutions exist. We can then label them by employing an additional integer number p, which is the radial order of the mode and corresponds to the number of field maxima along the radial axis of the microcylinder.

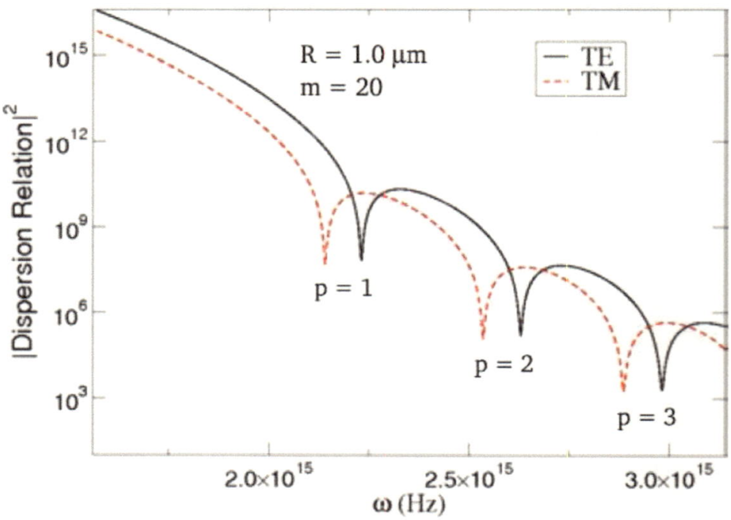

Fig. 2. Square modulus of the dispersion relations (5) versus angular frequency. The azimuthal symmetry of the modes is fixed (m=20): different function dips correspond to different radial order modes, as indicated.

It is interesting to note that higher p order modes have higher frequencies, as is shown in Fig. 2. This can be intuitively understood in terms of the geometrical picture of a WGM: a WGM is a mode confined in a microdisk by total internal reflections occurring at the dielectric/air interface and that, additionally, satisfies the round trip condition.
The resonance frequencies of the modes with p = 1 are then:

$$2\pi R = m\frac{\lambda}{n} = m\frac{2\pi c}{n\omega} \tag{7}$$

High p modes have their "center of mass" displaced towards the microdisk center, so that, for these modes, we can always use equation (7) but with a smaller "effective radius" R. As equation (7) suggests, once m is fixed, this results in a higher mode frequency.
If the resonance frequencies are known, expression (4) allows to obtain the radial function $\psi(\rho)$ for TM or TE modes. At this point, we can write the independent field component E_x or H_x, since the functions $\Theta(\theta)$ and $G(x)$ are already known.
Once E_x or H_x is found, the other field components can be directly obtained by using Maxwell's equations:

TM modes
$$\begin{cases} H_\rho = \dfrac{m}{\mu_0 \rho \tilde{\omega}} E_x \\ H_\theta = -\dfrac{j}{\mu_0 \tilde{\omega}} \dfrac{\partial E_x}{\partial \rho} \end{cases} \tag{8}$$

TE modes
$$\begin{cases} E_\rho = -\dfrac{m}{\varepsilon_0 \varepsilon \rho \tilde{\omega}} H_x \\ E_\theta = \dfrac{j}{\varepsilon_0 \varepsilon \tilde{\omega}} \dfrac{\partial H_x}{\partial \rho} \end{cases} \tag{9}$$

If the vertical part fulfills the condition

$$\int_{-\infty}^{+\infty} G^2(x) dx = 1 \tag{10}$$

then the constant N in equation (4) can be chosen in order to normalize the mode to the azimuthal power flow:

TM modes / TE modes
$$P_\theta = \begin{cases} +\dfrac{1}{2} \mathrm{Re}\left(\displaystyle\int_0^\infty E_x H_\rho^* d\rho \right) \\ -\dfrac{1}{2} \mathrm{Re}\left(\displaystyle\int_0^\infty E_\rho H_x^* d\rho \right) \end{cases} \tag{11}$$

2.2 Quality factor

The Q-factor of a resonance physically represents the number of optical cycles needed before its original energy decays by 1/e in the absence of further sourcing; this means that if U is the energy stored in the cavity, then we have:

$$\frac{dU}{dt} = -\frac{\omega}{Q} U \tag{12}$$

Since the term -dU/dt represents the dissipated power P_d, we find an alternative definition of Q:

$$Q = \omega \frac{U}{P_d} \tag{13}$$

On the other hand, Q can be written in the following form (Srinivasan, 2006):

$$Q = \omega \tau_{ph} = \frac{2\pi\, n_g L_{ph}}{\lambda} \tag{14}$$

where τ_{ph} is the photon lifetime, L_{ph} is the cavity decay length and n_g is the group index within the cavity.

Equation (14) is a useful relation because it allows to compare the losses of a microcylinder with those of other devices (e.g. a planar waveguide): in fact, for a planar waveguide it is customary to write the losses in terms of an inverse decay length α (in cm^{-1}). Once we know the resonance quality factor, we can use this equation to obtain L_{ph} and then express the losses in the form $\alpha = 1/L_{ph}$.

Until now, the only loss mechanism introduced for the microcylinder resonances was represented by the intrinsic radiation losses responsible for the finite value of Q_{WGM}. In

physical experiments, the situation is slightly more complex, and additional losses affect the overall Q-factor of a WGM.
Under the hypothesis that all loss factors are so small that their effects on the intra-cavity field can be treated independently, the overall quality factor can be written in the following form:

$$\frac{1}{Q}=\frac{1}{Q^{WGM}}+\frac{1}{Q^{mat}}+\frac{1}{Q^{cpl}}=\frac{1}{Q^{int}}+\frac{1}{Q^{cpl}} \tag{15}$$

Q^{cpl} represents the losses due to an eventual external coupling (see section 3), and Q^{mat} quantifies the losses due to bulk absorption. In the linear regime, this can be the case of free-carrier absorption, whereas, in the nonlinear regime, this term could include two-photon (or, in general, multi-photon) absorption. In the latter case, Q^{mat} will then depend on the field intensity circulating inside the cavity.
Both Q^{WGM} and Q^{mat} are intrinsic terms, whereas the last part of equation (15) describes the external coupling. In the next section, we will use the coupled mode theory for a thorough study of the evanescent coupling of a microcylinder and a bus waveguide or fiber; for the moment, the discussion is limited to a qualitative picture. Looking at Fig. 3, we can imagine to inject a given power into the fundamental mode of a single-mode waveguide sidecoupled to the microcylinder. In the region where the two structures almost meet, the exponential tail of the waveguide mode overlaps the WGM giving rise to an evanescent coupling.

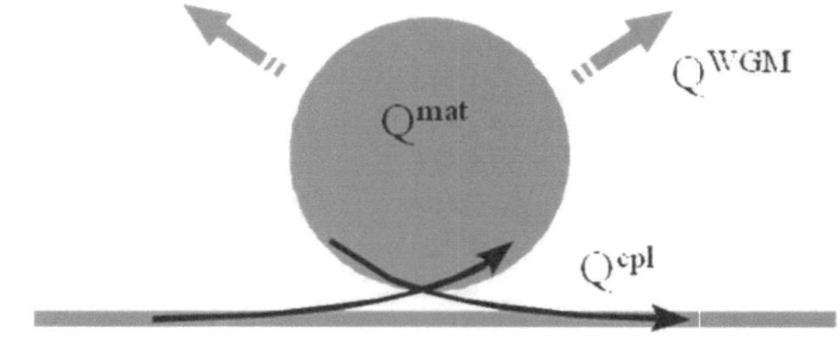

Fig. 3. Evanescent coupling scheme with a bus waveguide.

A final remark concerns the fact that the intrinsic quality factor Q^{int} can be reduced by additional contributions, e.g. the surface loss terms caused by surface scattering and surface absorption (Borselli et al., 2005). For this reason, we will denote with Q^{rad} (and not Q^{WGM}) the radiation losses.
Surface losses cannot always be neglected and become dominant in particular situations; moreover, they give rise to important phenomena like the lift of degeneracy for standingwave WGMs.

3. Three-wave mixing in semiconductor microcylinders

Microcavities are very promising for nonlinear optics applications, thanks to the high optical quality factors attainable with today's technology. For example, the group of J. D. Joannopoulos at MIT proved that high quality photonic crystal resonators can be very effective in obtaining low-power optical bistable switching (Soljačić et al., 2002), Second-Harmonic Generation (SHG), and in modifying the bulk nonlinear susceptibility through the Purcell effect (Soljačić et al., 2004; Bravo-Abad et al., 2007).

For nonlinear optics applications, the advantage of having a high-Q resonator is that its modes are stored in the cavity for many optical periods: this provides a considerable interaction time between modes and can be used to enhance parametric interactions.
WGM resonators are particularly well suited to attain high Q: for example, quality factors as high as $Q = 5 \times 10^6$ and $Q = 3.6 \times 10^5$ have been reported, at telecom wavelengths, for Si (Borselli et al., 2005) and AlGaAs (Srinivasan et al., 2005) microdisks, respectively.
In a DFG process, two pumps of frequencies ω_1 and ω_2 interact in order to generate a signal at the frequency difference $\omega_3 = \omega_1 - \omega_2$: in this way, energy conservation is ensured at photon level.
In this context, the exploitation of GaAs offers peculiar advantages with respect to other materials. Apart from having a wide transparency range, large refractive index, and a huge nonlinear coefficient, GaAs has in fact highly mature growth and fabrication technologies, and offers attracting possibilities in terms of optoelectronic integration and electrical pumping. On the other hand, due to its optical isotropy, GaAs-based nonlinear applications normally require technologically demanding phase-matching schemes (Levi et al., 2002). These are not necessary in the case of WGM resonators since, as theoretically demonstrated for a second harmonic generation process (Dumeige & Feron, 2006; Yang et al., 2007), the symmetry of a [100]-grown AlGaAs microdisk and the circular geometry of the cavity result in a periodic modulation of the effective nonlinear coefficient experienced by the interacting WGMs. This modulation can then be used to phase-match the pump and the generated fields without additional requirements.
The evanescent coupling between a semiconductor microcylinder and a waveguide is a way to excite two pump WGMs inside the microcavity. This technique has already been adopted in our laboratory for the characterization of GaAs microdisks.
In Fig. 4 we report the top view of a cylindrical cavity of radius R side-coupled to a bus waveguide used to inject two pump fields at ω_1 and ω_2. The intracavity generated field could be extracted by using a second waveguide, and the waveguide/microcavity distances can be chosen to optimize the injection/extraction efficiency.
The difference frequency generation in a triply resonant microcylinder can be described using the standard coupled mode theory.
The set of coupled mode theory equations describing this nonlinear process is (Haus, 1984):

$$\begin{cases} \dfrac{da_1}{dt} = j\omega_1 a_1 - \dfrac{a_1}{\tau_1^{tot}} + j\sqrt{\dfrac{2}{\tau_1^{tot}}}s_1 - s_1^{NL} \\ \dfrac{da_2}{dt} = j\omega_2 a_2 - \dfrac{a_2}{\tau_2^{tot}} + j\sqrt{\dfrac{2}{\tau_2^{tot}}}s_2 - s_2^{NL} \\ \dfrac{da_3}{dt} = j\omega_3 a_3 - \dfrac{a_3}{\tau_3^{tot}} - s_3^{NL} \end{cases} \quad (16)$$

For the i-th resonant mode ($i = 1, 2, 3$), a_i is the mode amplitude normalized to its energy, $\tau_i^{tot} = 2Q_i^{tot}/\omega_i$ is the total photon lifetime (including intrinsic and coupling losses). The terms s_i describe the external pumping, with $|s_i|^2 = P_i^{in}$ (P_i^{in} being the input power in the bus waveguide).
The third equation is slightly different since the WGM field at ω_3, which is generated inside the cavity, is not injected from the outside: its source is then constituted by the nonlinear

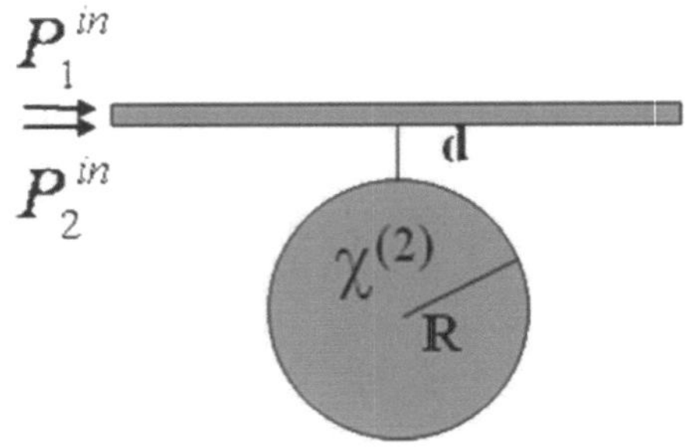

Fig. 4. Top view of a microcylinder coupled to an input waveguide.

term s_3^{NL}. For typical values like the ones we will see in the following, the pump depletion can be ignored, i.e. we can neglect the terms s_i^{NL} with i = 1,2.
In this way, putting $a_i = A_i e^{j\omega_i t}$ and looking for the steady state solution of the two pumps, we find:

$$|A_i|^2 = \frac{4}{\omega_i} \frac{Q_i^{cpl}}{\left(1 + Q_i^{cpl}/Q_i^{int}\right)^2} P_i^{in} \tag{17}$$

Where Q_i^{cpl} is the loss term due to the presence of the coupling to the waveguide, and Q_i^{int} the intrinsic quality factor, with $1/Q^{tot} = 1/Q^{int} + 1/Q^{cpl}$.
Equation (17) suggests that the power transfer from the waveguide to the cavity can be adjusted by changing the coupling losses, i.e. by properly varying the distance between waveguide and microcylinder and/or reducing the width of the waveguide. This transfer is maximized under critical coupling ($Q_i^{cpl} = Q_i^{int}$).
The power fed into the mode at ω_3 is:

$$P_3 = -\frac{j\omega_3}{4} \int \vec{E}^*(\omega_3) \cdot \vec{P}^{NL}(\omega_3) dV + c.c. \tag{18}$$

where c.c. denotes the complex conjugate, and $\vec{P}^{NL}$ is the nonlinear polarization given by:

$$P_i^{NL}(\omega_3) = \varepsilon_0 \sum_{jk} \chi_{ijk}^{(2)} E_j(\omega_1) E_k^*(\omega_2) \tag{19}$$

By using equations (18) and (19) we can rewrite s_3^{NL} in the form:

$$s_3^{NL} = -\frac{j\omega_3}{4} a_1 a_2^* I_{ov} \tag{20}$$

where I_{ov} is the nonlinear overlap integral between the WGMs:

$$I_{ov} = \varepsilon_0 \int_V \sum_{ijk} \chi_{ijk}^{(2)} E_i^*(\omega_3) E_j(\omega_1) E_k^*(\omega_2) dV \tag{21}$$

with V the cavity volume and $\chi^{(2)}$ the nonlinear tensor.
The $\bar{4}3m$ GaAs symmetry (Palik, 1999) and the growth axis in the [100] direction imply that the overlap integral differs from zero only when two of the three WGMs are TE polarized

and one is TM polarized. Moreover, the angular part of the integral in equation (20) can be readily calculated, resulting in the phase-matching condition $\Delta m = m_2 + m_3 - m_1 \pm 2 = 0$.
The $\pm$ 2 is due to the additional momentum provided by the periodic modulation of the $\chi^{(2)}$ coefficient that comes from the circular geometry of the cavity.
Looking for the steady state solution of the field at ω_3, and taking into account equation (17), we then find:

$$|A_3|^2 = \frac{(\omega_3/4)^2}{(1/\tau_3^{tot})^2} \prod_{i=1}^{2} \left[\frac{4}{\omega_i} \frac{Q_i^{cpl}}{(1 + Q_i^{cpl}/Q_i^{int})^2} \right] |I_{ov}|^2 P_1^{in} P_2^{in} \tag{22}$$

Therefore, if the difference-frequency mode is extracted with an additional waveguide, and under the hypothesis that the critical coupling condition is fulfilled for the three WGMs, the generated power is:

$$P_3^{cpl} = \frac{2|A_3|^2}{\tau_3^{cpl}} = \frac{1}{16} \frac{\omega_3}{\omega_1 \omega_2} Q_1^{int} Q_2^{int} Q_3^{int} |I_{ov}|^2 P_1^{in} P_2^{in} \tag{23}$$

On the other hand, if the intracavity-generated field is not coupled to any waveguide (it is simply radiated) and under the hypothesis of critical coupling for the two pumps, we find:

$$P_3^{rad} = \frac{2|A_3|^2}{\tau_3^{rad}} = \frac{1}{4} \frac{\omega_3}{\omega_1 \omega_2} \frac{1}{Q_3^{rad} (1/Q_3^{rad} + 1/Q_3^{mat})^2} Q_1^{int} Q_2^{int} |I_{ov}|^2 P_1^{in} P_2^{in} \tag{24}$$

We see that, in both cases, the non linear efficiency is directly related to the overlap between the three interacting fields and it is enhanced proportionally to the time the mode spend inside the resonator: higher Q-factors result in a longer interaction time between the fields in the nonlinear mixing.

4. Nonlinear GaAs Microcylinder for Terahertz Generation

4.1 Introduction

In the field of Terahertz spectroscopy there is a clear distinction between the broad-band time-domain spectroscopy (TDS) and the single-frequency (CW) spectroscopy. In TDS, the THz source is often a photoconductive dipole antenna excited by femtosecond lasers: by Fourier transforming the incident and transmitted optical pulses, it is possible to obtain the dispersion and absorption properties of the sample under investigation. This technique has proven powerful to study the far-infrared properties of various components, like dielectrics and semiconductors (Grischkowsky et al., 1990) or gases (Harde & Grischkowsky, 1991), and for imaging (Hu & Nuss, 1995).
However, besides requiring costly and often voluminous mode-locked lasers, the principal drawback of the THz TDS relies on its limited frequency resolution ($\Delta\nu \sim 5$ GHz) resulting from the limited time window ($\Delta t \sim 100$ ps), (Sakai, 2005).
On the other hand, narrow-band THz systems have found many applications in atmospheric and astronomical spectroscopy, where a high spectral resolution (1−100 MHz) is generally required (Siegel, 2002).
Among the large number of proposed CW-THz source schemes, it is worth mentioning at least two. The first one, known as photo-mixing, makes use of semi-insulating or

lowtemperature grown GaAs (Sakai, 2005). However, no significant progress in terms of output power has been demonstrated in CW photoconductive generation during the last few years, and the maximum output powers are in the 100 nW range.
The second CW scheme is the Quantum Cascade Laser (QCL) (Faist et al., 1994): in this case, the photons are emitted by electron relaxations between quantum well sub-bands. The original operating wavelength was $\lambda = 4.2$ μm and was extended in the THz region (Kohler et al., 2002). However, the main drawback of this kind of sources is that they are poorly tunable and only operate at cryogenic temperatures.
An alternative and interesting approach for the generation and amplification of new frequencies, both pulsed and CW, is based on second-order nonlinear processes: in this case, the first THz generation from ultrashort near-infrared pulses was demonstrated in bulk nonlinear crystals such as ZnSe and $LiNbO_3$ (Yajima & Takeuchi, 1970).
In 2006 Vodopyanov et al. demonstrated the generation of 0.9 to 3 THz radiation in periodically inverted GaAs, with optical to THz conversion efficiencies of 10^{-3} (Vodopyanov, 2006). With respect to terahertz generation in $LiNbO_3$ (Kawase et al., 2002), GaAs constitutes a privileged material choice, thanks to its large nonlinearity and inherently low losses at THz frequencies (~ 1 cm^{-1}). However, the periodically inversed GaAs sources are neither compact nor easy to use outside research laboratories, since they require bulky mode-locked pump sources. To avoid this technological complexity, it has been proposed to exploit the anomalous dispersion created by the phonon absorption band in GaAs to phase match a difference-frequency generation in the terahertz range (Berger & Sirtori, 2004).
In 2008 Vodopyanov and Avetisyan reported generation of terahertz radiation in a planar waveguide: using an optical parametric oscillator operating near 2 μm (with average powers of 250 and 750 mW for pump and idler), the THz output was centered near 2 THz and had 1 μW of average power (Vodopyanov & Avetisyan, 2008).
In the same year, Marandi et al. proposed a novel source of continuous-wave terahertz radiation based on difference frequency generation in GaAs crystal. This source is an integration of a dielectric slab and a metallic slit waveguide. They predicted an output power of 10.4 μW at 2 THz when the input infrared pumps have a power of 500 mW (Marandi et al., 2008).
In this section, we will present a CW, room-temperature THz source based on DFG from two near-IR WGMs in a high-quality-factor GaAs microcylinder: these pump modes are excited by the emission of quantum dots (QDs) embedded in the resonator.
The cavity, as sketched in Fig. 5, is a cylinder composed of a central GaAs layer sandwiched between two lower-index AlAs layers, capped on both sides by a metallic film (e.g. Au). This configuration provides both vertical dielectric confinement for the near-IR pump modes and plasmonic confinement for the THz mode. The design stems from two opposite

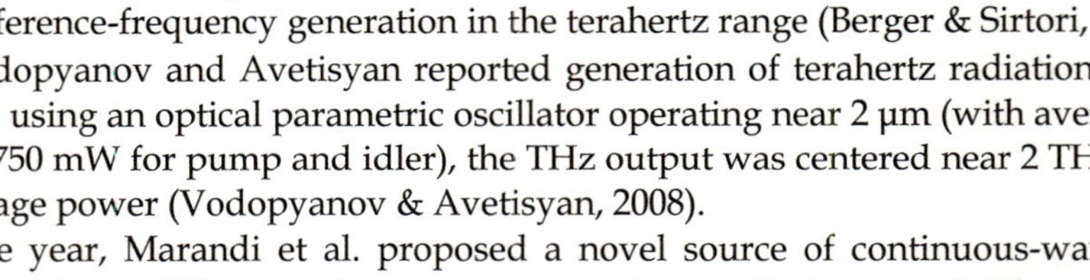

Fig. 5. Sketch of a GaAs/AlAs microcylinder.

requirements on the thickness of AlAs layers, aimed at increasing the DFG efficiency: maximize the overlap between the interacting modes, and prevent the exponential tails of the near-IR modes from reaching the metallic layers, thus avoiding detrimental absorption losses.

Fig. 6 shows an example of the pump and THz mode profile.

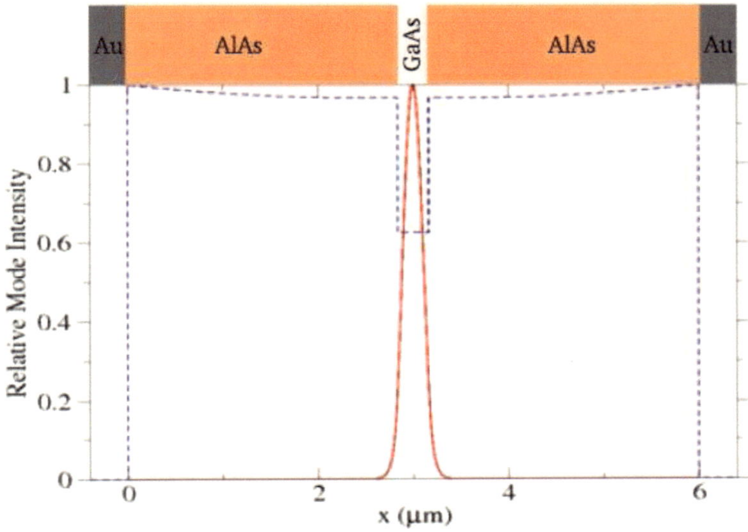

Fig. 6. Example of the vertical near-IR (solid line) and THz (dashed line) mode profiles. The wavelength are $\lambda = 0.9$ µm e $\lambda = 70.0$ µm for the IR e THz mode respectively.

The double metal cap allows to strongly confine the THz mode: with respect to a structure with just a top metallic mirror, where the THz mode would leak into the substrate, this allows to increase the overlap between the WGMs, thus improving the conversion efficiency.

In the horizontal plane, the light is guided by the bent dielectric/air interface, which gives rise to high-Q WGMs (Nowicki-Bringuier et al., 2007). The central GaAs layer contains one or more layers of self-assembled InAs quantum dots, which excite the two near-IR modes, and can be pumped either optically or electrically. The simultaneous lasing of these modes, without mode competition, can be obtained thanks to the inhomogeneously broadened gain curve of the QD ensemble, as observed for QDs in microdisks at temperatures as high as 300K (Srinivasan, 2005), and in microcylinders (Nowicki-Bringuier et al., 2007).

Fig. 7 shows the micro-photoluminescence (µPL) spectra of a 4 µm diameter pillar containing QDs reported in (Nowicki-Bringuier et al., 2007). The number next to each peak corresponds to the azimuthal number of a TE WGM excited by the QD ensemble emission. The figure also shows that increasing the pillar diameter results in a reduced free spectral range: if the structure diameter is big enough, it is possible to find two WGM whose frequency difference lies in the THz range.

In order to find the WGM spectrum of the cavity shown in Fig. 5, we can use the effective index method described in the previous sections: as demonstrated in (Nowicki-Bringuier et al., 2007), this approach gives an excellent approximation for micropillar WGMs.

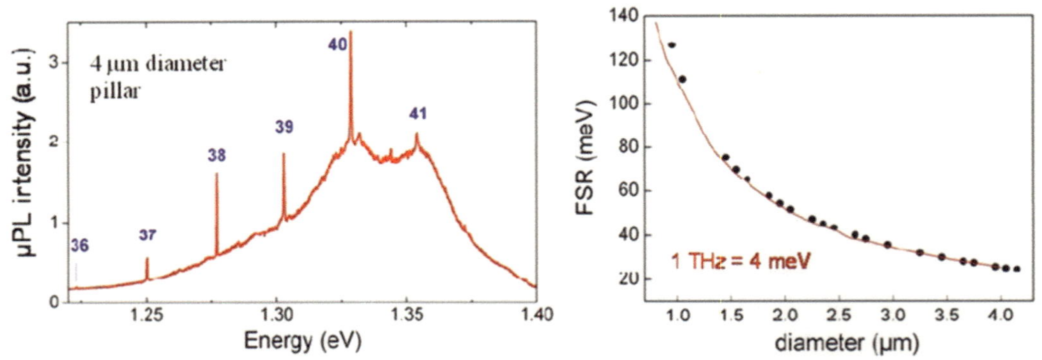

Fig. 7. Left: experimental µPL spectra measured at 4K on a 4 µm diameter pillar. Right: calculated (solid line) and observed (filled points) free spectral range versus diameter (Nowicki-Bringuier et al., 2007).

Applying the coupled mode theory to the present case, we obtain the following equation for the THz mode amplitude a_3:

$$\frac{da_3}{dt} = j\omega_3 a_3 - \left(\frac{1}{\tau_3^{rad}} + \frac{1}{\tau_3^{mat}}\right) a_3 + s_3^{NL} \qquad (25)$$

where τ_3^{rad} (τ_3^{mat}) represents the radiation (material absorption) limited photon lifetime. Again, the s_3^{NL} term represents the nonlinear polarization source, and it is given by (20).
As mentioned before, in order to generate the third mode, we have to fulfill two conditions:

1. two of the three WGMs must be TE polarized and one TM polarized;
2. the phasematching condition $\Delta m = m_2 + m_3 - m_1 \pm 2 = 0$ must hold.

If A_3 is the steady state solution of (25), the radiated THz power is:

$$P_3^{rad} = \frac{2|A_3|^2}{\tau_3^{rad}} = \frac{\omega_3}{4} \frac{Q_3^{rad}}{\left(1+Q_3^{rad}/Q_3^{mat}\right)^2} U_1 U_2 |I_{ov}|^2 \qquad (26)$$

with U_1 and U_2 the electromagnetic energy stored in the two pump WGMs and I_{ov} the nonlinear overlap integral given by (21).
This shows that the emitted THz power is proportional to the energy of the pump modes, and it can be increased by maximizing the overlap integral between the interacting WGMs.
As a final remark, we stress that the quality factor of the THz mode is mainly limited by intrinsic (radiation and material) losses. Conversely, intrinsic losses are extremely small for near-IR WGMs; therefore these modes will display experimentally quality factors that are limited by extrinsic losses, such as scattering by sidewall roughness (Srinivasan et al., 2005; Nowicki-Bringuier et al., 2007).

4.2 Numerical results

By numerically studying (Andronico et al., 2008) the structure of Fig. 5 with w = 0.325 µm and h = 6 µm, we find that a radius R = 40.6 µm allows to phase-match two pumps near 1 µm (λ_1 = 0.923 µm and λ_2 = 0.936 µm) and a THz WGM with λ_3 = 63.385 µm (i.e. ν_3 = 4.8 THz). The corresponding azimuthal numbers are m_1 = 917, m_2 = 913 and m_3 = 2. For the two pump modes, we took AlGaAs dispersion into account according to the Gehrsitzs model (Gehrsitzs et al., 2000).

Since the dipole of the fundamental transition in the InAs QDs is oriented in the microcylinder plane (Cortez et al., 2001), the only WGMs excited by the QDs are TE polarized. The THz WGM has then to be a TM mode.

Moreover, unlike quantum wells, the gain curve of QD ensembles is mostly broadened due to QD size fluctuations (inhomogeneous broadening). For InAs QDs in GaAs, the latter is 60-100 meV, and is centered around 1.3 eV (λ = 0.95 μm), (Nowicki-Bringuier et al., 2007). Such inhomogeneous broadening is thus much larger than the homogeneous broadening (10 meV at room-temperature (Cortez et al., 2001)): this allows to have different WGMs simultaneously lasing, with no mode competition (Siegman, 1986).

Under the hypothesis of Q = 10^5 for the two pump modes for AlGaAs microdisks with embedded QDs, we can make important statements for our source: 1) its estimated phasematching width, dictated by the finesse of the near-IR WGMs, is 3 GHz; 2) under the conservative assumption of extracting 1 mW (corresponding to a circulating power of 16 W) from each of the pump modes, the emitted THz power, calculated from equation (25), is expected to be about 1 μW.

It is also interesting to observe that, at these pump powers, two-photon absorption does not affect the performance of our device and can be safely neglected in the calculations.

Fig. 8 shows the far-field pattern of the source at room temperature obtained with a semianalytic method developed following (Heebner et al., 2007). The emission is concentrated at high angles, due to the strong diffraction experienced by the tightly confined THz mode.

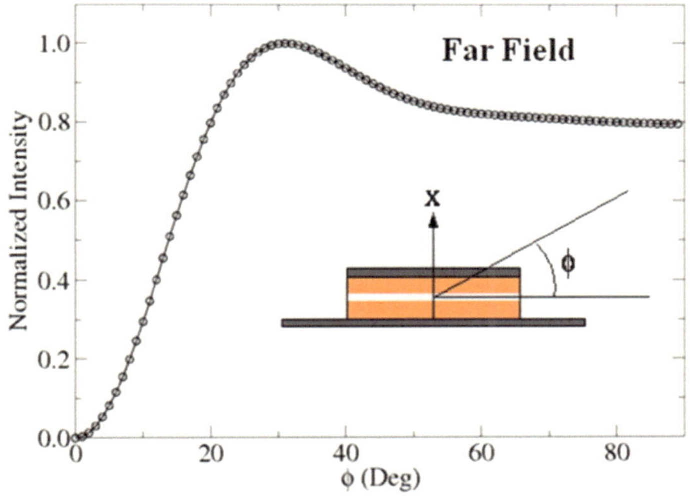

Fig. 8. Far Field pattern of the THz microcylinder source at room temperature, emitting at λ_3 = 63.4 μm. The inset shows the coordinate system used.

In Fig. 9 we report the effect of radius fabrication tolerance on the generated THz frequency, for three different temperatures: the slight THz frequency shift resulting from non-nominal fabrication is comparable to the phase-matching spectral width, and it is therefore negligible. Once the temperature has been chosen, each point in Fig. 14 corresponds to a phase-matched triplet with fixed azimuthal and radial numbers on each curve (different for each temperature).

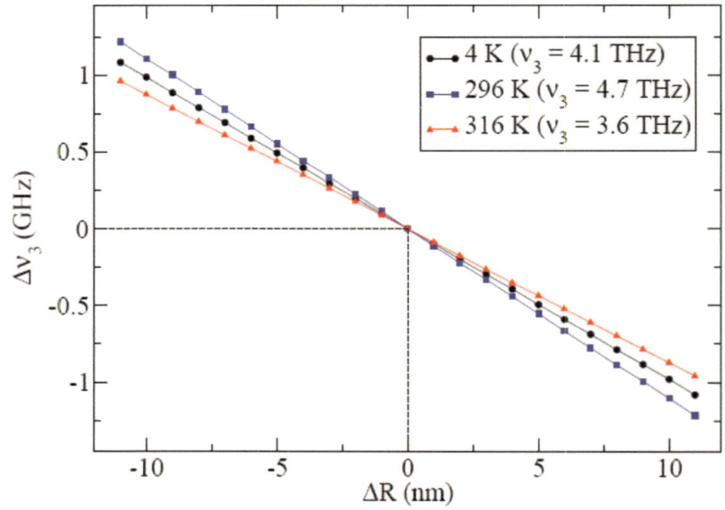

Fig. 9. Frequency deviation from nominal case ν_3 versus radius tolerance, for three different temperatures: 4K (circles), 296K (squares) and 316K (triangles).

A final remark concerns the wavelength range covered by this device: there are in fact two independent factors that contribute to it. As previously stated, at 300K, the homogeneous broadening of QDs is of the order of 10meV (Borri et al., 2001), which restricts the THz generation to frequencies $\nu_3 > 2.4$ THz. Emission at lower frequencies can be obtained by reducing the QDs homogeneous broadening, at the expense of a low-temperature operation. Conversely, the upper limit is set by the GaAs Rest-Strahlen band, i.e. $\nu_3 < 6$ THz.

In conclusion, this THz source is based on intracavity three-wave mixing between WGMs. As compared to the other THz sources today available, it could have noteworthy characteristics, such as: 1) room-temperature operation; 2) relatively high output power; 3) compactness; and 4) fabrication simplicity.

5. Perspectives

Being based on a cylindrical geometry, our THz source results in a compact and solid device, which provides a natural and elegant solution to combine the vertical confinement of the near-IR WGMs and THz mode. The two mirrors optimize the spatial overlap between the THz and the near-infrared modes, thus providing efficient conversion.

The presence of the QDs as active medium also allows, as an interesting perspective, to electrically pump the two near-IR modes. This can be accomplished by modifying the structure as shown in Fig. 10: the p-doped top and the n-doped bottom AlGaAs slabs allow the current flow. Besides their use as mirrors, the metallic layers are then exploited as external electrical contacts of the structure. Moreover, in order to selectively inject the current in the outer part of the cylinder where the near-infrared WGMs are located, a highly resistive region is defined by ion implantation in the central part of the cavity.

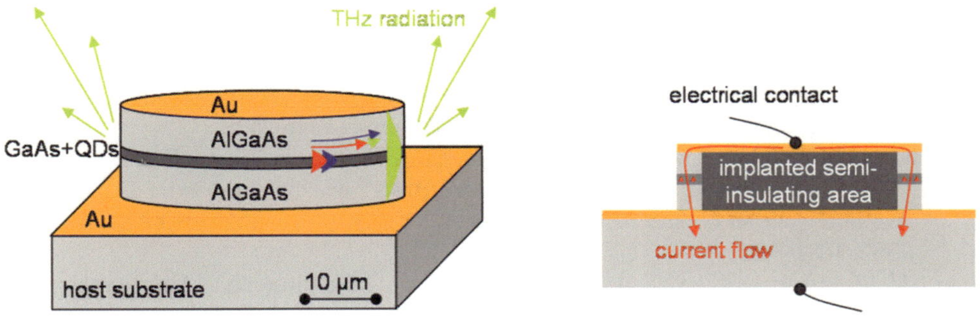

Fig. 10. Side view and cross section of microcylinder for electrical injection.

Compared to the standard air-suspended microdisk, the microcylinder geometry comes with a few crucial advantages, and it is well suited to achieve the high power lasing necessary to increase the efficiency of the nonlinear conversion process. Indeed, the heat sinking is significantly better than in air-suspended microdisks, where heating degrades the lasing properties (mode shifts, power saturation).

Moreover, although electrical pumping of microdisk lasers has been reported (Fujita et al. 2000), the lasing performances are strongly limited by the fact that the current is injected through the disk pedestal near the center of the microdisk. In the microcylinder geometry, the current can be selectively injected at the circumference of the GaAs active layer, which sustains the WGM modes.

Up to now, the lasing of WGMs in micropillars has been demonstrated only at low temperature and under optical pumping. Structures enabling THz confinement and a selective electrical injection at the edges of the microcylinder still need some technological progress, in order to minimize the threshold current, and optimize the lasing properties of the WGM modes under high injection conditions.

Although Q-factors in the 10^4 range have already been observed in similar structures (Nowicki-Bringuier et al., 2007), further improvement would ameliorate the lasing properties of the microcylinder, and increase the intensity of the near-IR light that is stored in the cavity.

To conclude, it is worth stressing that our source offers other interesting possibilities, as more long term developments: (a) the use of other semiconductors like the widely exploited InGaAsP/InP system; (b) a microcylinder with a single top mirror: to vertically confine the THz in the microcylinder a doped layer could be used below the AlAs layer, replacing the bottom mirror. This configuration is technologically simpler to realize; however, the THz mode would inevitably leak into the substrate, and we also expect a reduced efficiency due to a reduced overlap integral; (c) the incorporation of a quantum cascade structure: quantum cascade devices can simultaneously laser on two frequencies, without mode competition (Franz et al., 2007). This dual emission could then provide the two pumps needed in the DFG. Nevertheless, since the intersubband selection rules only allow TM polarized light, we could not use the [100] direction as growth axis, but we could replace it with the [111] direction (Berger & Sirtori, 2004); (d) a multi-spectral emission with lithographically defined arrays of microcylinders; (e) the fabrication of phased-array architectures by adjusting the distance between the microcylinders: this would allow to control the far-field pattern of the device; (f) optical pumping the device by approaching a fiber.

6. References

Andronico, A.; Claudon, J.; Gerard, J. M.; Berger, V. & Leo, G. (2008). Integrated Terahertz Source based on Three-Wave Mixing of Whispering-Gallery Modes. *Optics letters*, Vol. 33, No. 21, (October 2008) pp. 2416-2418, ISSN 0146-9592

Armaroli, A.; Morand, A.; Benech, P.; Bellanca, G. & Trillo, S. (2008). Three-Dimensional Analysis of Cylindrical Microresonators Based on the Aperiodic Fourier Modal Method. *Journal Optical Society America A*, Vol. 15, No. 8, (March 2008) pp. 667-675, ISSN 1084-7529

Berger, V. & Sirtori, C. (2004). Nonlinear Phase Matching in THz Semiconductor Waveguides. *Semiconductor Science and Technology*, Vol. 19, No. 8, (June 2004) pp. 964-970, ISSN 0268-1242

Borri, P.; Langbein, W.; Schneider, S.; Woggon, U.; Sellin, R. L.; Ouyang, D. & Bimberg, D. (2001). Ultralong Dephasing Time in InGaAs Quantum Dots. *Physical Review Letters*, Vol. 87, No. 15, (October 2001) pp. 157401.1–157401.4, ISSN 0031-9007

Borselli, M.; Johnson, T. J. & Painter, O. (2005). Beyond the Rayleigh Scattering Limit in High-Q Silicon Microdisk: Theory and Experiment. *Optics Express*, Vol. 13, No. 5, (March 2005) pp. 1515-1530, ISSN 1094-4087

Boyd, R. W. (2003). *Nonlinear Optics*, Academic Press, ISBN 0-12-121682-9, New York

Bravo-Abad, J.; Rodriguez, A.; Bermel, P.; Johnson, S. J.; Joannopoulos, J. D. & Soljačić, M. (2007). Enhanced Nonlinear Optics in Photonic-Crystal Microcavities. *Optics Express*, Vol. 15, No. 24, (November 2007) pp. 16161-16176, ISSN 1094-4087

Cortez, S.; Krebs, O.; Voisin, P. & Gerard, J. M. (2001). Polarization of the Interband Optical Dipole in InAs/GaAs Self-Organized Quantum Dots. *Physical Review B*, Vol. 63, No. 23, (May 2001) pp. 233306.1–233306.4, ISSN 0163-1829

Dumeige, Y. & Feron, P. (2006). Whispering-Gallery-Mode Analysis of Phase-Matched Doubly Resonant Second-Harmonic Generation. *Physical Review A*, Vol. 74, No. 6, (December 2006) pp. 063804.1-063804.7 ,ISSN 1050-2947

Faist, J.; Capasso, F.; Sivco, D. L.; Sirtori, C.; Hutchinson, A. L. & Cho, A. Y. (1994). Quantum Cascade Laser. *Science*, Vol. 264, No. 5158, (April 1994) pp. 553 556, ISSN 1095-9203

Franz, K. J. ; Wasserman, D. ; Hoffman, A. J.; Jangraw, D. C. ; Shiu, K.-T.; Forrest, S. R. & Gmachl, C. (2007). Evidence of Cascaded Emission in a Dual-Wavelength Quantum Cascade Laser. *Applied Physics Letters*, Vol. 90, No. 9, (February 2007) pp. 091104.1–091104.3, ISSN 0003-6951

Fujita, M.; Ushigome, R. & Baba, T. (2000). Continuous Wave Lasing in GaInAsP Microdisk Injection Laser with Threshold Current of 40 μA. *Electronics Letters*, Vol. 36, No. 9, (April 2000) pp. 790–791, ISSN 0013-5194

Gehrsitz S.; Reinhart, F. K.; Gourgon, C.; Herres, N.; Vonlanthen, A. & Sigg, H. (2000). The Refractive Index of $Al_xGa_{1-x}As$ below the Band Gap: Accurate Determination and Empirical Modeling. *Journal of Applied Physics*, Vol. 87, No. 11, (June 2000) pp. 7825-7837, ISSN 0021-8979

Grischkowsky, D.; Keiding, S.; Van Exter, M. & Fattinger, C. (1990). Far-Infrared Time-Domain Spectroscopy with Terahertz Beams of Dielectrics and Semiconductors. *Journal of Optical Society of America B*, Vol. 7, No. 10, (October 1990) pp. 2006–2015, ISSN 0740-3224

Haus, H. A. (1984). *Waves and Fields in Optoelectronics*. Prentice-Hall, ISBN 0-13-946053-5, New Jersey

Harde, H. & Grischkowsky, D. (1991). Coherent Transients Excited by Subpicosecond Pulses of Terahertz Radiation. *Journal of the Optical Society of America B*, Vol. 8, No. 8, (August 1991) pp. 1642-1651, ISSN 0740-3224

Heebner, J. E.; Bond, T. C. & Kallman, J. S. (2007). Generalized Formulation for Performance Degradations due to Bending and Edge Scattering Loss in Microdisk Resonators. *Optics Express*, Vol. 15, No. 8, (April 2007) pp. 4452-4473, ISSN 1094-4087

Hu, B. B. & Nuss, M. C. (1995). Imaging with THz Waves. *Optics Letters*, Vol. 20, No. 16, (August 1995) pp. 1716-1718, ISSN 0146-9592

Kawase, K.; Minamide, H.; Imai, K.; Shikata, J. & Ito, J. (2002). Injection- Seeded Terahertz-Wave Parametric Generator with Wide Tunability. *Applied Physics Letters*, Vol. 80, No. 2, (January 2002) pp. 195–197, ISSN 0003-6951

Kohler, R.; Tredicucci, A; Beltram, F.; Beere, H.; Linfield, E. H.; Davies, A. G.; Ritchie, D. A.; Iotti, R. C. & Rossi, F. (2002). Terahertz Semiconductor-Heterostructure Laser. *Nature*, Vol. 417, No. 6885, (May 2002) pp. 156-159, ISSN 0028-0836

Levi, O.; Pinguet, T. J.; Skauli, T.; Eyres, L. A.; Parameswaran, K. R.; Harris, J. S. J.; Fryer, M. M.; Kulp, T. J.; Bisson, S. E.; Gerard, B.; Lallier E. & Becouarn L. (2002). Difference Frequency Generation of 8 1m Radiation in Orientation-Patterned GaAs. *Optics Letters*, Vol. 32, No. 7, (December 2002) pp. 2091-2093, ISSN 0146-9592

Marandi, A.; Darcie, T. E. & So, P. P. M. (2008). Design of a Continuous-Wave Tunable Terahertz Source using Waveguide-Phase-Matched GaAs. *Optics Express*, Vol. 16, No. 14, (July 2008) pp. 10427-10433, ISSN 1094-4087

Mittleman, D. (2003). *Sensing with Terahertz Radiation*, Springer-Verlag, ISBN 3-540-43110-1, Berlin

Nowicki-Bringuier, Y.-R.; Claudon, J.; Bockler, C.; Reitzenstein, S.; Kamp, M.; Morand, A.; Forchel, A. & Gerard, J. M. (2007). High Q Whispering Gallery Modes in GaAs/AlAs Pillar Microcavities. *Optics Express*, Vol. 15, No. 25, (December 2007) 17291–17304, ISSN 1094-4087

Palik, E. D. (1998). *Handbook of optical constants of solids*, Academic Press, ISBN 0-12-544423-0, New York

Sakai, K. (2005). *Terahertz optoelectronics*, Springer-Verlag, ISBN 3-540-20013-4, Berlin

Siegel, P. H. (2002). Terahertz Technology. *IEEE Microwave Theory and Techniques Society*, Vol. 50, No. 3, (March 2002) pp.910–928, ISSN 0018–9480

Siegman, A. E. (1986). *Lasers*, University Science Books, ISBN 0-935702-11-3, Mill Valley

Soljačić, M.; Ibanescu, M.; Johnson, S. G.; Fink, Y. & Joannopoulos J. D. (2002). Optimal Bistable Switching in Nonlinear Photonic Crystals. *Physical Review E*, Vol. 66, No. 5, (November 2002) pp. 055601.1-0.55601.4, ISSN 1539-3755

Soljačić, M. & Joannopoulos, J. D. (2004). Enhancement of Nonlinear Effects Using Photonic Crystal. *Nature Materials*, Vol. 3, No. 4, (April 2004) pp. 211-219, ISSN 1476-1122

Srinivasan, K.; Borselli, M.; Johnson, T. J.; Barclay, P. E.; Painter, O; Stintz, A. & Krishna, S. (2005). Optical Loss and Lasing Characteristics of High-Quality-Factor AlGaAs Microdisk Resonators with Embedded Quantum Dots. *Applied Physics Letters*, Vol. 86, (April 2005) 151106: 1-3, ISSN 0003-6951

Srinivasan, K. (2006). Semiconductor Optical Microcavities for Chip-Based Cavity QED, Ph.D. thesis, California Institute of Technology.

Tamir, T. (1990). *Guided-Wave Optoelectronics*, Springer-Verlag, ISBN 3-540-52780-x, Berlin

Tonouchi, M. (2007). Cutting-edge terahertz technology. *Nature Photonics*, Vol. 1, No. 2, (February 2007) pp. 97-105, ISSN 1749-4893

Vodopyanov, K. L.; Feyer, M. M., Yu, X.; Harris, J. S.; Lee, Y.-S.; Hurlbut, W. C.; Kozlov, V. G.; Bliss, D. & Lynch, C. (2006). Terahertz-Wave Generation in Quasi Phase-Matched GaAs. *Applied Physics Letters*, Vol. 89, No. 14, (October 2006) pp. 141119.1-141119.3, ISSN 0003-6951

Vodopyanov, K. L. & Avetisyan, Yu. H. (2008). Optical Terahertz Wave Generation in a Planar GaAs Waveguide. *Optics Letters*, Vol. 33, No. 20, (October 2008), pp. 2314-2316, ISSN 0146-9592

Yajima, T. & Takeuchi, N. (1970). Far-Infrared Difference-Frequency Generation by Picosecond Laser Pulses. *Japanese Journal of Applied Physics*, Vol. 9, No. 11, (June 1970) pp. 1361–1371, ISSN 0021-4922

Yang, Z.; Chak, P.; Bristow, A. D.; Van Driel, H. M.; Iyer, R.; Aitchison, J. S.; Smirl, A. L.; Sipe, J. E. (2007). Enhanced Second-Harmonic Generation in AlGaAs Microring Resonators. *Optics Letters*, Vol. 32, No. 7, (April 2007) pp. 826-828, ISSN 0146-9592

10

Terahertz Time-Domain Spectroscopy of Metallic Particle Ensembles

Kenneth J. Chau
University of British Columbia
Canada

1. Introduction

The terahertz (THz) frequency range is the region of the electromagnetic spectrum between the microwave and optical bands spanning from 0.1 THz to 10 THz. Historically, electromagnetic radiation in this frequency range has been inaccessible due to the lack of widespread electronic or laser-based radiation sources. Electronic radiation sources such as crystal oscillators are generally confined to operate at frequencies below ~ 100 GHz, while laser radiation sources are generally confined to operate at frequencies above ~ 30 THz. In recent years, the development of femtosecond lasers and quantum electronics have enabled a wide range of implementations to both generate and detect THz radiation [14]. One of the earliest and most widespread techniques is THz time-domain spectroscopy. THz time-domain spectroscopy is based upon the generation of a broad-band, free-space THz transient, which is detected using a femtosecond pulse to sample the THz electric field in the time-domain. THz spectroscopic measurements are performed by illuminating materials with a THz pulse and measuring the pulse after reflection from or transmission through the material. The electromagnetic properties of the material are inferred from changes in the amplitude and phase of the measured electric field pulse relative that of the incident electric field pulse. THz time-domain spectroscopy has been applied in transmission mode to characterize the THz-frequency optical constants of dielectrics and superconductors [5, 8, 15, 16] and in reflection mode to characterize the reflection amplitude and associated phase change due to semiconductors such as InSb [6] and highly doped silicon [17, 18].

The implementation of THz time-domain spectroscopy requires that the reflected/ transmitted THz pulse undergo measurable transformation upon interacting with the material; spectroscopic measurements made in transmission mode require that the investigated material exhibit partial transparency to THz radiation (that is, some radiation must pass through the material), while spectroscopic measurements made in reflection mode require that the material exhibit partial reflectance to THz radiation (that is, the reflected radiation must be altered relative to the incident radiation). Highly reflective materials such as bulk metals are not amenable to THz time-domain spectroscopy in either transmission or reflection modes. Due to the large and negative real part of the relative permittivity of most metals at THz frequencies (where typically $Re[\underline{\varepsilon}(\omega) \sim -10^5]$), incident THz radiation penetrates only a short, subwavelength distance δ ~ 100nm into the surface of the metal and nearly all the incident electromagnetic energy is reflected. The high reflectivity of bulk metals at THz frequencies precludes transmission-based measurements, and the short penetration distance of THz electromagnetic radiation into bulk metal limits the amplitude and phase change observable in a reflection measurement.

While bulk metals (defined as materials composed of a continuous, conducting medium with physical dimensions much greater than the wavelength) are completely opaque to THz radiation, dense collections of subwavelength sized metallic particles have been shown to exhibit partial transparency at THz frequencies [2, 3]. This transparency is unexpected since the particles that constitute the ensemble are composed of a material that is opaque to THz radiation and the particles are densely packed in a manner which precludes direct THz propagation through the ensemble. The objective of this Chapter is twofold: 1) we will explore the physical mechanisms underlying the THz transparency of metallic particle ensembles through experimental evidence supported by simulation and 2) we will demonstrate the application of THz time-domain spectroscopy to study the effective optical constants of a metallic sample. First, the THz electromagnetic response of a single, isolated metallic particle is modeled using finite difference time-domain (FDTD) calculations of the Maxwell Equations. FDTD calculations are then applied to model THz electromagnetic wave interaction with a dense collection of metallic particles, where it is shown that THz electromagnetic propagation through the particle ensemble is mediated by near-field electromagnetic coupling between nearest-neighbor particles across the ensemble. The influences of the extent of the ensemble L and the particle size d on the THz transparency are experimentally tested using THz time-domain spectroscopy in transmission mode and the experimental evidence is compared with numerical simulations based on FDTD calculations. THz time-domain spectroscopy is then applied in transmission mode as a non-invasive, direct probe of the effective dielectric properties of a metallic particle ensemble. The sensitivity of this methodology for probing metallic media is tested by monitoring the properties of the ensemble during the liquid-solid phase transition of the metallic medium.

2. Single subwavelength metallic particle

The electromagnetic response of a single, subwavelength metallic particle excited by a THz electromagnetic wave is governed by two sequence of events: 1) the THz electromagnetic wave incident on the particle surface penetrates $\delta \sim 100$nm into the metal where it induces charge motion and subsequently, current density, and 2) a dipolar electric field, also known as a particle plasmon, is formed by the accumulation of negative and positive charge at opposite sides of the particle's surface. At the surface of the particle, the dipolar electric field induced by excitation of the particle is oriented normal to the particle surface and has a net orientation along the direction of the incident electric field.

To visualize the electric fields associated with the THz particle plasmon, the electromagnetic response of a single, isolated subwavelength metallic particle to electromagnetic wave excitation is studied using the FDTD method to solve the Maxwell equations in two dimensions. In the FDTD method, the material properties of each spatial grid point in the simulation space are independently specified, and the complete spatial and temporal evolution of the electric and magnetic fields are solved. Shown in Fig. 1 is a series of snapshots of the electric field amplitude distribution obtained from an FDTD calculation describing a THz electromagnetic pulse (with spectral contents centred at 0.6 THz and a 1 THz bandwidth) incident on a single copper particle having a diameter $d = 75\ \mu$m immersed in free-space. In the calculations, the THz pulse propagates upward from the bottom of the

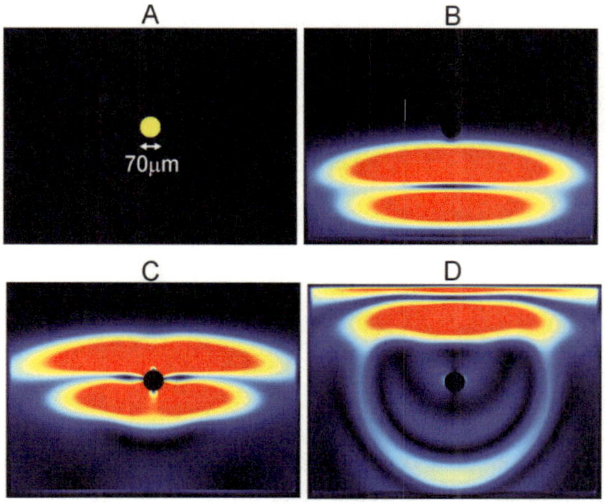

Fig. 1. Images of a FDTD calculation modeling single-cycle THz pulse excitation of an isolated 75 μm diameter copper particle (A) prior to excitation, (B) at 0.0 ps, (C) at 3.5 ps, and (D) at 8.5 ps.

images and is polarized in the plane of the images (transverse magnetic or TM). The images in Fig. 1B to 1D correspond to snapshots of the THz electric field magnitude at various times in the progression of the simulation. At t = 0.0 ps, the single-cycle polarized THz pulse propagates towards the subwavelength sized metallic particle, and when the THz pulse overtakes the particle at 3.5 ps, negligible THz electric field amplitude is present inside the particle, since the skin depth (penetration distance) of the THz electromagnetic wave is significantly less than the particle diameter. At 3.5 ps, the electric field amplitude can be conceptually divided into contributions from the external THz pulse and the electric field amplitude arising from the induced charges at the particle's surface. In this frame of the simulation, it is not possible to separate the external and induced-charge contributions to the total electric field amplitude. After the passage of the THz electric field pulse at 8.5 ps of the simulation, the electric field amplitude (Fig. 1D) and vector electric field (Fig. 2A) arising from the charges induced on the particle by the external electromagnetic wave can be visualized. At this snapshot after the THz pulse has propagated past the particle, the remnant electric field around the particle is confined to the surface and exhibits dipole-like signatures. Such a surface field is attributed to the excitation of charge oscillations on the particle oriented along the polarization of the external THz electric field. By taking the divergence of the electric field distribution, the charge density distribution associated with the dipolar electric fields can be obtained. As shown in Fig. 2B, the induced charge density illustrates dipolar charge induction by the incident THz pulse, where positive and negative charge density accumulate at opposing sides of the particle along the direction of the incident THz pulse polarization. The induced charge densities are coupled to an electromagnetic field confined to the surface of the particle. The dipolar electric field (highlighted in Fig. 3) associated with the induced charge density is strongest directly above the surface of the particle and decays exponentially within a distance of $\simeq$ 250 μm. This distance is less than the central wavelength of the THz pulse, λ = 500nm, indicating that the surface fields are confined to within a subwavelength region in the vicinity of the particle.

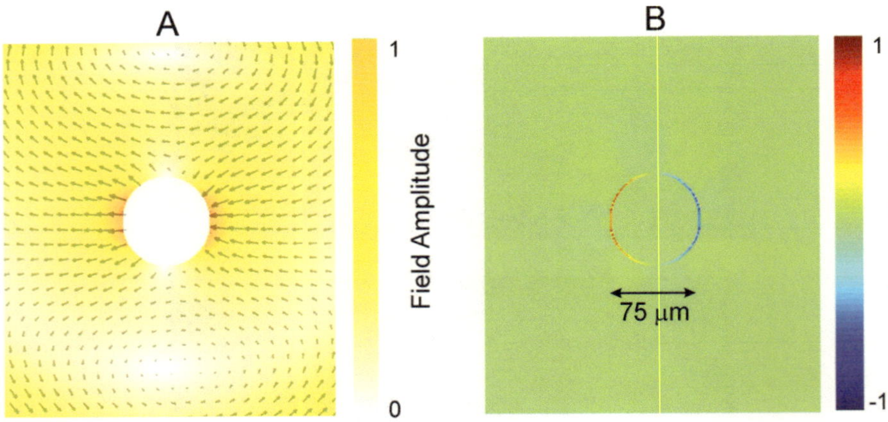

Fig. 2. (A) Vector plot of the electric field in the vicinity of a 75 μm copper particle after excitation by a single-cycle THz pulse at 8.5 ps of the simulation shown in Fig. 1. (B) illustrates the corresponding dipolar charge distribution at the surface of the particle.

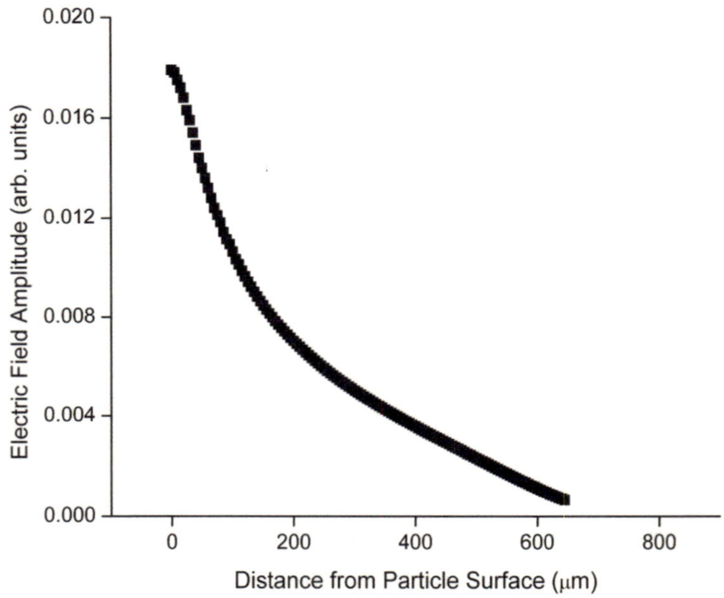

Fig. 3. Calculated amplitude of the electric field outside the surface of a 75- μmdiameter copper particle after excitation by a single-cycle THz pulse (which has propagated past the particle) with respect to the distance from the particle surface.

3. Ensemble of subwavelength metallic particles

In a collection of closely-spaced subwavelength metallic particles, electromagnetic interaction between the particles plays an important role in the overall electromagnetic properties of the ensemble. Since the particles are electromagnetically coupled, each particle is excited by the external electric field in addition to the field scattered from all the other particles. The complex interactions between metallic particles make it difficult to analytically

describe the electromagnetic properties of the ensemble. One common technique to determine the electromagnetic properties of subwavelength metallic particle collections is via effective medium approximations [1, 11]. The effective medium approximation replaces an inhomogeneous medium with a fictitious homogeneous effective medium which expresses the linear response of the whole inhomogeneous sample to an external electric field. Thus, rather than laboriously describing the microscopic interactions between the constituents, the entire heterogeneous medium is described by a single effective parameter.
Effective medium approximations have been employed to derive the homogeneous optical parameters of metallic clusters and metamaterials with subwavelength features [19]. The validity of effective medium approximations is governed by the quasi-static approximation, wherein the electric and displacement fields throughout the heterogeneous medium must be approximately uniform. To illustrate, consider a subwavelength metallic sphere having a diameter of d. The sphere is centred at $z = 0$ and illuminated by an electromagnetic plane-wave from free-space. For the field amplitude within the particle to be uniform, there must be minimal absorption over the particle dimension, or

$$x\mathrm{Im}[\sqrt{\underline{\epsilon}(\omega)}] \ll 1 \tag{1}$$

where $\mathrm{Im}[\sqrt{\underline{\epsilon}(\omega)}]$ is the imaginary part of the complex refractive index of the metal and $x = \pi d/\lambda$ is the size parameter. Similarly, there must be minimal spatial variation of the electromagnetic wave in the sphere, which implies that the wavelength inside the sphere is much greater than the particle size, or

$$x\mathrm{Re}[\sqrt{\underline{\epsilon}(\omega)}] \ll 1 \tag{2}$$

where $\mathrm{Re}[\sqrt{\underline{\epsilon}(\omega)}]$ is the real part of the refractive index of the metal. Combining the inequalities Eqs. 1 and 2 gives the condition in which the quasi-static regime is valid

$$x\left|\sqrt{\underline{\epsilon}(\omega)}\right| \ll 1 \tag{3}$$

Eq. 3 can be applied to test the applicability of field-averaging for micron-scale particles excited by electromagnetic waves at THz frequencies. Assuming a spherical copper particle with a diameter of 75 μm and a metal permittivity $\underline{\epsilon}_{Cu}(\omega)$ (1 THz) $\simeq -10^4 + -10^5$ excited by an electromagnetic wave with a wavelength of 300 μm (corresponding to a frequency of 1 THz)

$$x\left|\sqrt{\underline{\epsilon}_{\mathrm{Cu}}(\omega)}\right| = 248 > 1. \tag{4}$$

Thus, field averaging techniques to derive effective homogeneous parameters cannot be applied to describe the optical properties of micron-scale metallic particle at THz frequencies.
Since field-averaging cannot be used to effectively homogenize the THz electromagnetic response of the dense metallic particle ensembles, electromagnetic interactions within the ensemble are simulated rigorously using FDTD calculations of the Maxwell Equations in two dimensions. The structure used in the simulation is a randomly generated ensemble of copper particles in an air ambient that have a circular cross section with a diameter d = 75

μm. The particles are randomly packed to achieve a packing fraction p = 0.56 and the ensemble size is 5mm × 5mm, as depicted in Fig. 4A. In the simulations, a single-cycle THz pulse (with spectral contents centred at 0.6 THz and a 1 THz bandwidth) is normally incident on a flat face of the ensemble and the transmission through the opposing flat face of the ensemble is measured.

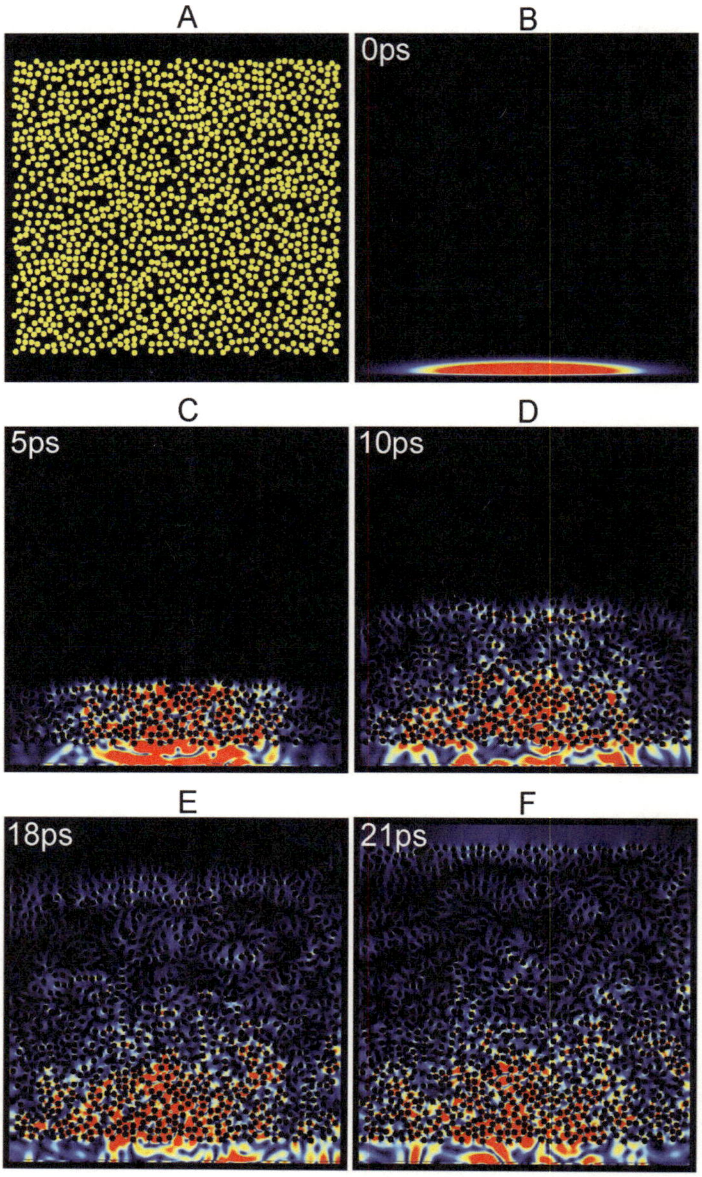

Fig. 4. (A) Simulation geometry in which an ensemble of 75- μm diameter copper particles with a volume fill fraction of 0.56 is excited by a TM-polarized THz electromagnetic pulse. Snapshots of the THz electric field magnitude within a random 5mm × 5mm ensemble of copper particles at times (B) 0 ps, (C) 5 ps, (D) 10 ps, (E) 18 ps, and (F) 21 ps.

Examination of the dynamics of the internal electric field amplitude obtained from the FDTD calculations shed insight into the mechanism underlying the propagation of THz electromagnetic energy through the particle ensembles. Figs. 4B to 4F depict snapshots of the spatio-temporal evolution of the electric field amplitude due to excitation of the L = 5mm ensemble by the THz pulse at representative times, t. At t = 0 ps, the polarized THz pulse is incident on the particle ensemble, and at t = 5 ps, the pulse couples into particle plasmons on the individual particles, evidenced by the high electric fields near the surfaces of the particles. The snapshots at time t = 10 ps and t = 18 ps show that significant electromagnetic energy is squeezed in the free-space gaps between the particles, while there is negligible field penetration into the individual particles. Collectively, the particles carry electromagnetic energy over the extent of the ensemble, evidenced by a wave-front which appears in the simulation as a large-electric-field-amplitude band progressing through the medium. By tracking the wavefront as it advances through the system, an electromagnetic energy velocity of $0.51c$ is measured. At t = 21 ps, this leading wave-front approaches the far boundary of the ensemble and radiates into free-space. The simulations demonstrate that a THz electromagnetic wave in a dense metallic particle ensemble is squeezed into the subwavelength-scale interstitial gaps of the metallic particle ensembles, yet can propagate through large, millimetre-scale distances and at the back face of the ensemble, radiate into free-space.

4. THz time-domain spectroscopy of metallic particles

In this section, the THz electromagnetic properties of an ensemble of subwavelength sized copper particles are studied using THz time-domain spectroscopy in transmission mode. The metallic particle ensemble consists of pure copper particles that are spherical in shape and nearly mono-dispersed in size, with a mean particle diameter d = 75 μm and a volume metal packing fraction (volume ratio of metal to the entire volume of the ensemble) p = 0.51 immersed in air. A scanning electron microscope image of a dispersed collection of the particles is shown in Fig. 5. The THz transmission through the particle ensemble is measured with the experimental configuration depicted in Fig. 6. Single-cycle, linearly polarized THz pulses, with spectral contents centred at 0.6 THz and a 1 THz bandwidth, are generated from a GaAs photoconductive switch excited with focused < 20 fs, 800 nm laser pulses supplied from a Ti:Sapphire laser at a repetition rate of 80MHz. The collimated THz beam is directed towards a sample cell, which is composed of THz-transparent polystyrene windows with variable separation distance L, that houses the metallic particle ensemble. The time-domain electric field transmission in addition to polarization of the transmission is measured to characterize electromagnetic wave transport through the medium. The on-axis THz electric field pulse transmitted through the ensemble is coherently detected via an optically gated 500 μm thick ⟨111⟩ ZnSe electrooptic crystal, and time-resolved information is obtained by varying the delay between the THz pulse and a sampling probe pulse.
Significant THz transmission through the particle ensemble is measured for sample thicknesses, L, up to 7.7 mm, where nearly 20% transmission is observed for the thinnest L = 0.6-mm ensemble. Fig.7 depicts the time-domain THz electric field pulses transmitted through particle ensembles over the range 0.6mm $\leq$ L $\leq$ 7.7 mm referenced to the

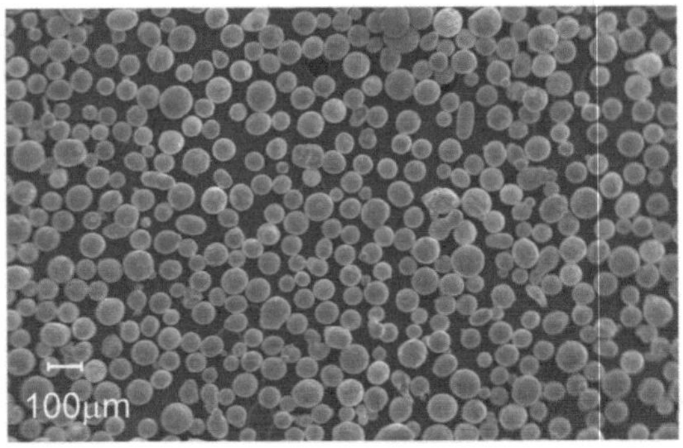

Fig. 5. Scanning electron microscope image of a dispersed collection of copper particles with an average diameter of 75 μm.

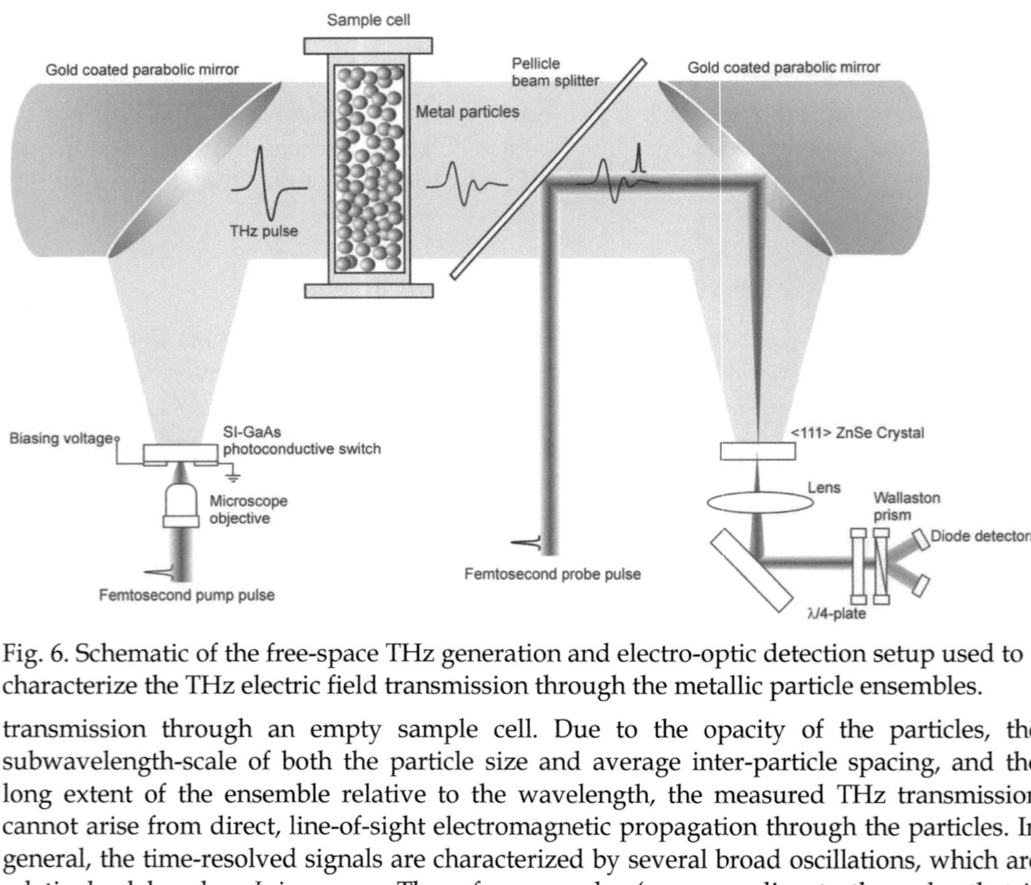

Fig. 6. Schematic of the free-space THz generation and electro-optic detection setup used to characterize the THz electric field transmission through the metallic particle ensembles.

transmission through an empty sample cell. Due to the opacity of the particles, the subwavelength-scale of both the particle size and average inter-particle spacing, and the long extent of the ensemble relative to the wavelength, the measured THz transmission cannot arise from direct, line-of-sight electromagnetic propagation through the particles. In general, the time-resolved signals are characterized by several broad oscillations, which are relatively delayed as L increases. The reference pulse (corresponding to the pulse that is incident on the ensembles) is localized in time (within ~ 1 ps); upon impulsive excitation of the sample, it requires a finite time for energy to propagate through the sample. To estimate the energy propagation velocity from one end to the sample to the other end, the relative delay of the transmitted field is measured. Here, the delay corresponds to the time difference between the centroid of the time-domain intensity distribution of the reference

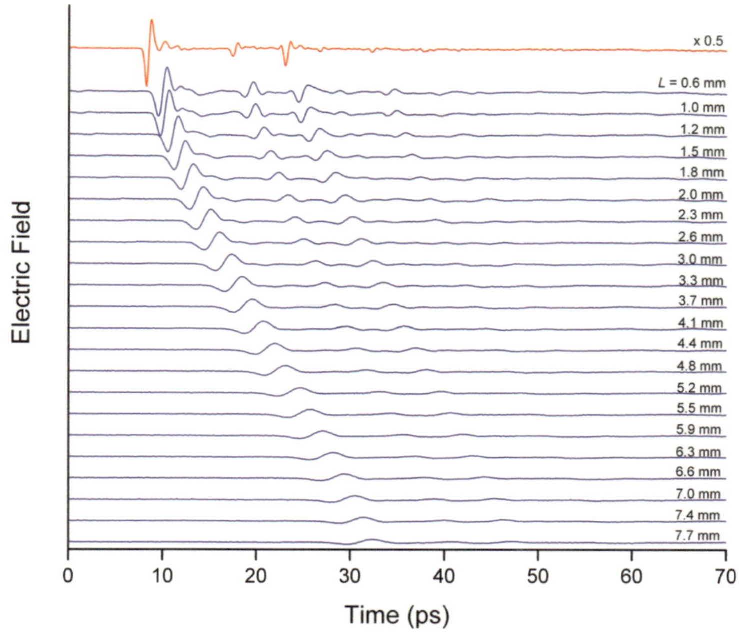

Fig. 7. Time-domain waveforms of the far-field THz transmission through ensembles of copper particles with lengths ranging from L = 0.6 mm to L = 7.7 mm.

pulse and the centroid of the time-domain intensity distribution of the transmitted pulse. Shown in Fig. 8 is the relative pulse delay as a function of sample thickness, referenced to an equivalent air path. The measured delay translates to an electromagnetic energy velocity of $0.51 \pm 0.01c$ or an effective refractive index of 2.0 ± 0.1. It should be noted that the effective macroscopic index reported here describes the overall response of the metallic particle ensemble to THz electromagnetic wave excitation, but is not derived from the effective medium approximation. As L increases from 0.6 to 7.7 mm, the durations of the transmitted electric field pulses are broadened from 2 ps to 6 ps. The pulse broadening, which increases with larger values of L, is indicative of a preferential amplitude reduction in the higher frequency components of the incident pulse. Due to the absence of significant intrinsic material resonances for bulk copper at THz frequencies, the preferential loss of higher frequency components likely originates from the scattering due to the extrinsic structural characteristics of the random metallic medium.

The experimentally measured relative delay of the THz transmission is compared with results obtained from FDTD calculations in Fig. 8. It should be noted that the packing fraction of the sample in the calculations (0.56) is larger than the experimentally measured packing fraction of the sample used in the experiments (0.51). Augmenting the packing fraction in the two-dimensional system in the simulations effectively increases the surface area of the particles, which more accurately accounts for non-radiative losses occurring at the surface of the three-dimensional particles used in the experiments. As shown in Fig. 8, the relative delay of the THz transmission obtained from FDTD calculations demonstrates excellent agreement with experimental observations. The linear increase in the relative delay with respect to the sample thickness indicates that the THz transmission is mediated by a phase accrual across the length of the ensemble. By increasing the length of the sample, the

THz electromagnetic wave interacts with a greater number of particles, which augments the delay in the measured transmission.

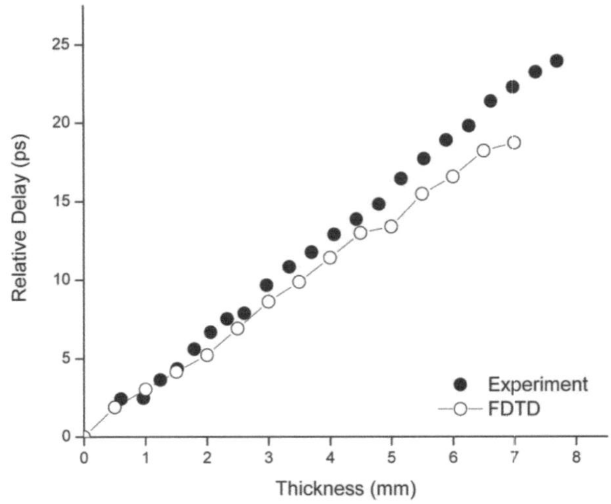

Fig. 8. The relative delay of the transmitted THz pulse through the copper ensembles with respect to the thickness of the ensemble obtained from experiment and simulation.

The polarization of the electric field transmitted through the particle ensemble provides further insights into the origin of the THz transmission. Comparison of the polarization of the transmitted THz pulse with the linear polarization of the incident THz pulse maps the degree of coherence of electromagnetic energy transport across the ensemble by onto a polarization change. A high correlation between the incident and transmitted polarizations would indicate a high degree of electromagnetic coupling between the incident and transmitted electric fields, whereas a low correlation would indicate a low degree of coupling. The transmitted electric field polarization is characterized by varying the angular orientation of the optical axis of the ⟨111⟩ ZnSe crystal electro-optic detector relative to the probe polarization. Fig. 9 illustrates polar plots of the intensity distribution of the free-space THz pulse incident onto the sample, in addition to those of the transmitted THz pulse through 2.2-mm thick and 7.7-mm thick ensembles of copper particles. The ensembles are composed of copper particles that have a mean diameter of $75 \pm 5\ \mu$m and a packing fraction of 0.51 ± 0.05. As highlighted in Fig. 9B, the THz electric field pulse transmitted through the 2.2-mm thick ensemble shows a high degree of polarization preservation of the incident horizontal, linear polarization of the incident pulse. As the sample thickness increases to 7.7mm (Fig. 9C), the transmission becomes more unpolarized. Polarization preservation of the transmission through the 2.2-mm thick ensemble indicates that at this thickness value, the THz transmission is predominantly mediated by coherent coupling across the ensemble. The diminishing polarization purity of the transmission as the sample thickness increases to 7.7 mm is attributed to augmented scattering of the THz electromagnetic wave, which randomizes the polarization and impairs the correlation between the incident and transmitted electric fields. From the data, it is inferred that the coherence length of electromagnetic transport across the ensemble, delineating a length scale below which the incident and transmitted electric fields are highly correlated, is on the order of several millimetres.

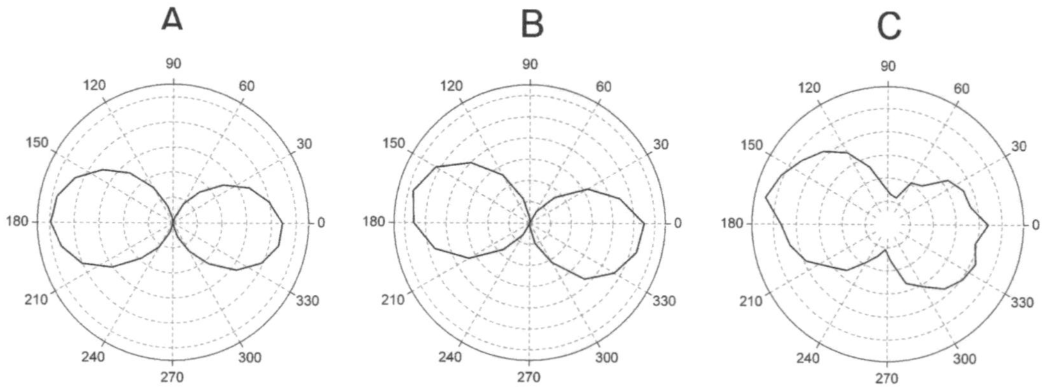

Fig. 9. Polar plot of the intensity distribution of (A) the incident THz pulse and the transmitted THz pulse through (B) a 2.2-mm thick and (C) a 7.7-mm thick ensemble of copper particles.

When a THz electromagnetic wave is incident on a subwavelength sized particle, a portion of the incident energy is coupled into the particle plasmon (as illustrated by simulation in Section 1) and a portion of the incident energy is scattered by the particle. Intuitively, the amount of electromagnetic energy "lost" to scattering from the metallic particle should be proportional to the cross-sectional area of the particle. As the particle cross sectional area increases, a larger portion of the incident electromagnetic energy is reflected and less electromagnetic energy is coupled into the localized particle plasmon at the surface of the particle. In the limit where the cross-sectional area of the particle is infinite, the incident electromagnetic wave encounters a bulk metallic surface and is completely reflected. To further investigate the origin of the THz transparency of metallic particle ensembles, the influence of the particle size on the THz transmission is studied. The relationship between the particle size of the ensemble and the THz transparency of the ensemble is studied using THz time-domain spectroscopy in transmission mode, where the THz electric field transmission is measured through several particle ensembles of fixed length (and fixed packing fraction) composed of copper spheres with average diameters of 85 ± 9 μm, 194 ± 9 μm, 250 ± 10 μm, 283 ± 8 μm, 372 ± 17 μm, 462 ± 17 μm, 560 ± 15 μm, and 670 ± 30 μm. In reporting the particle sizes, the nominal size corresponds to the average particle diameter and the error represents one standard deviation.

Shown in Fig.10 are the time-domain waveforms, the associated Fourier spectra, and the total integrated power of the transmission through copper particle ensembles where the particle sizes range from 85±9 μm to 670±30 μm. The thickness of the ensemble is kept constant at L = 3.0 mm throughout the experimentation. As the average particle size increases from 85 μm to 372 μm, the THz electric field transmission amplitude is dramatically reduced and is nearly zero for particles with diameters exceeding 462 μm. Associated with this attenuation is a shift in the central frequency of the transmission from 0.1 THz to 0.08 THz, indicating a preferential attenuation of the higher frequency components of the THz pulse (Fig. 10B). The preferential attenuation associated with the increasing particle size is due to the frequency-selective particle plasmon response of the particles. As the particle size increases, the shorter wavelength (higher frequency) components of the incident THz pulse cannot efficiently polarize the individual particles and do not excite the particle plasmon mode. As a result, the higher frequency components do not couple across the medium and are not radiated into the far-field.

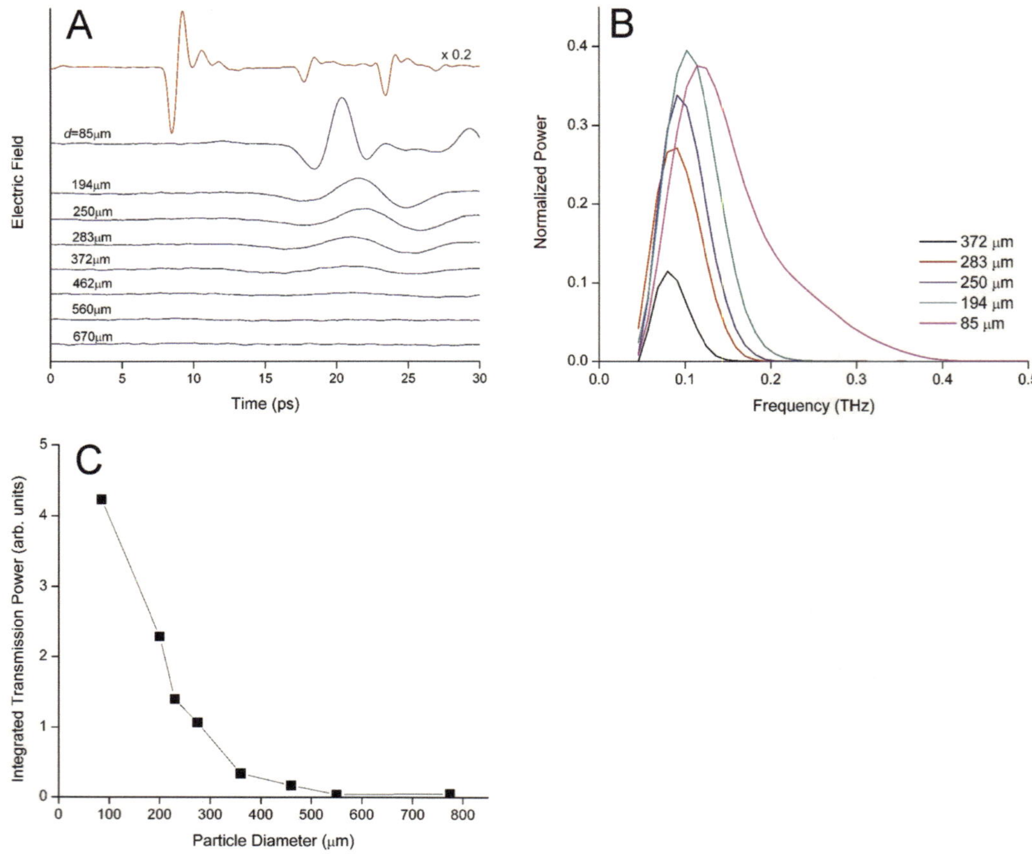

Fig. 10. (A) Time-domain waveforms of the far-field THz transmission through a 3.0-mm thick ensemble of copper particles with average particle diameters ranging from 85 μm to 670 μm. (B) Power spectra of the transmission through the particle ensembles normalized by the power spectrum of the incident THz pulse. (C) Total integrated transmitted power through the particle ensembles with respect to the average diameter of the particles that constitute the ensemble.

Fig.11 illustrates polar plots of the intensity distribution of the transmitted THz radiation through a $L = 3$ mm thick ensemble of densely-packed copper particles with mean diameters of 85 μm, 283 μm, and 372 μm. For the ensemble of 85- μm diameter particles, the transmission polarization preserves the incident linear polarization, indicating a high degree of electromagnetic coupling across the ensemble. The diminished polarization of the transmission through the ensemble of 283- μm diameter particles indicates reduced electromagnetic coupling. This observation coincides with a four-fold reduction in the transmitted intensity through the ensemble of 283- μm diameter particles relative to that of the 85- μm diameter particles.

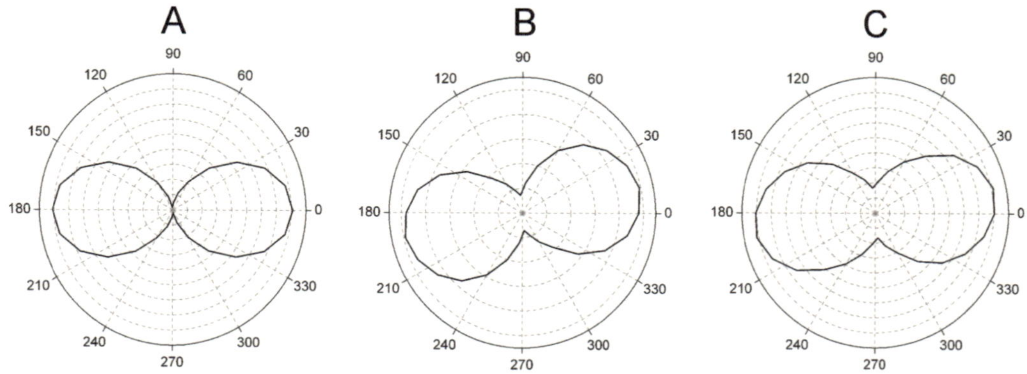

Fig. 11. Polar plot of the intensity distribution of the THz pulse transmitted through a 3.0-mm thick ensemble of copper particles with average diameters of (A)85 μm, (B)283 μm, and (C)372 μm.

The transmission through the ensemble of 372- μm diameter particles is nearly unpolarized, indicating that the transmitted energy is not coherently channeled across the extent of the ensemble. For the ensemble of 372- μm diameter particles, the transmitted intensity is almost fully extinguished.

The effect of the particle diameter on the polarizability of a single, isolated metallic particle (which is indicative of the degree of coupling to the particle plasmon mode of the particle) is illustrated via FDTD calculations of two situations in which a single, isolated metallic particle with a diameter of either 75 μm or 200 μm is excited by a THz electromagnetic pulse from free-space. The excitation pulse is a single-cycle THz transient centred at 0.6 THz with a 1 THz bandwidth, matching the THz pulses employed in previous experiments and simulations. In these simulations, the single-cycle THz pulse propagates upward toward the metallic particle. To map out the charge density induced by the external THz electric field pulse, the induced charge density distribution is calculated by taking the divergence of the vector displacement field distribution. Fig. 12 illustrates the instantaneous induced charge density distribution at the surface of the two particles after THz pulse excitation taken at the same time. For the 75- μm diameter particle, the THz electric field pulse induces a dipolar charge density distribution where conduction electrons at the surface of the two halves of

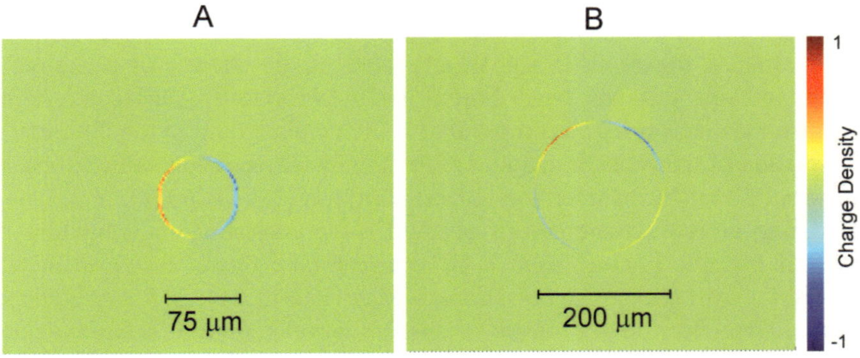

Fig. 12. FDTD calculation of the induced charge density distribution of a (A) 75- μm diameter particle and (B) 200- μm diameter particle after excitation by a single-cycle THz pulse in free-space.

the particle oscillate anti-parallel. As the particle size increases to 200 μm, the predominant polarization mode induced by the THz electric field pulse is quadrupolar. The magnitude of the charge density distribution for the larger particle is significantly weaker than the dipolar charge density distribution of the smaller particle. The peak charge density of the quadrupolar distribution is reduced to $\simeq$ 0.6 relative to the peak charge density of the dipolar distribution. The larger metallic particle is not efficiently polarized by the incident THz pulse, since a significant portion of the incident pulse is reflected by the larger particle and only the lower frequency components of the pulse can polarize the particle. In an ensemble of particles, this effect leads to a preferential reduction in the transmission of higher frequency components and an overall reduction in the total transmitted power.

5. Phase-transition THz spectroscopy of metallic particles

Terahertz time-domain spectroscopy is applied to study intrinsic, temperature-dependent phase transitions in a metallic particle ensemble. A phase transitions is defined as a transformation of a thermodynamic system from one phase to another. A distinguishing feature of phase transitions is an abrupt change in one or more physical properties of the material with a small change in a thermodynamic quantity such as temperature. For instance, when the specific energy of a metal is raised to the latent heat of fusion, the metal changes from the solid phase to the liquid phase. The microscopic mechanism for melting can be understood by considering the motions of ions in the solid and liquid states. Prior to melting, the ions that constitute the metal remain relatively fixed in the vicinity of their equilibrium positions. As the metal is heated above the melting temperature, the ions acquire enough energy leave their equilibrium positions and wander relatively large distances, resulting in a liquid state. Melting of solid metal is a typical example of a first order phase transition. First order phase transitions are those in which the substance releases or absorbs heat energy during the phase change. Since the energy cannot be absorbed or released instantaneously by the substance at the phase transition temperature, first order phase transitions are characterized by a mixed phase regime in which different phases of the medium coexist.

To date, metallic phase transitions are widely investigated using calorimetry techniques, such as AC- calorimetry [7]. A disadvantage of this method is that an invasive physical contact is required to accurately measure heat flow through the metallic sample. To overcome this constraint, several groups [9, 12, 13] have employed the photo-acoustic effect to non-invasively probe metallic phase transformations. In such experiments, phase transition modulates the acoustic signal generated at the surface of a sample when a surrounding ambient gas has been heated by a periodically modulated light beam. However, such mechanism requires a gas that is highly absorbing to the illuminating light, and interpretation of the acoustic signal is restricted by the complex nature of heat transfer between the solid metallic sample and surrounding gas [13]. Gallium is a unique metallic element existing at room temperature as solid α-Ga consisting of a mixture of stable molecular and metallic phases. Solid α-Ga is a complex phase described as a metallic molecular crystal with strong Ga_2 bonds and weaker intermolecular forces, whereas liquid gallium is more free-electron like [4]. At a free-space wavelength of 1.55 μm, the permittivity of liquid gallium has been estimated to be approximately 7 times larger than the permittivity of α-Ga [10]. Gallium possesses one of the lowest melting points of all metals T_m = 29.8 C°, which provides an ideal platform to study metallic solid-liquid phase transformation behavior via THz time-domain spectroscopy. Gallium particles are prepared by cooling bulk 99.99% gallium pellets to 77K and mechanically grinding the gallium pellets to achieve a powder having an average particle size of 109 ± 10 μm and a packing fraction of ~ 0.4 ± 0.1. In order to probe the phase transition of the gallium, THz time-domain spectroscopy in transmission mode is employed to monitor the effective optical properties of an ensemble of gallium particles as the temperature of the particles is raised past the melting point.

In the experimental setup, THz radiation is focused onto a polystyrene sample cell housing a L = 2.3mm thick collection of the random gallium particles. To examine the temperature-dependent THz transmissivity of the particles, the gallium particles are homogeneously heated, at a rate of 0.08 C°/min, from room temperature up to a temperature, T, of 38.2 C° (> T_m). Because the time over which the temperature increases is much longer than the heat diffusion time across the thin sample (< 1 s), it is ensured that the sample temperature is at equilibrium during the transmission measurements. The particles ensemble temperature is monitored (within ±0.1 C°) via a thermocouple inserted into the particle collection adjacent to the THz beam probing spot. During the measurements, both the beam spot size and location are kept fixed, thus ensuring that the THz radiation interacts with the same random realization of the particle ensemble throughout the temperature variation.

Melting is a thermal effect, and the temporal duration over which melting occurs is determined by the time over which heat can diffuse and equilibrate throughout the sample. The experiments are carefully designed and performed at an extremely slow heating rate (0.08 C°/min) in order to ensure that equilibrium conditions are established through the measurements. To quantify this condition, the heat diffusion times are estimated for both gallium metal (a lower bound) and air (an upper bound) through a distance of 2.3 mm corresponding to the sample thickness. Gallium has a thermal conductivity H_t = 40 $WK^{-1}m^{-1}$, a density u = 5910 kgm^{-3}, and a heat capacity C = 25.86 $Jmol^{-1} K^{-1}$. For air, H_t = 0.02$WK^{-1}m^{-1}$, u = 1.251 kgm^{-3} and C = 29.12 $Jmol^{-1} K^{-1}$. From these quantities, the thermal diffusivity is obtained from

$$D_t = H_t\, u^{-1}\, C^{-1}. \quad (5)$$

For a sample composed of gallium D_t = 1.9 × 10^{-5} m^2/s and for a sample composed of air D_t = 9 × 10^{-6} m^2/s. The characteristic diffusion time over a distance L is estimated by $t_{diff} = L^2/D_t$, yielding t_{diff} = 0.6 s for a sample composed entirely of gallium and t_{diff} = 0.28 s for a sample composed entirely of air. The sample used in the experiments is a mixture of air and gallium, and the characteristic heat diffusion time for the sample will lie between those bounds. To obtain an upper bound of time lagged thermal effects, we assume that it requires 0.6 s for heat to diffuse from one end of the sample to another. Over this time interval, a time-lagged temperature increase of 0.08 C°/min ×0.6 s = 0.0008 C°($\ll$ error in the temperature measurement) may develop across the sample. Since the time over which the temperature of the sample increases is much slower than the heat diffusion time across the sample thickness, it can be confidently concluded that the samples have reached thermal equilibrium as the THz time-domain spectroscopic measurements are taken. Fig. 13A illustrates the time-domain THz electric field waveforms, $E(t)$, transmitted through gallium particle collections measured at various temperatures. Notably, for temperatures below the melting point ($T < T_m$), the bipolar pulses transmitted through the particle collection all have an initial peak at a time t = 3.1 ps. The fact that the arrival delay, the amplitude, and the pulse shape of the transmitted pulses do not change throughout the temperature range 22.4 C° < T < 29.7 C° suggests an absence of phase transformation or any changes to the gallium metallic properties. However, once the temperature reaches the melting temperature of 29.9 C°, a temporal advancement (or early arrival) of the pulse peak by 0.3 ps provides evidence of the onset of a significant transformation in the electronic properties of the gallium particles. Although the pulse corresponding to $T = T_m = 29.9$ C° is temporally advanced, interestingly, the pulse shape remains unaltered at T_m. Further heating of the gallium particles from 29.9 C° to 38.2 C° induces striking pulse shape transformation where the pulse is attenuated and broadened in time.

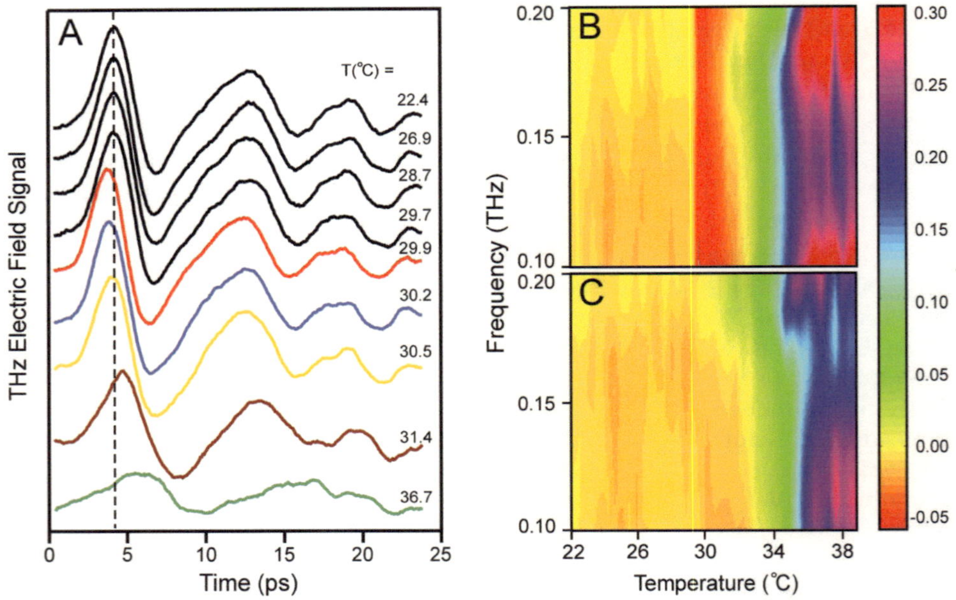

Fig. 13. (A) Experimental time-domain waveforms of THz pulses transmitted through 2.3-mmthick random gallium particle ensembles measured at various temperatures. The dashed line indicates the arrival time of the peak of the THz electric field pulse. Shown in (B) are the effective real refractive index change and (C) effective imaginary refractive index change versus temperature and frequency. The refractive indices are measured relative to the reference pulse transmitted through the sample at 21.2 C°.

Accompanying the temporal pulse shape trend with increasing temperature is a marked progressive delay and attenuation of $E(t)$. The pulse temporal shape, delay and amplitude trends for $T > T_m$ suggest conglomeration between adjacent, near-touching gallium particles. Because the THz transmission through the particle collections is mediated by nearest neighbor coupling between particles, conglomeration of the nearest-neighbor particles quenches radiation propagation mechanism. As the particles coalesce, the particles become larger and begin to exhibit metallic bulk-like electromagnetic properties, resulting in reduced transmission amplitude. Similarly, particle conglomeration results in a higher metal filling fraction, which increases the effective index of the particle ensemble and manifests as a temporal delay of the transmitted pulse.

To further explore the temperature-dependent evolution of the waveforms, the frequency-dependent relative effective refractive index inferred from the amplitude and phase of the transmitted THz electric fields relative to a reference THz electric field is analyzed. The effective real refractive index change

$$\Delta\mathrm{Re}[n] = \frac{\Phi(\omega) - \Phi_{ref}(\omega)}{kL} \tag{5}$$

and the effective imaginary refractive index change

$$\Delta\mathrm{Im}[n] = \frac{E(\omega)/E_{ref}(\omega)}{kL} \tag{6}$$

are obtained as a function of temperature, T. In these relations, $\Phi(\omega)$, $\Phi_{ref}(\omega)$, $E(\omega)$, and $E_{ref}(\omega)$ are the phase of the transmitted pulse, the phase of the reference pulse transmitted through the sample at T_{ref} = 21.2 C°, the amplitude of the transmitted pulse, and the amplitude of the reference pulse, respectively. Shown in Fig. 13B and 13C are $\Delta\mathrm{Re}[n]$ and $\Delta\mathrm{Im}[n]$ with respect to T over a frequency range between 0.1 THz and 0.2 THz (corresponding to the bandwidth of the transmitted pulse). As shown in the plot, there is negligible refractive index change between the temperature range 21.2 C° < T < T_m. At T_m = 29.9 C°, $\Delta\mathrm{Re}[n]$ decreases abruptly. As shown in Fig. 13B, this sharp discontinuity in $\Delta\mathrm{Re}[n]$ precisely at T_m is consistent over the entire transmission bandwidth. The abrupt, frequency-independent change in $\Delta\mathrm{Re}[n]$ suggests that the intrinsic electronic properties of gallium have been altered at T_m and is strongly indicative of metallic phase transformation. Interestingly, the onset of phase transition eludes detection in $\Delta\mathrm{Im}[n]$, as $\Delta\mathrm{Im}[n]$ remains approximately zero up to $T \simeq$ 30.5 C°. With further increase in the sample temperature above 30.5 C°, both $\Delta\mathrm{Re}[n]$ and $\Delta\mathrm{Im}[n]$ show large increases over the transmission bandwidth as a function of T. These significant increases in the complex effective refractive indices of the ensemble show that the particle ensemble becomes less transparent to the THz pulse for $T > T_m$ due to coalescing of nearest-neighbor particles. The strikingly different effective refractive index features for the range $T < T_m$, $T \simeq T_m$, and $T >$ 30.5 C° highlight three distinctive regimes where 1) the particles have not melted (constant $\Delta\mathrm{Re}[n]$ and $\Delta\mathrm{Im}[n]$), 2) the particles have melted but remain granular (discontinuity in $\Delta\mathrm{Re}[n]$, but constant $\Delta\mathrm{Im}[n]$), and 3) the particles have melted and are coalesced (large increases in both $\Delta\mathrm{Re}[n]$ and $\Delta\mathrm{Im}[n]$).

The temperature-dependent $\Delta\mathrm{Re}[n]$ and $\Delta\mathrm{Im}[n]$ trends at two frequencies, ω_1 = 0.1 THz and ω_2 = 0.2 THz are charted in Figs. 14A and 14B. As shown in Figure 14A, for 21.2 C° < T < 29.9 C°, $\Delta\mathrm{Re}[n]$ is nearly zero. Upon reaching T_m, the real part of the relative effective index exhibits a notably large, discontinuous jump of −0.06, indicative of an abrupt change in the intrinsic properties of gallium associated with metallic solid-liquid phase transformation. Above the melting temperature, $\Delta\mathrm{Re}[n]$ is strongly affected by conglomeration of the particles, which changes the underlying extrinsic microstructure of the ensemble. This extrinsic effect influences the effective index of the ensemble in a different way than the intrinsic metallic phase transition at T_m. For $T > T_m$, $\Delta\mathrm{Re}[n]$ increases from −0.06 to $\simeq$ 0.3 between 29.9 C° and 33.0 C° and beyond $T >$ 33.0 C°, is constant at $\simeq$ 0.3. Particle conglomeration occurring at $T > T_m$ increases the effective real refractive index of the ensemble, causing the arrival delay of the transmitted pulses. $\Delta\mathrm{Im}[n]$ exhibits similar overall trends as $\Delta\mathrm{Re}[n]$. Below the melting temperature, $\Delta\mathrm{Im}[n]$ shows negligible temperature dependence and is approximately zero. As shown in Fig. 14B, $\Delta\mathrm{Im}[n(\omega_2)]$ increases linearly for $T > T_m$ and saturates at 0.2 for $T >$ 33.0 C°. Such an increase in the imaginary effective refractive index reveals increased absorption or scattering losses within the ensemble due to particle melting and subsequent coalescing. In contrast to $\Delta\text{Im}[n(\omega_2)]$, $\Delta\text{Im}[n(\omega_1)]$ does not significantly increase from zero until the temperature exceeds 30.5 C° > T_m. The slightly different trends observed for $\Delta\text{Im}[n(\omega_1)]$ and $\Delta\text{Im}[n(\omega_2)]$ suggest that the higher frequency components of the pulse are more sensitive to particle conglomeration than the lower frequency components. Overall, the real and imaginary parts of the complex effective index of the sample exhibit high sensitivity to the solid-liquid phase transition of the gallium particles and subsequent melting and coalescing dynamics beyond T_m.

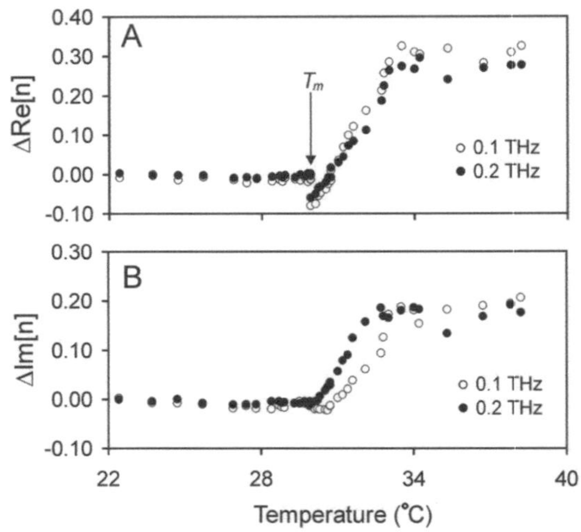

Fig. 14. Experimental effective (A) real refractive index change and (B) imaginary refractive index change at a frequency of 0.1 THz (empty circles) and 0.2 THz (filled circles) at various temperatures. The discontinuity in the effective real refractive index occurs at the gallium melting temperature, 29.8 C°.

The refractive index behavior for $T > T_m$ shows interesting particle conglomeration behavior of the gallium particles, where the particles begin to form interconnected networks. The experimental results show that coalescing does not occur concurrently with particle melting. To quantify the temperature where the particles begin to coalesce, the correlation function, $C(\tau) = \langle E(t+\tau)E_{ref}(t)\rangle$ is calculated, where $E(t+\tau)$ is the sample pulse (at a given temperature T) shifted by a time τ and $E_{ref}(t)$ is the reference pulse transmitted at reference temperature T_{ref}. It is noted that referencing the correlation function to the transmitted signal at T_{ref} cancels out the inherent spectral response of the setup since the spectral response of the system is fixed throughout the temperature variation. Because the only experimental variable is the sample temperature, changes in $C(\tau)$ as a function of T arise directly from temperature-dependent changes in the transmissivity of the gallium sample. As highlighted in the plot of the maximum correlation amplitude versus T [Fig. 15B], the transmitted pulse remains highly correlated even for T = 30.5 C° > T_m. Thus, at temperatures exceeding the melting transition, the extrinsic microstructure of the particle ensemble has not changed. However, at a coalescing temperature, T_c = 30.5 C°, there is a significant decrease in $C(\tau)$, marking the onset of particle conglomeration and transmission quenching. Because the particles must overcome their surface energy prior to liquefying, T_c is slightly higher than the bulk melting temperature. As shown in Fig. 15B, $C(\tau)$ decreases to 0.35 at 33.0 C°, and for T >

33.0 C°, the maximum correlation amplitude saturates and remains fixed. The experimental results reveal a narrow temperature range, $T_m < T < T_c$, where the individual particles have melted, yet the nearest-neighbor particles do not conglomerate.

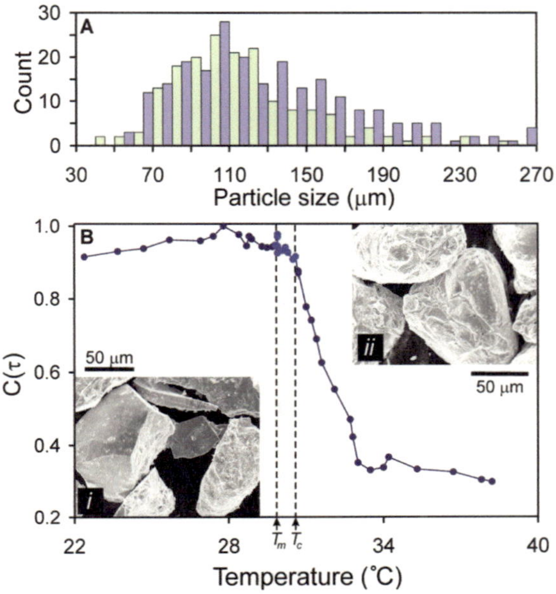

Fig. 15. (A) Size distribution of the gallium particles before melting (light bars) and after melting (dark bars). (B) shows the maximum correlation amplitude of transmitted pulses at various temperatures relative to the reference pulse at 21.2 C°. Insets are scanning electron microscope images of (i) the gallium particles prior to the heating cycle and (ii) the solidified particles after the heating cycle.

The gallium particle collection undergoes significant structural transformation over the heating cycle. After heating the particles above T_m and cooling back to room temperature, the nearest-neighbor particles have coalesced at small regions conjoining the particles, but overall, the ensemble retains a granular appearance and structure with no significant decrease in the total volume. The individual particles shapes are slightly distorted by the heating. As shown in the scanning electron microscope images in the insets of Fig. 15B, the particles prior to heating are characterized by sharp edges and flat faces. After cycling the temperature, the particles are rounded and have a rougher surface. Although heating induces shape change in the particles and coalescing between nearest-neighbor particles, the overall size distribution of the ensemble after heating is not significantly affected. As shown in Fig. 15A, the size distribution of the particles is nearly identical before and after heating. This further confirms that over the heating cycle, the particles do not fully conglomerate to form particles with augmented sizes. Rather, nearest-neighbor particles join at small sections of the particles that are in direct contact with each other.

6. Conclusion

THz time-domain spectroscopy has been employed to study the THz transparency of densely packed ensembles of subwavelength size metallic particles. Experimental investigations of the THz transmission with respect to the sample length and particle size, with supporting evidence from numerical simulations based on FDTD calculations, indicate that the transmission is mediated by coherent, near-field electromagnetic coupling between nearest-neighbor particles. Transmission-based THz spectroscopy is applied as a non-invasive probe to study the phase transition of a metallic particle sample.

7. References

[1] C. F. Bohren and D. R. Huffman. *Absorption and Scattering of Light by Small Particles*. John Wiley & Sons, 1983.

[2] K. J. Chau, G. D. Dice, and A. Y. Elezzabi. Coherent plasmonic enhanced terahertz transmission through random metallic media. *Phys. Rev. Lett.*, 94:173904, 2005.

[3] K. J. Chau and A. Y. Elezzabi. Terahertz transmission through ensembles of subwavelength-size metallic particles. *Phys. Rev. B*, 72:075110, 2005.

[4] X. G. Gong, G. L. Chiarotti, M. Parrinello, and E. Tosatti. α-gallium: A metallic molecular crystal. *Phys. Rev. B*, 43:14277–14280, 1991.

[5] D. Grischkowsky, S. Keiding, M. v. Exter, and C. Fattinger. Far-infrared time-domain spectroscopy with terahertz beams of dielectrics and semiconductors. *J. Opt. Soc. B*, 7:2006–2015, 1990.

[6] S. C. Howells and L. A. Schlie. Transient terahertz reflection spectroscopy of undoped InSb from 0.1 to 1.1 THz. *Appl. Phys. Lett.*, 69:550–552, 1996.

[7] M. Kano. Adiabatic calorimeter for the purpose of calorimetry in the solid, liquid and supercooled phases of metals. *J. Phys. E*, 22:907–912, 1989.

[8] M. Khazan, I. Wilke, and C. Stevens. Surface impedance of T1-2212 thin films at THz-frequencies. *IEEE Trans. Appl. Super-cond.*, 11:3537–3540, 2001.

[9] P. Korpiun and R. Tilgner. The photoacoustic effect at first-order phase transition. *J. Appl. Phys.*, 51:6115–6119, 1980.

[10] A. V. Krasavin, A. V. Zayats, and N. I. Zheludev. Active control of surface plasmon-polaritons. *J. Opt. A*, 7:S85–S89, 2005.

[11] U. Kreibig and M. Vollmer. *Optical Properties of Metal Clusters*. Springer-Verlag, 1995.

[12] F. A. McDonald and G. C. Wetsel Jr. Generalized theory of the photoacoustic effect. *J. Appl. Phys.*, 49:2313–2322, 1978.

[13] E. V. Meija-Uriarte, M. Nararrete, and M. Villagran-Muniz. Signal processing in photoacoustic detection of phase transitions by means of the autospectra correlation-based method: Evaluation with ceramic $BaTiO_3$. *Rev. Sci. Instrum.*, 75:2887–2891, 2004.

[14] D. Mittleman. *Sensing with Terahertz Radiation*. Springer, 2003.

[15] D. M. Mittleman, R. H. Jacobsen, R. Neelamani, R. G. Baraniuk, and M. C. Nuss. Gas sensing using terahertz timedomain spectroscopy. *Appl. Phys. B*, 67:379–390, 1998.

[16] S. Mujumdar, K. J. Chau, and A. Y. Elezzabi. Experimental and numerical investigation of terahertz transmission through strongly scattering sub-wavelength size spheres. *Appl. Phys. Lett.*, 85:6284–6286, 2004.

[17] S. Nashima, O. Morikawa, K. Takata, and M. Hangyo. Measurement of optical properties of highly doped silicon by terahertz time domain reflection spectroscopy. *Appl. Phys. Lett.*, 79:3923–3925, 2001.

[18] A. Pashkin, M. Kempa, H. Nemec, F. Kadlec, and P. Kuzel. Phase-sensitive time-domain terahertz reflection spectroscopy. *Rev. Sci. Intrum.*, 74:4711–4717, 2003.

[19] V. M. Shalaev. Electromagnetic properties of small-particle composites. *Phys. Rep.*, 272:61–137, 1996.

11

Applications of Tilted-Pulse-Front Excitation

József András Fülöp and János Hebling
University of Pécs, Department of Experimental Physics
Hungary

1. Introduction

Tilting the pump pulse front has been proposed for efficient phase-matched THz generation by optical rectification of femtosecond laser pulses in $LiNbO_3$ (Hebling et al., 2002). By using amplified Ti:sapphire laser systems for pumping, this technique has recently resulted in generation of near-single-cycle THz pulses with energies on the 10-μJ scale (Yeh et al., 2007, Stepanov et al., 2008). Such high-energy THz pulses have opened up the field of sub-picosecond THz nonlinear optics and spectroscopy (Gaal et al., 2006, Hebling et al., 2008a).

The method of tilted-pulse-front pumping (TPFP) was introduced as a synchronization technique between the optical pump pulse and the generated THz radiation. Synchronization was accomplished by matching the group velocity of the optical pump pulse to the phase velocity of the THz wave in a noncollinear propagation geometry. Originally, TPFP was introduced for synchronization of amplified and excitation pulses in so called traveling-wave laser amplifiers (Bor et al., 1983). By using such traveling-wave excitation (TWE) of laser materials, especially dye solutions, extremely high gain (10^9) and reduced amplified spontaneous emission could be obtained (Hebling et al., 1991).

Contrary to the case of TWE, when TPFP is used for THz generation by optical rectification, a wave-vector (momentum) conservation condition or, equivalently, a phase-matching condition has to be fulfilled. It was shown (Hebling et al., 2002), that such condition is automatically fulfilled if the synchronization (velocity matching) is accomplished. The reason is that in any tilted pulse front there is present an angular dispersion of the spectral components of the ultrashort light pulse and there is a unique connection between the tilt angle of the pulse front and the angular dispersion (Bor & Rácz, 1985, Martínez 1986, Hebling 1996).

Angular dispersion was introduced into the excitation beam of so called achromatic frequency doubler (Szabó & Bor, 1990, Martínez, 1989) and sum-frequency mixing (Hofmann et al., 1992) setups in order to achieve broadband frequency conversion and keeping the ultrashort pulse duration. It was pointed out that in non-collinear phase-matched optical parametric generators (OPG) and optical parametric amplifiers (OPA) tilted pulse fronts are expected (Di Trapani et al., 1995). TPFP was used in the non-collinear OPA (NOPA) producing sub-5-fs pulses (Kobayashi & Shirakawa, 2000). The different aspect of tilted pulse front and angular dispersion is usually not mentioned in these papers dealing with broadband frequency conversion.

It is well known that the bandwidth of parametric processes is connected to the relative group velocities of the interacting pulses (Harris, 1969). Phase matching to first order in frequency can also be formulated as matching the group velocities of (some of) the interacting pulses. In schemes utilizing angular dispersion for broadband frequency conversion it is important to consider the effect of angular dispersion on the group velocity for a precious connection between the Fourier-domain and the spatio-temporal descriptions.

In this Chapter, we give a comprehensive overview of the different types of applications relying on TPFP or angular dispersion with an emphasis on THz generation. The connection between pulse front tilt, group velocity and angular dispersion will be discussed for each type of application. The Chapter is organized as follows.

The introduction is followed by a discussion of the connection between pulse front tilt, group velocity and angular dispersion. The main part of the Chapter deals with the different types of applications. For the sake of simplicity we start with the applications relying on synchronization by tilting the pulse front. These include traveling-wave excitation of dye lasers, as well as possible future applications such as traveling-wave excitation of short-wavelength x-ray lasers, and ultrafast electron diffraction. Subsequently, applications based on achromatic phase matching for broadband frequency conversion will be discussed. Finally, high-field THz pulse generation by optical rectification of femtosecond laser pulses with tilted pulse front and its application to a new field of research, nonlinear THz optics and spectroscopy will be reviewed.

2. Pulse front tilt, group velocity, and angular dispersion

Tilting of the pulse front of picosecond pulses after traveling through a prism (Topp & Orner, 1975) or diffracting off a grating (Schiller & Alfano, 1980) was early recognized. Later, the following expression was deduced between the angular dispersion $\mathrm{d}\varepsilon/\mathrm{d}\lambda$ and the pulse front tilt γ created by the prism or the grating (Bor & Rácz, 1985):

$$\tan\gamma = -\bar{\lambda}\frac{\mathrm{d}\varepsilon}{\mathrm{d}\lambda}, \tag{1}$$

where γ is the tilt angle (the angle between the pulse front and the phase front, see Fig. 1), $\bar{\lambda}$ is the mean wavelength and $\mathrm{d}\varepsilon/\mathrm{d}\lambda$ is the angular dispersion. It was also shown that for a grating immersed in a material Eq. (1) has to be modified as (Szatmári et al., 1990):

$$\tan\gamma = -\frac{n}{n_{\mathrm{g}}}\bar{\lambda}\frac{\mathrm{d}\varepsilon}{\mathrm{d}\lambda}, \tag{2}$$

where n and n_{g} are the (phase) index of refraction and the group index of the material, respectively.

A device-independent derivation of Eqs. (1) and (2) is possible (Hebling, 1996) for ultrashort light pulses having large beam sizes. In this case the short pulse consists of plane-wave monochromatic components with different frequencies (wavelengths). Such a case is illustrated schematically in Fig. 2 assuming that the beam propagates, and has an angular dispersion in the x-z plane. Hence, the phases of the spectral components are independent of the third (y) coordinate, and the electric field of the components can be described as:

$$E_{\omega}(x,z,t) = E_{\omega}\cdot\sin(\omega t - k_x x - k_z z + \varphi_0) \tag{3}$$

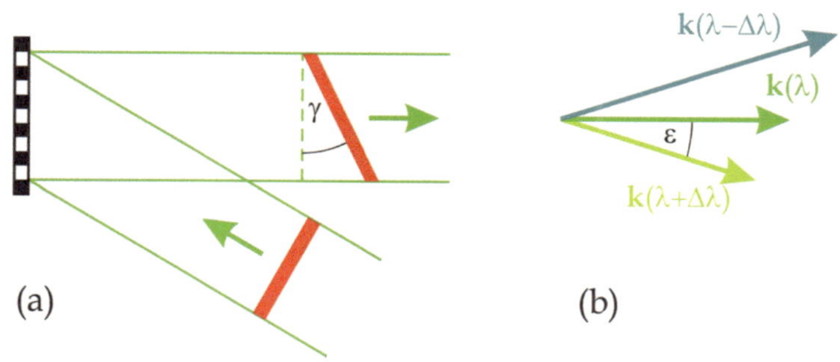

Fig. 1. (a) Pulse front tilt created by an optical element with angular dispersion (grating). (b) The corresponding angular dispersion of the wave vectors of the different spectral components of the ultrashort pulse.

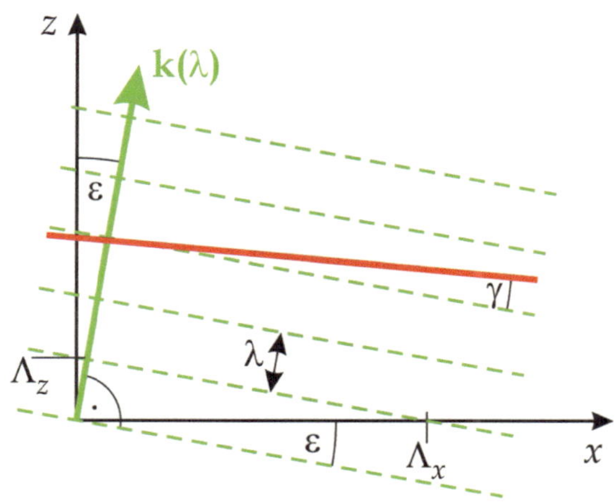

Fig. 2. Phase fronts (dashed lines) and pulse front (continuous line) for a light beam dispersed in the z–x plane. The phase front is indicated only for the mean wavelength. For different wavelengths the phase fronts are tilted relative to this. Positive angles are measured clockwise: $\varepsilon > 0$ and $\gamma < 0$.

Here $k_x = 2\pi/\Lambda_x = (2\pi/\lambda)\cdot\sin\varepsilon$ and $k_z = 2\pi/\Lambda_z = (2\pi/\lambda)\cdot\cos\varepsilon$ are the two components of the wave-vector, ε is the angle of propagation measured from the z-axis as shown in Fig. 2, and $\omega = 2\pi/\lambda$ is the angular frequency. Since for a phase front (a surface with constant phase) the argument of the sine function in Eq. (3) is constant, the slope of the phase front is given by:

$$m = -\frac{k_x}{k_z} = -\tan\varepsilon\ . \tag{4}$$

The pulse front at any time is the surface consisting of the points where the light intensity has a maximum. The intensity has maximum at points where the plane wave components with different frequencies have the same phase, i.e. where the derivative of the phase (the argument of the sine function in Eq. (3)) with respect to frequency is equal to zero. The result of such derivation is that the pulse front is plane with a slope of:

$$m_g = -\tan\gamma = -\tan\varepsilon + \frac{k_z}{\cos^2\varepsilon}\cdot\frac{\mathrm{d}\varepsilon}{\mathrm{d}k_z}, \tag{5}$$

where Eq. (4) was also used. If we choose the coordinate-system such that the main component propagates parallel to the *z*-axis, ε becomes equal to zero. With this choice, and by introducing the angular dispersion $\mathrm{d}\varepsilon/\mathrm{d}\lambda$ instead of $\mathrm{d}\varepsilon/\mathrm{d}k_z$, Eq. (5) reduces to Eq. (1). Since it was not necessary to suppose anything about the device which created the angular dispersion, with the above derivation we proved that the relation between the angular dispersion and the pulse front tilt as given in Eq. (1) is universal.

In order to prove the more general relationship given by Eq. (2) we suppose that the beam with angular dispersion propagates in a medium with wavelength (frequency) dependent index of refraction $n(\lambda)$. In this case the two components of the wave-vector of the plane wave, propagating in the direction determined by ε, are given by $k_x = 2\pi n(\lambda)/\Lambda_x = $ $= \left[2\pi n(\lambda)/\lambda\right]\cdot\sin\varepsilon$ and $k_z = 2\pi n(\lambda)/\Lambda_z = \left[2\pi n(\lambda)/\lambda\right]\cdot\cos\varepsilon$, respectively. By using these wave-vector components in the same derivation as above, one obtains Eq. (2) (Hebling, 1996). Again, since the derivation is independent of any device parameters, the relationship between the angular dispersion and the pulse front tilt as given by Eq. (2) is universal.

We can easily obtain (the reciprocal of) the group velocity of a short light pulse in the presence of angular dispersion. To this end we rewrite k_z by introducing the frequency instead of the wavelength as independent variable:

$$k_z = \frac{n(\omega)\omega}{c}\cos\varepsilon . \tag{6}$$

Since the group velocity is equal to the derivative of the angular frequency with respect to the wave vector (Main, 1978), we obtain for the reciprocal of the (sweep) group velocity along the *z* axis:

$$v_g^{-1} = \frac{\mathrm{d}k_z}{\mathrm{d}\omega} = \frac{1}{c}\left[n\cdot\cos\varepsilon + \omega\left(\cos\varepsilon\cdot\frac{\mathrm{d}n}{\mathrm{d}\omega} - \sin\varepsilon\cdot n\cdot\frac{\mathrm{d}\varepsilon}{\mathrm{d}\omega}\right)\right]. \tag{7}$$

Although this depends on the angular dispersion, the reciprocal of the group velocity is independent of it. Really, using $\varepsilon = 0$ in Eq. (7) results in the well known expression:

$$v_g^{-1} = \frac{1}{c}\left(n + \omega\cdot\frac{\mathrm{d}n}{\mathrm{d}\omega}\right) = \frac{n_g}{c}, \tag{8}$$

where $n_g = c / v_g$ is the usual group index.

It is important to notice that the frequency derivative of the reciprocal of the group velocity depends on the angular dispersion. The most important and well known implication of this is the working of pulse compressors consisting of prism or grating pairs. In such compressors angular dispersion is present in the light beam during the path between the two dispersive elements (prisms or gratings). The (negative) group delay dispersion (GDD) of such compressor is given as:

$$\mathrm{GDD} = l\cdot\frac{\mathrm{d}v_g^{-1}}{\mathrm{d}\omega}, \tag{9}$$

where l is the distance between the dispersive elements along the beam path. According to this, by taking the frequency derivative of the reciprocal of the group velocity as given by Eq. (7), substituting $\varepsilon = 0$ and multiplying by l result in the general expression for the GDD caused by propagation in the presence of angular dispersion:

$$\mathrm{GDD} = \frac{l}{c} \cdot \left[2 \cdot \frac{\mathrm{d}n}{\mathrm{d}\omega} + \omega \cdot \frac{\mathrm{d}^2 n}{\mathrm{d}\omega^2} - \omega \cdot n \cdot \left(\frac{\mathrm{d}\varepsilon}{\mathrm{d}\omega} \right)^2 \right], \tag{10}$$

in accordance with (Martínez et al., 1984). For the case of prism or grating compressors, with a large accuracy, $n \equiv 1$ and Eq. (10) simplifies to Eq. (10b), and the GDD is always negative:

$$\mathrm{GDD} = -\frac{l \cdot \omega}{c} \cdot \left(\frac{\mathrm{d}\varepsilon}{\mathrm{d}\omega} \right)^2 . \tag{10b}$$

When pulse front tilt (or equivalently, angular dispersion) is introduced into a beam of ultrashort pulses in order to achieve achromatic frequency conversion or synchronization of pump and generated pulses (see examples below), $n \not\equiv 1$ in the medium, and the full expression of Eq. (10) has to be considered. Since the first and second terms on the right hand side of Eq. (10) are usually positive, and the third term always negative, the effect of the angular dispersion and the material dispersion can sometimes compensate each other. If, however, a very large angular dispersion is needed the third term becomes much higher than the first two ones, and it causes a rapid change of the pulse length during propagation hindering efficient frequency conversion.

In the above derivation plane wave components with infinite transversal extension were assumed. For finite beam sizes a more complicated derivation (Martínez, 1986) is needed. According to this, the tilt angle changes with propagation distance and besides angular dispersion also spatial dispersion will be present. The spatio-temporal distortions in this case are described and investigated by an elegant theory (Akturk et al., 2005). According to numerical calculations, however, such distortions are usually not significant on a distance smaller than the beam size. This condition is typically fulfilled in frequency conversion processes of high-energy ultrashort pulses.

Finally, we have to recognize that a strong restriction was used in the above discussions, namely, an isotropic index of refraction was assumed. However, it is well known that in frequency conversion processes at least one of the beams involved has extraordinary polarization with a refractive index depending on propagation direction. (The only possible exception is quasi-phase-matching.) Because of this, it is essential to re-consider the above derivations. Let us first investigate again the group index! If we just apply the definition of the group index as it was introduced in Eq. (8) and take into account that besides the explicit frequency dependence of the refractive index, in the presence of angular dispersion an implicit dependence $n^*(\omega) = n(\omega, \varepsilon(\omega))$ can also be present, we obtain:

$$v_{\mathrm{g}}^{-1} = \frac{1}{c}\left(n^* + \omega \cdot \frac{\mathrm{d}n^*}{\mathrm{d}\omega} \right) = \frac{1}{c}\left(n + \omega \cdot \frac{\partial n}{\partial \omega} + \omega \cdot \frac{\partial n}{\partial \varepsilon} \cdot \frac{\mathrm{d}\varepsilon}{\mathrm{d}\omega} \right) = \frac{1}{c}\left(n_{\mathrm{g}} + \omega \cdot \frac{\partial n}{\partial \varepsilon} \cdot \frac{\mathrm{d}\varepsilon}{\mathrm{d}\omega} \right) = \frac{n_{\mathrm{g}}^*}{c} . \tag{11}$$

According to Eq. (11), in the presence of angular dispersion (which is in a plane containing the optical axis) an n_{g}^* modified group index is effective. Depending on the signs of the angular dependence of the refractive index and that of the angular dispersion, the effective group index can be either larger or smaller than the (usual, material) group index. Furthermore, the value of the effective group index and that of the group velocity can be set by adjusting the angular dispersion. (A more rigorous derivation staring from Eq. (7) results in the same expression as Eq. (11).) We note that even though extraordinary propagation of an angularly dispersed beam is a common situation in many types of frequency conversion schemes (see Section 4), not much attention has been payed previously to generalize the definition of group velocity to such a case.

Describing the angular dispersion with the frequency dependence of the angle instead of its wavelength dependence, the relation between the pulse front tilt angle and the angular dispersion becomes:

$$\tan\gamma = \frac{n}{n_{\mathrm{g}}}\bar{\omega}\frac{\mathrm{d}\varepsilon}{\mathrm{d}\omega}. \tag{2b}$$

By using a similar calculation as above, it is easy to show that for anisotropic materials Eq. (2b) is modified similarly to the modification of the group velocity:

$$\tan\gamma = \frac{n}{n_{\mathrm{g}}^{*}}\bar{\omega}\frac{\mathrm{d}\varepsilon}{\mathrm{d}\omega}. \tag{12}$$

Both the group velocity and the tangent of the tilt angle are changing by a factor of $n_{\mathrm{g}} / n_{\mathrm{g}}^{*}$ in presence of angular dispersion in anisotropic materials.

In conclusion, if in the beam of an ultrashort pulse angular dispersion is present, then necessarily a pulse front tilt is also present. Eq. (12) gives the relationship between the pulse front tilt and the angular dispersion. Furthermore, we have shown that contrary to the isotropic case, for anisotropic media the group velocity of the light pulse depends on the angular dispersion, too. Hence, the group velocity can be adjusted by adjusting the angular dispersion. This is utilized in many broadband frequency conversion schemes where the group velocities of interacting pulses are matched in such a way. We will show examples for this in Section 4.

3. Pulse front tilt for synchronization

In the first group of applications of TPFP the tilt of the intensity front is used to achieve the same sweep velocity of the pump pulse along a surface or along a volume close to a surface as the velocity of the generated excitation (i.e. amplified spontaneous emission, or surface polariton), or to the velocity of an other pulse (i.e. electron packet). In this group of applications angular dispersion is not an issue.

3.1 Traveling-wave excitation of lasers

Tilted-pulse-front excitation for generation of amplified spontaneous emission (ASE) pulses in dye solutions was introduced in 1983 by two groups independently (Polland et al., 1983, Bor et al., 1983). Both groups applied transversal pumping geometry, that is the pump pulses illuminated the long side of the dye cell perpendicularly, and the generated ASE pulse propagated along the surface in the pencil-like excited volume (see Fig. 3). The

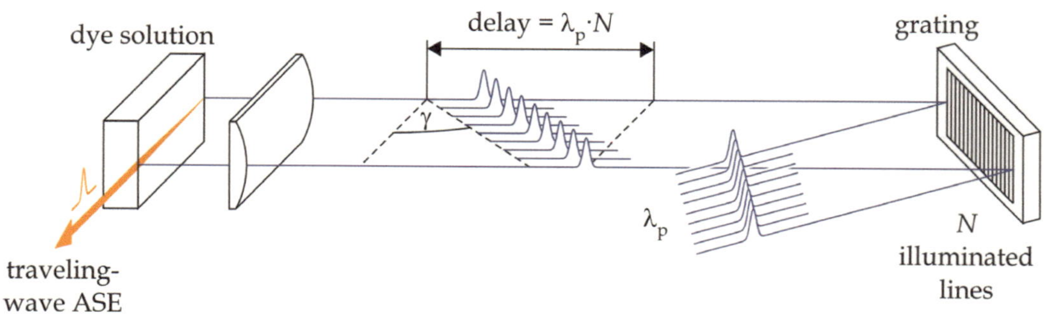

Fig. 3. Ultrashort (ps) light pulse generation by traveling-wave pumped ASE (Bor et al., 1983).

technique of diffracting the pump pulses off an optical grating was applied to create the tilted pulse front (Schiller & Alfano, 1980). The tilt angle γ was chosen to fulfill the condition $\tan\gamma = n_g$, where $n_g = c/v_g$ is the group refractive index of the dye solution at the mean ASE wavelength. In this way the pump pulse swept the surface of the dye solution with the group velocity of the ASE pulse inside the dye solution. This situation is called traveling-wave excitation(TWE). The exact temporal overlap between the pump pulse and the generated ASE pulse allowed an effective use of the pump energy even for dyes having excited state lifetimes of only a few ps. For example, 2% of the pump energy was converted into the energy of the ASE pulse for IR dyes that have less than 10^{-3} fluorescence quantum efficiency (Polland et al., 1983) due to the short lifetime dictated by fast non-radiative processes. The reason for the much higher energy conversion efficiency of the traveling-wave pumped amplifier as compared to the fluorescence quantum efficiency is that in the traveling-wave pumped amplifier the dye molecules are in excited state only for a short duration at every point of the amplifier because of the synchronism between the pump and the generated pulse. Another consequence of this is that the TWE resulted in two times shorter (6 ps) ASE pulse duration than that of the pump pulse (Bor et al., 1983). Using the TWE scheme in distributed feedback dye lasers resulted in sub-ps Fourier-transform limited pulses (Szabó et al. 1984).

It was also possible to use TWE efficiently with fs pump pulses (Hebling & Kuhl, 1989a, Hebling & Kuhl, 1989b, Klebniczki et al. 1990, Hebling et al., 1991). In these experiments TWE was achieved by non-perpendicular excitation and by the pulse front tilt introduced by a glass prism contacted to the dye cell, as shown in Fig. 4. By seeding the traveling-wave amplifier by white-light continuum pulses stretched in BK7 glass it was possible to generate about 100 fs long tunable pulses in the 640–680 nm range (Klebniczki et al., 1990). Furthermore, by seeding with attenuated continuous light of a Kr^+ laser, it was demonstrated, that the traveling-wave pumped dye amplifier can work as a gated amplifier with 100 fs gate window and as large as 10^9 gain (Hebling et al., 1991). According to the measurements as well as to model calculations for exact synchronisms the duration of the gate window is approximately equal to the pump pulse duration (somewhat shorter), while for different pump sweep and ASE pulse velocities it is equal to the difference of the times the pump and ASE pulses need to travel the length of the amplifier.

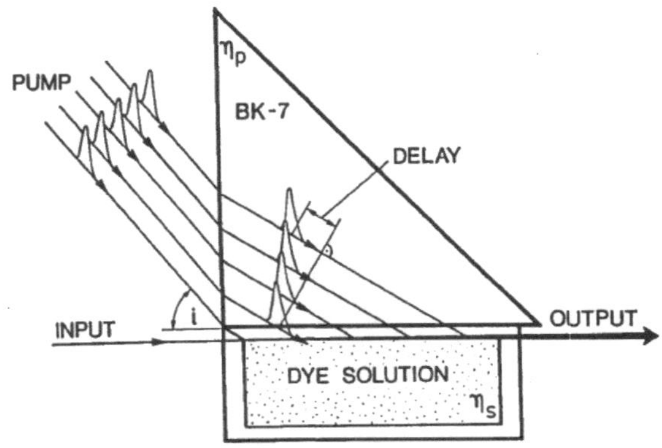

Fig. 4. Non-perpendicular TWE for ultrashort (fs) light pulse generation (Hebling et al., 1991).

3.2 Traveling-wave excitation of X-ray lasers

Laser action at extreme ultraviolet (XUV) and x-ray wavelengths is of great practical importance for many applications. For biological applications especially the so-called water window (4.4–2.2 nm) is of great interest. X-ray laser action can be achieved in highly ionized plasmas or in free electron lasers. Many plasma-based x-ray laser schemes use short laser pulses for creating the plasma and the population inversion required for x-ray lasing. In many cases reported so far, the pumping mechanism is either electron collisional excitation of neon-like or nickel-like ions, or inner-shell excitation or ionization processes (Daido, 2002, and references therein).

One of the main difficulties associated with x-ray lasing is the short lifetime of the excited states, which typically scales as λ^2 (Simon et al., 2005), where λ is the wavelength. In the 1 nm - 100 nm region, spontaneous transition rates correspond to 0.1 ps^{-1} - 10 ps^{-1} (Kapteyn, 1992). Processes with even shorter time constants, such as Auger decay of inner-shell vacancies with typical decay rates of 1 fs^{-1} - 10 fs^{-1}, can in many schemes impose further constraints on the required pumping rate (Kapteyn, 1992). Due to the short time available for the population inversion to build up, ultrashort pump pulses can be used to excite population inversion efficiently.

Due to the decreasing rate of stimulated emission with decreasing wavelength (Simon et al., 2005), in order to obtain a useful output level from an x-ray laser it is essential to provide population inversion over sufficient length, typically a few millimeters up to a few centimeters. This, together with the short amount of time available for population inversion requires using ultrashort pump pulses with a traveling-wave pumping geometry, which provides exact synchronization between the pump pulse and the generated x-ray pulse.

Sher et al. (Sher et al., 1987) have used grazing incidence excitation for nearly synchronous traveling-wave pumping of an XUV laser at 109 nm wavelength in Xe (Fig. 5). Later, a modified setup with grating-assisted traveling-wave geometry was used, where a grating pair introduced a small tilt of the pump pulse front for more exact pump-XUV synchronization (Barty, et al., 1988). However, in these early experiments a grooved target had to be used in order to compensate for the reduced pumping efficiency caused by the grazing incidence geometry. Kawachi et al. (Kawachi, et al., 2002) employed a quasi traveling-wave pumping scheme using a step mirror, which was installed in the line-focusing system to excite the nickel-like silver and tin x-ray lasers at the wavelengths of 13.9 and 12.0 nm.

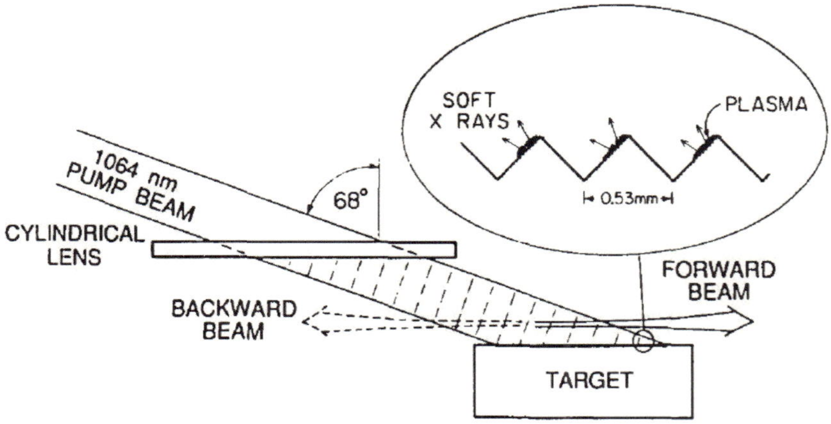

Fig. 5. Grazing incidence x-ray laser pumping scheme (Sher, et al., 1987).

On the way towards achieving x-ray lasing from laser produced plasmas with photon energies approaching the keV milestone, an important step, besides the investigation of fundamental physics concerned with the creation of population inversion, is to develop extremely accurate and controllable traveling-wave pumping systems (Daido, 2002). We would like to emphasize that the existing technique of tilted-pulse-front pumping (see Section 5) can provide the required tools for pump–x-ray synchronization with femtosecond accuracy. It also allows for working with perpendicular incidence of the pump beam onto the target. This gives higher effective pump intensity due to reduced pump spot size and due to improved energy coupling-in efficiency allowed by the reduced reflection coefficient from the plasma surface. A possible experimental setup is shown in Fig. 6. The pulse-front-tilting setup made up of a grating and a spherical focusing lens is extended by a concave cylindrical lens, which shifts the focus of the pump beam introduced by the spherical lens to the surface of the target by its defocusing effect in the direction perpendicular to the plane of the drawing. Thus, in the plane of the drawing a line focus is generated at the target. The pump intensity can also be increased by using demagnifying imaging.

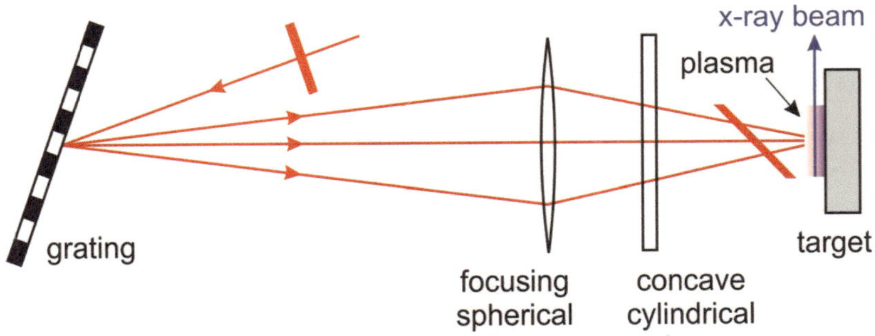

Fig. 6. Tilted-pulse-front pumping scheme for exactly synchronized excitation of an x-ray laser from laser-produced plasma.

3.3 Ultrafast electron diffraction

In recent years it became possible to achieve atomic-scale resolution simultaneously in time and space in revealing structures and dynamics. One important tool for such high-resolution studies is ultrafast electron diffraction and microscopy (Baum & Zewail, 2006). In these techniques an ultrashort (femtosecond) laser pulse is used to initiate a change in the sample. The dynamics is probed by an ultrafast electron pulse. Besides the temporal spread of the electron pulse due to space charge effect the difference in the group velocities of the optical and the electron pulses can impose an often more severe limitation on the temporal resolution (Fig. 7(a)). This mismatch in the propagation velocities along the sample becomes especially significant in case of ultrafast electron crystallography, where the electrons probe the sample at grazing incidence and the laser pulse triggering the dynamics has (nearly) perpendicular incidence (Fig. 7(b)).

Baum & Zewail (Baum & Zewail, 2006) proposed to use a traveling-wave type excitation scheme with tilted laser pulse front for exact synchronization with the probing electron pulse (Fig. 7(c–e)). They demonstrated a 25-fold reduction in time spread. They also discuss limitations in time resolution due to the possible curvature of the tilted pulse front. Furthermore, they proposed tilting the electron packet (generated by a tilted optical pulse illuminating a photocathode) to overcome space charge problems that are especially important in single-shot experiments.

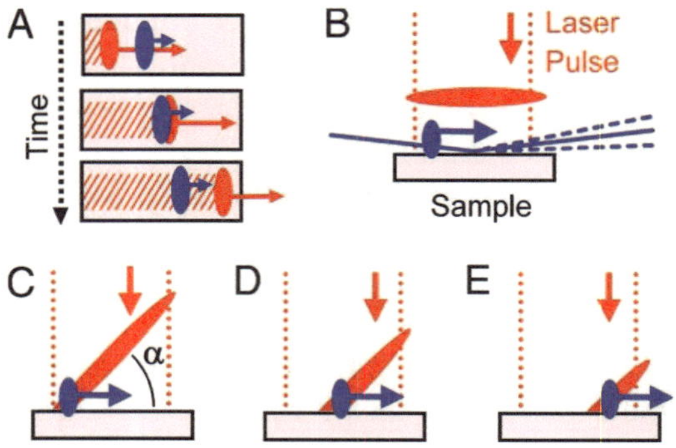

Fig. 7. (a) Group velocity mismatch in ultrafast electron diffraction. (b) Ultrafast electron crystallography without laser pulse front tilt. (c–e) Traveling-wave excitation of the sample by tilting the optical pump pulse front for synchronization with the slow (33% of *c*) electron pulse (Baum & Zewail, 2006). Copyright 2006 National Academy of Sciences, U.S.A.

4. Achromatic phase matching

In this part we consider various achromatic phase matching schemes used for frequency conversion or parametric amplification of ultrashort laser pulses. In this group of applications angular dispersion plays a key role, which affects the bandwidth of nonlinear processes. It was early recognized that achromatic phase matching is related to group velocity matching of the interacting pulses (Harris, 1969). An often neglected aspect in such schemes is the pulse front tilt linked to angular dispersion. The following discussion will include this aspect, too.

In frequency conversion of ultrashort laser pulses the bandwidth of the nonlinear material is of crucial importance. In nonlinear processes such as second- and third-harmonic generation, sum-frequency generation, optical parametric amplification, etc., broadband phase matching usually requires the use of thin nonlinear crystals. This, in turn, seriously limits the efficiency of frequency conversion. One way to overcome this limitation is to adopt the technique of achromatic phase matching to frequency conversion of ultrashort laser pulses.

Many different schemes have been proposed and used for various nonlinear optical processes with ultrashort laser pulses. The theoretical analysis of these schemes was carried out in most cases either in the Fourier domain (in terms of wave vectors and frequencies) or in the spatio-temporal domain. Little or no attention was paid to the connection of the two distinct descriptions. Even though descriptions in the two domains are equivalent, to consider the connection between them gives a more complete picture and can, in some cases, reveal new important features, which can be relevant for designing an experimental setup. As examples, in the following we will consider collinear and non-collinear achromatic second-harmonic generation (SHG), non-collinear achromatic sum-frequency generation (SFG), and NOPA with and without angular dispersion of the signal beam.

4.1 Collinear achromatic SHG

Achromatic phase matching in nonlinear frequency conversion was originally introduced for automatic, i.e. alignment-free phase matching in second harmonic generation (SHG) of tunable monochromatic laser sources (Saikan, 1976; see also references in Szabó & Bor, 1990). The technique relies on sending the pump beam at each wavelength through a birefringent nonlinear crystal under its respective phase-matching angle, which varies with wavelength. To this end, an element with appropriate angular dispersion was used in front of the crystal. In such a way the effective bandwidth of the nonlinear crystal could be increased considerably.

The femtosecond frequency doubler proposed independently by Szabó & Bor (Szabó & Bor, 1990; Szabó & Bor, 1994) and Martínez (Martínez, 1989) adopts the principle of automatic phase matching to ultrashort pulses. It utilizes the angular dispersion of gratings to achieve collinear achromatic phase matching in the nonlinear crystal and, subsequently, to eliminate the angular dispersion from the generated second-harmonic radiation (Fig. 8).

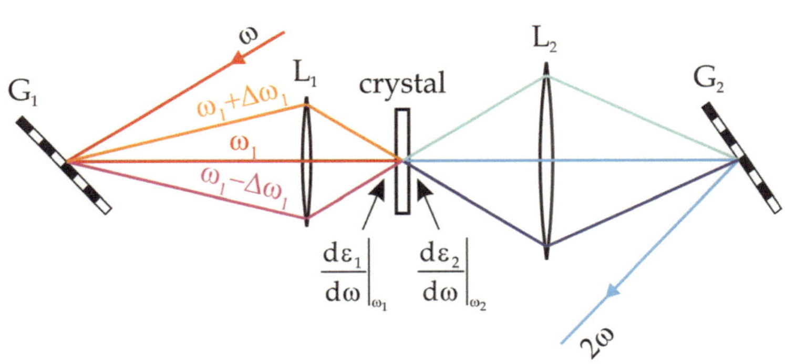

Fig. 8. Femtosecond frequency doubler based on collinear achromatic phase matching (Szabó & Bor, 1990). G_1, G_2: gratings, L_1, L_2: lenses.

Martínez (Martínez, 1989) gave a thorough analysis of the scheme in the Fourier-domain, without discussing the spatio-temporal implications. The achromaticity of the scheme relies on sending each fundamental frequency component into the crystal under its respective phase-matching angle. The phase matching condition at the fundamental and second harmonic central frequency components ω_1 and $\omega_2 = 2\omega_1$, respectively, reads as:

$$\mathbf{k}_2(\omega_2 = 2\omega_1) = 2\mathbf{k}_1(\omega_1), \tag{13}$$

where $\mathbf{k}_1$ and $\mathbf{k}_2$ are the fundamental and second harmonic wave vectors, respectively. Their magnitude is given by $k_i(\omega) = \omega_i \cdot n_i(\omega)/c$ with $i = 1, 2$. Eq. (13) implies that $n_1(\omega_1) = n_2(\omega_2)$. The condition for achromatic phase matching can be given as follows:

$$\mathbf{k}_2(\omega_2' = 2\omega_1 + \Delta\omega_1 + \Delta\omega_1') = \mathbf{k}_1(\omega_1 + \Delta\omega_1) + \mathbf{k}_1(\omega_1 + \Delta\omega_1'), \tag{14}$$

where $\Delta\omega_1$ and $\Delta\omega_1'$ are arbitrary detunings from the fundamental central frequency. Martínez showed that Eq. (14) implies

$$\left.\frac{dn_1^*}{d\omega}\right|_{\omega_1} = 2 \cdot \left.\frac{dn_2^*}{d\omega}\right|_{\omega_2}. \tag{15}$$

Please note that by writing n^* rather than simply n in Eq. (15), we allow that any of the two pulses may have extraordinary propagation. (In case of birefringent phase matching at least one of them must, in fact, be extraordinary.) In practical setups, such as that in Fig. 8, the required angular dispersion is usually matched only to first order by the dispersive optics and higher-order mismatch still imposes bandwidth limitation. Matching of angular dispersion to higher order was demonstrated with optimized setups and tunable cw lasers (Richman et al., 1998).

It is now straightforward to find the connection to the spatio-temporal description of the arrangement. According to Eq. (11), and the phase matching condition $n_1(\omega_1) = n_2(\omega_2)$ together with Eq. (15) imply that the fundamental and the second-harmonic group velocities are matched: $v_{g1} = v_{g2}$. On the other hand, it follows from the collinear geometry for each $(\omega_1', 2\omega_1')$ frequency pair (Fig. 8) that the angular dispersion of the fundamental is twice that of the second harmonic:

$$\left.\frac{d\varepsilon_1}{d\omega}\right|_{\omega_1} = 2 \cdot \left.\frac{d\varepsilon_2}{d\omega}\right|_{\omega_2}, \tag{16}$$

where $\varepsilon(\omega)$ is the angle between the optic axis and the wave vector of the respective frequency component. By using s, the matching of group velocities $v_{g1} = v_{g2}$, and Eqs. (12) and (16), one obtains that also the fundamental and second-harmonic pulse fronts are matched: $\gamma_1 = \gamma_2$.

In summary, we have shown that for the collinear SHG scheme (Fig. 8) achromatic phase matching to first order is equivalent to simultaneous group-velocity and pulse-front matching between the fundamental and the generated SHG pulses. Pulse-front matching allows the scheme to be used with large beam sizes.

4.2 NOPA

In recent years the technique of OPA (see e.g. Cerullo & De Silvestri, 2003, and references therein) has opened up a new path towards generating few-cycle laser pulses with unprecedented peak powers. Such pulses are crucial for the investigation of laser-driven strong-field phenomena, having become an accessible field of research with the advent of suitable laser systems.

In the OPA process phase-matching in a non-collinear geometry (NOPA) allows for extremely large amplification bandwidths, and thereby enables the amplification of broadband seed pulses using relatively narrowband pump pulses delivered by conventional laser amplification technology. Usually, a narrowband pump pulse is used, and the signal is stretched in time. The non-collinearity angle between signal and pump introduces an additional adjustable parameter besides the crystal orientation angle, which allows achieving phase matching to first order in signal frequency (achromatic phase matching). In such a setup the broadband signal beam has no angular dispersion while the idler beam generated in the amplification process has one. In such a scheme, achromatic phase matching is related to matching the (projected) group velocities of the signal and the idler pulses. Due to the long pulse durations, pulse front matching is usually not a practical issue.

Recently, the use of short (few-ps to sub-ps) pump pulses was proposed for generation of high-power few-cycle pulses (Fülöp et al., 2007, Major et al., 2009). In such a high-energy short-pulse-pumped NOPA pulse front matching between signal and pump is crucial to enable amplification with large beam sizes. This can be conveniently accomplished by introducing a small amount of angular dispersion into the pump (Kobayashi & Shirakawa, 2000, Fülöp et al., 2007).

As the above examples show, achromatic phase matching can also be achieved without appropriate pulse front matching in certain schemes. Therefore, in case of large beam sizes, which is the typical situation in high-power applications, it is important to consider both the Fourier domain as well as the spatio-temporal domain when designing frequency converting or OPA stages.

4.3 Non-collinear achromatic SFG

In NOPA schemes an additional degree of freedom can be provided by introducing angular dispersion into the signal beam. As we will show below, in such a scheme simultaneous group velocity and pulse front matching is achieved for all three interacting pulses in case of achromatic phase matching. We will discuss the scheme as achromatic SFG (the inverse process of OPA), which makes its symmetry with respect to signal and idler more obvious. The results are valid both for NOPA and SFG.

A special case of achromatic SFG is the non-collinear SHG scheme with two identical input beams proposed by Zhang et al. (Zhang et al., 1990), which utilizes the angular dispersion of prisms for achromatic phase matching. Each of the two incoming beams of the same central frequency ω_1 is dispersed in separate prisms to have angular dispersions equal in magnitudes but opposite in signs. The dispersed beams enter the nonlinear crystal under opposite angles of incidence. The emerging second-harmonic beam propagates along the angle bisector of the input beams and has no angular dispersion. Zhang et al. showed that simultaneous phase and group velocity matching is possible in such a scheme in BBO.

In the non-collinear achromatic SFG scheme shown in Fig. 9 two ultrashort pulses with central frequencies ω_1 and ω_2 are entering the nonlinear crystal under different angles α_1 and α_2, respectively. These non-collinearity angles are measured from the propagation direction of the generated SFG beam having the central frequency $\omega_3 = \omega_1 + \omega_2$. The incidence angles and the angular dispersions $\mathrm{d}\alpha_1 / \mathrm{d}\omega|_{\omega_1}$ and $\mathrm{d}\alpha_2 / \mathrm{d}\omega|_{\omega_2}$ of the two incoming beams are chosen such that the output SFG beam is generated with maximal bandwidth (achromatic phase matching) and without angular dispersion. The latter condition is important for practical applications of the scheme.

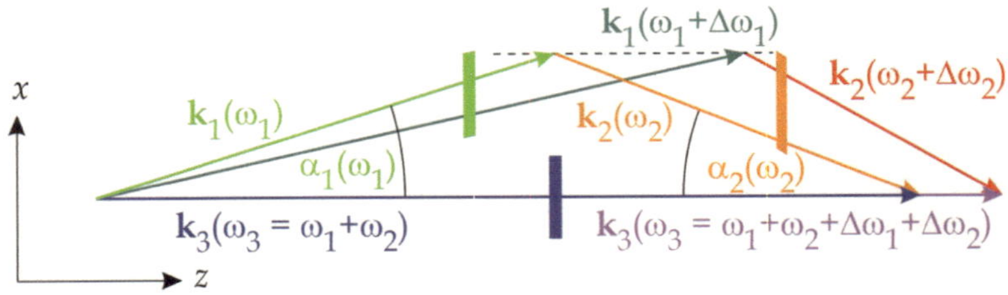

Fig. 9. Non-collinear achromatic SFG (or NOPA) scheme with angularly dispersed input beams. The corresponding pulse fronts are indicated by the vertical bars.

Let us first consider the scheme in the Fourier domain. The phase matching condition at the central frequencies can be described in terms of the wave vectors as

$$\mathbf{k}_3(\omega_3 = \omega_1 + \omega_2) = \mathbf{k}_1(\omega_1) + \mathbf{k}_2(\omega_2), \tag{17}$$

where $k_i(\omega) = |\mathbf{k}_i(\omega)| = \omega_i \cdot n_i(\omega)/c$ with $i = 1,2,3$, and n_i being the respective refractive index, which can be either ordinary or extraordinary. Depending on the type of phase matching one or more of the interacting beams propagate as extraordinary waves in the birefringent nonlinear medium. The condition for achromatic phase matching can be written as

$$\mathbf{k}_3(\omega_3' = \omega_1 + \omega_2 + \Delta\omega_1 + \Delta\omega_2) = \mathbf{k}_1(\omega_1 + \Delta\omega_1) + \mathbf{k}_2(\omega_2 + \Delta\omega_2), \tag{18}$$

where $\Delta\omega_1$ and $\Delta\omega_2$ are arbitrary detunings from the input central frequencies ω_1 and ω_2 within the bandwidth of the input pulses. The independent values of $\Delta\omega_1$ and $\Delta\omega_2$ reflect the fact that all possible combinations of the spectral components of the input pulses contribute to the SFG process. Since the generated SFG beam is not allowed to have any angular dispersion, the direction of $\mathbf{k}_3$ is along the *z*-axis for all sum-frequency components. Usually, in a practical setup it is sufficient to fulfill the achromatic phase matching condition, Eq. (18), only to first order in the frequency offsets $\Delta\omega_1$ and $\Delta\omega_2$. The non-collinearity angles α_1 and α_2, as well as the angular dispersions $\mathrm{d}\alpha_1/\mathrm{d}\omega|_{\omega_1}$ and $\mathrm{d}\alpha_2/\mathrm{d}\omega|_{\omega_2}$ can be found by decomposing Eq. (18) into components parallel with (*z*-axis) and perpendicular to (*x*-axis) $\mathbf{k}_3$, and by keeping only the zero- and first-order terms in the frequency offsets (first-order achromatic phase matching). Similar calculation was introduced by Martínez for collinear achromatic SHG (Martínez, 1989), as was discussed above in Section 4.1.

Let us now consider the spatio-temporal implications of Eq. (18), the achromatic phase matching condition. In the following we will outline the main results; details of the calculation will be given elsewhere (Fülöp & Hebling, to be published). The setup corresponding to the phase matching scheme shown in Fig. 9 is depicted in Fig. 10.

The terms first-order in $\Delta\omega_1$ and $\Delta\omega_2$ of the *x*-component of Eq. (18) imply the following relation between the incidence angles α_i and the corresponding angular dispersions $\mathrm{d}\alpha_i/\mathrm{d}\omega|_{\omega_i}$:

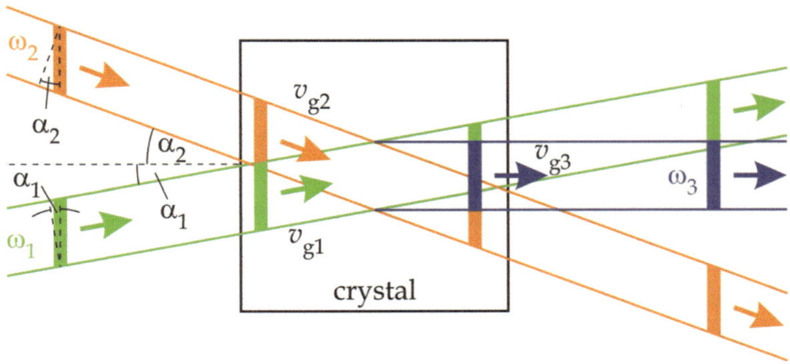

Fig. 10. Matching of the pulse fronts and the projected group velocities in the non-collinear achromatic SFG scheme. In case of NOPA the signal (ω_1) and pump (ω_3) beams are present at the input, and the idler (ω_2) is missing.

$$\tan(\alpha_i) = \frac{n_i}{n_{gi}^*}\omega_i \left.\frac{\mathrm{d}\alpha_i}{\mathrm{d}\omega}\right|_{\omega_i}, \quad (i=1,2). \tag{19}$$

Please note that it is the effective group index n_g^* which enters Eq. (19). By comparing Eq. (19) to Eq. (12) one can see that the incidence angles of the incoming beams are equal to their respective pulse front tilt angles γ: $\alpha_1 = \gamma_1$ and $\alpha_2 = \gamma_2$. Hence, the pulse fronts of both incoming beams are perpendicular to the propagation direction of the generated SFG beam (z-axis). This means that the pulse fronts of all three interacting pulses are matched to each other.

By expressing the angular dispersions from Eq. (19) and inserting them into the first-order z-component of Eq. (18), one obtains

$$v_{g1}\cos(\alpha_1) = v_{g2}\cos(\alpha_2) = v_{g3}. \tag{20}$$

Thus, the projections of the group velocities onto the z-axis are also matched. As shown in Fig. 10, both incoming pulses are propagating through the crystal with their own group velocities such that their pulse fronts are matched to each other as well as to that of the generated SFG pulse. In addition, the equal sweep velocity of all three pulses along the SFG propagation direction ensures that the generated SFG pulse has minimal pulse duration or, equivalently, maximal bandwidth. We also note that Eq. (20) together with the zero-order z-component of Eq. (18) lead to the following symmetric relation between the phase and group refractive indices:

$$\omega_1 n_1 n_{g1}^* + \omega_2 n_2 n_{g2}^* = \omega_3 n_3 n_{g3}^*. \tag{21}$$

In summary, we have shown that the achromatic phase matching condition for NOPA with angularly dispersed signal or for non-collinear SFG is equivalent to simultaneous pulse-front and group-velocity matching between all three interacting pulses. Hence, the scheme can be of interest for high-power (large beam cross section) applications, too.

We note that a calculation similar to that outlined above can be carried out for NOPA with a quasi-monochromatic pump (ω_3). The phase matching scheme of Fig. 9 can also be used here with a single $\mathbf{k}_3$ vector representing the pump. Our calculations show that simultaneous group velocity and pulse front matching between signal (ω_1) and idler (ω_2) can be achieved in case of achromatic phase matching. It is not clarified yet if achromatic phase matching also allows for a possible pulse front mismatch.

5. Generation of THz pulses by optical rectification

In the last two decades a new branch of science, the terahertz (THz) science emerged (Tonouchi, 2007, Lee, 2009). Usually the 0.1-10 THz range of the electromagnetic spectrum is considered as THz radiation, earlier referred to as the far-IR range. The reason of the new name is justified by the fact that nowadays very frequently time-domain terahertz spectroscopy (TDTS) (Grischkowsky, 1990) based on single-cycle THz pulses is used for investigations. In this method the temporal dependence of the electric field in the THz pulse is measured rather than the intensity envelope. Usually in TDTS setups a biased photoconductive antenna creates the THz pulses and an unbiased one is used for detection. Both are triggered by ultrashort laser pulses. While these devices are well suited for linear absorption and index of refraction measurements the energy of the THz pulses generated by a usual photoconductive antenna is not enough for creating nonlinear effects. Because of this, pump-probe measurements and other applications need THz pulse sources of much higher energy.

Optical rectification (OR) of ultrashort laser pulses (Hu et al., 1990) is a simple and effective method for high-energy THz pulse generation. Similarly to other nonlinear optical frequency conversion methods, a phase-matching condition needs to be fulfilled also in OR. For OR this requires the same group velocity of the pump pulse than the phase velocity of the generated THz radiation. Most frequently ZnTe is used as electro-optic crystal for OR, since in this material velocity matching is accomplished for the 800-nm pulses of Ti:sapphire lasers (Löffler et al., 2005, Blanchard et al., 2007). In this way 1.5-μJ single-cycle THz pulses were produced (Blanchard et al., 2007). However, ZnTe has more than two times smaller figure of merit for THz generation than $LiNbO_3$ (LN). Furthermore, THz absorption of free carriers generated by two-photon absorption of the pump can seriously limit the applicable pump intensity and the generation efficiency in ZnTe (Hoffmann et al., 2007, Blanchard et al., 2007). So LN is a much more promising material for THz generation by OR. However, collinear phase-matching is not possible in LN since the group velocity of the (near-IR) pump is more than two times larger then the THz phase velocity (Hebling et al., 2008b).

Tilted-pulse-front-excitation was suggested to achieve velocity matching for THz generation in LN (Hebling et al., 2002). The operation of this velocity matching method is obvious according to Fig. 11(a). If the intensity front of the excitation pulse is plane, the generated THz radiation will propagate perpendicularly to this plane. This means that for the case of tilted-pulse-front excitation not the group velocity of the pump has to be equal to the THz phase velocity, but the projection of the group velocity into the direction of the THz pulse radiation. So the velocity matching condition can be expressed as:

$$v_{g.p} \cdot \cos\gamma = v_{\mathrm{THz}}, \tag{22}$$

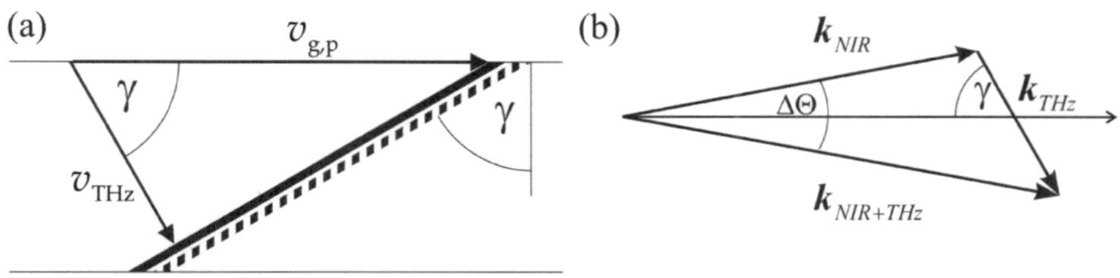

Fig. 11. Velocity matching using tilted-pulse-front excitation. (a) THz wave (bold line) generated in the LN crystal by the tilted intensity front of the pump pulse (dashed bold line) propagates perpendicularly to the THz phase front. (b) Wave-vector diagram for difference-frequency generation (Hebling et al., 2008b).

where $v_{g,p}$ is the group velocity of the pump, and v_{THz} is the phase velocity of the generated THz pulse. According to Eq. (22) an appropriate tilt angle γ can be chosen if the pump velocity is larger than the THz velocity. This is the case for LN.

The first experimental realization of THz-pulse generation by tilted-pulse-front-excitation (Stepanov et al., 2003) resulted in 30-pJ THz pulses for 2-μJ pump pulses. With the same pump energy it was possible to increase the THz energy to 100 pJ by using a better quality (Mg-doped stoichiometric) LN crystal (Hebling et al., 2004). A 200-times increase in the pump energy resulted in 2000 times larger THz energy (Stepanov et al., 2005). Further increasing the pump energy up to the 10-mJ range resulted in THz pulses with energies on the tens-of-μJ range (Yeh et al., 2007, Stepanov et al., 2008). Self-phase modulation of such high-energy THz pulses was observed inside the generating LN crystal (Hebling et al. 2008a). THz pulses with a few μJ energy at 1 kHz repetition rate were successfully used in THz pump-THz probe measurements (Hoffmann et al., 2009, Hebling et al., 2009). In low-bandgap n-type InSb (E_g=0.2 eV) an increase, while in Ge (E_g=0.7 eV) a decrease of the free carrier absorption was observed with sub-ps resolution. The increase in InSb was caused by impact ionization effect created by the electrons energized by the high field of the THz pulses. In Ge having higher bandgap impact ionization was not possible at the achieved field strength. The decrease of absorption was caused by the redistribution of the electrons in the conduction band (Mayer & Keilmann, 1986).

Above we used Fig. 11(a) to explain that for tilted-pulse-front excitation velocity matching is expressed by Eq. (22). On the other hand we know that an angular dispersion according to Eq. (12) is present if the pulse front is tilted and OR can be considered as the result of difference frequency generation (DFG) between individual frequency components of the broadband excitation pulse. In presence of angular dispersion non-collinear DFG occurs. The vector diagram demonstrating phase-matching for non-collinear DFG is depicted in Fig. 11(b). Using Eq. (12), it is easy to show (Hebling et al., 2002) that for small values of $\Delta\varepsilon$ the inclination angle of $\mathbf{k}_{THz}$ from the average propagation direction of the excitation pulse is the same γ as the pulse front tilt angle. Therefore the two pictures used in this section to describe THz pulse excitation by ultrashort pulses with tilted-pulse-front, based on velocity matching and on wave-vector conservation, respectively, predict the same propagation direction for the created THz pulse.

Since for LN there is more than a factor of two between the velocity of the excitation and the velocity of the THz radiation, the tilt angle has to be as large as 63°. According to Eq. (12) this implies a large angular dispersion, and according to Eq. (10) a large GDD. Because of this the duration of the pulse with tilted front is short only in a limited region of space. In order to deal with this problem in THz pulse generation a setup depicted in Fig. 12(a) is used for tilted-pulse-front excitation (Hebling et al. 2004). The necessary angular dispersion is introduced by the grating. The lens images the grating surface to the entrance aperture of the LN crystal. This reproduces the short initial pump pulse duration and high peak power in the image plane inside the crystal, which is essential for efficient THz generation.

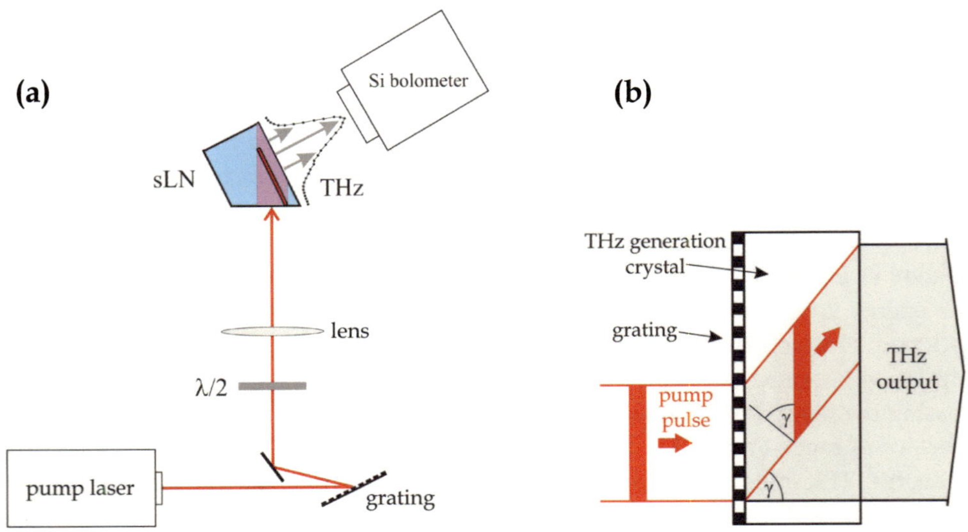

Fig. 12. (a) Experimental setup for THz generation by tilted-pulse-front excitation.
(b) Contact grating scheme.

An important advantage of the TPFP technique is its inherent scalability to higher THz energies. This can be accomplished simply by increasing the pump spot size and energy. However, our detailed analysis (Fülöp et al., 2009) of the pulse-front-tilting setup shows that aberrations caused by the imaging optics can introduce strong asymmetry of the THz beam profile and significant curvature of the THz wavefronts. Such distortions can limit application possibilities of a high-field THz source. In order to overcome the limitations imposed by the imaging optics in the pulse-front-tilting setup we have recently proposed (Pálfalvi et al., 2008) a compact scheme (see Fig. 12(b)), where the imaging optics is omitted and the grating is brought in contact with the crystal.

Besides LN also a few semiconductor materials, such as GaP, GaSe, GaAs, etc., are promising candidates for high-energy THz pulse generation in a contact-grating setup. Due to their lower bandgap as compared to that of LN, semiconductors have to be pumped at longer wavelengths, where only higher-order multiphoton absorption is effective and, as a consequence, higher pump intensities can be used (Fig. 13). In most cases TPFP is necessary for phase matching in semiconductors for longer-wavelength pumping. The required pulse-front-tilt angles are smaller (about 30° or below) than in case of LN, which allow for using larger material thicknesses for THz generation (due to smaller GDD, see above), thereby compensating for the smaller nonlinear coefficients of semiconductors (Fülöp et al., 2009). In addition, a smaller pulse-front-tilt angle makes the realization of a contact-grating setup technically less challenging. The absorption of LN in the THz range is rapidly increasing with increasing frequency above ca. 1 THz, which makes it less advantageous for generation of higher THz frequencies. The THz absorption of many semiconductors is smaller at higher THz frequencies than that of LN, and they can be used to efficiently generate THz radiation above 1 THz, provided that they are pumped at sufficiently long wavelengths to suppress free-carrier absorption.

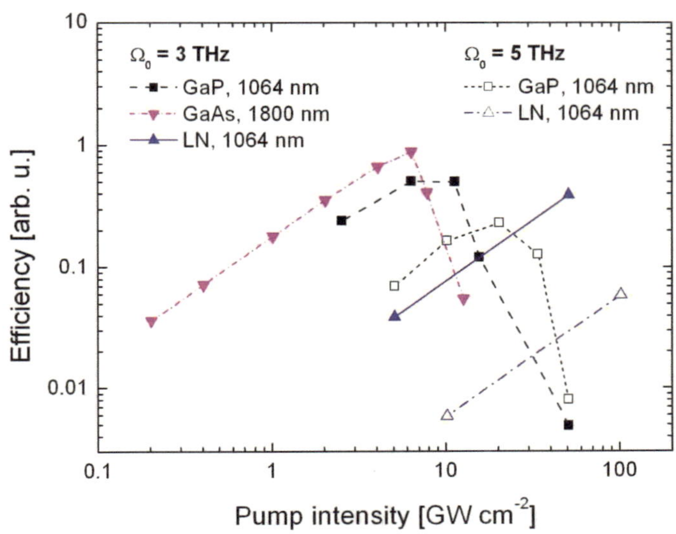

Fig. 13. Calculated maximal THz generation efficiencies in GaP, GaAs and LN for 3 THz and 5 THz phase matching frequencies. The pump wavelength is indicated for each c urve.

6. Conclusion

In summary, a survey on various applications of tilted-pulse-front excitation was given. We have started with considering the relation between pulse front tilt and angular dispersion. It was pointed out that the group velocity can, in general, depend on angular dispersion. Such dependence should be taken into account when considering a beam with angular dispersion propagating as extraordinary wave in a birefringent medium. To our knowledge, this fact was not explicitly mentioned in previous works.

Among the applications of TPFP, which provide an exact synchronization between the pump pulse and the generated excitation along the sample, TWE of visible and x-ray lasers, and ultrafast electron diffraction were briefly reviewed. We have proposed to use a modified pulse front tilting setup for pumping short-wavelength x-ray lasers.

A survey on various nonlinear optical schemes with achromatic phase matching was given including SHG, various types of NOPA and SFG. It was shown that for NOPA with angularly dispersed signal (and the corresponding SFG scheme) achromatic phase matching

is equivalent to simultaneous group velocity and pulse front matching. The importance of this feature to high-power applications was outlined.
Finally, the generation of intense ultrashort THz pulses by OR in LN using TPFP was reviewed, together with applications to ultrafast nonlinear THz spectroscopy. The potential of further upscaling the THz energy and field strength was assessed when using the contact grating scheme with semiconductor materials for OR pumped at IR wavelengths.

7. Acknowledgement

Financial support from Hungarian Scientific Research Fund (OTKA) grant numbers 76101 and 78262 is acknowledged.

8. References

Akturk, S.; Gu, X.; Gabolde, P. & Trebino, R. (2005). The general theory of first-order spatio-temporal distortions of Gaussian pulses and beams. *Opt. Express,* Vol. 13, No. 21, (October 2005) 8642-8660

Barty, C. P. J.; King, D. A.; Yin, G. Y.; Hahn, K. H.; Field, J. E.; Young, J. F. & Harris, S. E. (1988). 12.8-eV laser in neutral cesium. *Phys. Rev. Lett.,* Vol. 61, No. 19, (November 1988) 2201-2204

Baum, P. & Zewail, A. H. (2006). Breaking resolution limits in ultrafast electron diffraction and microscopy. *Proc. Natl. Acad. Sci.,* Vol. 103, No. 44, (October 2006) 16105-16110

Bor, Zs.; Szatmári, S. & Müller, A. (1983). Picosecond pulse shortening by travelling wave amplified spontaneous emission. *Appl. Phys. B,* Vol. 32, No. 3, (November 1983) 101-104

Bor, Zs. & Rácz, B. (1985). Group velocity dispersion in prisms and its application to pulse compression and travelling-wave excitation. *Opt. Commun.,* Vol. 54, No. 3, (June 1985) 165-170

Blanchard, F.; Razzari, L.; Bandulet, H.-C.; Sharma, G.; Morandotti, R., Kieffer, J.-C.; Ozaki, T.; Ried, M.; Tiedje, H. F.; Haugen, H. K. & Hegmann, F. A. (2007). Generation of 1.5 μJ single-cycle terahertz pulses by optical rectification from a large aperture ZnTe crystal. *Opt. Express,* Vol. 15, No. 20, (October 2007) 13212-13220

Cerullo, G. & De Silvestri, S. (2003). Ultrafast optical parametric amplifiers. *Rev. Sci. Instrum.,* Vol. 74, No. 1, (January 2003) 1-18

Daido, H. (2002). Review of soft x-ray laser researches and developments. *Rep. Prog. Phys.,* Vol. 65, (September 2002) 1513-1576

Di Trapani, P.; Andreoni, A.; Solcia, C.; Foggi, P.; Danielius, R.; Dubietis, A. & Piskarskas, A. (1995). Matching of group velocities in three-wave parametric interaction with femtosecond pulses and application to traveling-wave generators. *J. Opt. Soc. Am. B,* Vol. 12, No. 11, (1 November 1995) 2237-2244

Fülöp, J. A.; Major, Zs.; Henig, A; Kruber, S.; Weingartner, R.; Clausnitzer, T.; Kley, E.-B.; Tünnermann, A.; Pervak, V.; Apolonski, A.; Osterhoff, J.; Hörlein, R.; Krausz, F. & Karsch, S. (2007). Short-pulse optical parametric chirped-pulse amplification for the generation of high-power few-cycle pulses. *New J. Phys.,* Vol. 9, No. 12, (December 2007) 438

Fülöp, J. A.; Pálfalvi, L.; Almási, G. & Hebling, J. (2009). Design of high-energy terahertz sources based on optical rectification. In preparation

Gaal, P.; Reimann, K.; Woerner, M.; Elsaesser, T.; Hey, R. & Ploog, K. H. (2006). Nonlinear terahertz response of n-type GaAs. *Phys. Rev. Lett.*, Vol. 96, No. 18, (May 2006) 187402

Grischkowsky, D.; Keiding, S.; Exter, M. van & Fattinger, Ch. (1990). Far-infrared time-domain spectroscopy with terahertz beams of dielectrics and semiconductors. *J. Opt. Soc. Am. B*, Vol. 7, No. 10, (October 1990) 2006-2015

Harris, S. E. (1969). Tunable optical parametric oscillators. *Proc. IEEE*, Vol. 57, No. 12, (December 1969) 2096-2113

Hebling, J. & Kuhl, J. (1989a). Generation of femtosecond pulses by travelling-wave amplified spontaneous emission. *Opt. Lett.*, Vol. 14, No. 5, (March 1989) 278-280

Hebling, J. & Kuhl, J. (1989b). Generation of tunable femtosecond pulses by travelling wave amplification. *Opt. Commun.*, Vol. 73, No. 5 (November 1989) 375-379

Hebling, J.; Kuhl, J. & Klebniczki, J. (1991). Femtosecond optical gating with a traveling-wave amplifier. *J. Opt. Soc. Am. B*, Vol. 8, No. 5, (May 1991) 1089-1092

Hebling, J. (1996). Derivation of the pulse front tilt caused by angular dispersion. *Opt. Quantum Electron.*, Vol. 28, No. 12, (December 1996) 1759-1763

Hebling, J.; Almási, G.; Kozma, I. Z. & Kuhl, J. (2002).Velocity matching by pulse front tilting for large area THz-pulse generation. *Opt. Express*, Vol. 10, No. 21, (October 2002) 1161-1166

Hebling, J.; Stepanov, A.G.; Almási, G.; Bartal, B. & Kuhl, J. (2004). Tunable THz pulse generation by optical rectification of ultrashort laser pulses with tilted pulse fronts. *Appl. Phys. B*, Vol. 78, No. 5, (March 2004) 593–599

Hebling, J.; Yeh, K.-L.; Hoffmann, M. C. & Nelson, K. A. (2008a). High-power THz generation, THz nonlinear optics, and THz nonlinear spectroscopy. *IEEE J. Sel. Top. Quantum Electron.*, Vol. 14, No. 2, (March-April 2008) 345-353

Hebling, J.; Yeh, K.-L.; Hoffmann, M. C.; Bartal, B. & Nelson, K. A. (2008b). Generation of high-power terahertz pulses by tilted-pulse-front excitation and their application possibilities. *J. Opt. Soc. Am. B*, Vol. 25, No. 7, (July 2008) 6-19

Hebling, J.; Hoffmann, M. C.; Hwang, H. Y.; Yeh, K.-L. & Nelson, K. A. (2009). Comparison of Nonequilibrium Carrier Distribution in Ge, Si and GaAs observed by THz-Pump/THz-Probe Measurements. *Phys. Rev. B*, (submitted)

Hoffmann, M. C., Yeh, K.-L.; Hebling, J. & Nelson, K. A. (2007). Efficient terahertz generation by optical rectification at 1035 nm. *Opt. Express*, Vol. 15, No. 18, (September 2007) 11706-11713

Hoffmann, M. C.; Hebling, J.; Hwang, H. Y.; Yeh, K.-L. & Nelson, K. A. (2009). Impact ionization in InSb probed by terahertz pump-terahrtz probe spectroscopy. *Phys. Rev. B*, Vol. 79, No. 16 (April 2009) 161201 (R)

Hofmann, Th.; Mossavi, K.; Tittel, F. K. & Szabó, G. (1992). Spectrally compensated sum-frequency mixing scheme for generation of broadband radiation at 193 nm. *Opt. Lett.*, Vol. 17, No. 23, (1 December 1992) 1691-1693

Hu, B. B.; Zhang, X.-C. & Auston, D. H. (1990). Free-space radiation from electro-optical crystals. *Appl. Phys. Lett.*, Vol. 56, No. 6, (February 1990) 506-508

Kapteyn, H. C. (1992). Photoionization-pumped x-ray lasers using ultrashort-pulse excitation. *Appl. Opt.*, Vol. 31, No. 24, (August 1992) 4931-4939

Kawachi, T.; Kado, M.; Tanaka, M.; Sasaki, A.; Hasegawa, N.; Kilpio, A. V.; Namba, S.; Nagashima, K.; Lu, P.; Takahashi, K.; Tang, H.; Tai, R.; Kishimoto, M.; Koike, M.; Daido, H. & Kato, Y. (2002). Gain saturation of nickel-like silver and tin x-ray lasers by use of a tabletop pumping laser system. *Phys. Rev. A*, Vol. 66, No. 3, (September 2002) 033815

Klebniczki, J.; Hebling, J. & Kuhl, J. (1990). Generation of tunable femtosecond pulses in a traveling-wave amplifier. *Opt. Lett.*, Vol. 15, No. 23, (December 1990) 1368-1370

Kobayashi, T. & Shirakawa, A. (2000). Tunable visible and near-infrared pulse generation in a 5 fs regime. *Appl. Phys. B*, Vol. 70, Supplement 1, (June 2000) S239-S246

Kozma, I. Z.; Almási, G. & Hebling, J. (2003). Geometrical optical modeling of femtosecond setups having angular dispersion. *Appl. Phys. B*, Vol. 76, No. 3, (March 2003) 257-261

Lee, Y.-S. (2009). *Principles of Terahertz Science and Technology*, Springer Science+ Business Media, LLC, ISBN: 978-0-387-09539-4, e-ISBN: 978-0-387-09540-0, New York

Löffler, T.; Hahn, T.; Thomson, M.; Jacob, F. & Roskos, H. G. (2005). Large-area electro-optic ZnTe terahertz emitters. *Opt. Express*, Vol. 15, No. 9, (April 2005) 5353-5362

Main, I. G. (1978). *Vibrations and Waves in Physics*, Cambridge University Press, ISBN 0-521-216621-1, Cambridge

Major, Zs.; Trushin, S. A.; Ahmad, I; Siebold, M.; Wandt, C.; Klingebiel, S.; Wang, T.-J.; Fülöp, J. A.; Henig, A.; Kruber, S.; Weingartner, R.; Popp, A.; Osterhoff, J.; Hörlein, R.; Hein, J.; Pervak, V.; Apolonski, A.; Krausz, F. & Karsch S. (2009). Basic concepts and current status of the Petawatt Field Synthesizer –A new approach to ultrahigh field generation. *The Review of Laser Engineering*, Vol. 37, No. 6, (June 2009) 431-436

Martínez, O. E.; Gordon, J., P. & Fork, R., L. (1984). Negative group-velocity dispersion using refraction. *J. Opt. Soc. Am A*, Vol. 1, No. 10, (October 1984) 1003-1006

Martínez, O. E. (1986). Pulse distortions in tilted pulse schemes for ultrashort pulses. *Opt. Commun.*, Vol. 59, No. 3, (September 1986) 229-232

Martínez, O. E. (1989). Achromatic phase matching for second harmonic generation of femtosecond pulses. *IEEE J. Quantum Electron.*, Vol. 25, No. 12, (December 1989) 2464-2468

Mayer, A. & Keilmann, F. (1986). Far-infrared nonlinear optics. II. $\chi^{(3)}$ contributions from the dynamics of free carriers in semiconductors. *Phys. Rev. B*, Vol. 33, No. 10, (15 May 1986) 6962-6968

Pálfalvi, L.; Fülöp, J. A.; Almási, G. & Hebling, J. (2008). Novel setups for extremely high power single-cycle terahertz pulse generation by optical rectification. *Appl. Phys. Lett.*, Vol. 92, No. 17, (28 April 2008) 171107

Polland, H. J.; Elsaesser, T.; Seilmeier, A.; Kaiser, W.; Kussler, M.; Marx, N. J.; Sens, B. & Drexhage, K. H. (1983). Picosecond dye laser emission in the infrared between 1.4 and 1.8 μm. *Appl. Phys. B*, Vol. 32, No. 2, (October 1983) 53-57

Richman, B. A.; Bisson, S. E.; Trebino, R.; Sidick, E. & Jacobson, A. (1998). Efficient broadband second-harmonic generation by dispersive achromatic nonlinear conversion using only prisms. *Opt. Lett.*, Vol. 23, No. 7, (April 1998) 497-499

Saikan, S. (1976). Automatically tunable second-harmonic generation of dye lasers. *Opt. Commun.*, Vol. 18, No. 4, (September 1976) 439-443

Schiller, N. H. & Alfano, R. R. (1980). Picosecond characteristics of a spectrograph measured by a streak camera/video readout system. *Opt. Commun.*, Vol. 35, No. 3, (December 1980) 451-454

Sher, M. H.; Macklin, J. J.; Young, J. F. & Harris, S. E. (1987). Saturation of the Xe III 109-nm laser using traveling-wave laser-produced-plasma excitation. *Opt. Lett.*, Vol. 12, No. 11, (November 1987) 891-893

Simon, P.; Földes, I. B. & Szatmári, S. (2005). Laser-induced x-rays and x-ray lasers. *A kvantumoptika és –elektronika legújabb eredményei,* ELFT Atom-, Molekulafizikai és Kvatnumelektronikai Szakcsoport, Tavaszi Iskola 2005. május 31- június 3. Balatonfüred (in Hungarian)

Stepanov, A. G.; Hebling, J. & Kuhl, J. (2003). Efficient generation of subpicosecond terahertz radiation by phase-matched optical rectification using ultrashort laser pulses with tilted pulse fronts. *Appl. Phys. Lett.*, Vol. 83, No. 15, (October 2003) 3000-3002

Stepanov, A. G.; Kuhl, J.; Kozma, I. Z.; Riedle, E.; Almási, G. & Hebling, J. (2005). Scaling up the energy of THz pulses created by optical rectification. *Opt. Express,* Vol. 13, No. 15, (July 2005) 5762-5768

Stepanov, A. G.; Bonacina L.; Chekalin S. V. & Wolf, J.-P. (2008). Generation of 30 μJ single-cycle terahertz pulses at 100 Hz repetition rate by optical rectification. *Opt. Lett.* Vol. 33, No. 21, (November 2008) 2497-2499

Szabó, G.; Rácz, B.; Müller, A.; Nikolaus, B. & Bor, Zs. (1984). Travelling-wave-pumped ultrashort-pulse distributed-feedback dye laser. *Appl. Phys. B,* Vol. 34, No. 3, (March 1984) 145-147

Szabó, G. & Bor, Zs. (1990). Broadband frequency doubler for femtosecond pulses. *Appl. Phys. B,* Vol. 50, No. 1, (January 1990) 51-54

Szabó, G. & Bor, Zs. (1994). Frequency conversion of ultrashort pulses. *Appl. Phys. B,* Vol. 58, No. 3, (March 1994) 237-241

Szatmári, S.; Kuhnle, G. & Simon, P. (1990). Pulse compression and traveling wave excitation scheme using a single dispersive element. *Appl. Opt.,* Vol. 29, No. 36, (December 1990) 5372-5379

Tonouchi, M. (2007). Cutting-edge terahertz technology. *Nature Photonics,* Vol. 1, No. 2, (February 2007) 97-105

Topp, M. R. & Orner, G. C. (1975). Group dispersion effects in picosecond spectroscopy. *Opt. Commun.,* Vol. 13, No. 3, (March 1975) 276-281

Yeh, K.-L.; Hoffmann, M. C.; Hebling, J. & Nelson, K. A. (2007). Generation of 10 μJ ultrashort terahertz pulses by optical rectification. *Appl. Phys. Lett.*, Vol. 90, No. 17, (April 2007) 171121

Zhang, T. R.; Choo, H. R. & Downer, M. C. (1990). Phase and group velocity matching for second harmonic generation of femtosecond pulses. *Appl. Opt.*, Vol. 29, No. 27, (September 1990) 3927-3933

12

Applications of Effective Medium Theories in the Terahertz Regime

Maik Scheller[1], Christian Jansen[1], and Martin Koch[2]
[1]Institute for High-Frequency Technology, Technische Universität Braunschweig
Schleinitzstr. 22, 38106 Braunschweig
[2]Physics Department, Philipps-Universität Marburg
Renthof 5, 35032 Marburg
Germany

1. Introduction

In recent years, the investigation of composite material systems in the terahertz (THz) regime has drawn a considerable attention from a wide spectrum of scientific areas, for instance the fields of nano-science (Beard et al., 2002), (Hendry et al., 2006) and metamaterials (Levy et al., 2007). The interaction of terahertz waves with a composite system consisting of particles embedded in a host material as illustrated in Fig. 1 can be described by effective material properties and effective medium theories (EMTs) enabling the calculation of the resulting macroscopic permittivity ε_R.

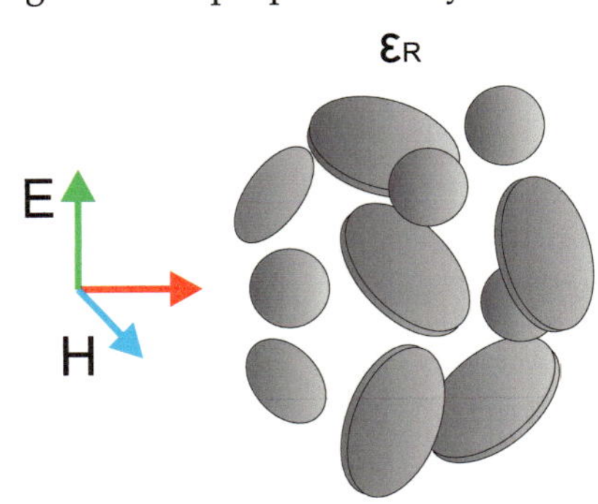

Fig. 1. The interaction between an electromagnetic wave and a composite system can be described by an effective permittivity ε_R.

If the particle size is much smaller than the wavelength of interest, as visualized in Fig. 2, scattering effects are negligible and quasi-static models suffice. Otherwise, scattering effects have to be taken into account.

In this book chapter we will review common quasi-static EMTs and their application to various composite material systems. The selection of theoretical models comprises the Landau-Lifshitz-Looyenga model, which is applicable to mixtures of arbitrarily shaped particles, the Polder-van-Santen theory, which explicitly considers the influence of the inclusions shape and orientation, the differential Bruggeman theory and a recent extension to the latter proposed by the authors.

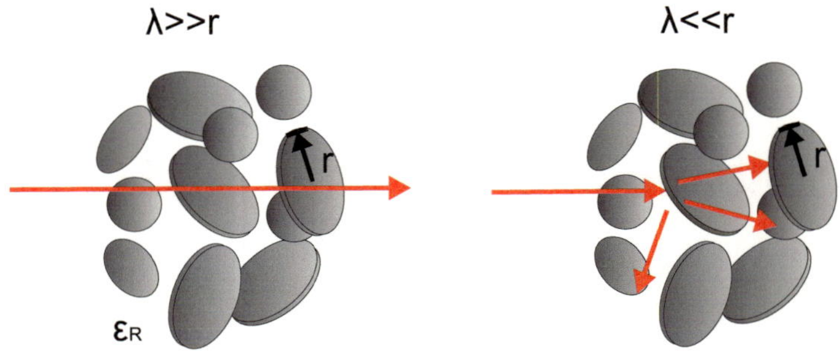

Fig. 2. In the case of small particles compared to the wavelength, quasi-static models suffices, otherwise scattering effects as to be taken into account.

The first application scenario that we will study is the characterisation of polymeric compounds. By adding microscopic particles to a polymeric host material, the resulting properties of the plastic like colour, material strength and flammability can be optimized. Moreover, the additives induce a change of the optical parameters of the mixture that can be studied with terahertz time domain spectroscopy (THz TDS). The resulting refractive index depends on the volumetric content and the dielectric constant of the additives as well as the particle shape. Due to the variety of commonly used additives, ranging from rod like glass fibres, over cellulose based fillers to spherical nanoparticles, polymeric compound systems are ideal to illuminate the applicability and limitations of the different EMTs.

Apart from the polymeric compounds, we will also discuss the usability of the EMTs to describe biological systems. As one example, the water content of plant leaves considerably effects their dielectric properties. Utilizing the EMTs allows for the determination of the water content of the plants with terahertz radiation.

In summary, the chapter will review a selection of effective medium theories and outline their applicability to various scientific problems in the terahertz regime. Additionally, a short overview on the THz time domain spectroscopy (TDS) which is employed to experimentally validate the models' predictions is presented.

2. Effective medium theories

The analysis of dielectric mixture systems, for instance particles embedded in a host material, is a problem of enormous complexity if every single particle is considered individually. Alternatively, the resulting macroscopic material parameter of the mixture can be derived which characterize the interaction between the material system and electromagnetic waves. To calculate this effective material parameter, effective medium theories (EMTs) can be employed. In this chapter, we will exemplarily present a selection of the most common quasi static EMTs which can directly be applied to the description of heterogeneous dielectrics in the THz range. Table 1 provides a basic overview of the characteristics of these models, which will be further described below.

2.1 Maxwell-Garnett

One of the first and probable the most well known EMT is the Maxwell-Garnett (MG) model (Maxwell-Garnett, 1904) which is based on analyzing the effective polarizability of spherical inclusions with the permittivity ε_p embedded in a vacuum environment as illustrated in Fig. 3.

Model	Volumetric content	Particle shape	Area of Application
Maxwell-Garnett	Low	spheres	Very low concentrations
Polder and van Santen	High	ellipsoidal	Ellipsoidal particles, anisotropic systems
Extended Bruggeman	High	ellipsoidal	High permittivity contrast, ellipsoidal particles, anisotropic systems
Landau, Lifshitz, Looyenga	Middle	arbitrary	Mixtures of irregular, unknown shaped particles
Complex Refractiv Index	middle	arbitrary	Mixtures with small permittivity contrast

Table 1. Overview of the EMTs mentioned in the text.

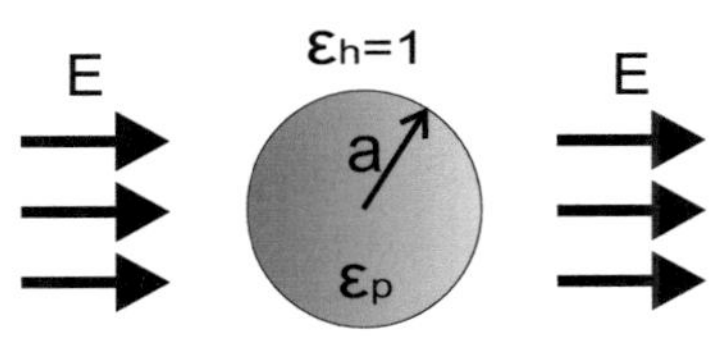

Fig. 3. To derive the MG model, the resulting polarizability of a single spherical particle is derived.

Following the basics of electrostatics the resulting polarizability α_p of a single spherical particle is given by (Jackson, 1999)

$$\alpha_p = 4\pi\varepsilon_0 \frac{\varepsilon_p - 1}{\varepsilon_p + 2} a^3 \tag{1}$$

where ε_0 is the permittivity of the vacuum and a is the radius of the particle. Now it is assumed, that the polarizability remains constant if multiple particles are present. Consequently the Clausius Mossoti relation (Kittel, 1995) that connects the relative permittivity ε_r of a material with the polarizability of a number of N microscopic particles

$$\frac{\varepsilon_r - 1}{\varepsilon_r + 2} = \frac{\sum_j N_j \alpha_j}{3\varepsilon_0} \tag{2}$$

can be exploited to calculate the effective permittivity ε_R of this inhomogeneous medium, where f_p is the volumetric content of the particles:

$$\frac{\varepsilon_R - 1}{\varepsilon_R + 2} = f_p \frac{\varepsilon_p - 1}{\varepsilon_p + 2} \tag{3}$$

If the particles are embedded in a host material with given permittivity ε_h, Eq. 3 changes into the MG equation:

$$\frac{\varepsilon_R - \varepsilon_h}{\varepsilon_R + 2\varepsilon_h} = f_p \frac{\varepsilon_p - \varepsilon_h}{\varepsilon_p + 2\varepsilon_h} \tag{4}$$

As can be seen from these deductions, the assumption is violated if a larger volumetric fraction of the medium is formed by the inclusions, since in this case the effective background permittivity changes. Thus, the model can be applied to very low concentrations only.

2.2 Polder and van Santen

Another approach with extended validity was derived by Polder and van Santen: Instead of employing the host ε_h in the calculation to derive Eq. 4, the effective dielectric constant ε_R is utilized. That way, the effect of the slightly increasing effective background permittivity can be taken into account. The equation

$$\frac{\varepsilon_R - \varepsilon_h}{3\varepsilon_R} = f_p \frac{\varepsilon_p - \varepsilon_h}{\varepsilon_p + 2\varepsilon_R} \tag{5}$$

results, which is known as the Böttcher equation (Böttcher, 1942). Despite this extension, the model is still restricted to spherical shaped inclusions. By including depolarization factors N in the deductions, it is possible to expand the validity to ellipsoidal particles. These factors can be calculated by the following equations (Kittel, 1995):

$$N_x = \frac{x_l y_l z_l}{2} \int_0^\infty \frac{du}{\left(x_l^2+u\right)\sqrt{(x_l^2+u)(y_l^2+u)(z_l^2+u)}} \tag{6}$$

$$N_x + N_y + N_z = 1 \tag{7}$$

The Fig. 4 shows the numerically calculated N_x values for different aspect ratios between the axis x and y in a) and the axis x and z in b).

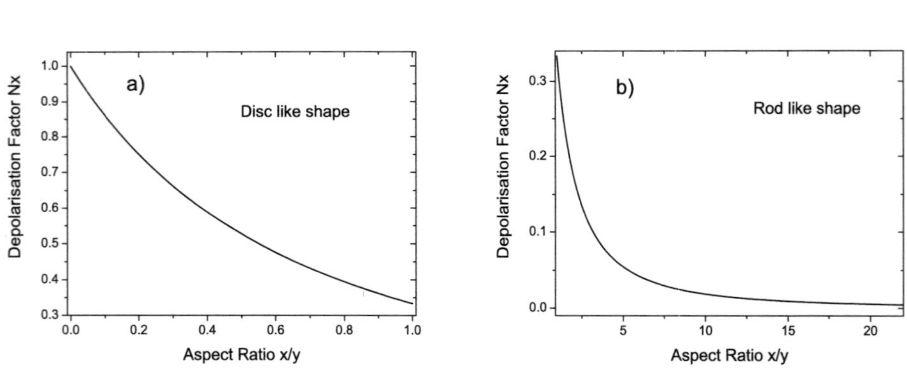

Fig. 4. Values of the depolarisation factor N_x as a function of the aspect ratio between the axis x and y in a) and the axis x and z in b).

In the case of ideal disc-like particles the aspect ratio x/y converges toward zero while the N_x value tends towards unity. For ideal rod like particles, the aspect ratio increase to infinity and N_x descends to zero. These shapes are illustrated together with the resulting depolarization factors in Fig. 5.

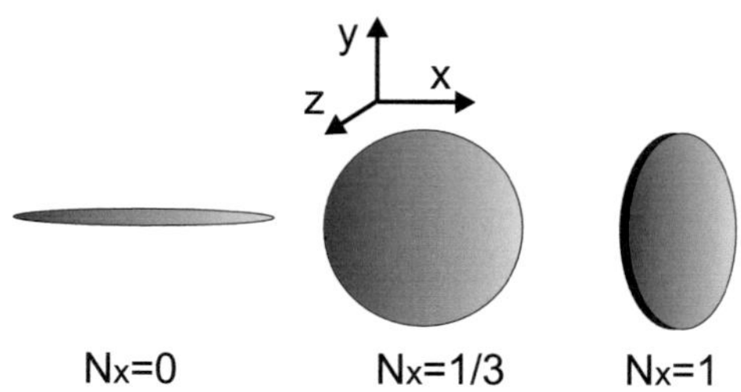

Fig. 5. Values of the deplarisation factor N_x for a) a rod b) a sphere and c) a disc

$$\frac{\varepsilon_R - \varepsilon_h}{3\varepsilon_R} = f_p \frac{\varepsilon_p - \varepsilon_h}{\varepsilon_p + 2\varepsilon_R} \tag{5}$$

Analogously to Eq. 5 the effective material parameter can be calculated by employing these factors which results in the Polder and van Santen (PvS) model (Polder & van Santen, 1946):

$$\varepsilon_R = \frac{\varepsilon_h}{1 - \frac{1}{3} f_p \left(\varepsilon_p - \varepsilon_h\right) \sum_i^3 \frac{1}{\varepsilon_R + (\varepsilon_p - \varepsilon_R) N_i}} \tag{8}$$

The special forms of the PvS model for ideal shapes, which are orientated isotropically in the mixture, are the following (Hale, 1976):

Spheres:

$$\frac{\varepsilon_R - \varepsilon_h}{3\varepsilon_R} = f_p \frac{\varepsilon_p - \varepsilon_h}{\varepsilon_p + 2\varepsilon_R} \tag{9}$$

Discs:

$$\frac{\varepsilon_R - \varepsilon_h}{\varepsilon_R + 2\varepsilon_p} = f_p \frac{\varepsilon_p - \varepsilon_h}{3\varepsilon_p} \tag{10}$$

Rods:

$$\frac{\varepsilon_R - \varepsilon_h}{5\varepsilon_R + \varepsilon_R} = f_p \frac{\varepsilon_p - \varepsilon_h}{3(\varepsilon_p + \varepsilon_R)} \tag{11}$$

As the Böttcher model is a special case of the PvS model, the Eq. 9 for spherical shaped particles equals the Böttcher equation Eq. 5.
Due to the consideration of the influence of the particles shape and the increasing background permittivity, the PvS model is widely applicable. Especially anisotropic mixture systems like orientated glass fibres can be described by this approach.

2.3 Bruggeman

While the PvS model well describes a variety of mixtures, a strong contrast in the permittivity between the mixture components still affects its validity. Here, a differential approach (Bruggeman, 1935) can be utilized. The Bruggeman theory makes use of a differential formulation of Eq. 4. After integration, the equation results:

$$1-f_p=\frac{\varepsilon_p-\varepsilon_R}{\varepsilon_p-\varepsilon_h}\sqrt[3]{\frac{\varepsilon_h}{\varepsilon_R}} \tag{12}$$

which is the basic form of the Bruggeman model. This basic form describes spherical particles embedded in a host where a large contrast in permittivity occurs.

By combining the two approaches (Bruggeman and PvS), more general forms of this model can be derived (Banhegyi 1986), (Scheller et al., 2009, a). The equation for this extended Bruggeman [EB] model in the general case, where one polarization factor is given by N, the other two by $1\text{-}N/2$ is:

$$f_p=1-\left(\frac{\varepsilon_h}{\varepsilon_R}\right)^{\left[\frac{-3N^2+3N}{3N+1}\right]}\left(\frac{\varepsilon_p-\varepsilon_R}{\varepsilon_p-\varepsilon_h}\right)\left(\frac{(1+3N)\varepsilon_p+(5-3N)\varepsilon_h}{(1+3N)\varepsilon_p+(5-3N)\varepsilon_R}\right)^{\left[\frac{12N-18N^2-2}{9N^2-12N-5}\right]} \tag{13}$$

For the case of isotopically orientated particles with ideal shapes the following set of equations results:

Spheres:

$$\frac{\varepsilon_R-\varepsilon_h}{3\varepsilon_R}=f_p\frac{\varepsilon_p-\varepsilon_h}{\varepsilon_p+2\varepsilon_R} \tag{14}$$

Discs:

$$1-f_p=\left(\frac{\varepsilon_p-\varepsilon_R}{\varepsilon_p-\varepsilon_h}\right)\left(\frac{2\varepsilon_p+\varepsilon_h}{2\varepsilon_p+\varepsilon_R}\right) \tag{15}$$

Rods:

$$1-f_p=\left(\frac{\varepsilon_p-\varepsilon_R}{\varepsilon_p-\varepsilon_h}\right)\left(\frac{5\varepsilon_h+\varepsilon_p}{5\varepsilon_R+\varepsilon_p}\right)^{\left[\frac{2}{5}\right]} \tag{16}$$

2.4 Landau, Lifshitz, Looyenga

Additionally, the Landau, Lifshitz, Looyenga (LLL) model (Looyenga 1965) makes use of a different assumption: Instead of taking the shape of the particles into account a virtual sphere is considered, which includes a given volumetric fraction of particles with unknown shape as illustrated in Fig. 3. By successively adding an infinitismal amount of particles, the effective permittivity increases slightly which can be described by a Taylor approximation. This procedure leads to the equation.

$$\sqrt[3]{\varepsilon_R}=f_p\sqrt[3]{\varepsilon_p}+(1-f_p)\sqrt[3]{\varepsilon_h} \tag{17}$$

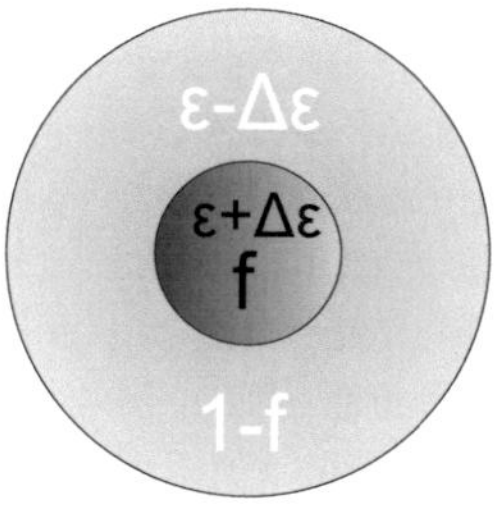

Fig. 3. The basic priniciple of the LLL model: A given volumetric fraction of particles are embedded in a virtual sphere. By differentially increasing the volumetric fraction f, a Taylor approximation can be utilized to calcuate the resulting permittivity ε_R of the system.

Here, no shape dependency is taken into account and thus, the model is favourably applicable to irregularly shaped particle mixtures (Nelson 2005).

2.5 Complex Refractive Index (CRI)

Besides from these deductive models, several more empirical approaches exist. The most common one is the Complex Refractive Index (CRI) model, that linearly connects the material parameter to the volumetric content resulting in the equation:

$$n = f_p n_p + \left(1 - f_p\right) n_h \tag{18}$$

This model was successfully applied to porous pressed plastics where irregularly shaped air gaps occur and a low permittivity contrast results (Nelson 1990).

3. Terahertz time domain spectroscopy

Terahertz time domain spectroscopy is a relatively young field of science. Apart from some early explorations (Kimmitt, 2003), for a long time the terahertz domain remained a most elusive region of the electromagnetic spectrum. This circumstance can be explained by the lack of suitable sources: while for long the high-frequency operation limit of electronic devices was found in the lower GHz regime, most optical emitters are not able to operate at the "low" THz frequencies.

Many researchers date the advent of nowadays THz science back to the upcoming of femtosecond laser systems (Moulton, 1985). Their short optical pulses could induce carrier dynamics on the timescale of a picosecond, which lead to the developement of different broadband THz sources (Kuebler et al., 2005).

Due to the scope of this chapter, the following section will only discuss one of many different ways to generate broadband terahertz radiation, namely the photoconductive switching pinoneered by Auston et al. in the 1970s (Auston, 1975) . For a more complete review of terahertz technology, its generation and its applications, the inclined reader is referred to the excellent articles of (Mittleman, 2003), (Sakai, 2005) and (Siegel, 2002).

In this section we will introduce a terahertz time domain spectrometer based on photoconductive switches driven by a Ti:Sa femtosectond laser. First, the single elements will be discussed followed by an explanation of the full spectroscopy system.

3.1 Ti:Sa femtosecond lasers

The centerpiece of a Ti:Sa femtosecond laser is a titan doped alumina crystal, which acts as the active medium inside a Fabry-Perot cavity. The crystal is driven by a diode pumped solid state laser, for example a neodymium-doped yttrium orthovanadate ($ND:YVO_4$) laser which emits at 532 nm and is commercially available with output powers exceeding 5W. This green light emission is well suited for pumping as it coincides with the absorption peak of the Ti:Sa crystal (see Fig. 6 right hand side). In this configuration a relaxation process inside the Ti:Sa leads to a monochromatic, continuous wave emission of the highest gain mode in the Ti:Sa gain region between 700 and 900 nm.

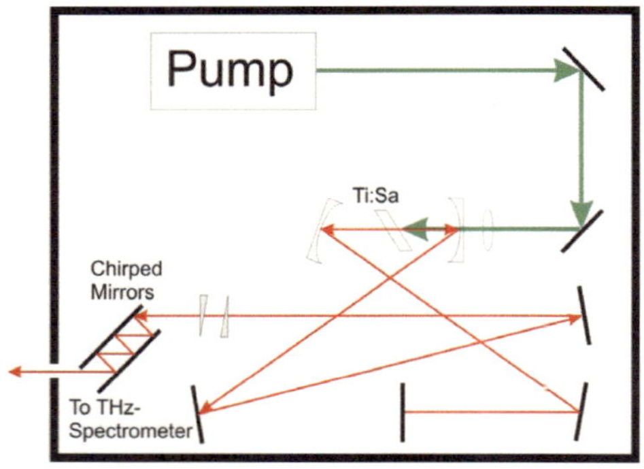

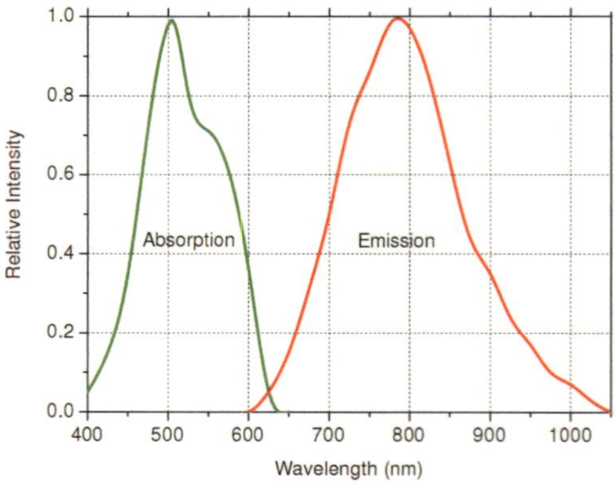

Fig. 6. The schematic setup of the femtosecond Ti:Sa laser system (left) and a sketch of the emission and absorption spectrum of the Ti:Sa crystal (right).

To obtain the desired femtosecond pulses, many modes inside the gain region have to be synchronously excited with a fixed phase relation - they have to be "mode locked". Often the Kerr-lens effect, also known as self focusing, is exploited to obtain this behaviour: For higher intensities, the laser pulse becomes strongly focused inside the crystal leading to a better overlap with the pump beam leading to an enhanced stimulated emission. Hence, pulsed emission becomes the favoured state of operation (Salin et al., 1991), (Piche & Salin, 1993). The mode locking can be induced by the artificial introduction of intensity fluctuations, e.g. by exciting the resonator end mirror with an mechanical impulse and the repition rate of the laser can be adjusted by selecting the appropriate resonator length.

The left hand side Fig. 6 shows a sketch of a Ti:Sa laser. The green pump light is focused into the Ti:Sa crystal mounted inside the resonator. After the out coupling mirror at one end of the cavity, a dispersion compensation system, consisting of chirped mirrors is located. This additional component becomes a necessity due to the ultra short nature of the optical pulses. With a typical 60 nm spectral bandwidth around the central wavelength of 800 nm, sub-30 fs pulses result. Such pulses are extremely broadened when transmitted through dispersive media, e.g. glass lenses or other optical components. The chirped mirrors have an anormal dispersive behaviour, pre-compensating for the normal dispersion inside the optical components after the laser, so that bandwidth limited pulses are obtained at the terahertz emitter and detector, respectively.

3.2 Photoconductive terahertz antennas

As previously mentioned, in this brief overview of terahertz spectroscopy we will focus on photoconductive antennas, also called Auston switches, as terahertz emitters and detectors (Auston, 1975), (Auston et al., 1984), (Smith et al., 1988). They consist of a semiconducting substrate with metal electrodes on top and a pre-collimating high resistivity silicon (HR-Si) lens mounted on the backside (Van Rudd & Mittleman, 2002). Though sharing the same basic structure, receiver and transmitter antenna differ in their requirements to the electrode geometry, substrate material and biasing voltage (Yano et al., 2005).

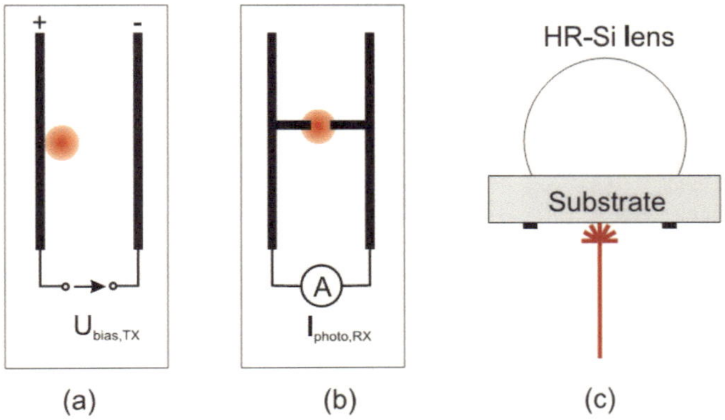

Fig. 7. Metallization structure of a stripline antenna (a) and a dipole antenna (b). The antenna mounted onto the collimating lens (c).

In case of the transmitter antenna, GaAs or low temperature grown GaAs (LT-GaAs) are commonly used as semiconducting substrates. The electrode geometry varies in different designs and can be custom-tailored to the application. A parallel strip line or a bowtie configuration is common. The electrodes are connected to a DC-bias voltage source and the laser spot is focused near the anode (Ralph & Grischkowsky, 1991). When a laser pulse hits the substrate, free carriers are generated and immediately separated in the bias field, giving rise to a photocurrent. The Drude model yields, that the electric THz field emitted, is directly proportional to the time derivative of the photocurrent. If the carrier scattering, relaxation and recombination times of the substrate are known, the behaviour of the transmitter can be accurately simulated (Jepsen et al., 1996).

The substrate of the receiver antenna is made of LT-GaAs which has arsenic clusters as carrier traps that ensure a short carrier lifetime. The electrodes usually have the form of a Hertzian dipole as illustrated in Fig. 7. The laser is focused into the gap between the electrodes. When the laser pulse hits the substrate, the generated carriers short the photoconductive gap and the electric field of the incoming THz pulse separates the free carriers, driving them towards the electrodes. Thus, a current can be detected which is a measure of the average strength of the electric THz field over the lifetime of the optically generated carriers. Due to the LT-GaAs substrate, the life time of the carriers is very short compared to the length of the terahertz pulse. Thus the approximation that only one point in time of the terahertz pulse is sampled can be made.

3.3 A terahertz time domain spectrometer

Now that we have discussed the single elements of a terahertz spectroscopy system (emitter, detector and femtosecond laser source) we shall investigate a complete terahertz time domain spectrometer as shown in Fig. 8.

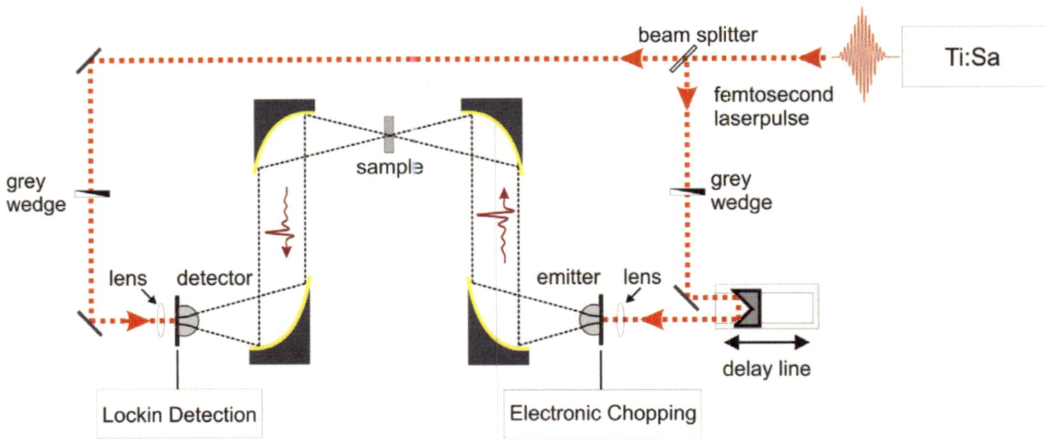

Fig. 8. Basic structure of a THz-TDS system.

The femtosecond laser pulse, generated by the Ti:Sa laser, is divided into an emitter and a detector arm inside a beam splitter. Grey wedges are used to set the desired power level and lenses focus the laser beams onto the photoconductive antennas. The terahertz radiation is guided by off-axis parabolic mirrors (OPMs). In order to analyze small-sized samples, the OPMs are used to create an intermediate focus, which is typically of the size of a few millimeters. In the emitter arm, a motorized delay line varies the time at which the laser pulse creates the terahertz radiation inside the photoconductive transmitter with respect to the time that the photoconductive receiver is gated. Thus, by varying the optical delay, the terahertz pulse is sampled step by step. A typical terahertz pulse with the corresponding Fourier spectrum obtained in such a spectrometer is depicted in Fig. 9.

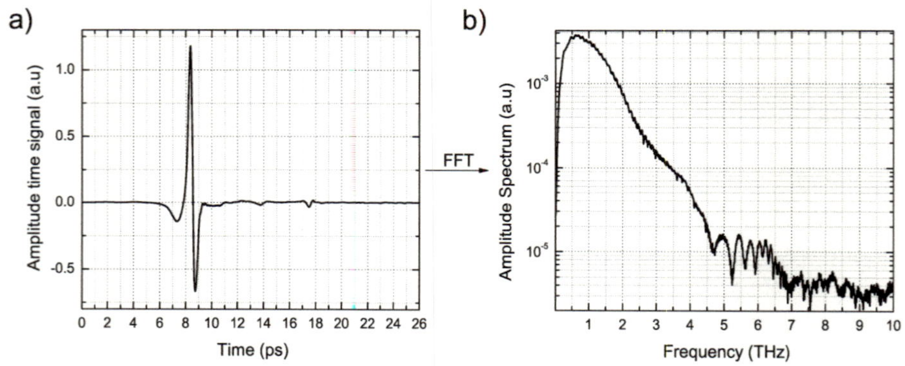

Fig. 9. Typical time domain signal (left) and the corresponding spectrum (right).

3.4 Material parameter extraction with terahertz time domain spectroscopy

Due to the coherent detection scheme of terahertz time domain spectroscopy, both phase and amplitude information of the electric field can be accessed. This circumstance enables the direct extraction of the complex refractive index $\underline{n} = n - i\kappa$ without the need for the Kramers-Kronig relations as required in the case of FIR spectroscopy. Here, n is the real part of the refractive index and κ the extinction coefficient. From these two measures the complex permittivity $\underline{\varepsilon_r} = \varepsilon_r' - i\varepsilon_r''$ with the real part $\varepsilon_r' = n^2 - \kappa^2$ and the imaginary part $\varepsilon_r'' = 2n\kappa$ as well as the absorption coefficient $\alpha = 2k_0\kappa$ can be determined. $k = \frac{\omega}{c_0}$ denotes the free space angular wave number, ω the angular frequency and c_0 the speed of light in free space.

The basic idea of material parameter extraction from THz TDS data is the comparison of a sample and a reference pulse, once with and once without the sample mounted in the terahertz beam. In most approaches the Fourier spectra of both pulses are calculated and a transfer function is defined as the complex quotient of the sample spectrum to the reference spectrum. Different algorithms for the data analysis were developed. Most recently a new approaches, which enables the simultaneous identification of the refractive index n, the absorption coefficient alpha and the sample thickness even in case of ultra thin samples in the sub 100 μm regime has been proposed (Scheller et al. 2009, b). As this approach was employed for the material parameter extraction of most datasets presented in this chapter we shall now briefly review its basic working principle.

The first step in the data extraction is the formulation of a general theoretic transfer function for the sample under investigation depending on the refractive index n, the absorption coefficient alpha, and the sample thickness L. In order to create such a transfer function the number of multiple reflections M which occur during the measured time window is determined in a preprocessing step which assumes an initial thickness L_0. The basic shape of the theoretical transfer function is given by

$$H(\omega) = A_0 \exp\left(-i\frac{\omega}{c_0}(\underline{n}-1)\cdot D\right) + A_1 \exp\left(-i\frac{\omega}{c_0}(3\underline{n}-1)\cdot D\right) + .. \qquad (18)$$

where A_i is given as the ith of M elements that are functions of the Fresnel coefficients. In a following step, an error function defined by the difference of the theoretical transfer function and the measured one is minimized which yields n and alpha as functions of the sample thickness L. To unambiguously derive n, alpha and the material thickness L, the Fabry-Perot oscillations superimposed to the measured material parameters have to be considered. Hence, an additional Fourier transform is applied to the frequency domain material parameters which transforms the superimposed Fabry-Perot oscillations to a discrete peak is the so called quasi space regime. Now the correct sample thickness as well as n and alpha can be determined by minimizing the peak amplitude completing the material parameter extraction.

As a demonstration of this technique, we analyse a 54.5 μm silicon wafer. If the correct thickness is chosen the peak values are minimized. Fig. 10 a) shows the refractive index n for different thicknesses over the frequency. For the correct thickness determined from the quasi space peak minimization (Fig. 10 b)), the Fabry Perot oscillations vanish from the material parameter spectra

The inclined reader finds a detailed discussion of this algorithm in (Scheller et al., 2009, b).

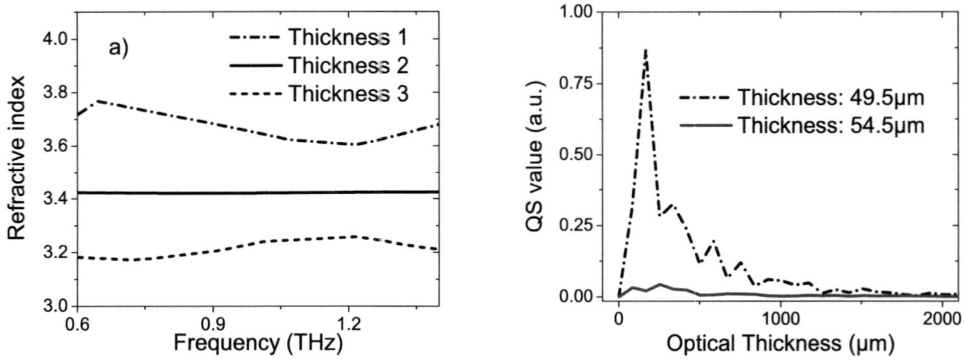

Fig. 10. The extracted refractive index for different thicknesses (a) and the corresponding QS values (b).

4. Application scenarios

4.1 Polymeric compounds

To evaluate the applicability of the different EMTs, we shall now investigate three additive concentration series of polymeric compounds, comparing measurement and simulation results of the refractive index at 1 THz (Scheller et al., 2009, a).. The additive particle sizes are less than 5 μm so that scattering effects should remain negligible. Hence, a quasi-static effective medium theory fully describes the dielectric behaviour of the composites at the wavelength of interest. The samples consist of injection-molded rods of 1 mm to 3 mm thickness. The additives differ both in shape and permittivity. After THz TDS measurements, the additive concentration was confirmed by combustion, determining the ignition residue content of the compounds. The measurement setup consists of a transmission terahertz time domain spectrometer as described in section 3.

The first series of samples comprises magnesium hydroxide $Mg(OH)_2$-filled low-density polyethylene (LLDPE) together with a compressed pellet of pure $Mg(OH)_2$, which was used to quantify the 100% additive content permittivity value. Utilizing the LLL model, this value was extrapolated from the measurement result of the porous pellet. Nelson et al. demonstrated that this procedure delivers accurate results for granular and powdered materials (Nelson, 2005). The $Mg(OH)_2$ particles have a hexagonal disc-like shape. Calculations employing the effective medium theories are compared to the measured data in Fig. 11. While the EB and the PvS model deliver good agreement with the measured data, the MG approaches show a significant discrepancy to the experimentally obtained results. This circumstance can be explained by the fact that the MG model only considers spherical particles while both the EB and the PvS model are shape-dependent. The predicted values based on the LLL theory, which does not consider any particle shape, cannot reach the accuracy of either the EB or the PvS calculations but performs better than the MG theory in case of this material system. The CRIs equation's prediction additionally exhibits a good agreement to measured data but lacks any physical motivation.

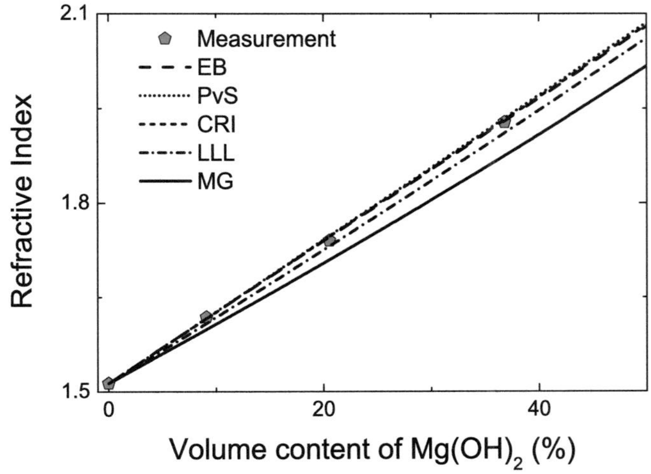

Fig. 11. The refractive index of the $Mg(OH)_2$ compound system for different volumetric contents of the additives compared to the EMT's predictions.

The second series consists of rutile titania (TiO_2) particles embedded in a polypropylene (PP) host matrix. In this case, the 100% additive volume content value was directly extracted from a rutile crystal. Both, measurements along the fast and the slow axis of the birefringent rutile titania crystal were conducted and a weighted average of the permittivity was determined. The obtained material parameters of the crystal are shown in Fig. 12.

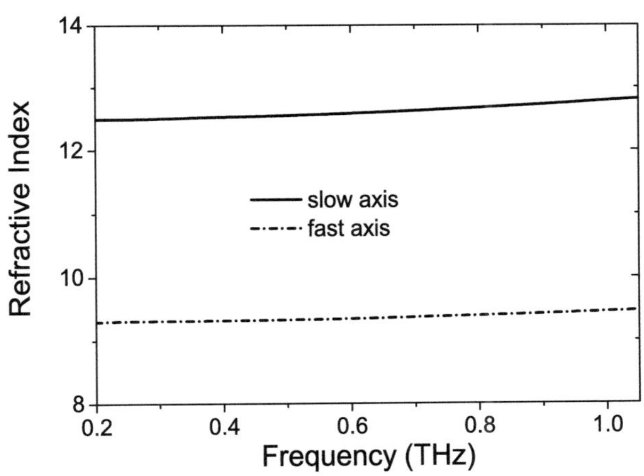

Fig. 12. The refractive index of the fast and the slow axis of a rutil crystal.

This step was necessary, as the LLL model, used for the extraction of the 100% additive values from the compressed pellets in case of the other material systems, is unable to properly describe the behaviour of the TiO_2-air mixture due to the large dielectric contrast.

As a Taylor approximation is contained in the LLL models derivation, valid predictions cannot be expected in this case. Apart, from the LLL and the CRI model, all other existing theories should be well suited to accurately describe the behaviour of this material system as long as only diluted mixtures are considered. An interesting aspect of this material system is the high permittivity contrast between the additive and the host combined with the spherical particle shape, which allows to clearly evaluate the capabilities of each approach.

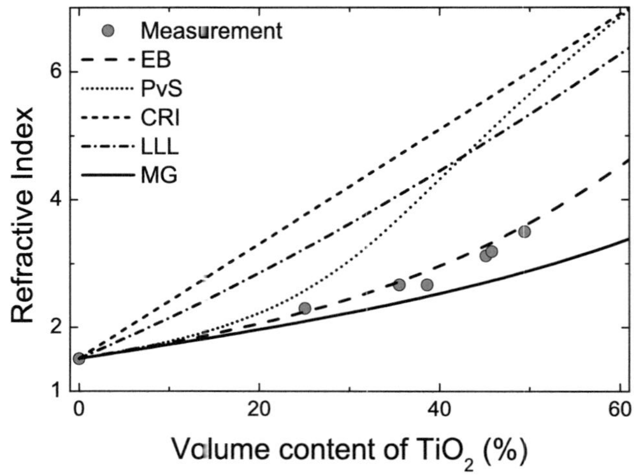

Fig. 13. The measured results of the TiO_2-PP mixture compared to the predictions of the different models.

The measured data together with the values predicted from the theories are shown in Fig. 13. As expected, all models, except the LLL and the CRI, deliver similar results as long as only low particle concentrations below 15 volume percent are considered. For higher concentrations, only the EB theory can still accurately model the mixture behaviour. While the MG theory delivers much smaller permittivity values compared to the measurements, the PvS model deviates in the opposite direction. The LLL and the CRI model do not yield a good estimation of the measured permittivity for any additive concentration due to the reasons discussed above.

The last series of samples which we will consider here comprises calcium carbonate ($CaCO_3$)-filled polypropylene samples. Analogous to the $Mg(OH)_2$-series, a compressed pure additive pellet of $CaCO_3$ was employed to extract the 100% additive content permittivity value. The $CaCO_3$ particles exhibit a cubic shape. Cubes are one of the few shapes that cannot be described as ellipsoids. In addition, the aspect ratio of the cuboids under investigation is not uniform. However, by considering a distribution of different aspect ratios, the PvS and the EB model should be able to reach a performance close to the modelling with the actual particle shape.

Calculations based on the physical models and the measured permittivity are illustrated in Fig. 14. For the EB and the PvS model, a rectangular distribution of aspect ratios between the longer and the shorter axis is assumed. While the EB and PvS model well describe the measured behaviour, closely followed by the LLL and the CRI approach, the MG theories suggest permittivity values that strongly deviate from the experimentally determined material parameters.

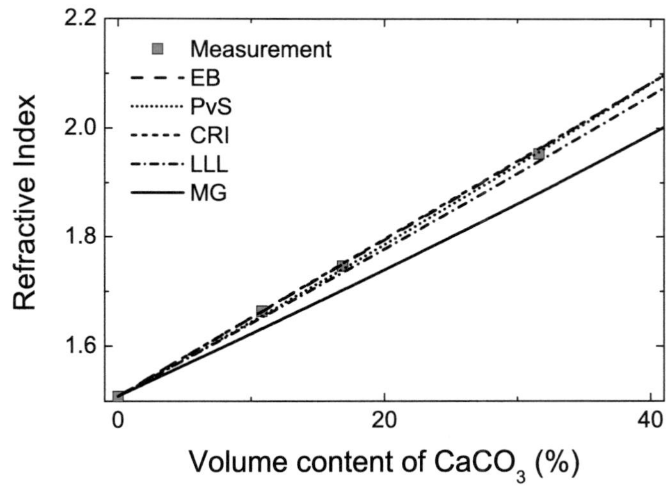

Fig. 14. The measured results of the CaCO3 system compared to the predictions of the different models.

These results indicate, that the PvS and the EB model offer most flexibilty and are sufficient for most applications. However, if the contrast in permittivity is small and the mixture consist of anisotropic shaped particles, the relation between the volumetric content and the refractive index exhibit a close to linear behaviour, allowing to utilize the CRI equation to estimate the resulting permittivity. The LLL model delivers good results for two of the three cases and is applicable if no high contrast in permittivity occurs and can be seen as a good choice if no information about the particle shape is known or if the mixture consists of irregularly shaped components.

4.2 Hydration monitoring of plant leaves

Hydration monitoring of leaves is of high importance for farmers and plant physiologists alike. It can provide valuable information about irrigation management and helps to control drought stress. THz radiation is ideally suited for studying the water content in leaf tissue due to the strong water induced absorption present at THz frequencies (Hadjiloucas et al., 1999). However, a physical modeling is required to create a connection between the measured data and the water content of the leaves.

A leaf can be described as a multi compound mixture, that mainly consists of the following three parts: air, water, and the solid plant material. Therefore, an EMT is in principal suitable to model the dielectric constant of this biological sample. Due to the irregular structure of the leave, the LLL model is suitable to solve this specific problem. However, LLL is just able to compute the effective permittivity of a compound consisting of two components so that an extension has to be made in order to match these enhanced requirements. Such a third order extension

$$\sqrt[3]{\varepsilon_R} = f_W \sqrt[3]{\varepsilon_W} + f_S \sqrt[3]{\varepsilon_S} + (1 - f_W - f_S) \; ,$$

where f_i is the concentration of the components is essential to account for the three parts of the leaf. The index W and S refers to the water and the solid plant material, respectively.
A series of fresh and water stressed Coffea arabica leaves was measured in a TDS setup by Jördens at coworkers (Jördens et al., 2009, a). Afterwards the leaves were detached from the plant, the thickness and mass was measured and the water was pressed out. The concentration of water, air, and solid plant material can be determined with the information of the thickness and the mass of each leaf before and after pressing.
The dielectric properties of the solid plant material were identified by TDS measurements on the pressed and dried leaves. The pure water permittivity was calculated by a dual Debye model analogue to (Liebe et al., 1991). As can be seen in Fig. 15, the water exhibits an intense anormal dispersion. Thus, a similar behavior is expected for the optical parameters of the leaf.
The measured refractive index and the absorption of two leaves, one fresh and one stressed, are shown in (Fig. 16) together with the model's predictions. As the refractive index of water exceeds the ones of the other mixtures components, the resulting refractive index of the leaves is higher for larger water content. Over the whole spectral range, a good agreement between the simulated and measured results is obtained, indicating the applicability of the THz technology in combination with EMTs to monitor the hydration status of plants

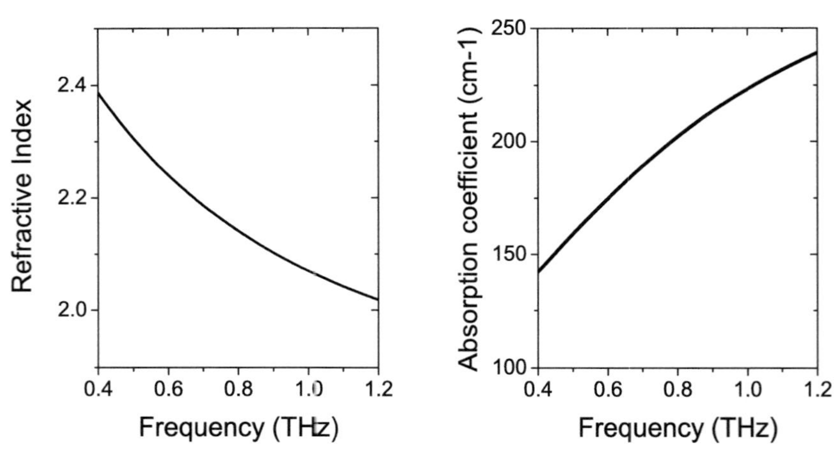

Fig. 15. The refractive index and the absorption coefficient of water at 20 °C calculated from the Debye model.

4.3 Isotropic material mixtures

In the case of the application scenarios described above anisotropic mixtures were considered. Yet, if the particles within the mixture exhibit an isotropic shape, an orientation of these will induce a macroscopic birefringence of the effective medium. As one example we will shortly examine a system of glass fibres filled HDPE. Due to the injection molding production process, a notable orientation of the fibres inside the direction of the mold flow is expected. Consequently, the refractive indices parallel and perpendicular to this direction will differ as a consequence of the isotropic material system.

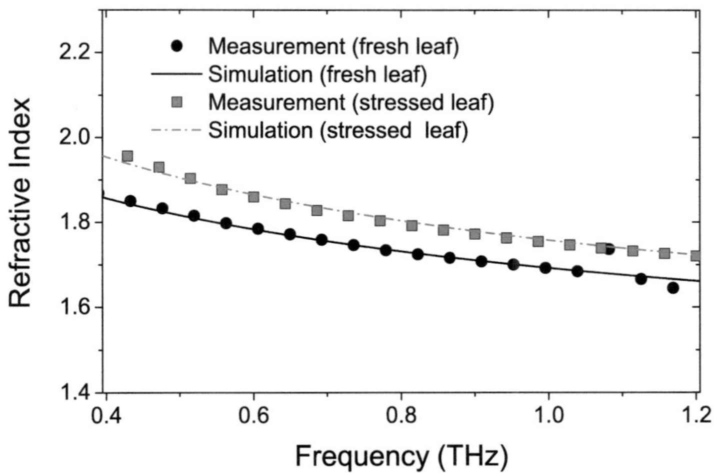

Fig. 16. Measured and simulated refractive index of a fresh and a stressed Coffea arabica leaf.

The PvS model was chosen to analyze the resulting effective medium, because the permittivity contrast between the HDPE and the borosilicate based fibres is small and the refractive indices along the different orientations can be calculated directly by Eq. 8, assuming rod like particles within the matrix. The resulting refractive indices are shown in Fig. 17 together with measured values, obtained from a 3 mm thick fibre enforced polymer.

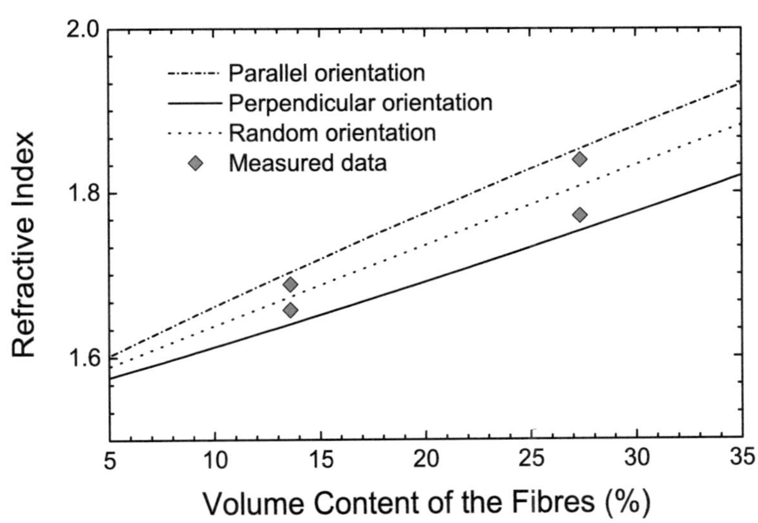

Fig. 17. The simulated refractive indices of the glass fibre - HDPE composite at 400GHz

A distinctive birefringence can be observed in the simulation. The measured data exhibit a slightly lower one. This discrepancy can be explained by a non uniform orientation of the particles. If only the majority of the fibres are orientated along the mold flow direction, the resulting birefringence will derivate from the maximum values. As a consequence, THz measurements can be utilized to determine the fibres content of the mixture and their degree of orientation simultaneously (Jördens et al., 2009, b).

5. Conclusion and outlook

In this chapter a short introduction into a selection of quasi static effective medium theories was presented. Their potential and restrictions were illuminated by a choice of representative application scenarios ranging from polymeric compounds to biological samples. However, the relatively long wavelengths of the terahertz waves enables for a wide applicability within this frequency range without the necessity of including volumetric scattering effects in the calculations. Therefore, an accurate analytical modelling of various problems can be achieved utilizing such EMTs. The Table 2 gives a short overview of exemplary application scenarios and the EMTs of choice.

Application Scenario		EMT of choice	
Mixtures with high contrast in permittivity, spherical particles		EB	
Mixtures with high contrast in permittivity, ellipsoidal particles		EB, PvS	
Mixtures with low contrast in permittivity, irregular or unknown shaped particles		LLL, CRI	
Mixtures of powders, porous pellets		LLL, CRI	
Mixtures consisting of isotropic components		EB, PvS	

Table 2. Overview of typical application scenarios for the EMTs mentioned in the text.

6. References

Auston, D. H. (1975). Picosecond optoelectronic switching and gating in silicon, *Applied* Physics Letters, AIP, 1975, 26, 101-103

Auston, D. H.; Cheung, K. P.; Smith, P. R. (1984). Picosecond photoconducting Hertzian dipoles, *Applied Physics Letters, AIP,* 1984, 45, 284-286

Banhegyi, G. (1986). Comparison of electrical mixture rules for composites. *Colloid Polym. Sci.,* Vol. 264, 1986, 1030–50

Beard, M. C; Turner, G.M.; Schmuttenmaer, C. A. (2002). Size-Dependent Photoconductivity in CdSe Nanoparticles as Measured by Time-Resolved Terahertz Spectroscopy. *Nano Letters,* Vol. 2 2002, 983-987

Böttcher, C. J. (1942). Zur Theorie der inneren elektrischen Feldstärke. *Physica,* Vol. 9, No. 10, 1942, 937-943

Bruggeman, D. (1935). Berechnung verschiedener physikalischer Konstanten von heterogenen Substanzen: I Dielektrizitätskonstanten und Leitfähigkeiten der Mischkörper aus isotropen Substanzen. *Annalen der Physik,* Vol. 24, 1935, 636-664

Hadjiloucas, S.; Karatzas, L. S.; Bowen, J. W. (1999). Measurements of leaf water content using terahertz radiation. *IEEE Transactions on microwave theory and techniques,* Vol. 47, No. 2, 1999, 142–149

Hendry, E.; Koeberg, M.; O'Regan B.; Bonn M. (2006). Local Field Effects on Electron Transport in Nanostructured TiO2 Revealed by Terahertz Spectroscopy. *Nano Letters,,* Vol. 6, 2006, 755-759

Jackson, J. D. (1999). *Classical electrodynamics,* Wiley & Sons, Inc, 047130932X, New York

Jepsen, P. U.; Jacobsen, R. H.; Keiding, S. R. (1996). Generation and detection of terahertz pulses from biased semiconductor antennas, *J. Opt. Soc. Am. B,* 1996, 13, 2424-2436

Jördens, C.; Scheller, M.; Breitenstein, B.; Selmar, D.; Koch, M. (2009, a). Evaluation of the Leaf Water Status by means of the Permittivity at Terahertz Frequencies. *Journal of Biological Physics,* 2009, 35, 255-264

Jördens, C.; Scheller, M.; Wietzke, S.; Romeike, D.; Jansen, C.; Zentgraf, T.; Wiesauer, K.; Koch, M. (2009, b). Terahertz spectroscopy to study the orientation of glass fibres in reinforced polymers. *Submitted to Composite Science and Technology,* 2009

Kimmitt, M. (2003). Restrahlen to T-Rays - 100 Years of Terahertz Radiation, *Journal of Biological Physics,* 2003, 29, 77-85

Kittel, C. (1995). *Introduction to solid state physics,* Wiley & Sons, Inc, 0471111813, New York

Kuebler, C.; Huber, R.; Leitenstorfer, A. (2005). Ultrabroadband terahertz pulses: generation and field-resolved detection, *Semiconductor Science and Technology,* 2005, 20, S128-S133

Levy, U.; Abashin, M.; Ikeda, K.; Krishnamoorthy, A.; Cunningham, J.; Fainman, Y. (2007). Local Field Effects on Electron Transport in Nanostructured TiO2 Revealed by Terahertz Spectroscopy. *Physical Review Letters,* Vol. 98, 2007, 243901

Liebe, H. J.; Hufford, G. A.; Manabe, T. (1991). A model for the complex permittivity of water at frequencies below 1 THz. *International Journal of Infrared and Millimeter Waves,* Vol. 12, No. 7, 1991, 659–675

Looyenga, H. (1965). Dielectric constants of heterogeneous mixtures. *Physica,* Vol. 31, 1965, 401-406

Maxwell-Garnett, J. C. (1904). Colours in metal glasses and in metallic films. *Philosophical Transactions of the Royal Society of London. Series A, Vol. 203, 1904,* 385-420

Mittleman (ed.), D. (2003). Sensing With Terahertz Radiation, *Springer-Verlag, Berlin,* 2003

Moulton, P. F. (1986). Spectroscopic and laser characteristics of Ti:Al2O3, *J. Opt. Soc. Am. B, OSA,* 1986, 3, 125-133

Nelson, S. O.; You, T. S. (1990). Relationships between microwave permittivities of solid and pulverised plastics. *J. Phys. D: Appl. Phys.,* Vol. 9, No. 10, 1990, 346-353

Nelson, S. O. (2005). Density-permittivity relationships for powdered and granular materials. *IEEE Transactions on Instrumentation and Measurement.,* Vol. 54, No. 5, 2005, 2033-2040

Piche, M.; Salin, F. (1993). Self-mode locking of solid-state lasers without apertures, *Opt. Lett., OSA,* 1993, 18, 1041

Polder, D.; van Santen, J. H. (1946). The effective permeability of mixtures of solids. *Physica,*Vol. No. 12, 5, 1946, 257-271

Ralph, S. E.; Grischkowsky, D. (1974). Trap-enhanced electric fields in semi-insulators: The role of electrical and optical carrier injection, *Applied Physics Letters, AIP,* 1991, 59, 1972-1974

Sakai (ed.), K. (2005). Terahertz Optoelectronics, *Springer Berlin,* 2005

Salin, F.; Squier, J.; Piché, M. (1991). Mode locking of Ti:Al2O3 lasers and self-focusing: a Gaussian approximation, *Opt. Lett., OSA,* 1991, 16, 1674-1676

Scheller, M.; Wietzke, S.; Jansen, C.; Koch, M. (2009, a). Modelling heterogeneous dielectric mixtures in the terahertz regime: a quasi-static effective medium theory. *J. Phys. D: Appl. Phys.,* Vol. 42, 2009, 065415

Scheller, M.; Jansen, C.; Koch, M. (2009, b). Analyzing sub-100-µm samples with transmission terahertz time domain spectroscopy. *Optics Communications,* Vol. 282, 2009, 1304-1306

Siegel, P. H. (2002). Terahertz technology, *IEEE Trans. on MTT,* 2002, 50, 910-928

Smith, P.; Auston, D.; Nuss, M. (1988). Subpicosecond photoconducting dipole antennas, *IEEE Journal of Quantum Electronics,* 1988, 24, 255-260

Van Rudd, J; Mittleman, D. M. (2002). Influence of substrate-lens design in terahertz time-domain spectroscopy, *Journal of the Optical Society of America B: Optical Physics, OSA,* 2002, 19, 319-329

Yano, R.; Gotoh, H.; Hirayama, Y.; Miyashita, S.; Kadoya, Y. & Hattori, T. (2005). Terahertz wave detection performance of photoconductive antennas: Role of antenna structure and gate pulse intensity, *Journal of Applied Physics, AIP,* 2005, 97, 103103

Nanoscale Optical Techniques and Applications

13

Local Electric Polarization Vector Detection

Kwang Geol Lee and DaiSik Kim
Seoul National University
Republic of Korea

1. Introduction

The function of nanophotonic devices, such as metamaterials for the visible range, photonic and plasmonic crystals, nanoscale waveguides, resonators, switches or optical antennas relies essentially on our ability to tailor and control electromagnetic fields on a sub-wavelength scale, in much the same way as the wavefunction of electronic nanostructures is localized on an atomic scale. Unlike matter waves, these localized electromagnetic waves are vectorial in nature, and their orientation and magnitude varies on a sub-wavelength scale. Therefore it is vital for a complete description of light in nano-scale devices to map the field vectors with subwavelength resolution. Experimental techniques aimed at probing the field intensity, e.g. aperture-less or aperture-based near-field microscopies, are rapidly improving, but the local orientation of the electromagnetic polarization vector, a fundamental property of the local electromagnetic field, could not be accessed experimentally.

Yet, the orientation of the field vector is a key quantity in many thousands of theoretical studies on nano-optics, nanophotonic devices and optical sciences in general. Evidently, the ability to experimentally probe electromagnetic field vectors with nanometer resolution is of fundamental importance for understanding and improving nano-optical devices and for reconciling experiment and theory in the bourgeoning field of nanophotonics.

In this chapter, we describe and demonstrate the local field polarization vector detection using the gold nanoparticle (GNP) attached tip as the local field scatterer acting as a nanometer-scaled polarizer. For a suitably small GNP, the far-field scattering is dominated by the electric dipole radiation. Dipole radiation conserve its polarization state into the far-field region enabling characterization of the dipole moment induced at the GNP by measuring the far-field polarization state. And also the dipole moment is determined by the local electric field via polarizability tensor of GNP. Therefore, by characterizing the polarizability tensor of GNP and the polarization state of far-field scattered light, the local electric field vector can be reconstructed. In doing that, the different scattering shape of GNP, the polarizability tensor, is carefully measured and considered to get a consistant result independently of the tip shape.

Mapping the local polarization vector in the near-field demands a careful consideration of the surface effect. For example, far field detected light is a complex mixture of the scattered light from GNP and its reflected light at the sample surface which interfere each other strongly depending on the polarization. This so called image dipole effect should be taken into account to correctly address the reconstructed local field vectors in near-field region. In the last section of this chaper, we discuss the limitations of our method and give suggestions to improve the functionality of the polarization vector microscopy.

2. Local field polarization vector detection on nanoscale

2.1 Polarizability tensor measurement of GNP attached tips

The metal particle has been used in a wide range of applications owing to its good photostability. For example, GNP is one of the excellent candidates as a tracking label in biosensing. It also can serve as an optical antenna to enhance the spontaneous emission rate of single molecules by using its plasmon resonance (Kühn et al., 2006). Moreover, a small GNP, less than 100 nm in diameter, is a good dipolar scatterer and it could be used as a nano-sized polarizer to detect the local electric field polarization vectors (Lee et al., 2007a). In the reconstruction of the local field vectors from the far-field detected signal radiated by the GNP, it is crucial to correctly and efficiently measure the polarizability tensor of the GNP. In this section, we present theoretically and experimentally how the polarizability tensor can be read out from two types of scattering measurement - rotational polarizer analyzer ellipsometry (RPAE) and rotational polarizer ellipsometry (RPE). By comparing two methods, the pros and cons of each method are discussed. Finally, we show the experimental results validating the model and the data process.

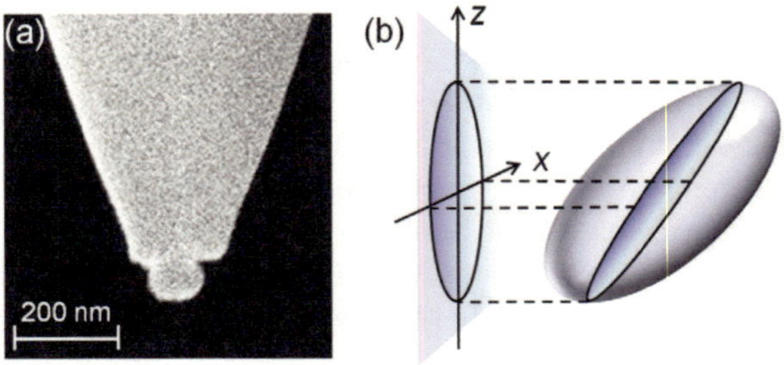

Fig. 1. (a) An SEM picture of a GNP attached tapered glass tip. (b) The projection of an arbitrary ellipsoid onto a plane makes an ellipse. In the same analogy, the geometrical shape of the polarizability tensor of an ellipsoidal GNP in a 2-dimensional plane can be regarded as an ellipse.

The GNP attached tips (Kalkbrenner et al., 2001) are prepared as following. Sharp glass fiber tips with the tip apex size of 100~200 nm are made by the chemical etching (hydrofluoric acid, 48%, 65 mins) or by using a mechanical puller machine. Then tips are chemically treated to generate a chemical layer sticky to the GNP. For the case of APTES (3-aminopropyltriethoxysilane), GNP is bonded to the glass fiber through the electro-static force, on the other hand, APTMS (3-aminopropyltrimethoxysilane) forms a layer which makes a covalent bonding to the GNP (Keating et al., 1999). This surface treated tip is scanned onto a cover slip or a slide glass where the GNP colloid is spin-coated on it. When the tip moves across a GNP, the GNP eventually sticks to the tip and moves together. A scanning electron microscope (SEM) image of the tip attached a 100 nm diameter sized GNP is shown in Fig. 1(a). The size variation of GNPs in the colloid is less 8% (from data sheet of BBinternational) and there is a small variation in shape also. Additional to these intrinsic morphological distributions, the attaching process may generate a deformation of the particle shape due to the squeezing forces applied between the tip and the sample surface. Therefore the geometrical shape of the GNP, in general, can be assumed as an arbitrarily oriented ellipsoid. Experimental data in the later part show that this assumption is quite well valid.

The polarizability tensor of the GNP is measured in a 2-dimensional plane (x-z plane) as depicted in Fig. 1(b). The projection of an arbitrary ellipsoid onto a 2-dimemsional plane makes an ellipse. The complete 3-dimensional polarizability tensor, the ellipsoid, can be obtained by performing this 2-dimensional measurement in three orthogonal directions and by combining them.

From now on, we describe how the polarizability tensor values are correctly read out from two scattering measurements - RPAE and RPE. In RPAE, the incident laser light of λ=780 nm is loosely focused at the tip end using a lens of 10 cm focal length while rotating the incident polarization using a half wave plate. The scattered light is polarization analyzed by rotating a linear polarizer (analyzer) in front of an APD (C-4777-01, Hamamatsu) in the direction of the incidence. The experimental schematic is shown in Fig. 2(a). For a suitably small GNP, below 100 nm in diameter, the far-field scattering is dominated by the electric dipole radiation. In that case, the polarizability tensor is described by a 2×2 matrix in a 2-dimensional plane. Here, we assume a simple case of the polarizability tensor ($\ddot{\alpha}$) which is symmetric to the laboratory frame with null values of off-diagonal terms, $\ddot{\alpha} = \begin{pmatrix} a & 0 \\ 0 & b \end{pmatrix}$. A general case of an arbitrarily oriented ellipse with non-zero off-diagonal terms will be discussed in later part. Fig. 2(b) shows the polar plots of the square rooted intensity of the scattered light as a function of the analyzer angle. Three polar plots are shown for the incident polarizations of 0 (black solid), 45 (black dashed), and 90 (black dotted) degrees from the x-axis of the laboratory frame. The incident and scattered lights are related through the following relations:

$$\vec{p} = \ddot{\alpha}\, \vec{E}_{inc} \tag{1}$$

$$\vec{E}_{dipole} = \frac{1}{4\pi\varepsilon_0}\left\{ k^2 (\vec{n} \times \vec{p}) \times \vec{n} \frac{e^{ikr}}{r} \right\} \tag{2}$$

,where $\vec{p}$, $\vec{E}_{inc}$, and $\vec{E}_{dipole}$ are induced dipole moment at GNP, incident electric field, and far-field radiated electric field by the induced dipole moment $\vec{p}$, respectively. Here, $\vec{n}$ is the detection position vector and to be $-\hat{y}$ in the configuration of Fig. 2. The maxima of the polar plot for every incident polarization (red dots) make an ellipse with the semi-major and the semi-minor axes to be *b* and *a* (*b*>*a*), respectively. From this, the original tensor value can be extracted.

In the next, RPE is applied in measuring the same polarizability tensor, $\ddot{\alpha} = \begin{pmatrix} a & 0 \\ 0 & b \end{pmatrix}$. Here, the incident beam polarization is also varied, but with no analyzer in front of the APD as shown in Fig. 2(a). The outer black curve in Fig. 2(c) shows the polar plot of the square rooted scattered light intensity as a function of the incident polarization. When the incident beam polarization is *θ*, the polarization angle of the scattered light is given as

$$\varphi = \tan^{-1}\left(\frac{b}{a} \tan\theta \right) \tag{3}$$

, and the square rooted intensity is calculated as,

$$r_1 = r_2 = \left(\frac{a^2 + b^2 \tan^2 \theta}{1 + \tan^2 \theta} \right)^{-\frac{1}{2}} . \tag{4}$$

The inner ellipse and the outer polar plot do not encounter each other except the maxima and the minima of the black curve which correspond to the major and the minor axes of the inner ellipse. By taking the four maxima and minima points of the outer polar plot, the inner ellipse - the polarizability tensor, can be determined. This RPE takes less measurement time compared to the RPAE, but it is less accurate since it has less number of data points. In addition, to apply RPE, it is required a prior study on whether the scatterer is a good dipolar scatterer, otherwise the measured data can be misleading.

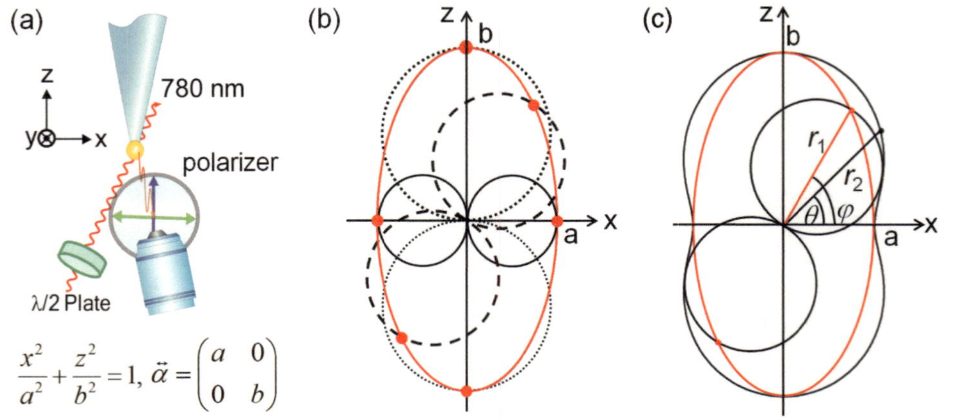

Fig. 2. RPAE vs RPE. (a) Schematics of RPAE (with polarizer) and RPE (without polarizer). A 780 nm plane wave is focused at the tip end part using a lens of 10 cm focal length, while rotating the incident polarization using a half wave plate. (b) RPAE. The scattered light is polarization analyzed rotating an analyzer (linear polarizer) in front of the detector. Three black polar plots show the square rooted scattered light intensities as a function of the analyzer angle, for cases of the incident polarization of 0 (solid), 45 (dashed), and 90 (dotted) degrees from the x-axis. The red ellipse is constructed by connecting the maxima of the polar plot of every incident polarization. (c) RPE. APD measures the total intensity of scattered light. The outer black polar plot shows the square rooted scattered light intensity as a function of the incident polarization. The inner red ellipse is determined by choosing the maxima and the minima of the polar plot (black curve) as the semi-major and the semi-minor axes of an ellipse (red). From (Lee et al., 2009). © 2006 The Korean Physical Society.

So far, we assumed a polarizability tensor of a simplified case - symmetric to the x and z axes. But in a general case, the off diagonal elements have non-zero values which implies that the major axis of the ellipse has a deviated angle from the laboratory axes. With β (≥1) defined as the length ratio between the major and the minor principal axes and ζ as the angle between the major axis of the scatterer and the laboratory frame x-axis, the polarizability tensor can be written as,

$$\ddot{\alpha} = \begin{pmatrix} \cos\zeta & -\sin\zeta \\ \sin\zeta & \cos\zeta \end{pmatrix} \begin{pmatrix} \beta & 0 \\ 0 & 1 \end{pmatrix} \begin{pmatrix} \cos\zeta & \sin\zeta \\ -\sin\zeta & \cos\zeta \end{pmatrix} = \begin{pmatrix} \beta\cos^2\zeta + \sin^2\zeta & (\beta-1)\sin\zeta\cos\zeta \\ (\beta-1)\sin\zeta\cos\zeta & \beta\sin^2\zeta + \cos^2\zeta \end{pmatrix} . \tag{5}$$

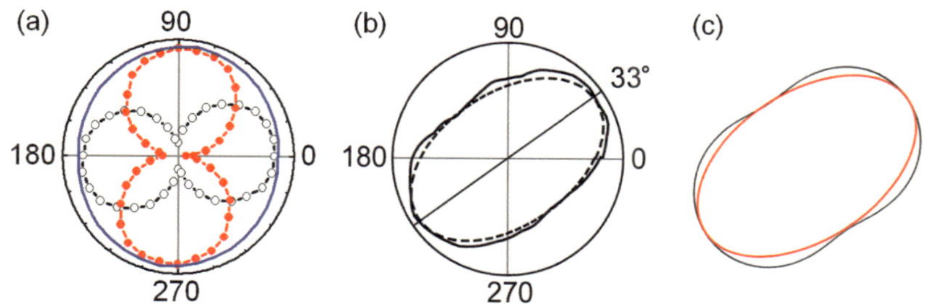

Fig. 3. (a) Two peanut shaped polar plots are measured applying RPAE. The incident polarization angle values are 0 (black open circles) and 90 (red filled circles) degrees. The blue curve is obtained applying RPE. (b) Another measurement applying RPE. Solid curve shows the measured data and the inner dashed ellipse is made by choosing the maxima and the minima of the polar plot as the semi-major and the semi-minor axes. (c) Red ellipse is same as the dashed one in (b), and the outer black curve shows the re-calculated polar plot from the ellipse parameters. From (Lee et al., 2009). © 2006 The Korean Physical Society.

Fig. 3 shows experimentally measured data of two GNP attached tips applying RPAE and RPE. Two peanut shaped polar plots in Fig. 3(a) show the square rooted scattered light intensity with the incident polarization directions parallel to the x-axis (black open circles) and the z-axis (red filled circles) of the laboratory frame. The dipolar radiation pattern of the polar plots shows that the attached GNP is a good dipolar scatterer. The outer blue curve is obtained applying RPE. It shows the polar plot of the square rooted scattered light intensity as a function of the incident polarization. From these measurements, values of β and ζ are determined as 1.05 and 80 degrees, respectively, and the resultant polarizability tensor in x-z plane is given as $\ddot{\alpha}_1 = \begin{pmatrix} 1 & 8\times10^{-3} \\ 8\times10^{-3} & 1.04 \end{pmatrix}$. In this case, the GNP shape is quite circular and the off diagonal terms are negligible compared to the diagonal terms. Note that with a small ellipticity (a≈b in Fig. 2(c)), the polar plot by RPE directly gives the ellipse of the polarizability tensor from Eqs. (3) and (4). The maxima of two polar plots by RPAE clearly touch the polar plot obtained from RPE demonstrating that two methods give a consistent result. Fig. 3(b) shows a scattering measurement of a different GNP functionalized tip obtained by applying RPE. Here, β and ζ values are 1.62 and 33 degrees, and the resulting polarizability tensor is given as $\ddot{\alpha}_2 = \begin{pmatrix} 1.208 & 0.257 \\ 0.257 & 1 \end{pmatrix}$. In this case, the scattering function of the GNP (inner dashed line of Fig. 3(b)) is quite elliptical and the major axis is rotated from the laboratory frame. To check the quality of the measured data, the polar plot is fitted with the determined $\ddot{\alpha}$ (black curve in Fig. 3(c)). The similar shape of the polar plots in Figs. 3(b) and 3(c) validates the model and the measurement.

To conclude this section, we demonstrate how the optical response of the gold nanoparticle attached tip, the polarizability tensor, can be measured from two different scattering measurements - RPAE and RPE. RPAE takes more time in determining the tensor values but generally it is more accurate. And also it gives the information of the dipole nature of the tip. In a comparison, RPE takes less time in measuring, but this method needs a prior knowledge whether the tip is a good dipolar scatterer. For our cases, the diameter of GNP attached to the tip is less than 100 nm, the electric dipole radiation is dominant in the far-field scattering process and two scattering measurements give a consistent result. The measured data is clearly well reconstructed by a simple calculation validating our analysis methods.

2.2 Polarization detection of light scattered off GNPs

A general elliptical polarization state of the local electric field at a fixed position $\vec{r}$ in the x-z plane can be written as (Born & Wolf, 1999):

$$\vec{E}_{Local}(\vec{r}) = (E_x, E_z) = (a_1 e^{i\omega t + i\delta_1}, a_2 e^{i\omega t + i\delta_2}), (a_1, a_2 > 0), \quad (6)$$

This field vector rotates at a frequency ω along the perimeter of an ellipse. A dipole scattering tip gives a scattered far-field $\vec{E}_S \propto \ddot{\alpha} \cdot \vec{E}_{Local}$ where $\ddot{\alpha}$ is the polarizability tensor of the scatterer (Eqs. (1), (2)). In determining of the polarization states of scattered light, we apply the RAE and the Stokes parameter measurement (Stokes, 1852).

Firstly, the polarization state of an arbitrarily shaped light is determined by the RAE method in which a linear polarizer, mounted inside the optical path of the scattered light and in front of the detector, is rotated by 360° in 10° steps. The detected field intensity passing through a polarizer is then given as

$$I \propto \left\langle \left| \vec{P} \cdot \vec{E}_S \right|^2 \right\rangle = \left\langle \left| \begin{pmatrix} \cos\varphi & \sin\varphi \end{pmatrix} \ddot{\alpha} \begin{pmatrix} E_{Local,x} \\ E_{Local,z} \end{pmatrix} \right|^2 \right\rangle, \quad (7)$$

where φ is the detecting polarizer angle from the x-axis and $\langle ... \rangle$ denotes a time average over many optical cycles.

The polar diagram $\sqrt{I(\varphi)}$ shown in Fig. 4(a), recorded by rotating the polarizer in 10° steps, allows us to determine the polarization state of the scattered light depicted as a red colored ellipse. The major axis angle of the ellipse corresponds to the detecting polarizer angle at which the measured intensity has its maximum and the major and minor axes lengths are proportional to the square-root of the maximum and minimum intensities, respectively. In this way, the shape of the polarization ellipse of the scattered field ($\vec{E}_S$) is reconstructed. One experimental polar diagram $\sqrt{I(\varphi)}$ is explicitly shown in Fig. 4(b): a gold nano-particle functionalized tip sits at a selected position and scatters a standing wave created by two counter-propagating evanescent waves on a prism surface. The corresponding ellipse is denoted in red color. In case of a highly elliptical polarization as in the standing wave which is our main interest in this section, we denote this ellipse with a double arrowed linear vector (red arrow) for a better visualization. Finally the polarization state of the local field $\vec{E}_{Local}$ is then reconstructed by a back-transformation $\ddot{\alpha}^{-1} \cdot \vec{E}_S$ (black arrow), for an example. The missing information is the sense of rotation and the absolute phase, i.e., the point on the ellipse at *t*=0, of the field vector. For a partially polarized light which contains certain amount of un-polarized light, it needs a careful analysis of data to be distinguishable from an elliptical polarization. Therefore, RAE is useful only for highly elliptical polarizations.

The Stokes parameter measurements can be applicable to address the missing information from RAE, such as sense of rotation and degree of polarization - the intensity of the polarized portion to the total intensity. The Stokes parameters are composed of 4 quantities

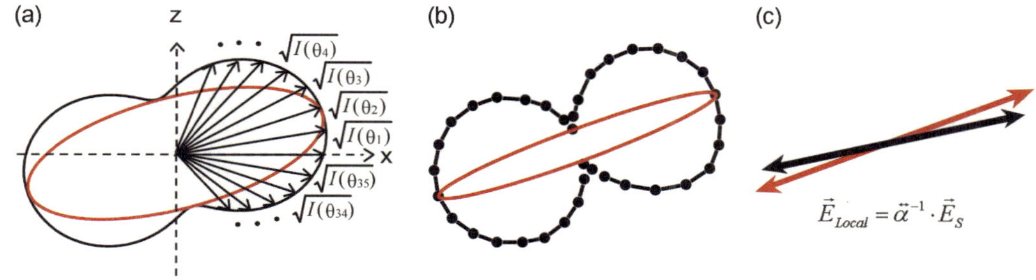

Fig. 4. (a) The outer-plot (black line) results from a polar plot of the squared-rooted intensities for every detecting polarizer angle. The angle (θ_{max}) of the measured intensity maximum corresponds to the major axis angle and the square-rooted maximum (minimum) intensity is proportional to the major (minor) axis length. (b) One such experimental polar plot of the scattered light at one selected position is shown as filled circles. The black line is a guide to the eye. The elliptical polarization state is reconstructed (inner red line). (c) The red arrow represents the long axis of the ellipse shown in (b). By back-transformation using the experimentally determined polarizability tensor of the scatterer, the local field vector is determined (black arrow). From (Lee et al., 2007c). © 2007 Optical Society of America.

(s_0 s_1 s_2 s_3) which can be measured by using a combination set of a phase retarder (λ/4) and a linear polarizer (Stokes, 1852; Born & Wolf, 1999),

$$
\begin{aligned}
s_0 &= I(0°,0) + I(90°,0) = \langle a_1^2 \rangle + \langle a_2^2 \rangle, \\
s_1 &= I(0°,0) - I(90°,0) = \langle a_1^2 \rangle - \langle a_2^2 \rangle = s_0 \cos 2\chi \cos 2\psi, \\
s_2 &= I(45°,0) - I(135°,0) = \langle a_1 a_2 \cos(\delta_1 - \delta_2) \rangle = s_0 \cos 2\chi \sin 2\psi, \\
s_3 &= I\left(45°,\frac{\pi}{2}\right) - I\left(135°,\frac{\pi}{2}\right) = \langle a_1 a_2 \sin(\delta_1 - \delta_2) \rangle = s_0 \sin 2\chi.
\end{aligned}
\tag{8}
$$

Here, $I(\theta, \varepsilon)$ represents the measured light intensity with the linear polarizer angle θ from the x-axis in the laboratory frame, when a phase retardation ε is given to the z-component relative to the x-component by a λ/4 plate. The bracket means the time average over many oscillation periods. χ and ψ are parameters of the ellipse shown in Fig. 5. The parameter s_0

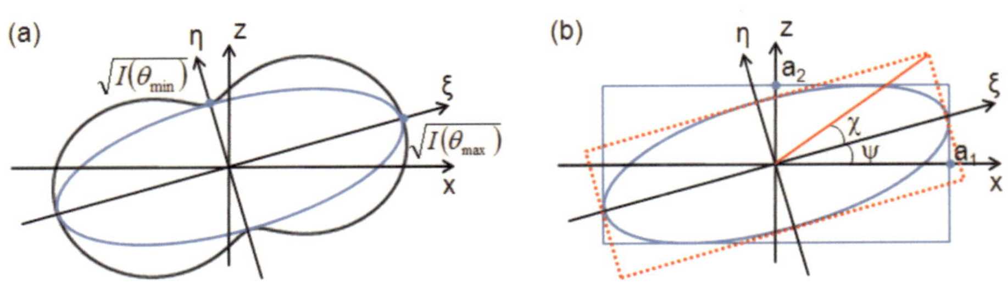

Fig. 5. RAE vs Stokes parameters. (a) The outer-plot (black line) results from a polar plot of the squared-rooted intensities for every detecting polarizer angle. (b) Parameters of an ellipse. Major axis angle ψ defines the orientation of an ellipse. The magnitude and the sign of angle χ characterize the ellipticity and the rotational sense, respectively.

represents the total intensity. The parameter s_1 determines whether the major axis is closer to the horizontal (x) or the vertical (z) axes. In the same analogy, the parameter s_2 tells whether the major axis is closer to the xz (45°) or $-xz$ (135°) directions. From these three parameters (s_0 s_1 s_2), the same amount of information can be derived comparing to the RAE, i. e., the values of ψ ($0 \leq \psi \leq \pi$) and $|\chi|$ ($-\pi/4 \leq \chi \leq \pi/4$). The final parameter s_3 represents the intensity difference between the right-handed polarization and the left-handed polarization - the sense of rotation (sign of χ). Additionally the Stokes parameters can define the degree of polarizaton, however, in our measurements, we use a monochromatic laser light as a light source and do not discuss this quantity in detail. One missing information of the phase can be determined by applying interferometric methods.

2.3 Reconstruction of local polarization vectors and tip shape independence

Due to a relatively simple way of picking up process, fabrication of the GNP funcitonalized tip is reliable and highly reproducible compared to other types nano-probes. Neverthless, the optical properties of a nano sized object are strongly dependent on its shape, size, and orientation. For an example, the polarization state of the scattered light is strongly dependent on the scattering function of this dipole scatterer, i.e., its polarizability tensor, it is important to characterize each tip carefully before the local electric field orientation is reconstructed.

To investigate how the effect of different tips can be corrected in the final determination of the local polarization vector, we prepared three tips attached with gold nanoparticles of different shapes and sizes. The corresponding polarizability tensor of each of these tips is measured as described in the section of 2.1. Using these tips we measured the polarization state of a standing wave generated on a prism surface. Our experimental setup is schematically depicted in Fig. 6(a). A p-polarized plane wave is guided into a prism and generates, with its reflected wave from the mirror at the other side of the prism, an evanescent standing wave on the prism surface, if the incident angle θ_i is set to be larger than the total internal reflection angle $\theta_c = \sin^{-1}\left(n_{air} / n_{prism}\right)$ given by the refractive indices of two media. For an evanescent standing wave, generated by two counter-propagating p-polarized beams of equal intensity, the field vector is given by:

$$\vec{E}(\vec{r}) = (E_x, 0, E_z) = E_0(\cos kx, 0, -\frac{k}{\kappa}\sin kx)e^{-\kappa z} \tag{9}$$

where E_0 is a constant magnitude. k and κ are related by the Helmholtz equation: $k^2 - \kappa^2 = (\frac{\omega}{c})^2$ and are determined by the angle of incidence and the index of refraction of the prism. In Fig. 6(b), theoretically calculated horizontal and vertical field intensities, $|E_x|^2$ and $|E_z|^2$, respectively, of this evanescent standing wave are presented with the corresponding field vectors of polarization shown in the upper part. For an incident angle of θ_i=60° and n_{prism}=1.51 at λ=780 nm, the peak vertical field intensity is about 2.25 times larger than its horizontal counterpart, and these two field components are spatially displaced with a 90° shift in their intensity profiles.

We scanned the prism surface along the x-direction and the polarization characteristics of photons scattered by these GNP attached tips are analyzed applying RAE method. The tip to sample distance was controlled to be constant using a shear force mode feedback system

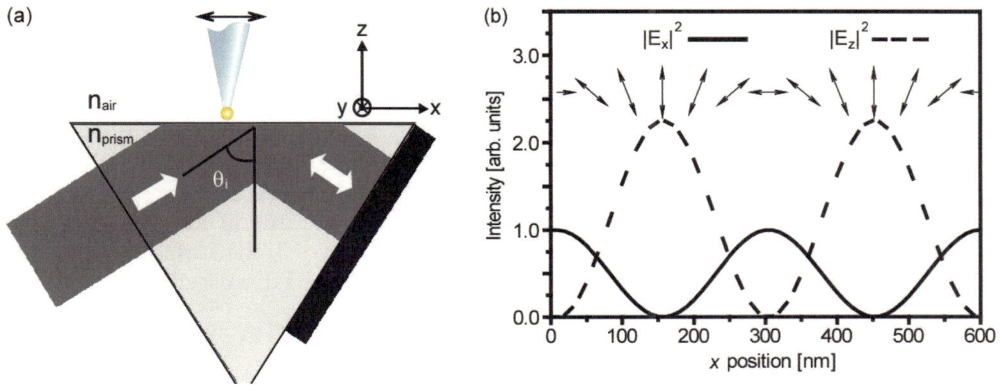

Fig. 6. (a) Experimental setup: A 780 nm cw-mode Ti:Sapphire laser enters at normal incidence into one side facet of an equilaterally shaped prism and is retro-reflected at the other side facet to generate an evanescent standing wave on the top surface. The gold nanoparticle attached tip scatters the local fields into far-field region. (b) Theoretically calculated local field components as a function of the scatterer position: vertical $|E_z|^2$ (dashed line) and horizontal $|E_x|^2$ (solid line) component, respectively. The corresponding local field vectors of polarization are presented at every position.

and the detection angle was set about 20° from the prism surface (-y axis) due to the experimental restrictions. The effects of the detection angle from the surface on the image contrast will be discussed later.

Fig. 7 shows the local field vectors of polarization obtained within a scan range of 600 nm on the prism surface obtained by using three different tips. The corresponding polarizability tensors are indicated above the vector plots. The results for Tip 1 and 2 are obtained with attached gold particles with a diameter of 200 nm and 100 nm, respectively. In these cases the effective polarizability tensors are close to the identity matrix, which means circular shape of GNPs. The bottom one is obtained with the tip introduced in Fig. 3(b). For all three tips attached with gold nanoparticles of different size and shape, the measured local polarization vectors show a good agreement with the theoretical prediction in Fig. 6(b), demonstrating the independence of the finally determined local polarization vector on the tip shape.

$$\overleftrightarrow{\alpha}_1=\begin{pmatrix} 1 & 0.014 \\ 0.014 & 1.075 \end{pmatrix} \quad \overleftrightarrow{\alpha}_2=\begin{pmatrix} 1 & 0.027 \\ 0.027 & 1.102 \end{pmatrix} \quad \overleftrightarrow{\alpha}_3=\begin{pmatrix} 1.208 & 0.257 \\ 0.257 & 1 \end{pmatrix}$$

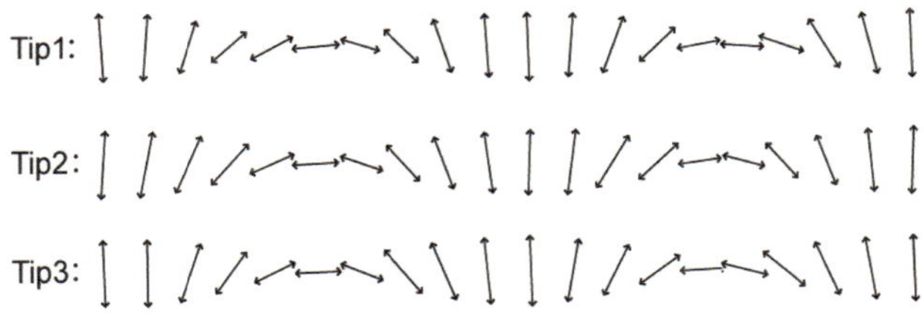

Fig. 7. Local field polarization vectors of the evanescent standing wave generated on the prism surface within a 600 nm scan range obtained by using three different gold-particle functionalized tips. The corresponding polarizability tensors are displayed above the scans. From (Lee et al., 2007c). © 2007 Optical Society of America.

Finally, Fig. 8 displays the theoretical and experimental vector field maps within a 600 nm × 300 nm scan area in x-z plane. The field vector rotates as we move along the x-direction and the electric lines of force are explicitly visualized. The reconstructed field polarization vectors match well with those expected for the evanescent surfaces waves unperturbed by the tip. Generally one may expect a certain perturbation of the local electric field by the field scatterer. The demonstrated ability to quantitatively map electric field vectors of local polarization in simple cases, such as the standing surface waves investigated here, will certainly be useful in obtaining a deeper understanding of the interaction between the tip scatterer and localized electric fields at surfaces. In the section of 3, the surface effects on the far-field detected light scattered from the near-field region will be discussed in more details.

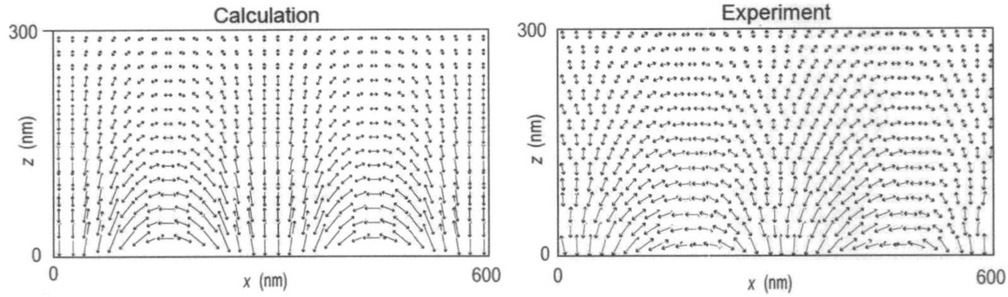

Fig. 8. Vector field plot of an x-z area of 600 nm by 300 nm of the theoretical (left) and experimental (right) results, respectively. From (Lee et al., 2007a). © 2007 Nature Publishing Group.

Before ending up this section, we need to check the validness of dipole approximation of GNPs when the measurements are carried out in the evanescent near-fields. With higher values of *k*-vector, the evanescent field is confined to the sample surface and exponentially decays to the direction normal to the surface. The evanescent field generated on the prism surface (BK7), with the incident angle of θ_i=60° as depicted in Fig 6, the decay constant in intensity is calculated as 147 nm. Then the far-field scattered field by a GNP of radius r is calculated in the Mie-scattering formalism (Chew, et al., 1979; Ganic et al., 2003).

$$E_{sc}(r)=\sum_{l=1}^{\infty}\sum_{m=-l}^{l}\{\frac{ic_0}{n_1^2\omega}\beta_E(l,m)\nabla\times\left[h_l(k_1r)X_{lm}(e_r)\right]+\beta_M(l,m)h_l(k_1r)X_{lm}(e_r)\} \quad (10)$$

Here, we do not include the effect of the glass tip shaft. The relative magnitude of the Mie-coefficient of electric and magnetic components for each radius is listed in Table 1. Upto GNP radius value of 100 nm, the electric dipole term dominates. For the case of r= 150, the magnetic and higher order terms significantly effect on the scattering signal and the dipole approximation cannot be applied anymore.

	r=50 nm	r=100 nm	r=150 nm
$\|a_1\|$	1	1	1
$\|a_2\|$	0.059	0.026	0.014
$\|b_1\|$	0.065	0.124	0.250
$\|b_1\|$	0.003	0.003	0.021

Table 1. Magnitude of two lowest orders of the Mie-coefficients a_l (electric) and b_l (magnetic).

2.4 Three dimensional expansion of local field polarization vector detection

Expanding the local polarization vector detection into a full 3-dimensional space, in principle, is straight forward by combining of 2-dimensional measurements in two orthogonal directions. As a target field, we chose a focused radial polarized light. The intense longitudinal field at the focus center of a radially polarized beam has attracted many attentions not only in a theoretical point of view but also in application respects such as confocal microscopy, optical data storage, and particle trapping and acceleration of particles. Generating good quality cylindrical vector beams, radially and azimuthally polarized beams, has been an intense research area itself. Several different methods are presented - interferometry, twisted liquid crystal, and laser mode controlling inside the cavity.

The interesting axis symmetric field distribution of the cylindrical beam at the focus stems from its axis symmetry of the field polarizations. The field configuration of the cylindrical beam has been demonstrated in theoretical works (Youngwoth & Brown, 2000), but it has been challenging to fully demonstrate it in experiment.

Experimental demonstration starts with a 3-dimensional tip characterization. Here, we adapted a slightly diffrent method to reduce down the total measuring time. Tip end is illuminated by loosely focused Ti:Sapphire laser beams in three orthogonal directions with various incident beam polarizations (Fig. 9(a)). The scattered electric field ($\vec{E}_{sca}$) is detected in (1 ±1 0), (1 0 0) and (0 -1 0) directions for each incident beam direction. With assuming the attached GNP as a dipolar scattering center, the incident and the scattered electric fields are related through the polarizability tensor $\ddot{\alpha}$ (Ellis & Dogariu, 2005):

$$\vec{E}_{sca} = \ddot{\alpha} \cdot \vec{E}_{inc} = \sum_{i,j=1}^{3} \alpha_{ij} E_{inc,j} \,. \tag{11}$$

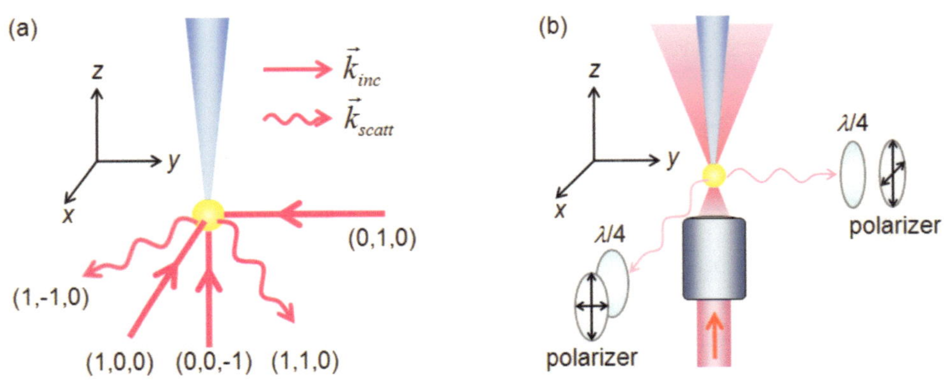

Fig. 9. (a) Three dimensional tip characterization. The tip end is illuminated by Ti:Sapphire laser beams in three orthogonal directions in sequence varying the incident beam polarization. The scattered electric field is detected in the direction of incidence and also in (1 ±1 0). (b) Polarization vector mapping of a focused radially polarized light. A radially polarized beam generated by using a radial polarization converter is focused by an objective. A GNP functionalized tip is scanned the focusing area in three dimensional space using a 3-axes nano positioner (Nano Cube, Physik Instrumente). The polarizaton state of the scattered light is determined by applying the RAE and measuring the Stokes parameters in two orthogonal axes.

From the pre-adjusted incident electric field and the measured scattered electric field polarization states, the polarizability tensor values are directly calculated from Eq. (11):

$$\ddot{\alpha} = \begin{pmatrix} 1.01 & 0.18 & -0.21 \\ 0.18 & 0.70 & 0.12 \\ -0.21 & 0.12 & 1 \end{pmatrix}. \qquad (12)$$

The radial polarization can be described as combination of Hermite-Gaussian modes:

$$\text{Radial polarization} = HG_{10}\hat{x} + HG_{01}\hat{y} \qquad (13)$$

The electric field at the focus in the Cartesian coordinate is given as follows (in air) (Youngworth & Brown, 2000; Novotny & Hecht, 2006):

$$E(\rho,\varphi,z) = \begin{pmatrix} E_x \\ E_y \\ E_z \end{pmatrix} = \frac{ikf^2}{2w_0} E_0 e^{-ikf} \begin{pmatrix} i(I_{11} - I_{12})\cos\varphi \\ i(I_{11} - I_{12})\sin\varphi \\ -4I_{10} \end{pmatrix},$$

where

$$\begin{aligned} I_{10} &= \int_0^{\theta\max} f_w(\theta)\sqrt{\cos\theta}\sin^3\theta\, J_0(k\rho\sin\theta)e^{ikz\cos\theta}d\theta \\ I_{11} &= \int_0^{\theta\max} f_w(\theta)\sqrt{\cos\theta}\sin^2\theta(1+3\cos\theta)\ J_1(k\rho\sin\theta)e^{ikz\cos\theta}d\theta \\ I_{12} &= \int_0^{\theta\max} f_w(\theta)\sqrt{\cos\theta}\sin^2\theta(1-\cos\theta)\ J_1(k\rho\sin\theta)e^{ikz\cos\theta}d\theta \\ f_w(\theta) &= \exp(-f^2\sin^2\theta / w_0^2)\ . \end{aligned} \qquad (14)$$

Here, J_n is the nth-order Bessel function and k is the wave vector of the incident beam. The focal length f, maximum focusing angle θ_{max} , and the incident beam radius w_0 are related as follows: $\frac{w_0}{f} = \sin\theta_{max} = NA$ (effective numerical aperture of the objective).

A radially polarized light is generated by using a radial polarization converter (Arcoptix) and focused by an objective (NA=0.39). A GNP functionalized tip scans the focus area and the scattered light is polarization analyzed in two orthogonal directions by applying the RAE and by measuring the Stokes parameters. The local polarization state of the focused light is reconstructed by performing the back transformation of the polarizability $\ddot{\alpha}$ obtained above in Eq. (12) to the scattered electric field.

Fig. 10 shows the local electric field components in the focus plane (z=0). Upper three intensity plots show the experimentally measured electric field components. As predicted by calculations as shown below, vertical field intensity is a maximum at the center of the focus. On the other hand, the x- and the y-components have intensity minima at the same spatial position. Combined image of (b) and (c) generates a donut shaped intensity distribution for the transversal field component (not shown). The NA value of the used oil immersion objective (n_{oil}=1.50~1.51) is 1.45 with full using the back aperture. The effective

NA value for this measurement performed in air side is chosen as 0.39 from the incident beam waist (w_0= 2 mm) and the back aperture radius of the objective (5 mm). Experiment and calculation agree well each other in the focused beam size and also in the relative intensity peak ratio between the transversal and the vertical components.

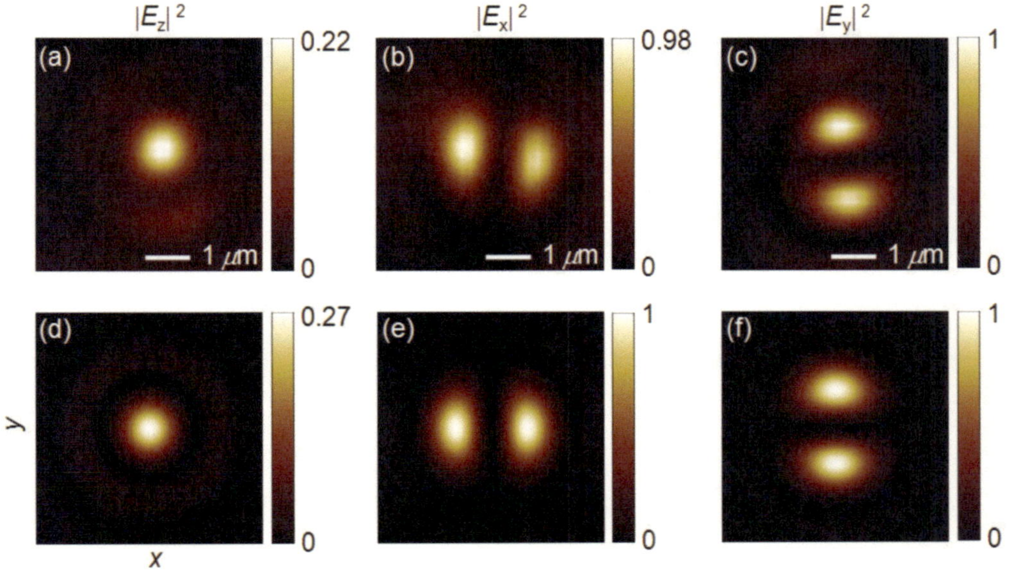

Fig. 10. (a-c) Experimentally measured field intensity distribution profiles for three orthogonal axes. (d-f) Numerically calculated field intensity distribution of the corresponding field component in the focus plane. From (Ahn et al., 2009). © 2009 Optical Society of America.

The 3-dimensional polarization vectors are shown in Fig. 11 determined from the RAE (a) and the Stokes parameters (b). They show quite complicated features, and the top and the side views of (b) are shown below in (c) and (d), respectively. In the top view (c), the polarization direction, the long axis of the ellipse, directs to the focus center. It tells that the transversal component still has a radial polarization state at the focus. However, due to a slight deviation of the beam axis from the z-axis, there are elliptical polarization states in the transversal field components unlike the calculations (Youngworth & Brown, 2000; Novotny & Hecht, 2006). Fig. 11(d) shows the side view (y=0) of (b) for several different tip height (z) values. Note that the vertical field amplitude is magnified by 5 times in this figure for a better visualization. The optical axis of the focused beam is slightly deviated from -x to x direction as it propagates from -z to z direction. It directly shows the imperfectness of the beam alignment together with the details of the focused radially polarized light.

In this section, a full 3-dimensional local polarization vector detection is demonstrated. This is achieved by performing the 2-dimensional polarization vector detection in two orthogonal directions and by combining them. Focused radially polarized light is a good target field due to an intense longitudinal field component at the focus center. The 3×3 polarizability tensor values of the GNP functionalized tip are also obtained by performing the scattering measurement in three orthogonal axes. The polarization vectors of a focused radially polarized light are mapped applying the RAE and the Stokes measurement.

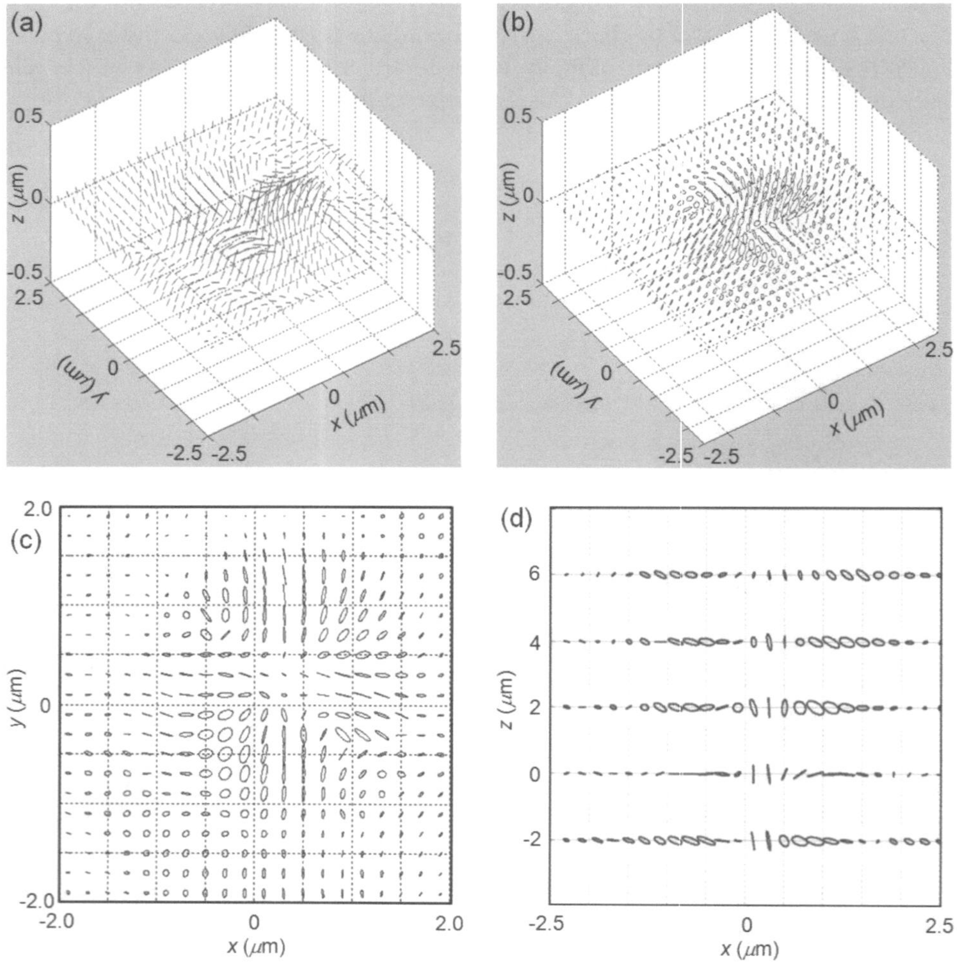

Fig. 11. Polarization vector mapping in the focus plane (z=0) by the (a) RAE and the (b) Stokes measurement. (c) Top view of (b). (d) Side view (y=0) of (b) for several tip height (z) values. The vertical field amplitude (E_z) is 5 times multiplied in (d) for a better visualization. From (Ahn et al., 2009). © 2009 Optical Society of America.

3. Sample surface effects on local field detection in near field region: image dipole effects

Unlike the light scattering by a tip in a homogeneous media, the scattered light in a near field region suffers significant modifications due to the existence of the surface. In the apertureless near-field scanning optical microscopy, the dipolar coupling between the real dipole at the tip apex and its image dipole induced at the sample surface has been widely applied in the analysis of the far-field scattered signals (Knoll & Keilmann, 2000; Raschke & Lieanu, 2003; Cvitkovic et al., 2007).

In this section, we systematically investigate the polarization dependent image dipole effects on the near-field polarization vector detection on a dielectric and a flat metal (Au) surfaces.

The experimental schematic is depicted in Fig. 12(a). A 780 nm cw Ti-sapphire laser is guided on one facet of the equilateral shaped prism (BK7) to generate a propagating evanescent wave at the air-prism interface. With the incident angle θ_i = 60° and the refractive index n_{prism} = 1.51 at the wavelength of 780 nm, the intensity ratio of $|I_z| / |I_x|$ and the skin depth into air are given as 2.25 and 147 *nm*, respectively, from Eq. (9). The evanescent field is then scattered by a GNP of radius 50 *nm* attached on a chemically etched optical fiber tip in a constant height mode (h~55nm). Tip was fixed at one selected x-position and the scattered light intensity is polarization analyzed. The relative intensity ratio of the vertical and the horizontal components of the local field at the GNP position is measured while varying the collection angle ϕ.

To account for the surface effects on the signal, we firstly consider the interference between the direct radiation from the GNP and its reflection from the sample surface to the detector by using a simple image dipole model ((i) in Fig. 12(b)). The reflected light from the surface can be considered as the radiation from the image dipole located at the opposite side to the interface. The relative strength of the radiation from the real and the image dipole is determined by the magnitude of the reflection coefficient ($R^{s(p)} = \left|R^{s(p)}\right| e^{i\varphi_{delay,s(p)}}$) of the s- and the p-polarized light at air-prism interface. The relative phase difference between the real and the image dipoles is determined by the argument of the reflection coefficient, $\varphi_{delay,(s,p)}$,

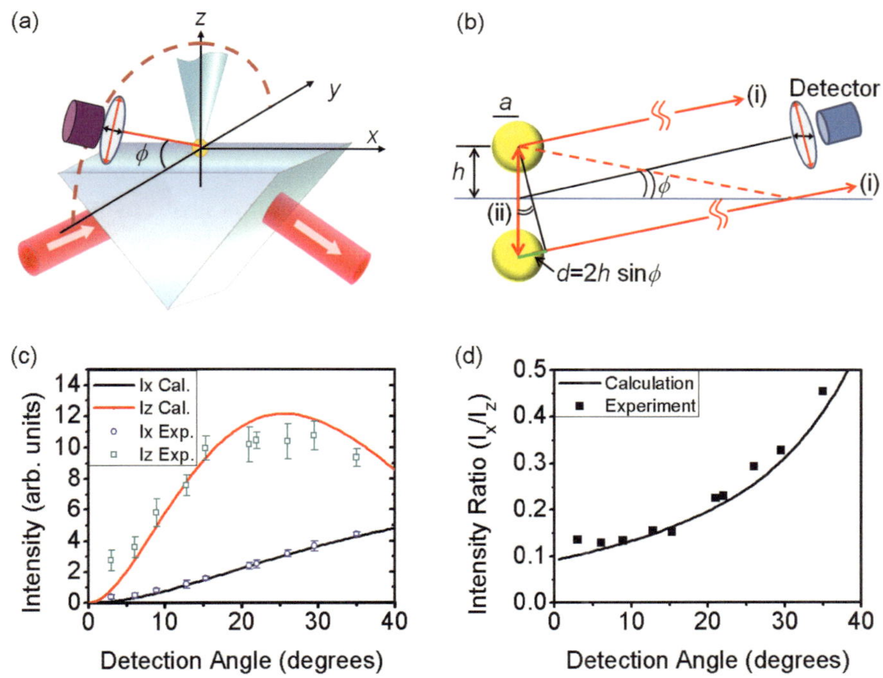

Fig. 12. Image dipole effects on a dielectric surface. (a) Experimental schematics. (b) The reflected light at the sample surface (dashed line) can be considered as the radiated field from the image dipole (i). The mutual interaction of the real (upper) and the image (below) dipoles modifies the radiation properties of their own (ii). (c-d) Relative intensities of the horizontal and the vertical field components of the progating evanescent wave on a prism surface. (c-d) from (Ahn et al., 2008). © 2008 Elsevier B.V.

as well as the phase difference caused by the optical path length difference $\varphi_{diff} = k_0 \cdot d$, where $d = 2\,h \sin\varphi$ is the path difference in Fig. 12(b). For analytical calculations, we applied the single dipole model (SDM) where the real (above the surface) and the image (below) dipoles are assumed to be point-like dipoles. We note that the reflection coefficient of the plane wave is used in this analysis because the scattered light is detected in far-field region. The effects of the higher nonlinear terms included in the effective polarizability change will be discussed in later part.

The signal intensity of the horizontal (*s*) and the vertical (*p*) dipoles in the SDM is written as

$$I_{s(p)} = \left|E_{s(p)-orig} + E_{s(p)-imag}\right|^2 = \left|E_{s(p)-orig}\right|^2 \left(1+|R_{s(p)}|^2 + 2|R_{s(p)}|\cos(\varphi_{delay,s(p)} + \varphi_{diff})\right). \quad (15)$$

Here, $\left|E_{s(p)-orig}\right|^2$ is determined by the relative time-integrated strength of the horizontal (*s*) and the vertical (*p*) field components of the propagating evanescent wave. For the vertical polarization case, i.e., *p*-polarization case, $\cos^2\phi$ term should be multiplied to Eq. (15) to compare with experimentally measured one because of that an oscillating dipole cannot radiate light in its oscillation direction. In a detailed explanation, the electric field at a detection position $\vec{r}$ radiated by a dipole moment $\vec{p}$ located at origin is described as follows (Jackson, 1998);

$$\vec{E}_{dipole} = \frac{1}{4\pi\varepsilon_0}\left\{k^2(\hat{d}\times\vec{p})\times\hat{d}\frac{e^{ikr}}{r} + \left[3\hat{d}(\hat{d}\cdot\vec{p}) - \vec{p}\right]\left(\frac{1}{r^3} - \frac{ik}{r^2}\right)e^{ikr}\right\}, \quad (16)$$

where k is the wave-vector and r is the distance from the dipole to the detector. With $\vec{p} = (0,0,p_z)$ and the detection position vector $\hat{d} = (0,-\cos\phi,\sin\phi)$, the radiated electric field in far-field region with consideration of the surface reflection is given by substituting Eq. (16) into Eq. (15).

$$\vec{E}_{detector} = \frac{k^2 p_z\left(1+|R_{s(p)}|^2 + 2|R_{s(p)}|\cos(\varphi_{delay,s(p)} + \varphi_{diff})\right)}{4\pi\varepsilon_0 r^2}\left(0,-\cos\phi\sin\varphi,\cos^2\phi\right) \quad (17)$$

Note that the vertical polarizer direction in Fig. 12(a) is differ from the z-axis in the laboratory frame but parallel to $\vec{E}_{detector}$ giving the measured intensity of the vertical component proportional to $\cos^2\phi$ instead of $\cos^4\phi$.

The relative intensities of the horizontal and the vertical field compoentns as a fuction of the detection angle are shown in Fig. 12(c-d). Simple analytical calculation well predicts the experiemtal result.

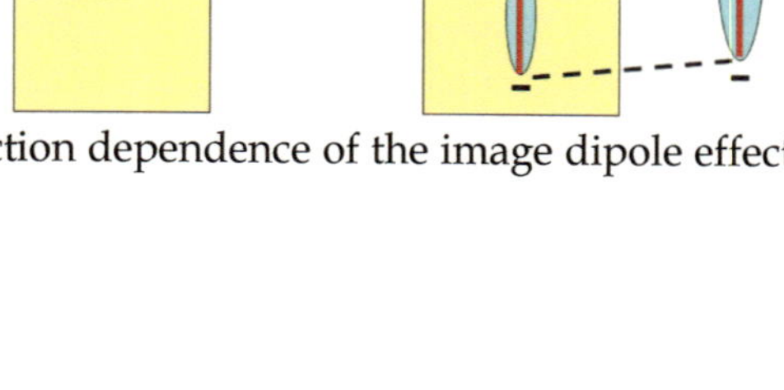

Fig. 13. Polarization direction dependence of the image dipole effects at the vicinity of metal surfaces.

Image dipole effects are more dramatic in metallic surface. Image dipole effects are highly dependent on the polarization direction, constructive (destructive) interference between real and image dipoles for the vertically (horizontally) aligned one in the vicinity of metal surfaces, respectively.

We use a propagating surface plasmon polaritons (SPP) as an excitation source. The well-characterized field profile of SPP and sufficiently reduced background noise by virtue of the evanescent nature of SPP make it possible to carry out quantitative studies of the image dipole effects on metallic surfaces. Fig. 14 shows our experimental schematics. A propagating SPP is generated at the slit position by impinging a beam of cw-mode Ti-Sapphire laser (wavelength λ_0=780 nm) at the back side of the sample. The incident polarization is adjusted perpendicular to the silt direction for the coupling of the incident light to the SPP. The thickness of the gold film and the slit width are chosen as 80 nm and 400 nm, respectively, to maximize the SPP coupling efficiency from the incident light (Kihm et al., 2008). A lens (focal length of 5 cm) focuses the excitation beam at the slit position to eliminate the position dependent interference between the directly transmitted light through the thin metal film and the propagating SPP at the tip position. The tip is fixed at one selected x-position at about 50 μm away from the slit exit to diminish unwanted backgrounds resulting from the deflected light at the tip shaft when the propagating light transmitted at the slit position touches the tip surface (Lee et al., 2007b).

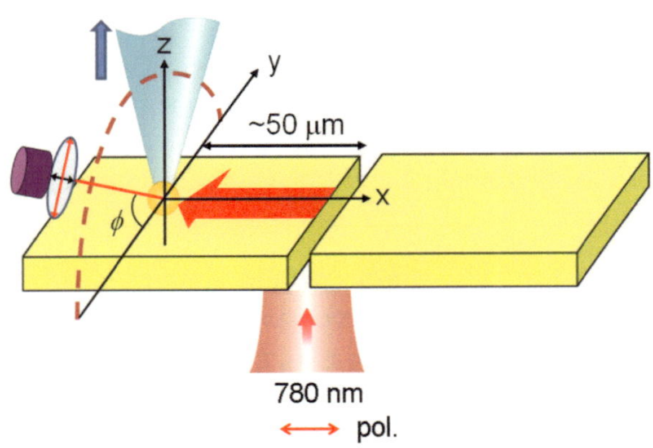

Fig. 14. Schematic diagram of the experimental setup. Tip is placed above a flat gold surface about 50 μm away from the slit position. A 780 nm cw Ti-Sapphire laser beam is incident from the bottom side of the sample to generate SPPs propagating in $\pm x$-direction on air-gold interface. This propagating SPP is scattered by the GNP functionalized tip and a linear analyzer is placed in front of the detector for the axis resolved detection. The tip-sample distance (h) is varied from near- to far-field region, and the detection angle (ϕ) between the sample surface and the detector position vector is also changed.

The electric field of the propagating SPP on a flat gold surface can be described as follows (Reather, 1988);

$$\vec{E}_{SPP}(\vec{r}) = (E_{SPP,x}, 0, E_{SPP,z}) = E_0\left(\cos(k_{SPP}x - \omega t), 0, -\frac{k_{SPP}}{\kappa}\sin(k_{SPP}x - \omega t)\right)e^{-\kappa z} \tag{18}$$

where E_0 is a constant amplitude, $\kappa = \mathrm{Im}\left[\frac{2\pi}{\lambda_0}\sqrt{\frac{\varepsilon_{air}^2}{\varepsilon_{air}+\varepsilon_{Au}}}\right]$ the reciprocal skin depth of the SPP into air (n_{air}=1), and $k_{SPP} = \frac{2\pi}{\lambda_0}\sqrt{\frac{\varepsilon_{air}\cdot\varepsilon_{Au}}{\varepsilon_{air}+\varepsilon_{Au}}}$ the wave-number of the SPP. As a dielectric constant of GNP, $\varepsilon_{\mathrm{Au}} \approx$-22.5+1.4$i$ of bulk gold at the wavelength λ_0=780 nm is used (Reather, 1988). The time integrated intensity ratio of the horizontal and the vertical field components of propagating SPP is determined by the dielectric constant of gold.

$$\frac{I_{SPP,z}}{I_{SPP,x}} = \frac{\left|E_{SPP,z}\right|^2}{\left|E_{SPP,x}\right|^2} = \left|\varepsilon_{Au}\right| \tag{19}$$

This propagating SPP induces the dipole moment at the GNP attached to the apex of an etched glass fiber. The scattered light intensity is measured while varying the tip-sample distance h and the detection angle ϕ between the sample surface and the detector position vector $\hat{d} = (0, -\cos\phi, \sin\phi)$. A long working objective lens (Mitutoyo M Plan Apo 10×) collects the scattered light and delivers it to an APD. A linear polarizer placed before the detector resolves the polarization direction of the scattered light.

Figure 15(a) shows the plot of the signal intensities versus the tip-sample distance (h) obtained with the detection polarizer oriented along the horizontal (black open circles) and vertical (red open circles) directions to the sample surface. The elliptical scattering shape of the GNP is taken into account by dividing the horizontal signal intensity with $(1.34)^2$. Here, the detection angle (ϕ) is 33°.

The signal intensity of the horizontal (s) and the vertical (p) dipoles in the SDM is given by Eq. (15). In this case, $\left|E_{s(p)-orig}\right|^2$ is originally determined by the relative time-integrated strength of the horizontal (s) and the vertical (p) field components of the propagating SPP as in Eq. (19). In addition to that, it is also considered the radiating property modifications of the GNP itself ($\left|E_{s(p)-orig}\right|^2$), caused by the reflected fields directly back to the GNP ((ii) in Fig. 12(b)). The radiated field from the real dipole (upper sphere in the Fig. 12(b)) influences the image dipole (below), and the resultant altered field of the image dipole modifies the real dipole again. This mutually repeating effect on the α_{eff} can be calculated in a self-consistent manner. In previous studies (Knoll & Keilmann, 2000; Raschke & Lieanu, 2003), the α_{eff} is calculated in the quasi-electrostatic limit assuming a small distance from the particle (or the tip apex of a metal tip) to the interface. The tip apex was considered as a point-like scattering center. In this study, in calculating α_{eff}, all terms of Eq. (16) are included for the case of a wider separation between the tip and the sample surface. Using the SDM and α_{eff} derived from it, we fail to reproduce the experimentally measured signals (dashed lines in Fig. 15(a)). The relative signal intensities of the vertical and the horizontal components are different in calculation and experiment. In addition, the lifted valley of the vertical polarization signal appeared in experiment at h~300 nm of the tip-sample distance cannot be recovered because the magnitude of the reflection coefficient $\left|R_{s(p)}\right|$ is close to unity for all polarization directions and detection angles at the gold-air interface (see Eq. 15).

In order to understand the origin of the observed deviation, we applied the coupled dipole method (CDM) (Martin et al., 1995; Novotny & Hecht, 2006) where the GNP of radius 50 nm is divided into approximately 500 identical sub-volumes, which act as point dipoles. In this calculation, all mutual interactions between sub-volumes including the reflected field from the sample surface are considered. In Fig. 15(a), the theoretical calculations with CDM (solid lines) are compared to the experimental data (open circles) for two orthogonal detection polarizer angle directions. The theory and the experiment are in excellent agreements to each other. Furthermore, the lifted non-zero value of the minimum at h~300 nm is clearly recovered by the theory.

The oscillation period $\frac{\lambda_0}{2\sin\phi}$ is determined by the condition of $\varphi_{diff} = k_0 \cdot d = 2\pi$ (Fig. 12(b)). In Fig. 15(b), the calculated values of oscillation period are plotted in solid line and experimentally measured values for three different detection angles are marked with open circles. For larger detection angles of ϕ=21° and 33° the values from the calculations and the experiments agree well to each other, but for ϕ=8.5° there is a relatively large discrepancy between them. This difference seems to result from the gradually confined numerical aperture (NA) of the collection objective. The used objective lens of NA=0.28 has the collection solid angle φ_s =16.3° (inset in Fig. 15(b)), which means that for a smaller detection angle $\phi < \varphi_s$, the lower part of the lens does not collect the signal, implying the bigger value of effective collection angle.

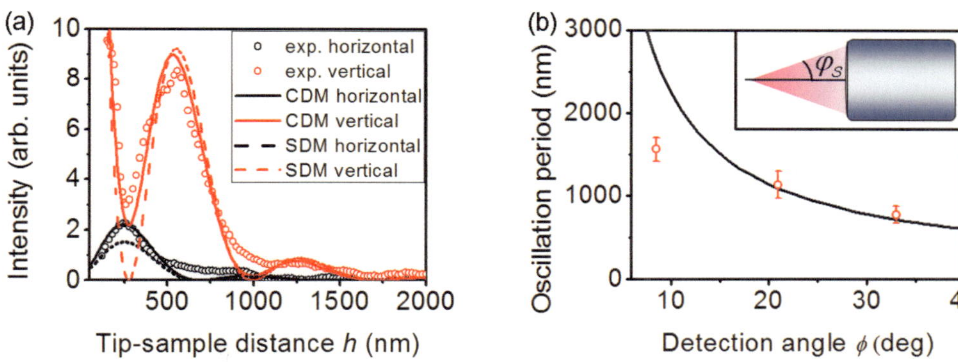

Fig. 15. (a) Experimentally measured tip-sample distance dependent signal intensity with the detection analyzer direction vertical (red) and horizontal (black) to the sample surface. Signal intensities calculated by applying SDM (dashed lines) and CDM in Green-function formalism (solid lines) are shown together with the experimental results (open circles). Here, ϕ=33°. (b) Oscillation periods from the calculation (solid line) and the experiment (open circles). Inset: the solid angle of the objective, φ_s=16.3°. From (Lee et al., 2008). © 2008 Optical Society of America.

In conclusions, we experimentally demonstrate how the image dipole modifies the far-field detected signal depending on its polarization direction to the dielectric and the metal surfaces. By using propagating evanescent optical wave and SPP as excitation sources, well characterized local dipoles are generated at the GNP. Contributions of dipoles aligned vertically and horizontally to the surface are completely separated from each other for a systematic analysis of the polarization dependent image dipole effects on the signal. Measured signals are fully explained by the Fabry-Perot like interference between the radiations from the GNP and from the image dipole induced at the flat gold surface, and by the finite size effects of the GNP.

We note that, in this study, flat surfaces to reflect the signal from the GNP are considered allowing a simple analysis of the detected signal. But, in real situations the sample may have a complex geometry and a more delicate treatment is required. To remove the complicate effects of the reflected light at the sample surface on the far-field detected signal, a confocal-like spatial masking technique can be applied to cut out lights radiated other than from the GNP, when the distance of the GNP from the sample surface is bigger than the diffraction limit of the collection lens. Within a shorter distance range, one should carefully consider the reflections at the sample surface to correctly account for the radiation from the GNP.

4. Problems and outlook

The scattered light by GNP has same frequency of the local electric field, therefore to correctly designate the polarization vector of the source field it is crucial to detect only the light scattered from GNP excluding any other background light. In our works, the source fields, the standing waves and outgoing waves from a slit structure, with a symmetry to the detection position, do not send light by themselves to the detector. However, due to the imperfection of the sample structure and the rough sample surface may generate background lights. And also the glass tip shaft scatters light into the detector. To study the background light effects on the determined polarization vector direction, we changed the amount of the background light to the detector and measured the changes of the local polarization vector direction. The experimental setup is same as depicted in Fig. 14. The tip was positioned at one position 50 μm away from the slit and the distance of the GNP from the sample surface was kept by few nm using a shear force feedback system. A spatial filter was placed in the image plane of the collection objective to cut out the background light other than from the GNP. The opening size of the spatial mask was increased allowing more background light to be detected. From the result in Fig. 15, in the vicinity of a flat gold surface, the polarization state of the scattered light from GNP is highly elliptical with the major axis of the ellipse standing vertically to the surface, along z-axis. With more background light, with wider opening the spatial mask, the major axis of the polarization ellipse deviates more from the vertical direction. The problem is that it is hardly possible to collect light only from the GNP for an arbitrary structured sample.

One possible way to make this local polarization vector detection method be applicable to an arbitrary situation including the self radiating samples is using the Stokes shifted light from nano-objects other than the GNP such as quantum dots and single molecules. For example, the absorption dipole moment of single molecules can be used to detemine the molecular orientation with carefully tailoring the excitation electric field (Betzig & Chichester, 1993). There are remaining questions how to implement such an object to a tip in a controllable way of the position and the orientation. Loosing the coherence in the Stokes shifted light is another disadvantage in determining the temporal phase. Searching and designing of new types probes to extend the functionality of the polarization vector microscope are currently underway.

Even though the scattering from the glass tip shaft can be much less than from the GNP due to the dielectric constant, nevertheless it should not be ignored since the whole volume of the tip shaft is much bigger than of the GNP. By making the tip shaft thinner, scattering from it can be reduced down. But making the tip shaft too much thin may prevent a good performance of the feedback system in controlling the tip position in the near-field region. In addition, it is needed to study the energy transfer between the tapered glass tip and the metallic structures attached to the end of it, depending on the material and on the structural shape.

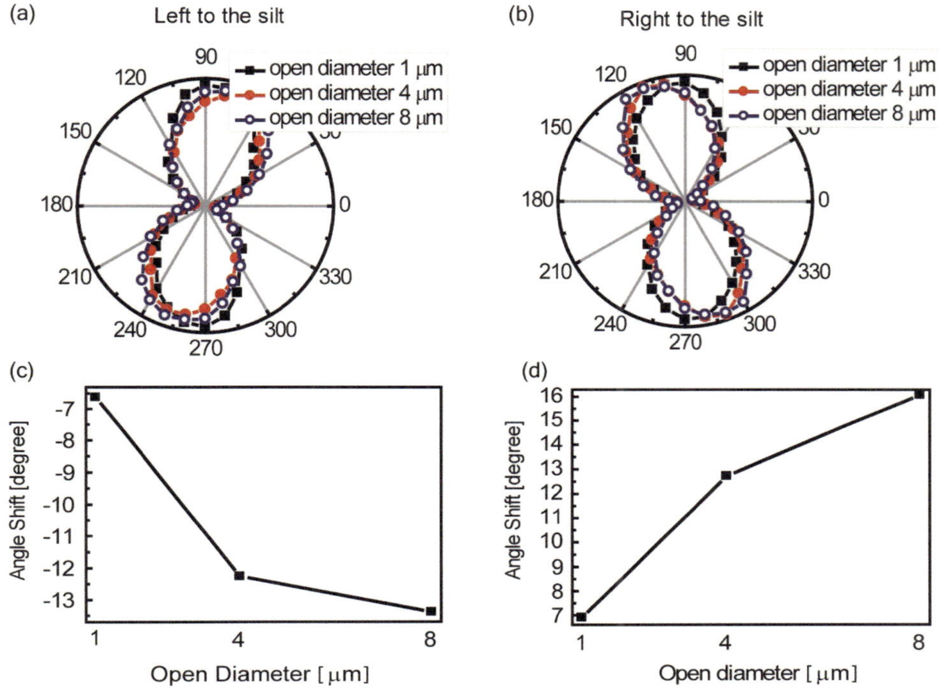

Fig. 16. Background light effect on the measured vector orientation. (a) Polar plot of the measured intensity. A gold nano-particle functionalized tip was located on a flat-metal region on the left-hand side of the slit about 50μm away from the slit position, where the vector field theoretically points towards the z-direction. When the iris opening of the spatial filter setup corresponds to a collection diameter of 8 microns, the experimental vector deviates from the z axis by minus 13.5 degrees. Closing the iris reduces this discrepancy to minus 7 degrees as shown in Figs. (a) and (c). (b) Moving the tip to the right-hand side of the slit and repeating the same experiments essentially gives the same results (Figs. (b) and (d)). From (Lee et al., 2007b). © 2007 Nature Publishing Group.

5. Conclusion

In this chapter, we demonstrate the local field polarization vector detection on nanoscale using the GNP attached tips as the local polarizer. This was enabled by means of a novel, scattering-type near-field microscopic technique combined with complete tip-characterization. For a suitably small GNP with the diameter less than 100 nm, the far-field scattering is dominated by the electric dipole radiation. Dipole radiation conserve its polarization state into the far-field region enabling characterization of the dipole moment induced at the GNP by measuring the far-field polarization state. And the dipole moment is determined by the local electric field via polarizability tensor of GNP. Therefore, by characterizing the polarizability tensor of GNP and the polarization state of far-field scattered light, the local electric field vector can be reconstructed. The polarizability tensor of the GNP is measured by two different scattering measurements-RPAE and RAE. And the polarization of the far-field scattered light is determined by applying RAE and Stokes measurements. By carefully considering the scattering shape of the GNP, the polarizability tensor, it is shown that the finally determined local polarization vectors are independent of the tip which confirms the reliability of our method. A full 3-dimensional polarization field mapping is demonstrated by detecting a focused radial polarization.

These results provide unprecedented images of light in nano-scale and demonstrate that the local field polarizatin vector mapping of light is indeed possible for nano-systems where spatially rapidly changing field orientations on a sub-wavelength scale are the rule rather than the exception and crucial for the functionality of novel nanophotonic devices. This method carry the potential for making the vector field mapping for nano-scale devices a common laboratory practice, instead of a conceptual one encountered only in theoretical realm, that can find wide applications in physics, engineering, chemistry and biology.

We also applied this method in studying the intricate surface effects when the scattering measruements are perfomed in the near-field. Supplementally, we discussed the limitations in our system and the possible ways to reconcile them to further improve the functionality of the system.

6. References

Ahn, J. S.; Kihm, H. W.; Kihm, J. E.; Kim, D. S. & Lee, K. G. (2009). 3-dimensional local field polarization vector mapping of a focused radially polarized beam using gold nanoparticle functionalized tips. *Opt. Express*, 17, 4, (FEB 2009), (2280-2286), ISSN 1094-4087

Ahn, K. J.; Lee, K. G. & Kim, D. S. (2008). Effect of dielectric interface on vector field mapping using gold nanoparticles as a local probe: Theory and experiment. *Opt. Commun.*, 281, (AUG 2008), (4136-4141), ISSN 0030-4018

Betzig, E. & Chichester, R. J. (1993). Single molecules observed by near-field scanning optical microscopy. *Science*, 262, (NOV 1993), (1422-1425), ISSN 0036-8075; online ISSN 1095-9203

Born, M. & Wolf, E. (1999). *Principles of Optics – 7th ed.*, Cambridge University Press, ISBN-10: 0521642221; ISBN-13: 978-0521642224, Cambridge, England

Chew, H.; Wang, D. –S. & Kerker, M. (1979). Elastic scattering of evanescent electromagnetic waves. *Appl. Opt.*, 18, (AUG 1979), (2679-2687), ISSN 0003-6935

Cvitkovic, A.; Ocelic, N. & Hillenbrand, R. (2007). Analytical model for quantitative prediction of material consrasts in scattering-type near-field optical microscopy. *Opt. Express*, 15, 14, (JUL 2007), (8550-8565), ISSN 1094-4087

Ellis, J. & Dogariu, A. (2005). Optical Polarimetry of Random Fields. *Phys. Rev. Lett.*, 95, (NOV 2005), (203905_1-4), ISSN 0031-9007

Ganic, D.; Gan, X. & Gu, M. (2003). Parametric study of three-dimensional near-field Mie scattering by dielectric particles. *Opt. commun.*, 216, (FEB 2003), (1-10), ISSN 0030-4018

Jackson, J. D. (1998). *Classical Electrodynamics*, John Wiley & Sons, Inc., ISBN-10: 047130932X; ISBN-13: 978-0471309321, New York

Kalkbrenner, T.; Ramstein, M.; Mlynek, J. & Sandoghdar, V. (2001). A single gold particle as a probe for apertureless scanning near-field optical microscopy. *J. Microsc.*, 202, (APR 2001), (72-76), ISSN 0022-2720

Keating, C. D.; Musick, M. D.; Keefe, M. H. & Natan, M. J. (1999). Kinetics and Thermodynamics of Au Colloid Monolayer Self-Assembly: Undergraduate Experiments in Surface and Nanomaterials Chemistry. J. *Chem. Ed. chem.wisc.edu.*, 76, (JUL 1999), (949), ISSN 0021-9584

Kihm, H. W.; Lee, K. G.; Kim, D. S.; Kang, J. H. & Park, Q-Han (2008). Control of surface plasmon generation efficiency by slit-width tuning. Appl. Phys. Lett., 92, (FEB 2008), (051115), ISSN 0003-6951

Knoll, B. & Keilmann, F. (2000). Enhanced dielectric contrast in scattering-type scanning near-field optical microscopy. *Opt. Commun.*, 182, (AUG 2000), (321-328), ISSN 0030-4018

Kühn, S.; Håkanson, U.; Rogobete, L. & Sandoghdar, V. (2006). Enhancement of Single-Molecule Fluorescence Using a Gold Nanoparticle as an Optical Nanoantenna. *Phys. Rev. Lett.*, 97, (JUL 2006), (017402), ISSN 0031-9007

Lee, K. G.; Kihm, H. W.; Kihm, J. E.; Choi, W. J.; Kim, H.; Ropers, C.; Park, D. J.; Yoon, Y. C.; Choi, S. B.; Woo, D. H.; Kim, J.; Lee, B.; Park, Q. H.; Lienau, C. & Kim, D. S. (2007a). Vector field microscopic imaging of light. *Nature Photon.*, 1, (JAN 2007a), (53-56), ISSN 1749-4885

Lee, K. G.; Kihm, H. W.; Kihm, J. E.; Choi, W. J.; Kim, H.; Ropers, C.; Park, D. J.; Yoon, Y. C.; Choi, S. B.; Woo, D. H.; Kim, J.; Lee, B.; Park, Q. H.; Lienau, C. & Kim, D. S. (2007b). On the concept of imaging nanoscale vector fields. *Nature Photon.*, 1, (MAY 2007b), (243-244), ISSN 1749-4885

Lee, K. G.; Kihm, H. W.; Ahn, K. J.; Ahn, J. S.; Suh, Y. D.; Lienau, C. & Kim, D. S. (2007c). Vector field mapping of local polarization using gold nanoparticle functionalized tips: independence of the tip shape. *Opt. Express*, 15, 23, (NOV 2007c), (14993-15001), ISSN 1094-4087

Lee, K. G.; Ahn, K. J.; Kihm, H. W.; Ahn, J. S.; Kim, T. K.; Hong, S.; Kim, Z. H. & Kim, D. S. (2008). Surface plasmon polariton detection discriminating the polarization reversal image dipole effects. *Opt. Express*, 16, 14, (JUL 2008), (10641-10649), ISSN 1094-4087

Lee, K. G.; Kihm, H. W. & Kim, D. S. (2009). Measurement of the Polarizability Tensor of Gold-nanoparticle-functionalized Tips. *J. Kor. Phys. Soc.*, (in press), ISSN 0374-4884

Martin, O. J. F.; Girard, C. & Dereux, A. (1995). Generalized Field Propagator for Electromagnetic Scattering and Light Confinement. *Phys. Rev. Lett.*, 74, (JAN 1995), (526-529), ISSN 0031-9007

Novotny, L. & Hecht, B. (2006). *Principles of Nano-Optics*, Cambridge University Press, ISBN-13: 9780521832243 ; ISBN-10: 0521832241, New York

Raschke, M. B. & Lienau, C. (2003). Apertureless near-field optical microscopy: Tip-sample coupling in elastic light scattering. *Appl. Phys. Lett.*, 83, 24, (DEC 2003), (5089-5091), ISSN 0003-6951

Reather, H. (1988). *Surface Polaritons on Smooth and Rough Surfaces and on. Gratings*, Springer-Verlag, ISBN-10: 0387173633; ISBN-13: 978-0387173634, Berlin

Stokes, G. G. (1852). *Trans. Cambridge Philos. Soc.*, 9, (399-416)

Youngworth, K. S. & Brown, T. G. (2000). Focusing of high numerical aperture cylindrical-vector beams. *Opt. Express*, 7, 2, (JUL 2000), (77-87), ISSN 1094-4087

14

Nanoimprint Lithography - Next Generation Nanopatterning Methods for Nanophotonics Fabrication

Jukka Viheriälä, Tapio Niemi, Juha Kontio and Markus Pessa
Optoelectronics Research Centre, Tampere University of Technology, Finland

1. Introduction

Nanophotonics is wide field covering many interesting applications branching from cutting edge science including plasmonics, metamaterials, cavity quantum electrodynamics in high-Q cavities all the way to applied sciences like silicon nanophotonics for on chip optical interconnections and single frequency semiconductor light sources. Most of the practical device demonstrations in these fields utilize nanopatterned surfaces. Applications require patterning of nanoscopic gratings, photonic crystals, waveguides and metal structures.

There are many wonderful demonstrations of nanotechnology-based lasers and other photonic components. However, difficult questions related to fabrication need to be addressed before these components enter any market. Demonstrations in the scientific literature have relied heavily on the use of direct writing lithography methods, such as electron beam lithography or focused ion beam lithography. These methods, although excellent for scientific studies, cannot be scaled up to allow cost effective production of nanophotonics. Lithography solutions developed for integrated circuits can produce extremely narrow linewidths and deliver high precision but are difficult to transfer to photonics fabrication. There exist many alternative lithography methods, but their scale-up to cost effective volume production is challenging.

Since the introduction of nanoimprint lithography (NIL) in 1995 (Chou et al. 1995), there has been widespread interest in the development of NIL for various applications. As early as in 2003 NIL had gained substantial support and was chosen as one of MIT's Technology Review's "10 Emerging Technologies That Will Change the World" (Technology Review 2003). The selection was justified by the fact that NIL can bridge the gap between lab level nanotechnology research and production level manufacturing requirements.

In this chapter, we briefly review the state-of-the-art lithography methods and introduce nanoimprint lithography (NIL), a very cost effective lithography method for nanophotonics applications. In section two we will introduce soft UV-NIL, an imprint method using soft and flexible stamps, as a method for patterning compound semiconductor optoelectronics. Finally, in section three, we highlight some NIL activities based on soft-UV-NIL.

1.1 Optical lithography

The era of microlithography started to develop in the 1970s, and was driven mainly by the development of integrated circuits (ICs). This industry created a need for high volume, perfect replication of ever smaller patterns on a substrate, at minimal costs. The main method to achieve this was, and still is, optical lithography. This branch of lithography utilizes templates, also known as photomasks, having transparent and opaque areas. Light is shone through the photomask on a substrate coated with a photosensitive thin film called photoresist. Photoresist areas that are exposed will transform to either soluble (positive photoresist) or nonsoluble (negative photoresist), depending on the chemistry of the photoresist. Light replicates patterns from photomask to the photosensitive film and further steps are taken to transfer the copied patterns to the substrate. In the early 1970s the required dimensions for the ICs were from 2 μm to 5 μm. Replication of these patterns was simply achieved by using mercury arc based UV-light and by bringing photomask and substrate in close proximity or into contact during the exposure. Systems based on this operating principle are still used today in microfabrication due their simplicity, relatively low cost, high throughput and good process quality. These systems, called UV-contact mask aligners, reach resolutions from a few micrometers to sub half micron level, depending on the exposure wavelength and the contact method. With fully automated systems the throughput can exceed 100 wafers per hour (wph) and reach an overlay accuracy of 0.25 μm (Suss 2009).

However, an ever increasing demand for lower linewidth has demanded more complex exposure systems. Nowadays the state-of-the-art systems in IC production reach 32 nm linewidths by using deep-UV ArF-light sources operating at 193 nm wavelength and expousing patterns using immersion scanners, phase shift masks and double exposure schemes. Exposure is based on an image reduction technique that projects the photomask onto the substrate and simultaneously reduces the size of the patterns many times. This allows the photomasks to be fabricated with looser tolerances than the final pattern. These systems are also very productive and able to pattern more than one hundred and fifty 300 mm wafers per hour and to reach better than 2.5 nm alignment between subsequent patterning steps (ASML 2009).

These exposure systems, reaching 45 nm or even 32 nm linewidths, cost tens of millions of euros, making acquisition and amortization of the instrument impossible unless very high volumes can be produced. For these reasons such instruments can only be owned by large IC-manufacturers. As the cost of optical lithography grows rapidly as linewidths get smaller, research and utilization of alternative techniques is tempting. Although the driving force in the development of lithography has been electronics, there are many other applications that benefit from effective nanofabrication methods.

The most interesting alternative lithography methods for repetitive nanofabrication, selected by the author, are interference and near field holographic lithography (Chapter 1.2), electron beam lithography utilizing one or multiple beams (Chapter 1.3), and nanoimprint lithography (Chapter 1.4).

1.2 Interference lithography

Interference lithography utilizes interference of two or several coherent beams that form an interference pattern on the substrate. Using photoresists similar to those used in optical lithography, this interference pattern can be transferred to the photoresist and subsequently to the other layers on the substrate. Near field holographic lithography is very similar to interference lithography. It uses a phase mask near the substrate to divide one beam into two diffracted beams propagating at different angles. These two beams interfere and generate diffraction patterns. Both methods can produce patterns cost effectively over large areas, but only allow exposure of periodical patterns whose pitch is limited by the exposure wavelength. However, state-of-the-art exposure tools produce good resolution. With a high index immersion fluid system a 32 nm half-pitch has been demonstrated using an exposure wavelength of 193 nm (French et al., 2005) and 12.5 nm half-pitch using an extreme ultraviolet light source emitting at 14.5 nm (Solak et al., 2007).

1.3 Electron beam lithography

Electron beam lithography (EBL) is traditionally based on a single beam of electrons focused on a small spot with a Gaussian shape, or on a beam of electrons that is cut down to the correct size and shape with an aperture. This beam is displaced with a magnetic field that is controlled with a computer. The beam exposes the electron beam sensitive material coated on the substrate in a similar fashion as photoresist is exposed in optical lithography. EBL allows replication of geometrical data structures from computer memory to the substrate. Therefore it is used to generate templates for other lithography techniques.

As the wavelength of the electron is very small, even a basic EBL system offers high resolution. At the same time, writing of large areas is very time consuming if the density of the patterns is high, the linewidth is narrow or the pattern geometry challenging. Registration of the patterns in EBL is not necessary nearly as good as the resolution of the system, because the substrate has to be moved over large distances during the exposure. This requires a fast and extremely accurate mechanical stage. High-end systems having a 10 nm registration accuracy and reasonable write speed can cost millions of euros. Even in high-end systems writing of a single wafer can take a very long time, and as a consequence exposure cost per wafer is substantial.

Electron beam lithography systems based on the use of multiple beams are being developed at present to tackle the throughput problem. These kinds of systems currently target prototyping or small volume manufacturing of IC circuitry, and first demo systems are being sold to customers. In 2008, for example, Mapper Lithography delivered the first systems having 13,000 beams to CEA-Leti and the Taiwan Semiconductor Manufacturing Company to be explored in 22 nm manufactures (Mapper 2009, Wieland et al. 2009). The KLA-Tencor Corporation and the Defense Advanced Research Projects Agency (DARPA) have launched a cost-shared program to develop high throughput EBL systems containing a million beams. The system is targeted to production of an astonishing five to seven densely patterned 300 mm wafers and up to forty sparsely patterned wafers per hour. The system is intended for the 45 nm node with extendibility to the 32 nm node and beyond (Petric et al. 2009). Although these systems produce unprecedented direct writing throughput it is expected that it will take a long time before these systems migrate into mainstream lithography due to the development status, complexity and cost of the instruments.

1.4 Nanoimprint lithography

Nanoimprint lithography (NIL) was introduced at 1995 by Stephen Chou (Chou et al. 1995). He demonstrated results from an experiment where a lab press was used to press together a patterned stamp, made from a SiO_2 coated Si-wafer, with a silicon substrate coated with a thermoplastic polymer (PMMA). Pillars having 25 nm diameters were successfully transferred from the template to the substrate. The process flow from the early paper is illustrated in figure 1. One can argue that Chou's method does not differ much from earlier imprint methods, i.e. those that were used to make compact disks, but the combination of a nanoscopic lateral scale and a thin residual layer allowing subsequent pattern transfer to the underlying layers differentiates Chou's work from others and defines NIL.

The NIL process is a mechanical replication process where surface reliefs from the template are embossed into a thin layer on the substrate. In principle, there are two versions of NIL. One is based on thermal embossing of thermoplastic polymers and the second is based on UV-curable polymers. Some special imprint chemistries require both temperature and UV-light (Schuster et al. 2009), but they are not very common. The NIL process and imprint instrument are conceptually very simple, but allow extremely good resolution and a relatively fast replication process. Compared to optical lithography it does not require extreme ultraviolet light sources and special optics, which increase the cost dramatically. In principle, NIL does not have any limitations in pattern geometry, therefore NIL can copy any patterns produced with EBL or by other techniques.

Thermal-NIL, as illustrated in figure 1, was the original version of NIL. It is based on the use of a thermoplastic polymer spin coated on the substrate. The thermoplastic polymer is heated above the glass transition point of the polymer, and the heated template is brought into contact with the polymer. Once the polymer has filled all the cavities of the template, the substrate and the template are cooled down and the template is separated from the substrate. A negative replica of the template is created on the polymer. In order to use imprinted polymer for pattern transfer to other layers on the substrate, polymer left on the indented areas has to be removed. This residual layer (see figure 1) originates from the fact that the flow of the polymers is not free of resistance.

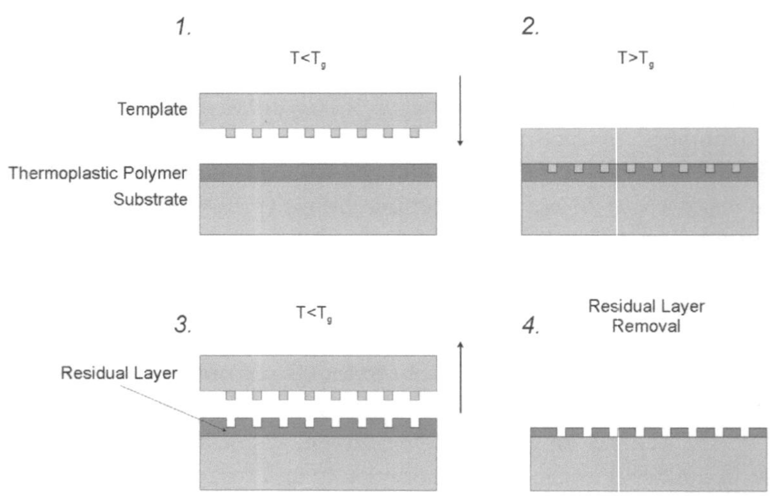

Fig. 1. Thermal NIL-process. Tg is the glass transition temperature of the thermoplastic polymer

Stefan's equation (Bird et al. 1977), describing the force needed to press two circular discs separated by a Newtonian fluid closer to each other, suggests that the imprint force is inversely proportional to the third power of the residual layer thickness. The equation states:

$$F = -\frac{3\pi R^4}{4h_0^3}\frac{dh}{dt}\eta_0.$$

In this equation F is the applied force, R is the disc radius, $2h_0$ is the separation between the discs, and η_0 is the viscosity of the fluid. This model implies that the displacement of fluid over large distances via thin channels requires a large force, or a small displacement rate. In particular, as the residual layer becomes very thin, the resistance grows rapidly, and it is no longer possible to displace all material within a finite time

NIL-process using UV-curable polymers is called UV-NIL. In this process, a layer of UV-curable fluid is spin coated on the substrate, the transparent template is brought into contact with the fluid, and cured using UV-light. The UV-curable layer must be exposed and cured through the template unless the substrate itself allows transmission of UV-Light. The UV-NIL process has some inherent advantages over Thermal NIL:

i. UV-NIL is a room temperature process, therefore time consuming heating and cooling cycles can be omitted.
ii. Room temperature processes eliminate the registration problems originating from the different coefficients of thermal expansion (CTE) of the substrate and the template.
iii. Typically fluids having very low viscosity (2 mPa·s to 50 mPa·s) can be used. UV-NIL therefore requires lower imprint pressures and shorter imprint cycles.

In addition to classifying NIL-processes by their curing properties, NIL-processes can also be classified by their strategy to cover large areas. In principle NIL can be applied on a whole substrate by using a template that is as large as the substrate. However, one small stamp can be used repetitively to cover large areas. These approaches are called "full field NIL" and "step and repeat NIL". Both of the methods have their advantages and disadvantages, as summarized in table 1.

In the case of UV-NIL, different processes can also be differentiated by the dispensing mechanism of the UV-curable polymer. The polymer can be dispensed as a uniform thin layer on the substrate by spin coating, or alternatively it can be dispensed as droplets[1] on pre-defined locations on the substrate by ink-jet or other means. Both methods have their advantages. Spin coating does not require any special equipment, and can deposit highly uniform layers with minimal investment. Droplet dispensing allows polymer to be delivered directly to the location where it is needed by adjustment of the droplet density, as illustrated in figure 2. Therefore the polymer does not need to flow over large distances, and throughput is improved in some cases. Droplet dispensing also helps when imprinting layers incorporating/containing local variations of nanopattern density, and consumes significantly less material than spincoating[2]. Droplets of strongly hydrophobic materials can be deposited on the substrate, whereas spin coating of these materials is challenging. Although this dispensing method is in many ways advantageous, it requires special instrumentation capable of first delivering potentially vast numbers of droplets with accurate volumes (from pico-litres to micro-litres depending on the droplet density) to the correct positions on the substrate, and then aligning the template correctly to the droplet pattern.

---

[1] Droplet dispensing is also known as step and flash imprint lithography (S-FIL) or jet and flash imprint lithography (J-FIL).

[2] A uniform 100 nm layer requires just 10 nl / cm^2 of polymer.

	Full field NIL	*Step and Repeat NIL*
Instrument complexity and cost	Very low to medium (1	Medium to very high
Throughput	Higher, since larger area is imprinted	Lower, since large area has to stitched from small areas
Cost of the Template	Very high. Depends on substrate size.	Medium. Depends on imprint field size.
Die size	Not limited	Limited by individual imprint field.
Within field pattern registration	Lower, due to larger area	Higher, due to smaller area
Over the substrate pattern registration	High, template dependent	Low to very high depending on instrument quality.
Overlay accuracy	Lower, limited by the pattern registration over large field	Higher since pattern must be aligned over smaller field
Requirement for planarity of the template and the substrate	Higher	Lower
Alignment accuracy	Medium	Medium to very high in advanced systems

Table 1. Comparison between Full field NIL and Step and Repeat NIL. (1 Instruments reaching extremely high throughput are also very expensive.

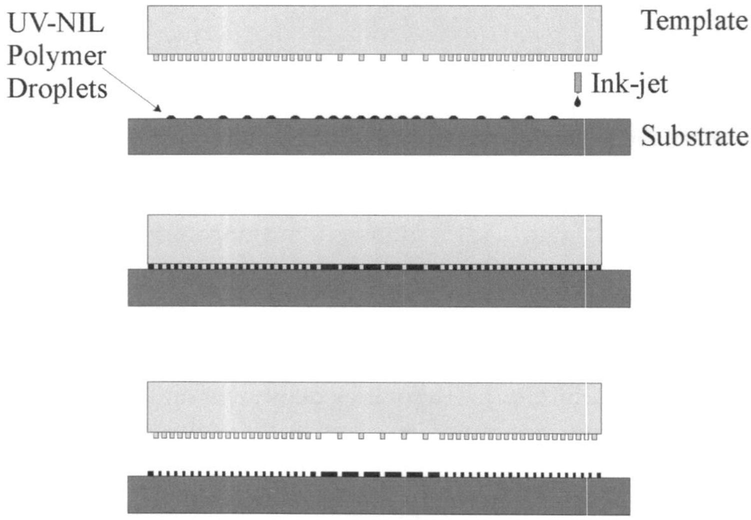

Fig. 2. Sequence for droplet dispense NIL. 1st Deposition of the UV-curable polymer using ink-jet or similar instrumentation. The density of the droplets can be chosen based on the template pattern volume. 2nd Imprint and UV-cure. 3rd Separation of the template.

1.41 State of the art in the NIL

The first published NIL paper already demonstrated a 25 nm pattern diameter and pattern transfer using the lift off technique (Chou et al. 1995). Later, a 10 nm pattern diameter was demonstrated by improving the resolution of the template (Chou et al. 1997). Both these records were based on the use of EBL-written templates that limited the resolution. The same group demonstrated a 6 nm linewidth using a template made from an epitaxially-grown GaAs/AlGaAs superlattice (Austin et al. 2005). However, as early as 2004 Hua had demonstrated linewidths as low as 2 nm using a carbon nanotube-based template (Hua et al. 2004). This demonstration can be considered a resolution record for imprint lithography. However, the patterns were not transferred to the underlying substrate.

The best reported throughput values for NIL systems are reported by Molecular Imprints Inc. For full field systems they have demonstrated throughputs as high as 180 wafers per hour using their HD2200 system designed for hard disk pilot production (Brooks et al. 2009). A state-of-the-art step and repeat system from Molecular Imprints Imprio 300 has demonstrated a throughput of 4 wafers per hour for 300 mm wafers using an imprint field of 26 mm × 32 mm. Although the current throughput record for the step-and-repeat system is rather modest compared to optical lithography systems, the company believes that the throughput can be further improved by clustering these low cost imprint heads together, and by improving their process further. In this way it would be possible to realise a high throughput at lower cost than competing optical lithography systems (Sreenivasan et al. 2009).

The alignment accuracy record for NIL is also set by the Molecular Imprints Imprio 300 system. The system has demonstrated a 3σ alignment accuracy of 16 nm using thick glass templates that have Moiré-alignment marks. The instrument can deform templates in situ by some nanometers in order to fine-adjust the overlay of the template to the substrate. They also utilize a feed-forward strategy that allows correction of the overlay based on the measurements of the previous overlays (Sreenivasan et al. 2009).

2. Nanoimprint lithography with soft stamp

Soft stamp UV-NIL is one of the most interesting nanoimprint methods, due its outstanding cost effectiveness. As the name implies this method utilises soft and sometimes flexible working stamps, replicated from the template. The softness of the stamp makes the imprint process more robust and economical. When the stamp is soft it can deform locally, for example, when a particle is trapped between the stamp and the target wafer. This improves the yield of the process, since with a soft stamp a small particle destroys only a small part of the wafer. Stamp softness also allows imprinting of a large area in a single step, while maintaining a uniform residual layer. High uniformity is possible because a soft and flexible stamp conforms to the overall nonflatness of the substrate. On the other hand, hard stamps can be typically used to imprint a maximum of one square inch area, in a single step. All the waviness of the substrate, within the imprint area, adds to the nonuniformity of the residual layer. This is important to realise, since it is not always possible to use sufficiently flat substrates. In particular, when the linewidth is narrow, the patterns height is small, and the residual layer must be highly uniform. The problem associated with hard stamps is illustrated in figure 3.

One of the best features of soft UV-NIL is that it can be applied with modern UV-mask aligners using special tooling. The mask aligner is, in any case, the work horse of microlithography processing, since it is often the best tool for patterning microscopic features. The increased cost of the mask aligner due to NIL-tooling is insignificant compared to any other feasible nanolithography solution. A manual mask aligner, equipped with NIL tooling, can produce from ten to twenty wafers per hour, and automated systems can reach even better throughputs. At the moment NIL tooling is available for EV Group, Suss Microtech and OAI-mask aligners. These tools typically allow submicron alignment accuracy and can be operated in manual, semi-automatic or fully automatic modes.

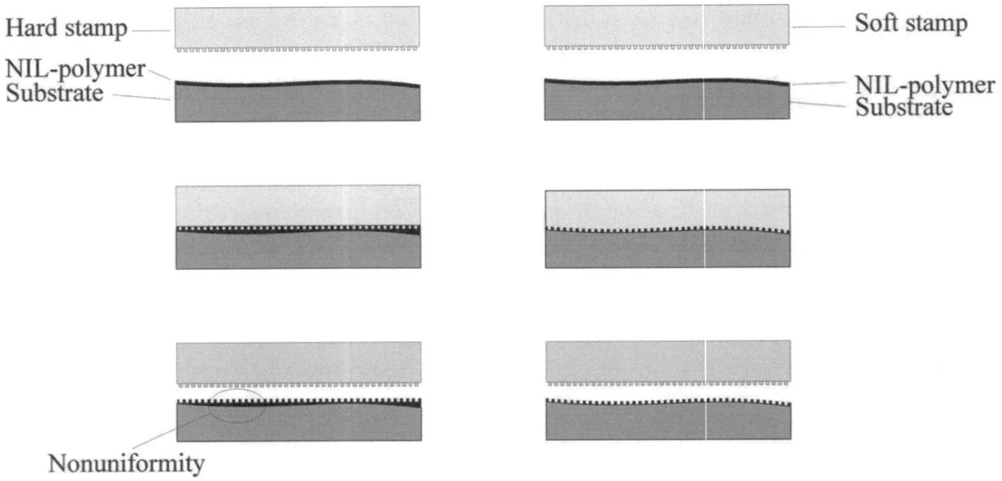

Fig. 3. Comparison between soft and hard stamp technology. On the left: Process flow for a hard stamp. It shows how nonuniformity of the wafer is transferred directly to the residual layer, since both stamp and substrate are rigid. On the right: Process flow for a soft stamp which can conform to the waviness of the wafer, thus improving uniformity of the residual layer.

2.1 Fabrication of large area templates

Based on experience, the most severe limitation for the potential of NIL, especially for NIL using full wafer sized stamps, is the cost of the mastering. When patterns are narrow and dense the exposure rate of any direct write system is slow. Therefore templates can be very expensive or non-feasible to fabricate. At the moment, the only generic method to circumvent this problem, and to make large area stamps, is the use of step and repeat lithography, since it allows replication of the small die to large areas. The benefit of optical lithography is that it uses reduction lenses that scale down the dimensions of the photomask. For this reason the, original template, the photomask, can be fabricated with looser tolerances for particle size, linewidth and registration of the patterns, than the final large area NIL template. On the other hand, step and repeat NIL allows narrower linewidths than optical lithography and NIL-steppers are more accessible than state-of-the-art optical lithography tools. Both lithography options are available from commercial vendors for fabrication of the large area templates.

We have studied use of NIL-templates fabricated from quartz (QZ) mask blanks, Si-wafers, silicon-on-insulator (SOI) wafers and Si-wafers with or without PECVD (plasma enhanced chemical vapor deposition) -deposited SiO_2-top layers. All material systems have a common

SiO_2 surface chemistry (thin surface oxide or bulk oxide) and for this reason they have similar adhesion properties. The main considerations for template fabrication are: etch depth uniformity, surface quality and flatness, etch profile and templates ruggedness. In many applications the quality of the EBL-tool and the processing have to be excellent in order to overcome all restrictions imposed by device physics.

When the linewidth of the patterns or the density of the patterns change over the template, etch depth uniformity issues may arise. These phenomena are referred to as reactive ion etching-lag and aspect ratio dependant etching. Dense and narrow features tend to etch more slowley than isolated and wide features. Both of the material systems, SiO_2 coated Si wafers and SOI-wafers, allow improved control of the etch depth by utilizing etch rate difference between Si and SiO_2. PECVD allows easy control of the etch depth but has higher surface roughness compared to SOI-wafers, which have to be bought from the wafer manufacturer with correct silicon layer thickness. On the other hand, SOI-wafers are almost atomically smooth, thus providing the best platform for large area template fabrication in terms of quality. The main inspection method for template cleanliness and quality is the scanning electron microscopy. Electrically conducting silicon templates are therefore preferred. However, QZ-template offers lowest CTE from these materials thus providing most stable pattern registration accuracy.

Pattern depth and sidewall profiles influence the adhesion between the template and the imprint resist. Obviously, shallow and smooth structures require less force since their surface areas are smaller. A positive etch profile has a lower adhesion force to NIL resist than a 90° profile, but it is easier to maintain a correct linewidth using a template having a 90° sidewall profile. Negative etch profiles have high adhesion forces and usually can't be replicated, although such profiles would be ideal for lift-off processing. However, replication of a slanted grating with opposite directions has been demonstrated, showing that even negative profiles can be replicated using certain material systems (Levola & Laakkonen, 2007).

2.2 Anti-adhesion treatment

Nearly all surfaces used in the template require a coating that makes it easier to separate template and the imprinted resist. There are maybe three factors that need to be evaluated when applying the anti adhesion layer. Obviously factor one is how it improves the demolding properties. Often this is evaluated simply by measuring the contact angel of wafer that is proportional to surface energy the surface. Surface having high contact angle and therefore being hydrophobic are believed to be ideal for NIL. However more correct way is to evaluate the force needed to separate template from the resist. Contact angle and fracture strength don't necessary correlate well since hydrophobic surface are not necessary chemically inert (Houle et al. 2007). Secondly coatings must be evaluated by their wear properties i.e. how many replicas you may make from the template before replacing the anti-adhesion coating. Finally, one has to evaluate how much coating distorts the dimensions of the patterns.

2.3 Soft stamps

Since the template is often very expensive it is common practice to fabricate a working stamp from the template. The replicated stamp can be used to make several imprints, therefore minimizing the risk of damage to the template. Stamps may also have different physical properties than the template. For example, an SOI-wafer is in many ways an ideal substrate for the template, as described earlier. However, it is nontransparent, and therefore unsuitable for UV-NIL applications. The replicated stamp might have also advantageous mechanical properties such as a soft surface and flexibility. The softness of the stamp may be tuned by the material selection (Schmid & Michel 2000) and by the geometry of the stamp, as illustrated by us in reference (Viheriälä et al, 2009). Finally, a working stamp might have advantageous chemical properties. For example, stamps made from polydimethylsiloxane (PDMS) have very hydrophobic surfaces and do therefore not require anti-adhesion treatment. Additionally, PDMS can absorb small amounts of solvent, which allows improved wetting of the stamp surface.

Perhaps the simplest of the soft stamps is a thick PDMS slab, typically made from the commercially available PDMS brand Sylgard 184, also known as soft-PDMS or s-PDMS (Dow Corning Inc., USA). The stamp is cast in a chamber formed between the template and an anti-adhesion treated glass (see stamp in figure 4a). The stamp can be used to replicate nanopatterns, but it does not offer high accuracy since PDMS has a high CTE, and the stamp does not have any supporting structure which would prevent the soft material from deforming laterally. More advanced geometries of the soft stamp are presented in figures 4b and 4c. In figure 4b the s-PDMS slab is bonded to glass. This improves the accuracy of the stamp, since different glasses have significantly lower CTEs (from 0.5 to 8 ppm/°C) and higher modulus than soft materials. For these reasons glass reduces lateral deformation caused by thermal expansion mismatch between stamp and a substrate or by other factors during imprint. Figure 4c improves the stamp concept of 4b since the nanopatterned layer of the stamp is composed of a different material than the soft compliant layer. The composite structure has the advantage that the pattern layer can now be made from harder material.

The stamp in figure 4d improves this concept further by introducing a thin layer of glass between the hard polymer material and the layer of s-PDMS. In this geometry, the thickness of the patterned polymer can be tuned nearly at will, and is backed by a thin glass plate (for example 50 μm thick). Since a glass layer is bonded close to the nanostructures, effectively anchoring the patterns, lateral deformation of the pattern is minimized. With this kind of stamp concept we have demonstrated registration reproducibility better than 8 ppm (80 nm/cm) from imprint-to-imprint over 3" substrate and showed that local lateral deformations are nearly non-existent (Viheriälä et. al., 2009b). We expect that three factors limited accuracy to this level; the most significant effect was the temperature variance of the clean room, which caused thermal expansion problems since glass and the substrate have different CTEs. Secondary error sources are in the overlay measurement accuracy based on use of an optical microscope, and variation in the imprint pressure. The backing glass in figure 4d is optional. When backing glass is used it eases stamp handling with some NIL-tools, but without the backing glass the stamp is fully flexible and can be easily peeled from the substrate after the imprint.

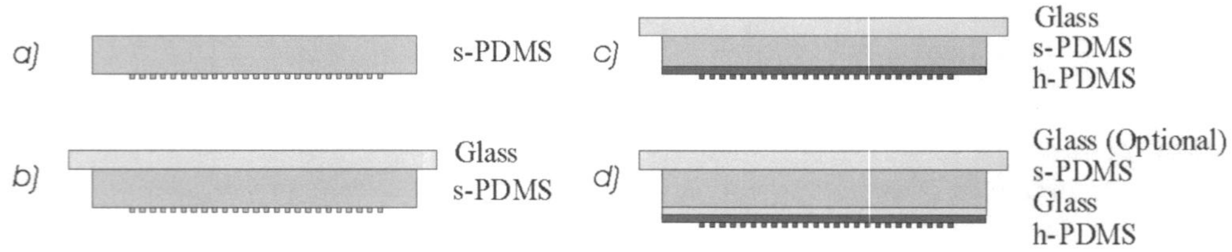

Fig. 4. Different stamp schemes.

Table 2 summarizes the main materials used in soft stamps. It should be noted that the mechanical properties of the materials may be altered to some degree by their processing conditions and mixing ratios.
Our group has used silicones Sylgard 184 (also known as s-PDMS, Dow Corning Inc., USA) and Optical Encapsulat 43 (Gelest Inc., USA) mainly as a compliant, non-patterned layer since these materials are relatively soft and durable. These materials can be spin coated onto the substrate and it is also possible to cast them into a mold. Spin coating offers good coating uniformity (>98%), but only about 30 μm layers can be coated in a single spin coat cycle. Thicker layers require multiple spin and cure cycles. Casting of thick, uniform layers is problematic even with a nearly perfectly flat mould. When we cased 2 mm thick slabs from these soft materials we observed pits and hills of several tens of microns in height. The lateral size of these features was in the sub-10 mm range. In particular pits in the compliant layer should be avoided since they will lead to the creation of areas that are not in contact during the imprint. Our observations concerning/regarding uniformity problems for compliant layers are in good agreement with reported values in reference (Glinsner et al, 2007).

	s-PDMS	**h-PDMS**	**Ormostamp**	**Optical Encapsulat 43**
Tensile modulus	1.8 MPa	8.2 MPa	650 MPa	Not known
Elongation at break (%)	160	7	Low	75-100
Hardness (MPa)	4.77	0.02	36	Not known
Durometer, Shore A	48	Not known	Not known	5-15
CTE (ppm/°C)	325	Not known, expected to be similar to s-PDMS	For similar materials 60-130	Similar to Sylgard 184
Curing	Temperature	Temperature	UV-Light	Temperature
Surface energy	26 mN/m	Similar to s-PDMS	High	Not relevant
Viscosity (mPa·s)	4575	Tunable	410-460	800-1500
Mixing	Two part (1:10)	Four part + solvent	Not needed	Two part (1:1)

Table 2. Materials used in stamp fabrication. Based on references Fuard et al. 2008, Choi & Rogers 2003, Gelest 2009, Dow Corning 2009, Klukowska et al. 2009, Micro Resist Technology 2009a and 2009b.

Sylgard 184 is widely used also as a patterned layer in the stamp and we also tested this approach. However, due to the excessive softness of the Sylgard 184 we often observed deformation of the nanopatterns during the imprint cycle, and pairing of neighboring patterns during stamp manufacturing. We believe that neighboring patterns stick together (pair), as illustrated in figure 5, because friction during the separation step generates surface charges. The generated electric field is several hundreds of volts/meter at one centimeter distance from the stamp, measured with a fieldmill-based static electric field meter. On the nanoscopic scale the field is nonuniform and causes forces strong enough to bend patterns together. Rather interestingly, as the patterns move the electric field changes, and the neighboring patterns can, under the right circumstances, change their pair. This can be observed through the optical microscope as a dynamically and chaotically changing pairing. Unfortunately, as the surface charge discharges over time, the patterns do not return to the non-paired situation, because PDMS surfaces kept in contact react chemically and are glued together. Pairing can be reduced by reduction of the aspect ratio, increase of the pattern spacing, and by using stiffer stamp materials (i.e. h-PDMS and Ormostamp). Long grating lines and tightly spaced pillars on PDMS are most prone to pairing with their neighbors. Larger patterns pair less likely than small ones, and softness of the stamp causes less deformations with microstructures than with nanostructures. For this reason Sylgard 184 is a rather popular material for micro-structuring of UV-polymers;on the nanoscale more rigid materials are required.

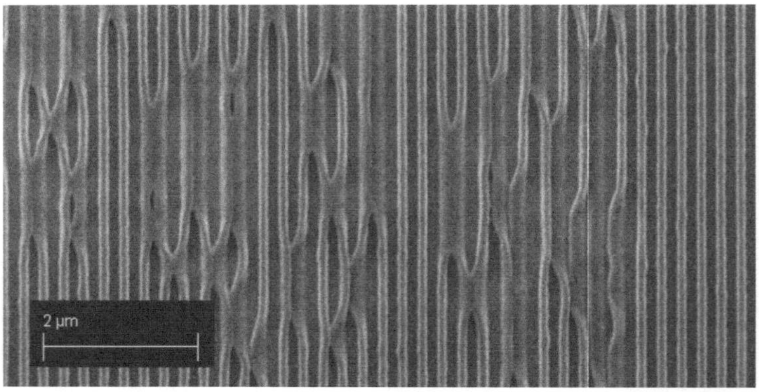

Fig. 5. Scanning electron microscope picture of pairing effect observed on soft stamp.

h-PDMS (aka hard-PDMS) was developed at IBM as early as 2000 (Schmid and Michel, 2000). They tried to formulate a better imprint material by trying different combinations of vinyl and hydrosilane end-linked polymers and vinyl and hydrosilane copolymers, with varying mass between cross-links and junction functionality. A nanoimprint resolution record of 2 nm (Hua et al. at 2004) was demonstrated using soft stamps based on h-PDMS. Based on Schmid's work and our studies we started to use a formulation according to table 3. Toluene was added to h-PMDS since it has very low viscosity (0.590 mPa·s) and a relatively suitable dipole moment. When toluene is mixed with h-PDMS prepolymer these properties improve h-PDMS's ability to fill all the nanocavities in the template (Kang et al. 2006, Koo et al. 2007). Toluene content in the h-PDMS can also be used to tailor the thickness of the spin coated h-PDMS,proved in our publication (Viheriälä et. al., 2009). Thickness control allows reduction of the stamp deformation in certain stamp geometries, as will be discussed later.

Ormostamp (Micro Resist Technology GmbH) is a recently developed UV-Curable inorganic-organic stamp material. It is significantly harder than h-PDMS, thus it has to be backed with soft material in order to realise robust full wafer imprinting. However, since it can be UV-cured, thermal mismatch problems observed when replicating thermally curable materials are eliminated. It is therefore clear that in applications requiring the highest overlay accuracy the best approach is to use UV-curable stamp materials. Unfortunately, not many of these materials are commercially available.

Amount	Brand name	Substance	Role of the substance
3.4 g	VDT-731 ABCR GmbH	Vinylmethylsiloxane-Dimethylsiloxane	Prepolymer
0.75 g	HMS-501 ABCR GmbH	Methylhydrosilane-Dimethylsiloxane	Copolymer
10mg	SIP6831.1 Gelest Inc.	Platinumdivinyltetramethyldisiloxane complex in xylene	Pt-catalyst
39mg	LA16645 Sigma-Aldrich Co.	2,4,6,8 - Tetramethyl - 2,4,6, 8 - tetravinylcyclotetrasiloxane	Inhibitor
For example 40 m%	Toluene	Methylbenzene	Thinner

Table 3. The h-PDMS recipe used by our group.

In many cases the softness of the stamp is a trade-off between process robustness against wafer non-ideality, and vertical deformation due to uneven load across the imprint field. A soft stamp improves the yield, since any possible particles deform only a small area of the imprint (see figure 6 on the left). On the other hand, the softness of the stamp complicates the process since it causes harmful bending under a locally varying load. This change of the load can be caused by the patterns in the stamp (see figure 6 on the right). The deformation can be compensated for by increasing the thickness of the resist (Viheriälä et al., 2009), as the resist layer (liquid) distributes the local pressure effectively over a large area. We have observed that low viscosity NIL-resist distributes pressure more efficiently. Although it is

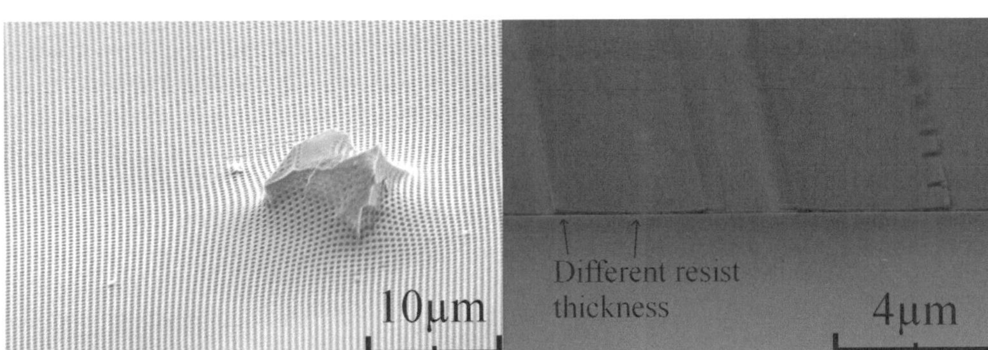

Fig. 6. Figure illustrates advantages and disadvantages of soft stamps. On the left: Softness has saved the imprint, since the pattern is only destroyed over a small area. On the right: The imprint pattern is vertically deformed, since the relatively large pattern (~3 μm linewidth) does not have enough mechanical support.

possible to imprint very high resolution imprints with this stamp (we have demonstrated 24 nm linewidth in Viheriälä et al., 2008), the *softness of the stamp limits the resolution of the transferred patterns in some cases*. Dense and small nanopatterns are relatively straightforward to imprint with a sub-10 nm residual layer, since the stamp load is uniform across the whole imprint field. However, if an imprint contains both wide and narrow patterns, isolated patterns, or if the density of patterns changes over the imprint field, the vertical deformation of the pattern layer must be compensated by a thick residual layer. When the thick residual layer is removed, with plasma etching, the smallest patterns might be washed away since during the residual layer removing linewidth may be reduced.

The stamp concept *d* in figure 4 can significantly reduce the unwanted *vertical* deformation of the stamp, compared with other soft stamps, since the thickness of the pattern layer can be tuned (Viheriälä et al., 2009). The stamp with a thin pattern layer exhibits smaller vertical deformation on the microscopic scale. The stamp with the thinner pattern layer is therefore effectively harder than the stamp with the thick layer, although they are made from the same materials. It is worth noting that although hardness of the stamp can be tuned on the microscopic scale by tuning the h-PDMS layer thickness, on the wafer level the stamp is still fully soft since a thin layer of glass backed by a very thick elastic layer deforms easily across wide (> 100 μm) lateral scale.

In addition to optimisation of the geometry of the stamp and the properties of the resist, vertical deformation can also be alleviated by load sequence and pattern layout. Obviously, low imprint pressure causes minimal deformation, but at the same time some force is required to overcome nonflatness of the substrate. We demonstrated in reference Viheriälä et al., 2009b, that by applying a dual sequence imprint process containing first a high pressure contact step and then a low pressure deformation release step, a better overall quality was attained compared to the traditional single step process.

Many nanophotonics devices already allow reduction of the deformation in the design phase. Isolated patterns, wide patterns and patterns having density variations are the most difficult to imprint. Interestingly, the situation is similar in dry etching or in chemical mechanical planarization, which may also suffer from similar layout restrictions although the physics behind the processes is rather different. However, often it is possible to design the device layout in a way that circumvents these problems by, for example, placing dummy patterns that increase pattern density without sacrificing device functionality. As an example we present in figure 7 two different ways to realize a nanopatterned waveguide. The figure on the left shows a straightforward way to realize the component. In this case the waveguide is isolated[3] and surrounded by an area having zero pattern density. The layout for the waveguide on the right corrects these problems. It is surrounded by a grating having a 50% pattern density, therefore consumption of the resist and pressure are more uniform across the imprint field. As a result the layout on the left exhibits as much as 3.4 times more vertical deformation compared to layout on the right under identical imprint conditions. The curves below the scanning electron microscope images show the surface profiles of the imprint, obtained by atomic force microscope.

[3] Spacing between parallel waveguides is 300 μm.

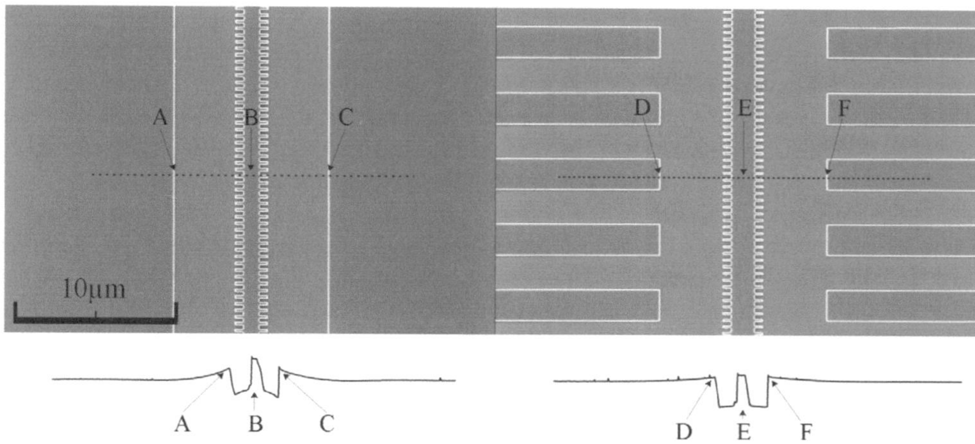

Fig. 7. Unoptimized pattern layout (left) versus more optimal (right). Both layouts can act as identical waveguides for distributed feedback laser diodes (DFB-LDs) but the pattern layout on the right is designed to cause less vertical deformation. Deformation of the imprint is illustrated on the surface curves below the electron microscope images. The dashed line on the electron microscope image represents the place from which the surface graph has been obtained. The letters indicate distinguishable pattern shapes, making it easier to compare graph and image.

3. NIL in nanophotonics applications

In chapter 3 we demonstrate the use of NIL in some applications. Chapter 3.1 demonstrates the first soft UV-NIL-based distributed feedback laser diodes (DFB-LDs) made using laterally coupled gratings. DFB-LDs emit a single longitudinal mode with narrow spectral linewidths and a low frequency chirp. These properties make them suitable for many applications, especially in optical telecommunications and optical spectroscopy, where they are used extensively. In chapter 3.2 we show how NIL can be used to make sharp metallic nanocones for controlling surface plasmons. These cones have many interesting properties for sensing and nonlinear optics, since they concentrate light on the tip of the cone, thus/thereby strongly enhancing the electric field. Chapter 3.3 illustrates the potential of NIL in a totally new class of functional optical fibres. We show the NIL can be used to pattern a functional element onto the facet of the fibre which alters the properties of light entering or exiting the fibre.

3.1 Distributed feedback laser diodes

Distributed feedback laser diodes (DFB-LDs) have a cavity consisting of a periodic structure, which forms a wavelength selective feedback mechanism. The periodic structure in DFB-LDs is normally a grating embedded within or at the side of the laser waveguide. The required period of the grating for lasers operating between 650 nm-1550 nm can be within the range of ~50 nm to 200 nm for first order gratings, and longer for higher order gratings. This resolution of these features is well within the reach of NIL.

The substrates used in the production of the DFB-LDs are relatively small (two or three inches in diameter), therefore patterning of the full wafer is possible with a single imprint.

However, the large area imprint requires a flexible stamp, because wafers are rarely completely flat since laser diodes, like many other optical components, are made on substrates that are not as uniform as large area prime grade silicon or glass substrates. The total thickness variance is regularly between 5 μm and 15 μm for GaAs and InP wafers (Sumitomo, 2009). A flexible stamp is also very easy to separate from the substrate, since it bends easily with minimal force. For this reason, the fragile substrate (typically GaAs, InP or GaSb) is not damaged. Softness of the stamp makes the imprint process more robust and economical as described in subsection 2. It is worth noting that even though the fabrication process of DFB-LDs requires narrow linewidths, patterns are not very sensitive to particles because the components are small and the waveguide uses only a small area of the chip.
We used laterally coupled gratings in our DFB-LDs. These components are based on a ridge waveguide laser diode having periodically corrugated ridge sidewall, as shown in figure 8. The corrugation acts as a grating. Light propagating below the ridge waveguide experiences small refractions caused by periodic perturbation of the effective refractive index of the waveguide. This generates distributed feedback.

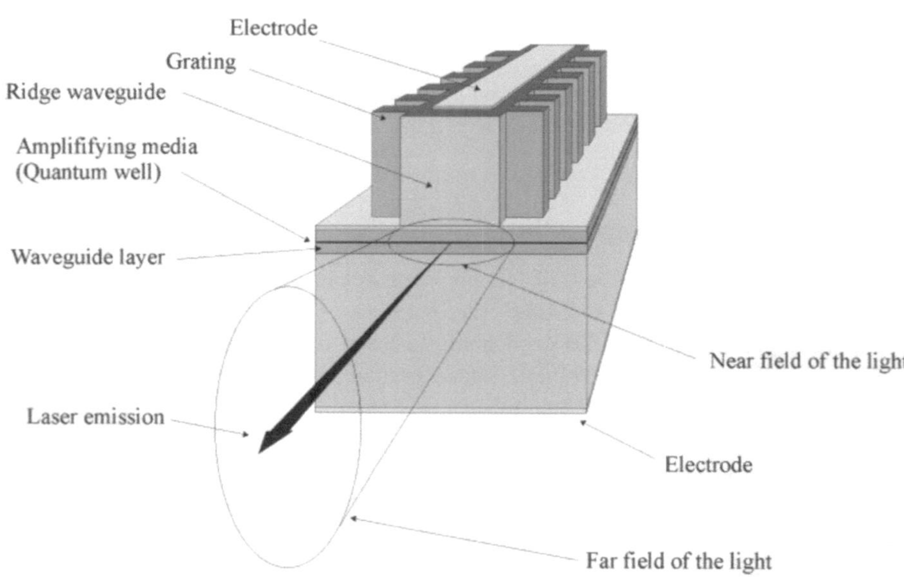

Fig. 8. Schematic operation principle of the laterally coupled distributed feedback laser diode.

Laterally coupled laser diodes are highly interesting in conventional applications (Abe et al. 1995), quantum cascade lasers (Williams et al. 2005 and Golga et al. 2005), terahertz generation (Pozzi et al. 2006) and photonic integrated circuits (Sorel et al. 2008). The main reasons for widespread interest towards this technology is that DFB-lasers based on laterally coupled gratings can be made without regrowth. Therefore, it can be applied to any compound semiconductor material system. Additionally, grating fabrication is only a slightly modified waveguide fabrication process, and therefore it is easily implemented on a photonic integrated circuit. It is also very easy to vary the dimensions of the waveguide and the gratings and thereby achieve complete control over the lasing mode. We show in figure 9 a DFB laser waveguide after it has been imprinted with NIL and the pattern has been transferred with dry etching to the semiconductor layers.

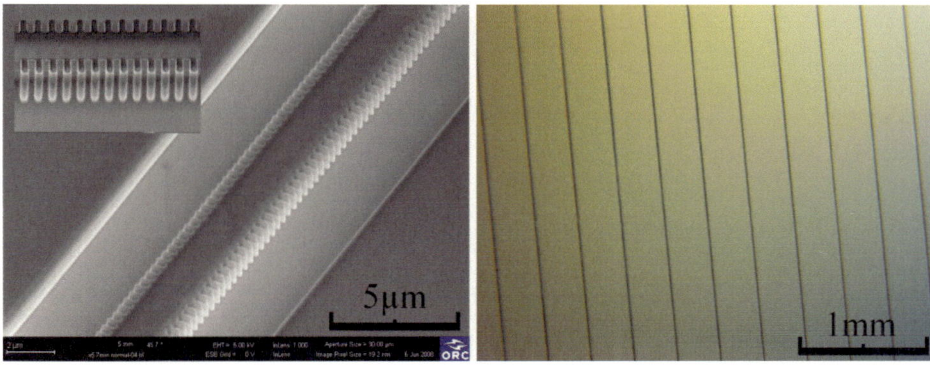

Fig. 9. On the left: Imprinted and etched waveguide for DFB-lasers. On the right: Wide area picture of a DFB-laser diode wafer after the imprint.

We have studied laser diodes operating at 975 nm and 894 nm wavelengths. The 975 nm laser diode was based on three InGaAs quantum wells embedded in a GaAs waveguide. The waveguide had an $Al_{0.6}Ga_{0.4}As$ cladding layer, and a heavily doped GaAs contact grown on top of the cladding. We used a third order grating period (~450 nm) to keep the aspect ratio of the etching at a reasonable level (around 7.5). These lasers exhibited a high, 50 dB, side-mode suppression-ratio near the gain-grating resonance, and a 40 dB side-mode suppression-ratio across the tuning area of 3 nm. The devices exhibited a wavelength tunability of 77 pm/°C. The Light-Current-Voltage relation and spectrum graph of the of one such device are shown in figure 10. The demonstrated laser diode is the first one fabricated with soft UV-NIL.

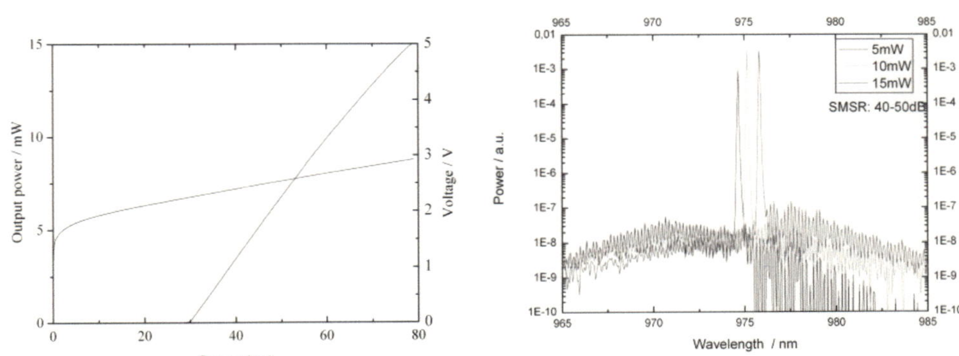

Fig. 10. On the left: Light-Current-Voltage behavior of the DFB laser diode showing threshold current of 30 mA and slope efficiency of 0.35 W/A. On the right: Spectrum of the device measured at 5 mW, 10 mW and 15 mW output power.

Our lasers operating at 894 nm are designed for pumping the D1 transition of Cs-atoms. They are based on a single GaInAs quantum well embedded in a GaInP-waveguide. The waveguide had an $Al_{0.7}Ga_{0.3}As$ cladding layer, and a heavily doped GaAs contact grown on top of the cladding. Grating periods of 418.6 nm and 421.4 nm produce resonances at 888 nm and 894 nm, respectively. Tunability of the laser is 89 pm/°C. The Light-Current-Voltage relation and spectrum graph of one of such is illustrated in figure 11.

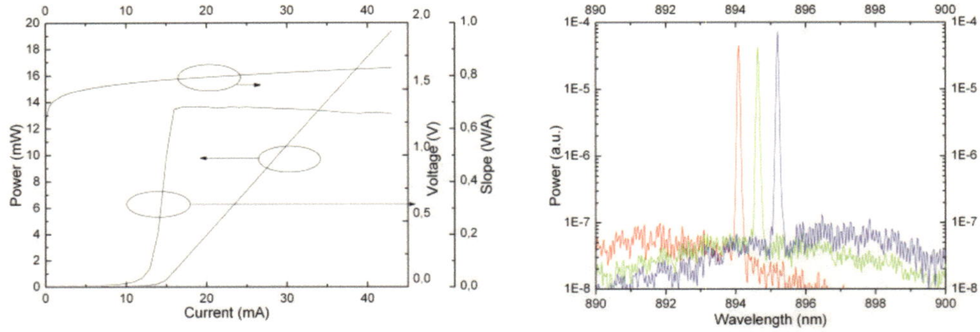

Fig. 11. On the left: Light-Current-Voltage behavior of the DFB laser diode showing threshold current of 15 mA and slope efficiency of 0.7 W/A. On the right: Spectrum of the device showing the tunability around the D1 transition of Cs-atoms.

3.2 Plasmonic nanostructures

In recent years metallic nanostructures have been under intense investigation in the field of nanophotonics as they enable the manipulation of light beyond the diffraction limit (Nature Photonics 2008). In particular sharp particles are particularry attractive, as they can produce highly localized electromagnetic fields due to a combination of plasmon resonances and the so-called lightning rod effect. Strong local fields enhance light-matter interactions and have various applications in tip-enhanced near-field microscopy, sensing, and nanofocusing of light.

The main challenge with these nanostructures is their fabrication, especially in large volumes. Electron beam lithography and focused ion beam (FIB) etching offer fast ways to producee plasmonic structures, but they have limitations in the large volume patterning needed for commercial applications. Here nanoimprint lithography has an advantage. It offers resolution on the sub 10-nm scale and also enables rapid fabrication on the wafer scale with low cost lithography equipment. The pattern can be replicated hundreds of times from the same stamp. NIL is also much less damaging to the substrate compared to FIB, an essential feature in patterning on top of compound semiconductor quantum well and dot structures.

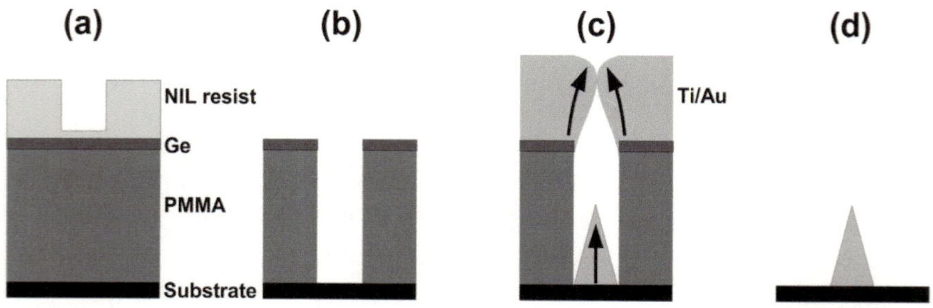

Fig. 12. The principle of nanocone fabrication by NIL.

Using nanoimprint lithography we have fabricated conical nanostructures, nanocones, with sharp tips and good uniformity (Fig. 13). In our tests we used a stamp with a 4 cm^2 pattern area for imprinting. The final wafer consisted of ~4,0 x 10^9 nanocones and the yield of the unoptimized process was 95 %. The principle of nanocone formation is similar to that used to fabricate Spindt-type field emitters (Fig. 11, Spindt et al. 1968). Although the fabrication process is quite simple and well-known in field emission applications, to the best of our knowledge it has not been exploited in plasmonic applications. We demonstrated that the nanocones lead to strongly localized electric fields which enhance nonlinear optical properties (Kontio et al. 2009a). The second-harmonic (SH) signal was enhanced by a factor of 150 compared to gold nanoparticles (half-cones) with the same period and base diameter, but without a sharp tip (Fig. 13). Evidently the strongly localized electromagnetic field of the fundamental beam enhances the SH signal. Possible application areas for metallic nanocones include tip probes, sensors and metamaterials. We have also fabricated nanocones from several different metals (Ag, Al, Au, Cr, Ge, Ni, Pt, and Ti) (Kontio et al. 2009b). The aspect ratio and overall quality strongly depends on the evaporated material.

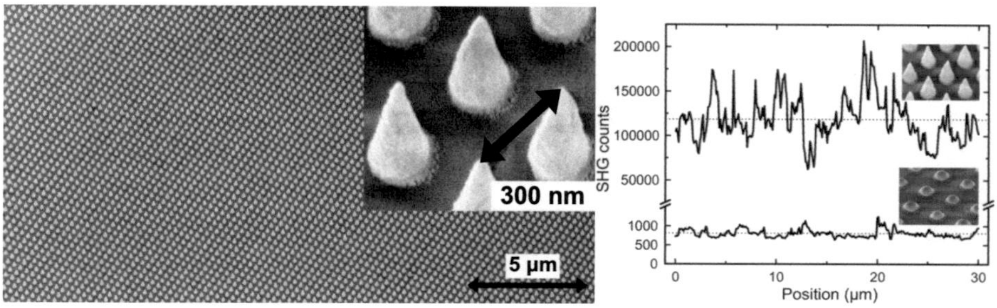

Fig. 13. On the left: A SEM image of an array of nanocones with a period of 300 nm, base diameter 130 nm, and height 290 nm. On the right: A line scan of the second-harmonic signal from the sharp nanocones and half-cones.

3.3 Patterned facets of optical fibres

Micro- and nanopattered surfaces of optical fibre can operate as various miniature optical elements. They can modify the propagation of light by diffracting, collimating, shaping, or focusing it. A properly designed optical element on the facet of an optical fibre improves the functionality of the fibre without compromising the compactness of an optical system. Miniaturized elements could subsequently be used for building miniature spectrometers, sensors, and other devices. However, until now suitable nano- and microfabrication methods that would allow efficient fabrication of such fibres have not existed.

So far, one simple optical element that can be prepared on the tip of a fibre is a lens. The lens may be made by grinding or melting the end of the fibre, or combining segments of fibres with different refractive index profiles (Shiraishi et al. 1997 and Yeh et al. 2004). More complex elements containing small features are made by micro- and nanopatterning using focused ion beam lithography or electron beam lithography (Giannini et al. 2000 and Schiappelli et al. 2003). These direct writing methods are expensive to deploy and capital investments are high. Moreover, their use for any small substrate, such as the facet of an optical fibre, is challenging.

We have demonstrated the world's first surface reliefs fabricated by NIL on the facet of a single fibre by (Viheriälä et al. 2007). The method utilized UV-curable polymer that was deposited on the facet by dip coating. Although dip coating delivers a rather non-repeatable quantity of polymer on the facet, due to the small size of the fibre it is possible to press excess low viscosity polymer away from the facet. We used polymer relief as the functional

element. This application only requires a simple imprint setup. The set-up is built built on an optical table, and includes a stamp holder and micromanipulator for bringing fibre and stamp into contact. A microscope was used to monitor the contact between the stamp and the fibre *in situ,* since excess contact force easily bends the fibre between the fibre chuck and the contact point. Polymer between the fibre and the stamp was cured with fibre-coupled UV-source delivering immense UV-intensity of 8 W/cm^2 Intensities this high cure the UV-NIL-polymer nearly instantaneously.
Using this simple set-up we patterned two sets of fibre facets. We used a standard single-mode fibre (Corning SMF-28). The first set of samples was patterned using a commercially available blazed grating with 830 lines / mm (Optometrics Corp). The second set of patterns consisted of holes with diameters of 250 nm, arranged in a square lattice with a period of 500 nm. The blazed grating was used in order to study the diffraction efficiency of the imprint. The grating efficiency was defined as the power of the first-order diffraction mode over the total light power in the modes. Efficiency versus wavelength graph is plotted in figure 14.

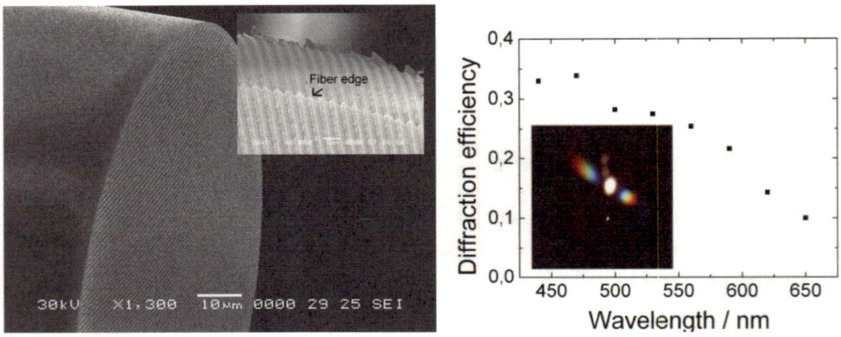

Fig. 14. On left: SEM image illustrating the facet of the optical fibre with the imprinted blazed grating. Insert: Close up near the fibre edge. On right: Graph of diffraction efficiency, and image from the output of the fibre when white light is launched into fibre.

We also demonstrated that nanopatterning of the fibre tip is possible. We used a stamp having 250 nm holes in a grid with a 500 nm period. The final structure showed good uniformity. The standard deviance for the diameter of the holes was below 7 nm, as analyzed from SEM images near the core of the fibre. We expect that that main mechanism causing this diameter deviation was the template having standard deviation of this magnitude. The very accurate replica obtained provides clear-cut evidence that UV-NIL can produce flawless sub-wavelength features on a small area fibre facet. In work published later, similar methods were also employed by other groups in order to fabricate fibre probes for on-wafer optical probing (Scheerlinck et al., 2008) and to make fibres with integrated surface enhanced Raman scattering sensors on their facet (Kostovski et al., 2009).

4. Conclusion

Nanophotonics is a rapidly growing field with great commercial potential. However, it is not yet clear how fabrication for a myriad of different applications can be scaled up. The electronics industry has developed its own fabrication methods largely around optical lithography but it is clear that the same model can not automatically be used for photonics fabrication. The field of nanophotonics is much more fragmented, less standardized, and requires different technical specifications than electronics. We expect that NIL will play an important role in the commercialization of many nanophotonics applications since it offers excellent cost effectiveness and requires relatively low capital investment. We argue that in many applications in particular UV-NIL based on soft working stamps is the best approach, since it offers perhaps the best cost effectiveness. However, like any technology soft UV-NIL has to be understood thoroughly before being applied to fabrication. We have underlined some of the keys issues one may encounter when UV-NIL, and especially soft UV-NIL, is applied and shown that, when NIL is mastered, it is possible to use it to demonstrate various imprinted components.

5. Acknowledgements

The authors wish to acknowledge financial support from the Finnish Funding Agency for Technology and Innovation within the projects Nanophotonics (161147-2) and Nano Extension (40149/08), the European Space Agency within the project ESA GSTP (21173/07/NL/PA), the EU within the FP7 project DeLight (224366) and the Academy of Finland in the project A-Plan (123109) and Lightcaviti (115428). Jukka Viheriälä also wishes to acknowledge the Ulla Tuominen Foundation, the Foundation for Financial and Technical Sciences, the Finnish Foundation for Technical Promotion, the Cultural Foundation and the National Graduate School in Materials Physics.

The authors also wish to thank MSc Tuomo Rytkönen, Mr Juha Tommila and Ms Milla-Riina Viljanen for their invaluable work with Nanoimprint Lithograpy, Mr Aki Wallenius and Mr Jarkko Telkkälä for their work with DFB-Laser diodes, and MSc Kimmo Harring for skillful preparation of optical coatings. Dr Charis Reith has had an important role in proofreading the English text. Without support from the epitaxy group - Dr Tomi Leinonen, MSc Lauri Toikkanen, MSc Teemu Hakkarainen and Ms Sanna Ranta - work with laser diodes would have been impossible. Optical design of the laser diodes was carried out by MSc Antti Laakso and Dr Mihail Dumitrescu. The authors acknowledge Dr Janne Simonen and Dr Mihail Dumitrescu as important forces in driving plasmonics and laser diode research forward. Finally we wish to acknowledge people that have prepared various NIL templates for our activities. Of these people we wish to especially acknowledge the University of Joensuu physics department : Prof Markku Kuittinen, Dr Hemmo Tuovinen, Dr Janne Laukkanen, MSc Kari Leinonen, MSc Ismo Vartiainen and XLith GmbH, AMO GmbH and Chalmers technical university.

6. References

Abe H., Ayling S. G., Marsh J. H., De La Rue R. M., and Roberts J. S. (1995). Single-mode operation of a surface grating distributed feedback GaAs-AlGaAs laser with variable-width waveguide, *IEEE Photon Technol. Lett.*, Vol. 7, No. 5, pp. 452–454

Austin M. D., Zhang W., Ge H., Wasserman D., Lyon S. A. and Chou S. Y. (2005). 6 nm half-pitch lines and 0.04 μm^2 static random access memory patterns by nanoimprint lithography, *Nanotechnology*, Vol. 16, pp. 1058-1061

ASML (2009). ASML, TWINSCAN™ NXT:1950i specifications

Bird R. B., Armstrong R. C. and Hassager O. (1977). *Dynamics of Polymer Liquids, Fluid Mechanics*, Vol. 1, John Wiley & Sons, 1977

Brooks C., Schmid G., Miller M., LaBrake D., Hofemann P., Resnick D., Sreenivasan S.V. (2009). Patterned Media Using Step and Flash Imprint Lithography, SPIE Advaned Lithography Conf., Alternative Lithographic Technologies, *Proc. of SPIE,* Vol. 7271, 72711L, San Jose, USA, 2009

Choi K.M. and Rogers J.A. (2003). A Photocurable Poly(dimethylsiloxane) Chemistry Designed for Soft Lithographic Molding and Printing in the Nanometer Regime, *J. Am. Chem. Soc.,* Vol. 125, No. 14, pp. 4060-4061

Chou S. Y., Krauss P. R., and Renstrom P. J. (1995). Imprint of sub-25 nm vias and trenches in polymers, *Appl. Phys. Lett.* Vol. 67, pp. 3114-3116

Chou S. Y., Krauss P. R., Zhang W.,Guo L. and Zhuang L. (1997). Sub-10nm imprint lithography and applications, J. Vac. Sci. Technology B, Vol. 15, pp. 2897-2904

Dow Corning (2009). Sylgard® 184 silicone elastomer kit, product datasheets, electronics

French R. H., Sewell H., Yang M. K., Peng S., McCafferty D. C., Qiu W., Wheland R. C., Lemon M. F., Markoya L. and Crawford M. K. (2005). Imaging of 32-nm 1:1 lines and spaces using 193-nm immersion interference lithography with second-generation immersion fluids to achieve a numerical aperture of 1.5 and a k_1 of 0.25, Journal of Micro/Nanolithography, MEMS, and MOEMS, Vol. 4, No. 3, 031103

Fuard D., Tzvetkova-Chevolleau T., Decossas S., Tracqui P. and Schiavone P. (2008), Optimization of poly-di-methyl-siloxane (PDMS) substrates for studying cellular adhesion and motility, *Microelectronic Engineering,* Vol. 85 No. 5-6, pp. 1289-1293

Gelest Inc. (2009). Silicone coatings, encapsulats, gels. Performance property guide for Dielectric, Optical, Thermal and Mechanical applications

Giannini G., Mastrogiacomo L., Cabrini S., Gerardino A., Nottola A., Foglietti V. and Cianci E. (2000). Microfabrication of diffractive optical elements onto the active region of optical fiber termination, in Proc. MICRO tec.2000.-VDE World Microtechnologies Congress, vol. 2, pp. 695-7, Berlin, Germany, 2000

Glinsner T., Plachetka U., Matthias T., Wimplinger M. and Lindner P. (2007). Soft UV-based nanoimprint lithography for large-area imprinting applications, Emerging Lithographic Technologies XI. *Proceedings of the SPIE,* Vol. 6517, pp. 651718

Golka S., Pflügl C., Schrenk W., and Strasser G. (2005). Quantum cascade lasers with lateral double-sided distributed feedback grating, *Appl. Phys. Lett.* Vol. 86, 111103

Houle F. A., Rettner C. T., Miller D. C., and Sooriyakumaran R. (2007). Antiadhesion considerations for UV nanoimprint lithography, *Appl. Phys. Lett.,* Vol. 90, 213103

Hua F., Sun Y., Gaur A., Meitl M. A., Bilhaut L., Rotkina L., Wang J., Geil P., Shim M. and Rogers J. A. (2004), Polymer Imprint Lithography with Molecular-Scale Resolution, *Nano Letters,* Vol. 4, No. 12, pp. 2467-2471

Kang H., Lee J., Park J. and Lee H. H. (2006). An improved method of preparing composite poly(dimethylsiloxane) moulds, *Nanotechnology,* Vol. 17, pp. 197–200

Anna Klukowska A., Kolander A., Bergmair I., Mühlberger M., Leichtfried H., Reuther F., Grützner G. and Schöftner R. (2009), Novel transparent hybrid polymer working stamp for UV-imprinting, *Microelectronic Engineering,* Vol. 86, No. 4-6, pp. 697-699

Kontio J.M, Husu H, Simonen J., Huttunen M.J, Tommila J., Pessa M., and Kauranen M. (2009a), Nanoimprint fabrication of gold nanocones with ~10 nm tips for enhanced optical interactions, *Optics Letters,* Vol. 34, pp. 1979–1981

Kontio J.M., Simonen J., Tommila J., and Pessa M. (2009b). Arrays of metallic nanocones fabricated by UV-nanoimprint lithography, Article in Press, *Microelectronic Engineering*

Koo N., Bender M., Plachetka U., Fuchs A., Wahlbrink T., Bolten J. and Kurz H. (2007). Improved mold fabrication for the definition of high quality nanopatterns by Soft UV-Nanoimprint lithography using diluted PDMS material, *Microelectronic Engineering*, Vol. 84, pp. 904-908,

Kostovski G., White D.J., Mitchell A., Austin M.W. and Stoddart P.R. (2009). Nanoimprinted optical fibres: Biotemplated nanostructures for SERS sensing, *Biosensors and Bioelectronics*, Vol 24, No. 5, pp. 1531-1535

Levola T. and Laakkonen P. (2007). Replicated slanted gratings with a high refractive index material for in and outcoupling of light, *Optics Express*, Vol. 15, pp. 2067-2074

Nature Photonics Issue "Antennas of Light" (2008), *Nature Photonics*, Vol. 4, pp. 199

Petric P., Bevis C., Brodie A., Carroll A., Cheung A., Grella L., McCord M., Percy H., Standiford K., and Zywno M. (2009). REBL Nanowriter: Reflective Electron Beam Lithography, Advanced Lithography, *Proc. of SPIE Vol.* 7271 727107-1, San Jose, USA, 2009

Pozzi F., De La Rue R. M., and Sorel M. (2006). Dual-Wavelength InAlGaAs–InP Laterally Coupled Distributed Feedback Laser, *IEEE Photon Technol. Lett.*, Vol. 18, No. 24, pp. 2563-2565,

Mapper Lithography (2009). Press Release 2008, http://www.mapperlithography.com/, Checked 1.6.2009

Micro Resist Technology (2009a). Preliminary process guidelines, contact person: Anna Klukowska

Micro Resist Technology (2009b). Ormocer®s (hybrid polymers) for micro optics, product flyer

Scheerlinck S., Taillaert D., Van Thourhout D., Baets R. (2008). Flexible metal grating based optical fiber probe for photonic integrated circuits, *Appl. Phys. Lett.*, Vol. 92 pp.031104

Schiappelli F., Prasciolu M., Cojoc D., Cabrini S. and Di Fabrizio E. (2003). Design and fabrication of lenses on the top of an optical fiber for efficient fiber-to-waveguide coupling by means of Focus Ion Beam (FIB) lithography, in *Dig. of Papers Microprocesses and Nanotechnology*, pp. 166-167., Int. Microprocesses and Nanotechnol. Conf., Tokyo, 2003

Schmid H. and Michel B. (2000). Siloxane Polymers for High-Resolution, High-Accuracy Soft Lithography. *Macromolecules*, Vol. 33 pp. 3042-3049

Schuster C., Reuther F., Kolander A., Gruetzner G. (2009). mr-NIL 6000LT - Epoxy-based curing resist for combined thermal and UV nanoimprint lithography below 50 °C , *Microelectronic Engineering*, Vol. 86, No. 4-6, pp. 722-725

Shiraishi K., Ohnuki H., Hiraguri N., Matsumura K., Ohishi I., Morichi H. and Kazami H. (1997). *J. Lightw. Technol.*, Vol. 15, 1997, pp. 356-363.

Sorel M., Strain M. and Mezosi G. (2008). Semiconductor ring lasers for ultra fast all-optical digital processing, European Semiconductor Laser Workshop 2008, Eindhoven, Netherlands, 9th - 20th September, 2008

Solak H. H., Ekinci Y., Käser P. and Park S. (2007). Photon-beam lithography reaches 12.5 nm half-pitch resolution, *J. Vac. Sci. Technol. B*, Vol. 25, No. 1, pp. 91-95

Spindt C.A. (1968). A Thin-Film Field-Emission Cathode, *Journal of Applied Physics*, Vol. 39, pp. 3504

Sreenivasan S.V., Schimaker P., Mokaberi B., Choi J., Perez J., Truskett V., Xu F., Lu X. (2009). Recent Developments in UV Nanoimprint Stepper Technology for Sub-30nm Half-Pitch Lithography, SPIE Advaned Lithography Conf., *Proc. of SPIE* Vol. 7271, San Jose, USA, February, 2009

Sumitomo (2009). Sumitomo Electric, GaAs and InP wafer flatness specifications, 19.1.2009. "http://www.sei.co.jp/sc/products_e/gaas/pdf/materials07_08.pdf"

Suss Microtech (2009). Spesfications for automatic and semiautomatic mask aligners

Technology review (2003), 10 Emerging Technologies That Will Change the World, *Technology review*, Vol. 38, Ferbruary 2003

Viheriälä J., Niemi T., Kontio J., Rytkönen T. and Pessa M. (2007). Fabrication of surface reliefs on facets of singlemode optical fibres using nanoimprint lithography, *Electronics letters*, Vol. 43, No. 3, pp. 150-152

Viheriälä J., Rytkönen T., Niemi T. and Pessa M. (2008). Narrow linewidth templates for nanoimprint lithography utilizing conformal deposition, *Nanotechnology, Vol.* 19, 015302

Viheriälä J., Tommila J., Leinonen T., Dumitrescu M., Toikkanen L., Niemi T. and Pessa M. (2009). Applications of UV-nanoimprint soft stamps in fabrication of single-frequency diode lasers, *Microelectronic Engineering*, Vol. 86, No. 3, pp. 321-324

Viheriälä J., Viljanen M.-R., Kontio J., Leinonen T., Tommila J., Dumitrescu M., Niemi T. and Pessa M. (2009b). Soft Stamp UV-Nanoimprint Lithography for Fabrication of Laser diodes, *Journal of Micro/Nanolithography, MEMS, and MOEMS*, Vol. 8, 033004

Wieland M.J., de Boer G., ten Berge G.F., Jager R., van de Peut T., Peijster J.J.M., Slot E., Steenbrink S.W.H.K., Teepen T.F., van Veen A.H.V., Kampherbeek B.J. (2009). MAPPER: High throughput maskless lithography, Alternative Lithographic Technologies, *Proc. of SPIE* Vol. 7271, 72710O, San Jose, USA, 2009

Williams B. S., Kumar S., John Q. H. and Reno L. (2005). Distributed-feedback terahertz quantum-cascade lasers with laterally corrugated metal waveguides, *Optics Letters*, Vol. 30 No. 21, pp. 2909-2911

Yeh S.-M., Lu Y.-K., Huang S.-Y., Lin H.-H., Hsieh C.-H., and Cheng W.-H (2004) A novel scheme of lensed fiber employing a quadrangular-pyramid-shaped fiber endface for coupling between high-power laser diodes and single-mode fibers, *J. Lightw. Technol.*, Vol. 22, No. 5, pp. 1374-1379.

15

Nanoscale Photodetector Array and Its Application to Near-Field Nano-Imaging

Boyang Liu[1], Ki Young Kim[1,2], and Seng-Tiong Ho[1]
[1]*Department of Electrical Engineering and Computer Science, Northwestern University*
[2]*Department of Physics, National Cheng Kung University*
[1]*USA*
[2]*Taiwan*

1. Introduction

Photodetector array have many applications, such as light detection and imaging. However, pixels in most photodetector array are micrometer scale or larger, which limits its application to relatively low spatial resolution detection. The possibility to realize photodetector array with pixel size at nanometer scale has become of great interest to various technologies. When the dimension of photodetector array's pixel is reduced to such a small scale, many new functions can be achieved. For example, with such a nanoscale photodetector array, it will enable us to image objects at a resolution better than that of conventional diffraction-limited imaging tool, for which the highest resolution that could be obtained is half of the illuminating light wavelength. Recently, many progresses on nano-scale photodetector array (NPD) have been made (Huang et al., 2001; Hayden et al., 2006; Yang et al., 2006; Maier et al., 2003), however most of them are based on nanotube technology and incapable of precisely controlling the position and configuration of detector array. It's desirable to have a photodetector array with nanoscale pixels while still having flexibility in device design and operation. In this chapter, we will present the research on such a photodetector array with nano-scale pixels based on dual side metal-semiconductor-metal (MSM) structure, including the design of NPD array, the simulation of NPD array's performance by finite difference time-domain (FDTD) method, the fabrication of NPD device, characterization of NPD array and the demonstration of nano-scale object imaging using the NPD array that fabricated.

2. Design of nanophotodetector array

The design of NPD array has a basic structure shown in Fig. 1. $In_{0.53}$ $Ga_{0.47}$ As ternary material is chosen as absorbing material for near-IR (1.0-1.6 μm) wavelength range detection. A dual side MSM structure is employed, where the semiconductor active material is sandwiched by the top and bottom electrode. The top and bottom electrode stripes are perpendicular to each other, which enables the pixel addressing by NPD array. Concerns and considerations for these configurations are described under the following categories: (1) selection of active material and structure; (2) considerations in choosing MSM structure.

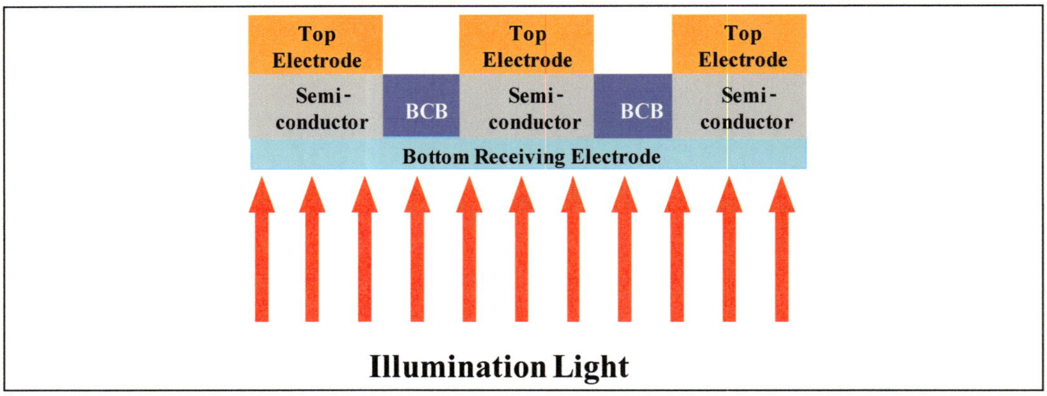

Fig. 1. The schematic of the novel NPD array design, where Benzocyclobutene (BCB) fills all areas between NPD array pixels.

2.1 Selection of active material and structures

Commonly, GaAs material is used for 0.8 μm wavelength detection and has demonstrated a good performance (Biyikli et al., 2001; Biyikli et al., 2004; Seo et al., 1992). For near-IR (1.0-1.6 μm) wavelength range, typically $In_{0.53}$ $Ga_{0.47}$ As, AlGaAs, InGaP and $In_{0.52}Al_{0.48}As$ are chosen as active material (Biyiklia et al., 2003; Chyi et al., 1994; DeCorby et al., 1997; Gao et al., 1994; Gao et al., 1995; Gao et al., 1997; Kim et al., 1998; Loualiche et al., 1990; Zhao et al., 2007). In practical application, $In_{0.53}$ $Ga_{0.47}$ As ternary compound is chosen as active material due to its band gap energy of 0.8 eV and lattice matching to InP material which has extensive applications in optical communications.

However, in near-IR region, the $In_{0.53}Ga_{0.47}As$ MSM photodetectors have not performed well. The primary difficulty is the low schottky barrier height (~0.2 eV) of commonly used Schottky contact metals on $In_{0.53}Ga_{0.47}As$ (Griem et al., 1990). Low barrier height results in excessive dark current and noise. One solution is to add an enhancement layer between the metal and $In_{0.53}Ga_{0.47}As$ absorbing layer to increase the Schottky barrier height. As a result, a Schottky barrier enhancement layer is used, i.e. digitally graded InAlAs/InGaAs super lattice (SL). The SL structure used is shown in Fig. 2. The graded super lattice consists of 3 periods of $In_{0.52}Al_{0.48}As$ and $In_{0.53}Ga_{0.47}As$, whereby the first period is composed of 7 nm of $In_{0.53}Ga_{0.47}As$ and 3 nm of $In_{0.52}Al_{0.48}As$, and the last period is reversed with 3 nm of $In_{0.53}Ga_{0.47}As$ and 7 nm of $In_{0.52}Al_{0.48}As$. The intermediary layer varies linearly between two endpoints in 2 nm increments. The graded SL structure is then capped with an additional 50nm i-$In_{0.52}Al_{0.48}As$ Schottky barrier enhancement layer.

2.2 Metal-Semiconductor-Metal structure

Metal-Semiconductor-Metal photodetector structure is equivalent to two Schottky diodes back to back, shown in Fig. 3. Their response is related to the current caused by the electron-hole pairs separated by the electric field in the depletion region of two Schottky diodes (Land et al., 1985). These devices usually have a simple planar design, often with interdigitated (IDT) fingers structure. The IDT MSM photodetector's respond speed is typically determined by the transit time rather than RC constant.

In practice, a dual side MSM structure is employed, where geometry with electrodes above and below a thin-layer of intrinsic semiconductor as active material is used for pixel

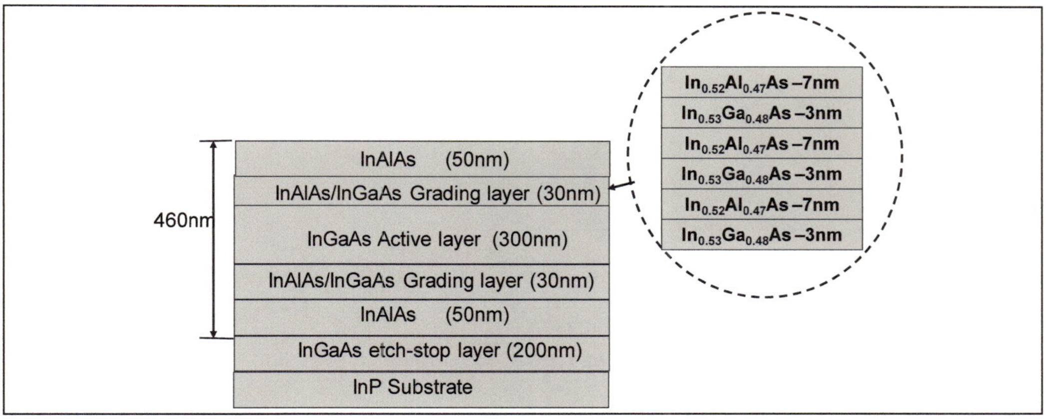

Fig. 2. Structure scheme of practical InGaAs active region with $In_{0.52}Al_{0.48}As$ Schottky barrier enhancement layer using digitally graded SL as transition layer.

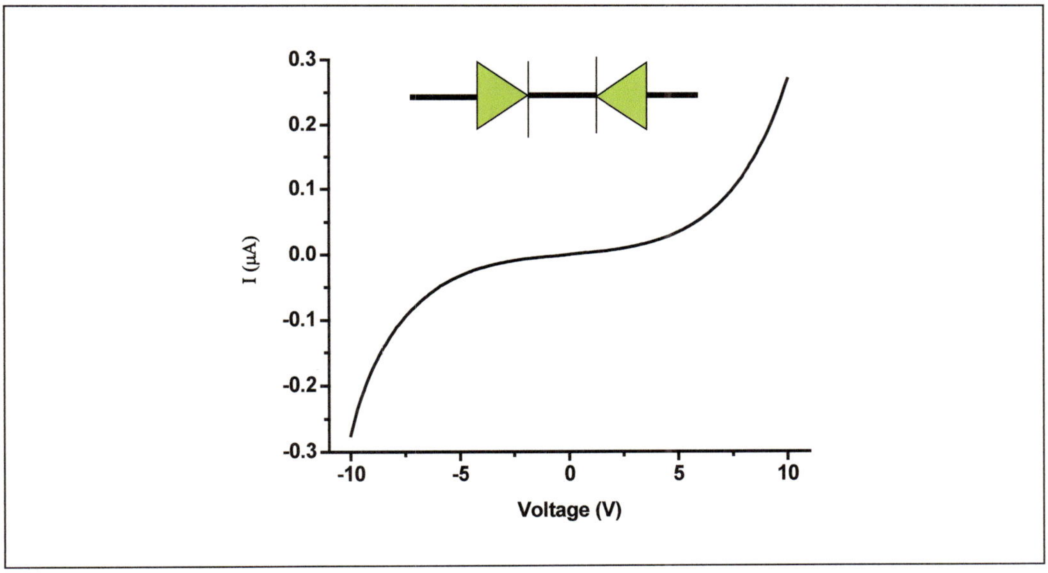

Fig. 3. MSM structure photodetector is equivalent to two Schottky diodes back to back.

addressing (shown in Fig. 4.). To detect a certain pixel, we will just connect to the right top and bottom electrode stripes. The detector where the top and bottom electrode stripe crosses will be the one that is sensed by the detector circuit. Thus, using M+N stripes, we can address M x N pixels.

3. Finite-difference time-domain simulation of nanophotodetector array

In this section, highest possible resolution obtainable with the proposed imaging array shown in Fig. 4 is investigated using the finite-difference time-domain method. The FDTD method (Yee 1966) is one of widely used numerical techniques in simulations for various optoelectronic and photonic devices. However, due to the lack of proper active semiconductor model for photonics applications, conventional FDTD simulations are yet

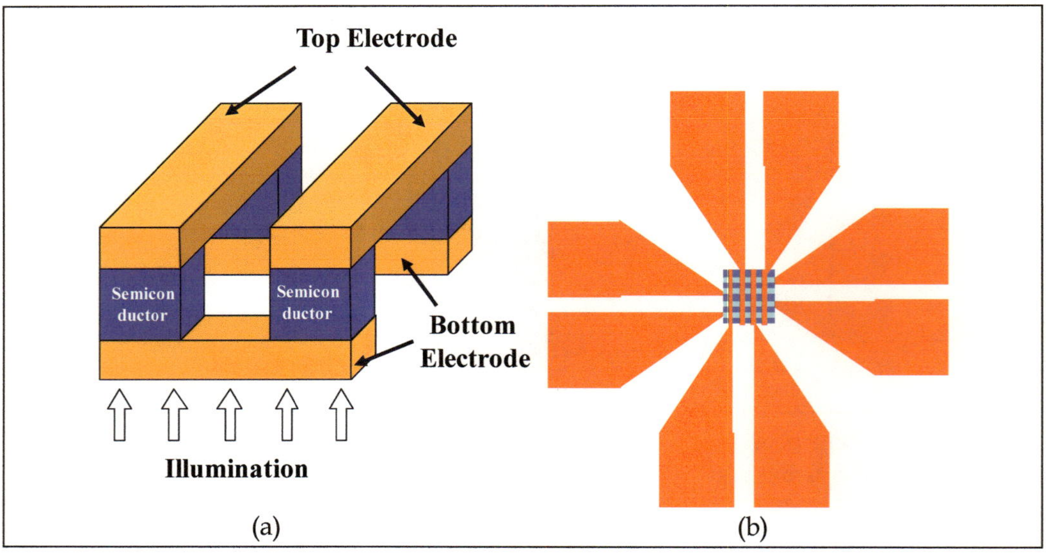

Fig. 4. (a) 3D schematic for channelized NPD array, where the front electrode stripes have a crossing direction to the back side electrodes, forming a matrix for pixel addressing; (b) Top view of a 4 x 4 NPD array.

simple enough while still taking into account the physics of semiconductor materials. A new idea using multi-level multi-electron (MLME) semiconductor FDTD model has been developed by Huang and Ho (Huang 2002; Huang & Ho 2006), in which multiple energy band levels to describe the essential characteristics of the semiconductor energy band structures have been incorporated. This simulation scheme has also been successfully utilized in various photonic device applications with active semiconductors including photodetector, photonic crystal fiber, photonic transistor, whispering gallery resonator, and so on (Kim *et al.* 2008; Khoo *et al.* 2008). In order to understand the working mechanism of NPD array as a nanoscale imaging device, a MLME FDTD method has been adopted in this Chapter.

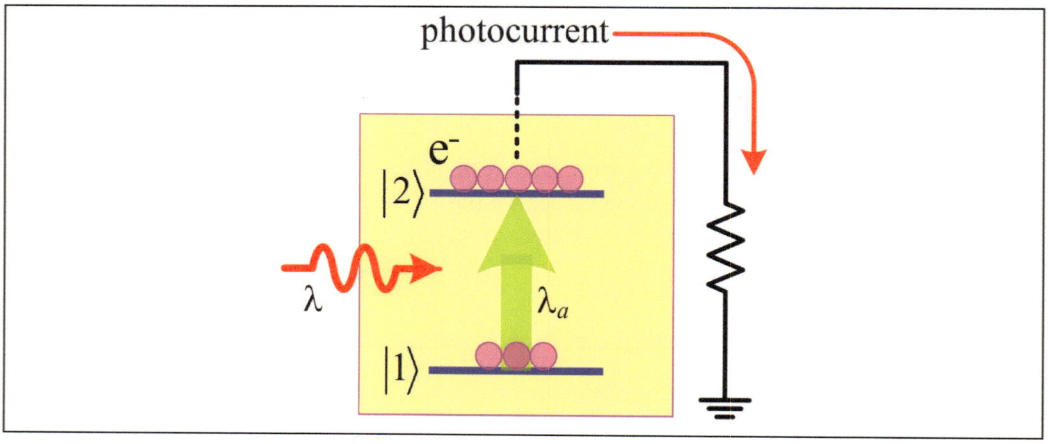

Fig. 5. Generation of photocurrent from photoelectrons in active semiconductor material. $\lambda=\lambda_a$=1550 nm is assumed, where λ and λ_a are the incident wavelength and resonant wavelength of the active semiconductor material, respectively.

In the FDTD simulations for the NPD array, photocurrent generated in semiconductor material by the incident light is one of key parameters in evaluating performance of a photodetector. The photocurrent generated in the active semiconductor medium is dependent on the incident (excitation) wavelength and the energy band gap structure of the semiconductor material. Conceptually, an active semiconductor material for photodetectors can be simplified as a medium with two different energy levels, in which the photocurrent can be calculated from the rate of excitation of ground state electrons from the ground level to the excitation level, which are subsequently retuned back to the ground level via external electric circuit as shown in Fig. 5. In Fig. 5, electrons in ground state (level 1) are excited and transited to excited state (level 2), when the incidence wavelength is matched to inherent resonance wavelength of an active semiconductor material. Here, we assumed $\lambda=\lambda_a$=1550nm, where λ and λ_a are incidence wavelength and resonance wavelength of an active semiconductor, respectively, which belongs to an optical telecommunication wavelength.

For conducting numerical simulations with the MLME FDTD code, the active semiconductor material needs to be spatially discretized as shown in Fig. 6, as well as other parts of the present device. The photocurrent generated in an active semiconductor material, which is composed of numerous FDTD pixels, can be quantitatively calculated from the following formula, which has been directly given based on the definition of the typical concept of electric current and photocurrent mechanism shown in Fig. 5.

$$I_{ph} = \frac{q}{t_{sim}} N = \frac{q}{t_{sim}} \left(\sum_{pixel} N_2 \right) \cdot N_{density} \cdot A \cdot h \,, \tag{1}$$

where $q = 1.6 \times 10^{-19} C$ is the electric charge, t_{sim} is the total time simulated, N_2 is the normalized number of electrons in the level 2 in a single FDTD cell, $N_{density}$ is the number of electrons per unit volume, A is the area of the FDTD pixel, and h is the height of the NPD pixel. Here, we set t_{sim}=1.0 ps, $N_{density}$=0.563x10^{22}/m^3, A=dx x dy=5 nm x 5 nm, and h=300 nm considering a dimension for fabrication.

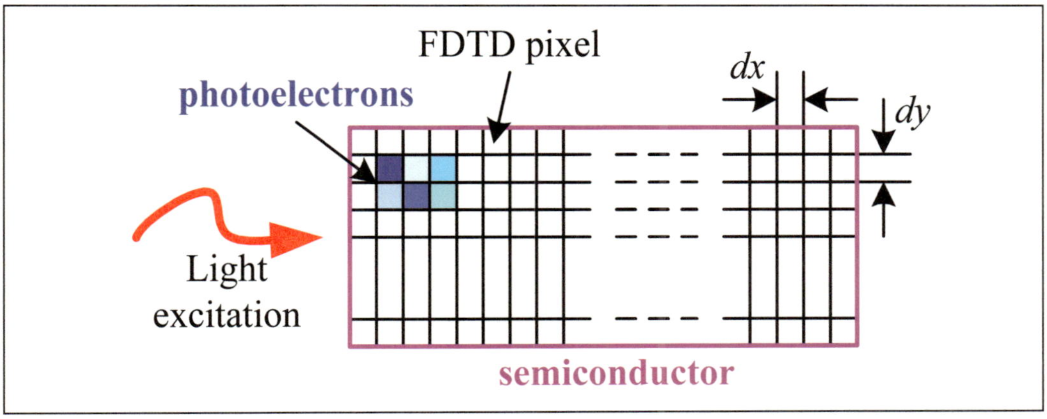

Fig. 6. Discretization of an active semiconductor material for the MLME FDTD method. Photoelectrons in each pixel are generated by light illumination. Photocurrents can be calculated with the photoelectrons.

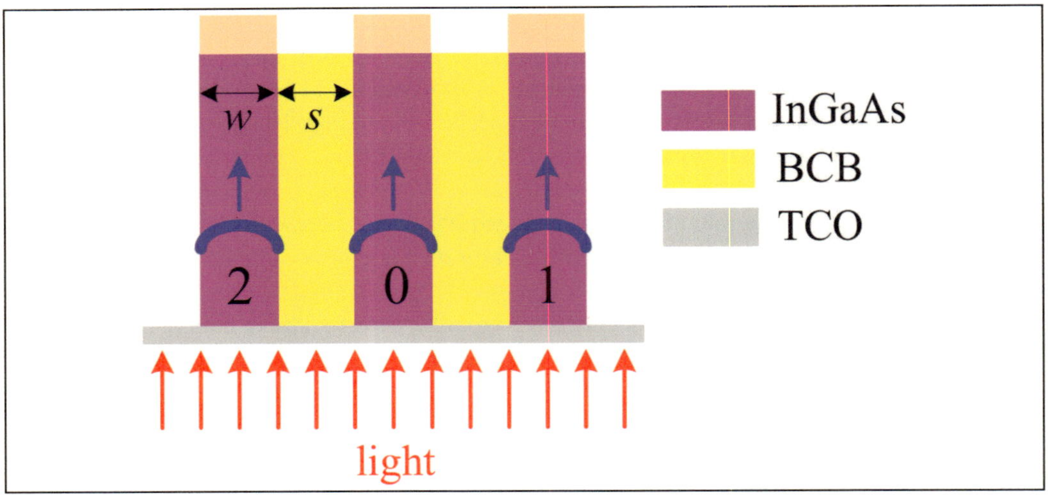

Fig. 7. Simplified two-dimensional schematic illustration of the NPD array.

InGaAs is used as active semiconductor regions of the NPD array. A protective material such as benzocyclobutene (BCB) is filled between pixels to support device structures and form cladding layer to each pixel. Top and bottom electrodes are placed at front and backside of the array for the purpose of photocurrent pickup to external electric circuit to detect the photocurrent generated in each pixel of the NPD array. Bottom electrode could be either a transparent conducting oxide (TCO) or a thin metal layer for light passing through. The refractive indexes of InGaAs, BCB, and air at 1550 nm are assumed to be 3.4, 1.5, and 1.0, respectively.

Fig. 7 illustrates a simplified two-dimensional schematic of the NPD array for clearer description of its working principles for a practical photocurrent pickup mechanism. The active semiconductor material slabs, where photoelectrons are generated by the incident light as described in Fig. 5 and 6, are separated by protective material with lower refractive index that forms cladding layer to each NPD pixel and supports the device structure mechanically. The active semiconductor layer is sandwiched between the top and bottom electrodes. Top and bottom electrodes are placed at front and backside of the array for the purpose of photocurrent pickup to external electric circuit to detect the photocurrent generated in each pixel of the NPD array. Very thin layer of metal layer or optically transparent conductor such as transparent conducting oxide (TCO) will be used for the bottom electrode, forming a matrix with top metal electrodes for pixel-array addressing as shown in Fig. 4. In our NPD design, the active semiconductor region is made from InGaAs with refractive index of 3.4 and surrounded by BCB with refractive index of 1.5. In working condition, the active semiconductor materials in NPD pixels get excited by the near-field point-like light source, which will cause an increase of active material's conductivity and increase the electric current of detection circuit. Therefore, the photocurrent generated by each pixel is the signal of the NPD imaging device. If we define the width of NPD pixel as w and spacing between two adjacent pixels as s, the $w+s$ would be the resolution of the NPD array imaging. In our study, the $1/e$ (~36.79%) resolution criterion is used for NPD imaging characterization. If we assume the photocurrent generated by the mth NPD pixel as 1.0, when the generated photocurrent by the (m+1)th or (m-1)th pixel is less than $1/e$, we say the mth pixel and the (m+1)th or (m-1)th pixel can be distinguishable from each other. For example, if the photocurrents generated in pixels 0 and 1 are 100 nA and 30 nA, respectively, we will have the ratio of 30% that is less than $1/e$. Then, these two pixels are distinguishable.

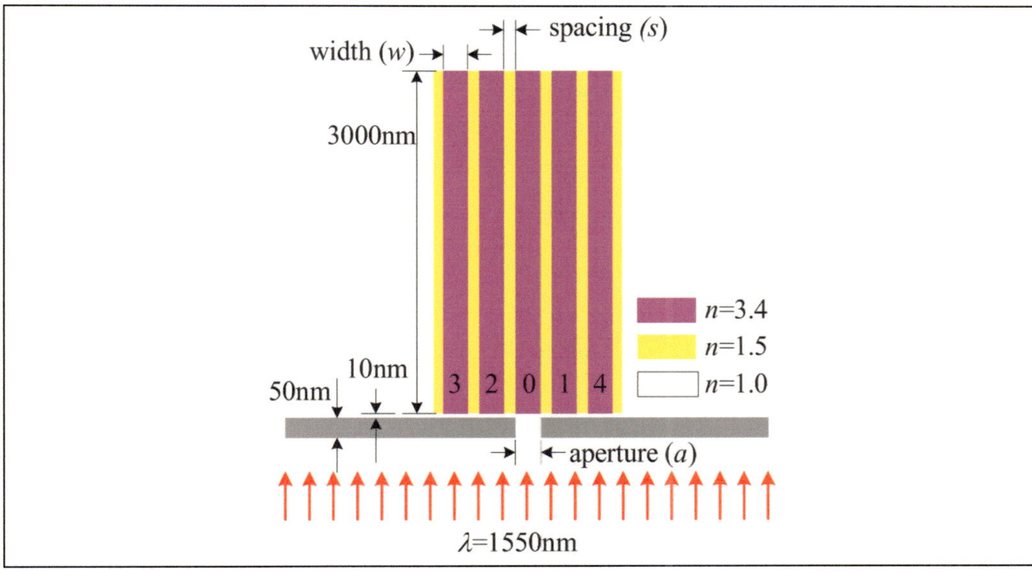

Fig. 8. The schematic of the NPD array with its dimensions for the FDTD simulation. Light to be detected is from the subwavelength metal slit. NPD pixels are labelled as 0, 1, 2, 3, and 4.

Fig. 8 shows a two-dimensional schematic illustration of the NPD array for our MLME FDTD simulations. The light of 1550 nm is incident from the bottom side. We assume a detector slab that is infinite in the direction perpendicular to the paper and the incident source has electric field polarization pointing along this infinite direction. The center-to center distance between the NPD pixels is $w+s$ with a width of the NPD pixel of w and an inter-pixel gap of s, as same as in Fig. 7. The length of the NPD pixels is set to be 3.0 μm to investigate the optical power coupling between pixels, although in practical fabrication, the length of NPD pixels is only a few hundred of nanometers. The semiconductor fingers (pixels) play an important role in detecting incident field, which are converted into photocurrent via the mechanism shown in Fig. 5.

The placement of the imaging device to immediately proximate distance from the near-field light source within a few nanometer orders is strongly required to avoid rapid fading-out of the near-field light. Since all practical NPD devices will work in the near-field region of the illuminating light, the distance between aperture and the NPD array is set to be 10 nm, which assures that central pixels are within the near-field region of the light from aperture. To generate the near-field point-like source, we block the incident plane wave of 1550 nm by a metal film with a small aperture having a small width (a). The front side of the center pixel of the NPD array has been placed at very near distance away from the aperture, thereby, the NPD array can pick up the light from the aperture having diffraction-limited subwavelength light.

The photocurrents from the NPD pixels are obtained to explore the resolution of this novel NPD device for subwavelength diffraction limited imaging. One limiting factor is the optical power coupling between adjacent detector pixels. The MLME FDTD simulation enables us to investigate such power coupling in the presence of absorbing media as well as the spatial distributions of electric field and photoelectron density.

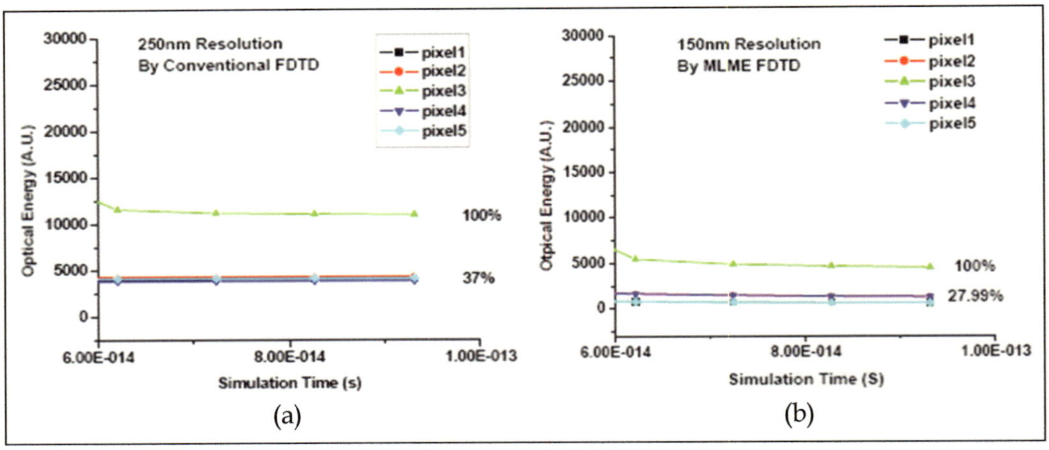

Fig. 9. (a) Simulation of NPD array by conventional FDTD method. The NPD pixel is 200nm wide with 50 nm spacing. The highest resolution is 250 nm for 1550 nm wavelength; (b) simulation of NPD array by MLME FDTD method, the NPD pixel is 100 nm wide with 50nm spacing. The highest resolution is 150 nm for 1550 nm wavelength.

In order to investigate the effect of the optical absorption in optical energy coupling between adjacent pixels in the NPD array, both conventional FDTD and MLME FDTD models are used and results for both cases are compared with each other in Fig. 9. In conventional FDTD simulation, only optical energy in each NPD pixel could be simulated, where the light propagates in dielectric NPD pixels and no interactions between light and NPD detection region are considered. Simulation shows a highest resolution of 250 nm, shown in Fig. 9(a). On the contrary, in MLME FDTD simulation, both the optical energy and photocurrent generated in each pixel could be simulated. Before simulation, the number density of active semiconductor material in MLME FDTD model is calibrated to match real property of InGaAs semiconductor to be used in experiments for 1550 nm wavelength. The calibrated active material loss of around $0.5/\mu m$ for a typical III-V semiconductor material, which is corresponding to a value $N_{density}=0.563 \times 10^{22}/m^3$ in eq. (1) as mentioned earlier. Shown in Fig. 9(b), MLME FDTD simulation shows a highest resolution of 150 nm, which is 100 nm (60%) higher than that by the conventional FDTD simulation. Compared with conventional FDTD model, the MLME FDTD simulation shows a better matching to the response of photosensitive material, which could be used to effectively simulate the photodetection process by the photodetectors.

In order to investigate the optical power coupling between NPD pixels, the average optical power in each pixel is calculated for the case of Fig. 9(b). Fig. 10(a) shows the electric field distributions, which indicates electric field is quasi-bounded by the center pixel (pixel 0) with subsequent coupling to the adjacent pixels (pixel 1, 2) and then to the next adjacent pixels (pixel 3, 4). Fig. 10(b) shows the corresponding photoelectron density of the whole NPD array from the electric field distributions of Fig. 10(a) with an arbitrary normalized linear scale, where most of the photocurrent is generated by the central pixel.

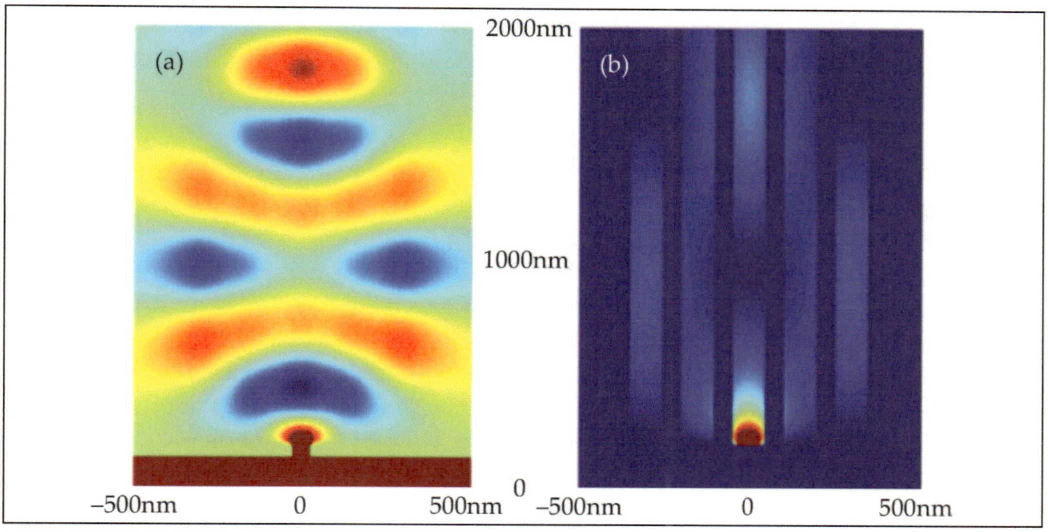

Fig. 10. (a) Electric field distributions and (b) corresponding normalized photoelectron density distributions obtained with MLME FDTD simulation in NPD array configuration. The dark red color indicates higher amplitude in arbitrary linear scale.

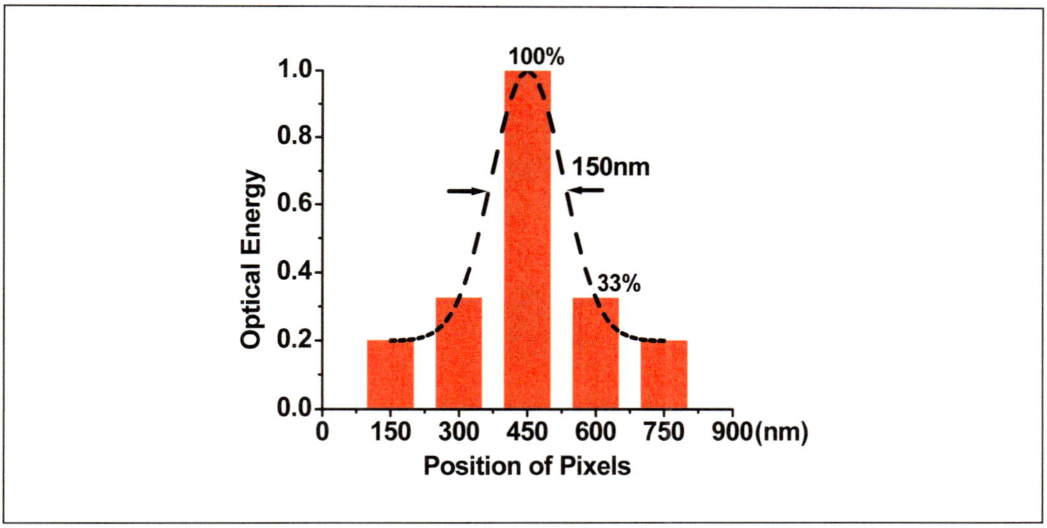

Fig. 11. Photocurrents generated in each NPD pixel.

Fig. 11 shows the photocurrents generated in each pixel from the spatial distribution of the photoelectron density profile of Fig. 10(b) by using eq. (1), where the photocurrent in each pixel for the 150 nm imaging resolution case, where the photocurrent generated in pixels adjacent to central pixel is less than 33% ($< 1/e$ criterion) of that in central pixel.

The photocurrent in each pixel has been normalized to that in the central pixel. The estimated spatial resolution for this NPD array geometry is about 150 nm, which corresponds to a resolution of $\lambda/10$. The resolution of 150 nm by NPD array corresponds to about $\lambda/10$ for near-IR wavelength and about 25 times higher than the diffraction limited conventional imaging system in terms of imaging area.

The achieved optical resolution is substantially below the subwavelength diffraction-limit of $\lambda/2$, which can be potentially applied to the observation of nano-scale moving objects or living cells.

4. Nanofabrication of the NPD Array

4.1 Fabrication of NPD array

Several techniques to fabricate such a nano-imaging device have been developed to realize the 3-dimentional structure of NPD array, including BCB wafer bonding technique and metal oxide sol-gel based nanoscale direct patterning technique (B. Liu et al., 2008a, b) . The pixel width and spacing of NPD array varies from 100 nm to 400 nm. A layer of Au/Ti (55 nm/5 nm) metal was deposited as receiving bottom electrodes. Up to 4×4 NPD array have been fabricated, where the smallest array pixel is as small as 100 nm wide with 100 nm spacing. Fig. 12(a) shows the top view of example 2×2 and 4×4 nanophotodetector array, where the bright electrodes are bottom receiving electrodes and the dark ones are front electrodes. An $In_{0.53}Ga_{0.47}As$ based super lattice structure with 460 nm thickness is sandwiched between the top and bottom electrode stripes. The Au/Ti top and bottom electrode stripes are perpendicular to each other and form an addressable pixel array. Fig. 12(b) shows the detection region of a 4×4 NPD array, the pixel width is 400 nm wide with 400 nm spacing.

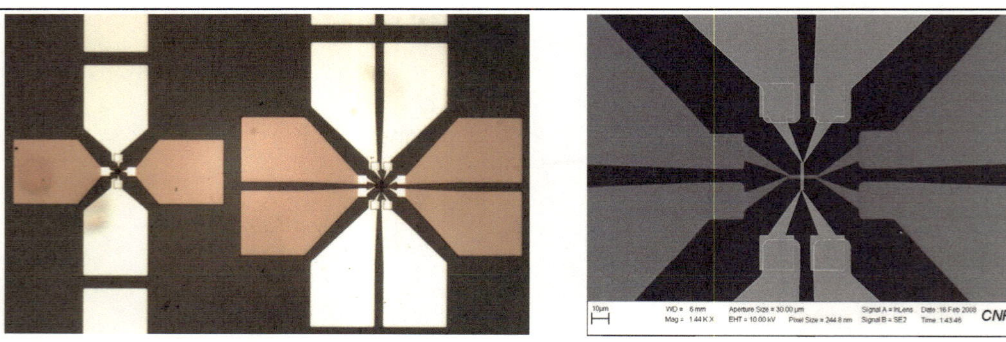

(a) (b)

Fig. 12. (a) The top view of 2×2 and 4×4 nanophotodetector array, where the dark electrodes are front electrodes and the bright ones are back electrodes; (b) The detection region of a 4×4 NPD array.

4.2 Device packaging of nanoscale photodetector array

Packaging is of great importance for the characterization of photonic devices, especially when the size of devices is down to the nanometer scale. Since each NPD device only has a size no more than 1 mm × 1 mm square, to successfully cleave the NPD array into individual pieces the thickness of each NPD array device has to be at least 4~5 times smaller than the width of each NPD array device. Otherwise, the cleaving machine has to cut the device wafer deep in order to make a successful cleaving, which will easily cause the damage of the tiny NPD device and more debris during cleaving. Therefore, before cleaving, the whole wafer that carries the NPD array devices were polished down to ~150 μm thick.

In addition to cleaving process, bonding is also very important to the NPD device's performance. The quality of bonding will have a great influence on the electrical performance and thermal conductivity of NPD device, especially for devices as small as a few hundred nanometers. In practice, a sliver paste is used to connect the NPD electrodes and extended electrodes. Fig. 13 shows the schematic of a bonded 4x4 NPD array using silver paste.

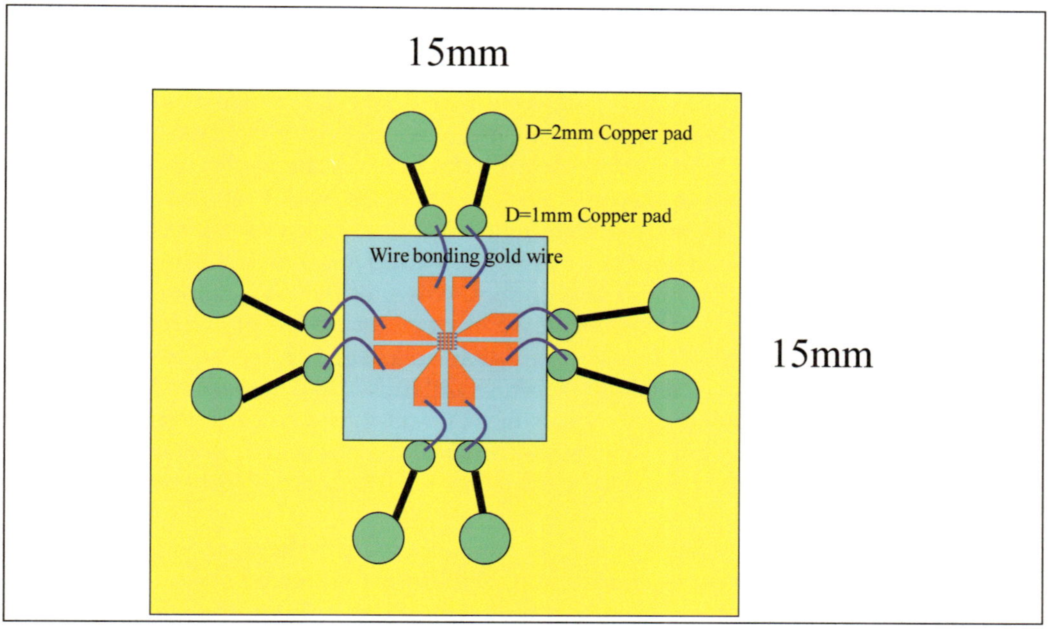

Fig. 13. The schematic of a bonded 4x4 NPD array.

5. Characterization and results

5.1 Photoresponse characterization

The electrical characterization of NPD array, which have pixel size of a few hundred nanometers, is different from micrometer scale MSM detectors. The primary reason is the small pixel size of NPD array. There are two main concerns on NPD characterization listed below:

(a) The conventional planar IDT MSM structure photodetector has tens of or even more pixels working together, which leads to a large detection area of hundreds of square micrometers or even larger. As a result, it could generate relative large photoresponse signals of typically around microamperes level. On the contrary, in order to enable the pixel addressing function, each pixel in NPD array has to work individually. Since the NPD pixel has a size of a few hundred nanometers and detection area is only a fraction of square micrometers, the generated photoresponse signal by NPD pixel is very small. For instance, the estimated signal of a single NPD array pixel could be as small as tens of picoamperes. However, the advantage of NPD array pixels with small size is its corresponding low dark current.

(b) Since the NPD array has pixel size of a few hundred nanometers, the corresponding detection area is only a fraction of square micrometers, which is already beyond the focusing limit of the optical objective lens used to focus the illuminating light onto the NPD array. The large spot size onto the pixel array will inevitably cause errors in estimations of input illuminating light power.

To overcome above difficulties in NPD characterization, a high sensitivity Stanford Research DSP 830 Lock-In Amplifier was used to directly measure photocurrent generated by NPD pixel. Before the measurement using the lock-in, the instrument was calibrated using a Newport 818-IR detector head. The SR 830 lock-in shows an accurate current measurement of 2 pA, which already reached the detection limit of Newport 818-IR detector head. Therefore, the calibration shows the measurement system has a detection capability of larger than 2 pA, which is high enough for the NPD array's low photoresponse.

For the focusing light spot issue, however, there is no ready instrument available for such small light spot focusing. One solution is to use near-field scanning optical microscope (NOSM) system, where the light coming out from the NSOM scanning tip could reach such small size. Nevertheless, the throughput of the light out of the NSOM tip is as low as 10^{-6} for a 50 nm wide aperture NSOM tip. Considering the light coupling efficiency of at most 20% into the NSOM fiber, the throughput of a NSOM tip with 50 nm wide aperture is only on the level of 10^{-7}, which means that for an input optical power of 100 mW, output through a 50 nm aperture NSOM tip is only a few nW. If we use a larger NSOM tip, which has the size comparable to our smallest NPD array detection area, i.e. 100 nm square, the strong diffraction by the small size of NOSM tip aperture will cause large amount of light scattered into NPD array pixel surrounding materials of BCB and then into detection area of NPD array. In practice, a large focusing light spot by an objective lens is used to characterize NPD pixel photoresponse with only one NPD pixel is biased. The focused light spot is estimated to be 2 μm in diameter.

The schematic of the NPD photoresponse measurement setup is shown in Fig. 14. A He-Ne laser is used for the alignment of whole setup. The IR light source will be introduced by a foldable mirror. A high reflection coating filter is used as a beam splitter to direct the illumination light to the NPD device through a 50X objective lens. An IR-CCD camera is put right on the top of the beam splitter to monitor the position of focused light spot. The NPD sample will be fixed onto a 6-dimension adjustable stage for a good alignment to the focused light spot from the objective lens. The bias is applied by a Keithley 2400 Digital Sourcemeter through two probe stations. Stanford Research DSP 830 Lock-In Amplifier is series-connected to the power supply and NPD detection circuit and works in current mode. The output signal from lock-in is input to a Tektronix TDS3032B 300MHz oscilloscope and then collected by a GIPB data collection card through Labview.

To characterize the nano-imaging performance of NPD array, another measurement scheme using near-field scanning optical microscope (NSOM) system is shown in Fig. 15. A Nanonics MultiView 400 Near-Field Scanning Optical Microscope system is used to characterize NPD pixel. Except for a conventional setup for NOSM measurement, a SR830 lock-in amplifier and a Keithley 2400 Sourcemeter are added into the system. Then the output photocurrent signal of NPD device from the lock-in is sent to an auxiliary port of NSOM system, and photocurrent generated by NPD array pixel at each scanning position of NSOM will be recorded along with the scanning position. As a result, the whole system is converted to a near-field scanning photocurrent microscopy system, which is capable of topographical imaging, NSOM imaging and photocurrent mapping simultaneously.

Fig. 16 shows the photo response characterization of sample pixel for a 2×2 channel NPD array versus reverse bias. The wavelength of illuminating light is 1310 nm. The measurement shows a photocurrent of ~735 nA at 3.3 V bias, where the corresponding dark

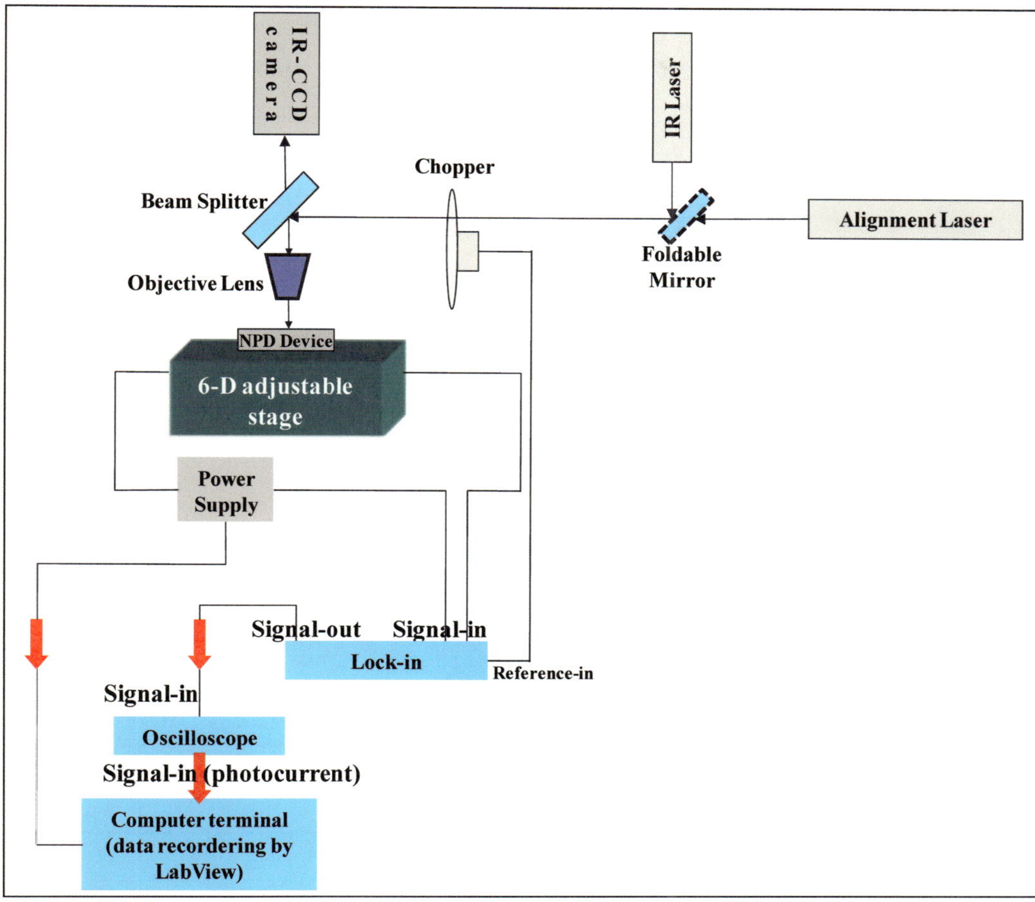

Fig. 14. The schematic of NPD photoresponse measurement setup.

current is ~0.483 nA and more than 1000 times lower than the registered photocurrent. The corresponding responsivity is ~0.28 A/W and quantum efficiency (η) is 26%. The primary reason of relatively low quantum efficiency is due to the blocking effect caused by Au/Ti thin metal electrode for the receiving bottom electrode. The loss of 1.31μm light through a 20 nm thick Au/Ti (10 nm/ 10 nm) by simulation is still as high as 70%. The inset of Fig. 16 shows the calculated responsivity of NPD pixel versus reverse bias, where the highest responsivity registered is 0.28 A/W at 3.3 V bias.

5.2 Near-field Scanning Photocurrent Microscopy (NSPM)

To realize NPD array's pixel addressable function, it's necessary to individually characterize each pixel's performance. However since each pixel of NPD array is nanometer scale, i.e. from 100 nm to 400 nm, it is difficult to characterize each pixel one by one. In this section, one way using NSOM system to qualitatively characterize NPD array and its nano-imaging performance is presented.

Actually, one popular application of NSOM system is to illuminate the scanned surface in near-field region of the light coming out of the NSOM tip. Such application is typically used

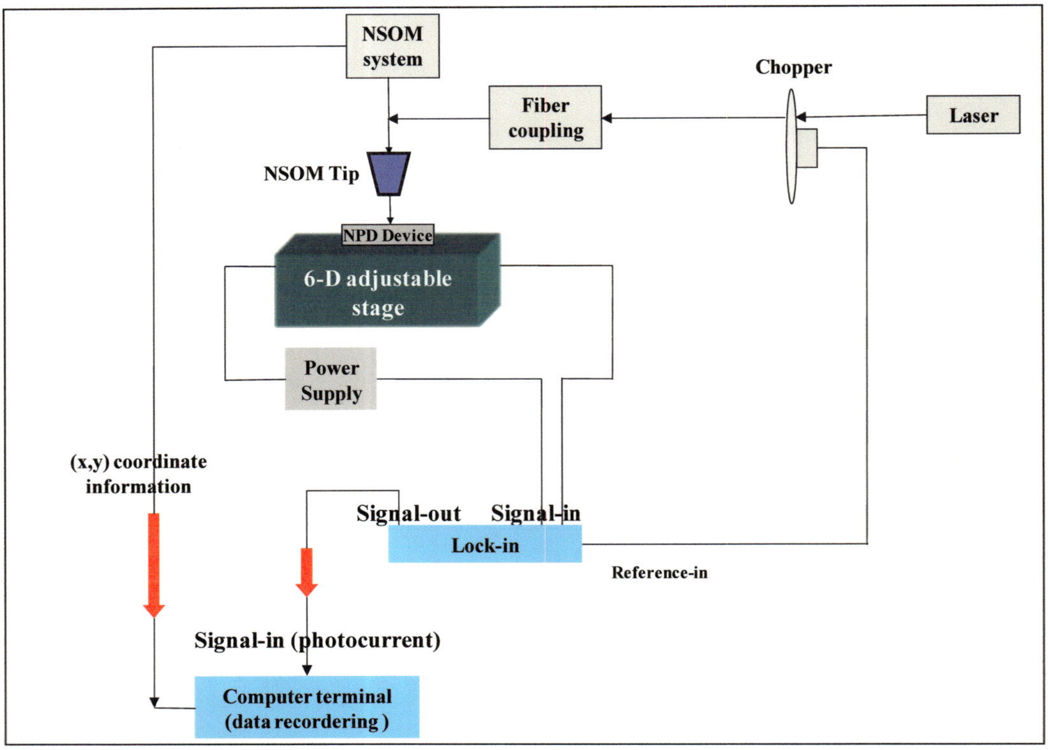

Fig. 15. The schematic of NPD characterization setup based on NSOM system for photocurrent mapping measurement.

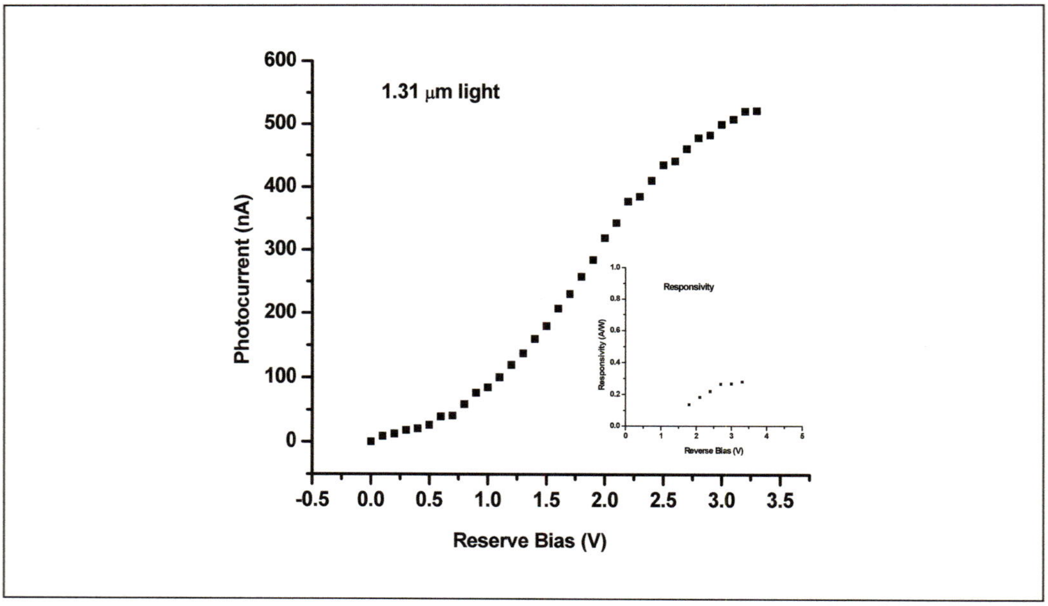

Fig. 16. Photocurrent of one 2×2 NPD pixel versus reverse bias. The pixel is 200 nm wide with 200 nm spacing. Inset, calculated responsivity of the NPD pixel at different bias.

in Biology to detect the fluorescence from single molecule or biological cell by illuminating them individually using NSOM tip. The region illuminated by NSOM tip only depends on the NOSM tip opening. Since NSOM probe tips' aperture generally varies from 50 nm to a few hundreds of nanometers, an illuminated region of 50 nm to a few hundred nanometer diameter circle is expected on the sample surface. To characterize NPD array using a NSOM system: (1) All electrical circuits should be connected and set up well; (2) A NSOM tip then scans the NPD array surface point by point. To obtain the image, the reflected or transmitted light is detected and recorded at each scanning point via a Photo Multiplier Tube carried in the NSOM instrument; (3) Simultaneously, the corresponding photocurrent generated by each NPD array pixel will also be recorded; (4) Imaging information and photocurrent at each scanning point are then integrated together using computer. As a result, one NSOM scan could generate both the physical imaging and the photocurrent map of the NPD array. By comparing the imaging and corresponding photocurrent map, one could characterize the effective pixel level resolution of the NPD array. The working mechanism of NSPM measurement is further shown in Fig. 17, where a 630 nm He-Ne laser is coupled to the NSOM tip to excite NPD array. In each measurement, one pair of NPD array electrodes, one from the top electrode stripes and the other from the bottom electrode stripes, is biased under 3 V. The NPD pixel at the crossing point of biased electrode stripes is the one to detect light signal from NSOM tip. The photocurrent generated by the biased NPD pixel is filtered and recorded while NSOM tip scanning the NPD array surface to generate the photocurrent map, which is also the image of NSOM tip by the NPD array.

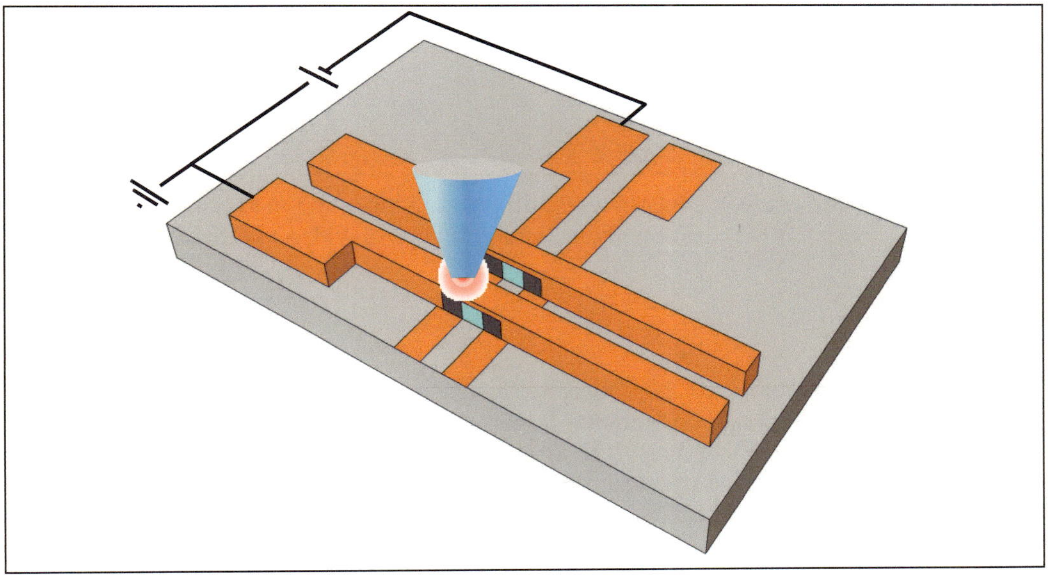

Fig. 17. The schematic of near-field scanning photocurrent mapping measurement of NPD device to characterize the nano-imaging performance by NPD array. The NPD pixel at the crossing point of biased electrode is the one to detect signal.

The NPD pixel under measurement is from a 2×2 array with 250 nm wide pixels and 150 nm spacing. The generated photocurrent map by the biased NPD pixel is shown in Fig. 18. The two pair of dashed lines represent the position of the 2×2 NPD array electrode stripes, where the crossing points are the four NPD pixels. The depth of the color represents the intensity of measured photocurrent generated by the biased NPD pixel, where the brighter the color the higher the photocurrent. The scanning area is 11 μm× 11 μm. The bright spot, shown in Fig. 18, in the 2D photocurrent map shows the image of the NSOM tip detected by the biased NPD array pixel. The measured FWHM diameter of the peak is ~ 390 nm. This means the photocurrent generated by one NPD pixel will drop to less than half of the peak value when the light source moves to the middle position between two adjacent NPD pixels, which indicates this two pixels are distinguishable and the test NPD array has an imaging resolution of 390 nm. Since the imaging resolution of an imaging device could be defined as the sum of one pixel and one spacing, as a nano-scale imaging device the NPD array measured will have an imaging resolution of 400 nm, which well matches our experimental results. As a result, a near-field nano-scale imaging using NPD array has been successfully demonstrated.

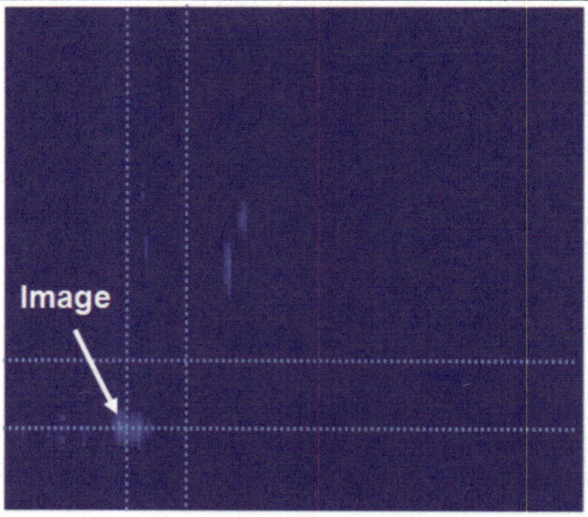

Fig. 18. 2D photocurrent map generated by NPD pixel using home-made NSPM system, where the bright spot is the position of the biased NPD pixel with a FWHM diameter of 390 nm.

6. Conclusion

In summary, a novel nanoscale photodetector array, which could offer new functionalities that cannot be achieved by conventional larger scale photodetectors, is presented in this chapter. The novel design of the NPD array is described as follows: (1) NPD array is designed to have a dimension with nanometer scale, where the smallest pixel is only 100 nm wide; (2) a vertical dual side MSM structure is used to enable pixel addressing; (3) the dimension of the NPD pixel detection area is specifically designed to form a wave-guiding structure to increase detection efficiency.

In addition, FDTD simulation of the NPD array is performed to simulate the photo detection and imaging performance of NPD array, where the photocurrent generated by the illuminated photodetector structure is investigated. The optical coupling effect between NPD pixels is explored by varying the width of NPD pixels and their spacing. Simulation results show that the MLME FDTD model has a good matching to the response of the photosensitive material, which could be used to effectively simulate the photodetection process by photodetectors. The smallest obtainable imaging resolution for 1550 nm wavelength is 150 nm.

Finally, with several newly developed fabrication techniques, NPD array with up to 4×4 array size are successfully realized. The smallest NPD array pixel realized is 100 nm wide with 100 nm spacing. The sample NPD array pixel shows a photo responsivity of ~0.28 A/W and quantum efficiency (η) of 26% for 1.31 μm wavelength light. Using near-field scanning optical microscope, we characterized the nano-imaging performance of the NPD array fabricated. An imaging resolution of ~390 nm, which is ~ 1/4 the designed wavelength of 1550 nm, by NPD array has been successfully demonstrated.

7. References

Biyiklia, N.; Kimukin, I.; Kartaloglu, T. & Aytur, O. (2003). High-speed solar-blind photodetectors with indium-tin-oxide Schottky contacts, *Applied Physics Letters*, Vol. 82, No. 14, Apr. 2003, 2344-2346, ISSN: 0003-6951

Biyikli, N.; Kimukin, I.; Butun, B.; Aytür, O. & Ozbay, E. (2004). ITO-Schottky photodiodes for high-performance detection in the UV-IR spectrum, *IEEE Journal of Selected Topics in Quantum Electronics*, Vol. 10, No. 4, Jul. 2004, 759-765, ISSN: 1077-260X

Biyikli, N.; Kimukin, I.; Aytur, O.; Gokkavas, M.; Unlu, M. S. & Ozbay, E. (2001). 45-GHz bandwidth-efficiency resonant-cavity-enhanced ITO-Schottky photodiodes, *IEEE IEEE Photonics Technology Letters*, Vol. 13, No.7, Jul. 2001, 705-707, ISSN: 1041-1135

Chyi, J.; Wei, T.; Hong, J.; Lin, W. & Tu, Y. (1994). Low dark current and high linearity InGaAs MSM photodetectors, *Electronics Letters*, Vol. 30, No. 4, Feb. 1994, 355-356, ISSN: 0013-5194

DeCorby, R. G.; MacDonald, R. I.; Sharma, R.; Gouin, F.; Noad, J. & Puetz, N. (1997). Frequency-domain demonstration of transit-time-limited, large area InGaP-InP-InGaAs MSM photodetectors, *IEEE Photonics Technology Letters*, Vol. 9, No. 7, Jul. 1997, 985-987, ISSN: 1041-1135

Gao, W.; Berger, P.; Zydzik, G.; O'Bryan, H.; Sivco, D. & Cho, A. (1997). $In_{0.53}Ga_{0.47}As$ MSM photodiodes with transparent CTO Schottky contacts and digital superlattice grading, *IEEE Transactions on Electron Devices,*, Vol. 44, No. 12, Dec. 1997, 2174-2179, ISSN: 0018-9383

Gao, W.; Berger, P.; Hunsperger, R.; Zydzik, G.; Rhodes, W.; O'Bryan, H.; Sivco, D. & Cho, A. (1995). Transparent and opaque Schottky contacts on undoped $In_{0.52}Ga_{0.48}As$ grown by molecular beam epitaxy, *Applied Physics Letters*, Vol. 66, No. 25, Jun. 1995, 3471-3473, ISSN: 0003-6951

Gao, W.; Khan, A.; Berger, P.; Hunsperger, R.; Zydzik, G.; O'Bryan, H.; Sivco, D. & Cho, A. (1994). $In_{0.53}Ga_{0.47}As$ metal-semiconductor-metal photodiodes with transparent cadmium tin oxide Schottky contacts, *Applied Physics Letters*, Vol. 65, No. 15, Oct 1994, 1930-1932, ISSN: 0003-6951

Griem, H. T.; Ray, S.; Freeman, J. L. & West, D. L. (1990). Long-wavelength (1.0–1.6 μm) $In_{0.52}Al_{0.48}As/In_{0.53}(Ga_xAl_{1-x})_{0.47}As/In_{0.53}Ga_{0.47}As$ metal semiconductor metal photodetector, *Applied Physics Letters.*, Vol. 56, No. 11, Mar 1990, 1067-1068, ISSN: 0003-6951

Huang, Y.; Duan, X.; Wei, Q. & Lieber, C. M. (2001). Directed assembly of one-dimensional nanostructures into functional networks, *Science*, Vol. 291, No. 5504, Jan. 2001, 630-633, ISSN: 0036-8075

Hayden, O.; Agarwal, R. & Lieber, M. C. (2006). Nanoscale avalanche photodiodes for highly sensitive and spatially resolved photon detection, *Nature Materials*, Vol. 5, May 2006, 352-354, ISSN: 1476-1122

Huang, Y. & Ho, S. –T. (2006). Computational model of solid-state, molecular, or atomic media for FDTD simulation based on a multi-level multi-electron system governed by Pauli exclusion and Fermi-Dirac thermalization with application to semiconductor photonics. *Optics Express*, Vol. 14, No. 8, Apr. 2006, 3569-3587, ISSN: 1094-4087

Khoo, E. H; Ho. S. –T.; Ahmed, I; Li, E. P. & Huang, Y. (2008). 3D Modeling of photonic devices using dynamic thermal electron quantum medium finite-differenr time-domain (DTEQM-FDTD) method. *2008 IEEE/LEOS International Conference on Optical MEMS and Nanophotonics*, Freiburg, Germany, Aug. 11-14, 2008.

Kim, S.; Mohseni, H.; Erdtmann, M.; Michel, E.; Jelen, C. & Razeghi, M. (1998). Growth and characterization of InGaAs/InGaP quantum dots for midinfrared photoconductive detector, *Applied Physics Letters*, Vol. 73, No. 7, Aug. 1998, 963-965, ISSN: 0003-6951

Kim, K. Y.; Liu, B.; Huang, Y. & Ho, S. –T. (2008). Simulation of photodetection using finite-difference time-domain method with application to near-field imaging based on nanoscle semiconductor photodetector array. *Optical and Quantum Electronics*, Vol. 40, No. 5-6, Apr. 2008, 343-347, ISSN: 0306-8919

Lang, J. & Herrick, H. (1985). Deep level impurities: a possible guide to prediction of band edge discontinuities in semiconductor heterojunction, *Physical Review Letters*, Vol. 55, No. 13, Sep. 1985, 1414-1417, ISSN: 0031-9007

Liu, B.; Huang, Y.; Xu, G. & Ho, S.T. (2008). Sub-10nm nanolithography and pattern transfer on semiconductor compounds using spin-coatable ZrO2 resist by inductively coupled plasma reactive ion etching (ICP-RIE), *Nanotechnology*, Vol. 19, Mar. 2008, 155303, ISSN: 0957-4484

Liu, B. & Ho, S.T. (2008). Sub-100nm nanolithography and pattern transfer on semiconductor compounds using sol-gel derived TiO2 resist, *Journal of The Electrochemical Society*, Vol. 155, No. 5, Mar. 2008, 57-60, ISSN: 0013-4651

Loualiche, S.; Corre, A.; Ginudi, A.; Henry, L.; Vaudry, C. & Clerot, F. (1990). Pseudomophic GalnP Schottky diode and MSM detector on InP, *Electronics Letters*, Vol. 26, No. 7, Mar. 1990, 487-488, ISSN: 0013-5194

Maier, S. A.; Kik, P. G.; Atwater, H. A.; Meltzer, S.; Harel, E.; Koel, B. E. & Requicha, A. A. G. (2003). Local detection of electromagnetic energy transport below the diffraction limit in metal nanoparticle plasmon waveguides, *Nature Materials*, Vol. 2, Apr. 2003, 229-232, ISSN: 1476-1122

Seo, J.; Ketterson, A.; Ballegeer, D.; Cheng, K.; Adesida, I.; Li, X. & Gessert, T. A (1992). Comparative study of metal-semicondutor-metal photodetectors on GaAs with Indium-Tin-Oxide and Ti/Au electrodes, *IEEE Photonics Technology Letters*, Vol. 8, No. 4, Aug. 1992, 888-890, ISSN: 1041-1135

Yang, C.; Barrelet, C. J.; Capasso, F. & Lieber, C. M. (2006). Single p-type/intrinsic/n-type silicon nanowires as nanoscale avalanche photodetectors, *Nano Letters*, Vol. 6, No. 12, Nov. 2006, 2929-2934, ISSN: 1530-6984

Yee, K. S. (1966). Numerical solution of initial boundary value problems involving Maxwell's equations in isotropic media. *IEEE Transactions on Antennas and Propagation*, Vol. 14, No. 3, May 1966, 802-807, ISSN: 0018-926X

Zhao, K.; Zhang, A.; Lo, Y. & Farr, W. (2007). InGaAs single photon avalanche detector with ultralow excess noise, *Applied Physics Letters*, Vol 91, 081107, Aug. 2007, ISSN: 0003-6951

16

Spontaneous and Stimulated Transitions in Impurity Dielectric Nanoparticles

K.K. Pukhov, Yu.V. Orlovskii and T.T. Basiev
General Physics Institute, Russian Academy of Sciences
Russia

1. Introduction

In recent years, great interest has been expressed by researchers in the optical properties exhibited by nanomaterials, including theoretical and experimental studies of the spontaneous lifetime of optical centers in nanosized samples (Christensen et al., 1982; Meltzer et al., 1999; Zakharchenya et al., 2003; Manoj Kumar et al., 2003; Vetrone et al., 2004; Chang-Kui Duan et al., 2005; Dolgaleva et al., 2007; Guokui Liu & Xueyuan Chen, 2007; Song & Tanner, 2008) A change in the spontaneous lifetime of optical centers (OCs) in nanoobjects as compared to bulk materials is of considerable interest for both fundamental physics and practical applications in the field of laser materials and phosphors. For example, the increased lifetime of a metastable level in a lasing medium makes it possible, by increasing the pump-pulse duration several times, to reduce the power and cost of the diode laser-pump source and superluminescence losses while keeping the output-radiation energy and power intact. The adequate theoretical interpretation of the experimental results is of primary importance at the current stage of investigations. It is of great interest to derive a formula describing the spontaneous decay rate of an excitation in a nanosized object and reveal its differences from the corresponding expression for the bulk sample.

The existence of spontaneous emission postulated in 1917 by Einstein in his quantum theory of the interaction between the equilibrium radiation and matter (Einstein, 1917). It is shown in this paper that the statistical equilibrium between matter and radiation can only be achieved if spontaneous emission exists together with the stimulated emission and absorption. The quantum-mechanical expression for the Einstein coefficient of spontaneous emission A equal to the probability of spontaneous emission from a two-level atom in a vacuum has been obtained by (Dirac, 1927; Dirac, 1982). In 1946, (Purcell, 1946) it is shown that the spontaneous emission probability can drastically increase if the radiating dipole is placed in a cavity (see also (Oraevskii, 1994; Milonni, 2007) and references therein). The inverse phenomenon, i.e., the inhibition of the spontaneous emission, can take place in three-dimensional periodic dielectric structures (Yablonovitch, 1987). Variations in the probability of the spontaneous emission from optical centers near the planar interface of the dielectrics have been the subject of active studies since the 1970s (Drexhage, 1970; Kuhn, 1970; Carnigia & Mandel, 1971; Tews, 1973; Morawitz & Philott, 1974; Agarwal, 1975; Lukosz & Kunz, 1977; Chance et al., 1978; Khosravi & Loudon, 1991; Barnes,1998).

Modifications of the spontaneous emission from OCs located in the vicinity of a metal mirror also has been analyzed (Amos & Barnes, 1997; Brueck et al., 2003). Chew (Chew,

1987; Chew, 1988) considered the modification of the spontaneous emission from an optical center inside and outside the dielectric sphere by modeling the optical center with an oscillating dipole. His analytical results were confirmed later by Fam Le Kien et al. (Fam Le Kien et al., 2000) devoted to the spontaneous emission from a two-level atom inside a dielectric sphere and by Glauber and Lewenstein (Glauber & Lewenstein, 1991). Klimov et al., 2001, considered the problem of the spontaneous emission from an atom near a prolate spheroid. The problem of the spontaneous emission from an atom in the vicinity of a triaxial nanosized ellipsoid is analyzed in a recent paper (Guzatov & Klimov, 2005). Of course, these short overviews far from being comprehensive.
It is well known now that spontaneous emission rate is not necessarily a fixed and an immutable property of optical centers but can be controlled.
The Chapter's central theme is the radiative characteristics of the small-radius optical centers (dopant ions of transition elements) in the subwavelength nanocrystals embedded in a dielectric medium. Our main aim is to provide answer the question "How the expressions derived for the radiative characteristics of optical centers in a bulk material should be modified upon changing over to a nanoobject?".
The rest of the Chapter is organized as follows. In Section 2 expression for the spontaneous radiative decay rate of OCs in a bulk crystal is presented and problem of the local field correction factor is briefly discussed. The expressions for the spontaneous radiative decay rate of OCs in the spherical nanocrystals are presented in Section 3. In Section 4 the expressions for the integrated emission and absorption cross - sections for spherical nanoparticles are given. The expressions for the spontaneous radiative decay rate of OCs in the ellipsoidal nanocrystals are presented in Section 5. Section 6 discusses the applicability of the Judd-Ofelt equation for nanoparticles. In Section 7 the experimental confirmation of the model for spontaneous radiative decay rates of rare-earth ions in the crystalline spherical nanoparticles of cubic structure embedded into different inert dielectric media is presented. The Chapter concludes in Section 8 showing directions for future research and conclusions.

2. Spontaneous radiative rate in a bulk crystal

The coupling between atom and the electric field in the dipole approximation is given by the electric-dipole interaction Hamiltonian

$$H_{\text{int}} = -\hat{\mathbf{d}}\,\hat{\mathbf{E}} \tag{2.1}$$

where $\hat{\mathbf{d}}$ is operator of the dipole moment and $\hat{\mathbf{E}}$ is the electric field operator, evaluated at the dipole position. In vacuum

$$\hat{\mathbf{E}}^{(vac)} = i\sum_{\boldsymbol{k},\sigma}\sqrt{\frac{2\pi\hbar\omega_k}{V}}\,e_{\boldsymbol{k},\sigma}\left[\mathrm{a}_{\boldsymbol{k},\sigma} - \mathrm{a}^{+}_{\boldsymbol{k},\sigma}\right] \tag{2.2}$$

where $\boldsymbol{k}$ is wave vector; σ denotes the state of polarization; $\mathrm{a}_{\boldsymbol{k},\sigma}$ and $\mathrm{a}^{+}_{\boldsymbol{k},\sigma}$ are the photon destruction and creation operators for field eigenmodes, which specified by indices ($\boldsymbol{k}$, σ); $e_{\boldsymbol{k},\sigma}$ is the polarization vector; V is the quantization volume. Photon frequency ω_k is connected with wave number $k = |\boldsymbol{k}|$ by the linear dispersion relation $\omega_k = c_0 k$ where c_0 is light velocity in vacuum. Fermi's golden rule leads to the following expression for the electric-dipole spontaneous emission rate in free space (Dirac, 1982):

$$A_0 = \left(4\pi^2 \omega |\mathbf{d}|^2 / 3\hbar \right) \rho_{vac}(\omega) . \tag{2.3}$$

Here ω is frequency of transition from excited atomic state i to lower-energy state j;

$$\rho_{vac}(\omega) = \omega^2 / \pi^2 c_0^3 \tag{2.4}$$

is the photon density of states in vacuum; $|\mathbf{d}|^2 = d_x^2 + d_y^2 + d_z^2$,where $d_a = \langle i | \hat{d}_a | j \rangle$ ($\alpha = x, y, z$) are the electronic matrix elements of the electric-dipole operator $\hat{\mathbf{d}}$ between the states i and j. The quantization of the electromagnetic field in a dielectric medium was first carried out by Ginzburg (Ginzburg, 1940). The macroscopic electric-field operator in a linear, isotropic, and homogeneous medium is given by

$$\hat{\mathrm{E}} = i \sum_{k,\sigma} \sqrt{\frac{2\pi\hbar\omega_k}{V\varepsilon}} e_{k,\sigma} \left[\mathrm{a}_{k,\sigma} - \mathrm{a}_{k,\sigma}^{+} \right] \tag{2.5}$$

with dielectric function $\varepsilon(\omega_k)$ and the dispersion relation $\omega_k = c_0 k/n$, where n is the refractive index of a dielectric. This dispersion relation results in changes in the photon density of states of a dielectric:

$$\rho(\omega) = n^3 \rho_{vac}(\omega) \tag{2.6}$$

Considering Eqs. (2.5) - (2.6) one can neglect the local-field effect for a moment and obtain for the electric-dipole spontaneous emission rate in the continual approximation (Nienhuis & Alkemade, 1976):

$$A = nA_0 \tag{2.7}$$

However, in general case the electric-dipole interaction Hamiltonian has the form

$$H_{\mathrm{int}} = -\hat{\mathbf{d}}\, \hat{\mathbf{E}}^{(loc)} , \tag{2.8}$$

where $\hat{\mathbf{E}}^{(loc)}$ is the local electric field operator acting at the position of the optical center. The local electric field in a crystal differs from the macroscopic electric field in a crystal. For this reason the expression for the electric-dipole spontaneous radiative rate of the small-radius optical centers in a bulk crystal is given by (Lax, 1952; Fowler & Dexter, 1962; Imbush & Kopelman, 1981)

$$A_{bulk} = n_{cr} (E^{(loc)} / E^{(cr)})^2 A_0 = n_{cr} f_L A_0 . \tag{2.9}$$

Here n_{cr} is the refractive index of a crystal; $E^{(loc)}$ and $E^{(cr)}$ are the strengths of the microscopic and macroscopic electric fields acting at the position of the optical center, respectively. Ratio

$f_L = (E^{(loc)} / E^{(cr)})^2$ is so called the local-field correction factor. In all existing local-field models f_L is a function of the crystal refractive index n_{cr} (see the comprehensive review of all currently available local-field models in paper (S. F. Wuister et al., 2004)); i.e., $(E^{(loc)} / E)^2 = f_L(n_{cr})$ and $f_L(1) = 1$. Most commonly used the local-field models are models of real cavity and virtual cavity (Rikken & Kessener, 1995). In case of an empty, real spherical cavity,

$$f_L(n) = \left[\frac{3n^2}{2n^2+1}\right]^2 . \tag{2.10}$$

In case of a virtual cavity (Lorentz model)

$$f_L(n) = \left[\frac{n^2+2}{3}\right]^2 . \tag{2.11}$$

The next sections will answer the question: How expression (2.9) for the electric-dipole spontaneous radiative rate of the small-radius optical centers in a bulk crystal should be modified upon changing over to a nanoobject?

3. Spontaneous radiative rate in the spherical nanocrystal

We shall refer to nanocomposite for dielectric nanocrystals embedded into different homogeneous dielectric media with refraction index n_{med}. The nanocrystals are assumed to be small enough compared with wavelength λ, but large compared with lattice constant a_L, so that the nanocrystals can be characterized by refraction index, which coincide with that of bulk crystal n_{cr}. The light wave propagates through the nanocomposite with amplitude E and a velocity c_0/ n_{eff}. The electric field E is macroscopic field averaged over volumes large enough as compared to the scales of inhomogeneities:

$$E = (1-x)E^{(med)} + x\, E^{(cr)} \tag{3.1}$$

where x is the volume fraction of nanocrystals in the medium (filling factor), $E^{(med)}$ and $E^{(cr)}$ are macroscopic fields in the dielectric medium and nanocrystals, respectively (Bohren & Huffman, 1998). So, expression (2.9) should be replaced by

$$A_{nano} = n_{eff} (E^{(loc)} / E)^2 A_0 . \tag{3.2}$$

After some obvious transformations of this expression we obtain (Pukhov et al., 2008; Basiev et al., 2008)

$$A_{nano} = n_{eff} (E^{(cr)} / E)^2 (E^{(loc)} / E^{(cr)})^2 A_0 , \tag{3.3}$$

or,

$$A_{nano} = n_{eff} f_N f_L(n_{cr}) A_0 , \tag{3.4}$$

where

$$f_N = (E^{(cr)} / E)^2 \tag{3.5}$$

is the correction factor that accounts for the difference between the macroscopic electric field $E^{(cr)}$ at the position of the optical center and the macroscopic electric field E in the nanocomposite. We assume here that the local-field correction factor is the same as in the bulk crystal because of macroscopic size of nanocrystals. This assumes that the microscopic surrounding of the optical centers is the same in a nanocrystal and in a bulk crystal. Of course, it is not valid for the optical centers located near the nanocrystal surface at distances smaller than perhaps ten lattice constant (Kittel, 2007). The arguments in support of the inference that the correction f_N differs from unity were clearly and thoroughly described by Yablonovitch et al. (Yablonovitch et al., 1988). Here, we will not repeat these arguments and note only that relationship (3.4) differs from the corresponding expression given by Yablonovitch et al. (Yablonovitch et al., 1988). The difference lies in the appearance of the factor f_L in relationship (3.4).
At last, we have (Pukhov et al., 2008; Basiev et al., 2008)

$$A_{nano} = n_{eff} f_N f_L(n_{cr}) A_0 = (n_{eff} / n_{cr}) f_N n_{cr} f_L(n_{cr}) A_0 = (n_{eff} / n_{cr}) f_N A_{bulk} \tag{3.6}$$

and for the A_{nano}/A_{bulk} get the following expression

$$A_{nano}/A_{bulk} = (n_{eff}/n_{cr}) f_N \,. \tag{3.7}$$

An important consequence of relationship (3.6) is that the ratio A_{nano}/A_{bulk} can be estimated without recourse to a particular local-field model. The problem of the theoretical determination of the ratio A_{nano}/A_{bulk} is reduced to the problem of determining the correction $f_N = (E^{(cr)}/E)^2$ (and, of course, to the problem of determining the effective refractive index n_{eff}).
Let us calculate the correction $f_N = (E^{(cr)}/E)^2$ for subwavelength spherical nanocrystals that have the radius R satisfying the condition $a_L \ll 2R \ll \lambda/2\pi$. The electrostatic approximation is applicable at this condition as it follows from the Lorenz-Mie solution to Maxwell's equations. In framework of the electrostatic approximation the electric field $\mathbf{E}^{(cr)}$ within a dielectric sphere placed in the external electric field $E^{(med)}$ is equal to (Landau & Lifshitz, 1984)

$$\mathbf{E}^{(cr)} = [3/(\varepsilon + 2)]\mathbf{E}^{(med)} \tag{3.8}$$

where $\varepsilon = \varepsilon_{cr}/\varepsilon_{med} = n_{cr}^2/n_{med}^2$ is relative permittivity.
On the lines of the Maxwell Garnett theory (Maxwell Garnett, 1904; Maxwell Garnett, 1906) we obtain

$$f_N^{spher} = \{3 / [2 + \varepsilon - x(\varepsilon - 1)]\}^2 \,. \tag{3.9}$$

So, the spontaneous emission rate of a two-level atom in the spherical nanoparticle is given by expression (Pukhov et al., 2008; Basiev et al., 2008)

$$A_{nano}^{spher}/A_{bulk}=\frac{n_{eff}}{n_{cr}}\left[\frac{3}{2+\varepsilon-x(\varepsilon-1)}\right]^2. \quad (3.10)$$

(Although for definiteness, we consider nanocrystals, all the inferences refer equally to nanoparticles from a dielectric material with the refractive index n_{cr}).
The Eq. (3.1) together with relation

$$P=(1-x)P^{(med)}+x\,P^{(cr)}, \quad (3.11)$$

where P, $P^{(med)}$ and $P^{(cr)}$ are average polarizations on nanocomposite, medium and nanocrysral, lead to well known Maxwell Garnett mixing rule for ε_{eff} (Maxwell Garnett, 1904; Maxwell Garnett, 1906):

$$\varepsilon_{eff}=n_{eff}^2=\varepsilon_{med}\left[1+\frac{3x\beta}{1-x\beta}\right], \quad (3.12)$$

where $\beta=(\varepsilon-1)/(\varepsilon+2)$. The Maxwell Garnett mixing rule predicts the effective permittivity ε_{eff} of a nanocomposite where homogeneous spheres of isotropic permittivity ε_{cr} dilutely mixed into isotropic medium with permittivity ε_{med} (see book (Bohren & Huffman, 1998) for details). As it can be seen from Eq. (3.10) and Eq. (3.12), the spontaneous emission rate in nanocomposite is enhanced for $\varepsilon<1$ and inhibited for $\varepsilon>1$.
From the expressions (3.10) and (3.12), for $x\to 1$, we obviously have the limiting case of $A_{nano}^{spher}\to A_{bulk}$. In the limit $x\to 0$, we obtain

$$A_{nano}^{spher}/A_{bulk}=\frac{n_{med}}{n_{cr}}\left[\frac{3}{2+\varepsilon}\right]^2. \quad (3.13)$$

The derived expression is consistent with both the result obtained by Yablonovitch et al. (Yablonovitch et al., 1988) and result derived by Chew (Chew, 1988) also without regard for the local-field effect. Thereby, formula of Eq. (3.10) yields the correct result for $x\to 1$ and fit the results of Refs. (Chew, 1988) and (Yablonovitch et al., 1988) for $x\to 0$. It is not yet clear whether this formula is applicable for the intermediate values of filling factors x as the experimental data are scarce.
It should be mentioned that in some papers (Meltzer et al., 1999; Zakharchenya et al., 2003; Manoj Kumar et al., 2003; Vetrone et al., 2004; Chang-Kui Duan et al., 2005; Dolgaleva et al., 2007; Liu et al, 2008) expression (2.9) for the spontaneous radiative rate in a bulk crystal is transformed into the formula for the decay rate of an optical center in a crystalline nanoparticles A_{nano} by direct replacing the refractive index of the crystal n_{cr} by the effective refractive index n_{eff} and the local-field correction $f_L(n_{cr})$ by the corresponding correction $f_L(n_{eff})$ with the use of a particular local-field model:

$$A_{nano}=n_{eff}f_L(n_{eff})A_0\,. \quad (3.14)$$

This leads to some arbitrariness in the interpretation of experimental data owing to the choice of the particular expression for the local-field correction $f_L(n)$ (this problem is discussed in the paper (Dolgaleva et al., 2007).

For the ratio between the excitation lifetimes of an optical center in a nanoparticle and a bulk crystal, expression (3.13) can be rearranged to give

$$\tau_{nano} / \tau_{bulk} = \frac{n_{cr}}{n_{med}} \left[\frac{2+\varepsilon}{3} \right]^2 \qquad (3.15)$$

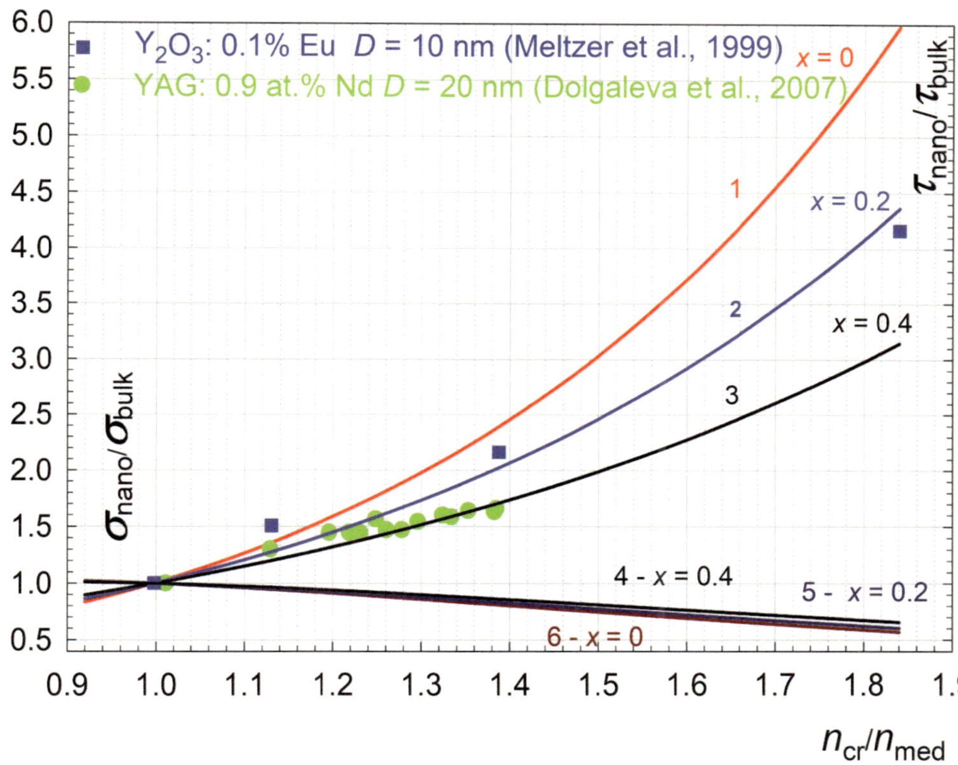

Fig. 1. (*1–3*) Theoretical dependences of the ratio τ_{nano}/τ_{bulk} (the right axis) on the ratio n_{cr}/n_{med} for crystalline matrices with volumefractions (*1*) $x \to 0$, (*2*) $x = 0.2$, and (*3*) $x = 0.4$. (*4–6*) Theoretical dependences of the ratio $\sigma_{nano}/\sigma_{bulk}$ (the left axis) on the ratio n_{cr}/n_{med} with volume fractions (*4*) $x = 0.4$, (*5*) $x = 0.2$, and (*6*) $x \to 0$. Dependences of the measured ratio of the decay time in a nanocrystal to the decay time in a bulk crystal τ_{nano}/τ_{bulk} on the ratio n_{cr}/n_{med} for the $^4F_{3/2}$ metastable level of Nd^{3+} ions in the YAG crystalline matrix ($n_{cr} = 1.82$) (Dolgaleva et al., 2007) (circles) and the $^5D_0$ metastable level of Eu^{3+} ions in the Y_2O_3 crystalline matrix ($n_{cr} = 1.84$) (Meltzer et al., 1999) (squares).

The calculations have demonstrated that, for $x = 0$ and $n_{med} = 1$ (one nanoparticle is suspended in air), the lifetime τ_{nano} of excitation of an optical center in a nanoparticle can increase as compared to the corresponding lifetime τ_{bulk} in a bulk crystal, for example, in Y_2O_3 (n_{cr}/n_{med} = 1.84) and YAG (n_{cr}/n_{med} = 1.82), by a factor of approximately 6 (Fig. 1, curve *1*). According to expressions (3.10) and (3.12) , an increase in the volume fraction x leads to a decrease in the ratio τ_{nano}/τ_{bulk} (Fig. 1, curves 2, 3).

4. Integrated emission and absorption cross - section

Apart from the lifetime of optical centers, the integrated emission and absorption cross - sections are important characteristics of laser materials. The integrated cross - section of the electric dipole emission in the band $i\rightarrow j$ for a bulk material can be represented in the form (Fowler & Dexter, 1962)

$$\sigma^{em}_{bulk}(i \rightarrow j) = A_{bulk}(i \rightarrow j) / [8\pi c_0 n^2_{cr} \bar{\nu}^2], \tag{4.1}$$

where $A_{bulk}(i \rightarrow j)$ is the probability of spontaneous decay in the channel $i\rightarrow j$ for a bulk crystal, $\bar{\nu}$ is the average energy of the transition $i\rightarrow j$ (in cm^{-1}).
It is evident that, in order to determine the integrated cross - section of the electric dipole emission in the band $i\rightarrow j$ for a nanocrystal, it is necessary to replace the probability of spontaneous decay $A_{bulk}(i \rightarrow j)$ for the bulk crystal by the probability of spontaneous decay $A_{nano}(i \rightarrow j)$ for the nanocrystal and the refractive index n_{cr} by the effective refractive index n_{eff} in relationship (4.1). As a result, we obtain (Pukhov et al., 2008; Basiev et al., 2008)

$$\sigma^{em}_{nano}(i \rightarrow j) = A_{nano}(i \rightarrow j)/[8\pi c_0 n^2_{eff} \bar{\nu}^2]. \tag{4.2}$$

After substituting the expression

$$A_{nano}(i \rightarrow j) = \frac{n_{eff}}{n_{cr}} f_N A_{bulk}(i \rightarrow j), \tag{4.3}$$

which was derived in much the same manner as expression (3.6) into relationship (4.2) , we find (Pukhov et al., 2008; Basiev et al., 2008)

$$\sigma^{em}_{nano} = \frac{n_{cr}}{n_{eff}} f_N \sigma^{em}_{bulk}. \tag{4.4}$$

The same relationship holds true for the integrated cross - section of the electric dipole absorption; i.e., the integrated cross - sections of electric dipole processes of all types are described by the expression (Pukhov et al., 2008; Basiev et al., 2008)

$$\sigma_{nano} = \frac{n_{cr}}{n_{eff}} f_N \sigma_{bulk}. \tag{4.5}$$

In the special case of spherical nanoparticles, substitution of relationship (3.9) for the correction f_N^{spher} into expression (4.5) gives

$$\sigma_{nano} = \frac{n_{cr}}{n_{eff}}\left[\frac{3}{2+\varepsilon - x(\varepsilon - 1)}\right]^2 \sigma_{bulk} \tag{4.6}$$

It is worth noting that the factor n_{cr}/n_{eff} in expressions (4.4) – (4.6) is the reciprocal of the factor n_{eff}/n_{cr}, which enters into the right-hand sides of relationships (3.7), (3.10), and (4.3). As a consequence, we have (Pukhov et al., 2008; Basiev et al., 2008)

$$\sigma_{nano}/\sigma_{bulk} = \left[\frac{n_{cr}}{n_{eff}}\right]^2 [A_{nano}/A_{bulk}] = \left[\frac{n_{cr}}{n_{eff}}\right]^2 [\tau_{bulk}/\tau_{nano}], \tag{4.7}$$

where $\tau_{\text{bulk}} = 1/A_{\text{bulk}}$ and $\tau_{\text{nano}} = 1/A_{\text{nano}}$.

It should be noted that, for example, at the refractive index $n_{cr} = 1.82$ (YAG), an increase in the time of radiative decay of optical centers in nanoparticles in an aerosol by a factor of 5 (as compared to that of the bulk crystal) results in a decrease in the corresponding emission cross - section by only 42% (Fig. 1, curves *1*, *6*). Moreover, the effect of an increase in the volume fraction x of nanoparticles on the decrease in the ratio $\sigma_{nano}/\sigma_{bulk}$ becomes considerably weaker as compared to that of the ratio τ_{nano}/τ_{bulk} (compare the changes in the ratios τ_{nano}/τ_{bulk} (Fig. 1, curves *1* - 3) and $\tau_{nano}/\sigma_{bulk}$ (Fig. 1, curves *4* – *6*)). An insignificant decrease in the pump absorption and emission cross - sections cannot bring about a negative effect on the laser medium, whereas a fivefold increase in the lifetime at the same pump power makes it possible to increase the product $\sigma_{\text{abs}}\,\tau$ and, therefore, inversion accumulated in a laser generator or an amplifier by a factor of 5. Therefore, an increase in the lifetime of metastable levels in laser media makes it possible to decrease the power and the cost of a diode laser pumping source and to reduce the superluminescence losses without changes in the energy and power of the output emission owing to a several fold increase in the time of pulse pumping.

5. Influence of the shape of samples: Ellipsoidal nanoparticles

5.1 An isolated ellipsoidal nanoparticle ($x \to 0$)

We now analyze the influence of the shape of nanoparticles on the decay rate of optical centers in subwavelength ellipsoidal dielectric nanoparticles as an example. The mathematical complication arising in the analysis of ellipsoidal nanoparticles lies in the fact that the electric field $E^{(cr)}$ inside an ellipsoidal dielectric nanoparticle placed in the external electric field $E^{(med)}$ is not parallel to the field $E^{(med)}$ (Landau & Lifshitz, 1984). Let us consider this problem in more detail. First, we shall restrict our consideration to a special case $x \to 0$ (an isolated ellipsoid). In this case $n_{eff} \to n_{med}$ and $E^{(med)} \to E$. The electric fields $E^{(cr)}$ and E are related by the linear relation $\mathbf{E}^{(cr)} = \hat{\mathbf{g}}\mathbf{E}$ (Landau & Lifshitz, 1984) with the tensor $\hat{\mathbf{g}}$ principal values given by

$$g_\alpha = 1/[1+(\varepsilon - 1)N_\alpha] \quad (\alpha = a, b, c) \tag{5.1}$$

where a, b, c are the principal axes of the ellipsoid and N_α are the depolarization factors

$(N_a + N_b + N_c = 1)$. Generally $g_a \neq g_b \neq g_c$.
The depolarization factors N_a are expressed through the elliptic integrals (Landau & Lifshitz, 1984):

$$N_\alpha = \frac{abc}{2} \int_0^\infty \frac{ds}{(s+\alpha^2)\sqrt{(s+a^2)(s+b^2)(s+c^2)}} \tag{5.2}$$

As a result, the Hamiltonian of interaction of electric field $\mathbf{E}^{(cr)}$ with dipole moment $\mathbf{d}$ takes the form (Pukhov, 2009)

$$H_{\text{int}} = -\hat{\mathbf{E}}^{(cr)}\hat{\mathbf{d}} = -\hat{\mathbf{d}}\hat{\mathbf{g}}\hat{\mathbf{E}} = -i\sum_{\boldsymbol{k},\sigma}\sqrt{\frac{2\pi\hbar\omega_k}{V\varepsilon_{med}}}\,\hat{\mathbf{d}}\hat{\mathbf{g}}e_{\boldsymbol{k},\sigma}\left[a_{\boldsymbol{k},\sigma} - a^+_{\boldsymbol{k},\sigma}\right] \tag{5.3}$$

The averaging of the quantity $\left|\mathbf{d}e_{\boldsymbol{k},\sigma}\right|^2$ over all orientations of the polarization vector $e_{\boldsymbol{k},\sigma}$ in the isotropic field results in the usual expression $|\mathbf{d}|^2/3 = \sum_\alpha |d_\alpha|^2/3$ (where d_α are the components of the transition dipole moment $\mathbf{d}$), whereas the corresponding averaging in the anisotropic field leads to the expression $\sum_\alpha g_\alpha^2 |d_\alpha|^2/3$.
As a result, instead of expression (3.13) for a sphere, we obtain the following relationship for an ellipsoid (Pukhov et al., 2008; Basiev et al., 2008; Pukhov, 2009):

$$A_{nano}^{ell}/A_{bulk} = \frac{n_{med}}{n_{cr}} \sum_{\alpha=a,b,c}\left[\frac{\gamma_\alpha}{1+(\varepsilon-1)N_\alpha}\right]^2 \tag{5.4}$$

where $\gamma_\alpha = |d_\alpha| / \sqrt{\sum_\alpha |d_\alpha|^2}$ are the direction cosines of the transition dipole moment in the principal axes *a*, *b*, and *c* of the ellipsoid. (In the case of sphere, $N_a = N_b = N_c = 1/3$, so that Eq. (5.4) reduces to Eq. (3.13) for a sphere.) This means that, now, the ratio A_{nano}/A_{bulk} depends on the dipole orientation with respect to the principal axes of the ellipsoid.
The anisotropy factor *K*, which is equal to the ratio of the probability of a transition in an ellipsoid A_{nano}^{ell} to the probability of a transition in a sphere A_{nano}^{spher} can be written in the form

$$K = \frac{A_{nano}^{ell}}{A_{nano}^{spher}} = \left[\frac{\varepsilon+2}{3}\right]^2 \sum_{\alpha=a,b,c}\left[\frac{\gamma_\alpha}{1+(\varepsilon-1)N_\alpha}\right]^2 \tag{5.5}$$

For a sphere, we have $Na = Nb = Nc = 1/3$ and, after the corresponding substitution and transformation, formula (5.5) leads to $K = 1$, as it must.

Ellipsoids of Revolution

From here on, we will consider only ellipsoids of revolution (where the c axis is the axial symmetry axis and the lengths of the semiaxes a and b are equal; i.e., $N_a = N_b$. In this case, expression (5.5) takes the form

$$K_{axial} = \left[\frac{\varepsilon+2}{3}\right]^2 \left[\frac{1-\gamma_c^2}{[1+(\varepsilon-1)N_a]^2} + \frac{\gamma_c^2}{[1+(\varepsilon-1)N_c]^2}\right], \tag{5.6}$$

where $N_a = N_b = (1 - N_c)/2$.

The elliptic integrals given by formula (5.2) are expressed through elementary functions for all ellipsoids of revolution (Landau & Lifshitz, 1984). For a prolate ellipsoid of revolution ($c > a = b$) with the eccentricity $e = \sqrt{1 - a^2/c^2}$, we have

$$N_c = \frac{1-e^2}{e^3}(\text{Arth}\,e - e), \tag{5.7}$$

$$N_a = N_b = (1 - N_c)/2. \tag{5.8}$$

When the ellipsoid is closely similar to a sphere ($e << 1$), the depolarization factors are approximately represented by the formulas (Landau & Lifshitz, 1984)

$$N_c = 1/3 - 2e^2/15, \tag{5.9}$$

$$N_a = N_b = 1/3 + e^2/15. \tag{5.10}$$

For an oblate ellipsoid of revolution ($c < a = b$) with the eccentricity $e = \sqrt{a^2/c^2 - 1}$, the depolarization factors are written as (Landau & Lifshitz, 1984)

$$N_c = \frac{1+e^2}{e^3}(e - \text{arctg}\,e), \tag{5.11}$$

$$N_a = N_b = (1 - N_c)/2. \tag{5.12}$$

For the eccentricity $e << 1$, the depolarization factors are given by the expressions (Landau & Lifshitz, 1984)

$$N_c = 1/3 + 2\,e^2/15, \tag{5.13}$$

$$N_a = N_b = 1/3 - e^2/15. \tag{5.14}$$

It can be seen from expression (5.6) that the dependence of the anisotropy parameter K_{axial} on the orientation of the transition dipole moment of the optical center is completely determined by the quantity γ_c, i.e., the projection of the transition dipole moment onto the axis of revolution c. For illustrative purposes, we will consider below several cases of the orientation of the transition dipole moment in long cylinders ($c>>a = b$) and thin disks ($c<<a = b$).

Case A

The transition dipole moment has nonzero components along the axis of revolution *c* and in the plane perpendicular to this axis ($\gamma_c^2 = (\gamma_a^2 + \gamma_b^2)/2 = 1/3$). The orientation of the crystallographic axes is chosen to be arbitrary with respect to the axes of the ellipsoid.

Cylindrical nanoparticles ($\gamma_c^2 = 1/3$). A dielectric cylinder is characterized by the depolarization factors $N_a = N_b = 1/2$ and $N_c = 0$. Setting $\gamma_c^2 = 1/3$ in relationship (5.6), we obtain

$$K_{cyl}(\varepsilon) = \left[\frac{\varepsilon + 2}{3(\varepsilon + 1)}\right]^2 \frac{\varepsilon^2 + 2\varepsilon + 9}{3}. \tag{5.15}$$

Disk-shaped nanoparticles ($\gamma_c^2 = 1/3$). A dielectric disk is characterized by the depolarization factors $N_a = N_b = 0$ and $N_c = 1$. Setting $\gamma_c^2 = 1/3$ in relationship (5.6), we find

$$K_{disk}(\varepsilon) = \left[\frac{\varepsilon + 2}{3}\right]^2 \frac{2\varepsilon^2 + 1}{3\varepsilon^2}. \tag{5.16}$$

In both variants, the function $K(\varepsilon)$ monotonically increases from the minimum at $\varepsilon = 1$. This indicates that the lifetime of optical centers in nonspherical nanocrystals is shortened as compared to their lifetime in spherical nanocrystals. It should be noted that, for Y_2O_3 nanocrystals in air ($\varepsilon = 3.4$), the lifetimes of optical centers in cylindrical and disk-shaped nanoparticles are shorter than that in spherical nanoparticles by factors of 1.5 and 2.2, respectively.

Case B

The transition dipole moment is perpendicular ($\gamma_c^2 = 0$) or parallel ($\gamma_c^2 = 1$) to the axis of revolution *c*.

Cylindrical nanoparticles ($\gamma_c^2 = 0$ or $\gamma_c^2 = 1$). For dipoles oriented perpendicular to the axis of revolution of the cylinder, we calculated $N_a = N_b = 1/2$, $N_c = 0$, and $\gamma_c^2 = 0$ in relationship (5.6) and derive the following expression (Fig. 2, curve *1*):

$$K_{cyl}^{\perp}(\varepsilon) = \left[\frac{2(\varepsilon + 2)}{3(\varepsilon + 1)}\right]^2. \tag{5.17}$$

For dipoles aligned parallel to the axis of revolution of the cylinder, we calculated $N_a = N_b = 1/2$, $N_c = 0$, and $\gamma_c^2 = 1$ in relationship (5.6) and obtain (Fig. 2, curve *2*)

$$K_{cyl}^{\parallel}(\varepsilon) = \left[\frac{\varepsilon + 2}{3}\right]^2. \tag{5.18}$$

Disk-shaped nanoparticles ($\gamma_c^2 = 0$ or $\gamma_c^2 = 1$). For dipoles oriented perpendicular to the axis of revolution of the disk, we calculated $N_a = N_b = 0$, $N_c = 1$, and $\gamma_c^2 = 0$ in relationship (5.6) and find (Fig. 2, curve *2*)

$$K_{disk}^{\perp}(\varepsilon) = \left[\frac{\varepsilon + 2}{3}\right]^2 . \quad (5.19)$$

It should be noted that, according to relationships (5.18) and (5.19), we have the equality $K_{disk}^{\perp}(\varepsilon) = K_{cyl}^{||}(\varepsilon)$.

For dipoles aligned parallel to the axis of revolution of the disk, we calculated $N_a = N_b = 0$, $N_c = 1$, and $\gamma_c^2 = 1$ in relationship (5.6) and derive (Fig. 2, curve *3*)

$$K_{disk}^{||}(\varepsilon) = \left[\frac{\varepsilon + 2}{3\varepsilon}\right]^2 . \quad (5.20)$$

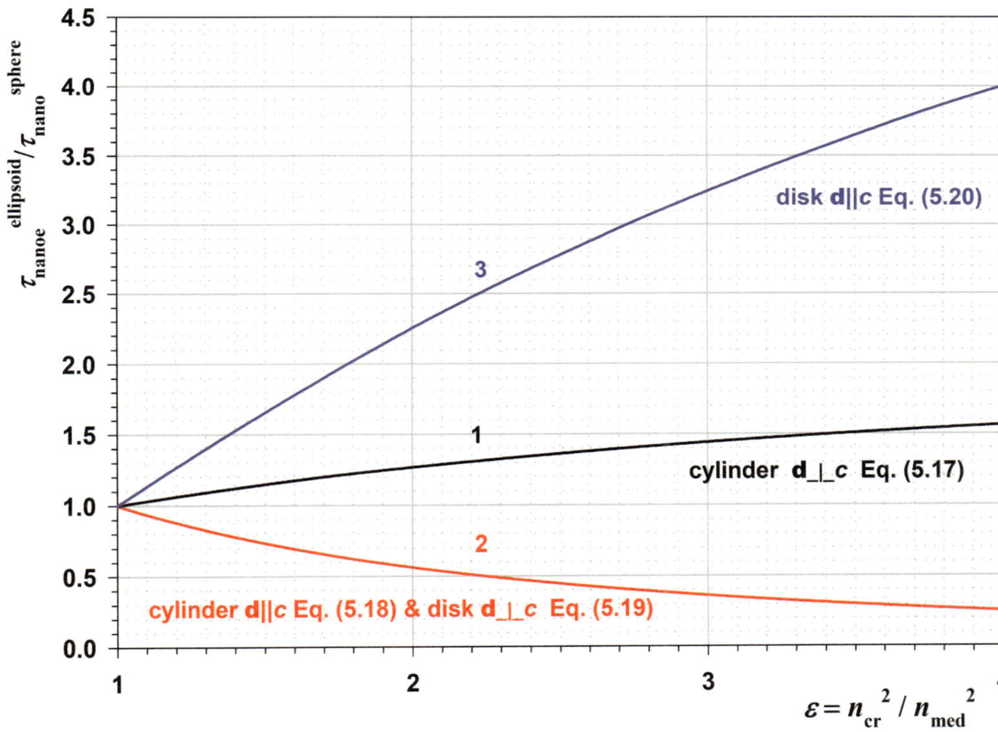

Fig. 2. Dependences of the ratio of the optical excitation lifetime in a nanoparticle in the form of an ellipsoid of revolution to the optical excitation lifetime in a nanosphere on the ratio $(n_{cr}/n_{med})^2$ at different ratios between the lengths of the *a*, *b*, and *c* axes and the directions of the dipole moment **d** with respect to the axes of the ellipsoid: (*1*) nanocylinder for $\mathbf{d} \perp c$, (*2*) nanocylinder for $\mathbf{d} \parallel c$ and nanodisk at $\mathbf{d} \perp c$, and (*3*) nanodisk for $\mathbf{d} \parallel c$.

Therefore, the radiative decay time of optical centers in nonspherical nanoparticles can vary over a wide range as compared to that in spherical nanoparticles. For example, the radiative decay time of optical centers in the Y_2O_3 matrix doped with rare-earth ions can increase by a factor of abuot 3.5 in disk-shaped nanocrystals with a transition dipole moment aligned parallel to the axis of revolution of the disk and, on the contrary, can decrease by a factor of about 3.3 in disk-shaped nanocrystals with a transition dipole moment oriented perpendicular to the axis of revolution of the disk or in cylindrical nanocrystals with a transition dipole moment directed parallel to the axis of revolution of the cylinder (Fig. 2, curves *2* and *3* at $\varepsilon = 3.39$, respectively).

The anisotropy factor *K* substantially complicates the interpretation of experimental data: in order to calculate the value of *K*, it is necessary to know the mutual orientation of the principal axes of the ellipsoid (*a*, *b*, and *c*) with respect to the crystallographic axes of the nanocrystal. Let us consider the simplest case in which the nanoparticle is a nanocrystal with cubic symmetry and the crystallographic axes of the nanocrystal are parallel to the principal axes of the ellipsoid. Owing to the cubic symmetry of the crystal, the appearance of optical centers with a transition dipole moment aligned parallel, say, to the [100] axis brings about the formation of an equal number of optical centers with transition dipole moments directed parallel to the [010] and [001] axes. This implies that the cubic crystal should contain optical centers of three types: optical centers of the first type are oriented parallel to the [100] direction ($\gamma_a^2= 1$, $\gamma_b^2 =\gamma_c^2 = 0$), optical centers of the second type are aligned parallel to the [010] direction ($\gamma_b^2 = 1$, $\gamma_a^2 = \gamma_c^2 = 0$), and optical centers of the third type are directed parallel to the [001] direction ($\gamma_a^2 = \gamma_b^2 = 0$, $\gamma_c^2 = 1$). Optical centers of all three types in a bulk crystal are equivalent and have the same lifetime, whereas a qualitatively different situation occurs with a nanocrystal. In the latter case, only the first two types of optical centers turn out to be equivalent (owing to the axial symmetry).

The luminescence decay kinetics of optical centers should be described by two exponential components (it is a manifestation of the "latent anisotropy" (Feofilov, 1961; Feofilov & Kaplyanskii, 1962) with a weight ratio of 2:1. In this case, the "fast and slow exponential components" should be characterized by lifetimes that are shorter and longer, respectively, than those observed in a nanosphere. For example, the luminescence kinetics of optical centers in the Y_2O_3: Yb^{3+} nanocylinder ($n_{cr} = 1.84$) in air should exhibit a nonexponential behavior (Fig. 3a, curve *1*) owing to the presence of two exponential components with equal weights and the same decay time τ_1= 6.4 ms (Fig. 3a, curve *2*) (the transition dipole moment is perpendicular to the axis of revolution of the cylinder) and one exponential component with the decay time τ_2=1.3 ms and a weight of 1/3 (Fig. 3a, curve *3*) (the transition dipole moment is parallel to the axis of revolution of the cylinder). For comparison, Fig. 3a also shows the luminescence decay curves of optical centers in a nanosphere with the decay time τ_{sphere} = 4.3 ms (Fig. 3a, curve *4*) and in a bulk crystal with τ_{bulk} = 0.72 ms (Fig. 3a, curve *5*). It can be seen from Fig. 3a that the radiative decay time of optical centers in the Y_2O_3:Yb^{3+} nanocylinder can increase by a factor of approximately 9 upon changing over from the bulk crystal to the cylindrical nanocrystal and by a factor of approximately 6 upon changing over to the spherical nanocrystal. A more pronounced nonexponential behavior is observed for the Y_2O_3:Yb^{3+} nanodisk (Fig. 3b, curve *1*). The fast and slow exponential components differ by a factor of approximately 12; in this case, we have τ_1= 1.3 ms with a weight of 2/3 (Fig. 3b, curve 2) (the transition dipole moment is perpendicular to the axis of revolution of the disk) and τ_2= 15.3 ms with a weight of 1/3 (Fig. 3b, curve *3*) (the transition dipole moment is

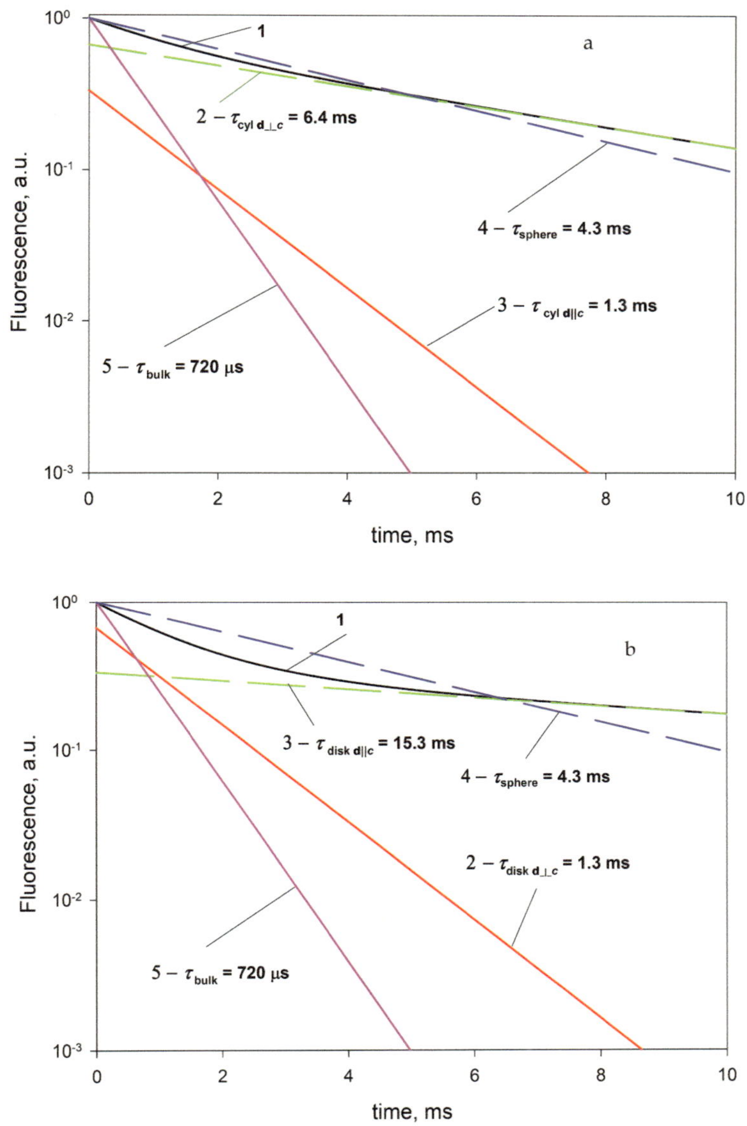

Fig. 3. (a) Luminescence kinetic curves for (*1*) cubic symmetry optical centers in the Y_2O_3:Yb^{3+} nanocylinder (n_{cr} =1.84) suspended in air for which the transition dipole moment **d** in equal proportion is parallel to one of the axes of the ellipsoid of revolution (*a*, *b*, or *c*); (*2*, *3*) optical centers with the transition dipole moments (*2*) **d**$\perp c$ and (*3*) **d** $\parallel c$; (*4*) the nanosphere; and (*5*) the bulk crystal
(b) Luminescence kinetic curves for (*1*) cubic symmetry optical centers in the Y_2O_3:Yb^{3+} nanodisk (n_{cr} = 1.84) suspended in air for which the transition dipole moment **d** in equal proportion is parallel to one of the axes of the ellipsoid of revolution (*a*, *b*, or *c*); (*2*, *3*) optical centers with the transition dipole moments (*2*) **d** $\perp c$ and (*3*) **d** $\parallel$ *c*; (*4*) the nanosphere; and (*5*) the bulk crystal.

parallel to the axis of revolution of the disk). It is clear that, in the absence of axial symmetry, the process of luminescence decay should involve three exponential components. The obtained theoretical inference that the kinetics of radiative decay of excited optical centers of the same type in cubic nanocrystals exhibits a strongly nonexponential behavior is confirmed by the experimental data available in the literature (Christensen et al., 1982).
It should be mentioned, that in a particular case of the cylinder nanocrystals with elliptical cross section (axes *a*, *b*) and an in-plane dipole moment ($d_c = 0$) we obtain from Eq. (5.4)

$$A_{nano}^{ell}/A_{bulk} = \frac{n_{med}}{n_{cr}}\left[\left(\frac{\gamma_a}{1+(\varepsilon-1)N_a}\right)^2+\left(\frac{\gamma_b}{1+(\varepsilon-1)N_b}\right)^2\right] \tag{5.21}$$

With the exception of the factor n_{med}/n_{cr} this expression coincides with result of Rogobete and Henkel for the cylinder nanocrystals with elliptical cross - section (see Eq.(A8) in Ref. (Rogobete&Henkel, 2004)). From our point of view the factor n_{med}/n_{cr} is important and cannot be dropped from Eq.(5.21).
It should also be noted that the known Judd–Ofelt parameters Ω_k (Krupke, 1966) for trivalent rare-earth ions in ellipsoidal nanocrystals should depend on the depolarization factors N_α and the orientation of the crystallographic axes with respect to the principal axes of the ellipsoid (see next Section 6).
A situation complicates when the spontaneous emission process involves several decay channels and each channel is determined by its own direction of the transition dipole moment. In this case, the branching ratio can also change for a nanocrystal. However, this circumstance can be used to obtain information on the shape of nanocrystals. The cases under consideration, of course, do not cover a wide variety of real optical centers in nanocrystals and, most likely, have only an illustrative character, because they have demonstrated the influence of the shape of nanoparticles on the spontaneous decay rate of optical centers. Nonetheless, the above analysis has shown that controlling the shape of samples is of primary importance for an adequate theoretical interpretation of experimental results. And visa versa, the deviation of the rate of kinetics of radiative decay from the theoretically predicted value for spherical samples has revealed morphological features of nanoparticles and can serve as an effective method for controlling their sphericity.

5.2 An ensemble of nanoellipsoids distributed in sizes and orientations ($x \neq 0$)

Let us now shortly consider a nanocomposite with the subwavelength ellipsoidal dielectric nanoparticles, which are identical in composition and shape but may be different in volume and orientation. We shall assume that all orientations are equally likely. Follow to Bohren and Huffman (Bohren & Huffman, 1998) we shall use equation

$$\mathbf{E} = (1-x)\mathbf{E}^{(med)} + x\left\langle \mathbf{E}^{(cr)} \right\rangle \tag{5.22}$$

where symbol <…> denotes averaging over orientations of ellipsoidal nanoparticles and the averaged fields $\mathrm{E}^{(med)}$ and $\langle \mathrm{E}^{(cr)} \rangle$ are connected by relation

$$\left\langle \mathbf{E}^{(cr)} \right\rangle = g\mathbf{E}^{(med)} \tag{5.23}$$

with

$$g = \sum_{\alpha} g_{\alpha} / 3 . \tag{5.24}$$

Eqs.(5.22)-(5.24) together with the corresponding equations for averaged polarizations give a generalization of Maxwell Garnett mixing rule for ε_{eff} (Bohren & Huffman, 1998):

$$\varepsilon_{eff} = \varepsilon_{med} \left[1 + \frac{xg(\varepsilon - 1)}{(1-x) + xg} \right]. \tag{5.25}$$

For spheres (g = 3/(ε+2)) formula (5.25) leads to Maxwell Garnett mixing rule (3.12), as it must.

It is follows from Eqs.(5.22)-(5.23) that

$$\left\langle \mathbf{E}^{(cr)} \right\rangle = \frac{g}{(1-x) + xg} \mathbf{E} \tag{5.26}$$

However, it is obvious that Eq. (5.26) is inappropriate for calculation of the decay rate of optical centers. With assumption that macroscopic electric field inside the specific ellipsoidal nanoparticle is given by

$$\mathbf{E}^{(cr)} = \frac{(1-x)\hat{\mathbf{g}} + xg}{(1-x) + xg} \mathbf{E}, \tag{5.27}$$

the Hamiltonian of interaction of electric field $\mathbf{E}^{(cr)}$ with dipole moment $\mathbf{d}$ takes the form

$$\begin{aligned} H_{\text{int}} &= -\hat{\mathbf{E}}^{(cr)}\hat{\mathbf{d}} = -\hat{\mathbf{d}}\left[(1-x)\hat{\mathbf{g}} + xg\right]\hat{\mathbf{E}} \Big/ \left[(1-x) + xg\right] = \\ &= -i\left[(1-x) + xg\right]^{-1} \sum_{\boldsymbol{k},\sigma} \sqrt{\frac{2\pi\hbar\omega_k}{V\varepsilon_{eff}}} \hat{\mathbf{d}} \left[(1-x)\hat{\mathbf{g}} + xg\right] \boldsymbol{e}_{\boldsymbol{k},\sigma} \left[\mathrm{a}_{\boldsymbol{k},\sigma} - \mathrm{a}^{+}_{\boldsymbol{k},\sigma} \right]. \end{aligned} \tag{5.28}$$

As a result, we obtain

$$\begin{aligned} & A^{ell}_{nano} / A_{bulk} = \\ & = \frac{n_{eff}}{n_{cr}} \left[(1-x) + xg\right]^{-2} \left[(1-x)^2 \sum_{\alpha=a,b,c} \gamma^2_{\alpha} g^2_{\alpha} + 2x(1-x)g \sum_{\alpha=a,b,c} \gamma^2_{\alpha} g_{\alpha} + x^2 g^2 \right]. \end{aligned} \tag{5.29}$$

We lead to expression (5.4) when $x \to 0$. It follows from Eq. (5.25) that $n_{eff} \to n_{med}$ either. In the limit $x \to 1$ ($n_{eff} \to n_{cr}$), we obviously have $A^{ell}_{nano} \to A_{bulk}$. Hence, Eq. (5.29) can be considered as a "matching" formula both yielding the correct result for the limiting cases $x \to 0$ and $x \to 1$. In the case of spherical nanoparticles, $N_a = N_b = N_c = 1/3$, $g = g_a = g_b = g_c = 3/(2+\varepsilon)$ so that Eq. (5.29) reduces to Eq. (3.10), as it must.

6. Is Judd-Ofelt equation valid for nanocrystal?

All above mentioned in Section 5 results are derived for a two-level atom. Now let us consider the multilevel systems, such as rare-earth (RE) ions. The Eq. (5.4) leads to following expression for the spontaneous emission rate of $J \to J'$ multilevel transition of RE ion in the ellipsoidal nanoparticles (Pukhov, 2009)

$$A_{nano}^{ell}(J,J')/A_{bulk}(J,J') = \frac{n_{med}}{n_{cr}}\left(\frac{3}{2+\varepsilon}\right)^2 S_{nano}^{ell}(J,J')/S_{bulk}(J,J') . \quad (6.1)$$

Here $S_{bulk}(J,J')$ is the line-strength of $J \to J'$ transition in a bulk crystal and $S_{nano}^{ell}(J,J')$ is "anisotropic line-strength" defined as

$$S_{nano}^{ell}(J,J') = \sum_{\alpha=a,b,c} h_\alpha^2 \gamma_\alpha^2(J,J') \quad (6.2)$$

where

$$h_\alpha = \frac{\varepsilon+2}{3} g_\alpha = \frac{\varepsilon+2}{3[1+(\varepsilon-1)N_\alpha]}, \quad (6.3)$$

$$\gamma_\alpha^2(J,J') = \sum_{BB'} \left|(B|D_\alpha|B')\right|^2 \quad (6.4)$$

(in Eqs. (6.2)-(6.4), the coordinate axes coincide with the principal axes of the ellipsoid a, b, c). Hereinafter we use Judd notations (Judd, 1962). B and B' denote Stark states of the J- and J'-manifolds, respectively. From Eqs. (6.2)-(6.4), we have $S_{nano}^{spher}(J,J') = S_{bulk}(J,J')$ for a sphere, as it must be because of invariance of the sum $\sum_{\alpha=x,y,z} \sum_{BB'} \left|(B|D_\alpha|B')\right|^2$ with respect to rotations of coordinate axes.

According to Judd-Ofelt theory (Judd, 1962; Ofelt, 1962), the linestrength for trivalent RE ions is given by equation

$$S_{bulk}(J,J') = \sum_{\lambda=2,4,6} \Omega_\lambda \left|(\kappa J\|U^{(\lambda)}\|\kappa' J')\right|^2 \quad (6.5)$$

where Ω_λ are Judd-Ofelt intensity parameters and $(\kappa J\|U^{(\lambda)}\|\kappa' J')$ is the reduced matrix element of the unit tensor operator $U^{(\lambda)}$ of rank λ that specifies the $J \to J'$ transition.

Is the Judd-Ofelt equation valid for $S_{nano}^{ell}(J,J')$?

It can be shown using Judd's results (Judd, 1962) that a sum $\sum_{BB'} \left|(B|D_\alpha|B')\right|^2$ ($\alpha = x, y, z$) can be expressed in terms of matrix elements $(\kappa J\|U^{(\lambda)}\|\kappa' J')$ in the arbitrary coordinate system. Namely,

$$\gamma_{\alpha}^{2}(J,J') = \sum_{BB'} \left|(B|D_{\alpha}|B')\right|^{2} = \sum_{\lambda} \Omega_{\alpha}^{(\lambda)} \left|(\kappa J \left\| U^{(\lambda)} \right\| \kappa' J')\right|^{2} \tag{6.6}$$

where

$$\Omega_{x}^{(\lambda)} = \frac{1}{2}\left[\Omega_{1,1}^{(\lambda)} + \Omega_{-1,-1}^{(\lambda)} - \Omega_{1,-1}^{(\lambda)} - \Omega_{-1,1}^{(\lambda)}\right] \tag{6.7a}$$

$$\Omega_{y}^{(\lambda)} = \frac{1}{2}\left[\Omega_{1,1}^{(\lambda)} + \Omega_{-1,-1}^{(\lambda)} + \Omega_{1,-1}^{(\lambda)} + \Omega_{-1,1}^{(\lambda)}\right] \tag{6.7b}$$

$$\Omega_{z}^{(\lambda)} = \Omega_{0,0}^{(\lambda)}. \tag{6.7c}$$

In Eqs. (6.7)

$$\Omega_{qq'}^{(\lambda)} = (2\lambda+1) \sum_{p,t,t'} \begin{pmatrix} 1 & \lambda & t \\ q & -p-q & p \end{pmatrix} \begin{pmatrix} 1 & \lambda & t' \\ q' & -p-q & p+q-q' \end{pmatrix} A_{tp} A_{t'p+q-q'}^{*} \Xi(t,\lambda)\Xi(t',\lambda) \tag{6.8}$$

(expressions for A_{tp} and $\Xi(t,\lambda)$ are given in Judd's paper (Judd, 1962)).

As a result we have that expression (6.2) for $S_{nano}^{ell}(J,J')$ takes the form (Pukhov, 2009)

$$S_{nano}^{ell}(J,J') = \sum_{\lambda} \Omega_{\lambda}^{ell} \left|(\kappa J \left\| U^{(\lambda)} \right\| \kappa' J')\right|^{2} \tag{6.9}$$

where intensity parameters are equal to

$$\Omega_{\lambda}^{ell} = \sum_{\alpha=a,b,c} h_{\alpha}^{2} \Omega_{\alpha}^{(\lambda)}. \tag{6.10}$$

Hence the expression for the spontaneous emission rate of $J \to J'$ transition of RE ion in the subwavelength ellipsoidal nanocrystals is of the same form as the Judd-Ofelt expression for a bulk crystal. However, the Judd–Ofelt intensity parameters for trivalent rare-earth ions in ellipsoidal nanocrystals should depend on the depolarization factors and the orientation of the crystallographic axes with respect to the principal axes of the ellipsoid. The intensity parameters for a sphere are equal to the bulk intensity parameters.

7. Comparsion of the theoretical data with the experimental data results

In this section, the ratios τ_{nano}/τ_{bulk} calculated from relationship (3.15) for spherical nanoparticles with a radius $2R << \lambda$ and a volume fraction $x \to 0$ are compared with the corresponding measured quantities taken from paper (Dolgaleva et al., 2007) for the $^4F_{3/2}$ level of Nd^{3+} ions in YAG nanocrystals and from paper (Meltzer et al., 1999) for the $^5D_0$ level of Eu^{3+} ions in Y_2O_3 nanocrystals in suspensions as a function of the ratio n_{cr}/n_{med} (Fig. 1).
In the case when nanoparticles have an identical shape (morphology) but are not spherical, instead of expression (3.15), it is necessary to use the general formula (5.4), which leads to the following relationship for the ratio between the decay times:

$$\frac{\tau_{nano}^{ell}}{\tau_{nano}^{spher}} = \left[\frac{3}{n_{cr}^2/n_{med}^2 + 2}\right]^2 \sum_{\alpha=a,b,c}\left[\frac{1+(n_{cr}^2/n_{med}^2 - 1)N_\alpha}{\gamma_\alpha}\right]^2 . \quad (7.1)$$

As can be seen from relationship (7.1), the smaller the ratio n_{cr}/n_{med}, i.e., the closer the refractive index of the nanoparticle n_{cr} to the refractive index of the surrounding medium n_{med}, the smaller the change in the rate of radiative relaxation of the rare-earth ion upon deviation of the shape of the nanoparticle from spherical, i.e., the weaker the effect of the nanoparticle morphology on the radiative relaxation rate. This tendency is clearly illustrated by the measured rates of radiative relaxation of the $^4F_{3/2}$ level of Nd^{3+} ions in YAG:(0.9 at % Nd^{3+}) nanoparticles with a diameter $2R$ = 20 nm and a small volume fraction ($x = 1.1 \times 10^{-3}$) in different immersion liquids (Fig. 1, circles). The size of the circles in Fig. 1 corresponds to the measurement error. It can be seen from Fig. 1 that the experimental points fit the theoretical curve fairly well at a ratio $n_{cr}/n_{med} < 1.18$. The deviation of the experimental data from the theoretical results becomes noticeable at a ratio $n_{cr}/n_{med} > 2$ and increases with an increase in its value. It should be remembered that, even at the late stage of the luminescence decay kinetics, the nonradiative relaxation due to the Nd–Nd concentration quenching can not be ignored at a Nd^{3+} content of 0.9 at % (Voron'ko et al., 1973). This can also result in a deviation of the experimental dependence τ_{nano}/τ_{bulk} versus n_{cr}/n_{med} from the theoretical curve obtained for YAG:Nd nanocrystals. In order to minimize the contribution of the nonradiative quenching to the luminescence kinetics, the lifetime, for example, in the YAG:Nd^{3+} crystal (τ_{bulk} = 255 μs (Basiev et al., 1974), was determined at a considerably lower concentration of Nd^{3+} ions as compared to that in the nanoparticles studied in (Dolgaleva et al., 2007).

Much better agreement with the theoretical curve, already at the ratio n_{cr}/n_{med} = 1.4, is observed for the decay time of the fluorescence associated with the $^5D_0$ level (not affected by quenching) of Eu^{3+} ions in Y_2O_3:(0.1%Eu^{3+}) nanoparticles with a diameter $2R$ = 7–10 nm (Fig. 1, squares) (Meltzer et al., 1999). The deviation observed at a ratio $n_{cr}/n_{med} \rightarrow n_{cr}$ can quite possibly be explained by the considerable volume fraction of the nanoparticles, their agglomeration, or deviation from sphericity.

8. Conclusion

Two principal physical reasons provide a basis for a change of spontaneous emission lifetime of an optical center, which moves from bulk crystal to nanocrystal. First, it is a change in photon density of states; second, it is a change of zero-point amplitude of electric-field modes, which are responsible for spontaneous emission. On this basis, an analytical expression is obtained for the electric-dipole radiative-decay rate of an excited optical center in an ellipsoidal dielectric nanoparticles (with sizes much less than the wavelength of light) embedded in a dielectric medium. It is found that the ratio of the decay rate A_{nano} of an excited optical center in the nanoparticle to the decay rate A_{bulk} of an excited optical center in the bulk sample is independent of the local-field correction and, therefore, of the adopted local-field model. The relation of the ratio A_{nano}/A_{bulk} with refractive indexes of the nanoparticles (n_{cr}) and the medium (n_{med}) and a volume fraction x of nanoparticles in the nanocomposite is established. The ratio of the absorption/emission cross-section $\sigma_{nano}/\sigma_{bulk}$ for spherical nanoparticles in dielectric medium to that in bulk crystal $\sigma_{nano}/\sigma_{bulk}$ is derived as well. Here, the functional dependence of A_{nano}/A_{bulk} and $\sigma_{nano}/\sigma_{bulk}$ ratios are found to be different. Strong increase in the radiative decay time in a nanoparticale in comparison with the bulk crystal gives rise to only a slight decrease in the corresponding cross - section for absorption and emission.

Comparison of the experimental and theoretical values of TR^{3+} metastable levels decay time τ in the YAG and Y_2O_3 spherical nanocrystals with those in bulk crystals proves the theory.

Thus, the above analysis of the theoretical expressions and the experimental results has demonstrated that the radiative characteristics of nanoparticles differ significantly from those of a bulk crystal. By varying the volume fraction x of nanoparticles in a suspension or an aerosol, the refractive index of the medium n_{med} surrounding the nanoparticles and their morphology it is possible to control the rates of spontaneous transitions and absorption and emission cross - sections of induced transitions. Thus, it is possible to increase $\sigma \times \tau$ product, an important laser parameter, several times and raises a population inversion in nanocomposite laser medium. By this method, one can control the laser properties of the nanocomposite materials and, thus, to design novel laser and luminescent media with improved characteristics.

The results obtained can be used in developing the fluorescence kinetic method for controlling the shape of nanoparticles and the degree of their agglomeration during subsequent applications to the synthesis of optical laser ceramic materials, control of the luminescence lifetime of nanophosphors, and observation of nanoagglomerates in organic and biological structures.

9. References

Agarwal, G. S. (1975). Quantum electrodynamics in the presence of dielectrics and conductors. IV. General theory for spontaneous emission in finite geometries. *Phys. Rev. A* , Vol.12, No.4, (1475-1497), ISSN 1050-2947

Amos, R. M. & Barnes, W. L. (1997). Modification of the spontaneous emission rate of Eu^{3+} ions close to a thin metal mirror. *Phys. Rev.* B, Vol.55, No.11,(7249- 7254), ISSN 1098-0121

Barnes, W. L. (1998). Fluorescence near interfaces: the role of photonic mode density. *J. Modern Optics*, Vol.45, (661-699) ISSN 0950-0340

Basiev, T. T., Dianov, E. M., Prokhorov, A. M. & Shcherbakov, I. A. (1974). On the Quantum Yield of Luminescence from the Nd^{3+} Metastable State in Silicate Glasses and $Y_3Al_5O_{12}$ Crystals. *Dokl. Akad. Nauk .SSSR.* Vol.216, No.2, (297-299), ISSN 0869-5652;[(1974). *Sov. Phys. Dokl.* Vol.19, No.5, (288-289), ISSN 0038-5689]

Basiev, T. T.; Orlovskii, Yu. V. & Pukhov, K. K. (2008). Spontaneous and induced emission in dielectric nanoparticles. *Nanotechnologies in Russia*, Vol.3, Nos.9-10, (551-559), ISSN 1995-0780 [Basiev, T. T.; Orlovskii, Yu. V. & Pukhov, K. K. (2008). *Rossiiskie nanotekhnologii*, Vol.3, Nos. 9–10, (66-73), ***ISSN*** 1992-7223]

Bohren, C. F.& Huffman D. R. (1998). *Absorption and scattering of light by small particles*, Wiley, ISBN 0-471-05772-X, New York

Brueck, S. R., Smagley, V. A. & Eliseev, P. G. (2003). Radiation from a dipole embedded in a multilayer slab. *Phys. Rev. E.* Vol.68, No.3, (036608 13pages), ISSN 1063-651X

Carnigia, C. K. & Mandel, L. (1971). Quantization of Evanescent Electromagnetic Waves. *Phys. Rev.* D, Vol.3, No.2, (280-296), ISSN 0556-2821

Chance, R. R., Prock, A. & Sibley, R. (1978). Molecular fluorescence and energy transfer near interfaces. *Adv. Chem. Phys.* Vol. 37, (1-65), ISSN 0065-2385

Chang-Kui Duan, Reid, M. F. & Zhongqing Wang (2005). Local field effects on the radiative lifetime of emitters in surrounding media: Virtual- or real-cavity model?. *Phys. Lett.* A, Vol.343, (474-480), ISSN 0375-9601

Christensen, H. P., Gabbe, D. R., & Jenssen, H. P. (1982). Fluorescence lifetimes for neodymium-doped yttrium aluminium garnet and yttrium oxide powders. *Phys. Rev.* B, Vol.25, No.3,(1467-1473), ISSN 1098-0121

Chew, H. (1987). Transition rates of atoms near spherical surfaces. *J. Chem. Phys.* Vol.87, No.2, (1355-1360), ISSN 0021-9606

Chew, H. (1988). Radiation and lifetimes of atoms inside dielectric particles. *Phys. Rev. A.* Vol.38, No.7, (3410-3416), ISSN 1050-2947

Dirac P.A.M. (1927). The Quantum Theory of the Emission and Absorption of Radiation. Proc. Roy. Soc., Vol. A114, No 243, (1-25) , Online ISSN: 1471-2946

Dirac, P. A. M., (1982). *The Principles of Quantum Mechanics,* Oxford Univ. Press, ISBN-10: 0198520115 London,

http://www.amazon.com/Principles-Quantum-Mechanics-International-Monographs/dp/0198520115#reader

Dolgaleva, K., Boyd R. W., & Milonni P. W., (2007). Influence of local-field effects on the radiative lifetime of liquid suspensions of Nd:YAG nanoparticles. *J. Opt. Soc. Am.* B, Vol.24, No.3, (516-521), ISSN 0740-3224

Drexhage, K. H. (1970). Influence of a dielectric interface on fluorescence decay time. *J. Lumin.* Vol.1–2, (693-701), ISSN 0022-2313.

Einstein, A. (1917). Zur Quantentheorie der Strahlung (On the Quantum Mechanics of Radiation). *Physik. Z.* Vol.18, (121-128), ISSN 0323-4479

Fam Le Kien, Nguyen Hong Quang, & K. Hakuta, (2000). Spontaneous emission from an atom inside a dielectric sphere. *Opt.Comm.* Vol.178, (151-164), ISSN 0030-4018

Feofilov, P. P. (1961) *The Physical Basis of Polarized Emission,* Consultants Bureau, ASIN: B0006AWJHG, New York.

Feofilov, P. P. & Kaplyanskii, A. A. (1962). Latent Optical Anisotropy of Cubic Crystals Containing Local Centers and Methods for Its Investigation, *Usp. Fiz. Nauk.* Vol.76, No.2, (201–237), ISSN 0042-1294 [(1962) *Sov. Phys.Usp.* Vol.5, No. 2, (79–104) ISSN 1063-7869

Fowler, W. B. & Dexter, D. L. (1962). Relation between Absorption and Emission Probabilities in Luminescent Centers in Ionic Solids. *Phys. Rev.* Vol.128, No.5, (2154-2165). ISSN 0031-899X

Ginzburg, V. L. (1940). The quantum theory of radiation of an electron uniformly moving in a medium. *J. Phys. USSR,* Vol. 2, No.6, (441-452) ISSN 0038-5697

Glauber, R.J. & Lewenstein, M. (1991). Quantum optics of dielectric media. *Phys. Rev. A.* Vol.43, No.1, (467-491), ISSN 1050-2947

Guokui Liu and Xueyuan Chen, (2007). *Spectroscopic properties of lanthanides in nanomaterials,* in Handbook on the Physics and Chemistry of Rare Earths, K. A. Gschneidner Jr., J. C. G. Bunzli, and V. K. Pecharsky, eds. (North-Holland, Amsterdam,), Vol.37, Chapter 233, p. 99-169. ISBN: 0-444-52144-5

Guzatov, D. V. & Klimov, V. V. (2005). Radiative decay engineering by triaxial nanoellipsoids. *Chem. Phys. Lett.* Vol.412, (341-346). ISSN 0009-2614

Imbush, G. F. & Kopelman, R. (1981). *Optical Spectroscopy of Electronic Centers in Solids. In: Laser Spectroscopy of Solids*, Ed. by E. W. W. Yen and P. M. Selzer (1-36) Springer, ISBN-10: 0387106383 ISBN-13: 978-0387106380, Berlin,

Judd, B. R. (1962). Optical absorption intensities of rare earth ions. *Phys. Rev.*, Vol.127, No.3, (750-761), ISSN 0031-899X

Khosravi, H. & Loudon, R. (1991). Vacuum Field Fluctuations and Spontaneous Emission in the Vicinity of a Dielectric Surface. *Proc. R. Soc. London.* A. Vol.433, (337-352). ISSN 1471-2946

Kittel C. (2007). *Introduction to solid state physics*, Wiley, ISBN 8126510455, New York

Klimov, V. V., Dyuklua, M. & Letokhov, V. S. (2001). Spontaneous emission of an atom in the presence of nanobodies. *Quantum electron.* Vol.31, No.7, (569-586), ISSN 1063-7818

Krupke, W. F. (1966). Optical absorption and fluorescence intensities in several rare-earth-doped Y_2O_3 and LaF_3 single crystals. *Phys. Rev.* Vol.145, No.1, (325-337), ISSN 0031-899X

Kuhn, H. (1970). Classical Aspects of Energy Transfer in Molecular Systems. *J. Chem. Phys.* Vol.53, (101-108), ISSN 0021-9606

Landau, L. D. & Lifshitz, E. M. (1984). *Electrodynamics of Continuous Media* (*Course of Theoretical Physics; v. 8*), Pergamon Press Ltd., ISBN 0 -08 - 030276 - 9, Oxford [Landau, L. D. & Lifshitz, E. M. (2003), *Elektrodinamika Sploshnykh Sred*, Fizmatlit, ISBN 5-9221-0123-4, Moscow]

Lax, M. (1952). The Franck-Condon principle and its application to crystals. J. Chem. Phys. Vol. 20, No 11, (1752- 1760), ISSN 0021-9606

Lukosz, W. & Kunz, R. E. (1977). Light emission by magnetic and electric dipoles close to a plane interface. *J.Opt. Soc. Am.* Vol.67, (1607-1615) 0740-3224

Manoj Kumar, G., Narayana Rao, D. & Agarwal, G. S. (2003). Measurement of Local Field Effects of the Host on the Lifetimes of Embedded Emitters. *Phys. Rev. Lett.*, Vol.91, No.20 (203903-1-203903-4), ISSN 0031-9007

Maxwell Garnett, J. C. (1904). Colours in metal glasses and in metallic films. *Philos. Trans. R. Soc. London.* Vol.A 203, (384-420), ISSN

Maxwell Garnett, J. C. (1906). Colours in metal glasses in metallic films and in metallic solutions. *Philos. Trans. R. Soc. London Ser.* Vol.*A* 205, (237-288), ISSN 0962-8436

Meltzer, R. S., Feofilov, S. P., Tissue, B. & Yuan, H. B. (1999). Dependence of fluorescence lifetimes of Y_2O_3:Eu^{3+} nanoparticles on the surrounding medium. *Phys Rev.* B, Vol.60, No.20, (R14012-R14015), ISSN 1098-0121

Milonni, P. W. (2007). Spontaneous emission between mirrors. *J. Mod. Opt.* Vol.54, No.13-15, (2115-2120), ISSN 0950-0340

Morawitz, H. & Philott, R. (1974). Coupling of an excited molecule to surface plasmons. *Phys. Rev. B*, Vol.10, No.12, (4863-4868), ISSN 1098-0121

Nienhuis, G. & Alkemade, C. Th. J. (1976).Atomic radiative transition probabilities in a continuous medium. *Physica (Amsterdam)*, Vol.81C (1), (181-188) ISSN 0921-4526

Ofelt, G. S. (1962). Intensities of crystal spectra of rare-earth ions. *J. Chem. Phys.*, Vol.37, No.3, (511-520), ISSN 0021-9606

Oraevskii, A. N. (1994). Spontaneous emission in a cavity *Usp. Fiz. Nauk.* Vol.164, No.4, (415-427), ISSN 0042-1294; [Oraevskii, A. N. (1994), Spontaneous emission in a cavity. *Phys.-Usp.* Vol.37, No. 4, (393-405), ISSN 1063-7869]

Pukhov, K. K.; Basiev, T.T. & Orlovskii, Yu. V. (2008). Spontaneous emission in dielectric nanoparticles. JETP Letters, Vol.88, No.1, (12-18), ISSN 0021-3640 [Pukhov K.K., Basiev T.T. & Orlovskii Yu.V., (2008) *Pis'ma v Zhurnal Éksperimental'noĭ i Teoreticheskoĭ Fiziki*, Vol. 88, No.1, (14 - 20), ISSN: 0370-274X]

Pukhov, K. K. (2009), Spontaneous emission in the ellipsoidal nanocrystals. *Journal of Rare Earths*, Vol. 27, No. 4, (637-640), ISSN 1002-0721

Purcell, E. M. (1946). Spontaneous emission probabilities at radio frequences. *Phys. Rev.* Vol.69, No.11-12 (681-681), ISSN 0031-899X

Rikken, G. L. J. A. & Kessener, Y. A. R. R. (1995). Local field effects and electric and magnetic dipole transitions in dielectrics. *Phys. Rev. Lett.*, Vol. 74, No.6, (880-883), ISSN 0031-9007

Rogobete L. & Henkel, C. (2004). Spontaneous emission in a subwavelength environment characterized by boundary integral equations. *Phys. Rev. A*, Vol. 70, No.6, (063815-1 - 063815-10), ISSN 1050-2947

Song, H. & Tanner P.A. (2008). *Luminescence properties of rare earth doped nanophosphors*, in *Doped Nanomaterials and Nanodevices*, Ed. Chen W., American Scientific Publishers, ISBN: 1-58883-110-8, California, USA

Tews, K. H., (1973). Zur Variation von Lumineszens-Lebensdauern(On the Variation of Luminescence Lifetimes.) *Annalen der Physik*. Vol. 29, No.2, (97-120) ISSN 0003-3804

Vetrone, F., Boyer, J.-C., Capobianco, J. A., et al., (2004). A Spectroscopic Investigation of Trivalent Lanthanide Doped Y_2O_3 Nanocrystals. *Nanotechnology*. Vol. 15, No.1, (75–81), ISSN 0957-4484

Voron'ko, Yu. K., Mamedov, T. G., Osiko, V. V., Timoshechkin, M. I., Shcherbakov, I. A.,(1973). Effect of donor-donor and donor-acceptor interactions on the decay kinetics on the metastable state of Nd^{3+} in crystas. *Zh. Éksp. Teor. Fiz.* Vol.65, No.3(9), (1141-1156). ISSN 0044-4510

Wuister, S. F., Celso de Mello Donega, & Meijerink, A. (2004). Local-field effects on the spontaneous emission rate of CdTe and CdSe quantum dots in dielectric media. *J. Chem. Phys.* Vol.121, No.9, (4310-4315), ISSN 0021-9606

Yablonovitch, E. (1987). Inhibited Spontaneous Emission in Solid-State Physics and Electronics. *Phys. Rev. Lett.* Vol.58, (2059-2062), ISSN 0031-9007

Yablonovitch, E., Gmitter, T. J. & Bhat, R. (1988). Inhibited and Enhanced Spontaneous Emission from Optically Thin AlGaAs/GaAs Double Heterostructures. *Phys. Rev. Lett.* Vol.61, No.22, (2546-2549). ISSN 0031-9007

Yongsheng Liu, Wenqin Luo, Renfu Li, et al, (2008). Optical Spectroscopy of Eu^{3+} Doped ZnO Nanocrystals. *J. Phys. Chem. C.* Vol.112, No.3, (686-694), ISSN 1932-7447

Zakharchenya, R. I., Kaplyanskii, A. A., Kulinkin, A. B. et al., (2003). Radiative transitions and spectral hole-burning in MgO: Cr^{3+} nanocrystals. *Phys. Solid State,* Vol.45, No.11, (2209-2212), ISSN 1063-7834; [(2003). Radiative transitions and spectral hole-burning in MgO: Cr^{3+} nanocrystals. *Fiz. Tverd. Tela*, Vol.45, No.11, (2104-2107), ISSN 0367-3294]

17

Photon-Number-Resolution at Telecom Wavelength with Superconducting Nanowires

Francesco Marsili[1,2], David Bitauld[2], Andrea Fiore[2],
Alessandro Gaggero[3], Francesco Mattioli[3], Roberto Leoni[3],
Aleksander Divochiy[4] and Gregory Gol'tsman[4]

[1]*Ecole Polytechnique Fédérale de Lausanne (EPFL), Institute of Photonics and Quantum Electronics (IPEQ), Station 3, CH-1015 Lausanne,*
[2]*COBRA Research Institute, Eindhoven University of Technology, P.O. Box 513, NL-5600MB Eindhoven,*
[3]*Istituto di Fotonica e Nanotecnologie (IFN), CNR, via Cineto Romano 42, 00156 Roma,*
[4]*Moscow State Pedagogical University (MSPU), Department of Physics, 119992 Moscow,*
[1]*Switzerland*
[2]*The Netherlands*
[3]*Italy*
[4]*Russian Federation*

1. Introduction

Optical detection involves converting an optical signal into an electrical signal. Most optical detectors are operated in a linear mode, i.e. the output signal is proportional to the incident photon flux. The main limitation in sensitivity of these linear detectors is the ability to extract a small signal from the amplifier noise in a given bandwidth. For example, avalanche photodiodes (APDs) commonly used in fiber-based optical communications have sensitivities of few hundreds of photons in a 100 ps detection window. When higher sensitivities are needed, single-photon detectors (SPDs) are often used, which operate in a strongly nonlinear mode. Indeed, a pulse containing more than one photon produces the same output signal as a single-photon pulse, which implies that it is not possible to directly measure the number n of photons in a pulse, if the pulse duration is smaller than the detector response time. However, photon number resolving (PNR) detectors are important in quantum communication, quantum information processing and quantum optics for two class of applications. In one case PNR detectors are needed to reconstruct the incoming photon number statistics by ensemble measurements. This is the case of the characterization of nonclassical light sources such as single photon (Yuan et al. 2002) or n-photon (Waks et al. 2004) state generators or of the detection of pulse splitting attacks in quantum cryptography (Brassard et al. 2000). In the second case PNR detectors are needed to perform a single-shot measurement of the photon number. Applications of this kind are linear-optics quantum computing (Knill et al. 2001), long distance quantum communication (which requires quantum repeaters (Sangouard et al. 2007)) and conditional-state preparation (Sliwa &

Banaszek 2003). Moreover, a linear detector with single-photon sensitivity can also be used for measuring a temporal waveform at extremely low light levels, e.g. in long-distance optical communications, fluorescence spectroscopy, and optical time-domain reflectometry.
Among the approaches proposed so far to PNR detection, detectors based on charge-integration or field-effect transistors (Fujiwara & Sasaki 2007; Gansen et al. 2007; Kardynal et al. 2007) are affected by long integration times, leading to bandwidths <1 MHz. Transition edge sensors (TES (Lita et al. 2008)) show extremely high (95%) detection efficiencies but they operate at 100 mK and show long response times (several hundreds of nanoseconds in the best case). Approaches based on photomultipliers (PMTs) (Zambra et al. 2004) and APDs, such as the visible light photon counter (VLPC) (Waks et al. 2003; Waks et al. 2004), 2D arrays of APDs (Yamamoto et al. 2006; Jiang et al. 2007) and time-multiplexed detectors (Achilles et al. 2003; Fitch et al. 2003) are not sensitive or are plagued by high dark count rate and long dead times in the telecommunication spectral windows. Arrays of single photon detectors SPDs additionally involve complex read-out schemes (Jiang et al. 2007) or separate contacts, amplification and discrimination (Dauler et al. 2007). We recently demonstrated an alternative approach (Divochiy et al. 2008; Marsili et al. 2009a), the Parallel Nanowire Detector (PND), which uses spatial multiplexing on a subwavelength scale to provide a single electrical output proportional to the photon number. The device presented significantly outperforms existing PNR detectors in terms of simplicity, sensitivity, speed, and multiplication noise (Divochiy et al. 2008). Here we present the working principle of the device (section 2), its fabrication process (section 3), the results of the optical characterization (section 4), an analysis of the device operation and corresponding design guidelines (section 5) and the first application of a PND to reconstruct an unknown incoming photon number statistics (section 6).

2. Photon number resolution principle

The structure of PNDs is the parallel connection of N superconducting nanowires (N-PND), each of which can be connected in series to a resistor R_0 (N-PND-R, Figure 1a). The detecting element is a 4-6 nm thick, 100 nm wide NbN wire folded in a meander pattern. Each section acts as a nanowire superconducting single photon detector (SSPD) (Verevkin et al. 2002). In SSPDs, if a superconducting nanowire is biased close to its critical current, the absorption of a photon causes the formation of a normal barrier across its cross section, so almost all the bias current is pushed to the external circuit. In PNDs, the currents from different sections can sum up on the external load, producing an output voltage pulse proportional to the number of photons absorbed.
The time evolution of the device after photon absorption can be simulated using the equivalent circuit of Figure 1b. Each section is modeled as the series connection of a switch which opens on the hotspot resistance R_{hs} for a time t_{hs}, simulating the absorption of a photon, of an inductance L_{kin}, accounting for kinetic inductance (Kadin) and of a resistor R_0. The device is connected through a bias T to the bias current source I and to the input resistance of the preamplifier R_{out}. The n firing sections, in red, all carry the same current I_f and the N-n still superconducting sections (unfiring), in blue, all carry the same current I_u. I_{out} is the current flowing through R_{out}. Let I_B be the bias current flowing through each section when the device is in the steady state. If a photon reaches the ith nanowire, it will cause the superconducting-normal transition with a probability $\eta_i=\eta(I_B/I_C^{(i)})$, where η is the current-dependent detection efficiency and $I_C^{(i)}$ is the critical current of the nanowire

(Verevkin et al. 2002) (the nanowires have different critical currents, being differently constricted (Kerman et al. 2007)). Because of the sudden increase in the resistance of firing nanowire, its current (I_f) is redistributed between the other N-1 unfiring branches and R_{out}. This argument yields that if n sections fire simultaneously (in a time interval much shorter than the current relaxation time), part of their currents sum up on the external load.

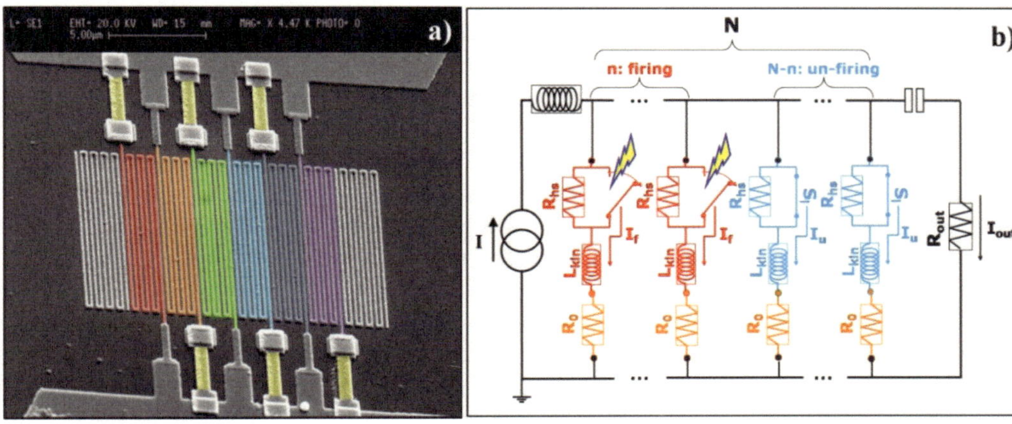

Fig. 1. (a) Scanning electron microscope (SEM) image of a PND with N=6 and series resistors (6-PND-R) fabricated on a 4nm thick NbN film on MgO. The nanowire width is w=100 nm, the meander fill factor is f=40%. The detector active area is Ad=10x10 µm². The devices are contacted through 70nm thick Au-Ti pads, patterned as a 50 Ω coplanar transmission line. The active nanowires (in color) of the PND-R are connected in series with Au-Pd resistors (in yellow). The floating meanders at the two edges of the PND-R pixel are included to correct for the proximity effect. (b) Circuit equivalent of a N-PND-R. The n firing sections, in red, all carry the same current I_f and the N-n still superconducting sections (unfiring), in blue, all carry the same current I_u. I_{out} is the current flowing through the input resistance R_{out} of the preamplifier. From (Marsili et al. 2009b).

The device shows PNR capability if the height of the current pulse through R_{out} for n firing stripes $\bar{I}_{out}^{(n)}$ is n times higher than the pulse for one $\bar{I}_{out}^{(1)}$, i.e. if the leakage current drained by each of the unfiring nanowires $\delta I_{lk}=I_u-I_B$ is negligible with respect to I_B. The leakage current is also undesirable because it lowers the signal available for amplification and temporary increases the current flowing through the still superconducting (unfiring) sections, eventually driving them normal. Consequently, δI_{lk} limits the maximum bias current allowed for the stable operation of the device and then the detection efficiencies of the sections. The leakage current depends on the ratio between the impedance of a section Z_S and R_{out} and it can be reduced by engineering the dimensions of the nanowire (thus its kinetic inductance) and of the series resistor (see sec. 5). The design without series resistors simplifies the fabrication process, but, as Z_S is lower, δI_{lk} significantly limits the detection efficiency of the device.

3. Fabrication

NbN films 3-4 nm thick were grown on sapphire (substrate temperature T_S=900°C (Gol'tsman et al. 2003; Gol'tsman et al. 2007)) or MgO (T_S=400°C (Marsili et al. 2008)) substrates by reactive magnetron sputtering in an argon–nitrogen gas mixture. Using an optimized sputtering technique, our NbN samples exhibited a superconducting transition temperature of T_C =10.5 K for 40-Å-thick films. The superconducting transition width was ΔT_C = 0.3 K.

Both the designs with and without the integrated bias resistors were implemented. Detector size ranges from 5x5 μm² to 10x10 μm² with the number of parallel branches varying from 4 to 14. The nanowires are 100 to 120 nm wide and the fill factor of the meander is 40 to 60%. The length of each nanowire ranges from 25 to 100 μm.

For the devices on MgO, the three nanolithography steps needed to fabricate the structure have been carried out by using an electron beam lithography (EBL) system equipped with a field emission gun (acceleration voltage 100 kV, 20 nm resolution). In the first step e-beam lithography is used to define pads (patterned as a 50 Ω coplanar transmission line) and alignment markers on a 450nm-thick polymethyl methacrylate (PMMA, a positive tone electronic resist) layer. The sample is then coated with a Ti-Au film (60 nm Au on 10nm Ti) deposited by e-gun evaporation, which is selectively removed by lift-off from un-patterned areas. In the second step, a 160nm thick hydrogen silsesquioxane (HSQ FOX-14, a negative tone electronic resist) mask is defined reproducing the meander pattern. The alignment between the different layers is performed using the markers deposited in the first lithography step. All the unwanted material, i.e. the material not covered by the HSQ mask and the Ti/Au film, is removed by using a fluorine based (CHF_3+SF_6+Ar) reactive ion etching (RIE). Finally, with the third step the series resistors (85nm AuPd alloy, 50%-each in weight), aligned with the two previous layers, are fabricated by lift off via a PMMA stencil mask. Our process is optimized to obtain both an excellent alignment between the different e-beam nanolithography steps (error of the order of 100 nm) and a nanowire with high width uniformity (less than 10% (Mattioli et al. 2007)).

Details on the fabrication process of the devices on sapphire can be found in (Gol'tsman et al. 2007).

4. Device optical characterization

In this section we present the results of the optical characterization of PNDs and PND-Rs (Divochiy et al. 2008), i.e. their speed performance (section 4.2), the proof of their PNR capability and their detection efficiency at λ=1.3 μm (section 4.3).

4.1 Measurement setup

Electrical and optical characterizations have been performed in a cryogenic probe station with an optical window and in cryogenic dipsticks. The bias current was supplied through the DC port of a 10MHz-4GHz bandwidth bias-T connected to a low noise voltage source in series with a bias resistor. The AC port of the bias-T was connected to the room-temperature, low-noise amplifiers. The amplified signal was fed either to a 1 GHz bandwidth single shot oscilloscope, a 40 GHz bandwidth sampling oscilloscope, or a 150MHz bandwidth counter for time resolved measurements and statistical analysis. The devices were optically tested using a fiber-pigtailed, gain-switched laser diode at 1.3 μm wavelength (100ps-long pulses, repetition rate 26 MHz), a mode-lock Ti:sapphire laser at 700 nm wavelength (40ps-long pulses, repetition rate 80 MHz), or an 850 nm GaAs pulsed laser (30 ps-long pulses, repetition rate 100 kHz).

In the cryogenic probe station (Janis) the devices were tested at a temperature T=5 K. Electrical contact was realized by a cooled 50 Ω microwave probe attached to a micromanipulator, and connected by a coaxial line to the room-temperature circuitry. The light was fed to the PNDs through a single-mode optical fiber coupled with a long working distance objective, allowing the illumination of a single detector.

In the cryogenic dipsticks the devices were tested at 4.2 K or 2 K. The light was sent through a single-mode optical fiber either put in direct contact and carefully aligned with the active area of a single device or coupled with a short focal length lens, placed far from the plane of the chip to ensure uniform illumination. The number of incident photons per device area was estimated with an error of 5 %.

Throughout the paper, the single photon detection efficiency of an *N*-PND ($\tilde{\eta}$) or of one of its sections (η) are defined with respect to the photon flux incident on the area covered by the device (active area A_d, typically 10 x 10 μm²) or by one section (A_d/N), respectively.

4.2 Speed performance

Figure 2.a shows a single-shot oscilloscope trace of the photoresponse of a 8.6x8 μm² 5-PND under laser illumination (λ=700 nm, 80 MHz repetition rate). Pulses with five different amplitudes can be observed, corresponding to the transition of one to five sections. The measured 80 MHz counting rate represents an improvement of three orders of magnitude over most of the PNR detectors at telecom wavelength (Rosenberg et al. 2005; Fujiwara & Sasaki 2007; Jiang et al. 2007), with the only exception of the SSPD array (Dauler et al. 2007).

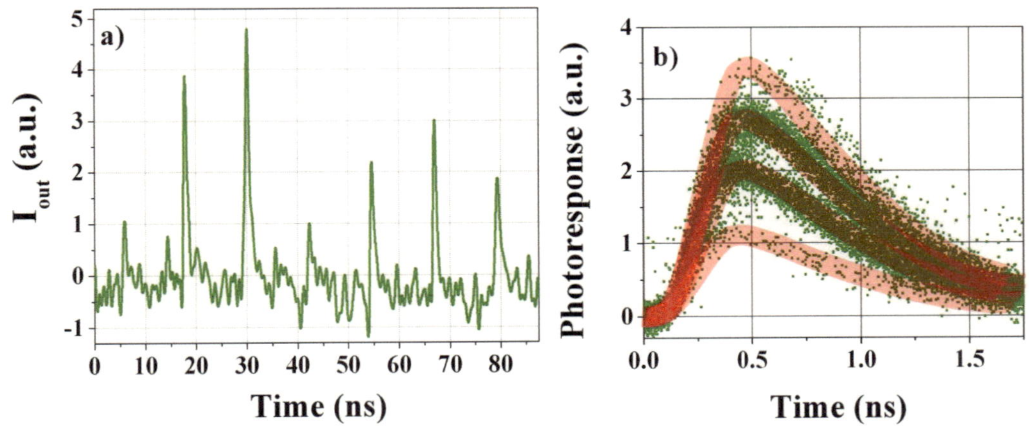

Fig. 2. a. Single-shot oscilloscope trace during photodetection by a 8.6x8 μm2 5-PND. The device was tested under uniform illumination in a cryogenic dipstick dipped in a liquid He bath at 4.2 K. The light pulses at 700 nm form a mode-locked Ti:sapphire laser had a repetition rate of 80 MHz. b. Photoresponse transients taken with a 40 GHz sampling oscilloscope while probing a 10x10 μm2 4-PND-R in the cryogenic probe station under illumination with 1.3 μm, 100ps-long pulses from a laser diode, at a repetition rate of 26MHz. The solid curves are guides to the eyes. From (Marsili et al. 2009b).

We investigated the temporal response of a 10x10 μm^2 4-PND-R probed with light at 1.3 µm wavelength using a 40 GHz sampling oscilloscope (Figure 2.b). All four possible amplitudes can be observed. The pulses show a full width at half maximum (FWHM) as low as 660ps. In a traditional 10x10 μm^2 SSPD, the pulse width would be of the order of 10 ns FWHM, so the recovery of the output current I_{out} through the amplifier input resistance is a factor $\sim 4^2$ faster (see section 5.3), which agrees with results reported by other groups (Gol'tsman et al. 2007; Tarkhov et al. 2008). As shown in section 5.3, the very attractive N^2 scaling rule for the output pulse duration unfortunately does not apply to the device recovery time.

4.3 Proof of PNR capability

Let an *N*-PND be probed with a light whose photon number probability distribution is **S**=[S(*m*)]=[s_m]. The probability distribution of the number of measured photons **Q**=[Q(*n*)]=[q_n] is related to **S** by the relation:

$$Q(n) = \sum_m P^N(n|m) \cdot S(m) \tag{1}$$

where $P^N(n|m)$ is the probability that *n* photons are detected when *m* are sent to the device.

To infer whether a PND is able to measure the number of incoming photons, it can be probed with a Poissonian distribution $S(m)=\mu^m \cdot \exp(-\mu)/m!$ (µ: mean photon number). The limited efficiency $\eta<1$ of the detector is equivalent to an optical loss, and reduces the mean photon number to: $\mu^*=\eta\mu$. In the regime $\mu^*<<1$, $S(m) \sim (\mu^*)^m/m!$, and for µ* low enough (1) can be written as:

$$Q(n) \sim P(n|n) \cdot S(n) \propto (\mu^*)^n / n! \qquad for\ \mu^* << 1 \tag{2}$$

Consequently, the probability Q(1) of detecting one photon is proportional to µ, Q(2) is proportional to μ^2, and so on.

A 10x10 μm^2 5-PND-R was tested with the coherent light of GaAs pulsed laser (λ=850 nm, 30 ps pulse width, 100kHz repetition rate), whose photon number distribution is close to a Poissonian. The photoresponse from the device was sent to a 150 MHz counter. The detection probabilities relative to one-, two- and three-photon absorption events are plotted for µ varying from 0.15 to 40 in Figure 3.a. As the mean single-photon detection efficiency $\tilde{\eta}$ of the device (defined with respect to the photon flux incident on the total active area covered by the device A_d) is a few percent (Figure 3.b) and µ is a few tens, the condition $\eta\mu=\mu^*<<1$ is verified and (2) is therefore valid. Indeed, the fittings clearly show that $Q(1) \propto \mu$, $Q(\mu,2) \propto \mu^2$ and $Q(\mu,3) \propto \mu^3$, which demonstrates the capability of the detector to resolve one, two and three photons simultaneously absorbed.

The device mean single-photon detection efficiency $\tilde{\eta}$ at λ=1.3 µm and the dark-counts rate DK were measured as a function of the bias current at T= 2.2 K (Figure 3.b). The lowest DK value measured was 0.15 Hz for $\tilde{\eta}$=2% (yielding a noise equivalent power (Miller et al. 2003) NEP=4.2x10^{-18} W/$Hz^{1/2}$), limited only by the room temperature background radiation coupling to the PND. This sensitivity outperforms most of the other approaches by one-two orders of magnitude (with the only exception of transition-edge sensors (Rosenberg et al. 2005), which require a much lower operating temperature).

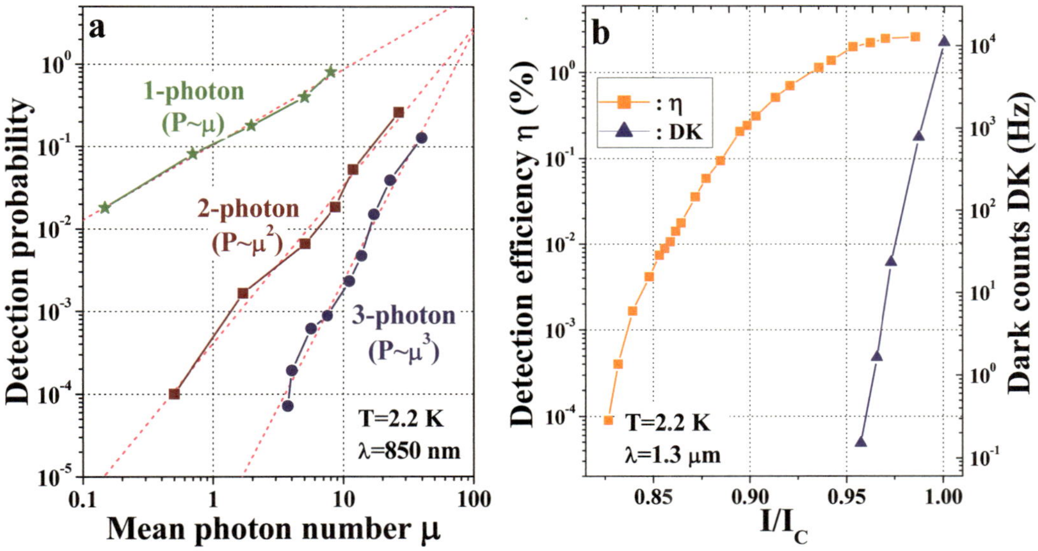

Fig. 3. a. Detection probabilities of a 10x10 μm² 5-PND-R relative to the one (stars), two (squares) and three-photon (circles) absorption events as a function of the mean photon number per pulse μ. device was mounted in cryogenic dipstick dipped in a liquid He bath at 2.2 K. A single-mode optical fiber was put in direct contact and aligned with the active area of the device. The power level was set with a variable fiber-based optical attenuator. b. Mean detection efficiency at 1.3 μm and dark-counts rate vs bias current of a 10x10 μm² 5-PND-R. The device was fiber-coupled and mounted on a cryogenic dipstick dipped in a liquid He bath at 2.2 K.

5. PND design

In this section, we provide a detailed analysis of the device operation and guidelines for the design of PNDs with optimized performance in terms of efficiency, speed and sensitivity (see also (Marsili et al. 2009b)).

The first step is to define the relevant parameter space. The width of the nanowire (w=100 nm) and the filling factor (f=50%) of the meander are fixed by technology, the thickness of the superconducting film (t=4nm) is the optimum value yielding the maximum device efficiency and the active area (A_d=10 x 10 μm²) is fixed by the size of the core of single mode fibers to which the device must be coupled. We consider single-pass geometries (no optical cavity), but the same guidelines can be applied to cavity devices with optimized absorption (Rosfjord et al. 2006). The parameters of the PND-R that can be used as free design variables are: the number of sections in parallel N, the value of the series resistor R_0 and the value of the inductance of each section L_0. The number of sections in parallel N can be chosen within a discrete set of values (N=2, 3, 4, 6, 7, 10, 17), which satisfy the constraints of w, f, size of the pixel and that the number of stripes in each sections is to be odd (we consider the geometry of Figure 1a). The value of L_0 is the sum of the kinetic inductance of each meander L_{kin} and of a series inductance which can be eventually added. L_{kin} is not a design parameter, as it is fixed by w, t, f, A_d and N. If no series inductors are added (bare devices, L_0=L_{kin}), the value of L_0 for each N is listed in Table 1.

N	L_0	SQ
2	225 nH	2500
3	153 nH	1700
4	117 nH	1300
6	81 nH	900
7	63 nH	700
10	45 nH	500
17	27 nH	300

Table 1. Inductance (L_0) and number of squares (SQ) of each section for all possible values of N. The width of the nanowires is w=100 nm, the thickness is t=4 nm. The kinetic inductance per square was estimated ($L_{kin}/\square$=90 pH) from the time constant of the exponential decay of the output current ($\tau_{out}=\tau_f=L_{kin}/R_{out}$, see sec. 5.3) for a standard 5x5μm² SSPD (Marsili et al. 2008).

An additional free parameter, relative to the read-out, is the impedance seen by the device on the RF section of the circuit R_{out}, which is 50 Ω (of the matched transmission line) in the actual measurement setup (see section 4.1), but which can be varied from zero to infinite introducing a cold preamplifier stage.

The target performance specifications are the single-photon detection efficiency (η), the signal to noise ratio (SNR) and the maximum repetition rate (speed), which must be optimized under the constraints that the operation of the device is stable and that it is possible to detect a certain maximum number of photons (n_{max}) dependent on the specific application.

This section is organized as follows. First we present the electrical equivalent model of the device developed to study its working principle and to define design guidelines (section 5.1). Then we define the figures of merit of the device performance in terms of efficiency (section 5.2), speed (section 5.3) and sensitivity (section 5.4) and we analyze their dependency on the design parameters (L_0, R_0, R_{out}, N).

5.1 Electrical model

Although a comprehensive description of PND operation should combine thermal and electrical modeling of the nanowires (Yang et al. 2007), it is possible to use a purely electrical model (see section 2 and Figure 1b) to make a reliable guess on how the device performance varies when moving in the parameter space (Marsili et al. 2009b).

In this model, the dependence of L_{kin} on the current flowing through the nanowire was disregarded, and it was assumed constant. Furthermore, it has been shown (Yang et al. 2007) that changing the values of the kinetic inductance of an SSPD or of a resistor connected in series to it results in a change of the hotspot resistance and of its lifetime, eventually causing the device to latch to the normal state. The simplified analysis presented here does not take into account these effects, and considers both R_{hs} and t_{hs} as constant (R_{hs}=5.5 kΩ, t_{hs}=250ps), and that device cannot latch. However, the results of this approach can still quantitatively predict the behavior of the device in the limit where the fastest time constant of the circuit τ_f (see section 5.3) is much higher than the hotspot lifetime ($\tau_f \gg t_{hs}$), and give a reasonable qualitative understanding of the main trends of variation of the performance of faster devices ($\tau_f \sim t_{hs}$).

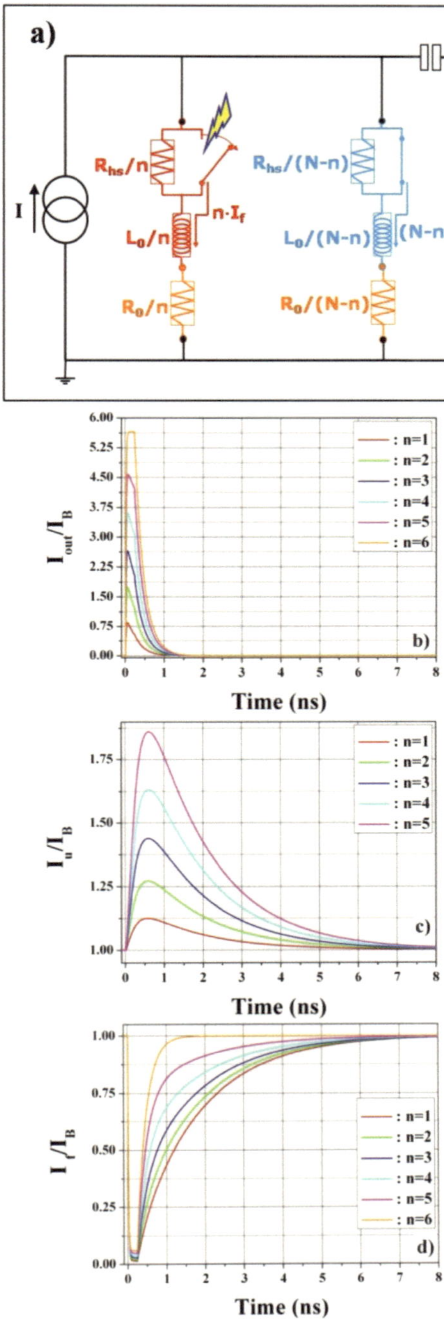

Fig. 4. (a) Simplified circuit of a *N*-PND-R, where the two sets of *n* firing and the *N*-*n* unfiring sections have been substituted by their Thévenin-equivalents. (b-d) Simulated time evolution of I_u (b), I_{out} (c) and I_f (d) for a 6-PND-R as *n* increases from 1 to 6. The parameters of the circuit are: L_0=L_{kin}=81 nH, R_0=50 Ω, R_{out}=50 Ω, R_{hs}=5.5 kΩ, and t_{hs}=250ps. From (Marsili et al. 2009b).

To gain a better insight on the circuit dynamics (see sec. 5.3) and to reduce the calculation time, the N+1 mesh circuit of Figure 1.b can be simplified to the three mesh circuit of Figure 4.a applying the Thévenin theorem on the n firing sections and on the remaining N-n still superconducting (unfiring) sections, separately. Figure 4.b to d show the simulation results for the time evolution of the currents flowing through R_{out} and through the unfiring (I_u) and firing (I_f) sections of a PND with 6 sections and integrated resistors (6-PND-R) and for the number of firing sections n ranging from 1 to 6. As n increases, the peak values of the output current ($I_{out,}$ Figure 4.b) and of the current through the unfiring sections (I_u, Figure 4.c) increase. The firing sections experience a large drop in their current (I_f, Figure 4.d), which is roughly independent on n. The observed temporal dynamics will be examined in the following sections.

5.2 Current redistribution and efficiency

Let $\overline{\delta I_{lk}}^{(n)}$ be the peak value of the leakage current drained by each of the still superconducting (unfiring) nanowires when n sections fire simultaneously. The stability requirement translates in the condition that for each unfiring section: $I_B + \overline{\delta I_{lk}}^{(n_{max})} \leq I_C$ (as the leakage current increases with n, $\overline{\delta I_{lk}}^{(n_{max})}$ represents the worst case). This limits the bias current and therefore the single-photon detection efficiency (η), which, for a certain nanowire geometry (i.e. w, t fixed), is a monotonically increasing function of I_B/I_C (Verevkin et al. 2002). For instance, to detect a single photon (at λ=1.3 µm, T=1.8K) in a section with an efficiency equal to 80% of the maximum value set by absorption (~32%, (Gol'tsman et al. 2007)), $\overline{\delta I_{lk}}^{(n_{max})}$ should be made ≤33% of I_B. Therefore the leakage current strongly affects the performance of the device and it is to be minimized, which makes it very important to understand its dependency from the design parameters: $\overline{\delta I_{lk}}^{(n)}(N, L_0, R_0, R_{out})$.

The leakage current can be investigated just in the case of n=1, as the design guidelines drawn from this analysis still apply to higher n (Marsili et al. 2009b). The dependency of $\overline{\delta I_{lk}}^{(1)}$ on N and L_0 at fixed R_0 and R_{out} (both equal to 50 Ω) is shown in Figure 5.a: an orange line highlights bare devices (L_0=L_{kin}, see Table 1) and the colored bars are relative to devices which respect the constraints on the geometry of the structure (L_0>L_{kin}), while the grey bars refer to purely theoretical devices which just show the general trend. For any N, the current redistribution increases with decreasing L_0, as the impedance of each section decreases. Keeping L_0 constant, $\overline{\delta I_{lk}}^{(1)}$ decreases with increasing N, as the current to be redistributed is fixed and the number of channels draining current increases. For this reason also the increase of redistribution with decreasing L_0 becomes weaker for high N.

The dependency of $\overline{\delta I_{lk}}^{(1)}$ on R_0 and R_{out} (for the same N, L_0) is very intuitive (Marsili et al. 2009b). Indeed, the redistribution decreases as R_0 increases because the impedance of each section increases with respect to the output resistance. For the same reason, $\overline{\delta I_{lk}}^{(1)}$ is strongly reduced when R_{out} is decreased.

In conclusion, the result of this simplified analysis is that, to minimize the leakage current and thus maximize the efficiency, N, L_0 and R_0 must be made as high as possible and R_{out} as low as possible. We note however that R_0 cannot be increased indefinitely to avoid that the nanowire latches to the hotspot plateau before I_B reaches I_C (Marsili et al. 2008).

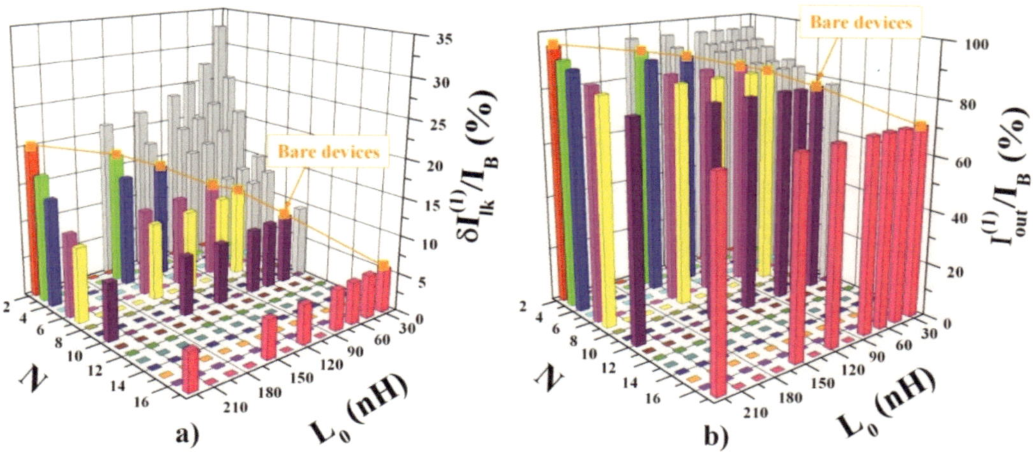

Fig. 5. Peak value of the leakage current $\overline{\delta I}_{lk}^{(1)}$ drained by each of the still superconducting (unfiring) nanowires (a) and of the output current $\overline{I}_{out}^{(1)}$ (b) when only one section fires plotted as a function of the number of sections in parallel N and of the value of the inductance of each section L_0. The leakage current and the output current are expressed in % of the bias current I_B because they are proportional to it. From (Marsili et al. 2009b).

5.3 Transient response and speed

Before proceeding to the analysis of the SNR and speed performances of the device, it is necessary to discuss the characteristic recovery times of the currents in the circuit.

The transient response of the simplified equivalent electrical circuit of the N-PND (Figure 4.a) to an excitation produced in the firing branch can be found analytically. Therefore, the transient response of the current through the firing sections I_f, through the unfiring sections I_u and through the output I_{out} after the nanowires become superconducting again ($t \geq t_{hs}$) can be written as:

$$\begin{cases} I_f \propto \frac{N-n}{N}\exp\left(-t/\tau_s\right)+\frac{n}{N}\exp\left(-t/\tau_f\right) \\ I_u \propto \exp\left(-t/\tau_s\right)-\exp\left(-t/\tau_f\right) \\ I_{out} \propto \exp\left(-t/\tau_f\right) \end{cases} \quad (3)$$

where $\tau_s = L_0/R_0$ and $\tau_f = L_0/(R_0+NR_{out})$ are the "slow" and the "fast" time constant of the circuit, respectively. This set of equations describes quantitatively the time evolution of the currents after the healing of the hotspot in the case $\tau_f >> t_{hs}$, and it provides a qualitative understanding of the recovery dynamics of the circuit for shorter τ_f.

The recovery transients ($t \geq t_{hs}$) of I_{out}, δI_{lk} and I_f for a 4-PND-R simulated with the circuit of figure Figure 4.a are shown in figure 6a, b, c, respectively (in blue) for different number of firing sections (n=1 to 4). As n increases from 1 to 4, the recoveries of I_{out} and δI_{lk} change only by a scale factor. On the other hand, the transient of I_f depends on n and becomes faster increasing n, as qualitatively predicted by the first of equations (3). Indeed, I_f consists in the

sum of a slow and a fast contribution, whose balance is controlled by the number of firing sections n. To prove the quantitative agreement with the analytical model in the limit $\tau_f >> t_{hs}$, the simulated transients of I_{out}, δI_{lk} and I_f have been fitted (figure 6a, b, c, respectively, in red) using the set of equations (3), and four fitting parameters (τ_s, τ_f, a time offset t_0 and a scaling factor K). The values of τ_s and τ_f obtained from the three fittings (of I_{out}, of δI_{lk} and of the whole set of four I_f for n=1,..., 4) closely agree with the values calculated from the analytical expressions presented above and the parameters of the circuit (τ_s^*=2.30 ns ,τ_f^*=460 ps).
To quantify the speed of the device, we take $f_0=(t_{reset})^{-1}$ as the maximum repetition frequency, where t_{reset} is the time that I_f needs to recover to 95% of the bias current after a detection event. According to the results presented above, which are in good agreement with experimental data (Figure 2.b), I_{out} decays exponentially with the same time constant for any n ($\tau_{out}=\tau_f$), which, for a bare N-PND, is N^2 times shorter than a normal SSPD of the same surface (Gol'tsman et al. 2007; Tarkhov et al. 2008). This however does not relate with the speed of the device. Indeed, t_{reset} is the time that the current through the firing sections I_f needs to rise back to its steady-state value ($I_f \sim I_B$). In the best case of $n=N$, I_f rises with the fast time constant τ_f, but in all other cases the slow contribution becomes more important as n decreases (see Figure 4.d and figure 6.c), until, for n=1, $I_f \sim [1-\exp(-t/\tau_s)]$. The speed performance of the device is then limited by the slow time constant ($t_{reset} \sim 3 \cdot \tau_s$), which means that an N-PND is only N times faster than a normal SSPD of the same surface, being as fast as a normal SSPD whose kinetic inductance is the same as one of the N section of the N-PND.

5.4 Signal to noise ratio

The peak value and the duration of the output current pulse are a function of the design parameters (see below and section 5.3, respectively). As the output pulse becomes faster, amplifiers with larger bandwidth are required and thus electrical noise become more important. To assess the possibility to discriminate the output pulse from the noise, we define the signal to noise ratio (SNR) as the ratio between the maximum of the output current $\bar{I}_{out}$ and the rms value of the noise-current at the preamplifier input I_n, SNR=$\bar{I}_{out}/\bar{I}_n$.
The peak value of the output current when n sections fire simultaneously (see Figure 4.b, relative to a 6-PND-R) can be written as:

$$I_{out}^{(n)} = n\left(I_B - I_f^{(n)*}\right) - (N-n)\,\delta I_{lk}^{(n)*}$$

where the starred values refer to the time $t=t^*$ when the output current peaks.
As n=1 represents the worst case, to evaluate the performance of the device in terms of the SNR, the dependency of $\bar{I}_{out}^{(1)}$ from the design parameters is investigated: $\bar{I}_{out}^{(1)}(N, L_0, R_0, R_{out})$.
The dependency of $\bar{I}_{out}^{(1)}$ on N and L_0 at fixed R_0 and R_{out} (both equal to 50 Ω) is shown in Figure 5.b. Inspecting the values of $\bar{I}_{out}^{(1)}$ and $\overline{\delta I}_{lk}^{(1)}$ for the same device in Figure 5, it becomes clear that they add up to a value well above to I_B, which is due to the fact that the output current and of the leakage current peak at two different times t* and t_{lk}, respectively (Figure 4.b, c). Furthermore, as $t_{lk}>t^*$, the output current is not significantly affected by redistribution, because I_{out} is maximum when δI_{lk} is still beginning to rise.

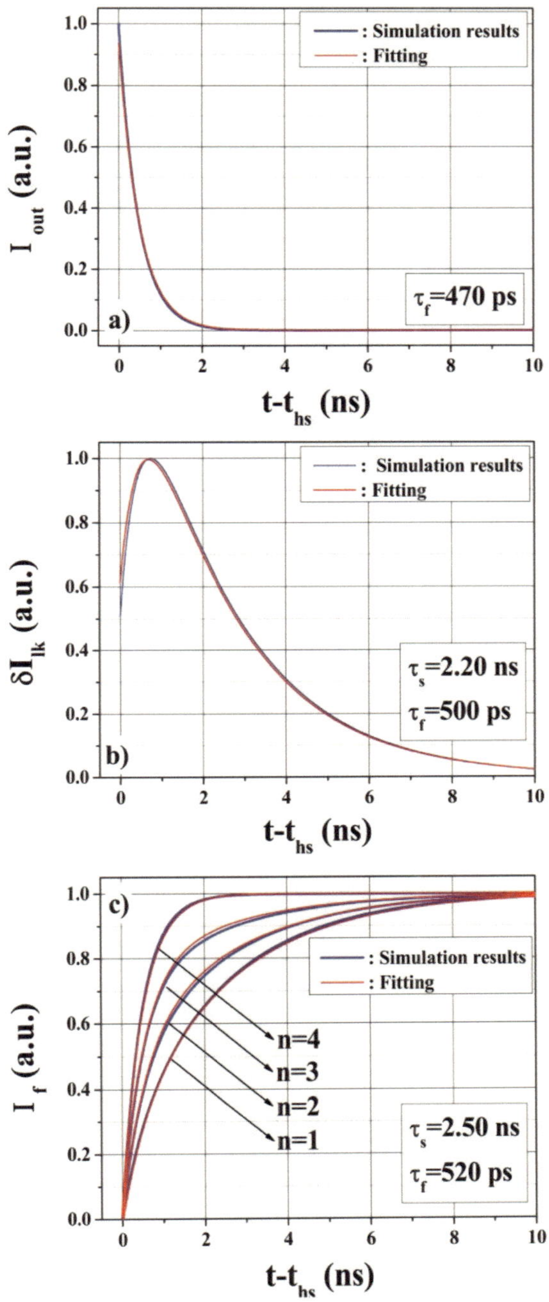

Fig. 6. Recovery transients ($t \geq t_{hs}$) of I_{out} (a), δI_{lk} (b), and I_f (c) for a 4-PND-R as n increases from 1 to 4. The simulated transients are in blue, the fitted curves are in red. The parameters of the circuit used for the simulations are: L_0=L_{kin}=117 nH, R_0=50 Ω, R_{out}=50 Ω, R_{hs}=5.5 kΩ, and t_{hs}=250ps. The three sets of curves are fitted by equations (3) (multiplied by K, and shifted by t_0), where the values of τ_s and τ_f are shown in the insets. From (Marsili et al. 2009b).

The expression for t_{lk} can be derived from (3): $t_{lk}= L_0/(N{\cdot}R_{out})\ln(1+N{\cdot}R_{out}/R_0)$, which means that increasing the device speed (decreasing L_0 or R_0, N or R_{out}) makes the redistribution faster and then $\bar{I}_{out}^{(1)}$ lower.

So, for any given N, $\bar{I}_{out}^{(1)}$ decreases (Figure 5.b) with decreasing L_0, both because $\overline{\delta I}_{lk}^{(1)}$ is higher and because t_{lk} is lower. Keeping L_0 constant, $\bar{I}_{out}^{(1)}$ decreases with increasing N because even though $\overline{\delta I}_{lk}^{(1)}$ decreases, the redistribution peaks earlier and the number of channels draining current increases.

The redistribution speed-up explains the dependency of $\bar{I}_{out}^{(1)}$ on R_0 (for the same N, L_0). Indeed, even though $\overline{\delta I}_{lk}^{(1)}$ decreases as R_0 increases (see section 5.2), the output current decreases due to the decrease of t_{lk}: $\delta I_{lk}^{(1)*}$ increases despite the decrease of the peak value of the leakage current. On the other hand, a decrease in R_{out} makes the redistribution much less effective, as t_{lk} decreases slower with decreasing R_{out} than with increasing R_0 (Marsili et al. 2009b).

In conclusion, to maximize the output current, N, R_0 and R_{out} must be minimized, while L_0 must be made as high as possible.

The rms value of noise-current at the preamplifier input I_n can be written as $I_n=\sqrt{S_n\Delta f}$, where S_n is the noise spectral power density of the preamplifier and Δf is the bandwidth of the output current I_{out}, which is estimated as $\Delta f=1/\tau_{out}$, where $\tau_{out}=\tau_f= L_0/(R_0+NR_{out})$ is the time constant of the exponential decay of I_{out} (see sec. 5.3). I_n is then a function of the parameters of the device and of the read-out through S_n and τ_f, and like I_{out} it is minimized minimizing N, R_0 and R_{out} and maximizing L_0.

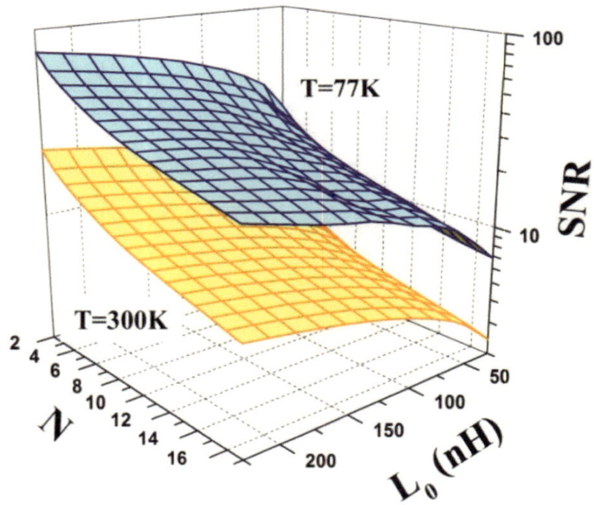

Fig. 7. SNR as a function of N and L_0 relative to commercially available cryogenic (77 K working temperature, in blue) and room-temperature amplifiers (in yellow). For the cryogenic amplifiers the following noise figures were used, relative to different -3 dB bandwidths: F=0.44 dB (Δf=0.1-4 GHz), F=1.3 dB (Δf=0.5-20 GHz), F=1.8 dB (Δf=0.5-40 GHz). For the room-temperature amplifiers: F=1.1 dB (Δf=0.1-4 GHz), F=2.13 dB (Δf=0.1-20 GHz), F=5 dB (Δf=0.1-40 GHz).

The same optimization criteria apply then naturally to the SNR. The dependence of the SNR from N and L_0 is shown in Figure 7 for cryogenic (77 K working temperature, in blue) and room-temperature amplifiers (in yellow). Amplifiers with different -3 dB bandwidths have been considered, depending on the bandwidth of the output current pulse that they were supposed to amplify. Depending on the amplifier bandwidth, noise figures of F=0.44 to 1.8 dB (F=1.1 to 5 dB) have been considered in the calculation of Sn for the room-temperature (cryogenic) amplifier. The input resistance is R_{out}=50Ω.
The main design guidelines which can be deduced from the analysis of sections 5.2 to 5.4 are summarized in Table 2. The type of dependency of $\overline{\delta I_{lk}}$, f_0, $\bar{I}_{out}$ and I_n from the design parameters (L_0, R_0, R_{out}, N) is indicated.

	L_0	R_0	R_{out}	N
δI_{lk}	↘	↘	↗	↘
f_0	↘	↗	--	--
I_{out}	↗	↘	↘	↘
I_n	↘	↗	↗	↗

Table 2. Dependency of the figures of merit from the design parameters: increasing with increasing the parameter (↗), decreasing with increasing the parameter (↘), independent (--).

6. Application to the measurement of photon number statistics

We wish to determine whether the PND can be used to measure an unknown photon number probability distribution **S** (see section 4.3). Indeed, the light statistics measured with a PND differ from the original one due to non-idealities such as the limited number of sections and limited and non-uniform efficiencies (η_i) of the different sections.
In this section, we present the modeling tools (section 6.1) used to fully characterize the device (section 6.2) (Marsili et al. 2009a) and to develop an algorithm to estimate the photon number statistics of an unknown light (section 6.3) (Marsili et al. 2009b).

6.1 Modeling and simulation

Equation (1) may be rewritten in a matrix form as $\mathbf{Q}=\mathbf{P^N}\cdot\mathbf{S}$, where $P^N=\left[P^N(n|m)\right]=\left[p^N_{nm}\right]$ is the matrix of the conditional probabilities of an N-PND. Assuming that the illumination of the device is uniform, the parallel connection of N nanowires can be considered equivalent to a balanced lossless N-port beam splitter, every channel terminating with a single photon detector (SPD) (Figure 8.a). Each incoming photon is then equally likely to reach one of the N SPDs (with a probability $1/N$). Each SPD can detect a photon with a probability η_i (i=1,..,N) different from all the others, and gives the same response for any number ($n\geq1$) of photons detected (Figure 8.b). The number of SPDs firing then gives the measured photon number. Two classes of terms in P^N can be calculated directly, the others being derived from these by a recursion relation (Divochiy et al. 2008; Marsili et al. 2009a). These terms are the probabilities $p^N_{m,m}$ that all the $m\leq N$ photons sent are detected and $p^N_{0,m}$ that no photons are detected when m are sent.

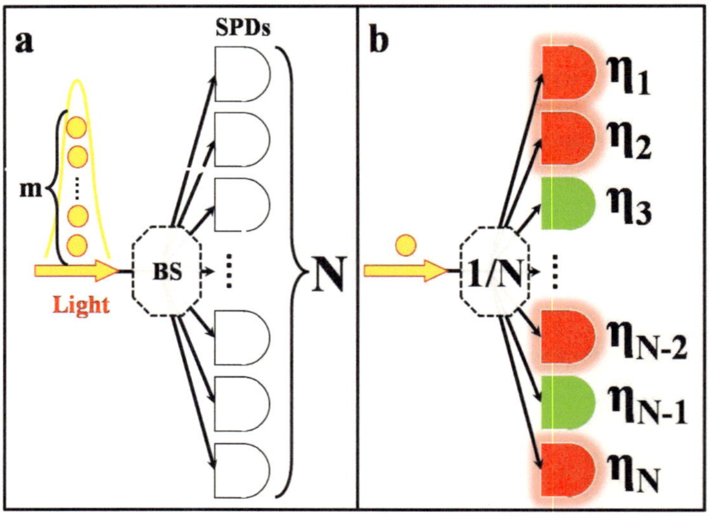

Fig. 8. a. Optical equivalent of an *N*-PND. b. k^{th} possible configuration of *n* firing (red) and *N*-*n* unfiring (green) sections. Each incoming photon is equally likely to reach one of the *N* SPDs (with a probability 1/*N*). Each SPD can detect a photon with a probability η_i (i=1,..,*N*) different from all the others, and it gives the same response for any number ($n \geq 1$) of photons detected. From (Marsili et al. 2009a).

In the case of zero detections, $p_{0,m}^{N}$ is given by:

$$p_{0,m}^{N} = \sum_{i_1=1,\dots,i_m=1}^{N} \left[\frac{1-\eta_{i_1}}{N} \cdot \dots \cdot \frac{1-\eta_{i_m}}{N} \right] \tag{4}$$

which assumes that a photon incident in the ith nanowire fails to be detected with an independent probability of $(1-\eta_i)$. The sum in (4) accounts for all the possible combinations when taking *m* elements in an ensemble of *N* with order and with repetition (permutations with repetitions). This because more than one photon can hit the same stripe (which gives the repetition), and the photons are considered distinguishable (which gives the order).

In the case that all the photons are detected, since *m* photons must reach *m* distinct nanowires:

$$p_{m,m}^{N} = \sum_{\substack{i_1=1,\dots,i_m=1 \\ i_p \neq i_q \ for \ p \neq q}}^{N} \left[\frac{\eta_{i_1}}{N} \cdot \dots \cdot \frac{\eta_{i_m}}{N} \right] \quad \text{for } m \leq N \tag{5}$$

The sum in (5) accounts for all the possible combinations when taking *m* elements in an ensemble of *N* with order and without repetition (permutations without repetitions). This because only one photon can hit the same stripe (which gives the non-repetition), and the photons are considered distinguishable (which gives the order).

The recursion relation for p_{nm}^{N} is:

$$p_{nm}^{N} = p_{n,m-1}^{N}\left[\frac{n}{N}+\frac{n!}{N!}\cdot\sum_{\substack{i_1=1,\dots,i_{N-n}=1\\ i_p\neq i_q\ for\ p\neq q}}^{N}\left(\frac{1-\eta_{i_1}}{N}+\dots+\frac{1-\eta_{i_{N-n}}}{N}\right)\right]+$$
$$+\,p_{n-1,m-1}^{N}\left[\frac{(n-1)!}{N!}\cdot\sum_{\substack{i_1=1,\dots,i_{N-(n-1)}=1\\ i_p\neq i_q\ for\ p\neq q}}^{N}\left(\frac{1-\eta_{i_1}}{N}+\dots+\frac{1-\eta_{i_{N-(n-1)}}}{N}\right)\right] \quad (6)$$

The first term on the right-hand side of (6) is the probability that n photons are detected when m-1 are sent, times the probability that the mth photon reaches one of the n nanowires already occupied (first term in the square brackets) or that it fails to be detected reaching one of the N-n unoccupied nanowires (second term in the square brackets). To clarify how the latter probability is derived, it is sufficient to consider a particular configuration k (see Figure 8.b) of n firing stripes (which have already detected a photon) and N-n unfiring stripes (still active). The probability that the incoming photon will not be detected when incident on any of the N-n unfiring stripes is then written as:

$$p_k = \frac{1-\eta_{i_1}}{N}+\dots+\frac{1-\eta_{i_{N-n}}}{N} \quad (7)$$

where $i_1 \dots i_{N-n}$ are the N-n stripes active in the kth configuration of (n)firing-(N-n)unfiring stripes considered. So a mean must be calculated on all the possible (n)firing-(N-n)unfiring configurations for the N stripes. Let C be the number of all these configurations. The mean is then calculated summing C terms of the type (7), and dividing by C: $1/C \cdot \sum_{k=1}^{C} p_k$. C is the number of permutations without repetitions of N-n elements in an ensemble of N, and it is given by the binomial coefficient: $C=N!/(N-(N-n))!=N!/n!$. The second term on the right-hand side of (6) is the probability that n-1 photons are detected when m-1 are sent times the probability that the mth photon reaches one of the N-(n-1) unoccupied nanowires and it is detected.

To prove the consistency of the analytical model, the probability distribution of the number of measured photons **Q** calculated from P^N by (1) was cross-checked with the $\mathbf{Q}_{MC}$ resulting from a Monte Carlo simulation (Marsili et al. 2009a). The input parameters of the simulation are the incoming photon number probability distribution **S**, the number of parallel stripes N, and the vector of the single-photon detection efficiencies of the different sections of the device $\boldsymbol{\eta}=[\eta_i]$.

6.2 Matrix of conditional probabilities

It has been shown (Achilles et al. 2004; Lee et al. 2004) that an unknown incoming photon number distribution **S** can be recovered if **Q** and $\mathbf{P}^N$ are known.

Let an N-PND be probed with a light whose photon number probability distribution is **S**, and its output be sampled H times. The result of the observation can be of N+1 different types (i.e. 0,..., N stripes firing), so an histogram of the H events can be constructed, which can be represented by a (N+1)-dimensional vector $\mathbf{r}=[r_i]$, where r_i is the number of runs in which the outcome was of the ith type. The expectation value of the statistics obtained from the histogram is $E[\mathbf{Q}_{ex}=\mathbf{r}/H]=\mathbf{Q}$. Considering equations (4) to (6), it is clear that the matrix of the conditional probabilities of a *N*-PND depends only on the vector of the *N* single-photon detection efficiencies of the different sections of the device $\eta=[\eta_i]$. The vector η can be determined from the statistics $\mathbf{Q}_{ex}$ measured when probing the device with a light of known statistics **S** as described in the following.

A 5-PND (A_d=8.6x8 μm^2) was tested with the coherent emission from a mode-locked Ti:sapphire laser under uniform illumination, whose photon number probability distribution is close to a Poissonian and could be fully characterized by the mean photon number μ with a power measurement. To determine $\mathbf{Q}_{ex}$, histograms of the photoresponse voltage peak V_{pk} were built for values of μ ranging from ~1 to ~100. The signal from the device was sent to the 1 GHz oscilloscope, which was triggered by the synchronization generated by the laser unit. The photoresponse was sampled for a gate time of 5ps, making the effect of dark counts negligible. The discrete probability distribution $\mathbf{Q}_{ex}$ was reconstructed from the continuous probability density $q(V_{pk})$ fitting the histograms to the sum of 6 Gaussian distributions (corresponding to the five possible pulse levels plus the zero level) and calculating their area (Figure 9).

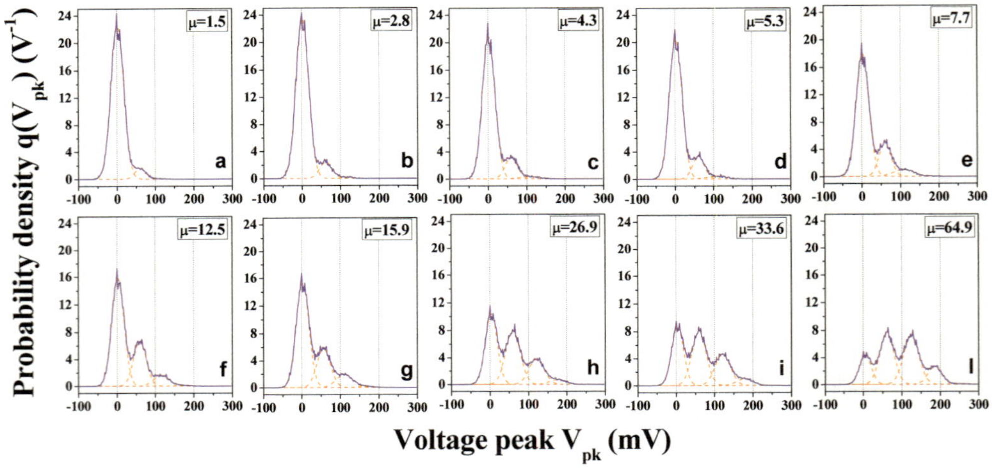

Fig. 9. Histograms of the photoresponse voltage peak. The 5-PND was tested in a dipstick dipped in a liquid He bath at 4.2 K. The light pulses at 700nm form the mode-lock Ti:sapphire laser were 40ps wide (after the propagation in the optical fiber) and the repetition rate was 80MHz. The average input photon number per pulse μ was set with a free space variable optical attenuator. Increasing μ, form 1.5 (a) to 64.9 (l), the shape of the histograms changes as the probability to observe higher response amplitudes increases. The solid lines are the experimental histograms. The dashed lines represent the fitted Gaussian distribution of each possible pulse level. From (Marsili et al. 2009a).

The probability distribution of the number of measured photons **Q** (expressed by(1)) was then fitted to the experimentally measured distribution $\mathbf{Q}_{ex}$ using η as a free parameter (Figure 10). The resulting η and matrix of the conditional probabilities are shown in Figure 11. The fitted efficiencies are rather uniform (2.9±0.5%), indicating a high-quality fabrication process. The value of η obtained from the fitting was then used as an input parameter of Monte Carlo simulations (see above) used to calculate $\mathbf{Q}_{MC}$ for each value of μ. The three sets of values for the photocount statistics of six levels are in good agreement over almost two orders of magnitude of μ, confirming the validity of the analytical model.

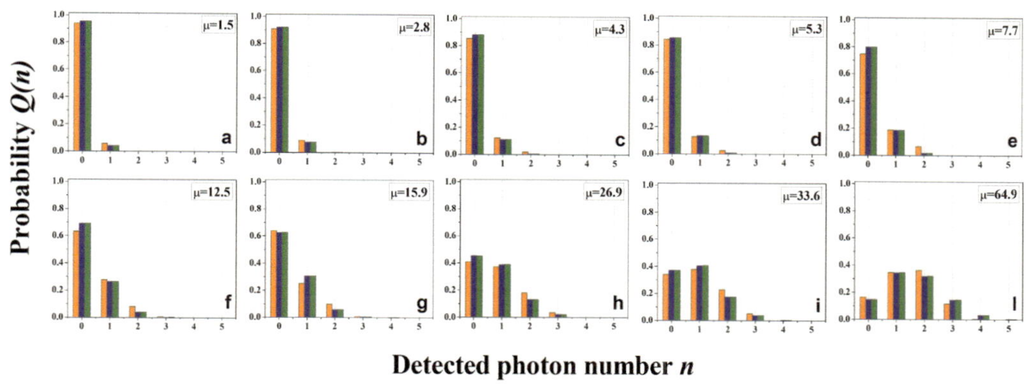

Fig. 10. Experimental (Q_{ex}), fitted (Q) and simulated (Q_{MC}) probability distributions of the number of measured photons. The experimental probability distribution Q_{ex} (orange bins) was estimated from the continuous probability density $q(V_{pk})$ of Figure 9. The 5-PND was probed with several incident mean photon numbers μ: 1.5, 2.8, 4.3, 5.3, 7.7, 12.5, 15.9, 26.9, 33.6, 64.9. The experimental values for Q_{ex} were then fitted (blue bins) using a genetic algorithm to recover the vector of detection efficiencies η. The value of η obtained was used to calculate Q_{MC} for each value of μ by Monte Carlo simulations (green bins). From (Marsili et al. 2009a).

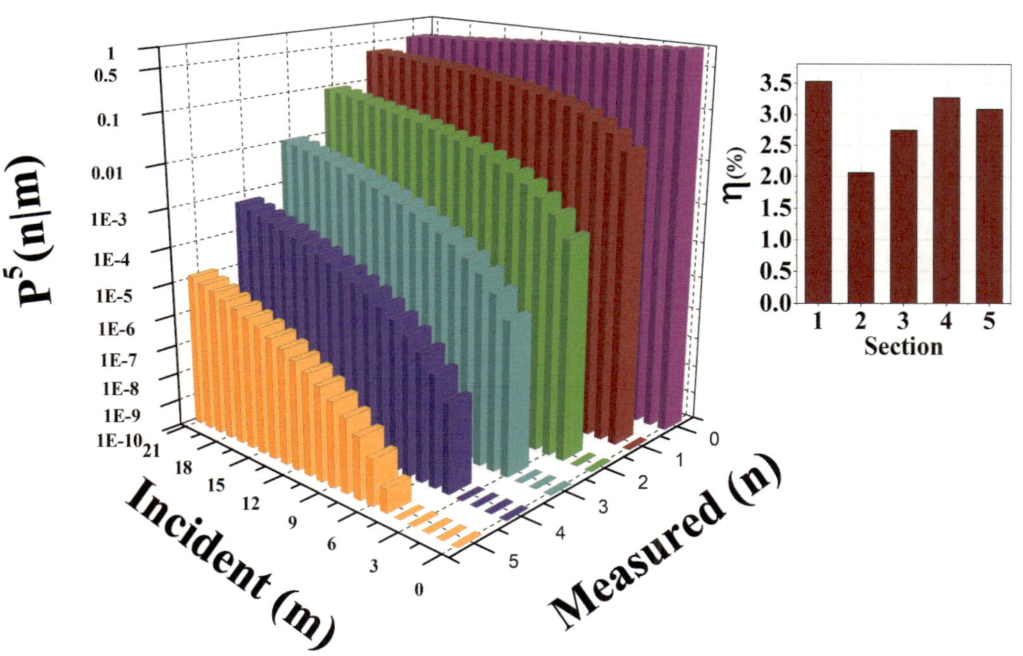

Fig. 11. Conditional probability matrix for a 8.6x8 μm² 5-PND (with no integrated series resistors), calculated from the vector η of the 5 single-photon detection efficiencies (relative to T=4.2 K, λ=700nm) of the different sections of the device (inset). From (Marsili et al. 2009b).

6.3 Maximum-Likelihood (ML) estimation

i. ML method

The $\mathbf{P}^N$ matrix provides a full description of the detector. Once $\mathbf{P}^N$ is known, several approaches can be used to reconstruct **S** from the histogram **r**. In the case no assumptions on the form of **S** are made, the maximum likelihood (ML) method is the most suitable, as it is the most efficient in solving this class of problems (Eadie et al.).

Let $\mathbf{R}= R_0,\dots, R_N$ be the random vector of the populations of the (N+1) different bins of the histogram after H observations. The joint probability density function $L(\mathbf{r}|\mathbf{Q})$ for the occurrence of the particular configuration $\mathbf{r}=r_0,\dots, r_N$ of **R** is called the likelihood function of **r** and it is given by (Eadie et al.):

$$L(\mathbf{r}|\mathbf{Q}) = H!\prod_{i=0}^{N}\frac{q_i^{r_i}}{r_i!} \tag{8}$$

where $\mathbf{Q}=[q_i]$ is the probability distribution of the number of measured photons, i.e. the vector of the probabilities to have an outcome in the bin i (i=0,…, N) in a single trial.

Considering **Q** as a function of **S** through (1), we can rewrite the likelihood function of the vector **r**, depending on the parameter **S**:

$$L(\mathbf{r}|\mathbf{S}) = H!\prod_{i=0}^{N}\frac{\left(\sum_m p_{i,m}^N s_m\right)^{r_i}}{r_i!} \tag{9}$$

which is then the probability of the occurrence of the particular histogram **r** when the incoming light has a certain statistics **S**.

As **r** is measured and then it is known, $L(\mathbf{r}|\mathbf{S})$ can be regarded as a function of **S** only, i.e. $L(\mathbf{r}|\mathbf{S})$ is the probability that a certain vector **S** is the incoming probability distribution when the histogram **r** is measured. The best estimate of the incoming statistics which produced the histogram **r** according to the ML method is the vector $\mathbf{S_e}$ which maximizes $L(\mathbf{r}|\mathbf{S})$, where **r** is treated as fixed. So, the estimation problem can in the end be reduced to a maximization problem.

ii. Description of the algorithm

For numerical calculations, it is necessary to limit the maximum number of incoming photons to m_{max} (in the following calculations, m_{max}=21). As **S** is a vector of probabilities, the maximization must be carried out under the constraints that the s_n are positive and that they add up to one. The positivity constraint can be satisfied changing variables: $s_n=\sigma_n^2$. Instead of L, we maximize the logarithm of L:

$$l(\Sigma) = \ln\left(L(\Sigma)\right) = \ln(C) + \sum_{i=0}^{N} r_i \ln\left(\sum_{m=0}^{m_{max}} p_{i,m}^N \sigma_m^2\right) \tag{10}$$

where $\Sigma=[\sigma_n]$ and C is a constant. The condition that the s_n add up to one can be taken into account using the Lagrange multipliers method: $F(\Sigma,\alpha)=l(\Sigma)-\alpha\left(\sum_{m=0}^{m_{max}}\sigma_m^2-1\right)$.

After developing (Banaszek 1998) the set of m_{max}+2 gradient equations $\nabla F(\Sigma,\alpha)=0$, we obtain that α=H and that the set of m_{max}+1 nonlinear equations to be solved respect to Σ is:

$$\sigma_l \left[\sum_{i=0}^{N} \frac{r_i p_{i,l}^N}{\sum_{m=0}^{m_{max}} p_{i,m}^N \sigma_m^2} - H \right] = 0 \tag{11}$$

for l=0,…, m_{max}. The set of equations (11) can be solved by standard numerical methods.

iii. ML reconstruction

To test the effectiveness of the reconstruction algorithm, a 8.6x8 μm² 5-PND was tested with the coherent emission from a mode-locked Ti:sapphire laser. $\mathbf{Q_{ex}}$ was determined as described in section 6.2.

The device was already characterized in terms of its conditional probability matrix $\mathbf{P^5}$ (Figure 11), so it was possible to carry out the ML estimation of the different incoming distributions with which the device was probed. Because of the bound on the number of incoming photons which is possible to represent in our algorithm (m_{max}=21) and as, for a coherent state, losses simply reduce the mean of the distribution, the ML estimation was performed considering $\mu^*=\mu/10$ and $\eta^*=10\eta$ (the efficiency of each section being lower than 10%).

Figure 12 shows the experimental probability distribution of the number of measured photons $\mathbf{Q_{ex}}$ obtained from the histograms measured when the incoming mean photon number is μ=1.5, 2.8, 4.3 photons/pulse (Figure 12.a, b, c respectively, in red), from which the incoming photon number distribution is reconstructed. The ML estimate of the incoming probability distribution $\mathbf{S_e}$ is plotted in Figure 12.d, e, f, (green), where it is compared to the real incoming probability distribution **S** (blue). The estimation is successful only for low photon fluxes (μ=1.5, 2.8 Figure 12.d, e) and it fails already for μ=4.3 (Figure 12.f). In Figure 12.a, b, c, $\mathbf{Q_{ex}}$ (red) is compared to the ones obtained from **S** and $\mathbf{S_e}$ through relation (1) (**Q**, $\mathbf{Q_e}$ in blue and green, respectively).

The main reasons why the reconstruction fails are not the low efficiencies of the sections of the PND or the spread in their values, but rather the limited counting capability (*N*=5) and a poor calibration of the detector, i.e. an imperfect knowledge of its real matrix of conditional probabilities. This assessment is supported by the following argument. If we generate $\mathbf{Q_{ex}}$ with a Monte Carlo simulation (see section 6.1) using the same η vector of Figure 11 and a Poissonian incoming photon number distributions and then we run the ML reconstruction algorithm (using the same $\mathbf{P^5}$, which this time describes perfectly the detector), **S** can be estimated up to much higher mean photon numbers (μ≥16, see Figure 13). Additional simulations will be needed to evaluate the performance of PNDs for the measurement of other, nonclassical photon number distributions. However, to alleviate this problem, a self-referencing measurement technique might be used (Achilles et al. 2006).

7. Discussion on the counting capability

Several factors may limit the counting capability (M_{max}) of a PNR detector.

One is the detection efficiency. From (5), assuming the detector saturation is negligible ($n \ll N$) and that all the branches are equal ($\eta_i=\eta$), the probability Q(*n*) of detecting *n* photons is proportional to η^n. In the PND tested $\tilde{\eta}$~2% at 1.3 μm (section 4.3), which we attribute to unoptimised film thickness and device design. This obviously prevents the application of the present device to *n*-photon states measurement for $n \gg 1$. Nevertheless, the η of SSPDs,

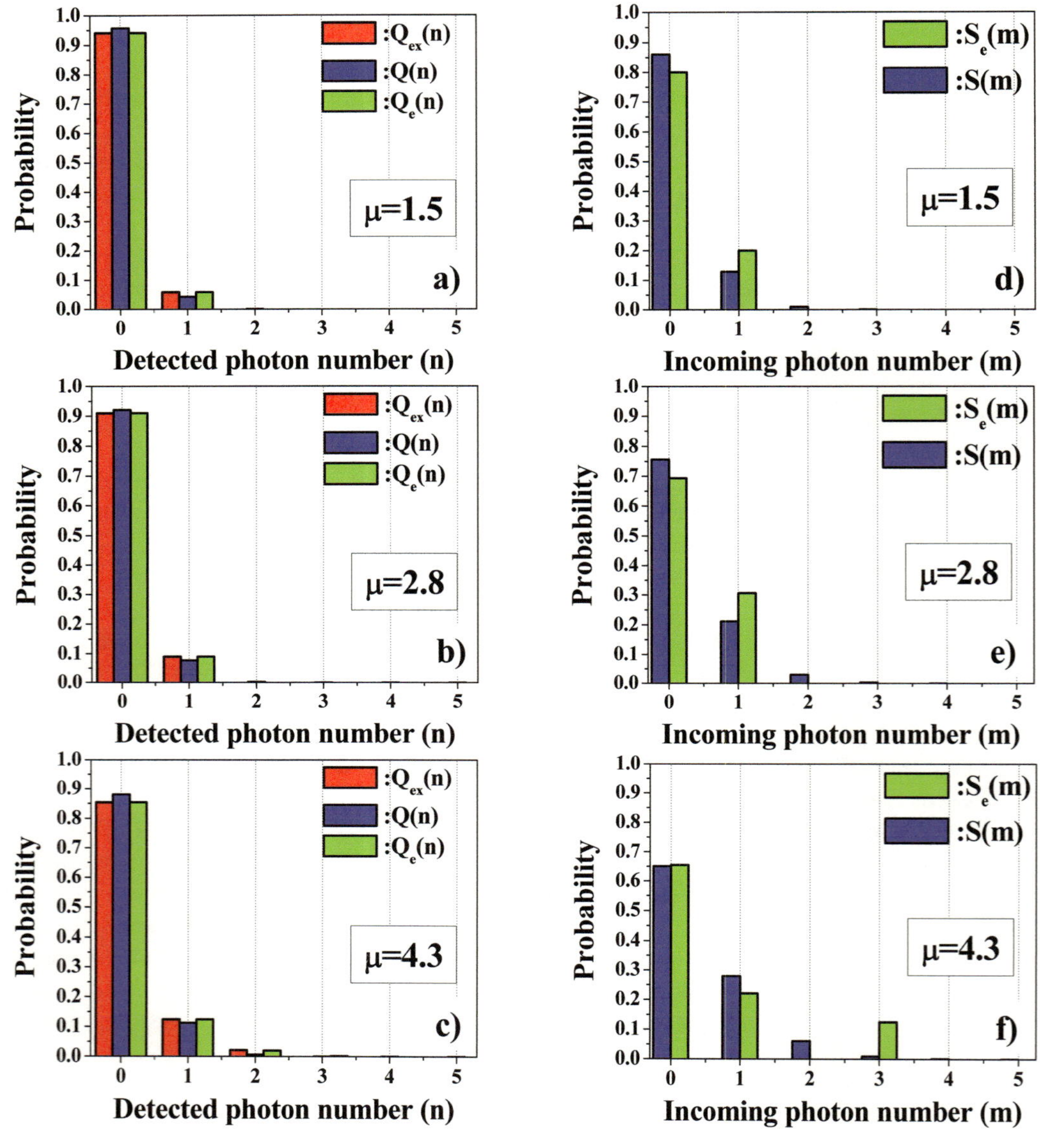

Fig. 12. a, b, c. Probability distribution of the number of measured photons obtained from experimental data Q_{ex} (red) from S (Q, in blue) and S_e (Q_e, in green) through relation (1), for μ=1.5, 2.8, 4.3 photons/pulse, respectively. d, e, f. Real incoming probability distribution S (blue) and its ML estimate S_e (green) for μ=1.5, 2.8, 4.3 photons/pulse, respectively. From (Marsili et al. 2009b).

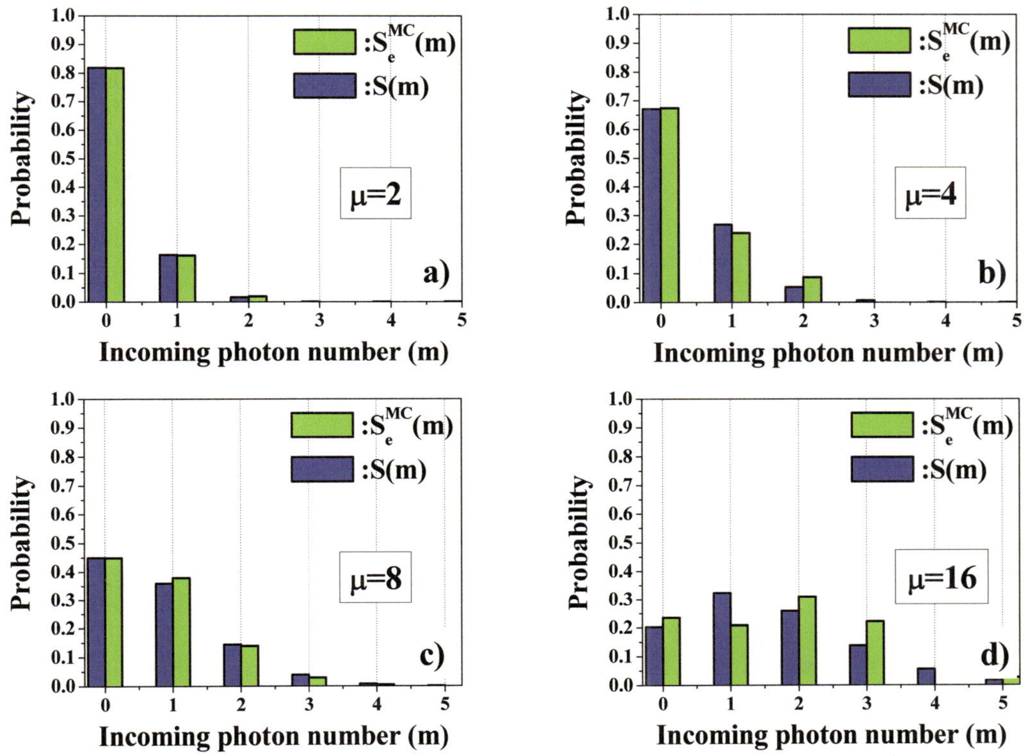

Fig. 13. Real incoming probability distribution S (blue) and its ML estimate $\mathbf{S}_e^{MC}$ (green), reconstructed from a $\mathbf{Q}_{ex}$ generated with a Monte Carlo simulation (see section 6.1) using the same η vector of Figure 11 for μ=2 (a), 4 (b), 8 (c), 16 (d) photons/pulse.

which are based on the same detection mechanism, can be increased up to ~60% (Rosfjord et al. 2006), and could potentially exceed 90% using optimized optical cavities. We also stress that uniform illumination of the wires is needed to achieve the optimum performance.

The second limitation is the intrinsic noise of the detector. As the currents from the sections of the PND are summed up to build the output, pulse height discrimination is used to achieve photon number resolution. This makes the noise performance of the device critical for its counting capability, as independent noisy signals are summed. Indeed, photon-number discrimination can be performed as long as the noise on the signal amplitude remains lower than the height of the one-photon pulse. The noise properties of any avalanche-based photon counting device are limited by its inner multiplication noise. In other avalanche PNR detectors (Waks et al. 2003; Waks et al. 2004; Zambra et al. 2004; Yamamoto et al. 2006; Fujiwara & Sasaki 2007; Gansen et al. 2007; Kardynal et al. 2007) the amplitude of the output signal is directly proportional to the number of carriers generated by single photon absorption events through a multiplication process which is intrinsically noisy. The noise on the multiplication gain is then completely transferred to the signal, which is then affected by a fluctuation of the same order. In contrast, with PNDs, the noisy avalanche carrier-multiplication process (Semenov et al. 2001) causes a fluctuation only in the resistance R_{hs} of the branch driven normal after the absorption of a photon and not in the

output current. Indeed, the amplitude of the photocurrent peak is determined by the partition between the fluctuating resistance R_{hs} of few kΩ and a resistance R_{out} almost 2 orders of magnitude lower, which is of fixed value. Comparing the broadening of the histogram peaks relative to different numbers of detected photons n (Figure 9), no multiplication noise buildup is observable, as the variance of the peak does not increase with n. The broadening of the peaks is then exclusively due to electrical noise originating from amplifiers and is not a fundamental property of the detector. To a good approximation the excess noise factor F (McIntyre 1966) of the PND is then close to unity and is not limiting M_{max}, which is not the case for most of the other approaches to PNR detection (Waks et al. 2003; Waks et al. 2004; Zambra et al. 2004; Rosenberg et al. 2005; Yamamoto et al. 2006; Fujiwara and Sasaki 2007; Gansen et al. 2007; Kardynal et al. 2007).
In PNDs, a third limitation to M_{max} arises from the leakage current δI_{lk}, which limits the bias current and therefore η. However, as discussed in section 5.2, this issue can be overcome with a careful design of the device.

8. Conclusions

A new PNR detector, the Parallel Nanowire Detector, has been demonstrated (Divochiy et al. 2008; Marsili et al. 2009a), which significantly outperforms existing approaches in terms of sensitivity, speed and multiplication noise in the telecommunication wavelength range. In particular, it provides a repetition rate (80 MHz) three orders of magnitude larger than any existing detector at telecom wavelength (Rosenberg et al. 2005; Fujiwara & Sasaki 2007; Jiang et al. 2007), and a sensitivity (NEP=4.2×10^{-18} W/Hz$^{1/2}$) one-two orders of magnitude better, with the exception of transition-edge sensors (Rosenberg et al. 2005) (which require a much lower operating temperature).
An electrical equivalent model of the device was developed to study its operation and to perform its design (Marsili et al. 2009b). In particular, we found that the leakage current significantly affects only the PND detection efficiency, while it has a marginal effect on its signal to noise ratio. To gain a better insight on the device dynamics, the (N+1)-mesh equivalent circuit of the N-PND was simplified and reduced to a three mesh circuit, so that the analytical expression of its transient response could be easily found. With this approach, we could predict a physical limit to the recovery time of the PND, which is slower than that previously estimated. Furthermore, the figures of merit of the device performance in terms of efficiency, speed and sensitivity ($\overline{\delta I_{lk}}$, f_0, SNR) were defined and their dependency on the design parameters (L_0, R_0, R_{out}, N) was analyzed.
To prove the suitability of the PND to reconstruct an unknown light statistics by ensemble measurements, we developed modeling tools to fully characterize the device and a maximum likelihood estimation algorithm (Marsili et al. 2009a; Marsili et al. 2009b). Testing a 5-PND with a Poissonian light we found that the reconstruction of the incoming photon number probability distribution to be successful only for low photon fluxes, most likely due to the limited counting capability (N=5) and the poor calibration (i.e. the imperfect knowledge of the real matrix of conditional probabilities) of the detector used, and not to its low detection efficiency ($\tilde{\eta} \sim 3\%$). Additional simulations will be needed to evaluate the performance of our detector for the measurement of other, nonclassical photon number distributions. Finally, despite the high sensitivity and speed of PNDs, their present performance in terms of detection efficiency (η=2% at λ=1.3μm) does not allow their application to single shot measurements, as required for linear-optics quantum computing (Knill et al. 2001), quantum repeaters (Sangouard et al. 2007) and conditional-state

preparation (Sliwa & Banaszek 2003). Nevertheless, the η of SPDs based on the same detection mechanism can be increased to ~60% (Rosfjord et al. 2006), and could potentially exceed 90% using optimized optical cavities.

9. Acknowledgments

This work was supported by the Swiss National Foundation through the "Professeur boursier" and NCCR Quantum Photonics programs, EU FP6 STREP "SINPHONIA" (contract number NMP4-CT-2005-16433), EU FP6 IP "QAP" (contract number 15848). The authors thank B. Deveaud-Plédran and F. Lévy for the useful discussions and B. Dwir, M. Benkahoul and H. Jotterand for technical support. A. Gaggero gratefully acknowledges a PhD fellowship at University of Roma TRE.

10. References

Achilles, D., et al. (2004). "Photon-number-resolving detection using time-multiplexing." *J. Mod. Opt.* 51(9-10): 1499-1515.

Achilles, D., et al. (2003). "Fiber-assisted detection with photon number resolution." *Opt. Lett.* 28(23): 2387-2389.

Achilles, D., et al. (2006). "Direct, Loss-Tolerant Characterization of Nonclassical Photon Statistics." *Phys. Rev. Lett.* 97(4): 043602.

Banaszek, K. (1998). "Maximum-likelihood estimation of photon-number distribution from homodyne statistics." *Phys. Rev. A* 57(6): 5013-5015.

Brassard, G., et al. (2000). "Limitations on practical quantum cryptography." *Physical Review Letters* 85(6): 1330-1333.

Dauler, E. A., et al. (2007). "Multi-element superconducting nanowire single-photon detector." *IEEE Trans. Appl. Supercond.* 17(2): 279-284.

Divochiy, A., et al. (2008). "Superconducting nanowire photon-number-resolving detector at telecommunication wavelengths." *Nature Photon.* 2: 302 - 306.

Eadie, W. T., et al. "Satistical Methods in Experimetal Physics." *in Statistical Methods in Experimental Physics, North-Holland, Amsterdam, 1971, Chap 8, and references therein.*

Fitch, M. J., et al. (2003). "Photon-number resolution using time-multiplexed single-photon detectors." *Phys. Rev. A* 68(4 B): 043814

Fujiwara, M. and M. Sasaki (2007). "Direct measurement of photon number statistics at telecom wavelengths using a charge integration photon detector." *Appl. Opt.* 46(16): 3069-3074.

Gansen, E. J., et al. (2007). "Photon-number-discriminating detection using a quantum-dot, optically gated, field-effect transistor." *Nature Photon.* 1: 585-588.

Gol'tsman, G., et al. (2007). "Middle-infrared to visible-light ultrafast superconducting single-photon detectors." *IEEE Trans. Appl. Supercond.* 17(2): 246-251.

Gol'tsman, G. N., et al. (2003). "Fabrication of nanostructured superconducting single-photon detectors." *IEEE Trans. Appl. Supercond.* 13(2 I): 192-195.

Jiang, L. A., et al. (2007). "Photon-number-resolving detector with 10 bits of resolution." *Phys. Rev. A* 75(6): 62325.

Kadin, A. M. *in Introduction to superconducting circuits, Wiley, New York, 1999, Chap 2.*

Kardynal, B. E., et al. (2007). "Photon number resolving detector based on a quantum dot field effect transistor." *Appl. Phys. Lett.* 90(18): 181114.

Kerman, A. J., et al. (2007). "Constriction-limited detection efficiency of superconducting nanowire single-photon detectors." *Appl. Phys. Lett.* 90(10): 101110.

Knill, E., et al. (2001). "A scheme for efficient quantum computation with linear optics." *Nature* 409(6816): 46-52.

Lee, H., et al. (2004). "Towards photostatistics from photon-number discriminating detectors." *J. Mod. Opt.* 51(9-10): 1517-1528.

Lita, A. E., et al. (2008). "Counting near-infrared single-photons with 95% efficiency." *Opt. Express* 16(5): 3032-3040.

Marsili, F., et al. (2009a). "Superconducting parallel nanowire detector with photon number resolving functionality." *J. Mod. Opt.* 56(2-3): 334-344.

Marsili, F., et al. (2008). "High efficiency NbN nanowire superconducting single photon detectors fabricated on MgO substrates from a low temperature process." *Opt. Express* 16(5): 3191-3196.

Marsili, F., et al. (2009b). "Physics and application of photon number resolving detectors based on superconducting parallel nanowires." *New J. Phys.* 11(4): 045022-045043.

Mattioli, F., et al. (2007). "Electrical characterization of superconducting single-photon detectors." *J. Appl. Phys.* 101(5): 054302.

McIntyre, R. J. (1966). "Multiplication noise in uniform avalanche diodes." *IEEE Trans. Electron Devices* 13(1): 164 - 168.

Miller, A. J., et al. (2003). "Demonstration of a low-noise near-infrared photon counter with multiphoton discrimination." *Appl. Phys. Lett.* 83(4): 791-793.

Rosenberg, D., et al. (2005). "Noise-free high-efficiency photon-number-resolving detectors." *Phys. Rev. A* 71(6): 1-4.

Rosfjord, K. M., et al. (2006). "Nanowire Single-photon detector with an integrated optical cavity and anti-reflection coating." *Opt. Express* 14(2): 527-534.

Sangouard, N., et al. (2007). "Long-distance entanglement distribution with single-photon sources." *Phys. Rev. A* 76(5): 050301-4.

Semenov, A. D., et al. (2001). "Quantum detection by current carrying superconducting film." *Physica C* 351(4): 349-356.

Sliwa, C. and K. Banaszek (2003). "Conditional preparation of maximal polarization entanglement." *Physical Review A* 67(3): 030101.

Tarkhov, M., et al. (2008). "Ultrafast reset time of superconducting single photon detectors." *Appl. Phys. Lett.* 92(24): 241112.

Verevkin, A., et al. (2002). "Detection efficiency of large-active-area NbN single-photon superconducting detectors in the ultraviolet to near-infrared range." *Appl. Phys. Lett.* 80(25): 4687.

Waks, E., et al. (2004). "Direct Observation of Nonclassical Photon Statistics in Parametric Down-Conversion." *Phys. Rev. Lett.* 92(11): 113602-1.

Waks, E., et al. (2003). "High-Efficiency Photon-Number Detection for Quantum Information Processing." *IEEE J. Sel. Top. Quant.* 9(6): 1502-1511.

Yamamoto, K., et al. (2006). "Development of Multi-Pixel Photon Counter (MPPC)." *IEEE Nuclear Science Symposium Conference Record, 2006* 2: 1094 - 1097.

Yang, J. K. W., et al. (2007). "Modeling the Electrical and Thermal Response of Superconducting Nanowire Single-Photon Detectors." *IEEE Trans. Appl. Supercond.* 17(2): 581-585.

Yuan, Z., et al. (2002). "Electrically driven single-photon source." *Science* 295(5552): 102-105.

Zambra, G., et al. (2004). "Counting photoelectrons in the response of a photomultiplier tube to single picosecond light pulses." *Rev. Sci. Instr.* 75(8): 2762-2765.

Optical Trapping and Manipulation

18

Optoelectronic Tweezers for the Manipulation of Cells, Microparticles, and Nanoparticles

Aaron T. Ohta[1], Pei-Yu Chiou[2], Arash Jamshidi[3], Hsan-Yin Hsu[3], Justin K. Valley[3], Steven L. Neale[3], and Ming C. Wu[3]
[1]*University of Hawaii at Manoa*
[2]*University of California, Los Angeles*
[3]*University of California, Berkeley*
USA

1. Introduction

Micro- and nanoparticle manipulation capabilities can benefit researchers in a wide variety of fields, from biological research to semiconductor physics. Optoelectronic tweezers (OET) is a novel technique that complements existing manipulation technologies. OET provides optically-controlled manipulation at the single-particle level. It is a dynamic, reconfigurable, minimally-invasive tool capable of massively parallel manipulation. In this chapter, the operating principle of OET will be explained, along with device design considerations and operational regimes. The capabilities of OET will be showcased in the context of applications in biological cell handling and micro- and nanoparticle assembly.

1.1 Methods of micro- and nanoparticle manipulation

Many research fields benefit from the ability to address particles in the micro- and nanoscale regimes. For example, biologists have traditionally studied cell behavior by observing the bulk response of a population of cells. However, it can be desirable to observe the behavior of a single cell. This information is useful for studying phenomena such as cell-cell interactions, cell signaling pathways, mutations or genetic damage among a population, or the differentiation of stem cells. In addition, the response of a single cell can be observed, instead of the average response of an entire population. Certain cells in a homogeneous population may exhibit behavior that deviates from the average response; single-cell observation may provide more insight into the cause of the deviant response. Thus, studies at the single-cell level are greatly facilitated by the ability to manipulate individual cells.

Non-biological particle manipulation is also of interest for engineering fields. Nanostructures that exhibit quantum-mechanical behavior are interesting to researchers studying nanoscale physics, or to engineers that are exploiting the unique properties of nanomaterials for improved devices. However, an obstacle to nanostructure research is the difficulty in addressing and assembling these extremely small particles.

Micro- and nanoscale manipulation can be achieved with a variety of forces, including mechanical, magnetic, fluidic, optical, and electrokinetic forces. These forces are controlled by a wide variety of devices. Perhaps the most intuitive devices are the mechanical manipulators, such as microgrippers for cells (Kim et al., 1992; Chronis & Lee, 2005) or atomic-force microscope tips for nanoparticle positioning (Avouris et al., 1999). However, non-contact methods are often preferred by biologists. In addition, it is difficult to scale up the number of mechanical manipulators in order to increase the parallel manipulation capabilities of the device.

Magnetic forces are not invasive, but this technique requires particles to be attached to magnetic beads, unless the particle has an intrinsic magnetic response. This limits the type of particles that can be manipulated by magnetic fields (Gosse & Croquette, 2002). Furthermore, tag-free protocols that do not require attaching cells to magnetic beads or other particles are more attractive to most biologists.

Microfluidic devices are another non-contact method of cellular manipulation (West et al., 2008). However, it is difficult to target a specific single cell in current microfluidic devices, although they are suitable for sorting heterogeneous cell populations.

Optical tweezers provides dynamic, flexible manipulation of specific single cells. Optical tweezers uses the gradient force of light to trap particles using a highly-focused laser beam (Ashkin et al., 1986), and has been used in many research applications, such as measuring cell motility (Maier et al., 2004), sorting colloidal particles (MacDonald et al., 2003), sorting cells (Paterson et al., 2005), and trapping and assembling nanowires (Pauzauskie et al., 2006). In addition, the basic single-beam optical trap has been extended to multiple traps from a single laser source using holographic imaging (Grier, 2003), evanescent-wave traps (Garces-Chavez et al., 2005), and plasmonic-enhanced traps (Miao & Lin, 2007; Grigorenko, et al., 2008).

The basic single-beam optical tweezers trap is formed when a laser beam is highly focused. Near the focal point, lateral trapping forces are created due to the gradient force of the light. Along the axial direction of the laser, a force due to radiation pressure also exists. Thus, the trap is only stable in three dimensions if the axial gradient force near the focal point is stronger than the forces from radiation pressure. A typical optical trap requires an intensity of approximately 10^7 mW/cm^2, which requires high-power lasers, and can damage some types of particles, including cells (Konig et al., 1996, Neuman et al., 1999; Mohanty et al., 2002). In addition, the generation of sufficient optical gradients requires high numerical aperture objective lenses to focus the laser; this results in a limited optical manipulation area, reducing parallel manipulation capabilities, especially for larger microscale particles such as cells.

1.2 Dielectrophoresis

Electrokinetic forces are capable of non-contact, non-invasive parallel manipulation. These forces include electrophoresis and dielectrophoresis. Electrophoresis is the movement of charged particles within an electric field, and is commonly used to separate DNA molecules by size. Dielectrophoretic manipulation is more attractive, as it is capable of acting on neutral particles as well as charged particles.

Dielectrophoresis (DEP) is the electrical analog of optical tweezers. This phenomena relies on the gradient force of an electric field rather than an optical field (Pohl, 1958). The induced dipole of a particle experiences a force due to the electric field gradient, resulting in dielectrophoretic force. Dielectrophoresis acts on a variety of particles, including non-conductive particles, conductive particles, particles that have a net electric charge, or particles that are charge-neutral. DEP forces are also capable of trapping nanoparticles

(Muller et al., 1996) through the use of microfabricated electrodes. DEP-based devices have been demonstrated in applications ranging from cell trapping and cell sorting (Gascoyne et al., 1997; Cheng et al., 1998; Huang et al., 2003; Gascoyne et al., 2004; Pethig et al., 2003) to carbon nanotube separation (Krupke et al., 2003) and nanowire assembly (Evoy et al., 2004). Electrode-based DEP devices have parallel manipulation capabilities, but the trapping patterns are fixed, and it is difficult to isolate a single particle of interest. This capability can be achieved by creating dynamic DEP cages via CMOS control (Fuchs et al., 2006). However, this device is limited by the pitch of the CMOS circuitry, and is currently limited to microscale manipulation.

2. Optoelectronic Tweezers (OET)

Optoelectronic tweezers (OET) enables the optically-controlled, parallel manipulation of single micro- and nanoscale particles. This device, first developed by Wu, et al. (Chiou et al., 2005) integrates the flexibility and control of optical manipulation with the parallel manipulation and sorting capabilities of dielectrophoresis, resulting in optically-induced dielectrophoresis. The OET device uses a photosensitive surface to allow an optical pattern to control the electric field landscape within an OET device. The resulting non-uniform electric field then generates a DEP force on particles. Unlike optical tweezers, the optical energy is not directly used for trapping, so much lower optical intensities can be used: 10 to 1000 mW/cm^2. These optical intensities can be achieved by a computer projector or an LED, allowing the creation of complex optical manipulation patterns (Ohta et al., 2007c). In addition, the optical gradient force is no longer required, relaxing the focusing requirements and allowing the use of low numerical aperture lenses. As a result, the effective manipulation area of OET can be over 500 times larger than that of a typical optical tweezers setup. Furthermore, since the manipulation patterns are controlled optically, OET retains the flexibility and dynamic control enjoyed by optical tweezers. Unlike electrode-based DEP, OET is capable of trapping a specific single particle from a larger population. In addition to the advantages presented by optically-controlled manipulation, OET also retains the properties of DEP. Different cells experience varying DEP forces, allowing the separation of cells based on the DEP response.

OET manipulation has been demonstrated on a variety of microparticles, including polystyrene beads (Chiou et al., 2005; Ohta et al., 2007a), *E. coli* bacteria (Chiou et al., 2004), red and white blood cells (Ohta et al., 2007c; Hwang et al., 2008a), HeLa cells (Ohta et al., 2007a; Neale et al., 2009a), Jurkat cells (Ohta et al., 2007a), yeast cells (Lu et al., 2005), and protozoa (Choi et al., 2008). In addition, OET is capable of manipulating semiconducting and metallic nanowires (Jamshidi et al., 2008).

The basic OET device consists of two electrodes (Figure 1a). The upper transparent electrode consists of a 100-nm-thick layer of indium tin oxide (ITO), a transparent conductive material, on a glass substrate. The lower photosensitive electrode typically consists of featureless layers of ITO and intrinsic amorphous silicon (a-Si) on a glass substrate. The two electrodes are separated by spacers, creating a chamber between the electrodes that is typically 100 μm in height. An aqueous solution containing the cells or particles under manipulation is introduced into the chamber between the electrodes. An electric field is created in the device by applying an ac bias across the top and bottom electrodes.

The OET device is placed under a microscope for observation of the particles under manipulation. Optical patterns are focused onto the photosensitive surface, typically using a microscope objective lens. The optical source is flexible. A low-power laser can be used, or an incoherent light source such as a halogen or mercury lamp, or a light-emitting diode (LED). Optical patterns are created using spatial light modulators, such as Texas Instrument's digital micromirror device (DMD), a MEMS mirror array, or Hamamatsu's liquid-crystal-based spatial light modulator. The optical source and the optical patterning system can also be integrated by focusing the output of a computer projector (Ohta et al., 2007c) or LCD display (Choi et al., 2007; Hwang et al., 2008a) onto the OET device.

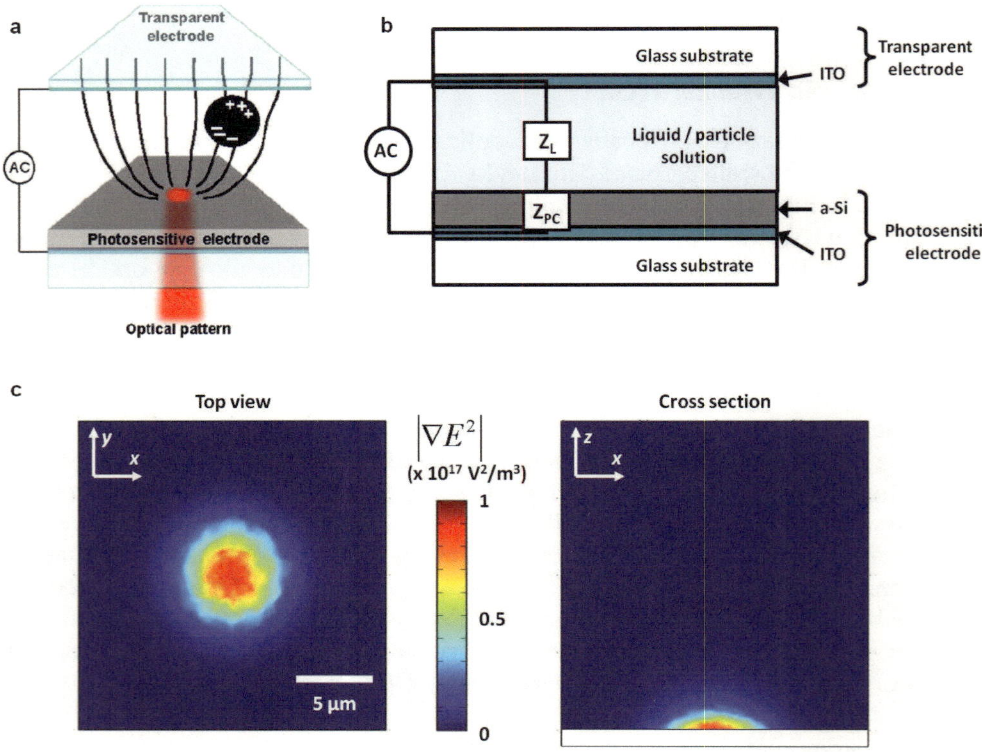

Fig. 1. The optoelectronic tweezers (OET) device. (a) Physical structure of the OET device, consisting of a transparent electrode and a photosensitive electrode. The illuminated areas create optically-defined electrodes on the photosensitive surface, creating electric field gradients that generate optically induced dielectrophoretic force. (b) Simplified equivalent circuit model. (c) Simulation of the gradient of the square of the electric field in the liquid layer of an OET device, which is proportional to the induced dielectrophoretic force. The areas of highest gradients correspond to a spot illuminated by an optical source.

2.1 Operating principle

The OET device relies on optically-induced dielectrophoresis to impart a force on the particles under manipulation. The dielectrophoretic force is controlled by optical patterns that are projected on the photosensitive surface of the device.

Amorphous silicon, the material used in the photosensitive electrodes of OET, has its conductivity modulated by the intensity of the light that is absorbed in the material. Under ambient lighting conditions, the a-Si film has a low conductivity. However, when illuminated with light, photogenerated charge carriers are produced in the a-Si. As a result, the conductivity of the a-Si layer increases by a few orders of magnitude. The a-Si layer thus has two states; a low-conductivity dark state, and high-conductivity illuminated state. These two states allow the a-Si layer to act as an optically-controlled and optically-defined electrode. The resolution of the optically-defined electrode has been shown to be a function of the size and intensity of the optical patterns (Ohta et al., 2007c; Neale et al., 2007).

The importance of the a-Si photoconductive layer can be seen by modeling the OET devices with a simplified equivalent circuit model (Figure 1b). In the dark state, the impedance of the a-Si, Z_{PC}, is larger than the impedance of the liquid layer, Z_L. Thus, in areas with no illumination, most of the applied ac voltage drops across the high impedance of the a-Si layer. However, in the illuminated areas, Z_{PC} becomes significantly lower, allowing a significant voltage drop to occur in the liquid layer. In this manner, an electric field gradient is set up between the illuminated and dark areas in the OET device (Figure 1c).

The electric field gradients in the OET and LOET devices result in an optically-induced dielectrophoretic force. This DEP force results from the interaction of a non-uniform electric field and the induced dipole of a particle within the electric field. The forces at each at end of the particle's dipole are unequal due to the non-uniform field, resulting in a net force. This force can be described by the following equation:

$$F = \overline{p} \cdot \nabla \overline{E} \tag{1}$$

where $\overline{p}$ is the dipole moment of the particle, and $\overline{E}$ is the first term of the Taylor series expansion of the electric field (Jones, 1995). If the particle is a homogeneous dielectric sphere (e.g. a polystyrene bead), then the time-averaged DEP force reduces to:

$$F_{DEP} = F_{OET} = 2\pi r^3 \varepsilon_m \mathrm{Re}[K(\omega)] \nabla E_{rms}^2 \tag{2}$$

where r is the particle radius, ε_m is the permittivity of the medium surrounding the particle, E_{rms} is the root-mean-square electric field strength, and $\mathrm{Re}[K(\omega)]$ is the real part of the Clausius-Mossotti factor, given by:

$$K(\omega) = \frac{\varepsilon_p^* - \varepsilon_m^*}{\varepsilon_p^* + 2\varepsilon_m^*}, \varepsilon_p^* = \varepsilon_p - j\frac{\sigma_p}{\omega}, \varepsilon_m^* = \varepsilon_m - j\frac{\sigma_m}{\omega} \tag{3}$$

where ε is the permittivity of the particle or medium (denoted by a subscript p or m), σ is the conductivity of the particle or medium, and ω is the angular frequency of the electric field (Jones, 1995). The magnitude of $\mathrm{Re}[K(\omega)]$ varies with frequency, resulting in a frequency-dependence of the dielectrophoretic force. Positive values of $\mathrm{Re}[K(\omega)]$ result in particle attraction to electric field maxima (positive DEP, referred to here as positive OET). For negative values of $\mathrm{Re}[K(\omega)]$, particles are repelled from field maxima (negative DEP, referred to here as negative OET). Applying an AC electric field thus allows the tuning of the type of DEP force induced on a particle, as well as negating any electrophoretic effects, or particle movement due to its surface charge.

If the particle under manipulation is non-homogeneous, like biological cells, Equation 3 must be modified. A single-shell model is typically used to determine the frequency-dependent Clausius-Mossotti factor for mammalian cells. The permittivity and conductivity of the cell membrane and interior are used to determine an effective complex permittivity that can then be used in Equation 3. This effective permittivity is given by:

$$\varepsilon_p^* = C_{mem}^* \frac{3r\varepsilon_{\text{int}}^*}{3\varepsilon_{\text{int}}^* + 3C_{mem}^* r} \qquad (4)$$

where $\varepsilon_{\text{int}}^*$ is the complex internal conductivity of the cell, assuming that the thickness of the cell membrane, *d*, is much less than the radius of the cell interior, *r* (Gascoyne et al., 1997). The membrane capacitance, C_{mem}^*, is given by:

$$C_{mem}^* = \frac{\varepsilon_{mem}}{d} - \frac{j\sigma_{mem}}{d} \qquad (5)$$

Many cell types are uniquely distinguishable by the real part of the Clausius-Mossotti factor. This enables the separation of different cell types using DEP force.

2.2 Operational modes of Optoelectronic Tweezers

In addition to optically-induced dielectrophoresis, several other effects can occur in the OET device (Valley et al., 2008). Most of these are undesirable parasitic effects, but some of these operational modes can be used to obtain expanded functionality in the OET device.

One of the parasitic effects in the OET device is electrolysis: the breakdown of water molecules into hydrogen and oxygen gas due to an electric current. This is undesirable, as the bubble formation can rupture cell membranes and interfere with OET operation. In addition, if the bubble formation process is violent enough, the a-Si film can crack, or pinholes can form. Electrolysis is more prevalent when using electric fields at a frequency of 1 kHz or less.

Another undesirable effect is electrothermal heating, as elevated temperatures can cause cell death or even boiling of the liquid media. The electrothermal heating is caused by phonon generation in the a-Si layer or joule heating in the a-Si and liquid layers. As a result, gradients in the electrical conductivity and permittivity of the liquid media arise, resulting in a liquid flow. However, these effects are only significant under high optical intensities (greater than 100 W/cm^2).

Light-induced AC electroosmosis (LACE) is another operational regime of the OET device, typically occurring at low bias frequencies of approximately 1 kHz. Unlike the other effects, LACE is a useful operating mode of the OET device. The LACE effect is an optically-patterned control of fluid flow in the OET device. The optical patterns control the zeta potential at the liquid-photoconductor interface as well as the electric field profile, creating localized fluid flows surrounding the illuminated areas. The LACE flows have been used to trap polystyrene beads with diameters of 2 μm down to 50 nm, λ-phage DNA molecules, and quantum dots (Chiou et al., 2008), and to separate 1-μm-diameter beads from 6-μm-diameter beads (Hwang & Park, 2009a).

A figure of merit has been developed to quantify the contribution of these other effects relative to the strength of the optically-induced DEP force. This is expressed as a ratio, β:

$$\beta = \frac{X_{DEP}}{X_{DEP} + X_{EXT} + \langle X_{BROWNIAN} \rangle} \qquad (6)$$

where X_{DEP}, X_{EXT}, and $\langle X_{BROWNIAN} \rangle$ represent the distance a particle travels in 1 s due to optically-induced DEP, all other external forces, and Brownian motion, respectively (Valley et al., 2008). If β is averaged over an area, a new figure of merit is obtained:

$$B = \frac{1}{A}\int_A \beta dxdy, \qquad x \in [-r, r], y \in [0, d] \tag{7}$$

where r is the maximum radius from the center of the optical pattern that particle perturbation is expected, d is the thickness of the liquid layer, and A is the area of integration, where $A = 2rd$. Thus, B represents the percentage of particle perturbation in the OET device due to optically-induced DEP. For applied bias frequencies of greater than 50 kHz and optical intensities of less than 100 W/cm^2 (typical OET operating parameters), optically-induced DEP is indeed the dominant force (Figure 2).

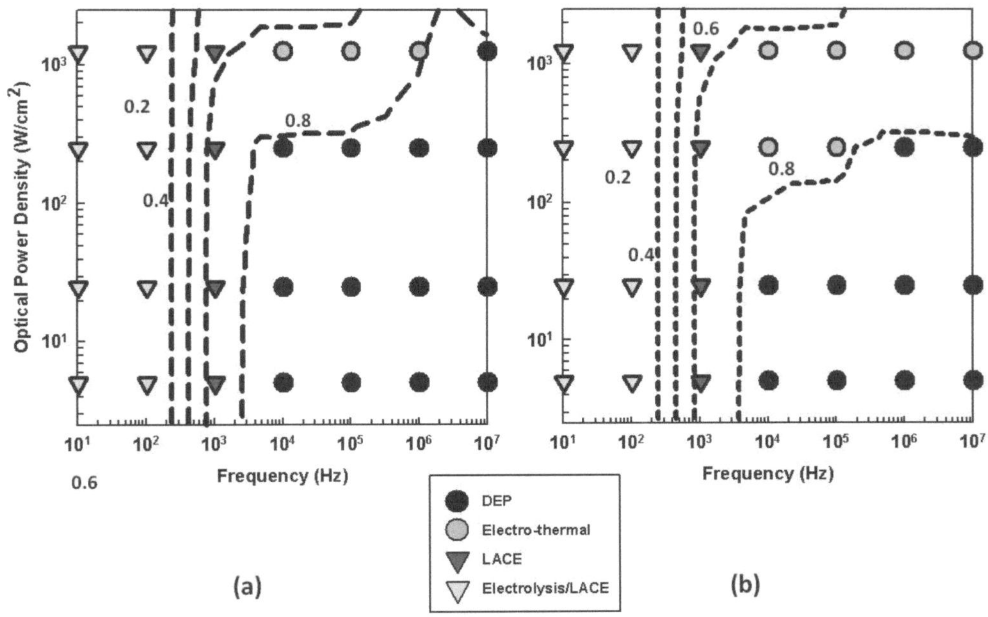

Fig. 2. Dominant effects in the OET device. Theoretical simulations are shown as dashed lines, and empirically-determined effects are shown as data points. The liquid solution has a conductivity of 1 mS/m. The applied voltage is at a frequency of 100 kHz, with amplitudes of (a) 20 Vpp and (b) 10 Vpp. Reprinted with permissions from Valley et al., 2008. Copyright 2008, IEEE.

2.3 Other Optoelectronic Tweezers devices

Thus far, the standard configuration of the OET device has been discussed. In addition to this standard configuration, there are other variants on the OET device, which are described here.

One modification is the substitution of alternate materials for the amorphous silicon photosensitive electrode. The amorphous silicon is deposited using plasma-enhanced vapor deposition (PECVD). If PECVD equipment is unavailable or infeasible, an alternative is to use photosensitive polymers in lieu of amorphous silicon. One suitable polymer blend material is a mixture of regioregular poly (3-hexylthiophene) (rr-P3HT) and phenyl-C61-butyric acid methyl ester (PCBM), forming a P3HT/PCBM blend, which is also used in plastic solar cells (Padinger et al., 2003). OET devices fabricated with P3HT/PCBM polymer films in place of amorphous silicon have similar functionality to standard OET devices (Wang et al., 2009).

Another alternate photosensitive material is cadmium sulphide (CdS), which is another photoconductive semiconductor. The bandgap of CdS is larger than that of a-Si, so OET made of CdS films are actuated using blue light (Higuchi et al., 2008). Functionality is also similar to the standard OET device, although a chemical process can be used for the CdS deposition, as opposed to PECVD for a-Si.

The OET device can also be reconfigured physically. One form is the use of amorphous-silicon-coated surfaces for both the upper and lower electrodes (Hwang et al., 2008b). This configuration is designed to reduce nonspecific particle adhesion when manipulation objects that experience negative OET forces, such as polystyrene beads. In the standard OET device, polystyrene beads can be levitated by the vertical component of the electric field gradient, bringing them into contact with the upper ITO-coated electrode. More than 50% of these particles adhere to the ITO-coated electrode, rendering them immovable by OET forces. By using a-Si for both electrodes, the forces due to the vertical electric field gradients are cancelled out, and the polystyrene beads remain at a metastable position that is approximately midway between the two electrode surfaces. In this version of the OET device, nonspecific adhesion is reduced to less than 20% of the particles (Hwang et al., 2008b). However, optical microscopy through the a-Si coated electrodes may present some viewing issues for certain types of particles, such as biological cells.

Another type of OET device combines the two discrete electrode surfaces onto a single substrate. This version of OET uses an interdigitated array of amorphous silicon electrodes to enable optical control of an electric field using a single substrate (Figure 3a). In this device, the electric field direction is primarily parallel to the surface of the substrate, so this configuration is known as lateral-field optoelectronic tweezers (LOET) (Ohta et al., 2007a) (Figure 3b). Using LOET, anisotropic particles such as microdisks (Tien et al., 2009) and nanowires (Ohta et al., 2007b; Ohta et al., 2008; Neale et al., 2009b) align with their long axis parallel to the surface of the LOET device, unlike in the conventional OET device (Jamshidi et al., 2008). In addition, the LOET device can be more easily integrated with other microdevices, as there is only one substrate.

Electrowetting-on-dielectric (EWOD) devices are an example of microsystems that can benefit from OET integration. EWOD devices can achieve complicated droplet-based manipulation (Cho et al., 2003; Srinivasan et al., 2004), but it is difficult to address objects within a droplet. OET functionality integrated with EWOD can create a platform capable of single-cell assays. In order to achieve OET integration with EWOD, an EWOD device and a LOET device were built on separate substrates, then sandwiched together to create a composite device that can perform droplet actuation as well as move cells within a droplet (Shah et al., 2009) (Figure 3c).

If the goal is droplet manipulation, another variant of the OET device can be used. The floating-electrode OET device is designed to manipulate aqueous droplets in oil for droplet-based assays (Park et al., 2008). In this device, electrodes are placed on top of an amorphous silicon layer, and the device is coated with a 15-μm-thick layer of polydimethylsiloxane (PDMS) polymer (Figure 3d). Optical patterns on the a-Si surface alter the applied electric field, producing a DEP force on the droplets suspended in oil. Movement of aqueous droplets was demonstrated at velocities of over 400 μm/s (Park et al., 2008).

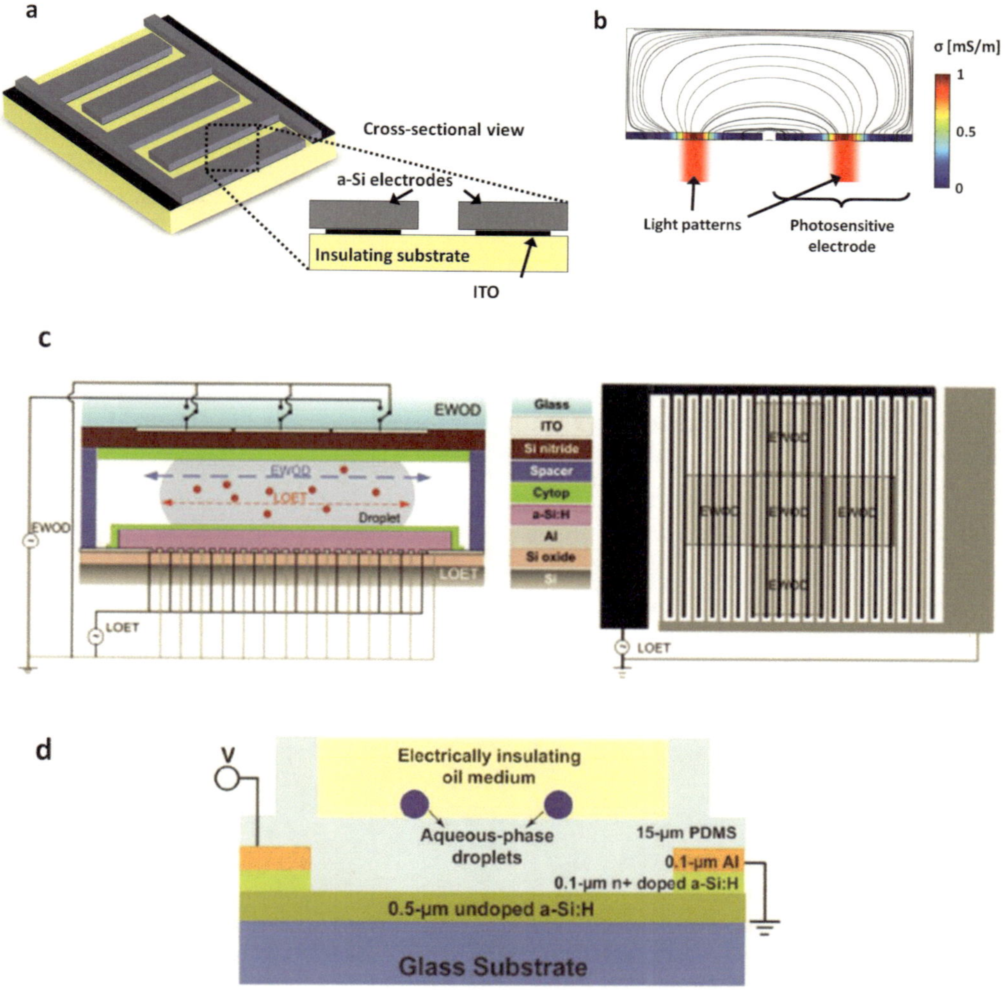

Fig. 3. Alternate OET configurations. (a) Lateral-field optoelectronic tweezers (LOET), which consist of an interdigitated array of a-Si electrodes. (b) Simulation of the electric field profile in the LOET device. The low-conductivity areas correspond to illuminated areas.
(c) Integrated EWOD-LOET device for simultaneous droplet and cell manipulation. The LOET device is fabricated on the lower substrate, and the EWOD electrodes are fabricated on the upper substrate. Reprinted with permission from Shah et al., 2009. Copyright 2009, Royal Society of Chemistry. (d) Floating-electrode OET for the manipulation of droplets in oil. Reprinted with permission from Park et al., 2008. Copyright 2008, American Institute of Physics.

3. Cell manipulation using Optoelectronic Tweezers

3.1 Single-cell trapping

The most straightforward biological application of optoelectronic tweezers is for the trapping and transport of cells. OET trapping of many cell types using a variety of optical patterns has been demonstrated, including *E. coli* bacteria (Chiou et al., 2004), red and white blood cells (Ohta et al., 2007c; Hwang et al., 2008a), HeLa cells (Ohta et al., 2007a; Neale et al., 2009a), Jurkat cells (Ohta et al., 2007a), yeast cells (Lu et al., 2005), and protozoa (Choi et al., 2008). One issue that arises when trapping mammalian cells is that of surface fouling, where cells nonspecifically adhere to the surface of OET devices. Cellular adhesion forces extend into the nanonewton range, whereas OET forces are typically tens to hundreds of piconewtons. Thus, once cells are fully adhered, it becomes difficult or impossible to actuate them by OET force.

In order to enable more reliable cellular trapping, OET devices were coated with a layer of poly(ethylene glycol) (PEG), a common antifouling polymer (Lau et al., 2009). The addition of the PEG coating resulted in 30 times less nonspecific adhesion as compared to uncoated OET devices, as tested with HeLa cells over the course of an hour. The PEG-coated devices allow reliable trapping and positioning of cells using OET manipulation. This was demonstrated by the formation of a 5 x 5 array of Jurkat cells on a PEG-coated OET device (Figure 3).

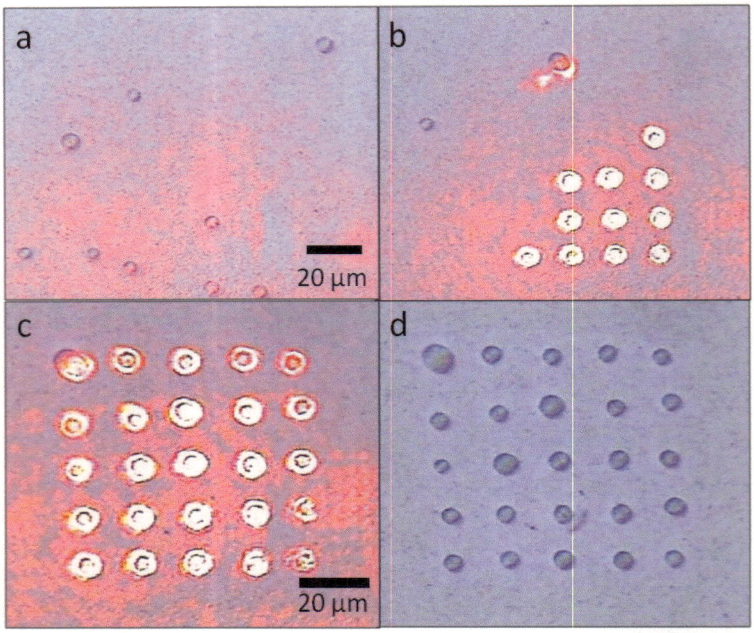

Fig. 3. Jurkat cell patterning on a PEG-coated OET device. (a) Initial random distribution of cells. (b) Formation of the array using OET manipulation. (c) Completed array. (d) Array with the optical manipulation pattern removed, for clarity. Reprinted with permission from Lau et al., 2009. Copyright 2009, Royal Society of Chemistry.

3.2 Cell separation

As described earlier, dielectrophoretic force is a function of the frequency-dependent electrical properties of the cells under manipulation. As different cell types exhibit dissimilar electrical properties, DEP can be used to sort between cell types, or even between widely varying cells of the same type (Gascoyne et al., 1997; Cheng et al., 1998; Huang et al., 2003; Gascoyne et al., 2004; Pethig et al., 2003). As OET uses DEP force, this capability can be used to selectively concentrate live human B cells from dead B cells, and to spatially discriminate a mixed population of Jurkat and HeLa cells.

In a live cell, the semi-permeable cell membrane allows a cell to maintain an ion differential between its interior and the surrounding liquid medium. In these OET experiments, cells are suspended in a low-conductivity isotonic buffer (8.5% sucrose and 0.3% dextrose), so the cells have internal conductivities greater than the liquid. However, once a cell dies, this ion differential is no longer maintained, and the conductivity of the cell interior becomes similar to that of the surrounding liquid. Thus, the Clausius-Mossotti factor is different for live and dead cells, assuming that the internal permittivity and conductivity of a dead cell is equal to that of the surrounding media, while all other parameters remain the same as a live cell (Figure 4a). These simulated results predict that for applied frequencies greater than approximately 60 kHz, live B cells will experience a positive OET force, while dead B cells will experience a negative OET force.

The difference in DEP response between live and dead B cells is used to selectively concentrate live B cells at an applied frequency of 120 kHz (Chiou et al. 2005). The selective collection pattern is a series of broken concentric rings (Figure 4b). As the concentric rings shrink, the live cells are focused to the center of the pattern by positive OET. In contrast, the dead cells experience negative OET, and slip through the gaps in the ring patterns. Dead B cells are verified by adding 0.4% Trypan blue dye, which is excluded by the live cells, and absorbed by the dead cells. In these images, the live cells appear clear, and the dead cells appear dark.

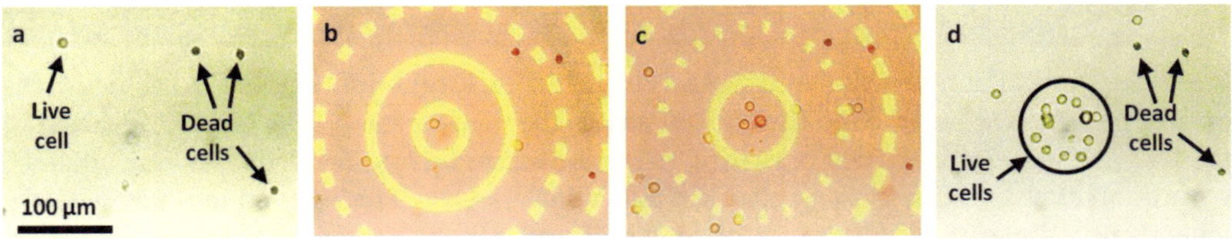

Fig. 4. Selective concentration of live and dead B cells. Reprinted with permission from Chiou et al., 2005. Copyright 2005, Nature Publishing Group.

OET force is also sensitive enough to distinguish between normal and abnormal live cells, which can improve the yield of *in vitro* fertilization procedures. This was demonstrated by the manipulation of oocytes cultured under standard condition and oocytes that had been kept in a nutrient-free solution for three days (Hwang et al., 2009b). By modelling oocytes using a protoplast structure, simulations predict that as cytoplasm conductivity decreases, the induced OET force also decreases. Hwang et al. hypothesize that this is the physical cause for different induced OET forces between the normal and nutrient-starved abnormal oocytes. It was observed that normal oocytes could be moved using OET patterns at approximately 12 µm/s, while the abnormal oocytes were unresponsive to OET manipulation (Hwang et al., 2009b).

In addition to differentiating between varying cells of the same type, OET force can be used to discriminate between different cell types. This ability is demonstrated through the spatial discrimination of live Jurkat and HeLa cells using OET.

In this experiment, cultured Jurkat cells were labeled with a green fluorescent dye. The labeled Jurkat cells and cultured HeLa cells are suspended in isotonic solution (Figure 5a). Culture media was added to adjust the conductivity of the cell solution to approximately 2 mS/m. Both cell types experience a positive OET force, and are attracted towards the optical manipulation patterns. However, the strength of the OET force varies according to the cell type, and also as a function of frequency (Ohta et al., 2007a). At an applied voltage of 10 Vpp at 100 kHz, sufficient variation in the OET force exists to differentiate between the two cell types using a scanning line optical pattern. A 15-µm-wide leading line and a 23-µm-wide trailing line are separated by ~40 µm, and are simultaneously scanned at a rate of 13 µm/s (Figure 5b). The leading line produces a weaker OET force than the thicker trailing line, as the manipulation velocity of cells exhibits a dependence on the width of the optical pattern (Ohta et al., 2007a; Ohta et al., 2007c). Thus, as the two lines are scanned across the OET device, the Jurkat cells, which experience a stronger OET force, are held by the leading line. The leading line does not produce sufficient force to transport the HeLa cells against the viscous drag, which are subsequently attracted to and transported by the trailing line. After the scan is completed, the cells retain a spatial separation equal to the spacing of the two scanning lines (Figure 5c). Fluorescent imaging is used to verify that the cells on the leading line pattern are the fluorescent-labeled Jurkat cells (Figure 5d).

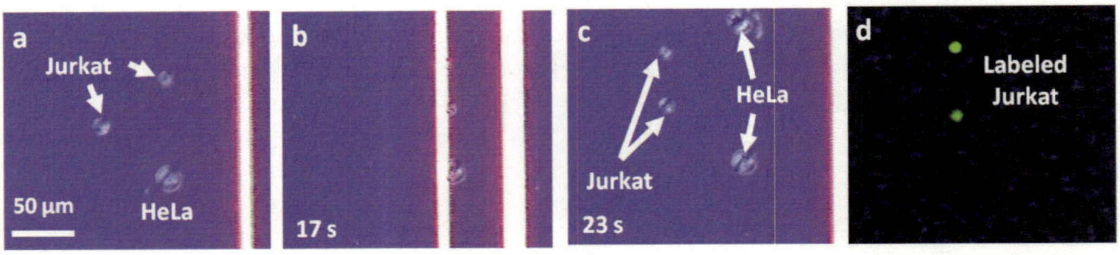

Fig. 5. Separation of live Jurkat and HeLa cells using OET. (a) Initial cell positions before the optical pattern is scanned from right to left across the field-of-view. (b) Cells are attracted to the leading line. The HeLa cell is starting to lag the scanning line. (c) Cells showing spatial separation after the scan is completed. An additional HeLa cell has moved into the field-of-view during the scan. (d) Fluorescent image of the cells in (c), verifying that the leading cells are the fluorescent-labeled Jurkat cells. The unlabeled HeLa cells do not appear in the fluorescent image. Reprinted with permissions from Ohta et al., 2007a. Copyright 2007, IEEE.

3.3 Light-induced electroporation

Other electric-field induced cellular operations can also be performed by the OET device. This includes electroporation, where electric pulses are used to temporarily open pores in the cellular membrane, allowing the introduction of molecules into cells. Currently, electroporation can be performed in a bulk procedure (Neumann et al., 1982) on thousands or millions of cells, restricting selectivity, or at the individual cell level (Lundqvistet al., 1998), in a time- and labor-intensive procedure. Although microfabricated devices have the potential to achieve electroporation at the single-cell level, with increased throughput (He et al., 2007; Khine et al., 2007; MacQueen et al., 2008; Yamada et al., 2008; Adamo & Jensen, 2008), this goal has not yet been achieved.

The OET device can be utilized as a platform to perform selective individual cell electroporation in parallel (Valley et al., 2009). The OET functionality allows single-cell resolution and selectivity of any cell within the OET manipulation area. In addition, the

electroporation can be performed in parallel, increasing throughput as compared to other single-cell electroporation techniques.
The parallel selective electroporation of HeLa cells was demonstrated using Propidium Iodide (PI) dye (Valley et al., 2009). The PI dye is normally membrane-impermeable. However, in the prescence of nucleic acids, the dye fluoresces red. Thus, a red fluorescent signal is evidence that cells have been successful electroporated, introducing PI dye into the cellular interior. Initially, OET manipulation is used to position the cells to be electroporated (Figure 6). Following this, two cells in the array corners are selectively illuminated by OET patterns, creating high electric field regions. An electroporation bias is briefly applied (1.5 kV/cm at 100 kHz for 5 s), electroporating only the illumated cells. This is verified by fluorescent microscopy, where only the illuminated cells show uptake and expression of the PI dye. A subsequent selective electroporation is performed to electroporate the remaining cells.

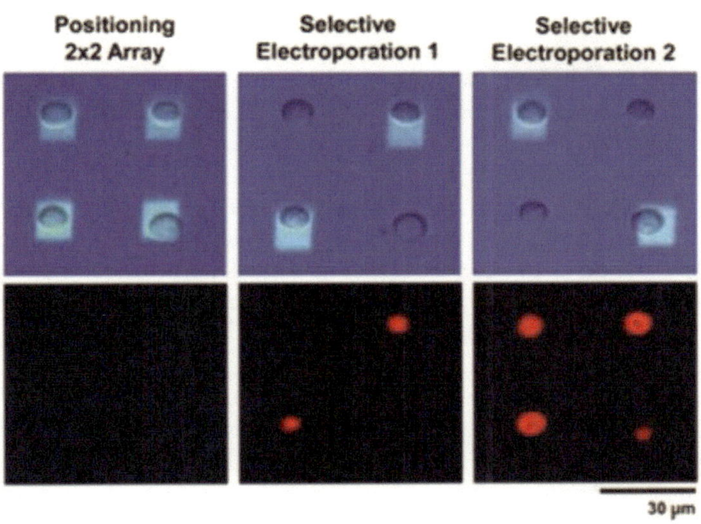

Fig. 6. Light-induced electroporation. Top row: Bright field microscopy of cells and optical patterns. Bottom row: Fluorescent microscopy showing PI dye. First, cells are arrayed using OET, which does not cause electroporation. Two cells on the diagonal are illuminated under the electroporation bias conditions (1.5 kV/cm). These cells fluoresce after uptaking PI dye (image taken 5 minutes following electroporation). Finally, the remaining two cells are electroporated, resulting in the fluorescence of all cells (image taken 5 minutes following electroporation). Reprinted with permission from Valley et al., 2009. Copyright 2009, Royal Society of Chemistry.

4. Micro- and nanoparticle assembly using Optoelectronic Tweezers

Optoelectronic tweezers is capable of assembling semiconducting and metallic micro- and nanoparticles for the creation of electronic and optoelectronic devices. This is demonstrated by the assembly of III-V semiconductor microdisk lasers on silicon for the integration of optical interconnects with CMOS circuits, and the assembly of semiconducting and metallic nanowires for nanowire-based devices.

4.1 Microdisk laser assembly on silicon

As the data transfer rates in computers increase, optical interconnects become attractive replacements for copper wiring. Silicon photonics provides an inexpensive method of integrating of CMOS electronics and optoelectronic components such as modulators and photodetectors for optical interconnects. However, on-chip optical sources are needed for most applications, necessitating the integration of semiconductor lasers with CMOS circuits on a silicon wafer. Silicon Raman lasers have been demonstrated (Boyraz & Jalali, 2004; Rong et al., 2005), but these require external optical pumps. Electrically-pumped lasers are only possible with compound semiconductor materials. Heteroepitaxy can grow compound semiconductor lasers directly on Si, but the growth temperature (> 400 °C) is usually too high for post-CMOS processing (Balakrishnan et al., 2006). To circumvent this issue, low temperature (300 °C) wafer bonding techniques have been used to integrate compound lasers on silicon wafers (Fang et al., 2006; Van Campenhout et al., 2007). However, integrating lasers on fully-processed CMOS wafers presents additional challenges as the silicon bonding surfaces are buried underneath up to ten layers of electrical interconnects. One approach is to build electrical interconnects and photonic circuits on separate Si wafers, bond the Si wafers, and then use flip-chip bonding to secure the III-V semiconductor materials on the Si wafers (Hattori et al., 2006). Another method to avoid any wafer bonding steps and enable simultaneous heterogeneous integration is an post-CMOS, optofluidic assembly process using LOET.

As the major axis of an anisotropic particle such as a microdisk aligns with the electric field lines, LOET must used to place the microdisks parallel to the surface of the device (standard OET would cause the disks to align with the major axis perpendicular to the OET surface). The microdisks used for the creation of on-chip lasers consist of an InGaAs/InGaAsP multiple-quantum-well structure, sandwiched by two larger-bandgap optical confinement layers, for a total thickness of 0.2 μm. Microdisks with diameters of 5 and 10 μm are fabricated from an InP epitaxial wafer, and suspended in ethanol for assembly by LOET.

The LOET electrodes create an optically-induced dielectrophoretic force, which is controlled by voltage applied across the electrodes and the position of optical patterns on the light-sensitive a-Si layer (Ohta et al., 2007a). The highest forces are in the illuminated areas near the electrode edges. Microdisks in solution are attracted to illuminated areas, and self-align in the gap between the electrodes (Figure 7). The optical patterns allow transportation of the microdisks along the length of the electrodes using an applied AC voltage of 1 to 10 Vpp at 200 kHz. Once the disks are aligned over a pedestal, the applied voltage is increased to 20 Vpp to hold the disks in place as the solution dries (Figure 7b, e). Ethanol is used to minimize surface tension forces during drying. After drying, the a-Si layer is removed by XeF_2 etching at 40 °C, so that the a-Si does not interfere with the optical mode of the microdisk. Subsequent SEM images show that the disks are aligned with an accuracy of approximately ± 0.25 μm (Figure 7c,f). This can be further improved by optimizing the optical imaging system.

Optical pumping of the assembled microdisk lasers show that at room temperature the 5- and 10-μm-diameter microdisks achieve single-mode lasing at wavelengths of 1558.7 nm and 1586 nm, at effective threshold pump powers of 0.34 mW and 1.0 mW, respectively (Tien et al., 2009). The threshold power for microdisks on the native InP wafer are similar to the threshold power of the assembled microdisks, indicating that the assembly process does not damage the disks.

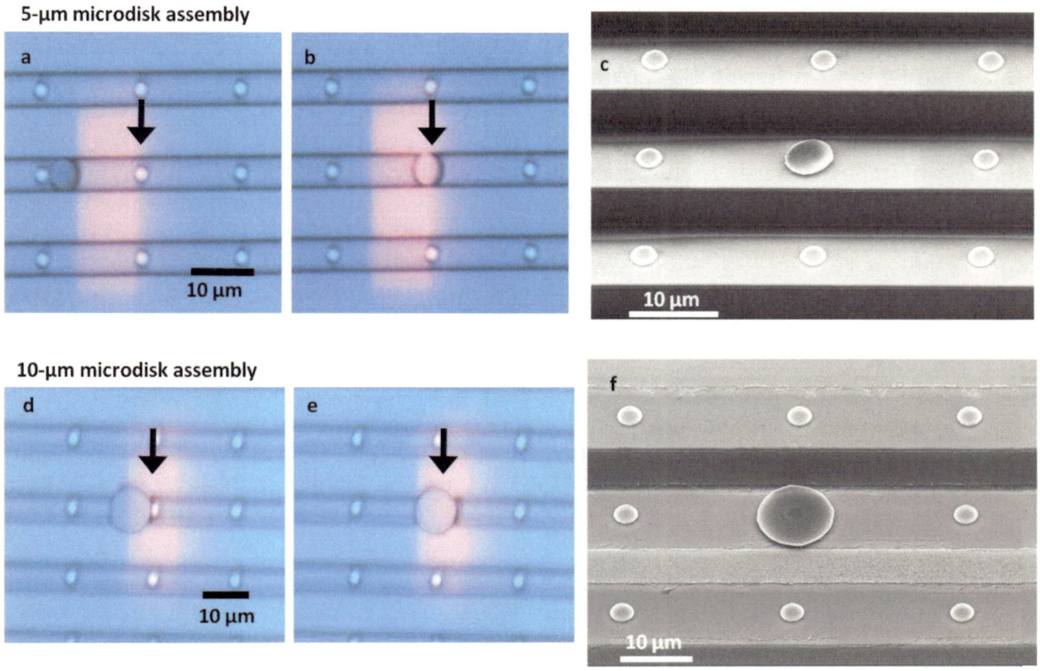

Fig. 7. III-V microdisk assembly. Assembly of a 5-µm-diameter microdisk (a, b) and a 10-µm-diameter microdisk (d, e) onto silicon pedestals. The target pedestals are indicated by arrows. (c, f) SEM images of the assembled microdisks.

4.2 Nanoparticle manipulation and assembly

Optoelectronic tweezers trapping works well for micron-scale objects, but the forces generated with OET typically are overcome by thermal (Brownian) fluctuations when particles are much smaller than 1 µm, as DEP force scales with r^3 (Equation 2). However, OET trapping is still possible when two dimensions of the particle are less than 100 nm, as long as the third dimension is on the order of micrometers. This anisotropy strengthens the particle dipole, increasing the DEP force. This effect is further enhanced for materials that are more polarizable than the liquid media, such as semiconductors or metals. As a result, OET is capable of exerting strong trapping forces both semiconducting and metallic nanowires with diameters as small as 20 nm (Jamshidi et al., 2008).

Nanowires experience a torque in addition to the DEP force, which aligns the long-axis of the nanowire with the electric field in the OET device. Therefore, the nanowire is aligned normal to the photoconductive electrode surface of the OET device. (If necessary, LOET can be used to manipulate nanowires parallel to the surface of the OET device [Ohta et al., 2007b; Ohta et al., 2008; Neale et al., 2009b]).

Once a nanowire is contained in an OET trap, the Brownian motion is significantly reduced. Without OET trapping, the nanowire covers an area of 28.9 µm^2; while in the OET trap, the nanowire is localized to an area of 0.22 µm^2, corresponding to a trap stiffness of 0.16 pN/µm (Jamshidi et al., 2008). Normalizing trap stiffness with respect to optical power results in 1.6×10^{-6} N/(m×mW), which is approximately 2 orders of magnitude larger than for an optical tweezers nanowire trap (Jamshidi et al., 2008). Maximum trapping speeds for an individual silicon nanowire with 100 nm diameter and 5 µm length approach 135 µm/s with a peak-to-peak trapping voltage of 20V, which is approximately 4 times the maximum speed achievable by optical tweezers (Pauzauskie et al., 2006) and is reached with 5-6 orders of magnitude less optical power density than optical tweezers.

In addition to the trapping of single nanowires, OET can be used to perform the parallel assembly of nanowires using real-time dynamic trapping. The formation of an individually-addressable 5×5 array of single silver nanowires has been demonstrated (Figure 8a). In addition, large-scale arrays of nanowires can be formed using appropriate optical patterns. The density of the trapped silver nanowires can also be tuned by varying the size of a rectangular trapping pattern in real-time (Figure 8b). The trapping density is ultimately limited by the strength of the dipole-dipole interactions between the nanowires.

Once the nanowires have been trapped, it is possible to preserve the position and orientation of the wires trapped with OET using a photocurable polymer solution such as PEGDA, enabling immobilization of nanowires in the polymer matrix within seconds by exposing the manipulation area to an ultra-violet source (Jamshidi et al., 2008). The ability to preserve the position and orientation of OET-trapped nanowires enables subsequent post-processing steps.

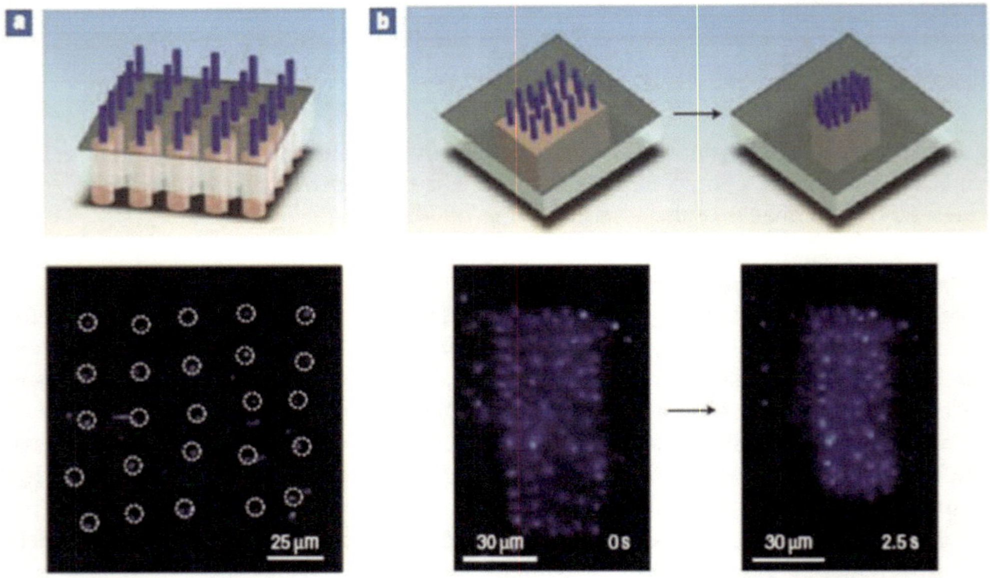

Fig. 8. Nanowire manipulation using OET. (a) Positioning silver nanowires into a 5 x 5 array. The top image shows a conceptual diagram; the bottom images show the microscope image, with the trapped nanowires circled for clarity. (b) Control of nanowire density by adjusting the size of the trapping pattern. More than 80 silver nanowires are concentrated from an area of approximately 3,000 µm^2 to 2,000 µm^2. Reprinted with permission from Jamshidi et al., 2008. Copyright 2008, Nature Publishing Group.

5. Future directions

New versions of the OET device are being developed. One promising device promises to improve OET manipulation of live cells. Currently, efficient OET operation requires cells to be suspended in a low-conductivity, isotonic media. This is beneficial for separating live and dead cells, as it results in OET forces of opposing polarities. However, isotonic media is not ideal for long-term cell viability or for fragile cells. In these cases, it is desirable to manipulate cells in their nature culture media. Unfortunately, typical cell culture media contains high concentrations of ions, resulting in an inefficient switching between the "on" and "off" states of the optically-defined OET electrodes. In order to observe OET actuation, prohibitively high optical intensities must be used.

One way to improve OET operation in high-conductivity cell culture media is to improve the illuminated photoconductivity of the OET device. This can be accomplished by replacing the a-Si material with a single-crystal silicon phototransistor structure. This phototransistor-based OET device (Ph-OET) manages to enhance the photoconductive response by more than 2 orders of magnitude as compared to a-Si, enabling OET operation in cell culture media (Hsu et al., 2009). The Ph-OET device provides more than 2 orders of magnitude of photoconductivity compared to a-Si under same illumination and bias, thanks to the high phototransistor gain and the high electron mobility of single crystalline silicon. The Ph-OET device can be actuated with an optical intensity of 1 W/cm^2. At these optical intensities, a-Si-based OET devices can only be actuated in low-conductivity media. Thus, the Ph-OET device promises to enable new applications of OET manipulation on live cells.

6. Summary

Optoelectronic tweezers and its variants are versatile devices for the manipulation of microparticles, nanoparticles, and cells. This chapter presented the theory and operation of the OET device and its variants, along with descriptions of various applications, including cell trapping and sorting, cell electroporation, microdisk laser assembly, and nanoparticle manipulation and assembly. Currently, the most promising fields for applications are biological research and micro- and nanoassembly. However, as OET is a relatively new tool, more applications are being developed that take advantage of the many capabilities of OET.

7. References

Adamo A. & Jensen, K. F. (2008). Microfluidic based single cell microinjection. *Lab on a Chip*, vol. 8, no. 8, 1258–1261.

Ashkin, A.; Dziedzic, J. M.; Bjorkholm, J. E. & Chu, S. (1986). Observation of a single-beam gradient force optical trap for dielectric particles. *Optics Letters*, vol. 11, no. 5, 288-290.

Avouris, P.; Hertel, T.; Martel, R.; Schmidt, T.; Shea, H. R. & Walkup, R. E. (1999). Carbon nanotubes: nanomechanics, manipulation, and electronic devices. *Applied Surface Science*, vol. 141, no. 3-4, 201-209.

Balakrishnan, G.; Jallipalli, A.; Rotella, P.; Huang, S.; Khoshakhlagh, A.; Amtout, A.; Krishna, S.; Dawson, L. R. & Huffaker, D. L. (2006). Room-temperature optically pumped (Al)GaSb vertical-cavity surface-emitting laser monolithically grown on an Si(100) substrate. *IEEE Journal of Selected Topics in Quantum Electronics*, vol. 12, no. 6, 1636-1641.

Boyraz O. & Jalali, B. (2004). Demonstration of a silicon Raman laser. *Optics Express*, vol. 12, no. 21, 5269-5273.

Cheng, J.; Sheldon, E. L.; Wu, L.; Heller, M. J. & O'Connell, J. P. (1998). Isolation of cultured cervical carcinoma cells mixed with peripheral blood cells on a bioelectronic chip. *Analytical Chemistry*, vol. 70, no. 11, 2321-2326.

Chiou, P. Y.; Wong, W.; Liao, J. C. & Wu, M. C. (2004). Cell addressing and trapping using novel optoelectronic tweezers, in *Proceedings of the 17th IEEE International Conference on Micro Electro Mechanical Systems (MEMS)*, pp. 21-24, Maastricht, Netherlands.

Chiou, P. Y.; Ohta, A. T. & Wu, M. C. (2005). Massively parallel manipulation of single cells and microparticles using optical images. *Nature*, vol. 436, no. 7049, 370-372.

Chiou, P. Y.; Ohta, A. T.; Jamshidi, A.; Hsu, H. Y. & Wu, M. C. (2008). Light-actuated ac electroosmosis for nanoparticle manipulation. *Journal of Microelectromechanical Systems*, vol. 17, no. 3, 525-531.

Cho, S. K.; Moon H. J. & Kim, C.-J. (2003). Creating, transporting, cutting, and merging liquid droplets by electrowetting-based actuation for digital microfluidic circuits. *Journal of Microelectromechanical Systems*, vol. 12, no. 1, pp. 70–80.

Choi, W.; Kim, S. H.; Jang, J. & Park, J. K. (2007). Lab-on-a-display: a new microparticle manipulation platform using a liquid crystal display (LCD). *Microfluidics and Nanofluidics*, vol. 3, no. 2, 217-225.

Choi, W.; Nam, S. W.; Hwang, H.; Park, S. & Park, J. K. (2008). Programmable manipulation of motile cells in optoelectronic tweezers using a grayscale image. *Applied Physics Letters*, vol. 93, no. 14, 143901-3.

Chronis, N. & Lee, L. P. (2005). Electrothermally activated SU-8 microgripper for single cell manipulation in solution. *Journal of Microelectromechanical Systems*, vol. 14, no. 4, 857-863.

Evoy, S.; DiLello, N.; Deshpande, V.; Narayanan, A.; Liu, H.; Riegelman, M.; Martin, B. R.; Hailer, B.; Bradley, J.-C.; Weiss, W.; Mayer, T. S.; Gogotsi, Y.; Bau, H. H.; Mallouk, T. E. & Raman, S. (2004). Dielectrophoretic assembly and integration of nanowire devices with functional CMOS operating circuitry. *Microelectronic Engineering*, vol. 75, no. 1, 31-42.

Fang, A. W.; Park, H.; Cohen, O.; Jones, R.; Paniccia, M. J. & Bowers, J. E. (2006). Electrically pumped hybrid AlGaInAs-silicon evanescent laser. *Optics Express*, vol. 14, no. 20, 9203-9210.

Fuchs, A. B.; Romani, A.; Freida, D.; Medoro, G.; Abonnenc, M.; Altomare, L.; Chartier, I.; Guergour, D.; Villiers, C.; Marche, P. N.; Tartagni, M.; Guerrieri, R.; Chatelain, F. & Manaresi, N. (2006). Electronic sorting and recovery of single live cells from microlitre sized samples. *Lab on a Chip*, vol. 6, no. 1, 121-126.

Garces-Chavez, V.; Dholakia, K. & Spalding, G. C. (2005). Extended-area optically induced organization of microparticies on a surface. *Applied Physics Letters*, vol. 86, no. 3, 031106-3.

Gascoyne, P. R. C.; Wang, X.-B.; Huang, Y. & Becker, F. F. (1997). Dielectrophoretic separation of cancer cells from blood. *IEEE Transactions on Industry Applications*, vol. 33, no. 3, 670-8.

Gascoyne, P. R. C. & Vykoukal, J. V. (2004). Dielectrophoresis-based sample handling in general-purpose programmable diagnostic instruments. *Proceedings of the IEEE*, vol. 92, no. 1, 22-42.

Grigorenko, A. N.; Roberts, N. W.; Dickinson, M. R. & Zhang, Y. (2008). Nanometric optical tweezers based on nanostructured substrates. *Nature Photonics*, vol. 2, no. 6, 365-370.

Gosse, C. & Croquette, V. (2002). Magnetic tweezers: Micromanipulation and force measurement at the molecular level. *Biophysical Journal*, vol. 82, no. 6, 3314-3329.

Grier, D. G. (2003). A revolution in optical manipulation. *Nature*, vol. 424, no. 6950, 810-816.

Hattori, H. T.; Seassal, C.; Touraille, E.; Rojo-Romeo, P.; Letartre, X.; Hollinger, G.; Viktorovitch, P.; Di Cioccio, L.; Zussy, M.; Melhaoui, L. E. & Fedeli, J. M. (2006). Heterogeneous integration of microdisk lasers on silicon strip waveguides for optical interconnects. *IEEE Photonics Technology Letters*, vol. 18, no. 1, 223-225.

He, H. Q.; Chang D. C. & Lee, Y. K. (2007). Using a micro electroporation chip to determine the optimal physical parameters in the uptake of biomolecules in HeLa cells. *Bioelectrochemistry*, vol. 70, 363–368.

Higuchi, Y.; Kusakabe, T.; Tanemura, T.; Sugano, K.; Tsuchiya, T. & Tabata, O. (2008). Manipulation system for nano/micro components integration via transportation and self-assembly, in *Proceedings of the IEEE 21st International Conference on Micro Electro Mechanical Systems (MEMS)*, pp. 836-839, Tucson, AZ, USA.

Hsu, H.-Y.; Ohta, A. T.; Chiou, P.-Y.; Jamshidi, A.; Neale, S. & Wu, M. C. (2009). Phototransistor-based optoelectronic tweezers for dynamic cell manipulation in cell culture media. *Lab on a Chip*, DOI: 10.1039/b906593h.

Huang, Y.; Yang, J. M.; Hopkins, P. J.; Kassegne, S.; Tirado, M.; Forster, A. H. & Reese, H. (2003). Separation of simulants of biological warfare agents from blood by a miniaturized dielectrophoresis device. *Biomedical Microdevices*, vol. 5, no. 3, 217-225.

Hwang, H.; Choi, Y. J.; Choi, W.; Kim, S. H.; Jang, J. & Park, J. K. (2008a). Interactive manipulation of blood cells using a lens-integrated liquid crystal display based optoelectronic tweezers system. *Electrophoresis*, vol. 29, no. 6, 1203-1212.

Hwang, H.; Oh, Y.; Kim, J. J.; Choi, W.; Park, J. K.; Kim, S. H. & Jang, J. (2008b). Reduction of nonspecific surface-particle interactions in optoelectronic tweezers. *Applied Physics Letters*, vol. 92, no. 2, 024108-3.

Hwang, H. & Park, J. K. (2009a). Rapid and selective concentration of microparticles in an optoelectrofluidic platform. *Lab on a Chip*, vol. 9, no. 2, 199-206.

Hwang, H.; Lee, D. H.; Choi, W. & Park, J. K. (2009b). Enhanced discrimination of normal oocytes using optically induced pulling-up dielectrophoretic force. *Biomicrofluidics*, vol. 3, no. 1, 014103-10.

Jamshidi, A.; Pauzauskie, P. J.; Schuck, P. J.; Ohta, A. T.; Chiou, P. Y.; Chou, J.; Yang, P. & Wu, M. C. (2008). Dynamic manipulation and separation of individual semiconducting and metallic nanowires. *Nature Photonics*, vol. 2, no. 2, 85-89.

Jones, T. B. *Electromechanics of Particles*. Cambridge: Cambridge University Press, 1995.

Khine, M.; Ionescu-Zanetti, C.; Blatz, A.; Wang L. P. & Lee, L. P. (2007). Single-cell electroporation arrays with real-time monitoring and feedback control. *Lab on a Chip*, vol. 7, no. 4, 457–462.

Kim, C.-J.; Pisano, A. P. & Muller, R. S. Silicon-Processed Overhanging Microgripper. *Journal of Microelectromechanical Systems*, vol. 1, no. 1, pp. 31-36.

Konig, K.; Liang, H.; Berns, M. W. & Tromberg, B. J. (1996). Cell damage in near-infrared multimode optical traps as a result of multiphoton absorption. *Optics* Letters, vol. 21, no. 14, 1090-1092.

Krupke, R.; Hennrich, F.; von Lohneysen, H. & Kappes, M. M. (2003). Separation of metallic from semiconducting single-walled carbon nanotubes. *Science,* vol. 301, no. 5631, 344-347.

Lau, A. N. K.; Ohta, A. T.; Phan, H. L.; Hsu, H.-Y.; Jamshidi, A.; Chiou, P.-Y. & Wu, M. C. (2009). Antifouling coatings for optoelectronic tweezers. *Lab on a Chip,* DOI: 10.1039/B907840A.

Lu, Y.-S.; Huang, Y.-P.; Yeh, J. A.; Lee, C. & Chang, Y.-H. (2005). Controllability of non-contact cell manipulation by image dielectrophoresis (iDEP). *Optical and Quantum Electronics*, vol. 37, no. 13-15, 1385-1395.

Lundqvist, J. A.; Sahlin, F. ; Aberg, M. A. I. ; Stromberg, A.; Eriksson, P. S. & Orwar, O. (1998). Altering the biochemical state of individual cultured cells and organelles with ultramicroelectrodes. *Proceedings of the National Academy of Sciences of the United States of America,* vol. 95, no. 18, 10356–10360.

MacDonald, M. P.; Spalding, G. C. & Dholakia, K. (2003). Microfluidic sorting in an optical lattice," *Nature,* vol. 426, no. 6965, 421-424.

MacQueen, L. A.; Buschmann M. D. & Wertheimer, M. R. (2008). Gene delivery by electroporation after dielectrophoretic positioning of cells in a non-uniform electric field. *Bioelectrochemistry,* vol. 72, no. 2, 141–148.

Maier, B.; Koomey, M. & Sheetz, M. P. (2004). A force-dependent switch reverses type IV pilus retraction. *Proceedings of the National Academy of Sciences of the United States of America,* vol. 101, no. 30, 10961-10966.

Miao, X. & Lin, L. Y. (2007). Trapping and manipulation of biological particles through a plasmonic platform. *IEEE Journal of Selected Topics in Quantum Electronics,* vol. 13, no. 6, 1655-1662.

Mohanty, S. K.; Rapp, A.; Monajembashi, S.; Gupta, P. K. & Greulich, K. O. (2002). Comet assay measurements of DNA damage in cells by laser microbeams and trapping beams with wavelengths spanning a range of 308 nm to 1064 nm. *Radiation Research,* vol. 157, no. 4, 378-385.

Muller, T.; Gerardino, A.; Schnelle, T.; Shirley, S. G.; Bordoni, F.; Gasperis, G. D.; Leoni, R. & Fuhr, G. (1996). Trapping of micrometre and sub-micrometre particles by high-frequency electric fields and hydrodynamic forces. *Journal of Physics D-Applied Physics,* vol. 29, no. 2, 340-349.

Neale, S. L.; Mazilu, M.; Wilson, J. I. B.; Dholakia, K. & Krauss, T. F. (2007). The resolution of optical traps created by light induced dielectrophoresis (LIDEP). *Optics Express,* vol. 15, no. 20, 12619-12626.

Neale, S. L.; Ohta, A. T.; Hsu, H.-Y.; Valley, J. K.; Jamshidi, A. & Wu, Ming C. (2009a). Trap profiles of projector based optoelectronic tweezers (OET) with HeLa cells. Optics Express, vol. 17, no. 7, 5232-5239.

Neale, S.L.; Fan, Z.; Ohta, A. T.; Jamshidi, A.; Valley, J. K.; Hsu, H.-Y.; Javey, A. & Wu, M. C. (2009b). Optofluidic assembly of red/blue/green semiconductor nanowires, in *Proceedings of the Conference on Lasers and Electro-Optics (CLEO),* Baltimore, MD, USA.

Neuman, K. C.; Chadd, E. H.; Liou, G. F.; Bergman, K. & Block, S. M. (1999). Characterization of photodamage to Escherichia coli in optical traps. *Biophysical Journal,* vol. 77, no. 5, 2856-2863.

Neumann, E.; Schaeferridder, M.; Wang, Y. & Hofschneider, P. H. (1982). Gene transfer into mouse lyoma cells by electroporation in high electric fields. *The EMBO Journal*, vol. 1, no. 7, 841–845.

Ohta, A. T.; Chiou, P.-Y.; Phan, H. L.; Sherwood, S. W.; Yang, J. M.; Lau, A. N. K.; Hsu, H.-Y.; Jamshidi, A. & Wu, M. C. (2007a). Optically Controlled Cell Discrimination and Trapping Using Optoelectronic Tweezers. *IEEE Journal of Selected Topics in Quantum Electronics*, vol. 13, no. 2, 235-243.

Ohta, A. T.; Jamshidi, A.; Pauzauskie, P. J.; Hsu, H.-Y.; Yang, P. & Wu, M. C. (2007b). Trapping and transport of silicon nanowires using lateral-field optoelectronic tweezers, in *Proceedings of the Conference on Lasers and Electro-Optics (CLEO)*, pp. 828-829, Baltimore, MD, USA.

Ohta, A. T.; Chiou, P. Y.; Han, T. H.; Liao, J. C.; Bhardwaj, U.; McCabe, E. R. B.; Yu, F.; Sun, R. & Wu, M. C. (2007c) Dynamic Cell and Microparticle Control via Optoelectronic Tweezers. *Journal of Microelectromechanical Systems*, vol. 16, no. 3, 491-499.

Ohta, A. T.; Neale, S.L.; Hsu, H.-Y.; Valley, J. K. & Wu, M. C. (2008). Parallel assembly of nanowires using lateral-field optoelectronic tweezers, in *Proceedings of the IEEE/LEOS International Conference on Optical MEMs and Nanophotonics*, pp. 7-8, Freiburg, Germany.

Padinger, F.; Rittberger, R. S. & Sariciftci, N. S. (2003). Effects of Postproduction Treatment on Plastic Solar Cells. *Advanced Functional Materials*, vol. 13, no. 1, pp. 85-88.

Park, S.; Pan, C.; Wu, T.-H.; Kloss, C.; Kalim, S.; Callahan, C. E.; Teitell, M. & Chiou, E. P. Y. (2008). Floating electrode optoelectronic tweezers: light-driven dielectrophoretic droplet manipulation in electrically insulating oil medium. *Applied Physics Letters*, vol. 92, no. 15, 151101-1-3.

Paterson, L.; Papagiakoumou, E.; Milne, G.; Garces-Chavez, V.; Tatarkova, S. A.; Sibbett, W.; Gunn-Moore, F. J.; Bryant, P. E.; Riches, A. C. & Dholakia, K. (2005). Light-induced cell separation in a tailored optical landscape. *Applied Physics Letters*, vol. 87, no. 12, 123901-3.

Pauzauskie, P. J.; Radenovic, A.; Trepagnier, E.; Shroff, H.; Yang, P. & Liphardt, J. (2006). Optical trapping and integration of semiconductor nanowire assemblies in water. *Nature Materials*, vol. 5, no. 2, 97-101.

Pethig, R.; Talary, M. S. & Lee, R. S. (2003). Enhancing traveling-wave dielectrophoresis with signal superposition. *IEEE Engineering in Medicine and Biology Magazine*, vol. 22, no. 6, 43-50.

Pohl, H. A. (1958). Some Effects of Nonuniform Fields on Dielectrics. *Journal of Applied Physics*, vol. 29, no. 8, 1182-1188.

Rong, H. S.; Jones, R.; Liu, A.; Cohen, O.; Hak, D.; Fang A. & Paniccia, M. (2005). A continuous-wave Raman silicon laser. *Nature*, vol. 433, no. 7027, 725-728.

Shah, G. J.; Ohta, A. T.; Chiou, E. P. Y.; Wu, M. C. & Kim, C.-J. (2009). EWOD-driven droplet microfluidic device integrated with optoelectronic tweezers as an automated platform for cellular isolation and analysis. *Lab on a Chip*, vol. 9, no. 12, 1732-1739.

Srinivasan, V.; Pamula, V. K. & Fair, R. B. (2004). An integrated digital microfluidic lab-on-a-chip for clinical diagnostics on human physiological fluids. *Lab on a Chip*, vol. 4, no. 4, pp. 310–315.

Tien, M.-C.; Ohta, A. T.; Yu, K.; Neale, S. L. & Wu, M. C. (2009). Heterogeneous integration of InGaAsP microdisk laser on a silicon platform using optofluidic assembly. *Applied Physics A: Materials Science & Processing*, vol. 95, no. 4, 967-972.

Valley, J. K.; Jamshidi, A.; Ohta, A. T.; Hsu, H.-Y. & Wu, M. C. (2008). Operational regimes and physics present in optoelectronic tweezers. *Journal of Microelectromechanical Systems*, vol. 17, no. 2, 342-50.

Valley, J. K.; Neale, S.; Hsu, H.-Y.; Ohta, A. T.; Jamshidi, A. & Wu, M. C. (2009). Parallel single-cell light-induced electroporation and dielectrophoretic manipulation. *Lab on a Chip*, vol. 9, no. 12, pp. 1714-1720.

Van Campenhout, J.; Rojo-Romeo, P.; Regreny, P.; Seassal, C.; Van Thourhout, D.; Verstuyft, S.; Di Cioccio, L.; Fedeli, J.-M.; Lagahe, C. & Baets, R. (2007). Electrically pumped InP-based microdisk lasers integrated with a nanophotonic silicon-on-insulator waveguide circuit. *Optics Express*, vol. 15, no. 11, 6744-6749.

Wang, W.; Lin, L.-H.; Guo, T.-F. & Lee, G.-B. (2009). Manipulation of Biosamples and Microparticles using Optical Images on Polymer Devices, in *Proceedings of the IEEE 22nd International Conference on Micro Electro Mechanical Systems (MEMS)*, pp. 415-418, Sorrento, Italy.

West, J.; Becker, M.; Tombrink, S. & Manz, A. (2008). Micro total analysis systems: Latest achievements. *Analytical Chemistry*, vol. 80, no. 12, 4403-4419.

Yamada, M.; Kobayashi, J.; Yamato, M.; Seki M. & Okano, T. (2008). Millisecond treatment of cells using microfluidic devices via two-step carrier-medium exchange. *Lab on a Chip*, vol. 8, no. 5, 772–778.

19

An Asymmetric Magneto-Optical Trap

Heung-Ryoul Noh[1] and Wonho Jhe[2]
Chonnam National University
Seoul National University
Korea

1. Introduction

Since the advent of the magneto-optical trap (MOT) in 1987 (Raab et al., 1987), it has been intensively studied and widely used as a pre-cooled atomic source for various experiments (Metcalf & van der Straten, 1999). In addition, there have been intensive studies on the MOT itself such as cold collisions (Walker & Feng, 1994), nonlinear optics (Tabosa et al., 1991), existence of sub-Doppler force (Wallace et al., 1994), or limit of density (Townsend et al., 1995). Nevertheless MOT itself is far from quantitative understanding and still keeps providing surprises as unexplored characteristics and applications are being developed. In the perspective of nonlinear dynamics in a MOT, there were several reports as follows: Sesko *et al.* observed several variations of atomic spatial distribution and abrupt change between the distributions when there exist laser beam misalignment, intensity imbalance or radiation trapping (Walker et al., 1990; Sesko et al., 1991). They explained the phenomena by optical torques exerted by the misaligned trapping lasers. Based on the studies of Sesko *et al.*, Bagnato *et al.* have observed the limit cycles and some abrupt changes of atomic spatial distributions (Bagnato et al., 1993; Dias Nunes et al., 1996). Recently, Wilkowski *et al.* found the instability phenomena in a MOT and explained them by means of shadow effect (Wilkowski et al., 2000; di Stefano et al, 2003). In addition, MOT exhibits very unique collective effects and critical behaviors when the number of atoms increases such as instability-induced pulsation (Labeyrie et al., 2006), and plasma oscillations of ultracold neutral plasma (Kulin et al., 2000).

In this article we present experimental and theoretical works on the applications of the magneto-optical trap by modifying the trap conditions, which is termed as an asymmetric magneto-optical trap (AMOT). This article is composed of three parts: In Sec. 2, we describe the parametric resonance achieved by the modulation of the trap laser intensities. When the modulation frequency is near twice the natural frequency of the trap and the modulation amplitude exceeds a threshold value, the parametric resonance can be excited; i.e., the trapped atoms are divided into two parts and oscillate in opposite directions. The various theoretical and experimental studies are presented. Section 3 is devoted to measurement of trap parameters by the method of parametric resonance and transient oscillation. By decreasing the modulation amplitude of the parametric excitation down to its threshold value one can measure the trap frequency. In the case of transient oscillation, the trap frequency and damping coefficient were obtained by measuring the trajectory of the atoms returning to the original trap center, after the applied uniform magnetic field, used for displacing the trap center, was turned off. In Sec. 4, we discusses adjustable trap where the trap-laser frequencies are unequal to one another. When the detuning of the laser along the symmetric axis is different from those of the transverse lasers, an array of several sub-Doppler traps with adjustable separations between the traps which are proportional to the relative laser-detuning difference.

2. Parametric resonance in a magneto-optical trap

The parametric resonances are ubiquitous phenomena: e.g., we can find them in vertically driven pendulum, fluid mechanics, ion trap such as a Paul trap or a Penning trap, optical dipole trap, and a lot of different fields (Landau & Lifshitz, 1976; Nayfeh & Moore, 1979). In an optical lattice or a dipole trap, the modulation of the potential depth was often used to measure the vibrational frequencies (Friebel et al., 1998). In a Paul trap, the observation of higher order resonance up to n = 10 has been reported (Razvi et al., 1998). Above all, our study is very analogous to parametrically-pumped electron oscillators in the Penning trap (Tan & Gabrielse, 1991; 1993; Tseng et al., 1999; Lapidus et al., 1999). The simple observed features of electrons in the cylindrical Penning trap are related to the cooperative behavior and nonlinear dynamics. The electron oscillators exhibit a rich and varied nonlinear dynamics which our systems also manifested.

In this section, we present the theoretical and experimental results of the parametrically driven MOT by using the modulation of the MOT potential, especially the cooling laser intensity (Kim et al., 2003; 2004). When we modulate the intensity of the cooling laser at about twice the resonant frequency of the trap, the atoms in the MOT are divided into two parts and oscillate reciprocally with the finite amplitude due to nonlinearity of the trap. The amplitude of oscillation and the phase with respect to the modulation of cooling laser are measured with the images of atomic cloud and compared to the theoretical calculations. Direct observation of the sub-Doppler trap in the parametrically driven MOT is also described (Kim et al., 2004).

2.1 Theory

The atomic motion in a parametrically-driven MOT can be described in terms of simple Doppler cooling theory. In the MOT the $\sigma^+(\sigma^-)$ polarized light propagates from the $-z(+z)$ axis, while the $\sigma^-(\sigma^+)$ polarized light propagates from the $-x(+x)$ and $-y(+y)$ axis. Also the inhomogeneous magnetic field is applied, which is given by

$$\vec{B} = b\left(-\frac{x}{2}\hat{x} - \frac{y}{2}\hat{y} + z\hat{z}\right),$$

where b is the magnetic field gradient for z-axis. The laser intensities are modulated to excite the parametric resonance. Although intensities of all laser beams are simultaneously modulated with the frequency ω, we can excite the resonance for one specific axis by controlling the parameters such as laser intensity or the detuning. In this study we only excite the parametric resonance for z-axis.

The atomic motion for z-axis can be described by the following equation for a two-level atom (Metcalf & van der Straten, 1999),

$$m\ddot{z} = F(z,\dot{z},t), \qquad (1)$$

with

$$F(z,\dot{z},t)=\frac{\hbar k\Gamma}{2}\left[\frac{s_0(1+h\cos\omega t)}{1+s_0(1+h\cos\omega t)+4(\delta-k\dot{z}-\frac{\mu_B b}{\hbar}z)^2/\Gamma^2}-\frac{s_0(1+h\cos\omega t)}{1+s_0(1+h\cos\omega t)+4(\delta+k\dot{z}+\frac{\mu_B b}{\hbar}z)^2/\Gamma^2}\right], \quad (2)$$

where k is the wave vector, μ_B is Bohr magneton, m is the mass of an atom, Γ is the decay rate of the excited state ($=2\pi\times5.9$ MHz), and $\delta(=\omega_L-\omega_A)$ is the detuning of the laser frequency ω_L relative to the atomic resonance frequency ω_A. Here $s_0(=I/I_s)$ is the unmodulated saturation parameters with I being the laser intensity of a laser beam for z-axis and I_s ($=1.62$ mW/cm^2) being the saturation intensity for a ^{85}Rb atom, and the term $(1+h\cos\omega t)$ describes the modulated laser intensity, where h is the modulation amplitude and ω is the modulation frequency.

In addition to the Doppler force in Eq. (2), we need to include the sub-Doppler force and the random force (Chang & Minogin, 2002). The sub-Doppler force describes the tight confinement in the vicinity of the origin of velocity or position and explains the bright and small trap at the origin. In the former part of Sec. 2, we ignore this force, since we do not consider the small trap and describes the motion with large amplitude. The random force which stems from the recoil of the emitted photon from atoms contributes to the determination of the width of trap and the transitions between the stable attractors. As far as the center of motion is concerned, however, we do not need to include this force. When we discuss the simulations about the width of oscillating clouds, we will include this force.

We present analytic solutions from the approximation of Eq. (1) in order to have concrete understanding of the system. When we expand the Eq. (1) up to third orders in position (z) and velocity ($\dot{z}$), Eq. (1) can be approximately written by

$$\ddot{z}+\beta\dot{z}+\omega_0^2(1+h\cos\omega t)z=-A_0\omega_0^2\left[z+\frac{\beta}{\omega_0^2}\dot{z}\right]^3, \quad (3)$$

where the natural frequency (ω_0), the damping coefficient (β), and the nonlinear coefficient (A_0) are given by

$$\omega_0=\sqrt{\frac{8k\mu_B b s_0(-\delta/\Gamma)}{m\left[1+4(\delta/\Gamma)^2\right]^2}}, \quad (4a)$$

$$\beta=\frac{\hbar k}{\mu_B b}\omega_0^2, \quad (4b)$$

$$A_0=8\left(\frac{\mu_B b}{\hbar\Gamma}\right)^2\frac{4(\delta/\Gamma)^2-1}{\left[4(\delta/\Gamma)^2+1\right]^2}, \quad (4c)$$

respectively. Here we have assumed that $h \ll 1$ and the damping is weak, that is, $\beta \ll \omega_0$. When we neglect the nonlinear term in Eq. (3), i.e., the term in the right hand side of Eq. (3), it becomes

$$\ddot{z} + \beta\dot{z} + \omega_0^2(1 + h\cos\omega t)z = 0, \tag{5}$$

which is a well-known Mathieu equation (Landau & Lifshitz, 1976; Nayfeh & Moore, 1979). The solutions of Mathieu equation exhibit the parametric resonance which occurs when the modulation frequency is $\omega = 2\omega_0/n$, where n is the integers. In practice, it is hardly possible to observe higher order resonances ($n > 1$). In the absence of damping of the system ($\beta = 0$), the parametric resonance occurs in the range

$$-\frac{1}{2}h\omega_0 < \omega - 2\omega_0 < \frac{1}{2}h\omega_0, \tag{6}$$

which is expressed as dotted lines in Fig. 1(a). Here the horizontal axis is the modulation amplitude (h), and the vertical axis is the modulation frequency normalized to the trap frequency (ω/ω_0). When the modulation frequency lies inside this region, the solution diverges and shows instability. If we take into account the damping, the parametric resonance range slightly changes from Eq. (6) to $\omega_1 < \omega < \omega_2$ as shown in Fig. 1(a) (solid curve), where the characteristic frequencies, ω_1 and ω_2, are given by

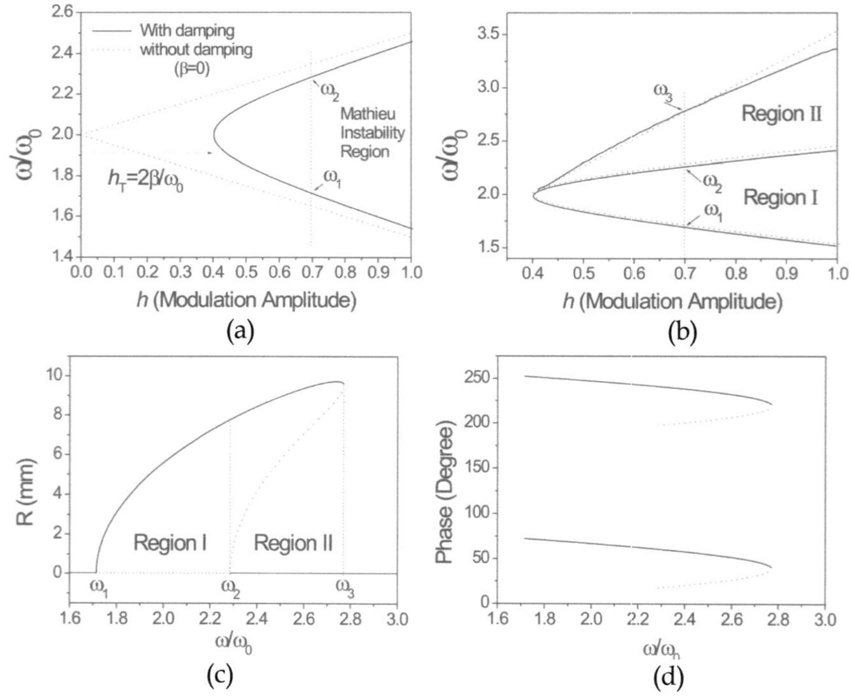

Fig. 1. (a) The instability region of the parametrically driven MOT. (b) The numerically calculated map from Eq. (1) as solid curves, while the result with the linearized Mathieu equation is plotted as a dotted line or curves. The calculated amplitude R (c) and the phase ψ (d) of the limit cycle at the modulation amplitude of $h = 0.7$.

$$\omega_{1(2)} = 2\omega_0 - (+)\frac{\omega_0}{2}\sqrt{h^2 - h_T^2}. \tag{7}$$

Here a threshold value (h_T) of the modulation amplitude above which the parametric resonance can occur is given by $h_T = 2\beta/\omega_0$.

Including the nonlinear terms, Eq. (3) can be analytically solved by means of usual averaging method (Nayfeh & Moore, 1979; Strogatz, 2001). In this case, the stability map is changed (Fig. 1(b)). In the steady-state regime in the rotating frame with the angular frequency $\omega/2$, the nontrivial solutions of Eq. (3) can be written by

$$z(t) = R(t)\cos\left[\frac{\omega}{2}t + \psi(t)\right], \tag{8}$$

where the amplitude $R(t)$ and the phase $\psi(t)$ satisfy the following equations:

$$\frac{dR}{dt} = -\frac{\beta}{2}R\left[1 + \frac{3}{16}(4 + h_T^2)A_0R^2 - \frac{h}{h_T}\sin 2\psi\right],$$

$$\frac{d\psi}{dt} = -\frac{\omega - 2\omega_0}{2} + \frac{h\omega_0}{4}\cos 2\psi + \frac{3}{32}(4 + h_T^2)A_0\omega_0R^2. \tag{9}$$

The solution in Eq. (8) is called a limit cycle motion, which comes from the reduction of the divergent solution to the finite one owing to the nonlinear term in Eq. (3).

The steady state solutions can be obtained by equating $dR/dt = 0$ and $d\psi/dt = 0$ in Eq. (9). We immediately notice that there exists a trivial solution $R = 0$. As well, we have the nontrivial steady-state solutions for Eq. (9) as

$$R^2 = \frac{16}{3A_0(4 + h_T^2)^2}\left(4(\eta - 2) - h_T^2 \pm \sqrt{h^2(4 + h_T^2) - 4h_T^2(\eta - 1)^2}\right),$$

$$\cos 2\psi = \frac{2}{h(4 + h_T^2)^2}\left(h_T^2(\eta - 1) \mp \sqrt{h^2(4 + h_T^2) - 4h_T^2(\eta - 1)^2}\right), \tag{10}$$

where the upper (lower) sign denotes the stable (unstable) solutions and $\eta(= \omega/\omega_0)$ is the normalized modulation frequency. The calculated results for the amplitude and phase are shown in Fig. 1(c) and 1(d), respectively, for $h = 0.7$. In Figs. 1(c) and 1(d), the stable (unstable) solutions are represented as solid (dotted) lines. We can see that there exist three characteristic frequencies (ω_1, ω_2 and ω_3) which characterize the existence of solutions. The frequencies ω_1 and ω_2 are already defined in Eq. (7) and the new frequency ω_3 is given by

$$\omega_3 = \omega_0\left(1 + \frac{h}{2h_T}\sqrt{4 + h_T^2}\right). \tag{11}$$

The frequency ω_3 does not exist for solutions of the Mathieu equation and arises when the nonlinear term is included. As shown in Fig. 1(b) (dotted line), the frequency ω_3 is linearly dependent on the modulation amplitude h.

In Figs. 1(c) and 1(d), for the frequency range $\omega_1 < \omega < \omega_2$ (Region I), as well as an unstable solution ($R = 0$), there exist two stable solutions. This is because for a given amplitude R, we have two solutions of the phase with the difference of 180° (Fig. 1(d)). When the nonlinear terms are neglected, the solution in this region exhibits instability. Due to the nonlinear terms, however, the solution ceases to diverge and have a finite value, which is called a limit cycle. When the frequency lies at the range $\omega_2 < \omega < \omega_3$ (Region II), we have three stable (solid line and curves) and two unstable solutions (dotted curves). Of three stable solutions, one is a stable attractor at the origin ($R = 0$) and the others are limit cycles with phase difference of 180°. In the other frequency region ($\omega < \omega_1$ or $\omega > \omega_3$), there exists one stable attractor at the origin. As can be seen in the experiment, while a limit cycle motion is presented as oscillating clouds with the frequency of $\omega/2$, the stable attractor at the origin is represented by a stationary trap.

When the modulation frequency increases, a trivial solution (stable attractor at the origin) converts to the limit cycle motion at the frequency $\omega = \omega_1$, which is called a super-critical Hopf bifurcation (Strogatz, 2001). In contrast, when the frequency decreases from the larger frequency region, the combination of a stable attractor at the origin and the limit cycle motions changes to the limit cycle motions alone at the frequency $\omega = \omega_2$, which is called a sub-critical Hopf bifurcation (Strogatz, 2001). In general we observe the hysteresis: when we increase the frequency across the frequency ω_2 we follow the nontrivial solution at Region II in Fig. 1(c). On the contrary, when we sweep down the frequency crossing the frequency ω_3, the solution is merely a trivial solution ($R = 0$). In reality, however, it is not possible to observe this normal hysteresis effect, which will be explained later. Comparison of numerical calculations of Eq. (1) with the results with the analytic solutions are shown in Fig. 1(b). In Fig. 1(b), the dotted curve denotes the results of Fig. 1(a). That is, they were obtained for the linearized Mathieu equation (Eq. (5)). Also the solid curve is the result of the numerical calculation. We can notice that ω_1 and ω_2 are slightly shifted towards smaller values.

The atomic motion can be described more clearly in the rotating frame with the angular velocity of $\omega/2$. In that frame, Eq. (9) can be transformed to the following equations:

$$\frac{dq_1}{dt} = -\frac{\beta}{2}q_1\left(1+q^2\right) + \frac{\omega_0}{2}q_2\left(\frac{h}{2}+\Omega-q^2\right),$$

$$\frac{dq_2}{dt} = -\frac{\beta}{2}q_2\left(1+q^2\right) + \frac{\omega_0}{2}q_1\left(\frac{h}{2}-\Omega+q^2\right), \tag{12}$$

where the dimensionless amplitudes are defined by $q_1 = q\cos\psi$ and $q_2 = q\sin\psi$, with

$$q^2 = q_1^2 + q_2^2 = \frac{3}{16}(4+h_T^2)A_0R^2, \quad \Omega = \frac{\omega - 2\omega_0}{\omega_0}.$$

The calculated stability regions in the rotating frame are plotted in Fig. 2(a) for $\omega/\omega_0 = 2.0$ (Region I in Fig. 1(c)) and Fig. 2(b) for $\omega/\omega_0 = 2.5$ (Region II in Fig. 1(c)). In Fig. 2 the stable (unstable) points are denoted by the filled circles (triangles). In Fig. 2(a), there are two stability regions, which are divided by two spiral-shaped regions. We have two stable points inside the regions and one unstable point at the border of two regions. An atom with initial condition in one specific region converges to a stable point which belongs to that region. In

Fig. 2(b), there exist three stable points and two unstable points. The regions are divided by three spiral-shaped regions. The behaviors of atomic motions are analogous to those in Fig. 2(a).

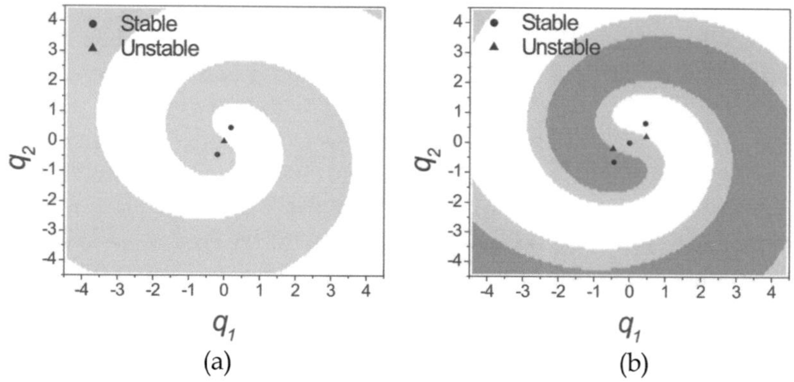

Fig. 2. The stability regions for the frequency Region I (a) and Region II (b).

When the modulation frequency corresponds to Region II in Fig. 1(b) or Fig. 1(c), in the experiment, the initial conditions of trapped atoms reside in the central spiral-shaped region in Fig. 2(a). Thus we only observe a stationary trap when the frequency is swept down across ω_3. On the contrary, when we sweep up the frequency crossing ω_2, we only observe the two limit cycle motions, because the initial conditions reside in the two large spirals. That is to say, we may observe hysteresis. However, in Region II, as will be seen in the next subsection, we can always observe the limit cycle motions as well as the central trap independent of the direction of the frequency sweeping. There are two reasons for the disappearance of the hysteresis effect. The first one is the fact that the atoms are always loaded from background atoms and lost due to the collisions with them. Thus the initial conditions cover the broad regions including three spiral-shaped regions. As a result, both the central trap and the limit cycle motions can be observed simultaneously in the experiment. The second reason is the existence of the diffusion. In the experiment there always exists diffusion which provokes atoms to make transition from the fixed point to the limit cycle motions. Therefore, although the frequency is swept downwards crossing ω_3, we can observe the limit cycles as well as the stationary trap. In reality, both reasons exist simultaneously, and contribute to the disappearance of the hysteresis effect.

2.2 Experimental and simulation results

We study parametric resonance of atoms in a standard vapor-cell MOT where the intensity of cooling laser is modulated. We used the first order diffracted laser beam by an acousto-optic modulator (AOM), which can vary the laser power in accordance with the applied voltage. The ^{85}Rb atoms in the MOT have three dimensional confinements with the natural frequency ω_0 ($\omega_0/\sqrt{2}$) for z (x or y)-axis, where ω_0 is defined in Eq. (4a). In order to observe the parametric resonance and limit cycle motion in the MOT, the s_0 should be very small ($<$ 0.1). In the experiment we excite parametric resonance for z-axis (the axis of anti-Helmholtz coils). For z-axis, $s_0 = 0.042$ (or 0.05), $b = 9$ G/cm, $\delta = -2.9\,\Gamma$, and $h = 0.9$ (or 0.7). Accordingly, $\omega_0 = 2\pi \times 31.5(34.3)$ s^{-1} and $h_T = 0.40$ (or 0.44). Thus the modulation amplitude should be greater than 0.40 (0.44) to excite parametric resonance, even if the modulation frequency is exactly twice the resonance frequency. On the other hand, the normalized laser intensity for x- or y-axis is about $5s_0$. Accordingly the corresponding threshold modulation amplitude is 0.9, which is not smaller than the modulation amplitude in the experiment. This is why only in the z-axis there occurs parametric resonance, although the cooling laser intensities for all three directions are modulated at the same frequency and modulation amplitude.

When the intensity of the cooling laser is modulated at about twice of its natural frequency, the atoms are divided into two parts and oscillate in opposite directions with the finite amplitudes. The typical photos of oscillating atomic clouds are shown in Fig. 3. The images of the cloud are taken at every 1 ms for 1/2400 s of exposure time. Here $s_0 = 0.042$, and the modulation frequency is 75 Hz ($\omega/\omega_0 = 2.38$) for (a) and 95 Hz ($\omega/\omega_0 = 3.02$) for (b). In all cases the modulation amplitude is 0.9. The modulation frequencies for the results in Figs. 3(a) and 3(b) belong to Region I and Region II in Fig. 1(b) or Fig. 1(c), respectively. As explained in the previous subsection, we have two stable points (limit cycles) in Region I (Fig. 3(a)) and three stable points (one fixed point and two limit cycles) in Region II (Fig. 3(b)). The bright spot in Fig. 3(b) is due to this sub-Doppler force (see Sec. 2.3). Figure 4 shows the series of signals measured for Region I (a) and Region II (b). The whole period is $2\pi/\omega$, which is the period of modulation signal and half the period of atomic oscillations, and the signals are equally separated in time. We can clearly see that the atomic clouds are divided into two parts and oscillate in opposite directions.

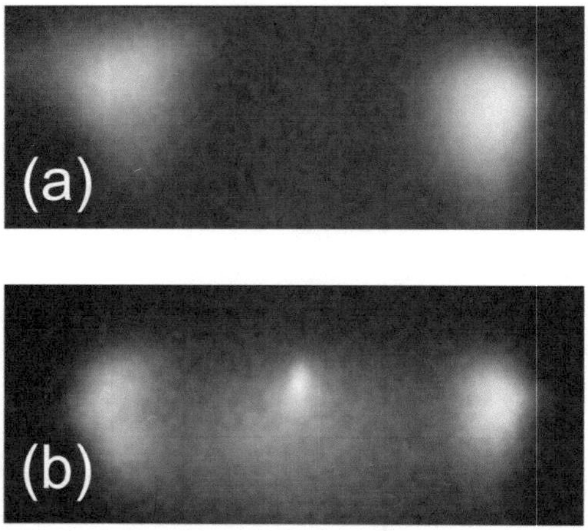

Fig. 3. The typical images of oscillating atomic clouds. The modulation frequency is 75 Hz for (a) and 95 Hz for (b).

The experimental results for the dependence of the amplitude and phase of limit cycles on the modulation frequency [amplitude] are presented in Fig. 5(a) and Fig. 5(b) [Fig. 5(c) and Fig. 5(d)], respectively. In Figs. 5(a) and (b), h = 0.7 and 0.9. Figs. 5(a) and (b), the solid [dotted] lines and curves are calculated results from the Eq. (1) while the filled squares [hollow circles] denote the experimental results at the modulation amplitude of h=0.9 [0.7]. In Figs. 5(c) and (d), the modulation frequency is $\omega = 2\pi \times 65$ Hz. The threshold amplitude (h_T) is 0.40, but we need at least h = 0.6 to observe the parametric resonance. In Fig. 5, we can clearly see the good agreement between the experiments and the calculations.

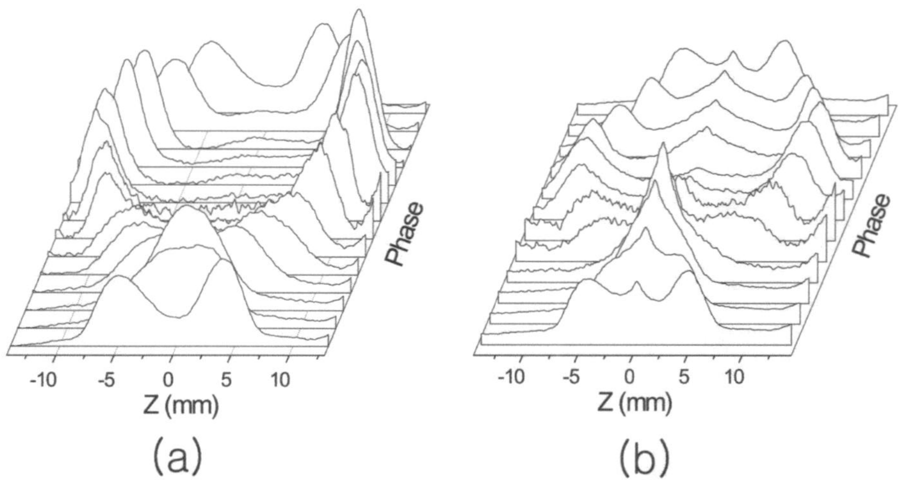

Fig. 4. Profiles of the oscillating clouds at the various phases with respect to the modulations at the frequency of Region I (a) and Region II (b).

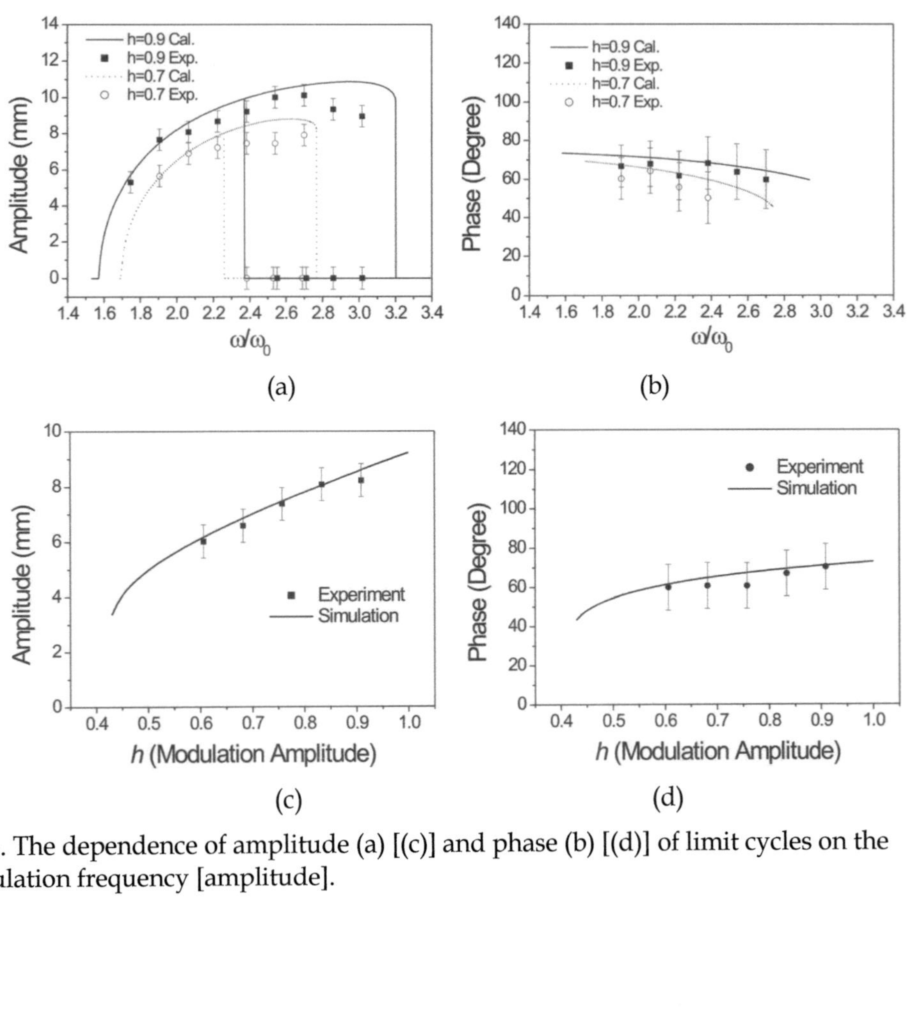

Fig. 5. The dependence of amplitude (a) [(c)] and phase (b) [(d)] of limit cycles on the modulation frequency [amplitude].

2.3 Observation of sub-Doppler trap

In this subsection we describe the direct observation of sub-Doppler part of the MOT through the parametric resonance (Kim et al., 2004). Typical experimental parameters are as follows: in the z-axis, $s_0 = 0.05$, $b = 9$ G/cm, the amplitude of modulation $h = 0.9$, and $\delta/\Gamma = -2.9$. Thus ω_{dop} is about $2\pi\times$ 34.3 Hz, and ω_{sub} (will be explained later) is $2\pi\times$ 460 Hz. The full width of MOT beams at the $e^{-1/2}$ intensity point is 2.5 cm in order to cover the whole range of atomic motion ($2R$). The total number of atoms in the unmodulated MOT is about 2×10^8 and that in the modulated MOT is about 8×10^7. Figure 6 shows the photos of parametrically excited atoms at the modulation frequency of 80 Hz (a) and 95 Hz (b). Both frequencies belong to region I and II in Fig. 1(c), respectively. We can easily notice that the broad and large trap in Fig. 6(b) is attributable to the third fixed point at the origin of the magnetic field with the Doppler theory. On the other hand, the narrow and sharp traps in Fig. 6(a) and (b) have a different origin, which can be very well understood by including the sub-Doppler cooling theory (Chang & Minogin, 2002; Jun et al, 1999).

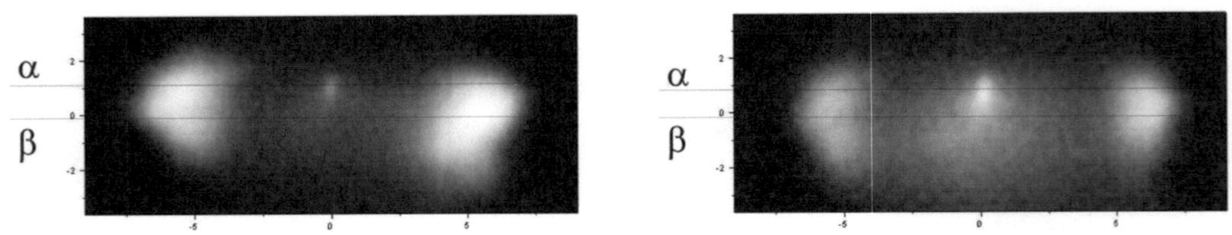

(a) (b)

Fig. 6. The typical appearance of (a) double and (b) triple fixed points in experiments. The modulation frequency is 80 Hz (a) and 95 Hz (b).

There exist two kinds of trap in a normal MOT: One is due to one photon process, that is, Doppler cooling theory, and the other is due to two photon process, which makes sub-Doppler temperature (Jun et al, 1999). Thus the force exerted on an atom is given by

$$F(z,v,t) = F_{\mathrm{dop}}(z,v,t) + F_{\mathrm{sub}}(z,v,t), \tag{13}$$

where the Doppler force ($F_{\mathrm{dop}}(z,v,t)$) is given in Eq. (2) and the sub-Doppler force for a $F = 1 \to F' = 2$ atomic transition line is analytically given by (Jun et al, 1999)

$$F_{\mathrm{sub}}(z,v,t) = f\left(\frac{kv}{\Gamma} + \frac{g_g \mu_B b z}{\hbar\Gamma}\right). \tag{14}$$

Here the function $f(x)$ is given by

$$f(x) = -\frac{120\hbar k\Gamma s(t)^2 \,|\,\delta/\Gamma\,|\, x}{17 s(t)^2\left[5 + 4(\delta/\Gamma)^2\right] + 2112\left[1 + 4(\delta/\Gamma)^2\right]^2 x^2},$$

with g_g the g–factor of the ground state and $s(t) = s_0(1 + h\cos\omega t)$. Fig. 7(a) shows the typical calculation of the total force and Fig. 7(b) shows the same force for the detailed region for $F = 1 \to F' = 2$ atomic transition line. We can see that the sub-Doppler force exists only near the origin. Although the sub-Doppler force for other transition lines can be calculated numerically (Walhout et al., 1992), we used Eq. (14) for the sub-Doppler force the transition $F = 3 \to F' = 4$ of ^{85}Rb atoms with $g_g = 1/3$. The trap frequency can be derived from Eq. (14) as

$$\omega_{\text{sub}} = \sqrt{\frac{P_1 \mu_B bk\,|\,\delta / \Gamma\,|}{m\left[1 + P_2(\delta/\Gamma)^2\right]}}\,. \tag{15}$$

with the coefficients $P_1 = 24/17$ and $P_2 = 4/5$ for 1→2, $P_1 = 6.84$ and $P_2 = 2.72$ for 2→3, and $P_1 = 12.5$ and $P_2 = 3.57$ for 3→4 transition line.

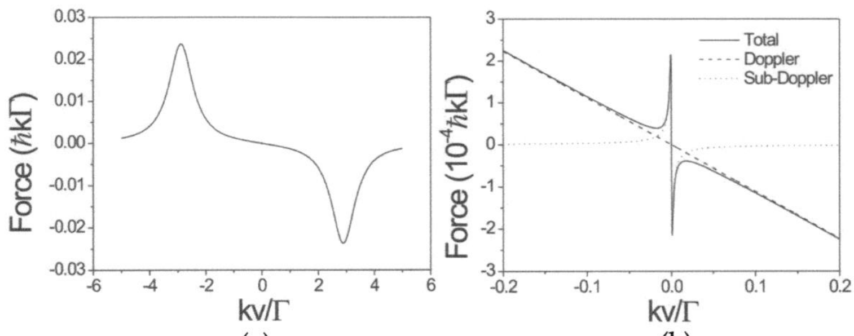

Fig. 7. The calculated total force in the enlarged (a) and detailed region (b).

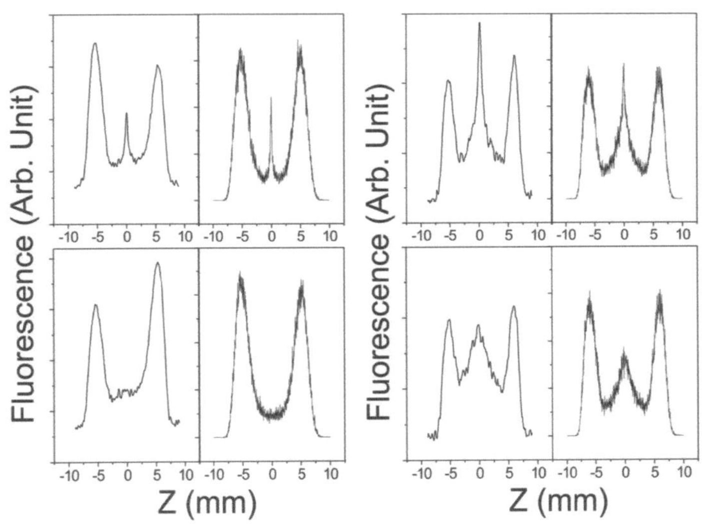

Fig. 8. Upper and Lower panels of (a)[(b)] is the profile of Fig. 6(a)[(b)] along α and β-line, respectively. In each panel, the left one is the measured profile and the right one is the simulation result.

The profiles of Fig. 6 are compared to the Monte-Calro simulations with and without considering the sub-Doppler force. Upper [Lower] panel of Figure 8(a) and 8(b) is the profile of Fig. 6(a) and 6(b), respectively along the α [β]-line. In the figures the left one is the experimental data and the right one is the simulation result. In Fig. 8(a), the experimental (simulation) result of width (twice the standard deviation in the Gaussian distribution) of broad peaks is approximately 2.1 mm (2.2 mm). Also the experimental (simulation) width of central peak is 0.45 mm (0.28 mm). In Fig 8(b), the experimental (simulation) width of oscillating broad peak, central broad peak, and the central sharp peak is 1.7 (1.9) mm, 5.6

(5.2) mm, and 0.68 (0.26) mm, respectively. We can see the good agreements between the experimental and simulation results. We used the combined force (Eq.(13)) in the simulation. In addition, we also take into account the random recoil force, f_r, from the spontaneously emitted photons. The random force is related with the momentum diffusion via the following equation:

$$< f_r(t) f_r(t') >= 2D_{zz}(t)\delta(t-t') , \tag{16}$$

where $D_{zz}(t)$ is the momentum diffusion coefficient for z-axis and the left-hand side of Eq. (16) denotes the average of product of the force at the different time over long time. The momentum diffusion coefficients for various transition lines are presented in Fig. 9(a) (Walhout et al., 1992), where we have assumed the unmodulated laser intensity, i.e., $s(t) = s_0$. In Fig. 9(a) the result for $F = 0 \rightarrow F' = 1$ corresponds to that of Doppler theory. The calculated results for the detailed region are shown in Fig. 9(b). Near the center of the velocity, there exists a sharp dip, which exhibits the feature of two-photon resonance, that is, sub-Doppler laser cooling theory. In the simulation we have used the calculated momentum diffusion coefficient in the sub-Doppler cooling theory for $F = 3 \rightarrow F' = 4$ transition line.

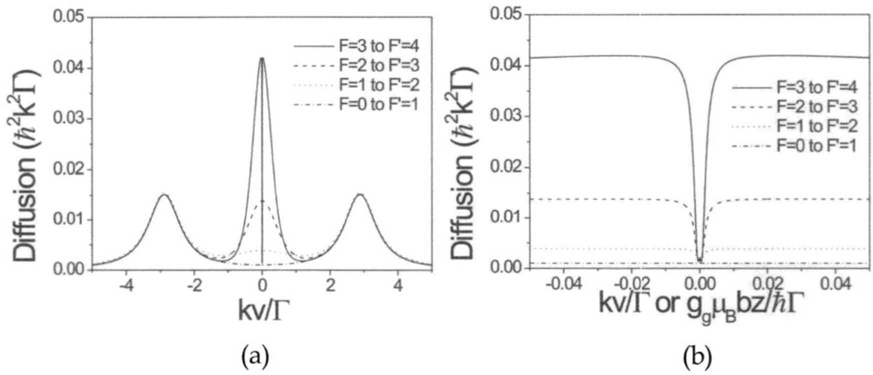

Fig. 9. (a) The calculated momentum diffusion coefficients for various atomic transition lines. (b) The detailed plots near the origin.

3. Measurement of trap parameters

3.1 Parametric resonance method

Since the spring constant is one of the basic parameters of a MOT, there has been much effort to measure the spring constant, inferred from the measured spatial profile and the temperature by means of the equi-partition theorem (Drewsen et al., 1994; Wallace et al., 1994), by using the beam imbalance (Wallace et al., 1994; Steane et al., 1992) or by using the imposed oscillating magnetic field (Kohns et al., 1993; Hope et al., 1993). Authors in Ref. (Xu et al., 2002) have used the method of free oscillation of atoms after rapid turning-off the constant pushing laser beam and study the three-dimensional Doppler cooling theory for two-level atoms. In Sec. 3, we present two methods to measure the trap parameters of the MOT such as the trap frequency and the damping coefficient. The first method, described in this subsection, is based on the parametric resonance described in the preceding section (Kim et al., 2005). The second is the method of transient oscillation of atomic trajectory (Kim et al., 2005), which is presented in the next subsection.

With a larger modulation amplitude, the parametric resonance occurs at a wider range of the modulation frequency. However, when the modulation amplitude decreases towards its threshold value, the amplitude becomes a Gaussian-like shaped function centered at twice the natural frequency. Therefore, one can accurately measure the trap frequency and thus the spring constant as well. In the experiment, the laser intensity in the *z*-axis was $I_z = 0.099$ mW/cm^2, while the intensities along the transverse directions were $I_x = I_y = 5I_z$. The modulation amplitude is decreased down to $h = 0.5$ to observe the trap frequency accurately. The full width of laser beams at the $e^{-1/2}$ intensity point is 2.5 cm to cover the whole atomic motion. When atoms are parametrically excited, the number of atoms decreases to about 8 $\times 10^7$. The laser detuning is about $\Delta = -2.3\Gamma$, where $\Gamma = 2\pi \times 5.9$ MHz is the decay rate of the excited state. In the experiment we vary the magnetic field gradient to study the dependence of the spring constant on the magnetic field gradient.

By varying the modulation amplitude, we measure the amplitudes of limit cycle motions as presented in Fig. 10(a). When h is large, two oscillating clouds with the phase difference of π can be clearly seen. As h decreases, two peaks approach each other and merge into a single cloud. In Fig. 10(a), the modulation frequency is $f = 84$ Hz. Figure 10(b) shows the typical experimental results for three values of the magnetic field gradients (b). For a given b, we vary the modulation frequency (f) and measure the width of single-merged cloud. Varying f, the amplitude has a peak at a certain frequency, which is approximately given by $f = 2f_0$.

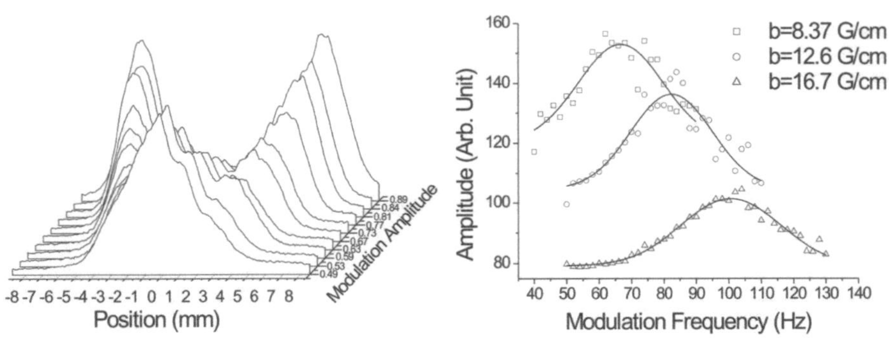

(a) (b)

Fig. 10. (a) The measured profiles of limit cycle motions at the various modulation amplitudes h ($f = 84$ Hz). (b) The measured profiles as functions of the modulation frequency for three magnetic filed gradients.

The measured trap frequencies (filled squares) for several magnetic field gradients are shown in Fig. 11(a). The curves in Fig. 11(a) are the theoretical results. The curves *A* and *C* are the results calculated by Eq. (4a) with the normalized saturation intensities $s_0 = I_z/I_s$ and $s_0 = I_z/(I_s)_{av}$, respectively. Here $(I_s)_{av} = 3.78$ mW/cm^2 is the averaged saturation intensity (Wallace et al., 1994). With the averaged saturation intensity, the result agrees well with experimental results. The curves *B* and *D* are the calculated results based on the rate equation model for $F_g = 0 \rightarrow F_e = 1$ and $F_g = 3 \rightarrow F_e = 4$ transition lines, respectively. We can see that the results of line *C* and *D* are almost equal and agree with the experimental results. The details of calculation using the rate equation model are presented in Ref. (Kim et al., 2005). The results for κ/b for the varying the intensities of $I_x + I_y$ are shown in Fig. 11(b), where $\kappa = m\omega_0^2$ is the spring constant. Since the spring constant is proportional to the magnetic field gradient (b), we have a constant value for all experimental data. As the laser intensity for the transverse directions increases, κ_0/b decreases. In Fig. 11(b), we find that the use of the averaged saturation intensity ($(I_s)_{av}$) is a good approximation at the broad range of the transverse laser intensities.

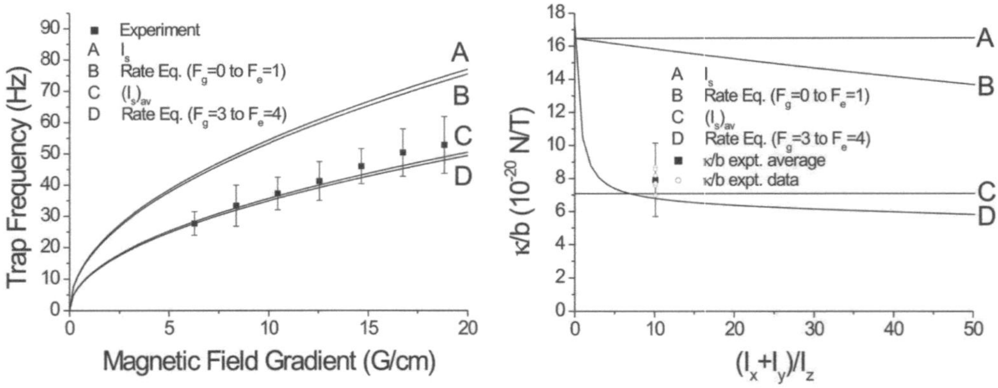

(a) (b)

Fig. 11. (a) The measured trap frequencies versus the the magnetic field gradients (filled squares). (b) The spring constants divided by the magnetic filed gradient.

3.2 Transient oscillation method

In this subsection, we present measurement of the trap frequency as well as the damping coefficient by detecting the temporal oscillatory behavior of the pushed atomic cloud as the magnetic field gradient or the laser detuning is varied (Kim et al., 2005). The atomic motion in the MOT is simply given by a damped harmonic oscillator model with the damping coefficient β and the trap frequency f_0. When a uniform magnetic field (B_z) is applied to the MOT, the position of the trap center is shifted by B_z/b, where b is the magnetic field gradient in the z-axis of the MOT. When the uniform magnetic field is suddenly turned off, the atomic cloud returns to the original trap center. In case of the underdamped motion, one can extract the trap parameters by measuring the trajectory of the released atomic cloud. The trajectory of the atomic cloud center is simply given by

$$z(t) = z_0 + A\exp(-\beta t/2)\left(\cos 2\pi f_0 t + \frac{\beta}{4\pi f_0}\sin 2\pi f_0 t\right), \qquad (17)$$

where $f_0 = \omega_0/(2\pi)$ is the trap frequency, β is the damping coefficient, z_0 is the equilibrium position, and A is the initial displacement from equilibrium.

Figure 12(a) shows the contour-plot of the typical absorption signals of the 16-channel photodiode array versus time (taken at 1/5000 s time interval), superposed by a curve corresponding to the maximum brightness. Here the vertical axis shows the position of the photodiode array and horizontal axis represents the time elapsed after switch off of the magnetic field. Figure 12(b) shows the same curve as in Fig. 12(a) and its fitted result using Eq. (17) for the region after elapse of one period, which is shown in the lower panel. From the fit presented in Fig. 12(b) one can obtain the trap frequency and the damping coefficient.

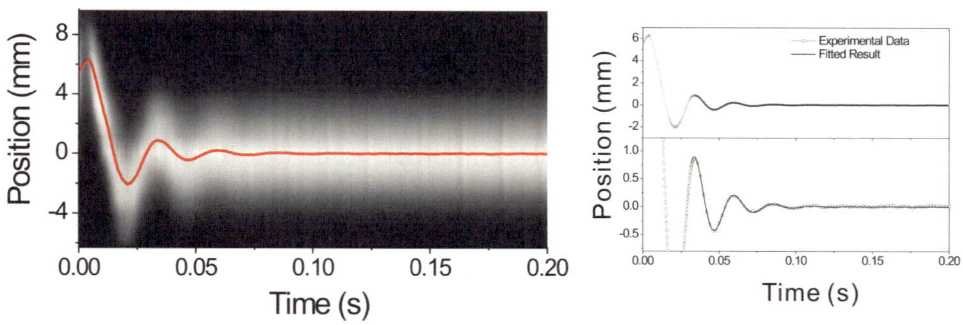

(a) (b)

Fig. 12. (a) The contour plot of the signals showing a typical oscillating motion of atomic cloud. (b) The same curve as in (a) and the fitted curve obtained by Eq. (17). The detailed plot after one period is shown in the lower panel.

We have measured the trap parameters by varying the laser intensity, detuning, and magnetic field gradient. The results are presented in Figs. 13. Fig. 13 shows the dependence of trap frequency (a)[(c)] and damping coefficient (b)[(d)] on the magnetic field gradient [detuning]. In Figs. 13(a) and (b), $\Delta = -2.71\Gamma$, and I_z = 0.10 mW/cm^2 (filled square), 0.13 mW/cm^2 (filled circle) and 0.17 mW/cm^2 (filled triangle), respectively. Note that the laser intensities in the transverse directions ($I_x = I_y$) are 0.62 mW/cm^2. The solid, dashed, and dashed-dotted lines in the figure are the calculated results from a theoretical model. In Figs. 13(c) and (d), I_z = 0.17 mW/cm^2, $I_x = I_y$ = 0.62 mW/cm^2, and b = 10 G/cm. Note that the solid line is obtained from the rate equation for the Doppler theory, whereas the dashed line is from the simple Doppler theory for two level atom where the saturation intensity is

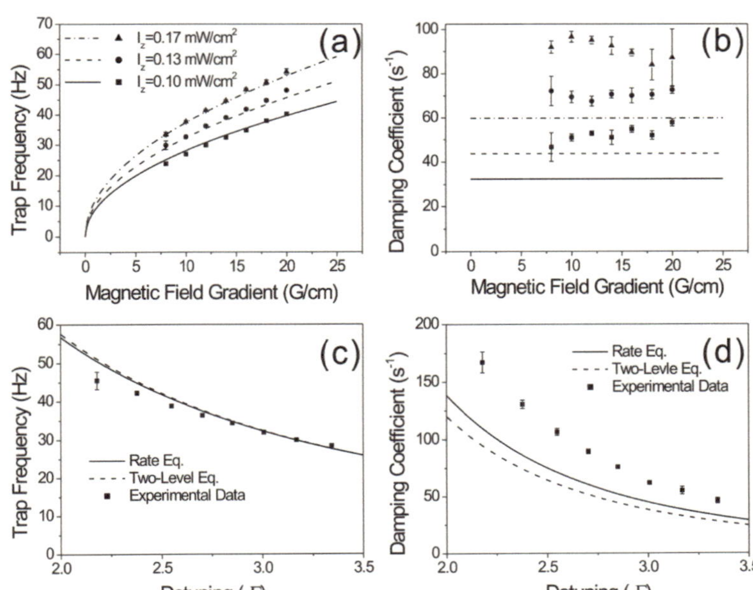

Fig. 13. The dependence of the trap frequency (a)[(c)] and damping coefficient (b)[(d)] on the magnetic field gradient [detuning].

substituted by the averaged value $(I_s)_{av}$ = 3.78 mW/cm². As shown in Fig. 13, the trap frequencies are in good agreement with the theoretical values. The damping coefficients, on the other hand, are about twice larger than the simple theoretical predictions. We provide a quantitative description of the theoretical model and explain the discrepancy found in the damping coefficient.

The summary of the data of Fig. 13 is presened in Fig. 14. The damping coefficient and the trap frequency are presented as a function of $s_0\delta/(1+4\delta^2)^2$ and $\sqrt{bs_0\delta}/(1+4\delta^2)$, respectively.

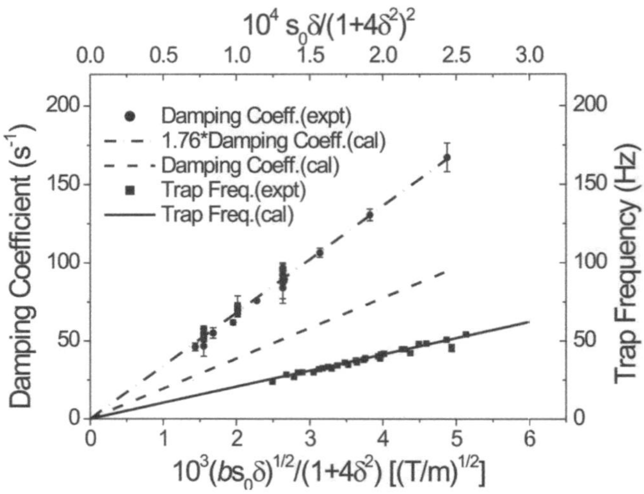

Fig. 14. The damping coefficient versus $s_0\delta/(1+4\delta^2)^2$ [filled circles, experimental data; dashed line, calculated results; dashed-dotted line, calculated results multiplied by 1.76] and the trap frequency versus $\sqrt{bs_0\delta}/(1+4\delta^2)$ [filled squares, experimental data; solid line, calculated results].

One can observe that the measured trap frequencies are in excellence agreement with the calculated results. On the other hand, one has to multiply the simply calculated damping coefficients by a factor 1.76 to fit the experimental data. We find that the discrepancy in the damping coefficients results from the existence of the sub-Doppler trap described in Sec. 2.3. In order to show that the existence of the sub-Doppler force affects the Doppler-cooling parameters, we have performed Monte-Carlo simulation with 1000 atoms. In the simulation, we used sub-Doppler forces and momentum diffusions described in Sec. 2.3. The results are presented in Fig. 15. Here we averaged the trajectories for 1000 atoms by using the same parameters as used in Fig. 12. We have varied the intensity (I) associated with F_{sub} without affecting the intensity for the Doppler force, and obtained the averaged trajectory, where I_{expt} =0.17 mW/cm² is the laser intensity used in the experiment [Fig. 13]. We then infer the damping coefficient and the trap frequency by fitting the averaged trajectory with Eq. (17). The fitted results for the damping coefficient and the trap frequency are shown in Fig. 15(b). While the trap frequency remains nearly constant, the damping coefficient increases with the intensity. Note that to obtain an increase of factor 1.76 as shown in Fig. 14, one should use I/I_{expt} = 1.6. The reason for the increase of the damping coefficient can be well explained qualitatively from the simulation.

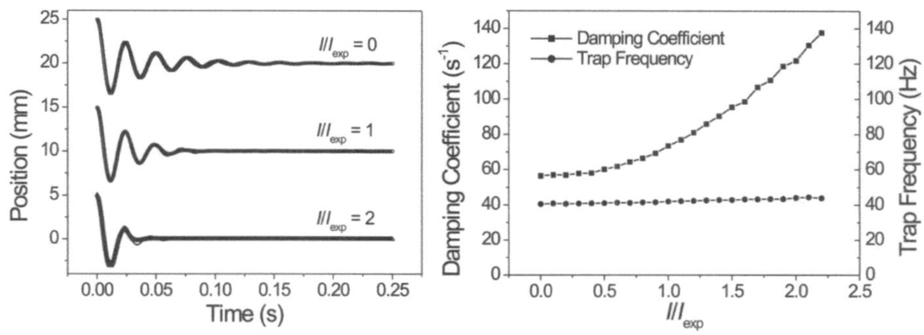

(a) (b)

Fig. 15. The Monte-Carlo simulation results. (a) The averaged trajectories for 1000 atoms together with the fitted curves obtained from Eq. (17). (b) The damping coefficient (filled square) and the trap frequency (filled circle) as a function of the laser intensity.

4. Adjustable magneto-optical trap

When the detuning and intensity of the longitudinal (z-axis) lasers along the symmetry axis of the anti-Helmholtz coil of the MOT are different from those of the transverse (x and y axis) lasers, one can realize an array of several sub-Doppler traps (SDTs) with adjustable separations between traps (Heo et al., 2007; Noh & Jhe, 2007). As shown in Fig. 16(a), it is similar to the conventional six-beam MOT, except that the detunings (δ_x and δ_y) and intensities (I_x and I_y) of the transverse lasers can be different from those of the longitudinal ones (δ_z and I_z). In the case of usual MOT, one obtains a usual Doppler trap superimposed with a tightly confined SDT at the MOT center, exhibiting bimodal velocity as well as spatial distributions (Dalibard, 1988; Townsend et al., 1995; Drewsen et al., 1994; Wallace et al., 1994; Kim et al., 2004). Under equal detunings but unequal intensities (I_x, $I_y \gg I_z$), which typically arise in the nonlinear dynamics study of nonadiabatically driven MOT (Kim et al., 2003; 2006), one still obtains the bimodal distribution. However, as the transverse-laser detuning δ_t ($\equiv \delta_x = \delta_y$) is different from the longitudinal one δ_z with the same configuration of laser intensity, the SDT at the center becomes suppressed with the usual Doppler trap still present. The existence of the central SDT, available at equal detunings, contributes not only to the lower atomic temperature but also to the larger damping coefficients than is expected

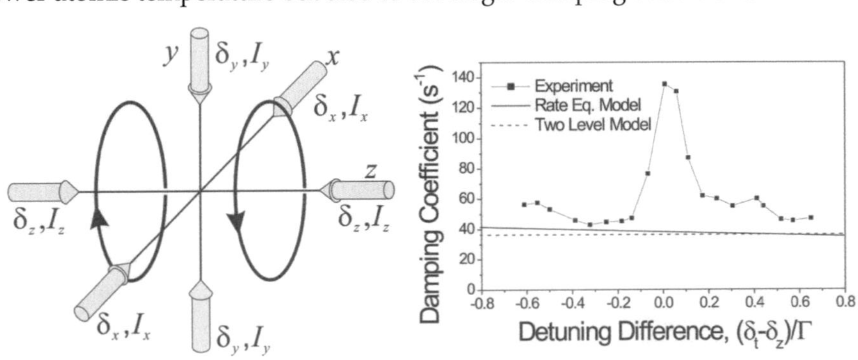

(a) (b)

Fig. 16. (a) Schematic of the asymmetric magneto-optical trap. (b) Measured damping coefficients versus normalized laser-detuning differences.

by the Doppler theory. In order to confirm the enhanced damping, we have measured the damping coefficients of MOT versus the laser detuning differences, $\delta_t - \delta_z$, by using the transient oscillation method described in Sec. 3.2 (Kim et al., 2005). As is shown in Fig. 16(b), one can observe a 'resonance' behaviour; the damping coefficient is suppressed by more than a factor of 2 and approaches the usual Doppler value at unequal detunings, which is directly associated with the disappearance of the central SDT.

When the transverse laser intensity is increased above a certain value at unequal detunings, we now observe the appearance of novel SDTs. In Fig. 17, the fluorescence images of the trapped atoms, obtained with $I_t \equiv I_x + I_y = 11.4I_z$ fixed, are presented for various values of $\delta_t - \delta_z$. The central peak, corresponding to the usual SDT, becomes weak when the detunings are different, as discussed in Fig. 16(b). However, the two side peaks, associated with the novel SDTs, are displaced symmetrically with respect to the MOT center, in proportion to $\delta_t - \delta_z$. In addition to these two adjustable side SDTs, there also exist another two weak SDTs located midway between each side SDT and the central one, which will be discussed later.

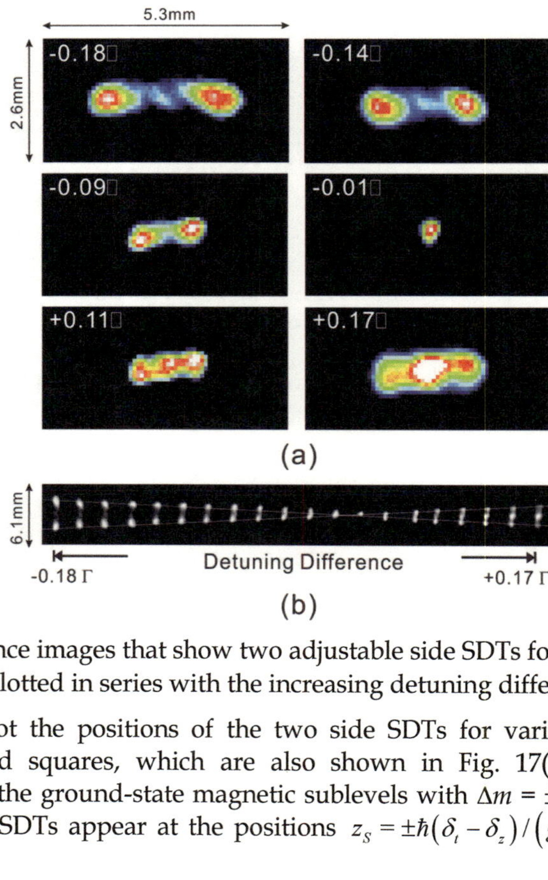

Fig. 17. (a) Fluorescence images that show two adjustable side SDTs for several values of $\delta_t - \delta_z$. (b) SDT pictures plotted in series with the increasing detuning differences.

In Fig. 18(a), we plot the positions of the two side SDTs for various values of $\delta_t - \delta_z$, represented by filled squares, which are also shown in Fig. 17(b). Attributed to the coherences between the ground-state magnetic sublevels with $\Delta m = \pm 1$ transitions (see Fig. 18(b)), the two side SDTs appear at the positions $z_S = \pm\hbar(\delta_t - \delta_z)/(g_g \mu_B b)$ and thus their separation satisfies,

$$\frac{\Delta z_S}{\Delta \nu} = \pm \frac{h}{\mu_B b g_g}, \tag{18}$$

where $\Delta\nu = (\delta_t - \delta_z)/(2\pi)$ and μ_B is the Bohr magneton. Since the ground-state g-factor is $g_g = 1/3$ for ^{85}Rb atoms and the magnetic field gradient is $b = 0.17$ T/m, the calculated value (solid line) is $\Delta z/\Delta\nu = 1.26$ mm/MHz, which agrees well with the experimental result of 1.25 (±0.12) mm/MHz, considering 10% error of position measurements. On the other hand, the two weak SDTs, resulting from the coherences due to $\Delta m = \pm 2$ transitions (refer to Fig. 18(b)), are located midway at $z_M = z_S/2$, as shown in Fig. 18(a) (open circles). The fitted result is 0.61 mm/MHz, which is almost half the value given by Eq. (18), in good agreement with the 'doubled' energy differences of the $\Delta m = \pm 2$ transitions with respect to the $\Delta m = \pm 1$ ones, responsible for the side SDTs.

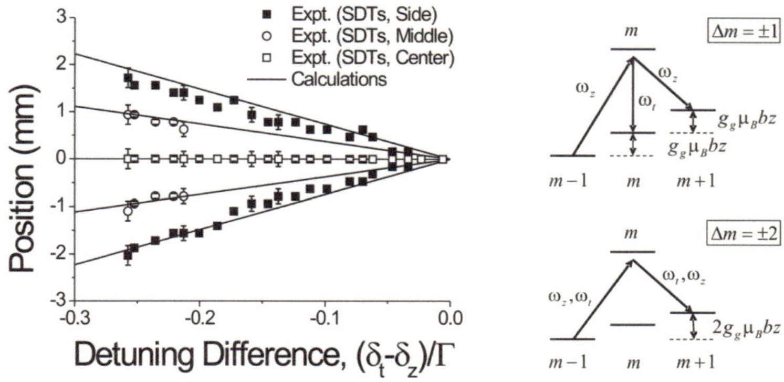

(a) (b)

Fig. 18. Measured positions of available SDTs versus negative detuning differences.

In order to have a qualitative understanding of the detuning-difference dependence, we have calculated the cooling and trapping forces in two dimension by using the optical Bloch equation approach (Dalibard, 1988; Chang & Minogin, 2002; Noh & Jhe, 2007). In Fig. 19(a), we present the calculated forces $F(z,v = 0)$ for $F_g = 3 \rightarrow F_e = 4$ atomic transition. In the presence of the transverse lasers, the ground-state sublevels with $\Delta m = \pm 1$ transitions can be coupled by a π photon from the transverse lasers in combination with a $\sigma^{\pm}$ photon from the longitudinal lasers (see Fig. 18(b)). As a result, for unequal detunings, there exists a position where the Zeeman shift compensates the laser-frequency difference, such that

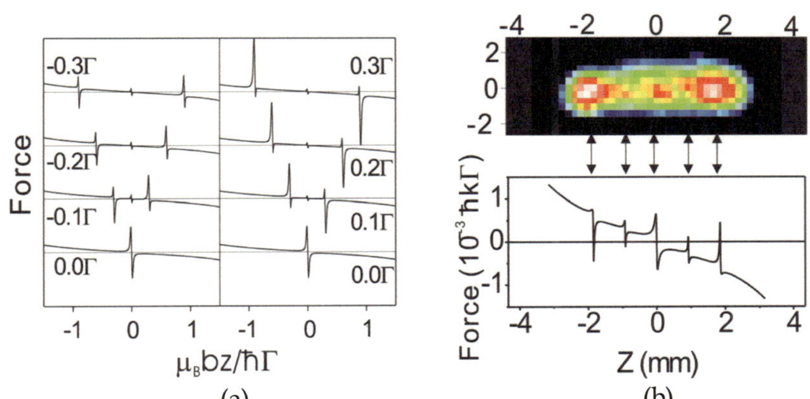

(a) (b)

Fig. 19. (a) Calculated forces $F(z,v = 0)$ for various detuning differences. The maximum forces at 0.3 Γ corresponds to $5 \times 10^{-3}\ \hbar k\Gamma$. Here $\delta_z = -2.7\Gamma$, $I_z = 0.11$ mW/cm^2, and $I_t = 5.6 I_z$. (b) Five SDTs, including two weak SDTs midway between the two side SDTs and the central one, for $\delta_t - \delta_z = -0.24\Gamma$.

$$\hbar\omega_z - \hbar\omega_t = \pm g_g \mu_B bz. \tag{19}$$

At this position, atoms can feel the sub-Doppler forces associated with the $\Delta m = \pm 1$ coherences and thus the novel SDT is obtained at two positions of $\pm\hbar\,(\delta_x - \delta_z)/(g_g\mu_B b)$, as confirmed in Fig. 18(a). As shown in Fig. 19(b), the two weak midway SDTs arise because the weak $\sigma^{\pm}$ photons, in addition to the dominant π ones, from the transverse lasers can contribute to the atomic coherences in the z-direction. Therefore, besides the $\Delta m = \pm 1$ transitions responsible for the side SDTs, the two-photon-assisted $\Delta m = \pm 2$ coherences (here, each $\sigma^{\pm}$ photon comes from the longitudinal and the transverse laser, as shown in Fig. 18(b)) can be generated, and atoms at the position z_M, satisfying the relation $\hbar\,\omega_t - \hbar\,\omega_z = \pm 2 g_g \mu_B b z_M$, feel this additional coherence. As a result, the midway SDTs can be obtained at $z_M = z_S/2$ (see Fig. 18(a)). The typically observed image and the calculated force are presented in Fig. 19(b).

5. Conclusions

In this article we have presented experimental and theoretical works on the asymmetric magneto-optical trap. In Sec. 2, we have studied parametric resonance in a magneto-optical trap. We have described a theoretical aspect of parametric resonance by the analytic and numerical methods. We also have measured the amplitude and phase of the limit cycle motions by changing the modulation frequency or the amplitude. We find that the results are in good agreement with the calculation results, which are based on simple Doppler cooling theory. In the final subsection we described direct observation of the sub-Doppler part of the MOT without the Doppler part by using the parametric resonance which. We compared the spatial profile of sub-Doppler trap with the Monte-Carlo simulation, and observed they are in good agreements.

In Sec. 3, we have presented two methods to measure the trap frequency: one is using parametric resonance and the other transient oscillation method. In the case of parametric resonance method, we could measure the trap frequency accurately by decreasing the modulation amplitude of the parametric excitation down to its threshold value. While only the trap frequency were able to be obtained by the parametric resonance method, we could obtain both the trap frequency and the damping coefficient by the transient oscillation method. We have made a quantitative study of the Doppler cooling theory in the MOT by measuring the trap parameters. We have found that the simple rate-equation model can accurately describe the experimental data of trap frequencies.

In Sec. 4, we have demonstrated the adjustable multiple traps in the MOT. When the laser detunings are different, the usual sub-Doppler force and the corresponding damping coefficient at the MOT center is greatly suppressed, whereas the novel sub-Doppler traps are generated and exist within a finite range of detuning differences. We have found that π and $\sigma^{\pm}$atomic transitions excited by the transverse lasers in the longitudinal direction are responsible for the strong side and the weak middle sub-Doppler traps, respectively. The adjustable array of sub-Doppler traps may be useful for controllable atom-interferometer-type experiments in atom optics or quantum optics.

The AMOT described in this article can be used for study of nonlinear dynamics using cold atoms such as critical phenomena far from equilibrium (Kim et al., 2006) or a nonlinear Duffing oscillation (Nayfeh & Moore, 1979; Strogatz, 2001).

6. Acknowledgement

This work was supported by the Korea Research Foundation Grant funded by the Korean Government (KRF-2008-313-C00355).

7. References

Bagnato, V. S., Marcassa, L. G., Oria, M., Surdutovich, G. I., Vitlina, R. & Zilio, S. C. (1993). Spatial distribution of atoms in a magneto-optical trap, *Phys. Rev. A* Vol. 48(No. 5): 3771–3775.

Chang, S. & Minogin, V. (2002). Density-matrix approach to dynamics of multilevel atoms in laser fields, *Phys. Rep.* Vol. 365(No. 2): 65–143.

Dalibard, J. (1998). Laser cooling of an optically thick gas: The simplest radiation pressure trap?, *Opt. Commun.* Vol. 68(No. 3): 203–208.

di Stefano, A., Fauquembergue, M., Verkerk, P. & Hennequin, D. (2003). Giant oscillations in a magneto-optical trap, *Phys. Rev. A* Vol. 67(No. 3): 033404-1–033404-4.

Dias Nunes, F., Silva, J. F., Zilio, S. C. & Bagnato, V. S. (1996). Influence of laser fluctuations and spontaneous emission on the ring-shaped atomic distribution in a magneto-optical trap, *Phys. Rev. A* Vol. 54(No. 3): 2271–2274.

Drewsen, M., Laurent, Ph., Nadir, A., Santarelli, G., Clairon, A., Castin, Y., Grinson, D. & Salomon, C. (1994). Investigation of sub-Doppler cooling effects in a cesium magnetooptical trap, *Appl. Phys. B* Vol. 59(No. 3): 283–298.

Friebel, S., D'Andrea, C., Walz, J., Weitz, M. & H¨ansch, T. W. (1998). CO_2-laser optical lattice with cold rubidium atoms, *Phys. Rev. A* Vol. 57(No. 1): R20–R23.

Heo, M. S., Kim, K., Lee, K. H., Yum, D., Shin, S., Kim, Y., Noh, H. R. & Jhe, W. (2007). Adjustable multiple sub-Doppler traps in an asymmetric magneto-optical trap, *Phys. Rev. A* Vol. 75(No. 2): 023409-1–023409-4.

Hope, A., Haubrich, D., Muller, G., Kaenders, W. G. & Meschede, D. (1993). Neutral cesium atoms in strong magnetic-quadrupole fields at sub-doppler temperatures, *Europhys. Lett.* Vol. 22(No. 9): 669–674.

Jun, J. W., Chang, S., Kwon, T. Y., Lee, H. S. & Minogin, V. G. (1999). Kinetic theory of the magneto-optical trap for multilevel atoms, *Phys. Rev. A* Vol. 60(No. 5): 3960–3972.

Kim, K., Noh, H. R., Yeon, Y. H. & Jhe,W. (2003). Observation of the Hopf bifurcation in parametrically driven trapped atoms, *Phys. Rev. A* Vol. 68(No. 3): 031403(R)-1–031403(R)-4.

Kim, K., Noh, H. R. & Jhe, W. (2004). Parametric resonance in an intensity-modulated magneto-optical trap, *Opt. Commun.* Vol. 236(No. 4-6): 349–361.

Kim, K., Noh, H. R., Ha, H. J. & Jhe, W. (2004). Direct observation of the sub-Doppler trap in a parametrically driven magneto-optical trap, *Phys. Rev. A* Vol. 69(No. 3): 033406-1–033406-5.

Kim, K., Noh, H. R. & Jhe, W. (2005). Measurements of trap parameters of a magneto-optical trap by parametric resonance, *Phys. Rev. A* Vol. 71(No. 3): 033413-1–033413-5.

Kim, K., Lee, K. H., Heo, M., Noh, H. R. & Jhe, W. (2005). Measurement of the trap properties of a magneto-optical trap by a transient oscillation method, *Phys. Rev. A* Vol. 71(No. 5): 053406-1–053406-5.

Kim, K., Heo, M. S., Lee, K. H., Jang, K., Noh, H. R., Kim, D. & Jhe, W. (2006). Spontaneous Symmetry Breaking of Population in a Nonadiabatically Driven Atomic Trap: An Ising-Class Phase Transition, *Phys. Rev. Lett.* Vol. 96(No. 15): 150601-1–150601-4.

Kohns, P., Buch, P., Suptitz, W., Csambal, C. & Ertmer, W. (1993). On-Line Measurement of Sub-Doppler Temperatures in a Rb Magneto-optical Trap-by-Trap Centre Oscillations, *Europhys. Lett.* Vol. 22(No. 7): 517–522.

Kulin, S., Killian, T. C., Bergeson, S. D. & Rolston, S. L. (2000). Plasma Oscillations and Expansion of an Ultracold Neutral Plasma, *Phys. Rev. Lett.* Vol. 85(No. 2): 318–321.

Labeyrie, G., Michaud, F. & Kaiser, R. (2006). Self-Sustained Oscillations in a Large Magneto-Optical Trap, *Phys. Rev. Lett.* Vol. 96(No. 2): 023003-1–023003-4.

Landau, L. D. & Lifshitz, E. M. (1976). *Mechanics*, Pergamon, London.

Lapidus, L. J., Enzer, D. & Gabrielse, G. (1999). Stochastic Phase Switching of a Parametrically Driven Electron in a Penning Trap, *Phys. Rev. Lett.* Vol. 83(No. 5): 899–902.

Metcalf, H. J. & van der Straten, P. (1999). *Laser Cooling and Trapping*, Springer, New York.

Nayfeh, N. H. & Moore, D. T. (1979). *Nonlinear Oscillations*, Wiley, New York.

Noh, H. R. & Jhe, W. (2007). Semiclassical theory of sub-Doppler forces in an asymmetric magneto-optical trap with unequal laser detunings, *Phys. Rev. A* Vol. 75(No. 5): 053411-1–053411-9.

Raab, E. L., Prentiss, M., Cable, A., Chu, S. & Pritchard, D. E. (1987). Trapping of Neutral Sodium Atoms with Radiation Pressure, *Phys. Rev. Lett.* Vol. 59(No. 23): 2631–2634.

Razvi, M. A. N., Chu, X. Z., Alheit, R., Werth, G. & Blümel, R. (1998). Fractional frequency collective parametric resonances of an ion cloud in a Paul trap, *Phys. Rev. A* Vol. 58(No. 1): R34–R37.

Sesko, D., Walker, T. & Wieman, C. (1991). Behavior of neutral atoms in a spontaneous force trap, *J. Opt. Soc. Am. B* Vol. 8(No. 5): 946–958.

Steane, A. M., Chowdhury, M. & Foot, C. J. (1992). Radiation force in the magneto-optical trap, *J. Opt. Soc. Am. B* Vol. 9(No. 12): 2142–2158.

Strogatz, S. H. (2001). *Nonlinear Dynamics and Chaos*, Perseus, New York.

Tabosa, J. W. R., Chen, G., Hu, Z., Lee, R. B. & Kimble, H. J. (1991). Nonlinear spectroscopy of cold atoms in a spontaneous-force optical trap, *Phys. Rev. Lett.* Vol. 66(No. 25): 3245–3248.

Tan, J. & Gabrielse, G. (1991). Synchronization of parametrically pumped electron oscillators with phase bistability, *Phys. Rev. Lett.* Vol. 67(No. 22): 3090–3093.

Tan, J. & Gabrielse, G. (1993). Parametrically pumped electron oscillators, *Phys. Rev. A* Vol. 48(No. 4): 3105–3121.

Townsend, C. G., Edwards, N. H., Cooper, C. J., Zetie, K. P., Foot, C. J., Steane, A. M., Szriftgiser, P., Perrin, H. & Dalibard. J. (1995). Phase-space density in the magneto-optical trap, *Phys. Rev. A* Vol. 52(No. 2): 1423–1440.

Walhout, M., Dalibard, J., Rolston, S. L. & Phillips, W. D. (1992). σ_+–σ_- Optical molasses in a longitudinal magnetic field, *J. Opt. Soc. Am. B* Vol. 9(No. 11): 1997–2007.

Tseng, C. H., Enzer, D., Gabrielse, G. & Walls, F. L. (1999). 1-bit memory using one electron: Parametric oscillations in a Penning trap, *Phys. Rev. A* Vol. 59(No. 3): 2094–2104.

Walker, T., Sesko, D. & Wieman, C. (1990). Collective behavior of optically trapped neutral atomstitle, *Phys. Rev. Lett.* Vol. 64(No. 4): 408–411.

Walker, T. & Feng, P. (1994). Measurements of Collisions Between Laser-Cooled Atoms, *Adv. At. Mol. Opt. Phys.* Vol. 34: 125–170.

Wallace, C. D., Dinneen, T. P., Tan, K. Y. N., Kumarakrishnan, A., Gould, P. L. & Javanainen, J. (1994). Measurements of temperature and spring constant in a magneto-optical trap, *J. Opt. Soc. Am. B* Vol. 11(No. 5): 703–711.

Wilkowski, D., Ringot, J., Hennequin, D. & Garreau, J. C. (2000). Instabilities in a Magnetooptical Trap: Noise-Induced Dynamics in an Atomic System, *Phys. Rev. Lett.* Vol. 85(No. 9): 1839–1842.

Xu, X., Loftus, T. H., Smith, M. J., Hall, J. L., Gallagher, A. & Ye, J. (2002). Dynamics in a two-level atom magneto-optical trap, *Phys. Rev. A* Vol. 66(No. 1): 011401-1–011401-4.

20

The Photonic Torque Microscope: Measuring Non-conservative Force-fields

Giovanni Volpe[1,2,3], Giorgio Volpe[1] and Giuseppe Pesce[4]
[1]*ICFO – The Institute of Photonic Sciences, Castelldefels (Barcelona),*
[2]*Max-Planck-Institut für Metallforschung, Stuttgart,*
[3]*Universität Stuttgart, Stuttgart,*
[4]*Università di Napoli "Federico II", Napoli,*
[1]*Spain*
[2,3]*Germany*
[4]*Italy*

1. Introduction

Over the last 20 years the advances of laser technology have permitted the development of an entire new field in optics: the field of *optical trapping and manipulation*. The focal spot of a highly focused laser beam can be used to confine and manipulate microscopic particles ranging from few tens of nanometres to few microns (Ashkin, 2000; Neuman & Block, 2004).

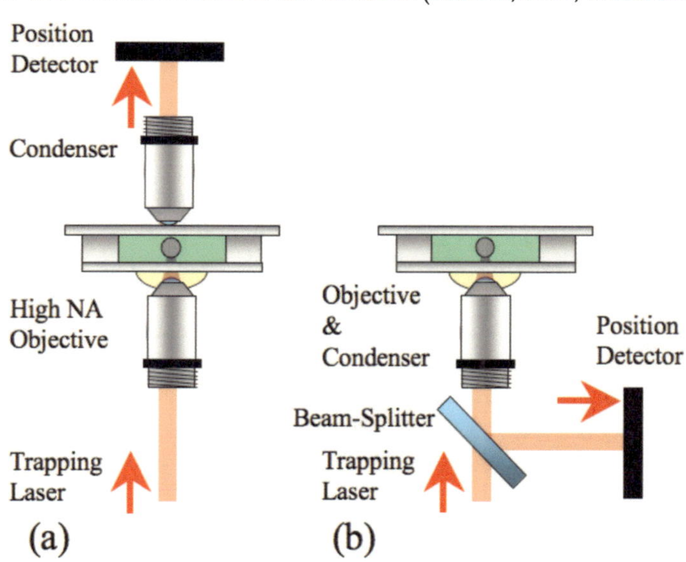

Fig. 1. PFM setups with detection using forward (a) and backward (b) scattered light.

Such an optical trap can detect and measure forces and torques in microscopic systems - a technique now known as photonic force microscope (PFM). This is a fundamental task in many areas, such as biophysics, colloidal physics and hydrodynamics of small systems.

The PFM was devised in 1993 (Ghislain & Webb, 1993). A typical PFM comprises an optical trap that holds a probe – a dielectric or metallic particle of micrometre size, which randomly moves due to Brownian motion in the potential well formed by the optical trap – and a position sensing system. The analysis of the thermal motion provides information about the local forces acting on the particle (Berg-Sørensen & Flyvbjerg, 2004). The PFM can measure forces in the range of femtonewtons to piconewtons. This range is well below the limits of techniques based on micro-fabricated mechanical cantilevers, such as the atomic force microscope (AFM).

However, an intrinsic limit of the PFM is that it can only deal with conservative force-fields, while it cannot measure the presence of a torque, which is typically associated with the presence of a non-conservative (or rotational) force-field.

In this Chapter, after taking a glance at the history of optical manipulation, we will briefly review the PFM and its applications. Then, we will discuss how the PFM can be enhanced to deal with non-conservative force-fields, leading to the photonic torque microscope (PTM) (Volpe & Petrov, 2006; Volpe et al., 2007a). We will also present a concrete analysis workflow to reconstruct the force-field from the experimental time-series of the probe position. Finally, we will present three experiments in which the PTM technique has been successfully applied:

1. *Characterization of singular points in microfluidic flows.* We applied the PTM to microrheology to characterize fluid fluxes around singular points of the fluid flow (Volpe et al., 2008).
2. *Detection of the torque carried by an optical beam with orbital angular momentum.* We used the PTM to measure the torque transferred to an optically trapped particle by a Laguerre-Gaussian beam (Volpe & Petrov, 2006).
3. *Quantitative measurement of non-conservative forces generated by an optical trap.* We used the PTM to quantify the contribution of non-conservative optical forces to the optical trapping (Pesce et al., 2009).

2. Brief history of optical manipulation

Optical trapping and manipulation did not exist before the invention of the laser in 1960 (Townes, 1999). It was already known from astronomy and from early experiments in optics that light had linear and angular momentum and, therefore, that it could exert radiation pressure and torques on physical objects. Indeed, light's ability to exert forces has been recognized at least since 1619, when Kepler's *De Cometis* described the deflection of comet tails by sunrays.

In the late XIX century Maxwell's theory of electromagnetism predicted that the light momentum flux was proportional to its intensity and could be transferred to illuminated objects, resulting in a radiation pressure pushing objects along the propagation direction of light.

Early exciting experiments were performed in order to verify Maxwell's predictions. Nichols and Hull (Nichols & Hull, 1901) and Lebedev (Lebedev, 1901) succeeded in detecting radiation pressure on macroscopic objects and absorbing gases. A few decades later, in 1936, Beth reported the experimental observation of the torque on a macroscopic object resulting from interaction with light (Beth, 1936): he observed the deflection of a quartz wave plate suspended from a thin quartz fibre when circularly polarized light passed through it. These effects were so small, however, that they were not easily detected. Quoting J. H. Poynting's presidential address to the British Physical Society in 1905, "a very short experience in attempting to measure these forces is sufficient to make one realize their extreme minuteness - a minuteness which appears to put them beyond consideration in terrestrial affairs." (Cited in Ref. (Ashkin, 2000))

Things changed with the invention of the laser in the 1960s (Townes, 1999). In 1970 Ashkin showed that it was possible to use the forces of radiation pressure to significantly affect the dynamics of transparent micrometre sized particles (Ashkin, 1970). He identified two basic light pressure forces: a *scattering force* in the direction of the incident beam and a *gradient force* in the direction of the intensity gradient of the beam. He showed experimentally that, using just these forces, a focused laser beam could accelerate, decelerate and even stably trap small micrometre sized particles.

Ashkin considered a beam of power P reflecting on a plane mirror: $P / h\nu$ photons per second strike the mirror, each carrying a momentum $h\nu / c$, where h is the Planck constant, ν is the light frequency and c the speed of light. If they are all reflected straight back, the total change in light momentum per second is $2 \cdot (P / h\nu) \cdot (h\nu / c) = 2P / c$, which, by conservation of momentum, implies that the mirror experiences an equal and opposite force in the direction of the light. This is the maximum force that one can extract from the light. Quoting Ashkin (Ashkin, 2000), "Suppose we have a laser and we focus our one watt to a small spot size of about a wavelength $\cong 1\mu m$, and let it hit a particle of diameter also of $1\mu m$. Treating the particle as a 100% reflecting mirror of density $\cong 1 gm / cm^3$, we get an acceleration of the small particle $= A = F / m = 10^{-3} dynes / 10^{-12} gm = 10^9 cm / sec^2$. Thus, $A \cong 10^6 g$, where $g \cong 10^3 cm / sec^2$, the acceleration of gravity. This is quite large and should give readily observable effects, so I tried a simple experiment. [...] It is surprising that this simple first experiment [...], intended only to show forward motion due to laser radiation pressure, ended up demonstrating not only this force but the existence of the transverse force component, particle guiding, particle separation, and stable 3D particle trapping."

In 1986, Ashkin and colleagues reported the first observation of what is now commonly referred to as an *optical trap* (Ashkin et al., 1986): a tightly focused beam of light capable of holding microscopic particles in three dimensions. One of Ashkin's co-authors, Steven Chu, would go on to use optical tweezing in his work on cooling and trapping atoms. This research earned Chu, together with Claude Cohen-Tannoudji and William Daniel Phillips, the 1997 Nobel Prize in Physics.

In the late 1980s, the new technology was applied to the biological sciences, starting by trapping tobacco mosaic viruses and *Escherichia coli* bacteria. In the early 1990s, Block, Bustamante and Spudich pioneered the use of optical trap *force spectroscopy*, an alternate name for PFM, to characterize the mechanical properties of biomolecules and biological motors (Block et al., 1990; Finer et al., 1994; Bustamante et al., 1994). Optical traps allowed these biophysicists to observe the forces and dynamics of nanoscale motors at the single-molecule level. Optical trap force spectroscopy has led to a deeper understanding of the nature of these force-generating molecules, which are ubiquitous in nature.

Optical tweezers have also proven useful in many other areas of physics, such as atom trapping (Metcalf & van der Straten, 1999) and statistical physics (Babic et al., 2005).

3. The photonic force microscope

One of the most prominent uses of optical tweezers is to measure tiny forces, in the order of 100s of femtonewtons to 10s of piconewtons. A typical PFM setup comprises an optical trap

to hold a probe - a dielectric or metallic particle of micrometer size - and a position sensing system. In the case of biophysical applications the probe is usually a small dielectric bead tethered to the cell or molecule under study. The probe randomly moves due to Brownian motion in the potential well formed by the optical trap. Near the centre of the trap, the restoring force is linear in the displacement. The stiffness of such harmonic potential can be calibrated using the three-dimensional position fluctuations. To measure an external force acting on the probe it suffices to measure the probe average position displacement under the action of such force and multiply it by the stiffness.
In order to understand the PFM it is necessary to discuss these three aspects:

1. the *optical forces* that act on the probe and produce the optical trap;
2. the *position detection*, which permits one to track the probe position with nanometre resolution and at kilohertz sampling rate;
3. the statistics of the *Brownian motion* of the probe in the trap, which are used in the calibration procedure.

3.1 Optical forces

It is well known from quantum mechanics that light carries a momentum: for a photon at wavelength λ the associated momentum is $p = h / \lambda$. For this reason, whenever an atom emits or absorbs a photon, its momentum changes according to Newton's laws. Similarly, an object will experience a force whenever a propagating light beam is refracted or reflected by its surface. However, in most situations this force is much smaller than other forces acting on macroscopic objects so that there is no noticeable effect and, therefore, can be neglected. The objects, for which this radiation pressure exerted by light starts to be significant, weigh less than $1\mu g$ and their size is below 10s of microns.
A focused laser beam acts as an attractive potential well for a particle. The equilibrium position lies near – but not exactly at – the focus. When the object is displaced from this equilibrium position, it experiences an attractive force towards it. In first approximation this restoring force is proportional to the displacement; in other words, optical tweezers force can generally be described by Hooke's law:

$$F_x = -k_x \left(x - x_0 \right), \tag{1}$$

where x is the particle's position, x_0 is the focus position and k_x is the spring constant of the optical trap along the x -direction, usually referred to as *trap stiffness*. In fact, an optical tweezers creates a three-dimensional potential well, which can be approximated by three independent harmonic oscillators, one for each of the x -, y- and z-directions. If the optics are well aligned, the x and y spring constants are roughly the same, while the z spring constant is typically smaller by a factor of 5 to 10.
Considering the ratio between the characteristic dimension L of the trapped object and the wavelength λ of the trapping light, three different trapping regimes can be defined:

1. the *Rayleigh regime*, when $L \ll \lambda$;
2. an *intermediate regime*, when L is comparable to λ;
3. the *geometrical optics regime*, when $L \gg \lambda$.

In Fig. 2 an overview of the kind of objects belonging to each of these regimes is presented, considering that the trapping wavelength is usually in the visible or near-infrared spectral region. In any of these regimes, the electromagnetic equations can be solved to evaluate the force acting on the object. However, this can be a cumbersome task. For the Rayleigh regime and geometrical optics regime approximate models have been developed. However, most of the objects that are normally trapped in optical manipulation experiments fall in the intermediate regime, where such approximations cannot be used. In particular, this is true for the probes usually used for the PFM: typically particles with diameter between 0.1 and 10 micrometres.

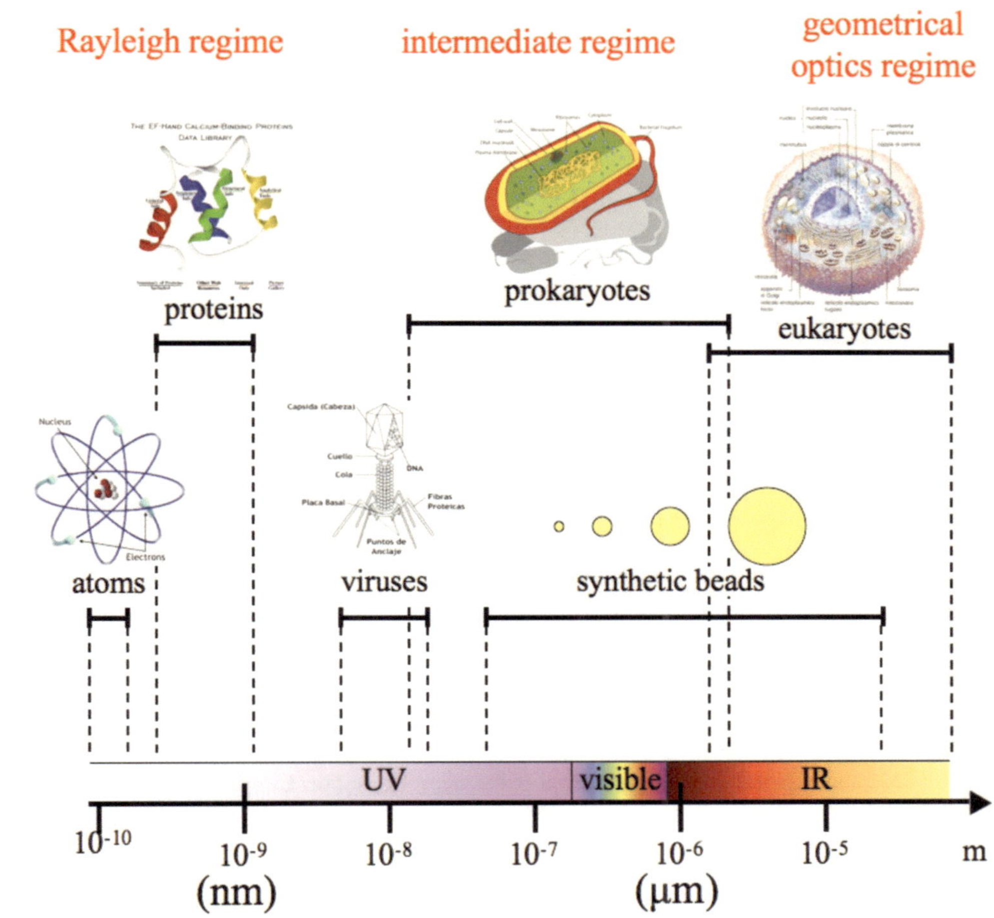

Fig. 2. Trapping regimes and objects that are typically optically manipulated: from cells to viruses in biophysical experiments, and from atoms to colloidal particles in experimental statistical physics. The wavelength of the trapping light is usually in the visible or near-infrared.

3.2 Position detection

The three-dimensional position of the probe is typically measured through the scattering of a light beam illuminating it. This can be the same beam used for trapping or an auxiliary beam.

Typically, position detection is achieved through the analysis of the interference of the forward-scattered (FS) light and unscattered (incident) light. A typical setup is shown in Fig.

1(a). The PFM with FS detection was extensively studied, for example, in Ref (Rohrbach & Stelzer, 2002).

In a number of experiments, however, geometrical constraints may prevent access to the FS light, forcing one to make use of the backward-scattered (BS) light instead. This occurs, for example, in biophysical applications where one of the two faces of a sample holder needs to be coated with some specific material or in plasmonics applications where a plasmon wave needs to be coupled to one of the faces of the holder (Volpe et al., 2006). A typical setup that uses the BS light is presented in Fig. 1(b). The PFM with BS detection has been studied theoretically in Ref. (Volpe et al., 2007b) and experimentally in Ref. (Huisstede et al., 2005).

Two types of photodetectors are typically used. The quadrant photodetector (QPD) works by measuring the intensity difference between the left-right and top-bottom sides of the detection plane. The position sensing detector (PSD) measures the position of the centroid of the collected intensity distribution, giving a more adequate response for non-Gaussian profiles. Note that high-speed video systems are also in use, but they do not achieve the acquisition rate available with photodetectors.

3.3 Brownian motion of an optically trapped particle

Assuming a very low Reynolds number regime (Happel & Brenner, 1983), the Brownian motion of the probe in the optical trap is described by a set of Langevin equations:

$$\gamma \mathbf{r}'(t) + \mathbf{K}\,\mathbf{r}(t) = \sqrt{2D}\,\gamma\,\mathbf{h}(t), \tag{2}$$

where $\mathbf{r}(t) = [x(t), y(t), z(t)]^T$ is the probe position, $\gamma = 6\pi R\eta$ its friction coefficient, R its radius, η the medium viscosity, $\mathbf{K}$ the stiffness matrix, $\sqrt{2D}\,\gamma[h_x(t), h_y(t), h_z(t)]^T$ a vector of independent white Gaussian random processes describing the Brownian forces, $D = k_B T / \gamma$ the diffusion coefficient, T the absolute temperature and k_B the Boltzmann constant. The orientation of the coordinate system can be chosen in such a way that the restoring forces are independent in the three directions, i.e. $\mathbf{K} = \mathrm{diag}(k_x, k_y, k_z)$. In such reference frame the stochastic differential Eqs. (2) are separated and, without loss of generality, the treatment can be restricted to the x-projection of the system.

When a constant and homogeneous external force $f_{ext,x}$ acting on the probe produces a shift in its equilibrium position in the trap, its value can be obtained as:

$$f_{ext,x} = k_x \langle x(t) \rangle, \tag{3}$$

where $\langle x(t) \rangle$ is the probe mean displacement from the equilibrium position.

There are several straightforward methods to experimentally measure the trap parameters – trap stiffness and conversion factor between voltage and length – and, therefore, the force exerted by the optical tweezers on an object, without the need for a theoretical reference model of the electromagnetic interaction between the particle and the laser beam. The most commonly employed ones are the *drag force method*, the *equipartition method*, the *potential analysis method* and the *power spectrum or correlation method* (Visscher et al., 1996; Berg-Sørensen & Flyvbjerg, 2004). The latter, in particular, is usually considered the most reliable one. Experimentally the trap stiffness can be found by fitting the autocorrelation function (ACF) of the Brownian motion in the trap obtained from the measurements to the theoretical one, which reads

$$r_{xx}(\tau) = \langle x(t+\tau)x^*(t)\rangle = \frac{k_B T}{k_x} e^{-\frac{k_x}{\gamma}|\tau|}. \quad (4)$$

4. The photonic torque microscope

The PFM measures a constant force acting on the probe. This implies that the force-field to be measured has to be invariable (homogeneous) on the scale of the Brownian motion of the trapped probe, i.e. in a range of 10s to 100s of nanometres depending on the trapping stiffness. In particular, as we will see, this condition implicates that the force-field must be conservative, excluding the possibility of a rotational component.

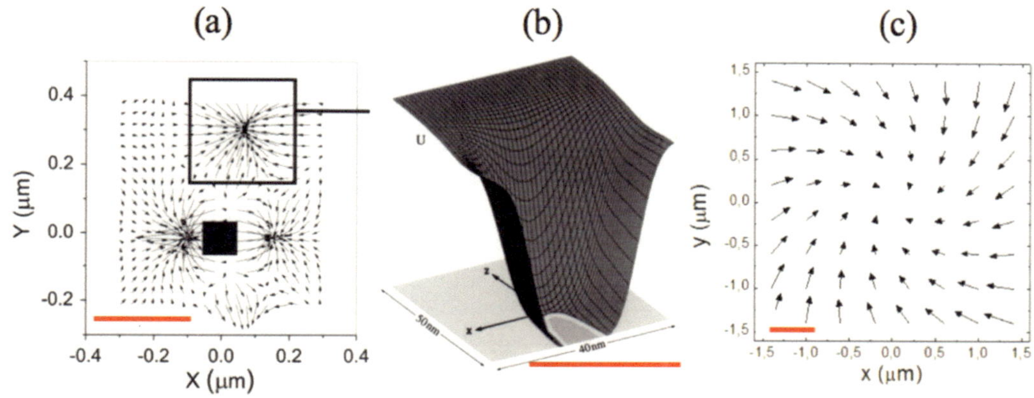

Fig. 3. Examples of physical systems that produce force-fields that cannot be correctly probed with a classical PFM, because they vary on the scale of the Brownian motion of the trapped probe (a possible range is indicated by the red bars): (a) forces produced by a surface plasmon polariton in the presence of a patterned surface on a 50*nm* radius dielectric particle (adapted from Ref. (Quidant et al., 2005)); (b) trapping potential for 10*nm* diameter dielectric particle near a 10*nm* wide gold tip in water illuminated by a 810*nm* monochromatic light beam (adapted from Ref. (Novotny et al., 1997)); and (c) force-field acting on a 500*nm* radius dielectric particle in the focal plane of a highly focused Laguerre-Gaussian beam (adapted from Ref. (Volpe & Petrov, 2006)).

However, there are cases where these assumptions are not fulfilled. The force-field can vary in the nanometre scale, for example, considering the radiation forces exerted on a dielectric particle by a patterned optical near-field landscape at an interface decorated with resonant gold nanostructures (Quidant et al., 2005) (Fig. 3 (a)), the nanoscale trapping that can be achieved near a laser-illuminated tip (Novotny et al., 1997) (Fig. 3(b)), the optical forces produced by a beam which carries orbital angular momentum (Volpe & Petrov, 2006) (Fig. 3(c)), or in the presence of fluid flows (Volpe et al., 2008). In order to deal with these cases, we need a deeper understanding of the Brownian motion of the optically trapped probe in the trapping potential.

In the following we will discuss the Brownian motion near an equilibrium point in a force-field and we will see how this permits us to develop a more powerful theory of the PFM: the Photonic Torque Microscope (PTM). Full details can be found in Ref. (Volpe et al., 2007a).

4.1 Brownian motion near an equilibrium position

In the presence of an external force-field $\mathbf{f}_{ext}(\mathbf{r}(t))$, Eq. (2) can be written in the form:

$$\gamma\,\mathbf{r}'(t)+\mathbf{f}(\mathbf{r}(t))=\sqrt{2D}\,\gamma\,\mathbf{h}(t), \tag{5}$$

where the total force acting on the probe $\mathbf{f}(\mathbf{r}(t))=\mathbf{f}_{ext}(\mathbf{r}(t))-\mathbf{K}\mathbf{r}(t)$ depends on the position of the probe itself, but does not vary over time.

The force $\mathbf{f}(\mathbf{r}(t))=\left[f_x(\mathbf{r}(t)),f_y(\mathbf{r}(t))\right]^T$ can be expanded in Taylor series up to the first order around an arbitrary point $\mathbf{r}_0$:

$$\mathbf{f}(\mathbf{r}(t))=\begin{bmatrix} f_x(\mathbf{r}_0) \\ f_y(\mathbf{r}_0)\end{bmatrix}+\begin{bmatrix} \dfrac{\partial f_x(\mathbf{r}_0)}{\partial x} & \dfrac{\partial f_x(\mathbf{r}_0)}{\partial y} \\ \dfrac{\partial f_y(\mathbf{r}_0)}{\partial x} & \dfrac{\partial f_y(\mathbf{r}_0)}{\partial y}\end{bmatrix}(\mathbf{r}(t)-\mathbf{r}_0)+o\left(\|\mathbf{r}(t)-\mathbf{r}_0\|\right), \tag{6}$$

where $\mathbf{f}_{\mathbf{r}_0}$ and $\mathbf{J}_{\mathbf{r}_0}$ are the zeroth-order and first-order expansion coefficients, i.e. the force-field value at the point $\mathbf{r}_0$ and the Jacobian of the force-field calculated in $\mathbf{r}_0$. In the following we will assume, without loss of generality, $\mathbf{r}_0=0$.

In a PFM the probe is optically trapped and, therefore, it diffuses due to Brownian motion in the total force-field (the sum of the optical trapping force and external force-fields). If $\mathbf{f}_{\mathbf{r}_0}\neq 0$, the probe experiences a shift in the direction of the force and, after a transient time has elapsed, the particle settles down in a new equilibrium position of the total force-field, such that $\mathbf{f}_{\mathbf{r}_0}=0$. As we have already seen, the measurement of this shift allows one to evaluate the homogeneous force acting on the probe in the standard PFM and, therefore, the zeroth order term of the Taylor expansion. In the following we will assume this to be null and study the statistics of the Brownian motion near the equilibrium point can be analyzed in order to reconstruct the force-field up to its first-order approximation.

4.2 Conservative and rotational components of the force-field

The first order approximation to Eq. (5) near an equilibrium point of the force-field, $\breve{\mathbf{r}}=0$, is:

$$\mathbf{r}'(t)=\gamma^{-1}\mathbf{J}_0\,\mathbf{r}(t)+\sqrt{2D}\,\mathbf{h}(t), \tag{7}$$

where $\mathbf{r}(t)=\left[x(t),y(t)\right]^T$, $\mathbf{h}(t)=\left[h_x(t),h_y(t)\right]^T$ and $\mathbf{J}_0$ is the Jacobian calculated at the equilibrium point.

According to the Helmholtz theorem, any force-field can be separated into its conservative (irrotational) and non-conservative (rotational or solenoidal) components. With simple algebraic passages, the Jacobian $\mathbf{J}_0$ can be written as the sum of two matrices:

$$\mathbf{J}_0=\mathbf{J}_c+\mathbf{J}_{nc}, \tag{8}$$

where

$$\mathbf{J}_c=\begin{bmatrix} \dfrac{\partial f_x(0)}{\partial x} & \dfrac{1}{2}\left(\dfrac{\partial f_x(0)}{\partial y}+\dfrac{\partial f_y(0)}{\partial x}\right) \\ \dfrac{1}{2}\left(\dfrac{\partial f_y(0)}{\partial x}+\dfrac{\partial f_x(0)}{\partial y}\right) & \dfrac{\partial f_y(0)}{\partial y}\end{bmatrix} \tag{9}$$

and

$$\mathbf{J}_{nc} = \begin{bmatrix} 0 & \frac{1}{2}\left(\frac{\partial f_x(0)}{\partial y} - \frac{\partial f_y(0)}{\partial x}\right) \\ \frac{1}{2}\left(\frac{\partial f_y(0)}{\partial x} - \frac{\partial f_x(0)}{\partial y}\right) & 0 \end{bmatrix}. \tag{10}$$

It is easy to show that $\mathbf{J}_c$ is the conservative component of the force-field and that $\mathbf{J}_{nc}$ is the rotational component.

The two components can be easily identified if the coordinate system is chosen such that $\frac{\partial f_x(0)}{\partial y} = -\frac{\partial f_y(0)}{\partial x}$. In this case, the Jacobian $\mathbf{J}_0$ normalized by the friction coefficient γ reads:

$$\gamma^{-1}\mathbf{J}_0 = \begin{bmatrix} -\phi_x & \Omega \\ -\Omega & -\phi_y \end{bmatrix}, \tag{11}$$

where $\phi_x = k_x / \gamma$, $\phi_y = k_y / \gamma$, $k_x = -\frac{\partial f_x(0)}{\partial x}$, $k_y = -\frac{\partial f_y(0)}{\partial y}$ and $\Omega = \gamma^{-1}\frac{\partial f_x(0)}{\partial y} = -\gamma^{-1}\frac{\partial f_y(0)}{\partial x}$.

In Eq. (11) the rotational component, which is invariant under a coordinate rotation, is represented by the non-diagonal terms of the matrix: Ω is the value of the constant angular velocity of the probe rotation around the z-axis due to the presence of the rotational force-field. The conservative component, instead, is represented by the diagonal terms of the Jacobian and is centrally symmetric with respect to the origin. Without loss of generality, it can be imposed that the stiffness of the trapping potential is higher along the x-axis, i.e. $k_x > k_y$ and, therefore, $\phi_x > \phi_y$.

4.3 Stability study

The conditions for the stability of the equilibrium point are

$$\begin{cases} \mathrm{Det}(\mathbf{J}_0) = \phi^2 - \Delta\phi^2 + \Omega^2 > 0 \\ \mathrm{Tr}(\mathbf{J}_0) = -2\phi < 0 \end{cases}, \tag{12}$$

where $\phi = (\phi_x + \phi_y)/2$ and $\Delta\phi = (\phi_x - \phi_y)/2$. The fundamental condition required to achieve the stability is $\phi > 0$. Assuming that this condition is satisfied, the behaviour of the optically trapped probe can be explored as a function of the parameters Ω / ϕ and $\Delta\phi / \phi$. The stability diagram is shown in Fig. 4(a).

The standard PFM corresponds to $\Delta\phi = 0$ and $\Omega = 0$. When a rotational term is added, i.e. $\Omega \neq 0$ and $\Delta\phi = 0$, the system remains stable. When there is no rotational contribution to the force-field ($\Omega = 0$) the equilibrium point becomes unstable as soon as $\Delta\phi \geq \phi$. This implicates that $\phi_y < 0$ and, therefore, the probe is not confined in the y-direction any more. In the presence of a rotational component ($\Omega \neq 0$) the stability region becomes larger; the equilibrium point now becomes unstable only for $\Delta\phi \geq \sqrt{\phi^2 - \Omega^2}$.

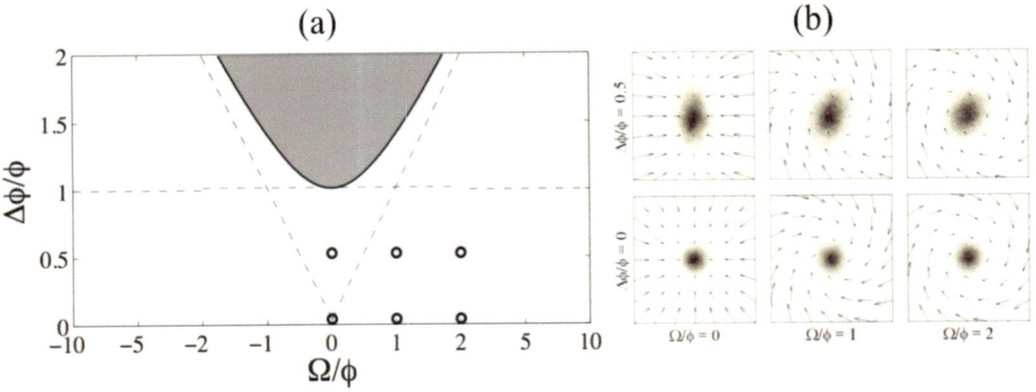

Fig. 4. (a) Stability diagram. Assuming $\varphi > 0$, the stability of the system is shown as a function of the parameters Ω / ϕ and $\Delta\phi / \phi$. The white region satisfies the stability conditions in Eq. (12). The dashed lines represent the $\Delta\phi = |\Omega|$ and $\Delta\phi = \phi$ curves. The dots represent the parameters that are further investigated in Figs. 4(b) and 5. (b) Brownian motion near an equilibrium point. The arrows show the force-field vectors for various values of the parameters Ω / ϕ and $\Delta\phi / \phi$. The shadowed areas show the probability distribution function (PDF) of the probe position in the corresponding force-field.

Some examples of possible force-fields are presented in Fig. 4(b). When $\Omega = 0$ the probe movement can be separated along two orthogonal directions. As the value of $\Delta\phi$ increases, the probability density function (PDF) of the probe position becomes more and more elliptical, until for $\Delta\phi \geq \phi$ the probe is confined only along the *x*-direction and the confinement along the *y*-direction is lost.

If $\Delta\phi = 0$, the increase in Ω induces a bending of the force-field lines and the probe movements along the x- and y-directions are not independent any more. For values of $\Omega \geq \phi$, the rotational component of the force-field becomes dominant over the conservative one. This is particularly clear when $\Delta\phi \neq 0$: the presence of a rotational component masks the asymmetry in the conservative one, since the PDF assumes a more rotationally symmetric shape.

4.4 The photonic torque microscope

The most powerful analysis method to characterize the stiffness of an optical trap is based on the study of the correlation functions - or, equivalently, of the power spectral density - of the probe position time-series. In order to derive the theory for the PTM, the correlation matrix for the general case of Eq. (5) will be first derived in the coordinate system considered in the previous section, where the conservative and rotational components are readily separated. Then, the same matrix will be given in a generic coordinate system and some invariant functions that are independent on its orientation will be identified.

Correlation matrix. The correlation matrix of the probe motion near an equilibrium position can be calculated from the solutions of Eq. (5). The full derivation is presented in Ref. (Volpe et al., 2007a). The correlation matrix results:

$$\begin{cases} r_{xx}(\Delta t) & = & D\frac{e^{-\phi|\Delta t|}}{\phi}\left[\left(\frac{\Omega^2-\alpha^2\Delta\phi^2}{\Omega^2-\Delta\phi^2}-\alpha^2\frac{\Delta\phi}{\phi}\right)C(\Delta t)-\alpha^2\frac{\Delta\phi}{\phi}\left(1-\frac{\Delta\phi}{\phi}\right)S(|\Delta t|)\right] \\ r_{yy}(\Delta t) & = & D\frac{e^{-\phi|\Delta t|}}{\phi}\left[\left(\frac{\Omega^2-\alpha^2\Delta\phi^2}{\Omega^2-\Delta\phi^2}+\alpha^2\frac{\Delta\phi}{\phi}\right)C(\Delta t)+\alpha^2\frac{\Delta\phi}{\phi}\left(1+\frac{\Delta\phi}{\phi}\right)S(|\Delta t|)\right] \\ r_{xy}(\Delta t) & = & D\frac{e^{-\phi|\Delta t|}}{\phi}\frac{\Omega}{\phi}\left[+S(\Delta t)+\alpha^2\frac{\Delta\phi}{\phi}\left(C(\Delta t)+S(|\Delta t|)\right)\right] \\ r_{yx}(\Delta t) & = & D\frac{e^{-\phi|\Delta t|}}{\phi}\frac{\Omega}{\phi}\left[-S(\Delta t)+\alpha^2\frac{\Delta\phi}{\phi}\left(C(\Delta t)+S(|\Delta t|)\right)\right] \end{cases}, \quad (13)$$

where

$$\alpha^2 = \frac{\phi^2}{\phi^2+\left(\Omega^2-\Delta\phi^2\right)} \quad (14)$$

is a dimensionless parameter,

$$C(t) = \begin{cases} \cos\left(\sqrt{|\Omega^2-\Delta\phi^2|}\,t\right) & \Omega^2 > \Delta\phi^2 \\ 1 & \Omega^2 = \Delta\phi^2 \\ \cosh\left(\sqrt{|\Omega^2-\Delta\phi^2|}\,t\right) & \Omega^2 < \Delta\phi^2 \end{cases} \quad (15)$$

and

$$S(t) = \begin{cases} \phi\frac{\sin\left(\sqrt{|\Omega^2-\Delta\phi^2|}\,t\right)}{\sqrt{|\Omega^2-\Delta\phi^2|}} & \Omega^2 > \Delta\phi^2 \\ \phi t & \Omega^2 = \Delta\phi^2 . \\ \phi\frac{\sinh\left(\sqrt{|\Omega^2-\Delta\phi^2|}\,t\right)}{\sqrt{|\Omega^2-\Delta\phi^2|}} & \Omega^2 < \Delta\phi^2 \end{cases} \quad (16)$$

In Fig. 5(a) these correlation functions are plotted for different ratios of the force-field conservative and rotational components.

For the case $\Delta\phi = 0$, the auto-correlation functions (ACFs) are $r_{xx}(\Delta t) = r_{yy}(\Delta t) = De^{-\phi|\Delta t|}\cos(\Omega\Delta t)/\phi$ and cross-correlation functions (CCFs) are $r_{xy}(\Delta t) = -r_{yx}(\Delta t) = De^{-\phi|\Delta t|}\sin(\Omega\Delta t)/\phi$. Their zeros are at $\Delta t = n\Omega/\pi$ and $\Delta t = (n+0.5)\Omega/\pi$ respectively, with n integer. However, when the rotational term is smaller than the conservative one ($\Omega < \phi$), the zeros are not distinguishable due to the rapid exponential decay of the correlation functions. As the rotational component becomes greater than the conservative one ($\Omega > \phi$), a first zero appears in the ACFs and CCFs and, as Ω increases even further, the number of oscillation grows. Eventually, for $\Omega >> \phi$ the sinusoidal component becomes dominant. The conservative component manifests itself as an exponential decay of the magnitude of the ACFs and CCFs.

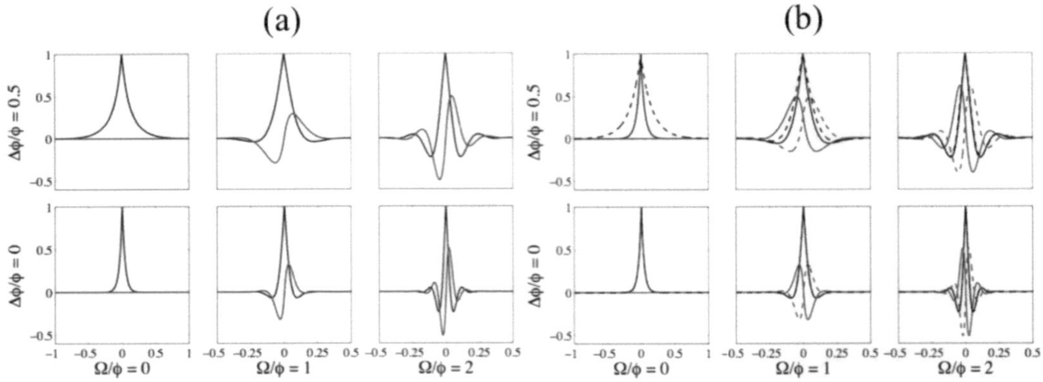

Fig. 5. (a) Autocorrelation and cross-correlation functions. Autocorrelation and cross-correlation functions for various values of the parameters Ω / ϕ and $\Delta\phi/\phi$: $r_{xx}(\Delta t)$ (black continuous line), $r_{yy}(\Delta t)$ (black dotted line), $r_{xy}(\Delta t)$ (blue continuous line) and $r_{yx}(\Delta t)$ (blue dotted line). (b) Invariant functions: $S_{ACF}(\Delta t)$ and $D_{CCF}(\Delta t)$. These functions, independent from the choice of the reference system, are presented for various values of the parameters Ω / ϕ and $\Delta\phi / \phi$: $S_{ACF}(\Delta t)$ (black line) and $D_{CCF}(\Delta t)$ (blue line).

When $\Omega = 0$, the movements of the probe along the x- and y-directions are independent. The ACFs are $r_{xx}(\Delta t) = De^{-\phi_x|\Delta t|}/\phi_x$ and $r_{yy}(\Delta t) = De^{-\phi_y|\Delta t|}/\phi_y$, while the CCFs are null, $r_{xy}(\Delta t) = r_{yx}(\Delta t) = 0$. In Fig. 4(a) this case is represented by the line $\Omega = 0$.
When both Ω and $\Delta\phi$ are zero, the ACFs are $r_{xx}(\Delta t) = r_{yy}(\Delta t) = De^{-\phi|\Delta t|}/\phi$ and the CCFs are null, i.e. $r_{xy}(\Delta t) = r_{yx}(\Delta t) = 0$. The corresponding force-field vectors point towards the centre and are rotationally symmetric.
When both Ω and $\Delta\phi$ are nonvanishing, the effective angular frequency that enters the correlation functions is given by $\sqrt{|\Omega^2 - \Delta\phi^2|}$. This shows that the difference in the stiffness coefficients along the x- and y-axes effectively influences the rotational term, if this is present. A limiting case is when $|\Omega| = \Delta\phi$. This case presents a pseudo-resonance between the rotational term and the stiffness difference.
Correlation matrix in a generic coordinate system. The expressions for the ACFs and CCFs in Eq. (13) were obtained in a specific coordinate system, where the conservative and rotational component of the force-field can be readily identified. However, the experimentally acquired time-series of the probe position required for the calculation of the ACFs and CCFs are usually given in a different coordinate system, rotated with respect to the one considered in the previous subsection. If a rotated coordinate system is introduced, such that

$$\mathbf{r}^{\theta}(t) = \mathbf{R}(\theta)\mathbf{r}(t), \tag{17}$$

where $\mathbf{r}^{\theta}(t) = \left[x^{\theta}(t), y^{\theta}(t)\right]^T$, $\mathbf{r}(t) = [x(t), y(t)]^T$ and $\mathbf{R}(\theta) = \begin{bmatrix} \cos\theta & -\sin\theta \\ \sin\theta & \cos\theta \end{bmatrix}$, the correlation functions in the new system are given by

$$\begin{bmatrix} r_{xx}^{\theta}(\Delta t) & r_{xy}^{\theta}(\Delta t) \\ r_{yx}^{\theta}(\Delta t) & r_{yy}^{\theta}(\Delta t) \end{bmatrix} = \begin{bmatrix} \cos\theta & -\sin\theta \\ \sin\theta & \cos\theta \end{bmatrix} \begin{bmatrix} r_{xx}(\Delta t) & r_{xy}(\Delta t) \\ r_{yx}(\Delta t) & r_{yy}(\Delta t) \end{bmatrix}, \tag{18}$$

which in general depend on the rotation angle θ.
However, it is remarkable that the difference of the two CCFs, $D_{CCF}(\Delta t) = r_{xy}^{\theta}(\Delta t) - r_{yx}^{\theta}(\Delta t)$, and the sum of the ACFs, $S_{ACF}(\Delta t) = r_{xx}^{\theta}(\Delta t) + r_{yy}^{\theta}(\Delta t)$, are invariant with respect to θ:

$$D_{CCF}(\Delta t) = 2D\frac{e^{-\phi|\Delta t|}}{\phi}\frac{\Omega}{\phi}S(\Delta t) \tag{19}$$

and

$$S_{ACF}(\Delta t) = 2D\frac{e^{-\phi|\Delta t|}}{\phi}\left[\left(\frac{\Omega^2 - \alpha^2\Delta\phi^2}{\Omega^2 - \Delta\phi^2} + \alpha^2\frac{\Delta\phi}{\phi}\right)C(\Delta t) + \alpha^2\frac{\Delta\phi^2}{\phi^2}S(|\Delta t|)\right]. \tag{20}$$

These functions are presented in Fig. 5(b).
Other two combinations of the correlation functions, which are also useful for the analysis of the experimental data, namely the sum of the CCFs, $S_{CCF}(\Delta t,\theta) = r_{xy}^{\theta}(\Delta t) + r_{yx}^{\theta}(\Delta t)$, and the difference of the ACFs, $D_{ACF}(\Delta t,\theta) = r_{xx}^{\theta}(\Delta t) - r_{yy}^{\theta}(\Delta t)$, depend on the choice of the reference frame:

$$S_{CCF}(\Delta t,\theta) = 2D\frac{e^{-\phi|\Delta t|}}{\phi}\alpha^2\frac{\Delta\phi^2}{\phi^2}\left(C(\Delta t) + S(|\Delta t|)\right)\left(\frac{\Omega}{\phi}\cos(2\theta) - \sin(2\theta)\right) \tag{21}$$

and

$$D_{ACF}(\Delta t,\theta) = -2D\frac{e^{-\phi|\Delta t|}}{\phi}\alpha^2\frac{\Delta\phi^2}{\phi^2}\left(C(\Delta t) + S(|\Delta t|)\right)\left(\frac{\Omega}{\phi}\sin(2\theta) + \cos(2\theta)\right). \tag{22}$$

In particular, they deliver information on the orientation θ of the coordinate system.

4.5 Torque detection using brownian fluctuations

We have seen that any force-field acting on a Brownian particle can be readily separated into its irrotational and rotational components. The last one, in particular, is completely defined by the value of the constant angular velocity Ω of the probe rotation around the z-axis. Such a rotation can be produced by the action of mechanical torque acting on the particle.
Once the value of Ω is known, the torque can be quantified. The constant angular velocity Ω results from a balance between the torque applied to the particle and the drag torque: $\tau_{\mathbf{drag}} = \mathbf{r} \times \mathbf{F}_{\mathbf{drag}} = \gamma\mathbf{r} \times \mathbf{v} = \gamma\mathbf{r} \times (\mathbf{r} \times \Omega)$, where $\mathbf{r}$ is the particle position and $\mathbf{v}$ is its linear velocity. Hence, the force acting on the particle from the torque source is given by $\mathbf{F} = \gamma\mathbf{r} \times \Omega$, which depends on the position of the particle. A time average of the torque exerted on the particle can then be expressed as

$$\langle\tau\rangle = \gamma\langle\mathbf{r} \times (\mathbf{r} \times \Omega)\rangle = \gamma\Omega\langle r^2\rangle \tag{23}$$

where $\langle r^2 \rangle$ is the mean square displacement of the sphere in the plane orthogonal to the torque.

With Eq. (23) we were able to measure torques in the range between 10s to 100s $fN\,\mu m$. The value of the measured torques (e.g. $4\,fN\,\mu m$ in Ref. (Volpe & Petrov, 2006)) is lower than the ones previously reported: $50\,fN\,\mu m$ for DNA twist elasticity (Bryant et al., 2003), $5\cdot 10^3\,fN\,\mu m$ for the movement of bacterial flagellar motors (Berry & Berg, 1997), $20\cdot 10^3\,fN\,\mu m$ for the transfer of orbital optical angular momentum (Volke-Sepulveda et al., 2002), or $5\cdot 10^2\,fN\,\mu m$ for the transfer of spin optical angular momentum (La Porta & Wang, 2004).

5. Data analysis workflow

The experimental position time-series need to be statistically analyzed in order to reconstruct all the parameters of the force-field, i.e. ϕ, $\Delta\phi$ and Ω, and the orientation of the coordinate system θ.

Supposing to have the probe position time-series in the experimental coordinate system $\mathbf{r}^e(t) = \left[x^e(t), y^e(t)\right]^T$, the data analysis procedure consists of three steps:

1. Evaluation of the parameters ϕ, $\Delta\phi$ and Ω;
2. Orientation of the coordinate system;
3. Reconstruction of the total force-field and subtraction of the trapping force-field to retrieve the external force-field under investigation.

In order to illustrate this method we proceed to analyze some numerically simulated data. The main steps of this analysis are presented in Fig. 6. In Fig. 6(a) the PDF is shown for the case of a probe in a force-field with the following parameters: $\phi = 37s^{-1}$, $\Delta\phi = 9.3s^{-1}$ (corresponding to $k_x = 43pN/\mu m$ and $k_y = 26pN/\mu m$), $\Omega = 0$ and $\theta = 30^\circ$. The PDF is ellipsoidal due to the difference of the stiffness along two orthogonal directions. In Fig. 6(b) the PDF for a force-field with the same ϕ, $\Delta\phi$ and orientation, but with $\Omega = 37s^{-1}$ is presented. The two time-series are chosen to have the same value of the parameters, except for Ω, in order to show not only how the method can obtain reliable estimates for the parameters, but also how it can distinguish between completely different physical situations, such as the absence or the presence of a non-conservative effect. The presence of the rotational component in the force-field produces two main effects. First, the PDF is more rotationally-symmetric and its main axes undergo a further rotation. Secondly, $D_{CCF}(\Delta t)$ is not null (Fig. 6(d)).

5.1 Estimation of the parameters

In order to evaluate the force-field parameters ϕ, $\Delta\phi$ and Ω the first step is to calculate the correlation matrix in the coordinate system where the experiment has been performed. $D_{CCF}(\Delta t)$ is invariant with respect to the choice of the reference system and it is different from zero only if $\Omega \neq 0$. The results are shown in Fig. 6(c) and Fig. 6(d) for the cases of the data shown in Fig. 6(a) and Fig. 6(b) respectively. The three aforementioned parameters can be found by fitting the experimental $D_{CCF}(\Delta t)$ to its theoretical shape.

When $\Omega = 0$, the $D_{CCF}(\Delta t)$ is null, as it can be seen also in Fig. 6(c), and, therefore, it cannot be used to find the two remaining parameters. For $\Omega = 0$, the other invariant function,

$S_{ACF}(\Delta t)$ can be used to evaluate ϕ and $\Delta\phi$. In general, $S_{ACF}(\Delta t)$ can also be used for the fitting of all the three parameters, but cannot give information on the sign of Ω, which must be retrieved from the sign of the slope at $\Delta t = 0$ of $D_{CCF}(\Delta t)$.

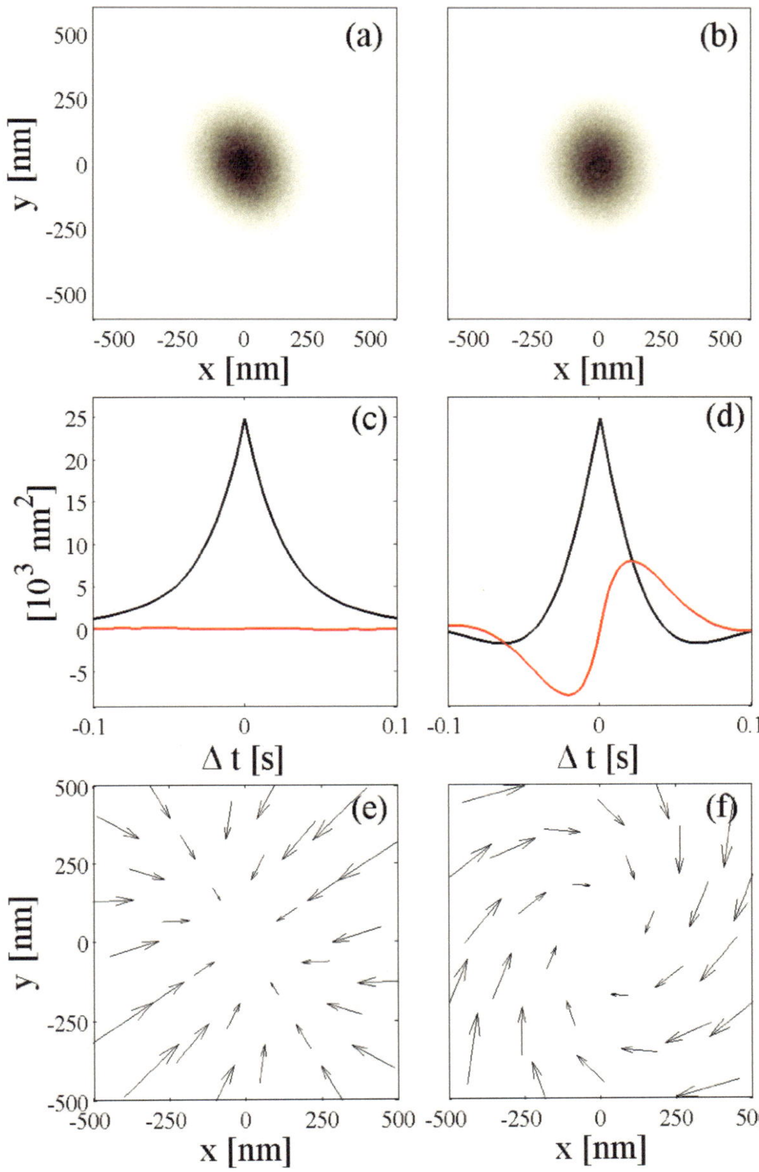

Fig. 6. Data analysis of numerically simulated time-series. (a-b) Probability density function for a Brownian particle under the influence of the force-field (simulated data $30s$ at $16kHz$); in (a) the force-field is purely conservative, while in (b) it has a rotational component. (c-d) Invariant function, $S_{CCF}(\Delta t)$ (black line) and $D_{CCF}(\Delta t)$ (red line) calculated from the simulated data and (e-f) reconstructed force-fields.

5.2 Orientation of the coordinate system

Although the values of the parameters ϕ, $\Delta\phi$ and Ω are now known, the directions of the force vectors are still missing. In order to retrieve the orientation of the experimental coordinate system, the orientation dependent functions $S_{CCF}(\Delta t,\theta)$ and $D_{ACF}(\Delta t,\theta)$ can be used. The best choice is to evaluate the two functions for $\Delta t = 0$, because the signal-to-noise ratio is highest at this point. The solution of this system delivers the value of the rotation angle θ:

$$\sin(2\theta) = \frac{\frac{\Omega}{\phi} D_{ACF}(0,\theta) - S_{CCF}(0,\theta)}{1 - 2D\frac{\alpha^2}{\phi}\frac{\Delta\phi}{\phi}\left(\frac{\Omega}{\phi}\right)^2}. \tag{24}$$

If $\Delta\phi = 0$, the value of θ is undetermined as a consequence of the PDF radial symmetry. In this case any orientation can be used. If $\Omega = 0$, the orientation of the coordinate system coincides with the axis of the PDF ellipsoid and, although Eq. (24) can still be used, the Principal Component Analysis (PCA) algorithm applied on the PDF is a convenient means to determine their directions.

5.3 Reconstruction of the force-field

Now everything is ready to reconstruct the unknown force-field acting on the probe around the equilibrium position. From the values of ϕ and $\Delta\phi$, the conservative forces acting on the probe result in $\mathbf{f}_c(x,y) = -(k_x x\mathbf{e}_x + k_y y\mathbf{e}_y)$ and, from the values of Ω, the rotational force is $\mathbf{f}_r(x,y) = \Omega(y\mathbf{e}_x - x\mathbf{e}_y)$. The total force-field is, therefore,

$$\mathbf{f}(x,y) = \mathbf{f}_c(x,y) + \mathbf{f}_r(x,y) = (-k_x x + \Omega y)\mathbf{e}_x + (-k_y y - \Omega x)\mathbf{e}_y \tag{25}$$

in the rotated coordinate system (Figs. 6(e) and 6(f)). Eq. (17) can be used to have the force-field in the experimental coordinate system. The unknown component can be easily reconstructed by subtraction of the known ones, such as the optical trapping force-field.

6. Applications: characterization of microscopic flows

The experimental characterization of fluid flows in micro-environments is important both from a fundamental point of view and from an applied one, since for many applications, such as lab-on-a-chip devices, it is required to assess the performance of microfluidic structures. Carrying out this kind of measurements can be extremely challenging. In particular, due to the small size of these environments, wall effects cannot be neglected. Additional difficulties arise studying biological fluids because of their complex rheological properties.

Following the data workflow presented in the previous section, the Brownian motion of an optically trapped polystyrene sphere in the presence of an external force-field generated by a fluid flow is analyzed (Volpe et al., 2008). Experimentally, two basic kinds of force-field –

namely a conservative force-field and a purely rotational one - are generated using solid spheres made of a birefringent material (Calcium Vaterite Crystals (CVC) spheres, radius $R = 1.5 \pm 0.2 \mu m$), which can be made spin through the transfer of light orbital angular momentum (Bishop et al., 2004).

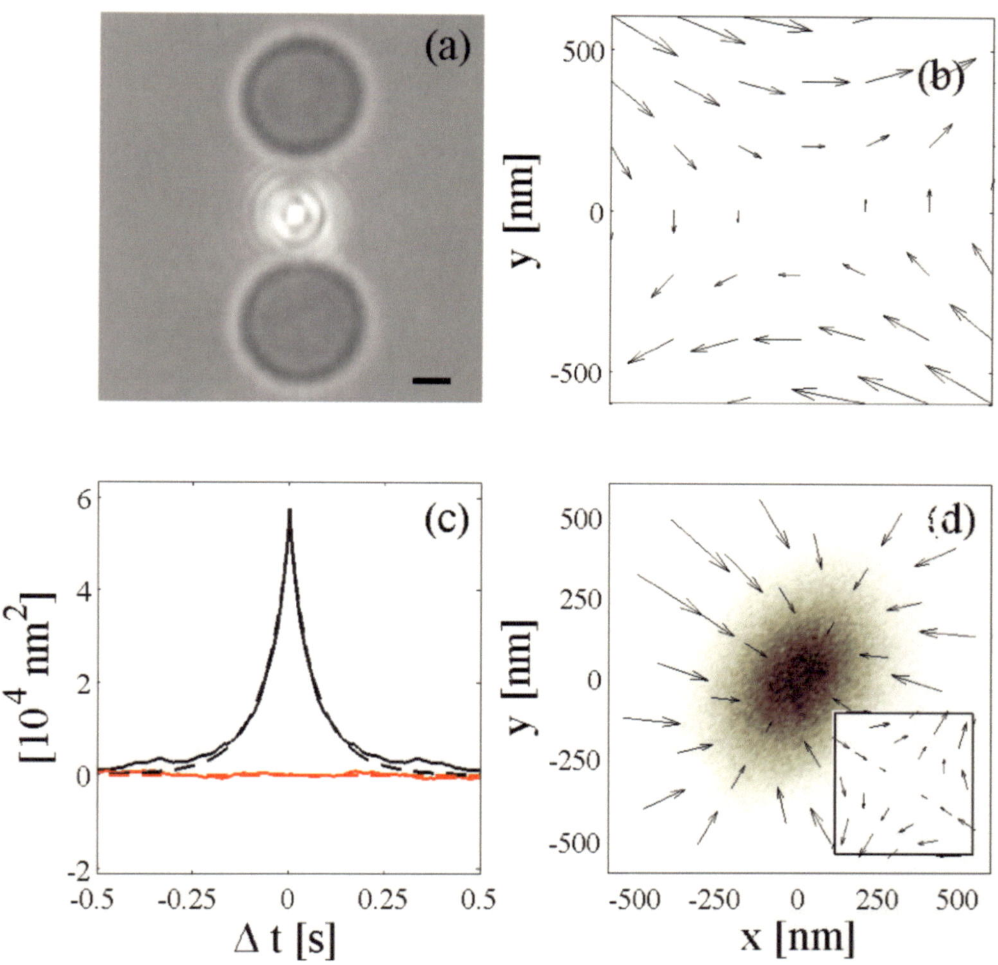

Fig. 7. Conservative force-field. (a) Experimental configuration with two spinning beads and (b) hydrodynamic component of the force-field (from hydrodynamic theory). (c) Experimental invariant functions $S_{ACF}(\Delta t)$ (black line) and $D_{CCF}(\Delta t)$ (red line) and their respective fitting to the theoretical shapes (dotted lines). (d) Experimental probability density function and reconstructed total force-field; inset: reconstructed hydrodynamic force-field.

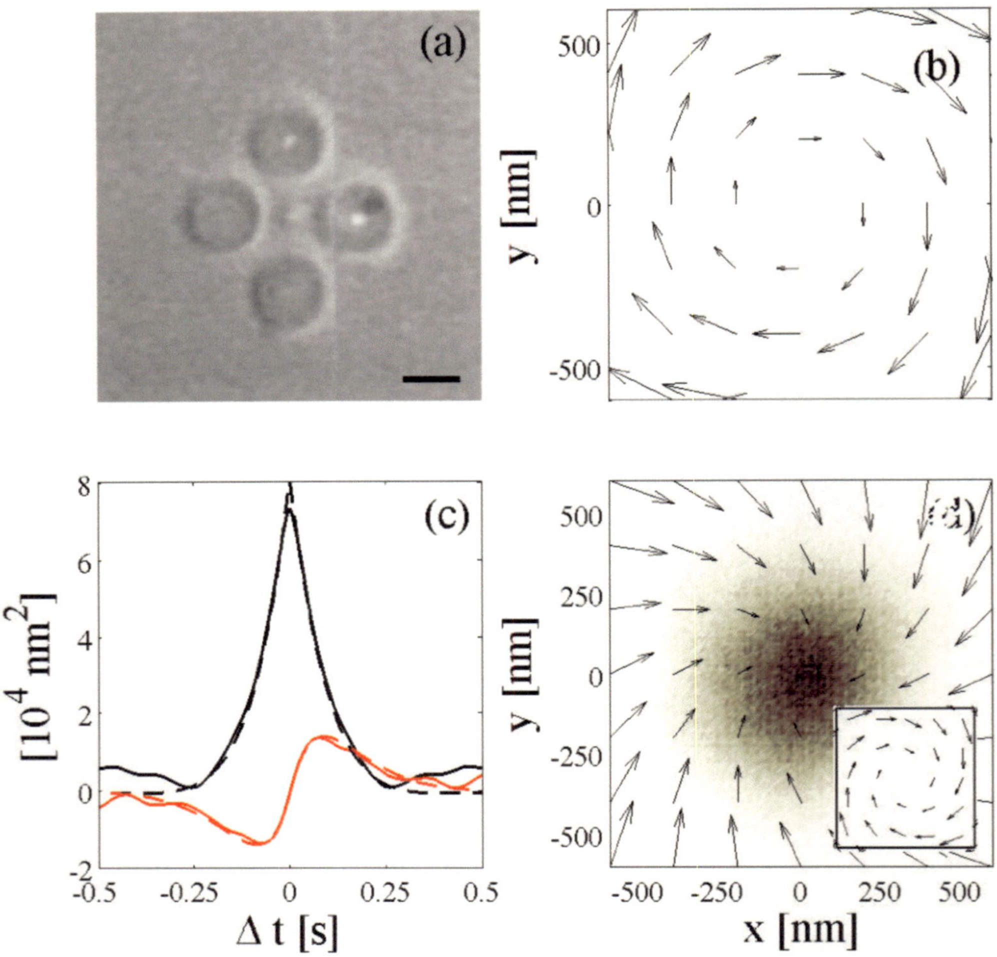

Fig. 8. Rotational force-field. (a) Experimental configuration with four spinning beads and (b) hydrodynamic component of the force-field (from hydrodynamic theory). (c) Experimental invariant functions $S_{ACF}(\Delta t)$ (black line) and $D_{CCF}(\Delta t)$ (red line) and their respective fitting to the theoretical shapes (dotted lines). (d) Experimental probability density function and reconstructed total force-field; inset: reconstructed hydrodynamic force-field.

6.1 Conservative force-field

In order to produce a conservative force-field, two CVCs are placed as shown in Fig. 7(a). In Fig. 7(b) the generated hydrodynamic force-field is presented as it is theoretically expected to be from hydrodynamic simulations. In Fig. 7(c), the invariant functions, $S_{ACF}(\Delta t)$ and $D_{CCF}(\Delta t)$ and their respective fitting to the theoretical shapes are presented. Since $D_{CCF}(\Delta t)$ is practically null, we see that in this case $\Omega = 0$, while the fitting to $S_{ACF}(\Delta t)$ allows one to find the values of $\phi = 18s^{-1}$ and $\Delta\phi = 6s^{-1}$. The value of the rotation of the coordinate system in this case is $\theta = 32^{\circ}$.

The total force-field can now be reconstructed: $k_x = 225 fN / \mu m$ and $k_y = 112 fN / \mu m$. This force-field is presented in Fig. 7(d). The hydrodynamic force-field can now be retrieved by subtracting the optical force-field ($k_{opt} = 118 fN / \mu m$ approximately constant in all directions), which can be measured in absence of rotation of the spinning particles (inset in Fig. 7(d)). This experimentally measured force-field corresponds very well to the theoretically predicted one (Fig. 7(b)).

6.2 Rotational force-field

In order to produce a rotational force-field, four CVCs are placed as shown in Fig. 11(a), which should theoretically produce the force-field presented in Fig. 8(b). In Fig. 8(c), the invariant functions, $S_{ACF}(\Delta t)$ and $D_{CCF}(\Delta t)$ and their respective fitting to the theoretical shapes are presented. Now $D_{CCF}(\Delta t)$ is not null any more and, therefore, it can be used to fit the three parameters: $\phi = 11s^{-1}$, $\Delta\phi \approx 0$ and $\Omega = 5 rad\, s^{-1}$. As already mentioned, $S_{ACF}(\Delta t)$ can be used for this purpose as well; however, using the latter, the sign of Ω stays undetermined.

The total force-field can now be reconstructed: $k_x \approx k_y = 100 fN / \mu m$. This force-field is presented in Fig. 8(d). The hydrodynamic force-field can be obtained by subtracting the optical force-field ($k_{opt} = 78 fN / \mu m$ approximately constant in all directions), which can be measured in absence of rotation of the spinning particles (inset in Fig. 8(d)). Again, this experimentally measured force-field corresponds very well to the theoretically predicted one (Fig. 8(b)).

7. Applications: measuring orbital angular momentum

A beam that carries an orbital angular momentum can induce a torque on the particle, which can be measured using the PTM. In the experiment fully presented in Ref. (Volpe & Petrov, 2006) the probe is trapped by a $785nm$ beam (Fig. 9(a)). We produced the Laguerre-Gaussian (LG) beam from a linearly polarized $532nm$ CW beam transformed by a holographic mask that generates a $l = 10$ order LG beam of power $P = 3mW$. The beam carries the flux of orbital angular momentum $\tau_{OAM} = lP / \omega = 7.5 \cdot 10^{-18} Nm$, where $\omega = 4 \cdot 10^{15} Hz$ is the light frequency. Only about 1% of the total flux of the orbital angular momentum is transferred to the dielectric sphere and only in the portion of the beam that overlaps the sphere. We estimate that in our experiments the beam transfers the torque $\tau_S = 7.5 \cdot 10^{-21} Nm$ to the sphere. A Dove prism can be inserted in the optical path to reverse the handedness of the helical phase front and therefore the sign of the orbital angular momentum without changing the direction or polarization of the beam.

In the absence of the $785nm$ optical trap (Fig. 9(b)) the $532nm$ beam moves the sphere towards the upper coverslip due to its radiation pressure and the gradient forces in the perpendicular plane do not allow the sphere to escape. In such conditions we can observe the clockwise and anti-clockwise rotation of the sphere depending on the handedness of the beam.

When the trapping $785nm$ beam power and therefore the trap stiffness are increased ($100\mu W, 16 fN/\mu m$), the sphere is more confined to the centre of the trapping beam (Fig.

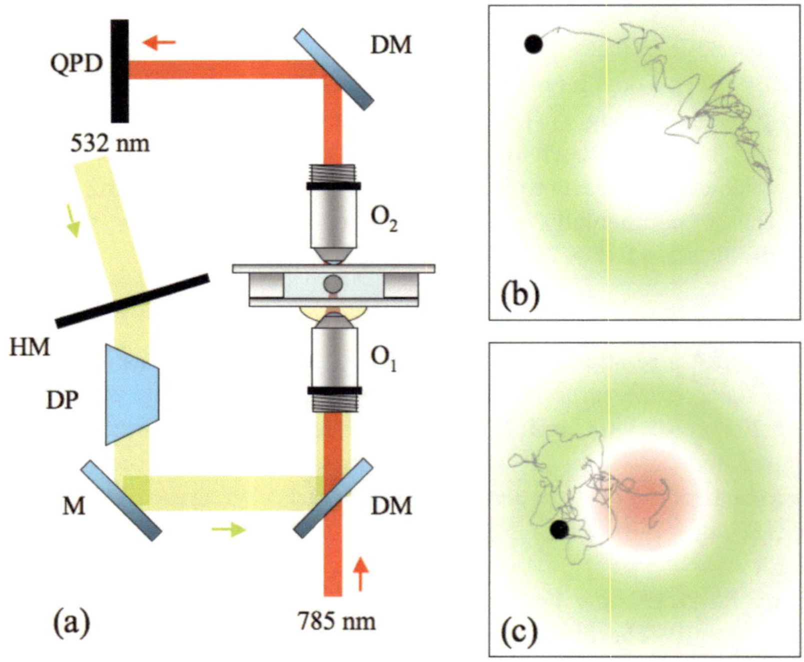

Fig. 9. Torque measurement setup. (a) Experimental setup: HM holographic mask, DP Dove prism, M mirror, DM dichroic mirror, O_1 focusing objective, O_2 collector objective, QPD quadrant photodetector. (b) Brownian motion of the sphere in the chamber when only the 532*nm* LG propagates. (c) Brownian motion of the sphere in the chamber when both the 532*nm* and the 785*nm* beams propagate.

9(c)) and does not display a rotational motion. We notice that in this case the Fourier analysis of the experimental traces does not show the presence of the torque existing in the system. However, the behaviour of the ACF and CCF functions near $\Delta t = 0$ (Fig. 10) unambiguously shows that the torque produced by the orbital angular momentum of the LG beam still affects the Brownian trajectories. Fitting these experimental functions to the theoretical ones and calculating the value of $\langle r^2 \rangle$ from the traces, we are able to measure the torque acting on the particle as $4.9 \pm 0.7 \cdot 10^{-21} Nm$.

8. Applications: non-conservative radiation forces in an optical trap

Back in 1992, Ashkin already pointed out that, in principle, scattering forces in optical tweezers do not conserve mechanical energy, and that this could have some measurable consequences (Ashkin, 1992). In particular, this nonconservative force would produce a dependence of the axial equilibrium position of a trapped microsphere as a function of its transverse position in the trapping beam (Fig. 11(a) and Fig. 10(c) of Ref. (Ashkin, 1992)); such prediction was first confirmed by Merenda and colleagues (Merenda et al., 2006). Recently, Roichman and colleagues (Roichman et al., 2008) have directly investigated the non-conservative component and discussed the implications that this might have for optical-tweezers-based experiments making use of the thermal fluctuations in the calibration procedure.

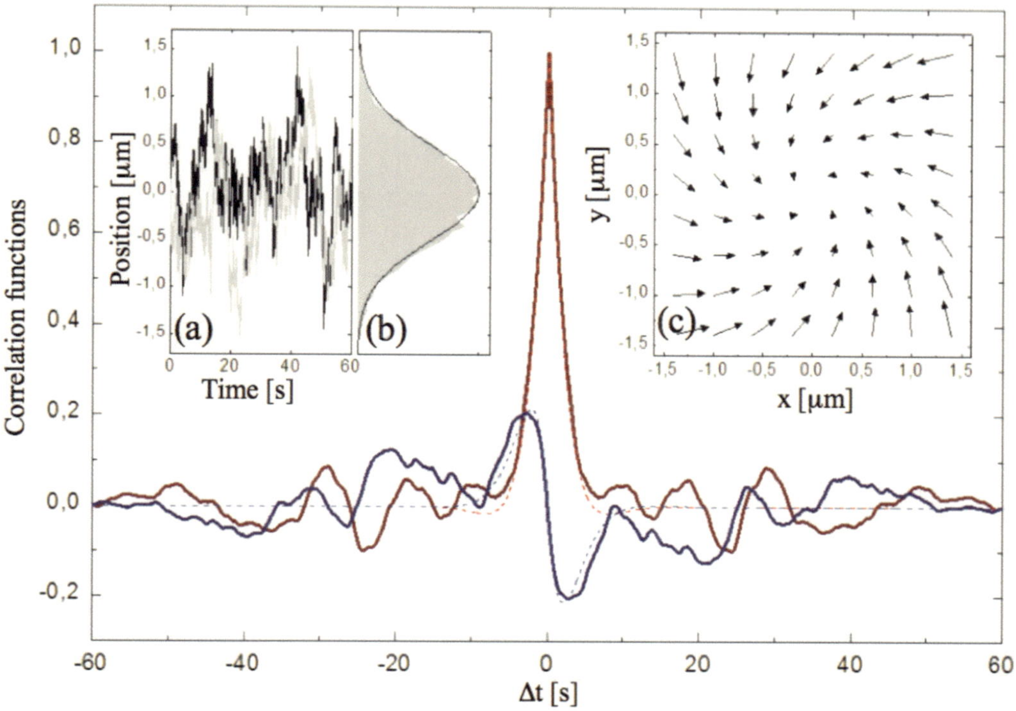

Fig. 10. Experimental ACF and CCF. Experimental ACF and CCF functions in presence of the torque induced by a LG beam with $l = +10$. The trap force constant k is high enough to confine the sphere. The continuous lines show the mean values obtained using five series of data acquisition (acquisition time $60s$, sampling rate $f_s = 1kHz$). The dotted lines show the fitting to the theoretical shape (the fitting was made on the central part of the curve for $\Delta t = [-2s, 2s]$). In the insets: (a), time traces for the x (black) and y (grey) coordinates; (b), histogram of the x coordinate and in black the fitting to a Gaussian distribution; and (c), vector force-field acting on the particle in the xy -plane.

Using the PTM we have evaluated the relative weight of the non-conservative component of the optical forces (Pesce et al., 2009). We used the PTM to analyze various optically trapped particles in different trapping conditions. The main result is that the non-conservative effects are effectively negligible and do not affect the standard calibration procedure, unless for extremely low-power trapping, far away from the trapping regimes usually used in experiments.

In Fig. 12, the ACFs and $D_{CCF}(\Delta t)$ are presented for a $0.45\mu m$ diameter particle held in an optical trap with an optical power at the sample of $6.0mW$. In order to have a non-vanishing $D_{CCF}(\Delta t)$, which is the signature of a rotational component of the force-field, the optical power at the sample needed to be reduced down to a few milliwatts. Notice that for this experiment the $D_{CCF}(\Delta t)$ is taken between the radial ($r = \sqrt{x^2 + y^2}$) and axial (z) coordinate. Furthermore, even under such low power a clearly non-vanishing $D_{CCF}(\Delta t)$ was

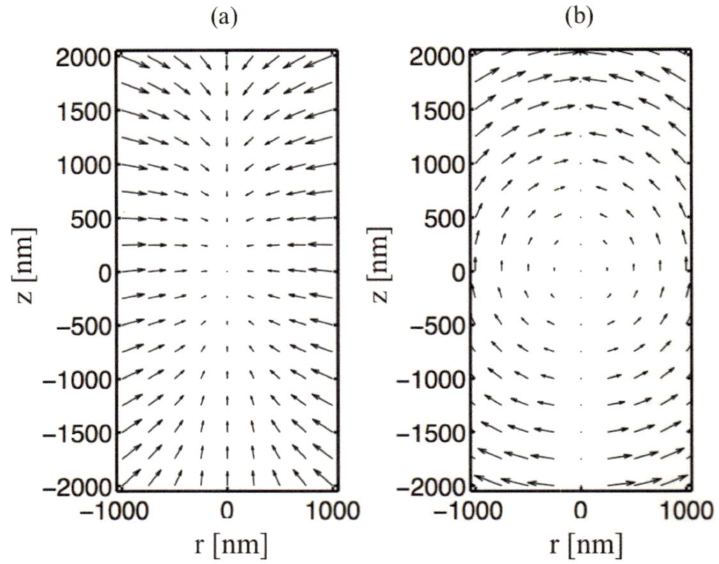

Fig. 11. Non-conservative radiation forces. (a) Force-field generated by an optical trap in the presence of a rotational component and (b) the rotational part of the force-field in the *rz*-plane. Note that for clarity of presentation the relative contribution of the non-conservative force-field has been exaggerated.

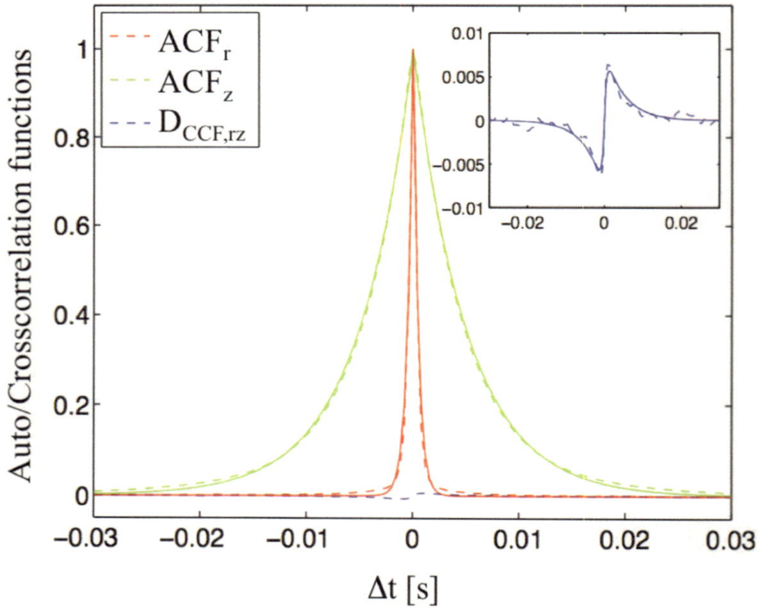

Fig. 12. Quantifying non-conservative radiation forces. ACF and $D_{CCF}(\Delta t)$ for a $0.45\,\mu m$ diameter colloidal particle optically trapped with a laser power of $6.0\,mW$ at the sample. Inset: Closeup of the $D_{CCF}(\Delta t)$. Dashed lines represent experimental data, while solid lines are the curves obtained from the fit.

only obtained when the acquisition time was increased up to $400s$. It is an important remark the fact that, due to the extremely low value of the rotational contribution to the total force-field, it is not immediately evident from the traces of the particle motion. Indeed the particle undergoes a random movement in the rz-plane, where it is not possible to distinguish the presence of a rotational component without the aid of a statistical analysis such as the one we propose.

9. Conclusion

The PTM technique can be applied to the detection of locally nonhomogeneous force-fields. This is achieved by analyzing the ACFs and CCFs of the probe position time series. We believe that this technique can help to gain new insights into microscale and molecular-scale phenomena. In these cases the presence of the Brownian motion is intrinsic and cannot be disregarded. Therefore this technique permits one to take advantage of the Brownian fluctuations of the probe in order to explore the force-field present in its surroundings. One of the most remarkable advantages of the technique we propose is that it can be implemented in all existing PFM setups and even on data acquired in the past. Indeed, it does not require changes to be made in the physical setup, but only to analyze the data in a new way. This method can even be applied to the study of the Brownian particle trajectories that can be obtained with techniques different from the PFM technique.

10. References

Ashkin, A. (1970). Acceleration and trapping of particles by radiation pressure. *Phys. Rev. Lett.*, 24, 156–159.

Ashkin, A (1992). Forces of a single-beam gradient laser trap on a dielectric sphere in the ray optics regime. *Biophys. J.*, 61, 569–582.

Ashkin, A. (2000). History of optical trapping and manipulation of small- neutral particle, atoms, and molecules. *IEEE J. Sel. Top. Quant. El.*, 6, 841–856.

Ashkin, A.; Dziedzic, J. M.; Bjorkholm, J. E. & Chu, S. (1986). Observation of a single-beam gradient force optical trap for dielectric particles, *Opt. Lett.*, 11, 288–290.

Babic, D.; Schmitt, C. & Bechinger, C. (2005). Colloids as model systems for problems in statistical physics. *Chaos*, 15, 026114.

Berg-Sørensen, K. & Flyvbjerg, H. (2004). Power spectrum analysis for optical tweezers, *Rev. Sci. Instrumen.*, 75, 594–612.

Berry, R. M. & Berg, H. C. (1997). Absence of a barrier to backwards rotation of the bacterial flagellar motor demonstrated with optical tweezers. *PNAS*, 94, 14433–14437.

Beth, R. A. (1936). Mechanical detection and measurement of the angular momentum of light. *Phys. Rev.*, 50, 115–125.

Bishop, A. I.; Nieminen, T. A.; Heckenberg, N. R. & Rubinsztein-Dunlop, H. (2004). Optical microrheology using rotating laser-trapped particles. *Phys. Rev. Lett.*, 92, 198104.

Block, S. M.; Goldstein, L. S. B. & Schnapp, B. J. (1990). Bead movement by single kinesin molecules studied with optical tweezers. *Nature*, 348, 348–352.

Bryant, Z.; Stone, M. D.; Gore, J.; Smith, S. B.; Cozzarelli, N. R. & Bustamante, C. (2003). Structural transitions and elasticity from torque measurements on DNA. *Nature*, 424, 338–341.

Bustamante, C.; Marko, J. F.; Siggia, E. D. & Smith, S. (1994). Entropic elasticity of lambda-phage DNA. *Science*, 265, 1599–1600.

Finer, J. T.; Simmons, R. M. & Spudich, J. A. (1994). Single myosin molecule mechanics: piconewton forces and nanometre steps. *Nature*, 368, 113–119.

Ghislain, L. P. & Webb, W. W. (1993). Scanning-force microscope based on an optical trap, *Opt. Lett.*, 18, 1678–1680.

Happel, J. & Brenner, H. (1983). Low Reynolds Number Hydrodynamics. Spinger, New York.

Huisstede, J. H. G.; van der Werf, K. O.; Bennink, M. L. & Subramaniam, V. (2005). Force detection in optical tweezers using backscattered light. *Opt. Express*, 13, 1113–1123.

La Porta, A. & Wang, M. D. (2004). Optical torque wrench: Angular trapping, rotation, and torque detection of quartz microparticles. *Phys. Rev. Lett.*, 92, 190801.

Lebedev, P. N. (1901). The experimental study of the pressure of the light. *Ann. Physik*, 6, 433–459.

Merenda, F.; Boer, G.; Rohner, J.; Delacrétaz, G. & Salathé, R.-P. (2006). Escape trajectories of single-beam optically trapped micro-particles in a transverse fluid flow. *Opt. Express*, 14, 1685–1699.

Metcalf, H. J. & van der Straten, P. (1999). Laser Cooling and Trapping. Springer, New York.

Neuman, K. C. & Block, S. M. (2004). Optical trapping, *Rev. Sci. Instrumen.*, 75, 2787–2809.

Nichols, E. F. & Hull, G. F. (1901). A preliminary communication on the pressure of heat and light radiation. *Phys. Rev.*, 13, 307–320.

Novotny, L.; Bian, R. X. & Xie, X. S. (1997). Theory of nanometric optical tweezers. *Phys. Rev. Lett.*, 79, 645–648.

Pesce, G.; Volpe, G.; De Luca, A. C.; Rusciano, G. & Volpe, G. (2009). Quantitative assessment of non-conservative radiation forces in an optical trap, *EPL*, 86, 38002.

Quidant, R.; Petrov, D. V. & Badenes, G. (2005). Radiation forces on a Rayleigh dielectric sphere in a patterned optical near field. *Opt. Lett.*, 30, 1009–1011.

Rohrbach A. & Stelzer, E. H. K. (2002). Three-dimensional position detection of optically trapped dielectric particles. *J. Appl. Phys.*, 91, 5474–5488.

Roichman, Y.; Sun, B.; Stolarski, A. & Grier, D. G. (2008). Influence of Nonconservative Optical Forces on the Dynamics of Optically Trapped Colloidal Spheres: The Fountain of Probability, *Phys. Rev. Lett.*, 101, 128301.

Townes, C. H. (1999). How The Laser Happened. Oxford University Press, Oxford, UK.

Visscher, K.; Gross, S. P. & Block, S. M. (1996). Construction of multiple-beam optical traps with nanometer-resolution position sensing. *IEEE J. Sel. Top. Quant. El.*, 2, 1066–1076.

Volke-Sepulveda, K.; Garcés-Chávez, V.; Chávez-Cerda, S. ; Arlt, J. & Dholakia, K. (2002). Orbital angular momentum of a high-order bessel light beam. *J. Opt. B*, 4, S82–S89.

Volpe, G. & Petrov, D. (2006). Torque Detection using Brownian Fluctuations, *Phys. Rev. Lett.*, 97, 210603.

Volpe, G.; Quidant, R.; Badenes, G. & Petrov, D. (2006). Surface plasmon radiation forces. *Phys Rev. Lett.*, 96, 238101.

Volpe, G.; Volpe, G. & Petrov, D. (2007a). Brownian motion in a non-homogeneous force field and photonic force microscope, *Phys. Rev. E*, 76, 061118.

Volpe, G.; Kozyreff, G. & Petrov, D. (2007b). Back-scattering position detection for photonic force microscopy, *J. Appl. Phys.*, 102, 084701.

Volpe, G.; Volpe, G. & Petrov, D. (2008). Singular point characterization in microscopic flows, *Phys. Rev. E*, 77, 037301.

21

Dynamics of a Kerr Nanoparticle in a Single Beam Optical Trap

Romeric Pobre[1] and Caesar Saloma[2]
[1]*Physics Department, CENSER, College of Science, De La Salle University-Manila*
[2]*National Institute of Physics, University of the Philippines-Diliman*
Philippines

1. Introduction

Single beam optical traps also known as optical tweezers, are versatile optical tools for controlling precisely the movement of optically-small particles. Single-beam trapping was first demonstrated with visible light (514 nm) in 1986 to capture and guide individual neutral (nonabsorbing) particles of various sizes (Ashkin *et. al.*, 1986). Optical traps were later used to orient and manipulate irregularly shaped microscopic objects such as viruses, cells, algae, organelles, and cytoplasmic filaments without apparent damage using an infrared light (1060 nm) beam (Ashkin, 1990). They were later deployed in a number of exciting investigations in microbiological systems such as chromosome manipulation (Liang et.al., 1993), sperm guidance in all optical *in vitro* fertilization (Clement-Sengewald *et.al.*,1996) and force measurements in molecular motors such single kinesin molecules (Svoboda and Block, 1994) and nucleic acid motor enzymes (Yim *et.al.*, 1995). More recently, optical tweezer has been used in single molecule diagnostics for DNA related experiments (Koch *et.al.*, 2002). By impaling the beads onto the microscope slide and increasing the laser power, it was tested that the bead could be "spot-welded" to the slide, leaving the DNA in a stretched state- a technique was used in preparing long strands of DNA for examination via optical microscopy.

Researchers continue to search for ways to the capability of optical traps to carry out multi-dimensional manipulation of particles of various geometrical shapes and optical sizes (Grier, 2003; Neuman & Block, 2004). Efforts in optical beam engineering were pursued to generate trapping beams with intensity distributions other than the diffraction-limited beam spot e.g. doughnut beam (He *et.al.*, 1995; Kuga *et.al.*, 1997), helical beam (Friese *et.al.*, 1998), Bessel beam (MacDonald *et.al.*, 2002). Multiple beam traps and other complex forms of optical landscapes were produced from a single primary beam using computer generated holograms (Liesener *et.al.*, 2000; Curtis *et.al.*, 2002; Curtis *et.al.*, 2003) and programmable spatial light modulators (Rodrigo *et.al.*, 2005; Rodrigo *et.al.*, 2005).

Knowing the relationship between characteristics of the optical trapping force and the magnitude of optical nonlinearity is an interesting subject matter that has only been lightly investigated. A theory that accurately explains the influence of nonlinearity on the behavior of nonlinear particles in an optical trap would significantly broaden the applications of optical traps since most materials including many proteins and organic molecules, exhibit considerable degrees of optical nonlinearity under appropriate excitation conditions (Lasky, 1997; Clays *et.al.*, 1993; Chemla & Zyss, 1987; Prasad & Williams, 1991; Nalwa & Miyata, 1997). One possible reason for the apparent scarcity of published studies on the matter is the difficulty in finding a suitable strategy for computing the intensity-dependent refractive index of the particle under illumination by a focused optical beam.

We have previously studied the dynamics of a particle in an optical trap that is produced by a single tightly focused continuous-wave (CW) Gaussian beam in the case when the refractive index n_2 of the particle is dependent on the intensity I (Kerr effect) of the interacting linearly polarized beam according to: $n_2 = n_2^{(0)} + n_2^{(1)}\mathbf{E}^*\mathbf{E}$, where $n_2^{(0)}$ and $n_2^{(1)}I$ are the linear and nonlinear components of n_2, respectively. We have calculated the (time-averaged) optical trapping force that is exerted by a focused TEM_{00} beam of optical wavelength λ on a non-absorbing mechanically-rigid Kerr particle of radius a in three different value ranges of the size parameter α: (1) $\alpha = 2\pi a/\lambda >>100$ geometric optics (Pobre & Saloma, 1997), (2) $\alpha \approx 100$ Mie scattering (Pobre & Saloma, 2002), and (3) $\alpha << 100$ Rayleigh scattering regime (Pobre & Saloma, 2006; Pobre & Saloma, 2008).

Here we continue our effort to understand the characteristics of the (time-averaged) optical trapping force $\mathbf{F_{trap}}$ that is exerted on a Kerr particle by a focused CW TEM_{00} beam in the case when $a \leq 50\lambda/\pi$. A nanometer-sized Kerr particle (bead) exhibits Brownian motion as a result of random collisions with the molecules in the surrounding liquid. The Brownian motion is no longer negligible and has to be into account in the trapping force analysis. The characteristics of the trapping force are determined as a function of particle position in the propagating focused beam, beam power and focus spot size, ω_0, a, and relative refractive index between the nanoparticle and its surrounding medium. The behavior of the optical trapping force is compared with that of a similarly-sized linear particle under the same illumination conditions.

The incident focused beam polarizes the non-magnetic Kerr nanoparticle ($a << \lambda$) and the electromagnetic (EM) field exerts a Lorentz force on each charge of the induced electric dipole (Kerker, 1969). We derive an expression for $\mathbf{F_{trap}}$ in terms of the intensity distribution and the nanoparticle polarizability $\alpha = \alpha(n_1, n_2)$, where n_2 and n_1 are the refractive index of the Kerr nanoparticle and surrounding medium, respectively. Optical trapping force ($\mathbf{F_{trap}}$) has two components, one that accounts for the contribution of the field gradient and the other from the light that is scattered by the particle. The two-component approach for computing the magnitude and direction of $\mathbf{F_{trap}}$ was previously used on linear dielectric nanoparticles in arbitrary electromagnetic fields (Rohrbach & Steltzer, 2001). We also mention that the calculation of the intensity distributions near Gaussian beam focus is corrected up to the fifth order (Barton & Alexander, 1989).

In the next section, we will show the equation of the motion of a Kerr nanoparticle near the focus of a single beam optical trap in a Brownian environment. Simulation results will be presented and discussed in detail for other sections.

2. Theoretical framework

A linearly polarized Gaussian beam (TEM_{00} mode) of wavelength λ, is focused via an objective lens of numerical aperture NA and allowed to propagate along the optical z-axis in a linear medium of refractive index n_1 (see Fig **1**). The beam radius ω_o at the geometrical focus (x = y = z = 0) is: $\omega_o = \lambda/(2NA)$.

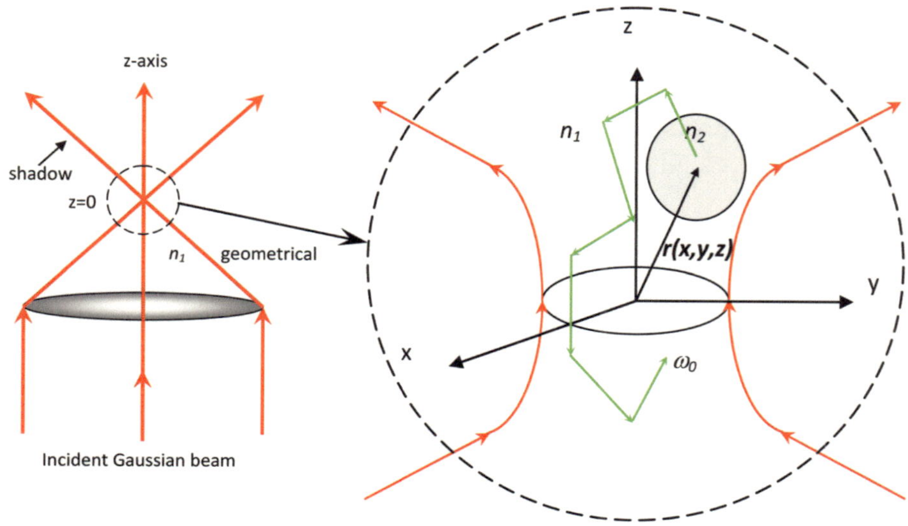

Fig. 1. Nonlinear nanoparticle of radius a and refractive index n_2 is located near the focal volume of a tightly-focused Gaussian beam of wavelength $\lambda >> a$ and beam focus radius ω_o. Gaussian beam propagates in a linear medium of index n_1. Nanoparticle center is located at r(x, y, z) from the geometrical focus at r(0, 0, 0). Enlarged figure in the focal volume shows Kerr nanoparticle undergoing Brownian motion near the focus.

The focused beam interacts with a Kerr particle of radius $a \leq 50\lambda/\pi$. The refractive index $n_2(\boldsymbol{r})$ of the Kerr particle is given by: $n_2(\boldsymbol{r}) = n_2^{(0)} + n_2^{(1)} I(\boldsymbol{r})$, where $I(\boldsymbol{r}) = \boldsymbol{E}^*(\boldsymbol{r})\boldsymbol{E}(r)$ is the beam intensity at particle center position $\boldsymbol{r} = \boldsymbol{r}(x, y, z)$ from the geometrical focus at $\mathbf{r} = 0$ which also serves as the origin of the Cartesian coordinate system. Throughout this paper, vector quantities represented in bold letters.

The thermal fluctuations in the surrounding medium (assumed to be water in the present case) become relevant when the particle size approaches the nanometer range. We consider a Kerr nanoparticle that is located at **r** above the reference focal point in the center of the beam waist ω_0 that is generated with a high NA oil-immersed objective lens of an inverted microscope - the focused beam propagates in the upward vertical direction (see inset **Fig. 1**). The dynamics of the Kerr nanoparticle as it undergoes thermal diffusion can be analyzed in the presence of three major forces: (1) Drag force, $\mathbf{F}_{drag}(d\mathbf{r}/dt) = \mathbf{F}_{drag}$, that is experienced when the particle is in motion, (2) Trapping force $\mathbf{F}_{trap}(\mathbf{r})$, which was derived in (Pobre & Saloma, 2006), and (3) time-dependent Brownian force $\mathbf{F}_{fluct}(t) = \mathbf{F}_{fluct}$, that arise from thermal motion of the molecules in the liquid. The Kerr nanoparticle experiences a net force $\mathbf{F}_{net}(\mathbf{r}, t) = \mathbf{F}_{net}$, that can be expressed in terms of the Langevin equation as:

$$\begin{aligned} \mathbf{F}_{net}(\mathbf{r},t) &= \mathbf{F}_{drag}(\dot{\mathbf{r}}) + \mathbf{F}_{trap}(\mathbf{r}) + \mathbf{F}_{fluct}(t) \\ m\ddot{\mathbf{r}} &= -\gamma\dot{\mathbf{r}} + \mathbf{F}_{trap}(\mathbf{r}) + \mathbf{F}_{fluct}(t) \end{aligned} \tag{1}$$

where: $\mathbf{F}_{drag} = -\gamma\, d\mathbf{r}/dt$, and γ is the drag coefficient of the surrounding liquid. According to Stokes law, $\gamma = 6\pi\eta a$, where η is the liquid viscosity. While the optical trapping force or optical trapping force, $\mathbf{F}_{trap}(\mathbf{r})$, on the Kerr nanoparticle was shown to be (Pobre & Saloma, 2006):

$$\mathbf{F}_{trap}(\mathbf{r}) = \left(\frac{2\pi n_1 a^3}{c}\right)\left(\frac{\left\{\frac{n_2^{(0)}+n_2^{(0)}I(r)}{n_1}\right\}^2-1}{\left\{\frac{n_2^{(0)}+n_2^{(0)}I(r)}{n_1}\right\}^2+2}\right)^2 \nabla I(r) + \left(\frac{8\pi n_1}{3c}\right)(|\mathbf{k}|a)^4 a^2 \left(\frac{\left\{\frac{n_2^{(0)}+n_2^{(0)}I(r)}{n_1}\right\}^2-1}{\left\{\frac{n_2^{(0)}+n_2^{(0)}I(r)}{n_1}\right\}^2+2}\right)^2 I(r) \quad (2)$$

Equation (2) reveals that $\mathbf{F}_{trap}$ consists of two components. The first component represents the gradient force and depends on the gradient of $I(\mathbf{r})$ and it is directed towards regions of increasing intensity values. The second component represents the contribution of the scattered light to $\mathbf{F}_{trap}$. The scattering force varies with I(**r**) and it is in the direction of the scattered field. Hence, the relative contribution of the scattering force to $\mathbf{F}_{trap}$ is weak for a particle that scatters light in an isotropic manner.

The Gaussian beam has a total beam power of P (Siegman, 1986) and its intensity distribution I(**r**) near the beam focus is calculated with corrections introduced up to the fifth-order (Barton & Alexander, 1989). Focusing with a high *NA* objective produces a relatively high beam intensity at $z = 0$, which decreases rapidly with increasing $|z|$ values. On the other hand, low *NA* objectives produce a slowly varying intensity distribution from z=0.

The molecules of the surrounding fluid affect significantly on the mobility of the Kerr nanoparticle since their sizes are comparable. As a result, the Kerr nanoparticle moves in a random manner between the molecules and exhibits the characteristics of a Brownian motion. The associated force can be generated via a white-noise simulation since it mimics the behavior of the naturally occurring thermal fluctuations of a fluid. The assumption holds when both the liquid and the Kerr nanopartilcle are non-resonant with λ. Localized (non-uniform) heating of the liquid is also minimized by keeping the average power of the focused beam low for example with a femtosecond laser source that is operated at high peak powers and relatively low repetition rate.

3. Optical trapping potential

As previously discussed, the Kerr nanoparticle of mass m and $2\pi a/\lambda \leq 100$ and $a << \lambda$, exhibits random (Brownian) motion in the liquid (Rohrbach & Steltzer, 2002; Singer *et.al.*, 2000). The thermal fluctuation probability increases with the temperature T of the liquid. To determine the dynamics of a Kerr nanoparticle near the focus of a single beam optical trap, we first determine the potential energy $V(\mathbf{r})$ of the optical trap near the beam focus, which can be characterized in terms of $\mathbf{F}_{\mathbf{trap}}$. The potential $V(\mathbf{r})$ as a function of the optical trapping force from all axes (in this case along the x, y, and z axes) is given by:

$$\begin{aligned} V(r) &= -\int_{\mathrm{r}_0}^{\mathrm{r}_\mathrm{f}} \mathbf{F}_{trap}(\mathbf{r})d\mathbf{r} \\ &= -\int_{\mathrm{x}_0}^{\mathrm{x}_\mathrm{f}} \mathbf{F}_{trap,\mathbf{x}}(\mathbf{r})d\mathrm{x} - \int_{\mathrm{y}_0}^{\mathrm{y}_\mathrm{f}} \mathbf{F}_{trap,\mathbf{y}}(\mathbf{r})d\mathrm{y} - \int_{\mathrm{z}_0}^{\mathrm{z}_\mathrm{f}} \mathbf{F}_{trap,\mathbf{z}}(\mathbf{r})d\mathrm{z} \end{aligned} \quad (3)$$

where: $\mathbf{F}_{trap,x}$, $\mathbf{F}_{trap,y}$ and $\mathbf{F}_{trap,z}$ are the Cartesian components of $\mathbf{F}_{trap}$, and $\mathbf{r}_0(x_0, y_0, z_0; t_0) = \mathbf{r}_0(t_0)$ and $\mathbf{r}_f(x_f, y_f, z_f; t_f) = \mathbf{r}_f(t_f)$ are the initial and final positions of the nanoparticle. For a nanoparticle in the focal volume of a Gaussian beam, V(**r**) can be approximated as a harmonic potential since the magnitude of $\mathbf{F}_{drag}$ is several orders larger than that of the inertial force. Equation (1) then describes an over-damped harmonic motion that is driven by time-dependent thermal fluctuations.

A nanoparticle at location **r**(t) in the optical trap has a potential energy V(**r**) and a kinetic energy $m|\mathbf{v}|^2/2$ where **v** = **v**(t) is the nanoparticle velocity. The probability that the Kerr nanoparticle is found at position **r**(t), is described by a probability density function $\Pi(\mathbf{r}) = \Pi_0 \exp[-V(\mathbf{r})/k_B T]$, where Π_0 is the initial probability density, T is the temperature of the surrounding medium, and k_B is the Boltzmann constant.

Figure 2 plots the potential energy (2a) of the optical trap and the corresponding time-dependent displacement trajectory (2b) of the Kerr nanoparticle (initial z position = 0.4 μm) along the optical z-axis assuming a zero initial velocity and a room temperature condition of 3.1 k_bT background energy of the surrounding medium. The trajectory (in blue trace) can be ascribed as overdamped oscillations of the Kerr nanoparticle that arise from the complex interplay of three forces indicated in the Langevin's differential equation. The oscillations

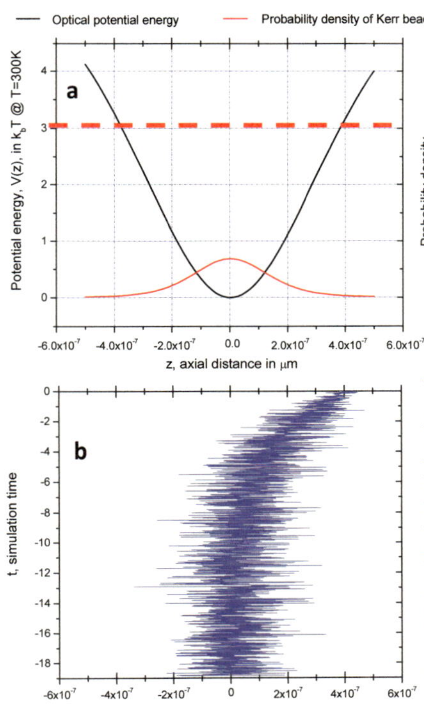

Fig. 2. **(a)** Potential energy and probability density function along the z-axis with trapping input parameters: z_o=0, p=100mW, a=30nm, *N.A.*=1.2, λ =1.064μm, n_1=1.33 , $n_2^{(0)}$=1.4, and $n_2^{(1)}$=1.8 x 10^{-12}m²/W. **(b)** Thermal diffusion of the Kerr nanoparticle along the z-axis with zero initial velocity at 0.4 μm with a 3.1 k_bT ambient energy (*T=300K*) of the surrounding water (in red dashed line).

are caused by random collisions between the Kerr nanoparticle and the relatively-large molecules. The narrower confinement of the Kerr nanoparticle indicates a stiffer potential trap that is contributed by the effects of the nonlinear interaction between the Kerr nanoparticle and the tightly focused Gaussian beam.

Figure 3 presents the three-dimensional (3D) plots of the trapping potential that is created by a focused beam (NA = 1.2) in the presence of a linear and a Kerr particle. The potential wells are steeper along the x-axis than along the z-axis since a high *NA* objective lens produces a focal volume that is relatively longer along the z-axis. The potential well associated with a Kerr nanoparticle is deeper than that of a linear nanospsphere.

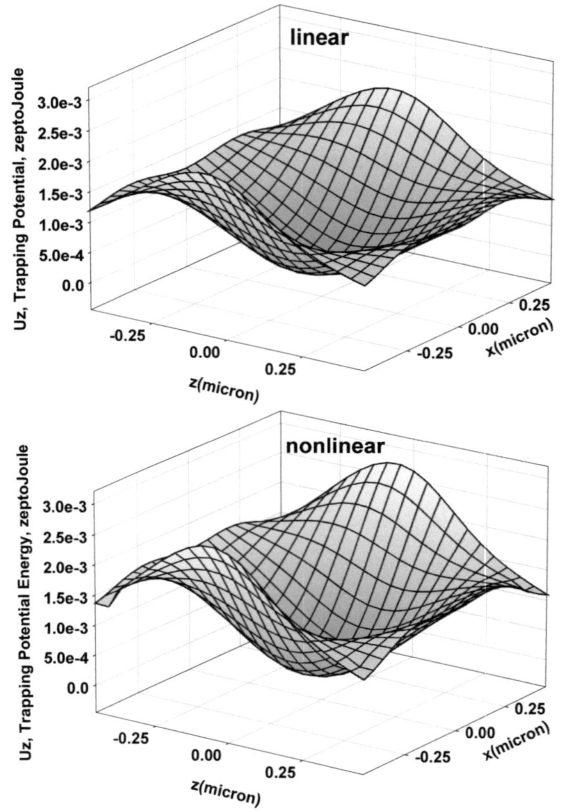

Fig. 3. Three-dimensional plot of the trapping potential energy along the transverse plane for both linear and nonlinear nanosphere as the focused laser beam propagates from left to right of the z-axis with the following trapping parameters: z_o=0, p=100mW, a=30nm, N.A.=1.2, λ =1.064um, n_1=1.33 , $n_2^{(0)}$=1.4, and $n_2^{(1)}$=1.8 x 10^{-12}m^2/W.

Under the same illumination conditions, a Kerr nanoparticle is captured more easily and held more stably in a single beam optical trap than a linear nanoparticle of the same size. A Kerr nanoparticle that is exhibiting Brownian motion is also confined within a much smaller volume of space around the beam focus as illustrated in 3D probability density of **figure 4**. The significant enhancement that is introduced by the Kerr nonlinearity could make the simpler single-beam optical trap into a viable alternative to multiple beam traps which are costly, less flexible and more difficult to operate.

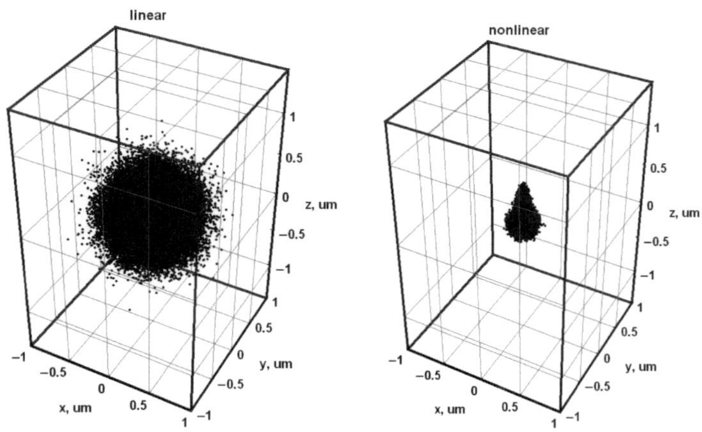

Fig. 4. Probability density distributions of linear and nonlinear (Kerr) nanospheres in a single-beam optical trap at T = 300K where t = 100,000 iterations, P = 100mW, a = 5 nm, NA = 1.2, λ = 1.064 μm, and n_1 = 1.33: a) Location probability distribution of linear ($n_2 = n_2^{(0)}$) and b) Kerr nanoparticle ($n_2^{(0)}$ = 1.4, $n_2^{(1)}$ = 1.8 x 10^{-12} m^2/W). Initially (t = 0), the nanoparticle is at rest at z = 0.5 μm.

4. Parametric analysis of the optical trapping force between linear and nonlinear (Kerr) nanoparticle

To better understand the underlying mechanism on how Kerr nonlinearity affects the trapping potential, let us perform a parametric analysis on how optical trapping force changes with typical trapping parameters on both linear and nonlinear (Kerr) nanoparticle.
The optical trapping force $\mathbf{F_{trap}(r)}$ that is described by Eq (2) was calculated using *Mathematica* Version 5.1 application program. **Figure 5a** presents the contour and 3D plots of $\mathbf{F_{trap}(r)}$ at different locations of the linear nanoparticle (n_2 = 1.4, a = 5 nm, λ = 1.062 μm, NA = 1.2) while **Figure 5b** shows the contour and 3D plots of $\mathbf{F_{trap}(r)}$ at different locations of the Kerr nanoparticle ($n_2^{(0)}$ = 1.4, $n_2^{(1)}$ = 1.8 x 10^{-11} m^2/W, a = 5 nm, λ = 1.062 μm, NA = 1.2). The $n_2^{(1)}$ value is taken from published measurements done with photopolymers which are materials that exhibit one of the strongest electro-optic Kerr effects (Nalwa & Miyata, 1997). Also shown is the contour plot of $\mathbf{F_{trap}(r)}$ for the case of a linear nanoparticle ($n_2^{(0)}$ = 1.4, a = 5 nm) of the same size.
For values of $z > 0$, $\mathbf{F_{trap}}$ is labeled negative (positive) when it pulls (pushes) the nanoparticle towards (away from) $\mathbf{r}$ = 0. For $z \leq 0$ the force is positive (negative) when it pushes (pulls) the nanoparticle towards (away from) the beam focus at $\mathbf{r}$ = 0. For both linear and nonlinear nanoparticles, the force characteristics are symmetric about the optical z-axis but asymmetric about the z = 0 plane. The asymmetry of the force is revealed only after the fifth-order correction is applied on the intensity distribution of the tightly focused Gaussian beam. The strongest force magnitude happens on the z-axis and it is 30% stronger in the case of the Kerr nanoparticle.
The stiffness of the optical trap may be determined by taking derivative of $\mathbf{F_{trap}(r)}$ with respect to $\mathbf{r}$. **Figure 6b** plots the stiffness at different locations of the Kerr nanoparticle. The stiffness distribution features a pair of minima at $r = (x^2 + y^2)^{1/2} \approx 0.1$ micron with a value of -25 x 10^{-12} N/m. Also presented in **Fig 6a** is the force stiffness distribution for the case of a

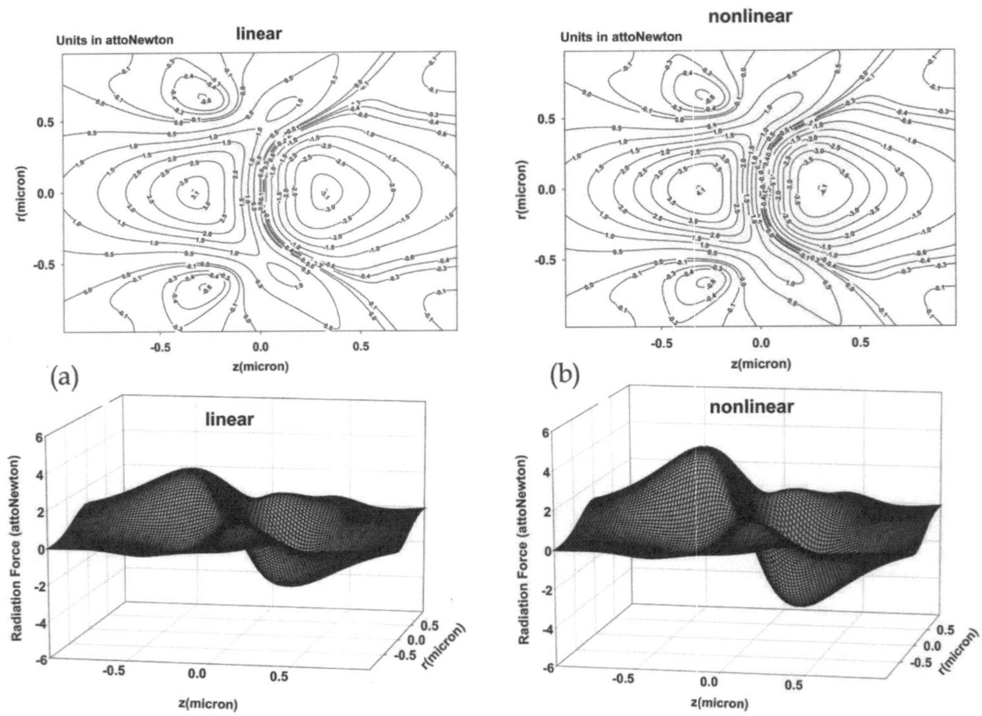

Fig. 5. Optical trapping force at different locations of both linear and Kerr nanoparticle ($n_2^{(0)}$ = 1.4, $n_2^{(1)}$ = 1.8 x 10^{-11} m^2/W) where r = $(x^2 + y^2)^{1/2}$. Parameter values common to both nanoparticles: P = 100 mW, a = 5 nm, NA = 1.2, λ = 1.064 microns, and n_1 = 1.33. The focused beam propagates from left to right direction. In all cases, F_{trap} = 0 at r(x, y, z) = 0.

linear nanoparticle exhibits a similar profile but a lower minimum value of -18 x 10^{-12} N/m at r ≈ 0.1 micron. The Kerr nanoparticle that is moving towards **r** = 0, experiences a trapping force that increases more rapidly than the one experienced by a linear nanoparticle of the same size. Once settled at **r** = 0, the Kerr nanoparticle is also more difficult to dislodge than its linear counterpart.

Figure 7a plots the behavior of $\mathbf{F_{trap}}$ at different axial locations of a linear nanoparticle (n_2 = 1.4) with a(nm) = 50, 70, 80, 90 and 100. In larger Kerr nanoparticles (a > 50 nm), the scattering force contribution becomes significant and the location of $\mathbf{F_{trap}(r)}$ = 0 shifts away from z = 0 and towards z > 0. Our results are consistent with those previously reported with linear dielectric nanoparticles (Rohrback and Steltzer, 2001; Wright *et.al.*, 1994).

Figure 7b plots the behavior of $\mathbf{F_{trap}(r)}$ at different axial locations of a bigger Kerr nanoparticle with a(nm) = 50, 70, 80, 90 and 100. The maximum strength of $\mathbf{F_{trap}(r)}$ increases with a. For a < 50 nm, $\mathbf{F_{trap}(r)}$ = 0 at z = 0 since $\mathbf{F_{trap}(r)}$ is contributed primarily by the gradient force. For larger Kerr nanoparticles, the relative contribution of the scattering force becomes more significant and the location where $\mathbf{F_{trap}(r)}$ = 0 is shifted away from z = 0 and towards the direction of beam propagation.

Figure 8a plots the behavior of $\mathbf{F}_{trap}\mathbf{(r)}$ as a function of the objective NA (0.4 ≤ NA ≤ 1.4) for a Kerr nanoparticle [$n_2^{(0)}$ = 1.4, $n_2^{(1)}$ = 1.8 x 10^{-11} m^2/W, P = 100 mW, a = 5 nm) that is located at r(0, 0, 0.5 micron). Also plotted is the behavior of $\mathbf{F}_{trap}\mathbf{(r)}$ with NA for a linear nanoparticle of the same size and initial beam location. Both the Kerr and the linear nanoparticle

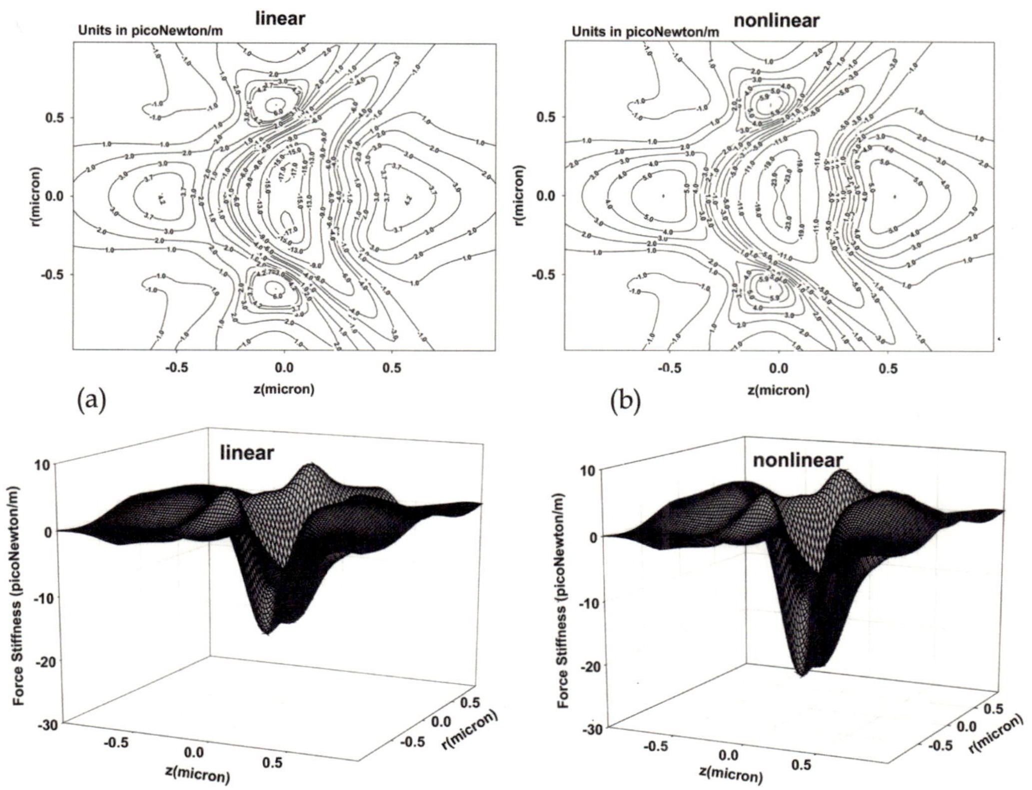

Fig. 6. Optical trapping force stiffness of optical trap at different locations of both linear ($n_2^{(0)} = 1.4$) and Kerr nanoparticle ($n_2^{(0)} = 1.4$, $n_2^{(1)} = 1.8 \times 10^{-12}$ m^2/W) where $r = (x^2 + y^2)^{1/2}$. Common parameter values: $P = 100$ mW, $a = 5$ nm, $NA = 1.2$, $\lambda = 1.064$ microns, and $n_1 = 1.33$.

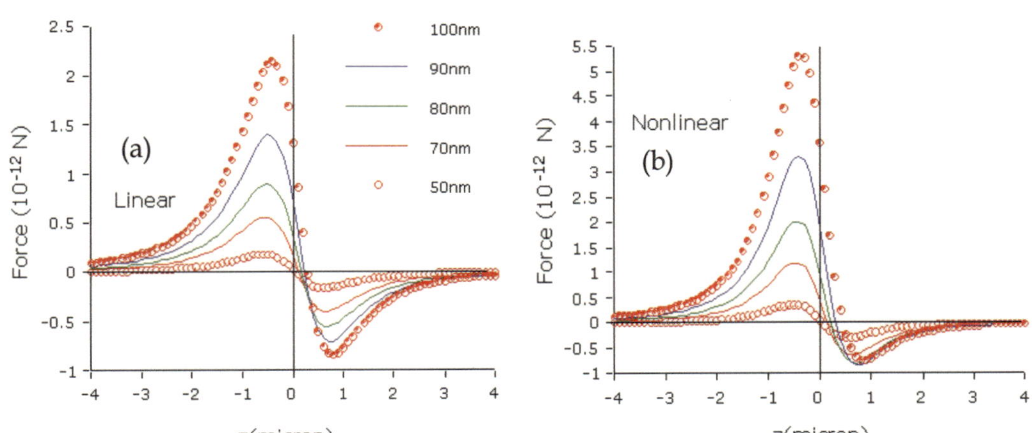

Fig. 7. Optical trapping force at different axial locations of: a) linear ($n_2 = n_2^{(0)} = 1.4$), and b) Kerr ($n_2^{(0)} = 1.4$, $n_2^{(1)} = 1.8 \times 10^{-12}$ m^2/W) nanoparticle of radius a(nm) = 50, 70, 80, 90 and 100. Common parameter values: $P = 100$ mW, $NA = 1.2$, $\lambda = 1.064$ microns, and $n_1 = 1.33$.

experience a trapping force that pulls them towards r = 0. The effect of the Kerr nonlinearity which is to increase the strength of $\mathbf{F}_{trap}(\mathbf{r})$, becomes more significant at *NA* > 1. At *NA* = 1.4, the force magnitude on the Kerr nanoparticle approximately twenty percent stronger than that experienced by the linear nanoparticle. The nonlinear effect is negligible in low *NA* focusing objectives (*NA* < 0.6). For the Kerr nanoparticle, the dependence of the force strength with *NA* is accurately described by a fourth order polynomial.

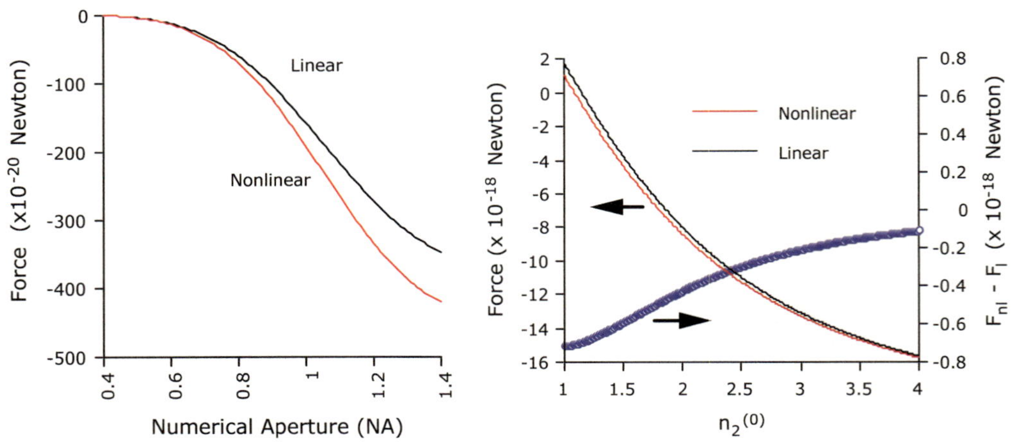

Fig. 8. (a) Optical trapping force on Kerr nanoparticle (solid line; $n_2^{(0)} = 1.4$, $n_2^{(1)} = 1.8 \times 10^{-12}$ m²/W) as a function of objective *NA*. Also plotted is the corresponding force (dotted line) on a linear nanoparticle with $n_2 = n_2^{(0)} = 1.4$. Common parameter values: z = 0.5 micron, P = 100 mW, a = 5 nm, λ = 1.064 microns, and n_1 = 1.33. For the Kerr nanoparticle the curve is accurately described by: $\mathbf{F}_{\mathbf{trap}} = 1311.65(NA)^4 - 3959.965(NA)^3 + 3656.265(NA)^2 - 1388.114NA + 188.105$. (b) Optical trapping force on non-absorbing Kerr (solid line; $n_2^{(1)} = 1.8 \times 10^{-12}$ m²/W) and linear ($n_2 = n_2^{(0)} = 1.4$) nanoparticle versus $n_2^{(0)}$, at λ = 1.064 microns. Also plotted (circles) is force difference ($F_{nl} - F_l$) as a function of $n_2^{(0)}$.

Figure 8b shows the behavior of $\mathbf{F}_{trap}(\mathbf{r})$ as a function of $n_2^{(0)}$ for a Kerr nanoparticle that is located at r(0, 0, 0.5 micron) in a surrounding medium with index n_1 = 1.33. Also plotted is the behavior of $\mathbf{F}_{trap}(\mathbf{r})$ with $n_2^{(0)}$ for a linear nanoparticle ($n_2 = n_2^{(0)}$) of the same size and initial beam location. The $\mathbf{F}_{trap}(\mathbf{r})$ profiles are similar for both the Kerr and linear nanoparticles. At $n_2^{(0)} > 1.1$, the Kerr nanoparticle experiences a negative (trapping) force that pulls it towards from r = 0. The trapping threshold is less than $n_2^{(0)} = 1.33$ because of the additional contribution of the nonlinear (Kerr) term $n_2^{(1)} I(r)$. For the linear nanoparticle, trapping is possible at a higher value of $n_2^{(0)} > 1.33$. Also plotted is the difference between the forces that are experienced by the two nanoparticles. The difference between the two trapping forces is highest near $n_2^{(0)} = 1$. The difference decreases with increasing $n_2^{(0)}$ value since the contribution of the Kerr term which has been held constant, becomes relatively small.

Figure 9 shows the dependence of $\mathbf{F}_{\mathbf{trap}}(\mathbf{r})$ with λ for a non-resonant Kerr nanoparticle at **r**(0, 0, 0.5 micron) in the range: $400 \le \lambda(\text{nm}) \le 1000$. Also presented is the behavior of $\mathbf{F}_{\mathbf{trap}}(\mathbf{r})$ for a linear nanoparticle of the same size and initial beam location. For a given *P* and *NA* value, the magnitude of $\mathbf{F}_{\mathbf{trap}}(\mathbf{r})$ increases nonlinearly with decreasing λ for both nanoparticles. However, the increase in the trapping force strength with λ is more rapid for the Kerr nanoparticle. In practice, λ is selected to avoid absorption by the nanoparticle and the surrounding liquid. Absorption could significantly heat up the nanoparticle and change its optical and mechanical properties. It can also lead to rapid evaporation of the surrounding liquid. In both cases, absorption reduces the efficiency of the optical trap.

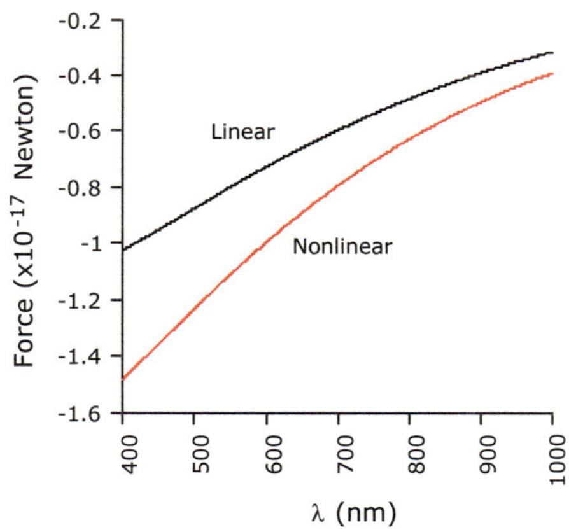

Fig. 9. Optical trapping force on non-absorbing Kerr nanoparticle (solid line, $n_2^{(0)} = 1.4$, $n_2^{(1)} = 1.8 \times 10^{-12}$ m^2/W) as a function of wavelength λ. Also plotted (dotted line) is the force on a linear nanoparticle ($n_2 = n_2^{(0)} = 1.4$). Other parameter values: P = 100 mW, a = 5 nm, NA = 1.2, z = 0.5 micron, and n_1 = 1.33.

Figure 10a plots the dependence of $\mathbf{F_{trap}(r)}$ with beam power P for a Kerr nanoparticle at r(0, 0, 0.5 micron). Also presented is the behavior of $\mathbf{F_{trap}(r)}$ for a linear nanoparticle of the same size and initial beam location. For the linear nanoparticle, the trapping force strength is directly proportional to P. For the Kerr nanoparticle, the trapping force strength increases more quickly (quadratically) with P for the Kerr nanoparticle. Figure 8b reveals that the force strength increases at a faster rate as the Kerr nanoparticle gets bigger.

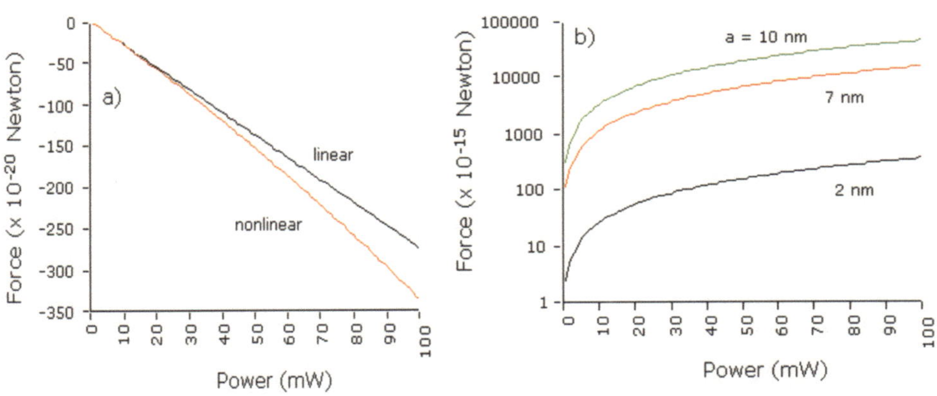

Fig. 10. (a) Optical trapping force on Kerr (filled circles; a = 5 nm) and linear nanoparticle (circles; a = 5 nm) as a function of beam power P, and (b) Force versus P for different radii of Kerr nanoparticle. Common parameter values: NA = 1.2, z = 0.5 micron), $n_2 = n_2^{(0)} = 1.4$, $n_2^{(1)} = 1.8 \times 10^{-12}$ m^2/W, and n_1 = 1.33. In (a) the force $\mathbf{F_{trap}}$ acting on the Kerr nanoparticle is accurately described by: $\mathbf{F_{trap}} = 0.006P^2 - 2.742P + 0.052$.

Figure 11, plots the trapping potential V(**r**) at different axial locations of a nanoparticle (a = 5 nm, λ = 1.064 microns, P = 100 mW) in the absence ($n_2 = n_2^{(0)}$) and presence ($n_2^{(1)}$ = 1.8 x 10^{-11} m^2/W) of Kerr nonlinearity. Compared to its linear counterpart, the Kerr nanoparticle is subjected to a V(**r**) that is significantly deeper and narrower. From the previous plots, **Figure 5a** and **5b** plot the movement of a linear and Kerr nanoparticle respectively, when they exhibit Browning motion in the single beam optical trap. The Kerr nanoparticle is released from rest at **r**(0, 0, 0.5 μm) and it is pulled towards **r** = 0 and confined to move within a region of about 0.2 μm radius, that is centered at **r** = 0 (Fig 5b). A linear nanoparticle that is released at **r**(0, 0, 0.5 μm) is constrained to move within a larger region of about 0.5 μm radius (Fig 5a).

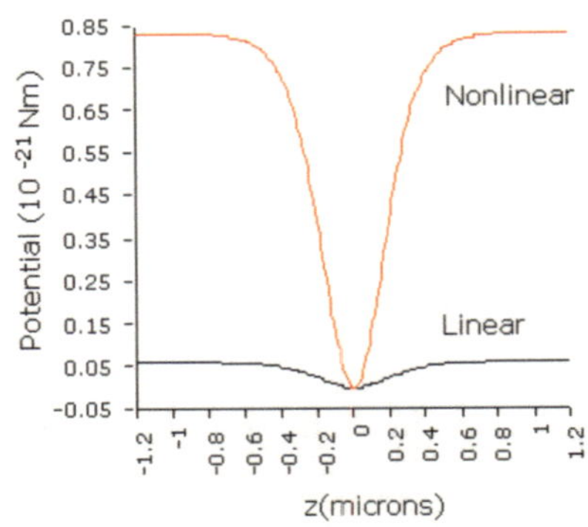

Fig. 11. Optical trapping potential in a single-beam optical trap at T = 300K where t = 100,000 iterations, P = 100mW, a = 5 nm, NA = 1.2, λ = 1.064 μm, and n_1 = 1.33: Kerr nanoparticle ($n_2^{(0)}$ = 1.4, $n_2^{(1)}$ = 1.8 x 10^{-12} m^2/W). Initially (t = 0), the nanoparticle is at rest at z = 0.5 μm.

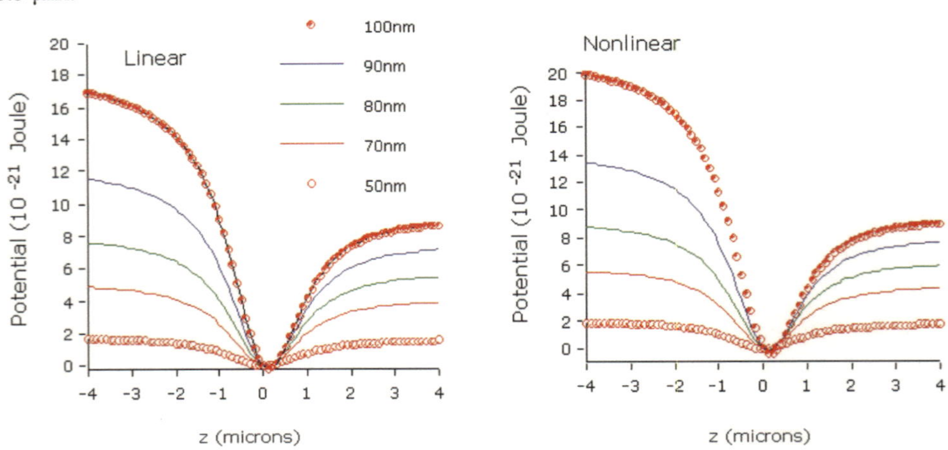

Fig. 12. Optical trapping potential at different axial locations of: a) linear ($n_2 = n_2^{(0)}$ = 1.4), and b) Kerr ($n_2^{(0)}$ = 1.4, $n_2^{(1)}$ = 1.8 x 10^{-12} m^2/W) nanoparticle of radius a(nm) = 50, 70, 80, 90 and 100. Common parameter values: P = 100 mW, NA = 1.2, λ = 1.064 microns, and n_1 = 1.33.

Figure 12a presents the V(z) profile at different z-locations of a linear nanoparticle with a(nm) = 50, 70, 80, 90 and 100. For comparison, **Figure 12b** shows the behavior of V(**r**) at different z-locations of a Kerr nanoparticle (λ = 1.064 microns, P = 100 mW, $n_2^{(1)} = 1.8 \times 10^{-11}$ m^2/W) with radius a(nm) = 50, 70, 80, 90 and 100. With respect to z = 0, the V(z) profile becomes asymmetric with increasing a for a > 50 nm. The distortion is caused by the increasing contribution of the scattering force to the net force $F_{trap}(r)$. It shifts the location of potential minimum away for z = 0, as well as lowers the escape threshold of a trapped Kerr nanoparticle in the direction of beam propagation. Optical trapping potential trends are similar except for differences in strengths, the V(z) profiles evolve in a similar manner with increasing nanoparticle size for both linear and nonlinear case.

5. Enhancement of single-beam optical trap due to Kerr nonlinearity

Our simulation results indicate that the performance of the single-beam optical trap is enhanced by the Kerr effect. For the same focused beam, a Kerr nanoparticle ($n_2^{(0)} = 1.4$, $n_2^{(1)} = 1.8 \times 10^{-11}$ m^2/W, P = 100 mW, NA = 1.2, λ = 1.064 microns, n_1 = 1.33) is subjected to a stronger trapping force than a linear nanoparticle (n_2 = 1.4,) of the same size (see Figs 7 - 10). The force magnitude increases rapidly as the nanoparticle approaches geometrical focus at **r** = 0 (Fig 8) especially along the optical z-axis. At the minimum of the trapping potential V(**r**), a Kerr nanoparticle encounters a higher escape threshold and therefore needs a greater amount of kinetic energy to escape from the optical trap (see Fig 11). Under the same illumination conditions, a Kerr nanoparticle that is exhibiting Brownian motion, is confined to move to within a much smaller region around **r** = 0, that a linear nanoparticle of the same size (see Fig 5).

The optical trapping force $\mathbf{F}_{trap}$ that is exerted on a Kerr nanoparticle with $a \leq 50$ nm = λ/21.3, is contributed primarily by the gradient force component. In such cases, $\mathbf{F}_{trap}$ = 0 at z = 0 (see Figs 8) and V(z) is symmetric about z = 0 (Figs 12 – 13). At a = 5 nm, we found that the maximum strength of the gradient force is about three orders of magnitude larger than that of the scattering force. The axial location where $\mathbf{F}_{trap}$ = 0 is shifted away from z = 0 and towards the general direction of the beam propagation, when the contribution of the scattering force component becomes comparable (see Fig 11). The corresponding V(z) profile becomes asymmetric with a lower escape barrier along the direction of beam propagation (see Fig 13). Such instances occur with larger Kerr nanoparticles ($a > \lambda/21.3$).

Except for differences in their relative magnitudes, the axial profiles of $\mathbf{F}_{trap}$ exhibit the similar characteristics with increasing nanoparticle size for the both the nonlinear and linear case. Our results indicate that the index increase that is introduced by the Kerr effect, does not affect the ability of a small Kerr nanoparticle ($a \leq \lambda/21.3$) to scatter light in an isotropic manner - the increase in n_2 is uniform distributed in the nanoparticle. The gradient force contribution to $\mathbf{F}_{trap}$ becomes significant when the non-absorbing nanoparticle scatters light in an anisotropic manner.

Figure 8a illustrates that the enhancement that is gained from the Kerr effect in trapping a non-resonant nanoparticle, is realized only with high NA focusing objectives (NA > 0.6). The strength of $\mathbf{F}_{trap}$ becomes stronger at shorter λ values (see Fig.10). The increase is faster for the Kerr nanoparticle due to the dependence of its refractive index with I(**r**) – the force strength increases quadratically with λ. We note that the strong dependence of $\mathbf{F}_{trap}$ with λ is not observed in larger Kerr nanoparticles especially in the regime of α > 100 and $a >> \lambda$) (Pobre & Saloma, 1997; Pobre & Saloma, 2002).

The optical trapping force increases rapidly with beam power P for the same NA and λ values (see Fig 10a). For a linear nanoparticle, the force strength is directly proportional to P.

For a Kerr nanoparticle, the relationship of the force strength with P is nonlinear - the Kerr effect permits the use of low power light sources that tend to be less costly to acquire and maintain. Trapping at low beam powers also minimizes the optical heating of the surrounding medium and even the nanoparticle itself. Reductions in unwanted thermal effects are vital in the manipulation and guidance of biological samples.

6. Summary and future prospects

We have analyzed the optical trapping force $\mathbf{F}_{trap}$ that is exerted on a Kerr nanoparticle by a focused Gaussian beam when $2\pi a/\lambda \leq 100$ and $a << \lambda$. The optical trapping mechanism consists of two dominant optical forces representing the contribution of the field gradient and that of the EM field that is scattered by the nanoparticle. The contributions of the two force components become comparable for nanoparticles with $a > \lambda/21.3$. The gradient force contribution is more dominant with smaller non-absorbing nanoparticles such that $\mathbf{F}_{trap} = 0$ at the beam focus $\mathbf{r} = 0$. The Brownian motion of the Kerr nanoparticle has an over-damped harmonic motion enveloped by white noise function due to thermal fluctuations generated by moving molecules defined by the background energy of 3.1 k_bT of the surrounding fluid. Confinement of the Kerr nanoparticle depends on the nonlinear refractive index of the nanoparticle as shown in the widths of the probability density of the Kerr nanoparticle.
Under the same illumination conditions, a Kerr nanoparticle is captured more easily and held more stably in a single beam optical trap than a linear nanoparticle of the same size. A Kerr nanoparticle that is exhibiting Brownian motion is also confined within a much smaller volume of space around the beam focus. The significant enhancement that is introduced by the Kerr nonlinearity could make the simpler single-beam optical trap into a viable alternative to multiple beam traps which are costly, less flexible and more difficult to operate.
Kerr nonlinearity enhances the performance of a single beam trap by increasing the magnitude of the trapping force. Its permits the trapping of nonlinear nanoparticles with $n_2^{(0)}$ values that are less than the index n_1 of the surrounding liquid and at lower NA values and optical beam powers. Low NA focusing objectives and low power laser sources are relatively inexpensive and are less likely to cause irreversible thermal damage on the sample and the surrounding medium.
Localized (non-uniform) heating of the liquid is also minimized if the average power of the focused beam is kept low using a femtosecond laser source with high peak powers and relatively low repetition rate.
Kerr nanoparticle can be an alternative probe handler when applied to photonic force microscope configuration for the imaging of hollow microbiological structures.

7. Acknowledgement

The authors are grateful for the financial support provided by University Research Coordination Office of De La Salle University (DLSU) and the University of the Philippines Diliman.

8. References

Ashkin, A.; Dziedzic, J.; Bjorkholm, J & Chu, S. (1986). Observation of a single-beam gradient force optical trap for dielectric particles. *Optics Letters*, Vol. 11, Issue 5, May 1, 1986, Pages 288 - 290, ISSN: 0146-9592.

Ashkin, A. (1997). Optical trapping and manipulation of neutral particles using lasers. *Proc. Natl. Acad. Sci. USA*, Vol 94, No. 10, May 13, 1997, Pages 4853 - 4860, ISSN: 0027-8424.

Barton, J., & Alexander, D. (1989). Fifth-order corrected electromagnetic field components for a fundamental Gaussian beam. *Journal of Applied Physics, 66(7)*, 2800-2802.

Clays, K., Hendrickx, E., Triest, M., Verbiest, T., Persoons, A., Dehu, C., & Brédas, J. (1993). Nonlinear optical properties of proteins measured by hyper-Rayleigh scattering in solution. *Science, 262(5138)*, 1419 - 1422.

Clement-Sengewald, A., Schütze, K., Ashkin, A., Palma, G., Kerlen, G., & Brem, G. (1996). Fertilization of bovine oocytes induced solely with combined laser microbeam and optical tweezers. *J. Assist. Reprod. Genet., 13(3)*, 259-265.

Curtis, J., & Grier, D. (2003). Structure of Optical Vortices. *Phys. Rev. Lett., 90(13)*, 133901.

Curtis, J., Koss, B., & Grier, D. (2002). Dynamic holographic optical tweezers. *Optics Communications, 207(1-6)*, 169-175.

Friese, M., Nieminen, T., Heckenberg, N., & Rubinsztein-Dunlop, H. (1998). Optical alignment and spinning of laser-trapped microscopic particles. *Nature, 394(6691)*, 348-350.

Gambogi, W., Gerstadt, W., Mackara, S., & Weber, A. (1991). Holographic transmission elements using photopolymer films. *Computer and Optically Generated Holographic Optics, 4th in a Series, Proc. SPIE, 1555*, 256–267.

Grier, D. (2003). A revolution in optical manipulation. *Nature, 424(6950)*, 810-816.

He, H., Friese, M., Heckenberg, N., & Rubinsztein-Dunlop, H. (1995). Direct Observation of Transfer of Angular Momentum to Absorptive Particles from a Laser Beam with a Phase Singularity. *Physical Review Letters, 75*(5), 826 - 829.

Kerker, M. (1997). *Scattering of Light & Other Electromagnetic Radiation*. Toronto: Academic Press.

Koch, S., Shundrovsky, A., Jantzen, B., & Wang, M. (2002). Probing protein-DNA interactions by unzipping a single DNA double helix. *Biophys J, 83*(2), 1098-1105.

Kuga, T., Torii, Y., Shiokawa, N., Hirano, T., Shimizu, Y., & Sasada, H. (1997). Novel Optical Trap of Atoms with a Doughnut Beam. *Physical Review Letters, 78*(25), 4713-4716.

Lasky, L. (1997). Cell adhesion: How integrins are activated. *Nature, 390*, 15-17.

Liang, H., Wright, W., Cheng, S., He, W., & Berns, M. (1993). Micromanipulation of chromosomes in PTK2 cells using laser microsurgery (optical scalpel) in combination with laser-induced optical force (optical tweezers). *Exp Cell Res, 204*(1), 110-120.

Liesener, J., Reicherter, M., Haist, T., & Tiziani, H. (2000). Multi-functional optical tweezers using computer-generated holograms. *Optics Communications, 185(1-3)*, 77-82.

Macdonald, M., Paterson, L., Volke-Sepulveda, K., Arlt, J., Sibbett, W., & Dholakia, K. (2002). Creation and Manipulation of Three-Dimensional Optically Trapped Structures. *Science, 296(5570)*, 1101-1103.

Nalwa, H., & Miyata, S. (1997). *Nonlinear optics of organic molecules and polymers* . Boca Raton, Fla.: CRC Press.

Neuman, K., & Block, S. (2004). Optical trapping. *Review of Scientific Instruments, 75*(9), 2787-2809.

Pobre, R., & Saloma, C. (2008). Thermal diffusion of Kerr nanobead under a tigthly focused laser beam. *IFMBE Proceedings World Congress on Medical Physics and Biomedical*

Engineering 2006, Vol. 14 Part 5, SI Kim and TS Suh, editors (Springer, Berlin, 2008), pp 321-325.

Pobre, R., & Saloma, C. (2006). Radiation force exerted on nanometer size non-resonant Kerr particle by a tightly focused Gaussian beam. *Opt. Commun., 267*(2), 295-304.

Pobre, R., & Saloma, C. (2002). Radiation forces on nonlinear microsphere by a tightly focused Gaussian beam. *Appl. Opt., 41*, 7694-7701.

Pobre, R., & Saloma, C. (1997). Single Gaussian beam interaction with a Kerr microsphere: characteristics of the radiation force. *Appl. Opt., 36*(*15*), 3515-3520.

Prasad, P. N., & Williams, D. J. (1991). *Introduction to Nonlinear Optical Effects in Molecules and Polymers*. New York: Wiley-Interscience.

Rodrigo, P., Daria, V., & Glückstad, J. (2005). Four-dimensional optical manipulation of colloidal particles. *Appl. Phys. Lett., 86*(*7*), 74103.

Rodrigo, P., Daria, V., & Glückstad, J. (2005). Dynamically reconfigurable optical lattices. *Opt. Express, 13*(*5*), 1384-1394.

Rohrbach, A., & Stelzer. (2001). Optical trapping of dielectric particles in arbitrary fields. *Appl. Opt., 18*(*4*), 839–853.

Rohrbach, A., & Stelzer, E. (2002). Trapping Forces, Force Constants, and Potential Depths for Dielectric Spheres in the Presence of Spherical Aberrations. *Appl. Opt., 41*(*13*), 2494-2507.

Siegman, A. E. (1986). *Lasers*. Sausalito, CA: University Science Books.

Singer, W., Bernet, S., Hecker, N., & Ritsch-Marte, M. (2000). Three-dimensional force calibration of optical tweezers. *J. Mod. Opt. , 47*(*14/15*), 2921-2931.

Svoboda, K., & Block, S. (1994). Biological applications of optical forces. *Annu Rev Biophys Biomol Struct, 23*, 247-285.

Wright, W., Sonek, G., & Berns, M. (1994). Parametric study of the forces on microspheres held by optical tweezers. *Appl. Opt., 33*(*9*), 1735-1748.

Yin, H., Wang, M., Svoboda, K., Landick, R., Block, S., & Gelles, J. (1995). Transcription against an applied force. *Science, 270*(*5242*), 1653-1657.

Zyss, D. &., (eds), J., & Chemla. (1987). *Nonlinear Optical Properties of Organic Molecules and Crystals, Volumes* 1. London: Academic Pr.